Alternatives for Dermal Toxicity Testing

Chantra Eskes • Erwin van Vliet
Howard I. Maibach
Editors

Alternatives for Dermal Toxicity Testing

Springer

Editors
Chantra Eskes
SeCAM
Magliaso
Switzerland

Erwin van Vliet
Innovitox Consulting & Services
Houten
The Netherlands

Howard I. Maibach
Department of Dermatology
University of California School
of Medicine
San Francisco
California
USA

ISBN 978-3-319-50351-6 ISBN 978-3-319-50353-0 (eBook)
DOI 10.1007/978-3-319-50353-0

Library of Congress Control Number: 2017955816

Printed on acid-free paper

This Springer imprint is published by Springer Nature
The registered company is Springer International Publishing AG
The registered company address is: Gewerbestrasse 11, 6330 Cham, Switzerland

Foreword

This substantial compilation of manuscripts provides an important and comprehensive collection of papers by world-renowned scientists covering the literature on alternatives for dermal toxicity testing.

Historically, dermal testing was initially thought of as one of the more difficult *in vitro* methods. The physiological basis of dermal toxicity is very complex and involves many different cell types and pathways for sensitivity, irritation, and corrosion. Yet surprisingly, dermal toxicity is one of the earliest areas of *in vitro* toxicity to provide useful human cell-based systems.

Initial toxicity assay developments were seen as simple (quick) approaches to commercial human skin systems that were being developed for treating burn patients. A few companies learned the hard way that *in vitro* toxicology was no simpler than using those cultured skin systems as skin grafts. After several years, they all went out of business. Several scientists who understood the complexity, however, focused on developing human skin models for the sole purpose of *in vitro* toxicity. These models, simple at first, became more standardized and more complex and provided a better matrix for testing.

The Johns Hopkins Center for Alternatives to Animal Testing (CAAT) was founded in 1981 specifically to develop *in vitro* methods for hazard evaluation and safety testing of cosmetic products (see [1]). One aspect of the research program, identified as Program Projects, was the coordination of several projects within a selected topic to develop a better understanding of mechanisms responsible for a toxic event.

The Avon Program Project

Avon funded CAAT from the first grant (from the Cosmetic, Toiletry, and Fragrance Association (CTFA)) and then continued independently funding the center. After a few years, Avon, in the person of Yale Gressel, asked if CAAT could take on a larger project—developing an *in vitro* assay to predict skin sensitization.

We approached the problem by inviting about eight laboratories working on various aspects of skin biology to present to their "competitors and colleagues." They were asked how they would approach the issue and what aspects they saw as the most important. At first, the discomfort was obvious: "Will what I share be used by my competitors?" As the day progressed, however, it became clear that each lab would be focusing on different aspects of the problem. We invited five individuals to submit grant applications with the provision that, if approved, up to three applications would be funded.

The funded project teams would get together twice yearly in a roll-up-your-sleeves discussion about their progress and how to proceed. The attendees at these "lab" meetings were the participants along with other experts from Hopkins, the government, and Avon. And they were wonderful meetings. At almost every meeting a person from one of the sectors would ask a question and the response from another sector would be, "That is a great question—I would have never thought of it." In essence, the corporate and government scientists wanted to know how to use the information generated and the academics wanted to better understand the mechanisms involved.

The project lasted nine years, and the science it generated formed the basis of our understanding of mechanisms of skin sensitization. This project was summarized by Craig Elmets [2].

"By all measures it was a very successful project, characterized by identification of many of the interleukins, cytokine pathways, and the recognition that keratinocytes play an important role in sensitization." (As quoted from [1])

Toxicity Testing in the Twenty-First Century

The NAS report, *Toxicity Testing in the 21st Century: A Vision and a Strategy*, was a seminal moment in the development of *in vitro* assays [3]. This report had undergone external review and I was one of the external reviewers.

The major conclusions of the study included the following:

1. Animal studies are time-consuming and expensive.
2. There is a lack of predictability of animal studies as they relate to humans.
3. We should be using human cells in culture.
4. We should explore systems biology and pathways and mechanisms of toxicity.

This publication was, and is, a major advancement in *in vitro* toxicology, alternatives, and risk assessment. It created major new research approaches and opportunities. It provided an important source of encouragement for the development of alternative toxicological methodologies and stimulated what is now recognized as a scientific revolution.

Human Cell in Culture

As the *in vitro* toxicology field began to develop, animal cells, mainly from rats and mice, were being used, as human cell culture was essentially not available. When CAAT was founded, Leon Golberg (1982) emphasized that human cell cultures would be the key to developing *in vitro* methods for risk assessment that would be accepted for decision making. How correct he was. As a result of this realization, CAAT, from the very first round of grants, funded research to advance the science of human cell culture. A number of contributors to this volume were funded by CAAT. A summary of many aspects of human cell culture can be found in Bressler et al. [4].

Skin

The skin represents the largest organ of the human body. The ability to understand how drugs and chemicals penetrate the skin and how they may adversely affect the health of skin is important for protecting consumers from undesired effects. Excised human skin sections from cadavers have been used extensively to understand the dermal penetration of drugs and cosmetics. And for more than 30 years, the scientific community has devoted much time developing monolayer cultures of cells and more recently has focused on 3D reconstituted human skin models.

Alternatives for Dermal Toxicity Testing editors Chantra Eskes, Erwin van Vliet, and Howard Maibach have compiled an excellent, important, and comprehensive book that is necessary for anyone in the field—from beginner students to highly acclaimed senior researchers.

The book contains six sections: irritation, corrosion, sensitization, UV-induced effects, genotoxicity, and a concluding section with three papers exploring integrated strategies and high-throughput systems.

I believe that every commercial model is covered, in depth, with adequate information to assist one in identifying the best model for their studies. The volume is an invaluable resource.

The editors should be congratulated for identifying essentially most, if not all, of the contributors in this field and synthesizing a highly readable and important reference publication.

Alan M. Goldberg, PhD
Center for Alternatives to Animal Testing
Departmental of Environmental Health and Engineering
Bloomberg School of Public Health, Global Food Ethics
Johns Hopkins University
Baltimore, MD 21205, USA

References

1. Goldberg AM. A history of the Johns Hopkins Center for alternatives to animal testing, the first 28 years (1981–2009). In: Applied in vitro toxicology. Vol 1. Mary Ann Liebert Publishers; 2015. p. 99–108.
2. Elmets C. The AVON Program Project. A report of progress. In Vitro Toxicol. 1996;9:223.
3. NAS. Toxicity testing in the 21st century—a vision and a strategy. NAS; 2007.
4. Bressler J, Bader J, Goldberg A. Alternatives to conventional toxicology testing. In: McQueen CA, editor. Comprehensive toxicology, vol. 3. Oxford: Academic; 2010. p. 247–259.

Preface

Dermal toxicity is one of the pioneer areas in which alternative methods to the use of animal testing have gained scientific, industrial, and regulatory acceptance. Over two decades have passed since the publication in 1994 of Mary Ann Liebert's book on *In Vitro* Skin Toxicology (Rougier A., Goldberg A.M., and Maibach H.I. Eds.). Since then, several alternative methods for dermal toxicity have been optimized, scientifically validated, and gained international regulatory acceptance. In some cases it is already possible to fully replace the regulatory animal test, such as for skin irritation and corrosion, by using, e.g., Integrated Approaches to Testing and Assessment (IATAs). In other cases, such as for skin sensitization, it is possible to partially replace the regulatory animal test with *in chemico* and *in vitro* test methods that address key events of the adverse outcome pathway (AOP) leading to allergic contact dermatitis. Furthermore, the use of human *in vitro* models in the area of skin irritation and the use of defined approaches (DA) for skin sensitization testing (i.e., which combine, e.g., *in chemico* and *in vitro* test methods) have shown comparable if not better correlations to human data than the regulatory animal tests.

In view of the considerable progress made, this book aims at providing up-to--date comprehensive information on the most advanced alternative test methods available for the assessment of dermal toxicity with particular emphasis on the areas of skin irritation, skin corrosion, skin sensitization, UV-induced effects, and skin genotoxicity. For each test method, a description of the currently available protocol is given including highlights of its critical steps, applicability, limitations, potential role, and use within testing approaches and correlation with the traditional animal data and, when available, also human data. Furthermore, the book addresses exploratory areas that may be of relevance for the future of dermal toxicity safety testing, including the use of human progenitor skin cells, integration of *in vitro* and clinical methodologies, and application of high-throughput screening techniques.

The editors warmly acknowledge all authors that contributed to make the project of this book a reality and Springer for their great support and belief in the project. Albeit attempting to be comprehensive, new and/or additional methods and authors that could not be involved in this book will be invited to contribute to the next editions to come, for which any comments and/or suggestions are welcomed.

Magliaso, Switzerland — Chantra Eskes
Houten, The Netherlands — Erwin van Vliet
San Francisco, CA, USA — Howard Maibach

Contents

Contributors

Pierre Aeby Consultant, Marly, Switzerland

Shaheda Ahmed Alcyomics, Ltd., Newcastle-upon-Tyne, UK

Haematological Sciences, Institute of Cellular Medicine, Newcastle University, Newcastle-upon-Tyne, UK

Nathalie Alépée L'Oréal Research and Innovation, Aulnay-sous-Bois Cedex, France

Lee Ann Applegate Department of Musculoskeletal Medicine, Plastic, Reconstructive and Hand Surgery, Unit of Regenerative Therapy, University Hospital of Lausanne, Epalinges, Switzerland

Takao Ashikaga Shiseido, Co., Ltd., Tokyo, Japan

Claude Auriault ImmunoSearch, Grasse, France

Marcela Contador Baptista R&D Department, Grupo Boticário, São José dos Pinhais, PR, Brazil

David A. Basketter DABMEB Consultancy Ltd, Sharnbrook, UK

Caroline Bauch Cyprotex, Kalamazoo, MI, USA

Carolina Gomes Benevenuto Faculty of Pharmaceutical Sciences of Ribeirão Preto, University of São Paulo, São Paulo, Brazil

Elodie Boitel ImmunoSearch, Grasse, France

Carl Borrebaeck Department of Immunotechnology, Lund University, Lund, Sweden

Rico Brendtke Department Tissue Engineering and Regenerative Medicine (TERM), University Hospital Wuerzburg, Wuerzburg, Germany

Senetics Healthcare Group GmbH & Co. KG, Erlangen, Germany

Carla Abdo Brohem R&D Department, Grupo Boticário, São José dos Pinhais, PR, Brazil

Petra Budde Proteome Sciences PLC, Cobham, UK

Anthony de Buys Roessingh Department of Pediatric Surgery, University Hospital of Lausanne, Lausanne, Switzerland

Andrezza di Pietro Micali Canavez R&D Department, Grupo Boticário, São José dos Pinhais, PR, Brazil

Emanuela Corsini Laboratory of Toxicology, Department of Pharmacological and Biomolecular Sciences, Università degli Studi di Milano, Milan, Italy

José Cotovio L'Oréal Research and Innovation, Aulnay-sous-Bois Cedex, France

Françoise Cottrez ImmunoSearch, Grasse, France

Maria Teresa Cruz Centre for Neuroscience and Cell Biology, University of Coimbra, Coimbra, Portugal

Faculty of Pharmacy, University of Coimbra, Coimbra, Portugal

Faculty of Pharmacy, Center for Neuroscience and Cell Biology, University of Coimbra, Coimbra, Portugal

Anne Dickinson Alcyomics, Ltd., Newcastle-upon-Tyne, UK

Haematological Sciences, Institute of Cellular Medicine, Newcastle University, Newcastle-upon-Tyne, UK

Solène Dreyfuss L'Oréal Research & Innovation, Aulnay-sous-Bois Cedex, France

Oliver Engelking CellSystems GmbH, Troisdorf, Germany

Chantra Eskes SeCAM Services & Consultation on Alternative Methods, Magliaso, Switzerland

Philipp R. Esser Allergy Research Group, Department of Dermatology, Faculty of Medicine, University of Freiburg, Freiburg, Germany

Isabel Ferreira Centre for Neuroscience and Cell Biology, University of Coimbra, Coimbra, Portugal

Leslie M. Foertsch The Procter & Gamble Company, Mason Business Center, Cincinnati, OH, USA

Horst W. Fuchs CellSystems GmbH, Troisdorf, Germany

Valentina Galbiati Laboratory of Toxicology, Department of Pharmacological and Biomolecular Sciences, Università degli Studi di Milano, Milan, Italy

Lorena Rigo Gaspar Faculty of Pharmaceutical Sciences of Ribeirão Preto, University of São Paulo, São Paulo, Brazil

G. Frank Gerberick The Procter and Gamble Company, Central Product Safety, Mason, OH, USA

Susan Gibbs Department of Dermatology, VU University Medical Centre, Amsterdam, The Netherlands

Department of Oral Cell Biology, Academic Centre for Dentistry Amsterdam (ACTA), Amsterdam, The Netherlands

Marie-Hélène Grandidier L'Oréal Research and Innovation, Aulnay-sous-Bois Cedex, France

Florian Groeber Translational Center Wuerzburg 'Regenerative Therapies in Oncology and Musculoskeletal Diseases', Fraunhofer Institute for Interfacial Engineering and Biotechnology (IGB), Wuerzburg, Germany

Hervé Groux ImmunoSearch, Grasse, France

Robert Guest Envigo Research Limited, Shardlow, Derbyshire, UK

Jan Hansmann Department Tissue Engineering and Regenerative Medicine (TERM), University Hospital Wuerzburg, Wuerzburg, Germany

Translational Center Wuerzburg 'Regenerative Therapies in Oncology and Musculoskeletal Diseases', Fraunhofer Institute for Interfacial Engineering and Biotechnology (IGB), Wuerzburg, Germany

Rosette Van Den Heuvel Unit of Environmental Risk and Health, VITO NV, Antwerp, Belgium

Nathalie Hirt-Burri Department of Musculoskeletal Medicine, Plastic, Reconstructive and Hand Surgery, Unit of Regenerative Therapy, University Hospital of Lausanne, Epalinges, Switzerland

Markus Hofmann Swiss Federal Office of Public Health, Bern, Switzerland

Jef Hooyberghs Unit of Environmental Risk and Health, VITO NV, Antwerp, Belgium

Theoretical Physics, Hasselt University, Hasselt, Belgium

Kazuhiro Hosoi Research & Development Center, Santen Pharmaceutical, Co., Ltd., Osaka, Japan

Paris Jafari Department of Musculoskeletal Medicine, Plastic, Reconstructive and Hand Surgery, Unit of Regenerative Therapy, University Hospital of Lausanne, Epalinges, Switzerland

Henrik Johansson SenzaGen AB, Lund, Sweden

Helena Kandarova MatTek In Vitro Life Science Laboratories, Bratislava, Slovakia

Masakazu Katoh Japan Tissue Engineering Co. Ltd., Gamagori City, Aichi, Japan

Camila Martins Kawakami Faculty of Pharmaceutical Sciences of Ribeirão Preto, University of São Paulo, São Paulo, Brazil

Donald Keller Cyprotex, Kalamazoo, MI, USA

Petra S. Kern The Procter & Gamble Company, Beijing Innovation Center, Beijing, China

Ian Kimber Faculty of Life Sciences, University of Manchester, Manchester, UK

Hajime Kojima National Institute of Health Sciences (NIHS), Tokyo, Japan

Odivania Kruger R&D Department, Grupo Boticário, São José dos Pinhais, PR, Brazil

Jochen Kühnl Beiersdorf AG, Hamburg, Germany

Nathalie Lambrechts Unit of Environmental Risk and Health, VITO NV, Antwerp, Belgium

Robert Landsiedel BASF SE, Experimental Toxicology and Ecology, Ludwigshafen, Germany

Manfred Liebsch "Centre of Alternative Methods to Animal Experiments—ZEBET" at the BfR, Berlin, Germany

Malin Lindstedt Department of Immunotechnology, Lund University, Lund, Sweden

Marcio Lorencini R&D Department, Grupo Boticário, São José dos Pinhais, PR, Brazil

Stefan F. Martin Allergy Research Group, Department of Dermatology, Faculty of Medicine, University of Freiburg, Freiburg, Germany

João Demétrio Martins Centre for Neuroscience and Cell Biology, University of Coimbra, Coimbra, Portugal

Faculty of Pharmacy, University of Coimbra, Coimbra, Portugal

Annette Mehling BASF Personal Care and Nutrition GmbH, Düsseldorf, Germany

Laurent Nardelli L'Oréal Research and Innovation, Aulnay-sous-Bois Cedex, France

Andreas Natch Givaudan Suisse SA, Duebendorf, Switzerland

Inge Nelissen Unit of Environmental Risk and Health, VITO NV, Antwerp, Belgium

Caroline Radoski Neumann R&D Department, Grupo Boticário, São José dos Pinhais, PR, Brazil

Bruno Miguel Neves Department of Chemistry, Mass Spectrometry Centre, QOPNA, University of Aveiro, Aveiro, Portugal

Centre for Neuroscience and Cell Biology, University of Coimbra, Coimbra, Portugal

Faculty of Pharmacy, University of Coimbra, Coimbra, Portugal

Satomi Onoue The School of Pharmaceutical Sciences, University of Shizuoka, Shizuoka Prefecture, Shizuoka, Japan

Prakash Patel Cyprotex, Kalamazoo, MI, USA

Stefan Pfuhler Procter and Gamble, Mason Business Centre, Mason, OH, USA

Ian Pike Proteome Sciences PLC, Cobham, UK

Cécile Piroird L'Oréal Research and Innovation, Aulnay-sous-Bois Cedex, France

Wassim Raffoul Department of Musculoskeletal Medicine, Plastic, Reconstructive and Hand Surgery, Unit of Regenerative Therapy, University Hospital of Lausanne, Epalinges, Switzerland

Tzutzuy Ramirez BASF SE, Ludwigshafen, Germany

Kerstin Reisinger Henkel Beauty Care, Düsseldorf, Germany

Hendrik Reuter Beiersdorf AG, Hamburg, Germany

Erwin L. Roggen 3Rs Management and Consulting ApS, Kongens Lyngby, Denmark

Hitoshi Sakaguchi Kao Corporation, Tokyo, Japan

Corinne Scaletta Department of Musculoskeletal Medicine, Plastic, Reconstructive and Hand Surgery, Unit of Regenerative Therapy, University Hospital of Lausanne, Epalinges, Switzerland

Andreas Schepky Beiersdorf AG, Hamburg, Germany

Freia Schmid Translational Center Wuerzburg 'Regenerative Therapies in Oncology and Musculoskeletal Diseases', Fraunhofer Institute for Interfacial Engineering and Biotechnology (IGB), Wuerzburg, Germany

Greet Schoeters Unit of Environmental Risk and Health, VITO NV, Antwerp, Belgium

Department of Biomedical Sciences, University of Antwerp, Antwerp, Belgium

Dept of Environmental Medicine, Institute of Public Health, University of Southern Denmark, Odense, Denmark

Desirée Cigaran Schuck R&D Department, Grupo Boticário, São José dos Pinhais, PR, Brazil

Ana Silva Centre for Neuroscience and Cell Biology, University of Coimbra, Coimbra, Portugal

Talita M.T.P. Silveira R&D Department, Grupo Boticário, São José dos Pinhais, PR, Brazil

Sander W. Spiekstra Department of Dermatology, VU University Medical Centre, Amsterdam, The Netherlands

Carine Tornier EPISKIN™, 4 rue Alexander Fleming, 69366 Lyon, France

John A. Troutman The Procter & Gamble Company, Mason Business Center, Cincinnati, OH, USA

Rich Ulmer InVitro International, Placentia, CA, USA

Natalia de Albuquerque Vita R&D Department, Grupo Boticário, São José dos Pinhais, PR, Brazil

Heike Walles Department Tissue Engineering and Regenerative Medicine (TERM), University Hospital Wuerzburg, Wuerzburg, Germany

Translational Center Wuerzburg 'Regenerative Therapies in Oncology and Musculoskeletal Diseases', Fraunhofer Institute for Interfacial Engineering and Biotechnology (IGB), Wuerzburg, Germany

Amy Wang InVitro International, Placentia, CA, USA

Xiao Nong Wang Haematological Sciences, Institute of Cellular Medicine, Newcastle University, Newcastle-upon-Tyne, UK

Ana Cristina Weihermann R&D Department, Grupo Boticário, São José dos Pinhais, PR, Brazil

Dirk Weisensee CellSystems GmbH, Troisdorf, Germany

Bart De Wever ATERA SAS, Nice, France

Hilda Witters Unit of Environmental Risk and Health, VITO NV, Antwerp, Belgium

Kathrin Stephanie Zeller Department of Immunotechnology, Lund University, Lund, Sweden

Hans-Dieter Zucht Proteome Sciences PLC, Cobham, UK

Part I

Skin Irritation

Overview on Current Status of Alternative Methods and Testing Approaches for Skin Irritation Testing

1

Chantra Eskes and Markus Hofmann

1.1 Background

If the animal *in vivo* study has been originally used to classify for potential skin corrosion and skin irritation hazard effects (such as the OECD Test Guideline 404 [1] originally adopted in 1981), the area of skin corrosion and irritation represents one of the pioneering areas in which a number of alternative methods have been validated and internationally adopted since 2000 (and 2004) for skin corrosion and since 2009 (and 2010) for skin irritation by the EU (and by the OECD respectively).

In order to replace or minimize to the extent possible the use of *in vivo* animal testing, current internationally agreed approaches (UN, OECD and EU) recommend the use of integrated approaches and strategies for the assessment of skin irritation and corrosion effects, such as the Integrated Approach for Testing and Assessment (IATA) endorsed by OECD member countries [2]. These approaches recommend considering all existing information sources, and conducting a weigh-of-evidence evaluation before performing prospective testing first on alternative test methods, and only as a last resort on animals. Depending upon regulatory requirements, some geographical regions already allow the use of alternative methods for skin irritation and corrosion testing as full replacement of the animal testing, as it is the case in the European Union (EU).

In the EU, a number of legislations indeed call for the use of alternative methods to animal toxicological testing. The EU Cosmetics Regulation [3] prohibits animal testing of finished products since 2004 and of cosmetic ingredients since 2009, reinforced by a marketing ban of cosmetics finished products tested on animals since 2004 and for cosmetics containing ingredients tested on animals since 2013 [3]. Furthermore,

C. Eskes (✉)
SeCAM Services & Consultation on Alternative Methods, Magliaso, Switzerland
e-mail: chantra.eskes@secam-ce.eu

M. Hofmann
Swiss Federal Office of Public Health, Bern, Switzerland

C. Eskes et al. (eds.), *Alternatives for Dermal Toxicity Testing*,
DOI 10.1007/978-3-319-50353-0_1

the EU regulation on the Registration, Evaluation, Authorisation and Restriction of Chemicals (REACH; [4, 5]), requires that *in vitro* testing is conducted by OECD member countries for skin corrosion and irritation unless the test chemical falls outside of the applicability domain of the available *in vitro* methods or the results obtained from such methods do not allow a conclusive decision on (non-)classification and risk assessment. The EU regulation on Classification, Labelling and Packaging of substances and mixtures (EU CLP; [6, 7]), which implemented the Globally Harmonized System for classification and labelling of substances and mixtures in the European Union, encourages the use of tiered weight-of-evidence evaluations, and makes use of information from *in vitro* testing in its tiered classification approach for skin corrosion and irritation. Finally, the EU Directive on the protection of animals used for scientific purposes [8] states that (article 13(1)) "*Member States shall ensure that a procedure is not carried out if another method or testing strategy for obtaining the result sought, not entailing the use of a live animal, is recognised under the legislation of the Union*".

1.2 Classification for Skin Irritation Hazard

The UN has published in 2003 the Globally Harmonized System (GHS) for classification and labelling to favour harmonized classification of hazards across the world, which is now in its 6th revision [9]. This classification system was still then based on the traditional *in vivo* animal test adopted within the OECD Test Guideline 404 [1] originally developed by Draize and co-workers [10]. Since validation studies on alternative methods for skin irritation testing have used the animal test as the reference test method, a description of this classification system is given here.

Skin irritation is defined *in vivo* as "*the production of reversible damage of the skin following the application of a test substance for up to 4 hours*" [2, 7, 9]. One main irritant category is defined by the UN GHS classification system, i.e., Category 2, as described in Table 1.1. However, an additional optional category for mild irritants (i.e., Category 3) is also defined for those authorities wanting to have more than one skin irritant category.

In the European Union, the UN GHS classification and labelling system has been implemented by means of the EU CLP regulation (1272/2008; [6, 7]). It replaced from December 2010 the EU Dangerous Substances Directive establishing the former EU classification system for substances (EU DSD; [11]), and from 2015 the EU Dangerous Preparation Directive establishing classification criteria for mixtures (EU DPD; [12]). The EU CLP is equivalent to the UN GHS as shown in Table 1.1, but makes use of a single category (Category 2) only, whereas the mild irritant category 3 is not required. Substances falling in the UN GHS category 3, require No Category classification under the EU CLP.

Figure 1.1 provides with a comparison of the criteria applied for skin irritation classification according to the UN GHS, EU CLP and EU DSD classification systems for skin irritation [6, 7, 9, 11]. In addition to the cut-offs shown in Fig. 1.1, the three classification systems also consider a substance irritant if effects persist at the

Table 1.1 UN GHS skin irritation category(ies)

Categories	Criteria[a]
Irritant Category 2	(1) Mean value of ≥ 2.3 and ≤ 4.0 for erythema/eschar or for oedema in at least 2 of 3 tested animals from gradings at 24, 48 and 72 h after patch removal or, if reactions are delayed, from grades on three consecutive days after the onset of skin reactions; or (2) Inflammation that persists to the end of the observation period normally 14 days in at least two animals, particularly taking into account alopecia (limited area), hyperkeratosis, hyperplasia, and scaling; or (3) In some cases where there is pronounced variability of response among animals, with very definite positive effects related to chemical exposure in a single animal but less than the criteria above
Optional mild irritant Category 3	Mean value of ≥1.5 and < 2.3 for erythema/eschar or for oedema from gradings in at least 2 of 3 tested animals from grades at 24, 48 and 72 h or, if reactions are delayed, from grades on three consecutive days after the onset of skin reactions (when not included in the irritant category above)

[a]Grading criteria are understood as described in the OECD Test Guideline 404 [1]

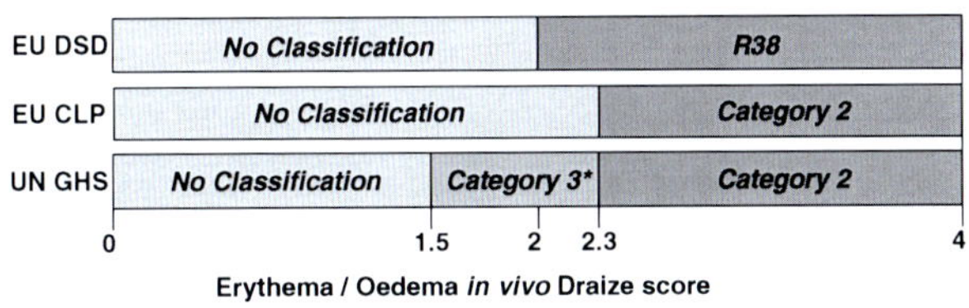

Fig. 1.1 Erythema/oedema Draize score ranges defining EU DSD, EU CLP and UN GHS classification of skin irritation. Scores refer to the mean value from gradings at 24, 48 and 72 h observed in at least two out of three animals (or as required in case of more than three animals). *Category 3 is an optional category available for those authorities wanting to have more than one skin irritant category

end of the observation period (day14) in two or more test animals, and other effects such as hyperplasia, scaling, discoloration, fissures, scabs and alopecia.

1.3 Integrated Approaches for Testing and Assessment (IATA)

Current internationally agreed approaches (OECD, EU and UN) recommend the use of integrated approaches and strategies for the assessment of skin irritation and corrosion effects. In particular, the OECD published in 2014 the first Guidance Document (GD No. 203) on an IATA adopted at an international level by OECD member countries for skin corrosion and irritation [2]. The IATA aims at hazard identification of the skin corrosion or irritation potential of chemicals (or the absence thereof) and to provide adequate information for classification and labelling according to the UN GHS classification system.

The IATA is divided in three major parts including as Part 1 the use of existing information, physico-chemical properties and non-testing methods, as Part 2 a weigh-of-evidence evaluation, and as Part 3 the conduct of prospective testing. The possible individual information sources integrating the IATA have been grouped into eight Modules according to the type of information provided, which can be used in one or more Parts of the IATA as described in Table 1.2. The strengths and limitations as well as the potential role and contribution of each Module and their individual components in the IATA for skin irritation and corrosion are described within the OECD GD 203 [2] with the purpose of minimizing the use of animals to the extent possible, whilst ensuring human safety. Furthermore, a schematic outline of the IATA for skin corrosion and irritation classification and labelling is presented in Fig. 1.2.

Table 1.2 Parts and modules of the IATA for skin corrosion and irritation (extract from [2])

Part[a]	Module	Data
Part 1 (existing information, physico-chemical properties and non-testing methods)	1	Existing human data – Non-standardised human data on local skin effects – Human Patch Test (HPT)
	2	*In vivo* skin irritation and corrosion data (OECD TG 404)
	3	*In vitro* skin corrosion data – OECD TG 430 – OECD TG 431 – OECD TG 435
	4	*In vivo* skin irritation data (OECD TG 439)
	5	Other *in vivo* and *in vitro* data – *In vitro* skin corrosion or irritation data from test methods not adopted by the OECD – Other *in vivo* and *in vitro* dermal toxicity data
	6	Physico-chemical properties (existing, measured or estimated) such as pH, acid/alkaline reserve
	7	Non-testing methods for substances: (Q)SAR, read-across, grouping and prediction systems; for mixtures: bridging principles and theory of additivity
Part 2 (WoE analysis)	8	Phases and elements of Weight of evidence (WoE) approaches
Part 3 (additional testing)	(5b)	Other *in vivo* and/or *in vitro* dermal toxicity testing (if required by other regulations)
	(3)	*In vitro* skin corrosion testing
	(4)	*In vitro* skin irritation testing
	(5a)	*In vitro* skin irritation testing in test method not adopted by the OECD
	(2)	*In vivo* skin irritation and corrosion testing

[a]While the three Parts are considered as a sequence, the order of Modules 1–7 of Part 1 might be arranged as appropriate

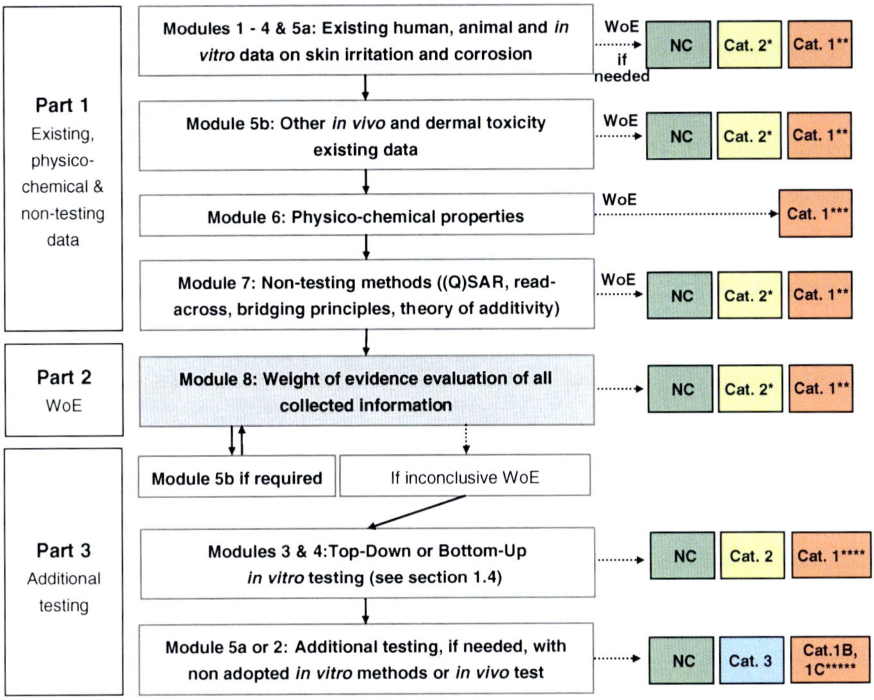

Fig. 1.2 Schematic overview of the IATA for skin irritation and corrosion based on the recommendations from the OECD GD 203 [2]. *Cat. 1* corrosive to skin, *Cat. 2* irritating to skin, *NC* no category. *Including optional Cat. 3, as applicable. **Including corrosive sub-categories 1A, 1B and 1C, as applicable. ***If corrosive sub-categorisation is required an appropriate *in vitro* skin corrosion test needs to be conducted. **** Possibilities to sub-categorise depends on the specific test method used: *OECD TG 435* allows for the discrimination between Sub-cat. 1A, Sub-cat. 1B and Sub-cat. 1C but with a limited applicability domain; *OECD TG 431* allows for the discrimination between Sub-cat. 1A and the combined Sub-cat. 1B-and-1C but does not permit the discrimination between sub-categories 1B and 1C; *OECD TG 430* only allows the identification of corrosives into a single category without sub-categorisation, i.e., Cat. 1. *****If outside of the applicability domain of OECD TG 435

While the three Parts are considered as a sequence, Modules 1–7 of Part 1 might be arranged as appropriate. Ideally, the IATA should be universally applicable to ensure human safety, whilst making maximum use of existing data, being resource efficient and minimising or eliminating the requirement for animal experiments.

Under *Part 1* of the IATA (*existing, physico-chemical & non-testing data*), existing and available information is retrieved from literature and databases and other reliable sources for Modules 1–5, while under Module 6 on physico-chemical properties, primarily the pH and the acidic/alkaline reserve are considered, and under Module 7 non-testing methods are considered. Whilst the retrieval of existing information for Modules 1–5a directly relate to skin corrosion and irritation, Module 5b requires a different search for other *in vitro* and *in vivo* dermal toxicity studies.

The collected information from Part 1 is then evaluated in a weight of evidence (WoE) approach in *Part 2 (WoE)*. While a WoE approach implies the weighing of each available piece of information on a case by case basis, the modules included in the IATA differ a priori with respect to their intrinsic weight e.g. based on considerations of relevance relating to the species of interest or biological and mechanistic aspects. Typically, the relative a priori weights of the modules can be expected to be as follows, based on regulatory acceptance of data when it is of equal quality (note that the following relative a priori weights are indicative only and depend on the quality of the individual data in each specific case):

- Reliable existing human data (in particular HPT data – Module 1b) would be expected to carry the highest weight;
- Followed by, with equal weights, *in vivo* rabbit skin corrosion/irritation data (Module 2) and *in vitro* skin corrosion or irritation data (Modules 3 and 4);
- Non-testing methods (Module 7), non-standard *in vivo* or *in vitro* and other dermal toxicity data (Module 5) and physico-chemical information (Module 6) would typically carry less intrinsic weight.

If the WoE is conclusive, decision for C&L can be conducted accordingly. However, if the WoE evaluation is inconclusive regarding the skin irritation and corrosion potential, other *in vivo* or *in vitro* dermal toxicity tests (Module 5b) for which data are still not available but may be needed to satisfy other regulatory requirements, shall be conducted. Once available, these additional test results should be incorporated into a new WoE analysis. If the WoE is still inconclusive or no other *in vivo* or *in vitro* dermal toxicity tests need to be conducted, all available information from the WoE should then be considered to formulate a hypothesis of the most likely skin corrosion or skin irritation potential of the chemical.

This hypothesis will then guide the sequence of *in vitro* prospective testing of *Part 3 (additional testing)* in either a top-down or a bottom-up approach. The top-down approach is to be used when available information suggests that the substance has a high likelihood of being irritant or corrosive to the skin, starting with an *in vitro* method for identification of skin corrosion (Module 3) followed eventually by an *in vitro* method for identification of skin irritation (Module 4). On the other hand, the bottom-up approach is to be used only when available information suggests that the substance has a high likelihood to not be irritant to the skin, starting with an *in vitro* method for identification of skin irritation (Module 4), followed eventually by an *in vitro* method for identification of skin corrosion (Module 3).

If additional testing is still required to satisfy specific requirements, the Guidance Document suggests that other *in vitro* skin irritation or corrosion test methods not yet adopted by the OECD are used that may resolve specific optional- or sub- categorisation issues (e.g., Cat. 3 for mild irritancy or resolving between sub-categories 1B and 1C in case the test chemical is outside of the applicability domain of OECD TG 435). Animal testing should be used only as a last resort when (1) discrimination between optional sub-categories 1B and 1C for chemicals outside of the applicability domain of OECD TG 435 is required, (2) discrimination of optional Cat. 3 from No Cat. is required, or (3) the test chemical cannot be tested with the *in vitro* test methods currently adopted by the OECD due to limitations or non-applicability.

The IATA is considered applicable to both substances and mixtures, although it is acknowledged that there is a different amount of information available on the applicability of the modules of this IATA to mixtures and that such applicability may depend on the information available in each specific case to be assessed. Indeed, with the exception of OECD TG 435, for which a number of tested mixtures ($n = 152$) were part of the validation dataset [13], only limited information is available in the public domain on the testing of mixtures with test methods falling under OECD TGs 430, 431 and 439 [2]. Despite the limited information available on mixtures, the test methods falling within these three TGs (430, 431 and 439) are currently considered to be applicable to the testing of mixtures as an extension of their applicability to substances. However, if new information becomes available, this should be taken into account, in combination with the existing evidence, to evaluate the usefulness of a test method to assess mixtures. In cases where evidence can be demonstrated on the non-applicability of the Test Guideline to a specific category of mixtures, the Test Guideline should not be used for that specific category of mixtures. Similar care should be taken in case specific chemical classes or physico-chemical properties are found not to be applicable to the current Test Guidelines (e.g., gases, aerosols, specific pH ranges, etc.).

1.4 *In Vitro* Prospective Testing

Although no single *in vitro* test method can cover across the full range of skin corrosion and irritation responses from the traditional Draize *in vivo* regulatory test [1, 14], currently validated and adopted *in vitro* methods can replace the Draize *in vivo* test when combined within a tiered testing strategy or depending on the outcome of the testing [2]. In cases where the weight-of-evidence assessment indicates a need for prospective testing, all available existing information should be used to formulate a hypothesis of the most likely skin irritation/corrosion potential of the chemical. This hypothesis and the regulatory context under which a decision must be taken will then guide the choice of test methods to be used and the sequence of the prospective *in vitro* testing in either a top-down or a bottom-up approach. Figure 1.3 provides a schematic overview of the construction of the top-down and bottom up *in vitro* testing strategies as recommended by the OECD GD 203 and within the EU [2, 15].

When all available collected information and the WoE assessment result in a high a-priori probability of the test chemical to be an irritant or a corrosive, the *top-down approach* should be used, starting with an *in vitro* method for the identification of skin corrosion hazard followed, in case the test chemical is identified as not being corrosive, by an *in vitro* method for the identification of skin irritation hazard.

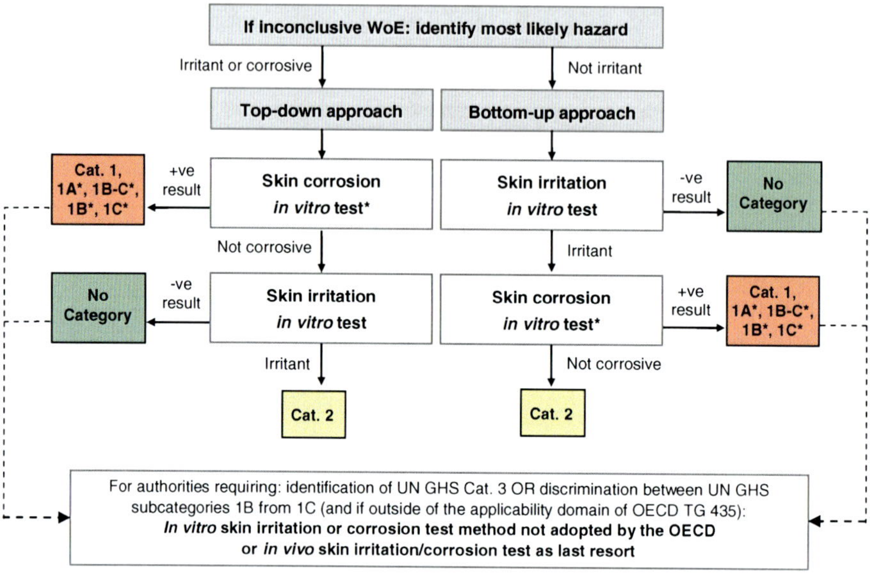

Fig. 1.3 Schematic overview of the top-down and bottom-up *in vitro* testing strategies [2, 15]. Cat. 1: Corrosive to skin; Cat. 2: Irritating to skin.* Corrosive sub-categories applicable as follows: *OECD TG 435* allows for the discrimination between Sub-cat. 1A, Sub-cat. 1B and Sub-cat. 1C but with a limited applicability domain; *OECD TG 431* allows for the discrimination between Sub-cat. 1A from Sub-cat. 1B-and-1C but does not permit the discrimination between sub-categories 1B and 1C. *OECD TG 430* only allows the identification of corrosives into a single category without sub-categorisation, i.e., Cat. 1

Conversely, when all available collected information and the WoE assessment result in a high a-priori probability of the test chemical not being an irritant to skin, *the bottom-up approach* should be used, starting with an *in vitro* method for identification of skin irritation followed, in case the test chemical is identified as being irritant, by an *in vitro* method for identification of skin corrosion. An example on the use of such approaches has been described using the SkinEthic™ RHE model, in which high accuracy values were reported using either a bottom-up or a top-down approach [16].

An overview of the validated and regulatory adopted *in vitro* test methods to be used within the bottom-up and top-down approaches for both skin corrosion and skin irritation hazard classification is given in Table 1.3. These methods have been validated according to internationally agreed principles [17], and adopted by the OECD since 2004 for skin corrosion and since 2010 for skin irritation. As a consequence they fall under the OECD international Mutual Acceptance of Data (MAD), in which test data generated in any OECD member country in accordance with these OECD Test Guidelines and following the Principles of Good Laboratory Practice (GLP) should be accepted in other OECD member countries for

Table 1.3 Overview of the validated and adopted *in vitro* methods available for skin corrosion and skin irritation regulatory testing, their purposes, application and limitations

Purpose	Test method	Application and limitations
Identification of skin corrosives Positive results lead to skin corrosion classification Negative results lead to no classification as corrosive	*OECD TG 431/EU B.40bis* Reconstructed human epidermis (RHE) test method - EPISKIN™ Standard Model (SM) - EpiDerm™ Skin Corrosion Test (SCT) - SkinEthic™ RHE - epiCS® (previously named EST-1000)	Applicable to substances and mixtures[a]. Allows identification of corrosives (GHS cat. 1), and discrimination between subcategory 1A from subcategories 1B-and-1C. The test guideline does not allow discrimination between skin corrosive subcategory 1B and subcategory 1C. An EPISKIN™ prediction model exists for distinguishing GHS subcategory 1B from 1C but its validity could not be evaluated due to the limited set of well-known *in vivo* corrosive subcategory 1C chemicals. The test guideline is not designed to provide information on skin irritation, and is not applicable to gases and aerosols Results obtained with test chemicals presenting non-specific interactions with MTT ≥ 50% should be taken with caution when OD is used as measurement for cell viability. This may be circumvented for coloured interference in case HPLC/UPLC is used as alternative measurement
	OECD TG 430/EU B.40 Transcutaneous Electrical resistance (TER) test method	Applicable to substances and mixtures[a]. Allows identification of corrosives (GHS cat. 1). The test guideline does not allow to distinguish the three GHS subcategories (1A, 1B and 1C). It is not designed to provide information on skin irritation, and is not applicable to gases and aerosols. Finally, the TER test method may be considered as an animal test in some countries.
	OECD TG 435 Membrane barrier test – Corrositex®	Applicable to substances and mixtures[a]. Allows identification of corrosives (GHS cat. 1) and sub-categorisation into the three GHS subcategories (1A, 1B and 1C). In EU, the method was not adopted in legislation as considered valid for the limited applicability domain of acids, bases and their derivatives. The test guideline is not designed to provide information on skin irritation, and is not applicable to gases and aerosols. Test chemicals not causing detectable changes in the chemical detection system cannot be tested.

(continued)

Table 1.3 (continued)

Purpose	Test method	Application and limitations
Identification of skin irritants Negative results lead to no classification[b] Positive results lead to skin irritation Cat. 2 classification if negative result with the skin corrosion test	*OECD TG 439 / EU B.46* Reconstructed human epidermis (RHE) test method EPISKIN™ Skin Irritation Test (SIT) EpiDerm™ SIT SkinEthic™ SIT[42bis] LabCyte EPI-MODEL24 SIT	Applicable to substances and mixtures[a]. Allows identification of skin irritants according to GHS Cat. 2, in case the test chemical is found to be non-corrosive. Furthermore, for countries not adopting the optional GHS Cat. 3 such as in the EU, the method also allows identification of non-classified substances. The test guideline is not designed to provide information on skin corrosion nor on mild irritants (optional GHS cat. 3), and is not applicable to gases and aerosols. Results obtained with test chemicals presenting non-specific interactions with MTT $\geq$ 50% should be taken with caution when OD is used as measurement for cell viability. This may be circumvented for coloured interference in case HPLC/UPLC is used as alternative measurement.

[a]Before use of the test method on a mixture for generating data for intended regulatory purposes, it should be considered whether, and if so why, it may provide adequate results for that purpose. Such considerations are not needed, when there is a regulatory requirement for testing of the mixture

[b]Classification according to EU CLP

assessment purposes and other uses relating to the protection of human health and the environment.

When limitations and domain of the validated and adopted *in vitro* tests are adequately considered, these tests can provide sufficient information for the decision on potential of the substance to cause skin irritation and/or corrosion. In case of *in vitro* skin corrosion testing, the most appropriate OECD TG for the test chemical and the specific purpose should be chosen. In particular, the applicability domain and the ability of the test methods to provide information on sub-categorisation may play an important role in the choice of test method to be used.

In the EU, only *in vitro* testing should be conducted for substances manufactured or imported in quantities between 1 tonne and 10 tonnes per year. In contrast, for substances manufactured or imported in quantities of ≥10 tonnes per year, "a*n in vivo study for skin corrosion/irritation shall be considered only if the in vitro studies (…) are not applicable, or the results of these studies are not adequate for classification and risk assessment*" [5]. As a consequence, no *in vivo* testing should be conducted in cases where the substance falls under the scope of the adopted *in vitro* test methods performed and there are no substance-specific limitations to using those tests [15]. Furthermore, the *in vivo* testing may be waived if an adaptation is formulated according to Annex XI to the REACH Regulation [15].

1.5 *In Vitro* Alternative Methods for Skin Irritation

1.5.1 Scientific Validation

The *in vitro* assays proposed for skin irritation have initially undergone pre-validation and optimization studies [18–21] that led to a formal validation study [22, 23] and the endorsement of the scientific validity of the EPISKIN™ Skin Irritation Test (SIT) [24]. This statement was based on the former EU DSD classification system, which was then still in place in the EU [11]. Following this validation study, two Reconstructed human Epidermis models were considered to be 'similar' tests to the validated test method EPISKIN™ SIT, i.e. the EpiDerm™ Skin Irritation Test modified protocol and the SkinEthic™ RHE test method, as they met the requirements of the performance standards as defined by EURL-ECVAM for *in vitro* skin irritation testing and had sufficient accuracy and reliability for predicting of skin irritating and non-skin irritating test chemicals as compared to the validated EPISKIN™ assay [25–29]. Following a review by ESAC, both test methods were also endorsed to be scientific valid based on the former EU DSD classification system [30].

With the implementation in 2008 of the UN GHS Classification system in the EU by means of the EU CLP Regulation [6], the performances of all three test methods (EPISKIN™ SIT, EpiDerm™ SIT modified protocol and SkinEthic™ RHE) were re-evaluated to take into account the change in the cut-off value for the classification of skin irritants (shifted from a cut-off of 2 for the EU DSD [11] to a cut-off of 2.3 for the EU CLP/UN GHS Cat.2 [6, 7], see Fig. 1.1). Results from the three test methods were considered to be satisfactory so that the statements relating to their scientific validity continued to be accurate and were extended to the EU CLP/UN GHS classification system [31].

The OECD TG 439 on "*In vitro* Skin Irritation: Reconstructed Human Epidermis Model" was then adopted in 2010 [32] and revised in 2013 to include a fourth test method, the Labcyte EPI-MODEL 24SIT, considered to be scientific valid also for having met the established performance standard criteria [33–37].

As a consequence, four commercially available RhE test methods currently comply with the OECD TG 439 for *in vitro* skin irritation regulatory testing [32]. These are:

- EPISKIN™ SIT [38], validated following an ECVAM prospective validation study [24, 31],
- EpiDerm™ EPI-200-SIT [39], validated for having met the established performance standards [30, 31],
- SkinEthic™ RHE SIT[42bis] [40], validated for having met the established performance standards [30, 31], and
- Labcyte EPI-MODEL 24SIT [41], validated for having met the established performance standards [37].

In case additional similar or modified test methods are developed, before they can be used for regulatory testing they should be evaluated to determine their similarity, reliability and predictive capacity using the Performance Standards defined in the OECD Guidance Document No. 220 [42].

1.5.2 Principles

The three-dimensional RhE models are comprised of non-transformed human-derived epidermal keratinocytes cells which have been cultured in an air-liquid interface to form a multilayered, highly differentiated model of the human epidermis. They consist of organised basal, spinous and granular layers, and a multilayered *stratum corneum* containing intercellular lamellar lipid layers representing main lipid classes analogous to those found *in vivo*. The *in vitro* RhE models represent therefore the target organ of the species of interest.

Chemical-induced skin irritation, manifested by erythema and oedema, is the result of a cascade of events beginning with penetration of the chemicals through the *stratum corneum* where they may damage the underlying layers of keratinocytes and other skin cells. The damaged cells may either release inflammatory mediators or induce an inflammatory cascade which also acts on the cells in the dermis, particularly the stromal and endothelial cells of the blood vessels. It is the dilation and increased permeability of the endothelial cells that produce the observed erythema and oedema [43].

The RhE-based test methods (in the absence of any vascularisation in the *in vitro* test system) measure the initiating events in the cascade of skin irritation, e.g. cell/tissue damage, using cell viability as readout. Test chemicals are applied topically to the RhE models and cell viability is measured by enzymatic conversion of the vital dye MTT into a blue formazan salt that is quantitatively measured after extraction from tissues. Irritant chemicals are identified by their ability to decrease cell viability below defined threshold levels (i.e. ≤50%, for UN GHS category 2 irritants).

If the main endpoint considered in the regulatory adopted RhE models is the cell viability assessed by the reduction of MTT, the release of IL-1α in the EPISKIN™ SIT was considered as a useful adjunct to the MTT assay as it has the potential to increase the sensitivity of the test without reducing its specificity. This endpoint may be used to confirm negative results obtained with the MTT endpoint [24].

1.5.3 Applicability and Limitations

The reconstructed human epidermis tests falling under the OECD TG 439 can be used for the hazard identification of UN GHS Cat. 2 irritant chemicals (substances and mixtures), when test results are supported by a non-corrosive outcome (see Sect. 4). In member countries or regions that do not adopt the optional UN GHS Category 3 (mild irritants), such as in the EU, test chemicals that produce cell viabilities above the defined threshold level (i.e. >50%), are identified as not requiring classification. Therefore, depending on the regulatory framework and the classification system in use, the OECD TG 439 may be used to determine the skin irritancy of chemicals either as a stand-alone replacement test for *in vivo* skin irritation testing or as a partial replacement test within a testing strategy [2].

A limitation of the OECD TG 439 is that it does not allow the classification of chemicals to the optional UN GHS Category 3 (mild irritants). Furthermore, the OECD TG 439 does not provide adequate information on skin corrosion. For this purpose other *in vitro* methods such as OECD TG 430, 431 or 435 may be used that specifically address the identification of skin corrosion hazard. For a full evaluation of local skin effects after a single dermal exposure, the use of the IATA for skin corrosion and irritation should be considered [2], in which *in vitro* testing for skin corrosion and skin irritation should be conducted before considering testing in living animals (see Sect. 3).

The OECD TG 439 is applicable to mixtures and substances as well as to liquids (aqueous or non-aqueous), semi-solids, solids (soluble or insoluble in water) and waxes. However, before using the test methods falling within the OECD TG 439 on a mixture for generating data for intended regulatory purposes, it should be considered whether, and if so why, it may provide adequate results for that purpose. Such considerations are not needed, when there is a regulatory requirement for testing the mixture. Due to the fact that mixtures cover a wide spectrum of categories and composition, and that only limited information is currently available on the testing of mixtures, in cases where evidence can be demonstrated on the non-applicability of the OECD TG 439 to a specific category of mixtures (e.g. following a strategy as proposed by [44]), the TG should not be used for that specific category of mixtures. Similar care should be taken in case specific chemical classes or physico-chemical properties are found not to be applicable to the current Test Guideline.

Finally, the OECD TG 439 does not allow testing of gases and aerosols. Furthermore, test chemicals absorbing light in the same range as MTT formazan and test chemicals that are able to directly reduce the vital dye MTT (to MTT formazan), may interfere with the tissue viability measurements and require the use of adapted controls for corrections. The type of adapted controls required will vary depending on the type of interference produced by the test chemical and the procedure used to measure MTT formazan (see [32]). Results for test chemicals producing non specific interactions with MTT $\geq 50\%$ of the negative control should be taken with caution when OD is used as means of measurement. However, the use of HPLC/UPLC spectrophotometry as an alternative means of measuring the MTT formazan offers the possibility of evaluating the skin irritation potential of strongly coloured test chemicals that could interfere with the standard OD measurements as well as with visual observations in the *in vivo* animal testing [45, 46].

1.6 Comparison to the *In Vivo* Test Method

Morphologically, the adopted *in vitro* Reconstructed human Epidermis models, which make use of 3D tissues are closer to the human epidermis as compared to the rabbit skin. Although these models do not present all functional complexity that exist *in vivo* (including the dermis and its components such as hair follicules, subaceous glands, nerve and immune cells, which could play a role in the mechanisms of skin irritation), the *in vitro* reconstructed human epidermis were found to have similar profiles of phase I and II enzymatic activities as compared to the human skin such as the low expression and function levels of phase I enzymes, and measurable activity of some phase II enzymes [47]. Furthermore, *in vitro* reconstructed human epidermis models using multiple endpoint analyses were shown to have good correlation with the results of the human patch test [48]. In particular, the RhE test methods were found to better predict the effects on humans than the rabbit test [49]. Out of 16 chemicals classified as irritants in the rabbit, only five substances were irritating to the human skin. The concordance of the rabbit test with the 4 h Human Patch Test (HPT) was only of 56% (n=25 overall chemicals), whereas the concordance obtained between the RhE test methods and the HPT was of 76% (n=25) for the EpiDerm™ RhE model and of 70% (n=23) for the EpiSkin™ RhE model [49]. Such findings were confirmed by Basketter and co-authors who showed that the rabbit skin irritation test largely over-predicts human responses to chemicals [50]. In their study, the authors show that out of 81 substances found to have HPT data, about 50% were classified as irritating based on the rabbit skin test whereas with the 4 h HPT test less than 20% were identified as acutely irritant to human skin.

The main endpoint considered in the OECD TG 439 is cell viability, based on the principle that irritant chemicals are able to penetrate the *stratum corneum* by diffusion and are cytotoxic to the cells in the underlying layers. The *in vitro* test methods cover mainly the initial mechanisms of skin irritation occurring in the *in vivo* test (Fig. 1.4). However, the evaluation of the release of Interleukin 1α, considered to be a useful adjunct to the MTT assay to increase the sensitivity of the EpiSkin™ SIT assay without reducing specificity [24], could give additional insight on the release of inflammatory mediators that may act in the subsequent mechanistic cascade of events occurring during skin irritation reactions.

A summary of the major components of the regulatory *in vivo* and *in vitro* reconstructed human epidermis methods for skin irritation is shown in Table 1.4. The

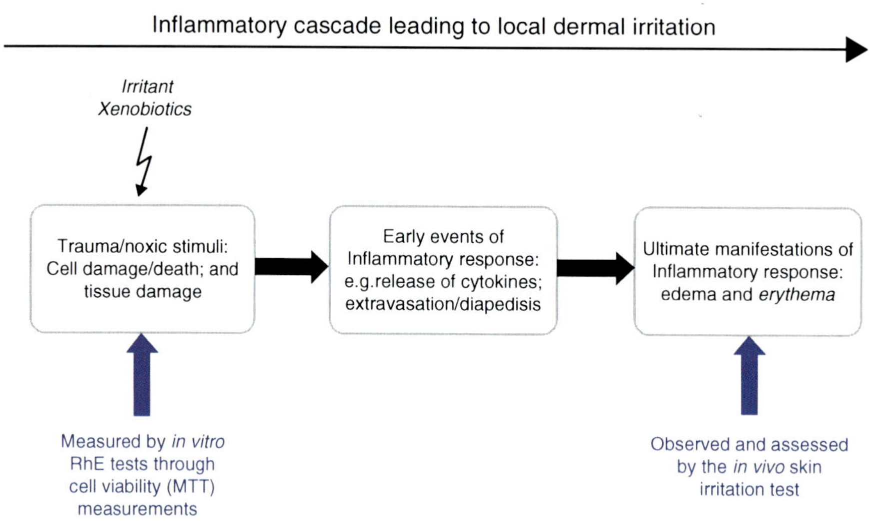

Fig. 1.4 Extract from the OECD Guidance Document 137 [28]. Schematic representation of the inflammatory cascade leading to local acute dermal irritation

Table 1.4 Comparison of the principal method components of the regulatory accepted *in vivo* and *in vitro* tests for skin irritation

	In vivo test for skin irritation (OECD TG 404)	*In vitro* reconstructed human epidermis (RhE) test methods (OECD TG 439)
Model used	Albino rabbit.	Three-dimensional reconstructed human epidermis, consisting of organized basal, spinous and granular layers, and a multilayered *stratum corneum*. Surface of tissue models: 0.3 cm^2 for Labcyte EPI-MODEL 24SIT; 0.38 cm^2 for Episkin™-SIT; 0.63 cm^2 for EpiDerm™ 200-SIT and 0.5 cm^2 for SkinEthic™ SIT.
Number of replicates	2 to 3 animals based on severity of effects.	At least 3 replicates for each test chemical.
Dose and application of test chemical	0.5 ml (liquids) or 0.5 g (solids) applied to ~6 cm^2 of skin and covered with a gauze patch (~83.3 µl or mg/cm^2). Solids might be moisten to ensure good skin contact.	Liquids: 10 to 30 µl (26 to 83 µl/cm^2 depending on model). Solids: 10 to 25 mg (26 to 83 mg/cm^2 depending on model). Tissues should be moisten prior to solid application to ensure good contact with the RhE.
Controls	Potential influence of the vehicle on irritation of the skin by the test chemical should be minimal, if any.	Negative control: water or PBS. Positive control: 5% aqueous SDS.
Exposure time	4 h.	15–60 min depending on the model.
Washing	At the end of exposure time to remove test chemical.	At the end of exposure time to remove test chemical.
Post-treatment incubation time	If no corrosive effects seen, the animal is observed up to 14 days.	After washing, the exposure time is followed by a post-treatment incubation time of 42 h to allow for recovery from weak cytotoxic effects as well as for appearance of clear cytotoxic effects.
Endpoint assessed	– Grading of skin reactions (erythema, edema). – Other reactions such as: defatting of skin, clinical signs of toxicity and body weight, persistence of alopecia, hyperkeratosis, hyperplasia and scaling. – Histopathology may be carried out in case of equivocal responses	Cell viability based on the premise that irritant chemicals are able to penetrate the *stratum corneum* by diffusion and are cytotoxic to the cells in the underlying layers. Use of HPLC/UPLC spectrophotometry allows evaluating strongly coloured test chemicals. Inflammatory mediators such as Interleukin 1alpha may be a useful adjunct to the MTT assay to increase sensitivity of the assay.
Interpretation of results	Method of reference for the hazard identification of: – UN GHS Cat. 1 (skin corrosion) including Sub-categ ories 1A, 1B and 1C, – UN GHS Cat. 2 (skin irritation), – UN GHS No Category.	Can be used for hazard identification of: – UN GHS Cat. 2 if supported by corrosive negative results, and – No-Category in countries not adopting the optional GHS Cat. 3 (mild irritant.s)

(continued)

Table 1.4 (continued)

	In vivo test for skin irritation (OECD TG 404)	*In vitro* reconstructed human epidermis (RhE) test methods (OECD TG 439)
Limitations	– Overpredicts human responses. – Coloured chemicals may interfere with observations. – May be variable between laboratories. – Does not assess repetitive low-dose exposure. – Has the potential to cause discomfort or pain to laboratory animals.	– Not designed to distinguish the optional GHS Cat. 3 (mild irritants), corrosive chemicals, gases and aerosols. – Results obtained with test chemicals presenting non-specific interactions with MTT ≥ 50% should be taken with caution when OD is used as measurement for cell viability. This may be circumvented for coloured interfering test chemicals with the use of HPLC/UPLC as an alternative measurement.

doses applied *in vitro* (26 to 47.6 ml or mg/cm^2) are generally smaller with respect to those applied *in vivo* (~83.3 ml or mg/cm^2), with the exception of the LabCyte EPI-MODEL24 SIT, which makes use of similar doses (83.3 ml or mg/cm^2). Furthermore, the exposure times used by the adopted *in vitro* RhE assays are in general shorter as compared to those used *in vivo* (15–60 min *in vitro* versus 4 h *in vivo*). Similarly the post-treatment time is shorter *in vitro* with respect to the *in vivo* test (42 h versus 14 days). Finally, the ability of the RhE test methods to detect skin irritants classified *in vivo* on the basis of persistence only could not be assessed during the validation study due to the poor availability of such test chemicals in around 5000 screened chemicals from the industrial commerce [22]. However, the need to identify such scarce occurring test chemicals classified based on persistence only may be questionable. Furthermore, these differences might be compensated by the more simple structure of the skin components involved in skin irritation reactions present in the *in vitro* models with respect to the *in vivo* situation.

Unlike the *in vivo* test, the *in vitro* assays make systematically use of positive and negative controls to check for the functionality of the test method. In addition, the OECD TG 439 recommend ensuring the technical proficiency of the assays, by the laboratory, prior to the routine use of the *in vitro* assays by testing a list of recommended proficiency chemicals.

In the EU, where the optional UN GHS Category 3 (mild irritants) is not implemented, the adopted *in vitro* assays for skin irritation can be used as a stand-alone assay for identifying test chemicals not requiring classification for skin irritation in case of a negative result. In the case of a positive result, they can be used for the hazard identification of UN GHS Cat. 2 irritant chemicals (substances and mixtures), when test results are supported by a separate non-corrosive outcome (based on e.g., OECD TG 430, 431 or 435). OECD TG 439 does however not allow classifying test chemicals in the optional GHS Cat. 3 as mild irritants, nor does it provide adequate information on skin corrosion.

1.7 Other *In Vitro* Test Methods for Skin Irritation Testing

A number of similar RhE models to the ones already adopted have been developed. Among those three models have undergone a catch-up multi-laboratory validation study based on the Performance Standards (PS) as defined in the OECD Guidance Document No. 220 [42]. These are:

- The commercially available epiCS® SIT model which underwent a positive independent peer-review (see Chap. 6),
- An open-source RhE model (OS-Rep), having an openly accessible protocol for tissue production, that underwent a PS-based validation study in which each participating laboratory made use of their in-house generated OS-Rep to assess the set of PS reference chemicals [51, 52], and
- The commercially available Skin +® RhE test system produced by Sterlab and commercialized by ATERA, which also underwent a PS-based validation study.

Other similar models developed include the Leiden human epidermal (LHE) model that showed similar skin irritation results with the 20 reference chemicals to those reported for the validated skin models [53], a model based on human skin obtained from surgery [54], and a viable human full thickness skin model [55].

In addition, new assays based on the measurement of parameters other than cell viability are being developed that show promise to distinguish not only irritants from non irritants but also to determine the skin irritancy potential of chemicals. For example, the IRR-IS assay exploits the quantitative analysis of expression profiles of relevant genes and is proposed to contribute to the discrimination of non-irritants (No Cat.), mild-irritants (Cat. 3) and irritants (Cat. 2) as shown in a study evaluating gene expression changes in the validated EpiSkin™ test system in response to chemical exposure [56]. Furthermore, use of biomarkers such as IL-1α, IL-1RA, IL-8 and MTT in a reconstructed epidermis model was shown to determine the skin irritant potency of chemicals in addition to distinguishing irritants from non irritants [57]. Other endpoints investigated include the use of proteomics [58] and toxicogenomics [59, 60].

Finally, attempts have also been made to develop an innervated *in vitro* model of human skin including sensory neurons derived from embryonic rat dorsal root ganglion as neural components [61]. The aim was to integrate the sensory neuronal components which are usually present in the skin and may play a role *in vivo* in the production of neurogenic inflammation leading to sensory irritation and pain [62].

References

1. OECD. OECD guidelines for the testing of chemicals no. 404. Acute dermal irritation/corrosion. Paris, France: organisation for economic cooperation and development. Originally adopted in 1981. 2015. Available at: http://www.oecd-ilibrary.org/environment/oecd-guidelines-for-the-testing-of-chemicals-section-4-health-effects_20745788.

2. OECD. Guidance document no. 203: Integrated Approaches to Testing and Assessment for Skin Irritation/Corrosion. Environment, health and safety publications, series on testing and assessment. Organisation for economic cooperation and development, Paris. 2014. Available at: http://www.oecd.org/env/ehs/testing/series-testing-assessment-publications-number.htm.
3. EC. Regulation 1223/2009 of the European Parliament and of the council of 30 November 2009 on cosmetic products. Off J Eur Union. 2009;L342:59–209.
4. EC. Regulation 1907/2006 of the European Parliament and of the council of 18 December 2006 concerning the Registration, evaluation, Authorisation and Restriction of chemicals (REACH), establishing a European chemicals agency, amending Directive 1999/45/EC and repealing council regulation 793/93 and Commission regulation 1488/94 as well as council Directive 76/769/EEC and Commission directives 91/155/EEC, 93/67/EEC, 93/105/EC and 2000/21/EC. Off J Eur Union L. 2006;396:1–849.
5. EC. Regulation 2016/863 amending annexes VII and VIII to regulation 1907/2006 of the European Parliament and of the council on the Registration, evaluation, Authorisation and Restriction of chemicals (REACH) as regards skin corrosion/irritation, serious eye damage/eye irritation and acute toxicity. Off J Eur Union L. 2016;144:27–31.
6. EC. Regulation 1272/2008 of the European Parliament and of the council of 16 December 2008 on classification, labelling and packaging of substances and mixtures, amending and repealing directives 67/548/EEC and 1999/45/EC, and amending regulation 1907/2006. Off J Eur Union. 2008;L353:1–1355.
7. EC. Regulation 2016/918 amending, for the purposes of its adaptation to technical and scientific progress, regulation 1272/2008 of the European Parliament and of the council on classification, labelling and packaging of substances and mixtures. Off J Eur Union L. 2016;156:1–103.
8. EC. Directive 2010/63/EU of the European Parliament and of the council of 22 September 2010 on the protection of animals used for scientific purposes. Off J Eur Union. 2010;L276:33.
9. United Nations (UN). Globally harmonized system of classification and labelling of chemicals (GHS). Sixth revised edition. Part 3: health hazards–chapter 3.2 skin corrosion/irritation. New York, USA, and Geneva, Switzerland: United Nations; 2015. p. 125–136. Available at: http://www.unece.org/fileadmin/DAM/trans/danger/publi/ghs/ghs_rev06/English/ST-SG-AC10-30-Rev6e.pdf.
10. Draize JH, Woodard G, Calvery HO. Methods for the study of irritation and toxicity of substances applied topically to the skin and mucous membranes. J Pharmacol Exp Ther. 1944;82:377–90.
11. EC. Commission Directive 2001/59/EC of 6 august 2001 adapting to technical progress for the 28th time council Directive 67/548/EEC on the approximation of laws, regulations and administrative provisions relating to the classification, packaging and labelling of dangerous substances. Off J Eur Commun L. 2001;225:1–333.
12. EC. Directive 1999/45/EC of the European Parliament and of the council of 31 May 1999 concerning the approximation of laws, regulations and administrative provisions of the member states relating to the classification, packaging and labelling of dangerous preparations. Off J Eur Commun L. 1999;200:1–68.
13. NIH. Corrositex®: an in vitro test method for assessing dermal corrosivity potential of chemicals, NIH Publication No. 99-4495. Research Triangle Park, NC: NIEHS; 1999.
14. EC. Regulation 440/2008 laying down test methods pursuant to regulation 1907/2006 of the European Parliament and of the council on the Registration, evaluation, Authorisation and Restriction of chemicals (REACH). Off J Eur Union. 2008;L142:1–739.
15. ECHA. Guidance on information requirements and chemical safety assessment. Chapter R.7a: endpoint specific guidance. Version 4.0. Subchapter R.7.2. Skin corrosion/irritation, serious eye damage/eye irritation and respiratory tract corrosion/irritation; 2015. p168–99. Available at: http://echa.europa.eu/documents/10162/13632/information_requirements_r7a_en.pdf.
16. Alépée N, Grandidier MH, Tornier C, Cotovio J. An integrated testing strategy for in vitro skin corrosion and irritation assessment using SkinEthic™ reconstructed human epidermis. Toxicol In Vitro. 2015;29:179–1792.
17. OECD. Guidance document no. 34: validation and international acceptance of new or updated test methods for hazard assessment. environmental health and safety monograph series on

testing and assessment. 2005. Available at: http://www.oecd.org/env/ehs/testing/series-testing-assessment-publications-number.htm.

18. Cotovio J, Grandidier M-H, Portes P, Roguet R, Rubinstenn G. The in vitro acute skin irritation of chemicals: optimisation of the EPISKIN prediction model within the framework of the ECVAM validation process. ATLA. 2005;33:329–49.
19. Fentem JH, Briggs D, Chesné C, Elliott GR, Harbell JW, Heylings JR, Portes P, Roguet R, Van De Sandt JJM, Botham PA. A prevalidation study on in vitro tests for acute skin irritation: results and evaluation by the management team. Toxicol In Vitro. 2001;15:57–93.
20. Portes P, Grandidier MH, Cohen C, Roguet R. Refinement of the EPISKIN protocol for the assessment of acute skin irritation of chemicals: follow-up to the ECVAM prevalidation study. Toxicol In Vitro. 2002;16:765–70.
21. Zuang V, Balls M, Botham PA, Coquette A, Corsini E, Curren RD, Elliott GR, Fentem JH, Heylings JR, Liebsch M, Medina J, Roguet R, van de Sandt H, Wiemann C, Worth AP. Follow-up to the ECVAM prevalidation study on in vitro tests for acute skin irritation. ECVAM skin irritation task force report 2. ATLA. 2002;30:109–29.
22. Eskes C, Cole T, Hoffmann S, Worth A, Cockshott A, Gerner I, Zuang V. The ECVAM international validation study on in vitro tests for acute skin irritation: selection of test chemicals. ATLA. 2007;35:603–19.
23. Spielmann H, Hoffmann S, Liebsch M, Botham P, Fentem JH, Eskes C, Roguet R, Cotovio J, Cole T, Worth A, Heylings J, Jones P, Robles C, Kandárová H, Gamer A, Remmele M, Curren R, Raabe H, Cockshott A, Gerner I, Zuang V. The ECVAM international validation study on in vitro tests for acute skin irritation: report on the validity of the EPISKIN and EpiDerm assays and on the skin integrity function test. ATLA. 2007;35:559–601.
24. ESAC Statement on the validity of in vitro tests for skin irritation. Available at: http://ecvam.jrc.ec.europa.eu/ under "Publications", "ESAC statements". 2007.
25. Alépée N, Tornier C, Rober C, Amsellem C, Roux MH, Doucet O, Pachot J, Méloni M, de Brugerolle de Fraissinette A. A catch-u validation study on reconstructed human epidermis (SkinEthic RHE) for full replacement of the Draize skin irritation test. Toxicol In Vitro. 2010;24:257–66.
26. Kandárová H, Hayden P, Klausner M, Kubilus J, Sheasgreen J. An in vitro skin irritation test (SIT) using the EpiDerm reconstructed human epidermal (RHE) model. J Vis Exp. 2009;29 doi:10.3791/1366.
27. Kandárová H, Hayden P, Klausner M, Kubilus J, Kearney P, Sheasgreen J. In vitro skin irritation testing: improving the sensitivity of the EpiDerm skin irritation test protocol. ATLA. 2009;37:671–89.
28. OECD. Guidance document no. 137: Explanatory background document to the OECD test guideline on in vitro skin irritation testing. Environment, health, and safety publications, series on testing and assessment. organisation for economic cooperation and development, Paris. 2010. Available at: http://www.oecd.org/env/ehs/testing/series-testing-assessment-publications-number.htm.
29. Tornier C, Amsellem C, de Brugerolle de Fraissinette A, Alépée N. Assessment of the optimized SkinEthic reconstructed human epidermis (RHE) 42bis skin irritation protocol over 39 substances. Toxicol In Vitro. 2010;24:245–56.
30. ESAC. Statement on the scientific validity of in vitro tests for skin irriation testing (SkinEthic and modified EpiDerm). Available at: http://ecvam.jrc.ec.europa.eu/under "Publications", "ESAC statements". 2008
31. ESAC. Statement on the performance under UN GHS of three in vitro assays for skin irritation testing and the adaptation of the reference chemicals and defined accuracy values of the ECVAM skin irritation performance standards. Available at: http://ecvam.jrc.ec.europa.eu/ under "Publications", "ESAC statements". 2009.
32. OECD. OECD guidelines for the testing of chemicals no. 439. In vitro skin irritation: reconstructed human epidermis test method. Paris, France: Organisation for economic

cooperation and development. Originally adopted in 2010. 2015. Available at: http://www.oecd-ilibrary.org/environment/oecd-guidelines-for-the-testing-of-chemicals-section-4-health-effects_20745788.
33. Katoh M, Hamajima F, Ogasawara T, Hata K. Assessment of human epidermal model LabCyte EPI-MODEL for in vitro skin irritation testing according to European Centre for the Validation of alternative methods (ECVAM)-validated protocol. J Toxicol Sci. 2009;34:327–34.
34. Katoh M, Hata K. Refinement of LabCyte EPI-MODEL24 skin irritation test method for adaptation to the requirements of OECD test guideline 439. AATEX. 2011;16:111–22.
35. Kojima H, Ando Y, Idehara K, Katoh M, Kosaka T, Miyaoka E, Shinoda S, Suzuki T, Yamaguchi Y, Yoshimura I, Yuasa A, Watanabe Y, Omori T. Validation study of the in vitro skin irritation test with the LabCyte EPI-MODEL24. ATLA. 2012;40:33–50.
36. Kojima H, Katoh M, Shinoda S, Hagiwara S, Suzuki T, Izumi R, Yamaguchi Y, Nakamura M, Kasahawa T, Shibai A. A catchu-up validation study of an in vitro skin irritation test method using reconstructed human epidermis LabCyte EPI-MODEL24. J Appl Toxicol. 2014;34:766–74.
37. OECD. Guidance document no. 155: Peer review report of validation of the skin irritation test using LabCyte EPI-MODEL24. Environment, health and safety publications, series on testing and assessment, organisation for economic cooperation and development, Paris. 2011. Available at: http://www.oecd.org/env/ehs/testing/series-testing-assessment-publications-number.htm.
38. EpiSkin™ SIT. SOP Version 1.8. In: ECVAM skin irritation validation study: validation of the EpiSkin™ test method 15 min–42 hours for the prediction of acute skin irritation of chemicals; 2009. Available at: https://eurl-ecvam.jrc.ec.europa.eu/validation-regulatory-acceptance/docs-skin-irritation-1/DOC11-SOP%20EpiSkin_revised_MTT.pdf.
39. EpiDerm™ EPI-200-SIT (2009). SOP version MK-24-007-0023. Protocol for: in vitro EpiDerm™ skin irritation test (EPI-200-SIT), for use with MatTek Corporation's reconstructed human epidermal model EpiDerm (EPI-200). Available at: https://eurl-ecvam.jrc.ec.europa.eu/validation-regulatory-acceptance/docs-skin-irritation-1/DOC13-SOP%20modified%20EpiDerm%20MTT.pdf.
40. SkinEthic™ RHE SIT-42bis. SOP version 2.0. skinethic skin irritation test-42bis test method for the prediction of acute skin irritation of chemicals: 42 minutes application + 42 hours post-incubation; 2009. Available at: https://eurl-ecvam.jrc.ec.europa.eu/validation-regulatory-acceptance/docs-skin-irritation-1/DOC14-SOP%20SkinEthic%20MTT.pdf.
41. Labcyte EPI-MODEL 24SIT. EPI-MODEL24 SIT SOP, Version 8.3, skin irritation test using the reconstructed human model "LabCyte EPI-MODEL24"; 2011. Available at: http://www.jpte.co.jp/english/business/LabCyte/Testprotocol.pdf.
42. OECD. Document no. 220: performance standards for the assessment of proposed similar or modified in vitro reconstructed human epidermis (RHE) test methods for skin irritation testing as described in TG 439. OECD environmental health and safety publications, Series on testing and assessment; 2015c. Available at: http://www.oecd.org/env/ehs/testing/series-testing-assessment-publications-number.htm.
43. Welss T, Basketter DA, Schroder KR. In vitro skin irritation: facts and future. State of the art review of mechanisms and models. Toxicol In Vitro. 2004;18:231–43.
44. Eskes C, Detappe V, Koeter H, Kreysa J, Liebsch M, Zuang V, Amcoff P, Barroso J, Cotovio J, Guest R, Hermann M, Hoffmann S, Masson P, Alepee N, Arce LA, Bruschweiler B, Catone T, Cihak R, Clouzeau J, D'Abrosca F, Delveaux C, Derouette JP, Engelking O, Facchini D, Frohlicher M, Hofmann M, Hopf N, Molinari J, Oberli A, Ott M, Peter R, Sa-Rocha VM, Schenk D, Tomicic C, Vanparys P, Verdon B, Wallenhorst T, Winkler GC, Depallens O. Regulatory assessment of in vitro skin corrosion & irritation data within the european framework: workshop recommendations. Regul Toxicol Pharmacol. 2012;62:393–403.
45. Alépée N, Barroso J, De Smedt A, De Wever B, Hibatallah J, Klaric M, Mewes KR, Millet M, Pfannenbecker U, Tailhardat M, Templier M, McNamee P. Use of HPLC/UPLC-spectrophotometry for detection of formazan in in vitro reconstructed human tissue (RhT)-based test methods employing the MTT-reduction assay to expand their applicability to strongly coloured test chemicals. Toxicol In Vitro. 2015;29:741–61.

46. Alépée N, Hibatallah J, Klaric M, Mewes KR, Pfannenbecker U, McNamee P. Assessment of cosmetic ingredients in the in vitro reconstructed human epidermis test method EpiSkin™ using HPLC/UPLC-spectrophotometry in the MTT-reduction assay. Toxicol In Vitro. 2016;33:105–17.
47. Hewitt NJ, Edwards RJ, Fritsche E, Goebel C, Aeby P, Scheel J, Reisinger K, Ouédraogo G, Duche D, Eilstein J, Latil A, Kenny J, Moore C, Kuehnl J, Barroso J, Fautz R, Pfuhler S. Use of human in vitro skin models for accurate and ethical risk assessment: metabolic considerations. Toxicol Sci. 2013;133:209–17.
48. Welss T, Matthies W, Schroeder KR. Compatibility testing in vitro: a comparison with in vivo patch test data. Int J Cosmet Sci. 2007;29:143.
49. Jírová D, Basketter D, Liebsch M, Bendová H, Kejlová K, Marriott M, Kandárová H. Comparison of human skin irritation patch test data with in vitro skin irritation assays and animal data. Contact Dermatitis. 2010;62:109–16.
50. Basketter D, Jírová D, Kandárová H. Review of skin irritation/corrosion hazards on the basis of human data: a regulatory perspective. Interdiscip Toxicol. 2012;5:98–104.
51. Groeber F, Schober L, Schmid FF, Traube A, Kolbus-Hernandez S, Daton K, Hoffmann S, Petersohn D, Schäfer-Korting M, Walles H, Mewes KR. Catch-up validation study of an in vitro skin irritation test method based on an open source reconstructed epidermis (phase II). Toxicol In Vitro. 2016;36:254–61.
52. Mewes KR, Fischer A, Zöller NN, Laubach V, Bernd A, Jacobs A, van Rompay A, Liebsch M, Pirow R, Petersohn D. Catch-up validation study of an in vitro skin irritation test method based on an open source reconstructed epidermis (phase I). Toxicol In Vitro. 2016;36:238–53.
53. El Ghalbzouri A, Siamari R, Willemze R, Ponec M. Leiden reconstructed human epidermal model as a tool for the evaluation of the skin corrosion and irritation potential according to the ECVAM guidelines. Toxicol In Vitro. 2008;22:1311–20.
54. Miles A, Berthet A, Hopf NB, Gilliet M, Raffoul W, Vernez D, Spring P. A new alternative method for testing skin irritation using a human skin model: a pilot study. Toxicol In Vitro. 2014;28:240–7.
55. Lönnqvist S, Briheim K, Kratz G. Non-occlusive topical exposure of human skin in vitro as model for cytotoxicity testing of irritant compounds. Toxicol Mech Methods. 2016;26:82–7.
56. Groux H., Lelievre D., Cottrez F., Auriault C., Alépée N., Meunier J.R., Cotovio J.. Evaluation of IRR-IS®, an EpiSkin™ based model for quantifying chemical irritation potency. Oral Communication N° 1.5, ESTIV2012, 17th October 2012, Lisbon, Portugal. In: Poster presented at 51st Annual Meeting of the Society of Toxicology, March 11–15, 2012, San Francisco, USA. The Toxicologist; 2012p. 286. Available at: http://www.toxicology.org/AI/Pub/Tox/2012Tox.pdf.
57. Spiekstra SW, Dos Santos GG, Scheper RJ, Gibbs S. Potential method to determine irritant potency in vitro - comparison of two reconstructed epidermal culture models with different barrier competency. Toxicol In Vitro. 2009;23:349–55.
58. Zhang Q, Dai T, Zhang L, Zhang M, Xiao X, Hu H, Zou P, Liu X, Xiang Q, Su Z, Huang Y, He QY. Identification of potential biomarkers for predicting acute dermal irritation by proteomic analysis. J Appl Toxicol. 2011;31:762–72.
59. Borlon C, Godard P, Eskes C, Hartung T, Zuang V, Toussaint O. The usefulness of toxicogenomics for predicting acute skin irritation on in vitro reconstructed human epidermis. Toxicology. 2007;30:157–66.
60. Niwa M, Nagai K, Oike H, Kobori M. Evaluation of the skin irritation using a DNA microarray on a reconstructed human epidermal model. Biol Pharm Bull. 2009;32:203–8.
61. Khammo N, Ogilvie J, Clothier RH. Development of an innervated model of human skin. ATLA. 2007;35:487–91.
62. Garle MJ, Fry JR. Sensory nerves, neurogenic inflammation and pain: missing components of alternative irritation strategies? A review and a potential strategy. ATLA. 2003;31:295–316.

2 Skin Irritation Hazard of Chemicals Assessed by the EpiSkin™ *In Vitro* Test Method

Nathalie Alépée, Marie Hélène Grandidier, and José Cotovio

2.1 Principle of the Test Method and Scientific Basis

Information about a chemical's potential to cause skin irritation is required by international regulations and testing guidelines for the safety assessment of chemicals and mixtures (REACH, EU CLP, Cosmetics Directive, [1, 2]). Until the last decade, the rabbit Draize dermal irritation test was the method traditionally used for this purpose [3, 4]. However, this animal test has major drawbacks such as different physiological characteristics as compared to human skin and the lack of reproducibility [5, 6]. Since the 1980s, the European Commission has advocated reducing the use of laboratory animals in safety testing as soon as scientifically valid alternative methods are available (Council Directive on the protection of animals used for scientific purposes 86/609/EEC revised as 2010/63/EU; [7, 8]). The 7th Amendment to the Cosmetics Directive (Directive 2003/15/EC taken up by Regulation 1223/2009) went even further and implemented a complete ban on animal testing for finished cosmetics products from 2004, and for cosmetic ingredients from 2009, for all human-health-related effects (EC 2009, [2]). An important step in worldwide harmonization was the adoption of the Globally Harmonized System (GHS) of Classification and Labelling of Chemicals [9]. Within the EU, UN GHS came into force in 2009 via the legislation referred to as the Classification, Labelling and Packaging System (CLP; European GHS Regulation (EC) No. 1272/2008) and is an integral part of REACH [10].

Skin irritation refers to the production of reversible damage to the skin following the application of a test substance. Chemical-induced skin irritation, manifested by erythema and oedema, is the result of a cascade of events beginning with

N. Alépée (✉) • M.H. Grandidier • J. Cotovio
L'Oréal Research and Innovation, 1, Avenue Eugène Schueller, 93600 Aulnay-sous-Bois Cedex, France
e-mail: nalepee@rd.loreal.com

C. Eskes et al. (eds.), *Alternatives for Dermal Toxicity Testing*,
DOI 10.1007/978-3-319-50353-0_2

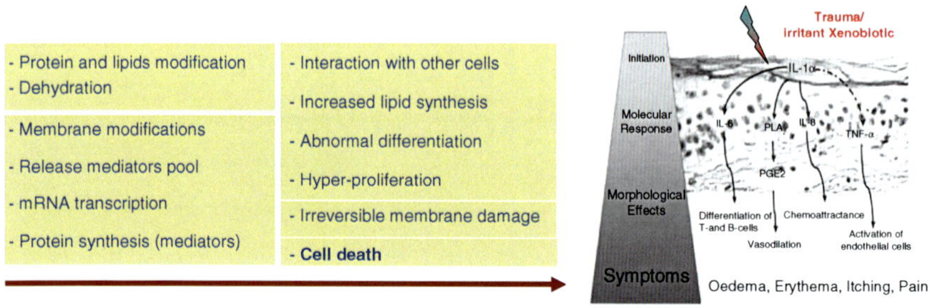

Fig. 2.1 Scientific basis of skin irritation [41, 42]

penetration of the stratum corneum and damage to the underlying layers of keratinocytes (Fig. 2.1). Irritation is initially manifested by redness (erythema), vesicles, serous exudates, serous scabs (scars) and various degrees of swelling (oedema). Over time, other reactions may be manifested, such as small areas of alopecia, hyperkeratosis, hyperplasia and scaling. Histopathology might be useful in discerning responses. In most cases, inflammation is well developed within the first 72 h of observation, commonly leading to the use of grades at 24, 48 and 72 h to evaluate irritancy potential. In some cases, as with defatting agents and certain petroleum-distillate-containing products, inflammatory responses may be delayed [11]. Stressed, damaged or dying keratinocytes release mediators that initiate an inflammatory reaction, which acts on the cells in the dermis, particularly the stromal and endothelial cells. It is the dilation and increased permeability of the endothelial cells that produce the observed erythema and oedema *in vivo*.

Several validated *in vitro* methods for skin irritation have been adopted by the Organisation for Economic Co-operation and Development (OECD) and by the European Union (EU) in the last decade [12]. These reconstructed human epidermis (RhE)-based test methods, including the adopted EpiSkin™ skin irritation (SIT) test method, measure the initiating events in the cascade irritation (i.e. cell and tissue damage measured through decreased tissue viability *in vitro*). Four commercially available RhE models have been endorsed as scientifically valid to be used within the framework of the Test Guideline 439 (OECD TG 439). Furthermore, EU test guideline B.46 considers the RhE model a stand-alone replacement for the assessment of acute dermal irritation test within a tiered testing strategy and/or in a weight-of-evidence approach. The use of human-derived, non-transformed epidermis keratinocytes as cell source and the use of representative tissue and cytoarchitecture closely mimic the biochemical and physiological properties of the upper parts of the human skin, i.e. the epidermis. The EpiSkin™ model is constructed by culturing the keratinocytes at the air-liquid interface to form a multilayered, highly differentiated model of the human epidermis. It consists of organized basal, spinous and granular layers and a multilayered stratum corneum containing intercellular lamellar lipid layers representing main lipid classes analogous to those found *in vivo*. Test chemicals are applied topically to the EpiSkin™ model and exposed for 15 min. Cell viability is measured after a 42 h post-treatment incubation period by dehydrogenase conversion of the vital dye MTT into a blue formazan salt that is quantitatively measured after extraction from tissues [13]. Irritant chemicals are identified by their ability to decrease tissue viability below the defined threshold level.

Many chemicals are known to induce contact dermatitis in humans, and a number are capable of doing so after a single exposure. Such primary irritants show great diversity with respect to chemical structure, molecular weight, polarity and binding capacity. They cause damage to the cellular components of the skin in a variety of different ways, including denaturation of epidermal keratins, the removal of surface lipids and water-holding substances, damage to cell membranes and direct cytotoxic effects.

2.2 Current Validation Status

An EURL ECVAM Skin Irritation Validation Study (SIVS) was conducted on the EpiSkin™ test method from 2003 to 2007. The SIVS was a prospective validation study involving the blind testing of 58 test substances representing a wide spectrum of chemical functionalities and the full range of dermal irritancy [14]. The goal of the study was to assess whether the *in vitro* test method would correctly predict *in vivo* classifications according to the former EU Dangerous Substance Directive (DSD) classification scheme, "R38" and "no label" (i.e. non-irritant). Following an independent peer review, the EURL ECVAM Scientific Advisory Committee (ESAC) endorsed the scientific validity of the EpiSkin™ SIT test method in 2007 as a replacement for the rabbit skin irritation method [15]. Furthermore, the ESAC endorsed that the original ESAC statement relating to the scientific validity of the test method remained valid and was extended in the context of the EU CLP/UN GHS classification system [16]. The EpiSkin™ test method using the MTT endpoint was therefore validated as a potential stand-alone method, capable of reliably distinguishing non-irritant (non-classified) from irritant chemicals according to the former (EU DSD) and recently implemented Globally Harmonized Systems (GHS) – EU CLP classifications. EpiSkin™ (as a 'reference method') was also used to specify the EURL ECVAM skin irritation performance standards with regard to the defined accuracy values. The test method is included in the EU Test Method B46 and accepted in the *in vitro* skin irritation OECD Test Guideline 439 adopted since 2010 [13].

2.3 Performance and Applicability of the Test Method

2.3.1 Reproducibility

Two types of reproducibility were evaluated for the EpiSkin™ test method: one obtained by testing the same chemicals over time in a single laboratory (within-laboratory reproducibility, WLR) and the other by testing the same chemicals in different laboratories (between-laboratory reproducibility, BLR). WLR was calculated as the percentage of chemicals for which 100% concordant classifications were obtained in the three valid runs performed. BLR was calculated as the percentage of chemicals for which 100% concordant classifications were obtained between laboratories.

Table 2.1 Within-laboratory and between-laboratory reproducibility of the EpiSkin™ skin irritation test method

	EURL ECVAM SIVS laboratories			
Reproducibility	L'Oréal	Sanofi	Unilever	BLR
58 SIVS chemicals (I vs. NI)	94.5% (52/55)	90.9% (50/55)	94.6% (53/56)	89.5% (51/57)

WLR calculated on the basis of data extracted from Spielmann et al. [14] for the three valid runs in L'Oréal, Sanofi and Unilever laboratories. BLR calculated on the basis of all median classification per laboratory combined

During the international SIVS study conducted under the supervision of the EURL ECVAM, 58 coded chemicals were tested in three laboratories (L'Oréal, Sanofi and Unilever). During phase 1, 100% agreement was obtained. For phase two, the same concordant classification was observed between the three valid runs for 158 out of 169 items (93.5%) for the three laboratories when considering irritants versus non-irritants [14] (Table 2.1). Therefore, the same prediction was observed for 44 out of 50 chemicals (88%) in each triplicate experiment and in all three laboratories when considering irritants versus non-irritants [14]. In only six cases (12%) did one laboratory gives results which were consistently in a different classification category to those from the other laboratories, with the 2-isopropyl-2-isobutyl-1,3-dimethoxypropane (*in vivo* UN GHS Category 2) being under-predicted in 2 out of 3 laboratories. Five chemicals (3-mercaptohexanol; 4-methylthio-benzaldehyde; bis[(1-methylimidazol)-(2-ethyl-hexanoate)], zinc complex and 2 mixtures of isomers) known to be UN GHS non-classified were over-predicted as UN GHS Category 2 in at least one laboratory.

Importantly, the test method showed acceptable reproducibility both within (>90%) and between (>85%) laboratories during its validation.

2.3.2 Predictive Capacity

The predictive capacity of the EpiSkin™ test method was originally validated in reference to the cut-off value for hazard categorization as used in the EU Dangerous Substance Directive, EU DSD (ESAC statements, [15]). As a result of the implementation of the UN GHS for Classification and Labelling in the EU from 2008 onwards through the EU Classification, Labelling and Packaging Directive (1272/2008), the cut-off value for distinguishing irritants from non-irritants shifted from an *in vivo* score of 2.0 to 2.3. A re-evaluation of the original results on predictive capacity taking this shift of the cut-off value into consideration was performed [5]. While the specificity of the EpiSkin™ method decreased from 81.8% (previous EU DSD system) to 71.1% (EU CLP), the test sensitivity increased from 72.0% (previous EU DSD system) to 84.6% (EU CLP) (Table 2.2). The original ESAC statement relating to the scientific validity of the test method therefore remains accurate and, with regard to its use in the context of decisions of classification, is now extended to the CLP system [5, 16].

Table 2.2 Predictive capacity of the EpiSkin™ skin irritation test method considering the SIVS (58 chemicals, valid runs) and L'Oréal data for a set of 74 chemicals alone or in combination with the data obtained in the EURL ECVAM SIVS (all 122 chemicals), based on weighted data

	SIVS (EU DSD)		SIVS (EU CLP)		L'Oréal only (median) (EU CLP)		Overall (EU CLP)	
Predictive capacity	Predictions/total	%	Predictions/total	%	Predictions/total	%	Total (substances)	%
Specificity (Non-classified correct predictions)	27/33	81.8%	32/45	71.1%	24/29	82.6%	n = 65	75.4%
Sensitivity (Irritant correct predictions)	18/25	72.0%	11/13	84.6%	45/46	97.7%	n = 57	94.7%
Accuracy (Category 2 vs. Non-classified)	45/58	77.6%	43/58	74.1%	68/74	91.9%	n = 122	84.4%

The predictive capacity of EpiSkin™ was further calculated considering the data obtained by L'Oréal, alone or in combination with the data obtained in the EURL ECVAM SIVS. Thus, different chemicals ended up with a different number of independent classifications used for calculating predictive capacity, i.e. ranging from 12 classifications (nine experiments from EURL ECVAM SIVS and three experiments from L'Oréal) to at least 3 classifications (chemicals that were not part of EURL ECVAM SIVS). To prevent different chemicals from weighing differently in the calculation of predictive capacity, a weighted calculation was used to reflect the real predictive capacity of the test method.

The EpiSkin™ test method showed a specificity of 75.4% considering the 65 tested *in vivo* non-irritant chemicals (Table 2.2). The EpiSkin™ test method showed a sensitivity of 94.7% (Table 2.2). Three chemicals (out of 57 tested irritants) had false-negative predictions, i.e. the *in vivo* Category 2 methyl palmitate, di-n-propyl disulphide and 2-isopropyl-2-isobutyl-1,3-dimethoxypropane. However, it is important to note that methyl palmitate, with an *in vivo* Draize score of 3.0, is known to be non-irritant to humans based on the 4-h human patch test (1/29 positive reactions) [17]. Similar observations were made with the di-n-propyl disulphide, with 6/30 positive reactions suggesting an over-prediction of *in vivo* Draize assay versus human effects [17, 18]. Some chemicals were over-predicted (e.g. phenethyl bromide; trans-cinnamaldehyde; 10-undecenoic acid; 4-methylthio-benzaldehyde; di-limonene; 1.6-dibromohexane; 1-bromo-4-chlorobutane; 3-mercaptohexanol; 3,4-dimethyl-1H-pyrazole; 3-chloro-4-fluoronitrobenzene; 4-methylthio-benzaldehyde; tri-isobutyl phosphate; eugenol; 2,4-xylidine) with some of them correctly categorized according to the former EU DSD classification. Erring on the side of caution for consumers, many of them are also misclassified in the other three validated test methods.

The overall accuracy of the EpiSkin™ test method to distinguish between irritant and non-irritant chemicals was 84.4% (103/122).

2.3.3 Applications and Limitations

The EpiSkin™ skin irritation test method is applicable to all mono-substances or multicomponent test substances that are solids, liquids, semi-solids, soluble or insoluble in water. Gases and aerosols cannot be evaluated (although this is true for all models included in OECD TG 439) [13].

The test method has been shown to be applicable to a broad range of chemicals covering relevant ranges of chemical classes (fragrances, dyes, preservatives, actives, UV filter and non-cosmetics ingredients), reaction mechanisms and physico-chemical properties. However, volatility may considerably reduce the amount of chemical in contact with the epidermis as in a human exposure. In addition, dyes and other strongly coloured chemicals may impair the scoring of effects. Using the alternative HPLC/UPLC-spectrophotometry as an endpoint measurement instead of the validated optical density (OD) measurement allows the determination of the cell viability for strongly coloured chemicals enhancing the applicability domain to this chemical type (see section on Possible Protocol Adaptation). Finally, the test method is applicable to mixtures, although only limited information on the testing of mixtures is available.

2.3.4 Comparison to Human Data

In vitro reconstructed human-based test methods for identification of skin irritation have been the subject of validation and acceptance endorsement as a full replacement for the *in vivo* rabbit Draize dermal irritation. However, an alternative strategy and the associated protocol for the use of human volunteers to identify skin irritation have been described [19–21]. A total of 81 substances tested according to the aforementioned four-hour human patch test protocol were found and collated into a dataset together with their existing *in vivo* classifications published in the literature [17, 22, 23]. Jírová, et al. [24] compared human patch test data with *in vitro* and animal data and found that only five substances were human skin irritants out of 16 materials classified as skin irritants in the rabbit test. The authors concluded that such results confirm observations that rabbits over-predict skin effects in humans. When considering the EpiSkin™ test method, predictivity towards humans or rabbits was evaluated on the basis of 25 test chemicals. Of the ten chemicals classified as irritants in rabbits, only six chemicals were found to be significantly irritating to human skin (Table 2.3). Concordance between the human epidermis model was the

Table 2.3 Summary table of *in vivo* and *in vitro* results

Chemical name	CAS number	*In vivo* EU CLP/ UN GHS class	4-h human patch test	*In vitro* EU CLP/ UN GHS class
Isopropyl myristate	110-27-0	NC	NC	NC
Methyl laurate	111-82-0	NC	NC	NC
Linalyl acetate	115-95-7	NC	NC	NC
Benzyl salicylate	118-58-1	NC	NC	NC
Isopropyl palmitate	142-91-6	NC	NC	NC
Dodecanoic acid	143-07-7	NC	NC	NC
Dipropylene glycol	25,265-71-8	NC	NC	NC
Heptyl butyrate	5870-93-9	NC	NC	NC
Hexyl salicylate	6259-76-3	NC	NC	NC
Isopropanol	67-63-0	NC	NC	NC
Water	7732-18-5	NC	NC	NC
Naphthaleneacetic acid	86-87-3	NC	NC	NC
10-Undecenoic acid	112-38-9	NC	NC	Cat 2
1-Bromo-4-chlorobutane	6940-78-9	NC	NC	Cat 2
Eugenol	97-53-0	NC	NC	Cat 2
Methyl palmitate	112-39-0	Cat 2	NC	NC
Di-n-propyl disulphide	629-19-6	Cat 2	NC	NC
1-Decanol	112-30-1	Cat 2	NC	Cat 2
Butyl methacrylate	97-88-1	Cat 2	NC	Cat 2
1-Bromohexane	111-25-1	Cat 2	I	Cat 2
Heptanal	111-71-7	Cat 2	I	Cat 2
Nonanoic acid	112-05-0	Cat 2	I	Cat 2
Octanoic acid	124-07-2	Cat 2	I	Cat 2
Sodium lauryl sulphate (20% aq.)	151-21-3	Cat 2	I	Cat 2
Hydrogenated tallow amine	61,788-45-2	Cat 2	I	Cat 2

Cat. category, *NC* no category

same for the human and rabbit data, i.e. 80% (Table 2.3). Concordance between the rabbit test and the 4-h HPT was 84% for the same dataset. Consistently false-positive results (also true for almost all reconstructed human epidermis test methods included in the OECD *in vitro* tests) were observed, indicating that the test method errs on the side of caution for the safety of consumers.

2.4 Brief Description of the Protocol

Test chemicals are applied topically to the three-dimensional epidermal model, comprised of human-derived epidermal keratinocytes which have been cultured to form a multilayered highly differentiated model. EpiSkin™ tissue is produced in accordance with the quality standard ISO 9001, ensuring traceability and reproducibility of the epidermal tissues. The reproducibility of each batch is checked by histological analysis, taking into account the general organization, the stratification of the epidermis, the nucleation of the basal layer, the size of the intercellular spaces, the adhesion of the basal layer to the support, the quantity of granular cells and the thickness of the horny layer. In addition, the reproducibility of the response of each EpiSkin™ batch is tested against a reference irritant, sodium dodecyl sulphate (SDS), for which the acceptability range of the model is 1 mg/mL $\geq IC_{50} \geq$ 3 mg/mL.

Three epidermis units are treated per test chemical for 15 min at room temperature. Concurrently to the test chemical, SDS 5% and PBS-treated epidermis are used as positive and negative controls, respectively. Exposure to the test chemical is terminated by rinsing with phosphate-buffered saline (PBS). The epidermis is then incubated at 37 °C for 42 additional hours (Fig. 2.2). Cell viability determination is based on cellular mitochondrial dehydrogenase activity, measured by MTT [3-(4,5-Dimethylthiazol-2-yl)-2,5-diphenyltetrazolium bromide, Thiazolyl blue; CAS number 298-93-1] reduction and conversion into blue formazan salt that is quantified after extraction from tissues [25]. The viability is assessed by incubating the tissues for 3 h with a MTT solution in a 12-well plate (0.3 mg/mL; 2 mL per well). The formazan precipitate is then extracted using acidified isopropanol (0.5 mL) and quantified spectrophotometrically at 570 nm using 96-well plates (200 μL/well). For each treated tissue, the viability is expressed as a % relative to negative control tissues (mean). Irritant chemicals are identified by their ability to decrease cell viability below a defined threshold level (i.e., ≤50%, for UN GHS Category 2). Details are provided in the SOP [26].

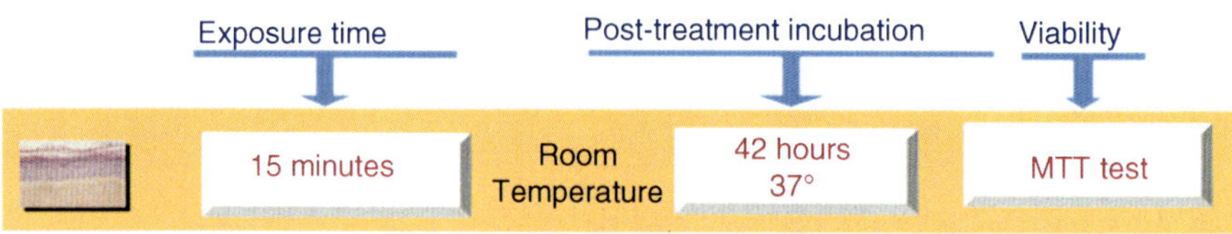

Fig. 2.2 Main steps of the EpiSkin™ skin irritation test method

2.5 Role in a Testing Strategy

The EpiSkin™ test method is able to identify UN GHS/EU CLP Category 2 (Cat. 2) and No Cat. chemicals and can thus serve as a stand-alone skin irritation method for non-irritants in countries where the optional UN GHS Category 3 is not implemented, e.g. in the European Union. If EpiSkin™ test method results show Cat. 2, an *in vitro* skin corrosion test—if not performed beforehand—is required to determine the final classification: Cat. 2 (irritant) or Cat. 1(A, B or C) (corrosive) [27]. Indeed, for a full evaluation of local skin effects after a single dermal exposure, the OECD Guidance Document No. 203 on an Integrated Approach to Testing and Assessment (IATA) should be consulted [28]. In particular, the OECD IATA introduces the option to prospectively test the *in vitro* skin irritation/corrosion potential of a chemical using either a top-down approach (an *in vitro* skin corrosion test followed by an *in vitro* skin irritation test) or a bottom-up approach (an *in vitro* skin irritation followed by an *in vitro* skin corrosion test) (Fig. 2.3).

Combining skin irritation and skin corrosion *in vitro* evaluations on EpiSkin™ tissues, the two approaches can be used to fully evaluate the local skin effects after chemical exposure [29]. This approach is feasible due to the difference in exposure times between the *in vitro* skin irritation and the *in vitro* skin corrosion tests. While the former has an exposure time of 15 min (followed by a 42-h post-exposure incubation) and a unified classification cut-off at 50% tissue viability (see Brief Description, above), the latter has a maximum exposure time of 4 h (and no post-exposure incubation time) (see Chap. 8) and classification cut-offs at 35% tissue viability [29]. An evaluation was conducted with the EpiSkin™ skin irritation and skin corrosion test methods on 87 test chemicals and demonstrated similar outcomes using both testing strategies (bottom-up or top-down) and their relevance for hazard assessment (manuscript pending).

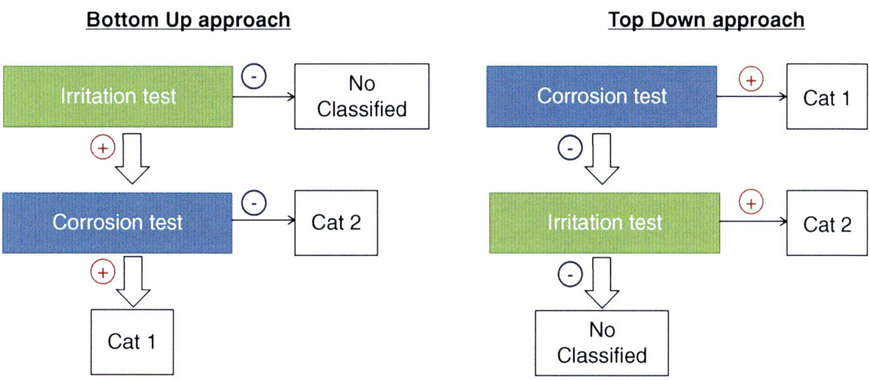

Fig. 2.3 Bottom-up and top-down testing strategies

2.6 Perspectives from the Test Developer

2.6.1 Critical Steps in the Protocol

The critical steps of the standardized operating procedure include the following:

- The absence of air bubble under the epidermis should be verified at each step.
- All test substances should be tested alone in separate plate.
- Ensure the entire surface of the epidermis is covered with the test chemical.
- For viscous and sticky chemicals, a curved flat spatula should be used or the test chemical shall be weighed directly on the nylon mesh (pretesting the compatibility of the test chemical with the nylon mesh should be considered).
- Tissues should be rinsed thoroughly.
- Always make use of a freshly prepared MTT solution (preparation to be used within 3 h).
- The epidermis should be gently detached from the matrix and turned with the epidermis topical side against the matrix before formazan extraction is conducted with isopropanol acid.

2.6.2 Possible Protocol Adaptations

Protocol adaptions could be applied to the validated EpiSkin™ skin irritation test method to either increase its applicability domain to strongly colourants or its sensitivity as described below.

The skin irritation potential of a test chemical is determined by measuring tissue viability using the photometric MTT-reduction assay. A known limitation of this assay is the possible interference of direct MTT reducers or strongly coloured test chemicals with measurement of formazan by absorbance (optical density). In this case, test chemicals that act directly on MTT (e.g. MTT reducer) or which have a colour that absorb at the same wavelength as MTT require the use of adapted controls as described in the test method SOP. For strongly coloured chemicals, when the evaluation is not possible by absorbance, Cosmetics Europe has evaluated the use of HPLC/UPLC-spectrophotometry as an alternative formazan measurement system. Using the approach recommended by the Federal Drug Administration (FDA) guidance for validation of bioanalytical methods [30], 26 chemicals were tested in the EpiSkin™ skin irritation test method. Results support that (1) formazan measurement by HPLC/UPLC-spectrophotometry and OD gave almost identical tissue viabilities for test chemicals exhibiting neither colour interference nor direct MTT reduction and (2) HPLC/UPLC-spectrophotometry can measure formazan for strongly coloured test chemicals when this is not possible by absorbance only [31]. Cosmetics Europe has undertaken a second study that focuses on evaluation of chemicals with functionalities relevant to cosmetic products. Such chemicals were primarily identified from the [32] memorandum (addendum) on the *in vitro* test EpiSkin™ for skin irritation testing. Fifty test items were evaluated in which both standard photometry and HPLC/UPLC-spectrophotometry were used for endpoint detection. The results obtained in this study (1) provide further support for within-laboratory reproducibility of HPLC/UPLC-spectrophotometry for measurement of formazan, (2) demonstrate, through use of a case study with Basazol C Blue pr. 8056, that HPLC/UPLC-spectrophotometry enables determination of an *in vitro* classification even when this is not possible using standard photometry and (3) addresses the question raised by SCCS in their 2010 memorandum (addendum) to consider an endpoint detection system not involving optical density quantification in *in vitro* reconstructed human epidermis skin irritation test methods [33]. The HPLC/UPLC technique has been adopted within OECD TG 439 [13] and by the authorities [34].

Furthermore, to improve the MTT viability-based prediction model, the release of a membrane damage marker, cytokines IL-1, was also investigated as a possible protocol adaption. In response to physical or chemical stress, keratinocytes produce and release inflammatory cytokines interleukins [IL-1α, tumour necrosis factor α (TNF-a)], chemotactic cytokines [IL-8, interferon, e.g. induced protein 10 (IP-10)], growth-promoting factor [IL-6, IL-7, IL-15, granulocyte/macrophage colony-stimulating factor GM-CSF)], transforming growth factor [TGF], cytokines regulating humoral versus cellular immunity [IL-10, IL-12] and other signalling factors, which rapidly generate cutaneous inflammation, suggesting that measurement of such keratinocyte responses may allow the evaluation of toxicological properties of chemicals in order to identify irritants [35, 36]. As an additional measure of skin irritation, release of inflammatory mediators (e.g. interleukin-1 alpha) may therefore be considered [37]. If the cytotoxic effect is absent or weak, a quantifiable amount of inflammatory mediators is released by the epidermis and may be used in a tiered approach to increase the sensitivity of the test. For EpiSkin™ tissues showing a cell viability >50%, the amount of IL-1α released into the tissue culture medium at the end of the 42-h post-treatment incubation period was measured in the medium (immediately or frozen) using ELISA (Roguet and Cotovió [38]). The test chemical might be considered to be an irritant if the viability is >50% and the amount of IL-1α release is >9.18 IU/mL or the viability only is ≤50%. This endpoint was found as a useful adjunct in the case of the EpiSkin™ test method, potentially increasing the sensitivity of the assay.

2.6.3 Challenges and Opportunities

Challenges and opportunities might be seen in the context of the assessment of specific categories of ingredients (e.g. mixtures and vegetal extracts) as well as for the identification of the optional UN GHS Category 3 (mild irritants) using the EpiSkin™ test method.

Regarding its applicability, the EpiSkin™ skin irritation test method is appropriate for the evaluation of mono- and multicomponent substances as well as mixtures. However, only limited information is available in the public domain on the testing of mixtures with test methods falling under OECD TG 439 [12, 39]. The applicability of the test method for the assessment of mixtures may depend on the types and categories of products tested and/or the *in vitro* test method protocol used. It is therefore not possible to generalize the applicability based on the types of mixtures assessed [12]. Furthermore, it is not possible to define criteria on the amount of evidence needed to demonstrate the applicability of an adopted *in vitro* assay to test mixtures, as this may depend on the availability of *in vivo* (animal and/or human) data, as well as on the variety, category and type of mixture evaluated. Further investigations would be beneficial due to the limited information reported.

According to the EU CLP classification, the EpiSkin™ test method can be used as a stand-alone skin irritation replacement test method to distinguish Category 2 (Cat. 2, irritant) from not classified (no category, non-irritant) chemicals. UN GHS foresees one category for irritant chemicals (Cat. 2) but allows the use of a further optional category (Cat. 3) to classify substances with intermediate irritancy potency (mild irritants) with *in vivo* scores of between 1.5 and 2.3. Using the EpiSkin™ tissues, a protocol based on the measurement of parameters other than cell viability has been developed. The IRR-IS assay—exploiting quantitative analysis of expression profiles of relevant genes—appears to be a promising methodology to contribute to the determination of skin irritancy potential, i.e. the discrimination of non-irritants, mild irritants and irritants as shown in a study evaluating gene expression changes in the validated EpiSkin™ test system in response to chemical exposure [40]. So before embarking on animal testing to generate information on UN GHS/EU CLP Cat. 3 chemicals to satisfy the requirements of authorities implementing this category, the use and/or generation of data from this approach should be considered. Nevertheless, further investigations would be beneficial due to the limited information reported in the literature.

2.7 Conclusion

The EpiSkin™ skin irritation test method evaluates the production of reversible damage to the skin following the application of a test substance. Upon contact with skin irritants, the tissue viability of EpiSkin™ decreased *in vitro*. According to the EU CLP classification, the EpiSkin™ test method can be used as a stand-alone skin irritation replacement test method to distinguish Category 2 (irritant) from non-classified (no category, non-irritant) chemicals. The reliability (transferability, intra- and inter-reproducibility) and the relevance were formally demonstrated, leading to the regulatory acceptance of reconstructed human epidermis in OECD Test Guidelines 439 and Guidance document 203.

References

1. EC. REGULATION (EC) No. 1907/2006 of the European Parliament and of the Council of 18 December 2006 concerning the Registration, Evaluation, Authorisation and Restriction of chemicals (REACH), establishing a European Chemicals Agency, amending Directive 1999/45/EC and repealing Council Regulation (EEC) No. 793/93 and Commission Regulation

(EC) No. 1488/94 as well as Council Directive 76/769/EEC and Commission Directives 91/155/EEC, 93/67/EEC, 93/105/EC and 2000/21/EC. Off J Eur Union. 2006;L396:1–1355.

2. EC. Regulation (EC) No. 1223/2009 of the European Parliament and of the Council of 30 November 2009 on cosmetic products. Off J Eur Union. 2009;L342:59–209. http://eur-lex.europa.eu/legal-content/EN/TXT/?uri=CELEX:32009R1223
3. Draize JH, Woodard G, Calvery HO. Methods for the study of irritation and toxicity of substances applied directly on the skin and mucous membranes. J Pharmacol Exp Ther. 1944;82:377–90.
4. OECD. Guideline for the testing of chemicals, No. 431. *In vitro* skin corrosion: human skin model test, 8pp. Paris, France: OECD. 2015. Available at http://www.oecd-ilibrary.org/environment/oecd-guidelines-for-the-testing-of-chemicals-section-4-health-effects_20745788.
5. OECD. Explanatory background document to the OECD draft Test Guideline on in vitro skin irritation testing. Published in OECD Series on Testing and Assessment, No. 137, OECD, Paris. 2010. Available at http://www.oecd.org/officialdocuments/displaydocumentpdf/?cote=env/jm/mono(2010)36&doclanguage=en.
6. Weil CS, Scala RA. Study of intra-and interlaboratory variability in the results of rabbit eye and skin irritation tests. Toxicol Appl Pharmacol. 1971;19:276–360.
7. EC. Regulation (EC). Council Directive 86/609/EEC of 24 November 1986 on the approximation of laws, regulations and administrative provisions of the Member States regarding the protection of animals used for experimental and other scientific purposes. Off J Eur Union. 1986;L358:0001–28.
8. EC. Regulation (EC). Directive 2010/63/EU of the European Parliament and of the Council of 22 September 2010 on the protection of animals used for scientific purposes Text with EEA relevance. Off J Eur Union. 2010;L276:33–79.
9. United Nations (UN). Globally harmonized system of classification and labelling of chemicals (GHS). 6th revised ed; 2015. UN New York and Geneva, 2013
10. EC. REGULATION (EC) No. 1272/2008 of the European Parliament and of the Council of 16 December 2008 on classification, labelling and packaging of substances and mixtures, amending and repealing Directives 67/548/EEC and 1999/45/EC, and amending Regulation (EC) No. 1907/2006. Off J Eur Union. 2008;L353:1–1355.
11. Emmett EA. Toxic responses of the skin. In: Klaassen CD, Amdur MO, Doull J, editors. Casarett and Dull's toxicology. New York, NY, USA: Macmillan; 1986. p. 412–31.
12. Eskes C, Detappe V, Koëter H, Kreysa J, Liebsch M, Zuang V, Amcoff P, Barroso J, Cotovio J, Guest R, Hermann M, Hoffmann S, Masson P, Alépée N, Arce LA, Brüschweiler B, Catone T, Cihak R, Clouzeau J, D'Abrosca F, Delveaux C, Derouette JP, Engelking O, Facchini D, Fröhlicher M, Hofmann M, Hopf N, Molinari J, Oberli A, Ott M, Peter R, Sá-Rocha VM, Schenk D, Tomicic C, Vanparys P, Verdon B, Wallenhorst T, Winkler GC, Depallens O. Regulatory assessment of in vitro skin corrosion and irritation data within the European framework: Workshop recommendations. Regul Toxicol Pharmacol. 2012;62:393–403.
13. OECD. *In vitro* skin irritation: reconstructed human epidermis test method. OECD guideline for the testing of chemicals No. 439, OECD, Paris. Originally adopted in 2010. 2015. Available at http://www.oecd-ilibrary.org/environment/oecd-guidelines-for-the-testing-of-chemicals-section-4-health-effects_20745788.
14. Spielmann H, Hoffmann S, Liebsch M, Botham P, Fentem J, Eskes C, Roguet R, Cotovio J, Cole T, Worth A, Heylings J, Jones P, Robles C, Kandárová H, Gamer A, Remmele M, Curren R, Raabe H, Cockshott A, Gerner I, Zuang V. The ECVAM international validation study on in vitro tests for acute skin irritation: Report on the validity of the EPISKIN and EpiDerm assays and on the skin integrity function test. ATLA. 2007;35:559–601.
15. EURL ECVAM Statement on the validity of in vitro tests for skin irritation, issued by the ECVAM Scientific Advisory Committee (ESAC26), 27 Apr 2007. Available at http://ihcp.jrc.ec.europa.eu/our_activities/alt-animal-testing.
16. EURL-ECVAM Statement on the "Performance under UN GHS of three in vitro assays for skin irritation testing and the adaptation of the Reference Chemicals and Defined Accuracy Values of the ECVAM skin irritation Performance Standards", issued by the ECVAM Scientific Advisory Committee (ESAC30), 9 Apr 2009. Available at http://ihcp.jrc.ec.europa.eu/our_activities/alt-animal-testing.

17. Basketter DA, York M, McFadden JP, Robinson MK. Determination of skin irritation potential in the human 4-h patch test. Contact Dermatitis. 2004;51:1–4.
18. Jirova D, Liebsch M, Basketter D, Spiller E, Kejlova K, Bendova H, Marriott M, Kandarova H. Comparison of human skin irritation and photo-irritation patch test data with cellular *in vitro* assays and animal *in vivo* data. AATEX. 2007;14(Special Issue):359–65.
19. Basketter DA, Whittle E, Griffiths HA, York M. The identification and classification of skin irritation hazard by human patch test. Food Chem Toxicol. 1994;32:769–75.
20. Basketter DA, Whittle E, Chamberlain M. Identification of irritation and corrosion hazards to skin: an alternative strategy to animal testing. Food Chem Toxicol. 1994;32:539–42.
21. Basketter DA, Chamberlain M, Griffiths HA, York M. The classification of skin irritants by human patch test. Food Chem Toxicol. 1997;35:845–52.
22. Basketter DA, Jirova J, Kandarova H. Review of skin irritation/corrosion Hazards on the basis of human data: a regulatory perspective. Interdiscip Toxicol. 2012;5(2):98–104.
23. Robinson MK, McFadden JP, Basketter DA. Validity and ethics of the human 4 hour patch test as an alternative method to assess acute skin irritation potential. Contact Dermatitis. 2001;45:1–12.
24. Jirova D, Basketter D, Liebsch M, Bendova H, Keilova K, Marriott M, Kandarova H. Comparison of human skin irritation patch test data with in vitro skin irritation assays and animal data. Contact Dermatitis. 2010;6:109–16.
25. Mosmann T. Rapid colorimetric assay for cellular growth and survival: application to proliferation and cytotoxicity assays. J Immunol Methods. 1983;65:55–63.
26. EpiSkin™ SOP, Version 1.8, ECVAM skin irritation validation study: validation of the EpiSkin™ test method 15 min–42 hours for the prediction of acute skin irritation of chemicals. 2009. Available at http://ihcp.jrc.ec.europa.eu/our_activities/alt-animal-testing.
27. OECD. *In vitro* skin corrosion: reconstructed human epidermis (RHE) test method. OECD guideline for the testing of chemicals. Originally adopted in 2004, OECD, Paris. 2016. Available at http://www.oecd-ilibrary.org/environment/oecd-guidelines-for-the-testing-of-chemicals-section-4-health-effects_20745788.
28. OECD. Guidance document on integrated approaches to testing and assessment of skin irritation/corrosion, series on testing and assessment, No. 203, OECD, Paris. 2014. Available at http://www.oecd.org/env/ehs/testing/seriesontestingandassessmenttestingforhumanhealth.htm.
29. Alépée N, Grandidier MH, Cotovio J. Sub-categorisation of skin corrosive chemicals by the EpiSkin™ reconstructed human epidermis skin corrosion test method according to UN GHS: Revision of OECD Test Guideline 431. Toxicol *In Vitro*. 2014;28:131–45.
30. FDA. Guidance for industry: bioanalytical method validation. U.S. Department of Health and Human Services, Food and Drug Administration. May 2001. Available at http://www.fda.gov/cder/guidance/index.htm.
31. Alépée N, Barroso J, De Smedt A, De Wever B, Hibatallah J, Klaric M, Mewes KR, Millet M, Pfannenbecker U, Tailhardat M, Templier M, McNamee P. Use of HPLC/UPLC-spectrophotometry for detection of formazan in in vitro reconstructed human tissue (RhT)-based test methods employing the MTT-reduction assay to expand their applicability to strongly coloured test chemicals. Toxicol *In Vitro*. 2015;29:741–61.
32. Scientific Committee on Consumer Safety (SCCS), 2010. Memorandum (addendum) on the in vitro test EPISKIN™ for skin irritation testing. SCCS/1392/10SCCS/1392/10. Available at http://ec.europa.eu/health/scientific_committees/consumer_safety/docs/sccss_003.pdf.
33. Alépée N, Hibatallah J, Klaric M, Mewes KR, Pfannenbecker U, McNamee P. Assessment of cosmetic ingredients in the *in vitro* reconstructed human epidermis test method EpiSkin™ using HPLC/UPLC-spectrophotometry in the MTT-reduction assay. Toxicol *In Vitro*. 2016;33:105–17.
34. Scientific Committee on Consumer Safety (SCCS). The sccs notes of guidance for the testing of cosmetic ingredients and their safety evaluation 9th revision. SCCS/1564/15 Revised version of 25 Apr 2016. Available at http://ec.europa.eu/health/scientific_committees/consumer_safety/docs/sccs_o_190.pdf.

35. Nickoloff BJ, Turka LA. Keratinocytes: key immunocytes of the integument. Am J Pathol. 1993;143:325–31.
36. Williams IR, Kupper TS. Immunity at the surface: Homeostatic mechanisms of the skin immune system. Life Sci. 1996;58:1485–507.
37. Cohen C, Selvi-Bignon C, Barbuer A, Rougier A, Lacheretz F, Roguet R. Measurement of pro inflammatory mediators production by cultured keratinocytes: a predictive assessment of cutaneous irritation. In Alternative Methods in Toxicology, Vol. 10. In: Rougier A, Goldberg AM, Maibach HI, editors. Vitro Skin Toxicology. New York, NY, USA: Mary Ann Liebert; 1994. p. 83–96.
38. Roguet, R., and Cotoviő, J.. Interleukin 1 alpha (IL-1α). Rationale and use. Role in inflammation/irritation processes. Measurement conditions specificities. Document provided on 30 March 2007 for consideration by the SIVS Management Team and Peer Review Panel. 2007
39. Molinari J, Eskes C, Andres E, Remoué N, Sá-Rocha VM, Hurtado SP. Improved procedures for in vitro skin irritation testing of sticky and greasy natural botanicals. Toxicol *In Vitro*. 2013;27(1):441–50.
40. Groux H, Lelievre D, Cottrez F, Auriault C, Alépée N, Meunier JR, Cotovio J. Evaluation of IRR-IS®, an EpiSkin™ based model for quantifying chemical irritation potency. Oral Communication No. 1.5, ESTIV 2012, 17th Oct 2012, Lisbon, Portugal. In: Poster presented at 51st Annual Meeting of the Society of Toxicology, Mar 11–15, 2012, San Francisco, USA. The Toxicologist, p. 286. 2012. Available at http://www.toxicology.org/AI/Pub/Tox/2012Tox.pdf.
41. Boelsma E, Ponec M. Basics (guidelines) on cell culture testing for topical/dermatological drugs/products and cosmetics with regard to efficacy and safety of the preparations. In: Dermatopharmacology of topical preparations, vol. 2000. Heidelberg: Springer; 2000. p. 37–57.
42. Welss T, Basketter DA, Schröder KR. *In vitro* skin irritation: facts and future. State of the art review of mechanisms and models. Toxicol *In Vitro*. 2004;18(3):231–43.

3 The EpiDerm™ Skin Irritation Test (EpiDerm™ SIT)

Helena Kandarova and Manfred Liebsch

3.1 Principle of the Test Method and Scientific Basis

Skin irritation is defined *in vivo* as the reversible damage to the skin following the application of a test chemical for up to 4 h [as defined by the United Nations (UN) Globally Harmonized System of Classification and Labeling of Chemicals (GHS)] [1]. The potential of chemicals to induce skin irritation (hazard) is an important consideration in establishing procedures for the safe handling, packing and transport of chemicals.

In vivo, skin irritation is determined using a modification of the Draize rabbit skin irritation test, as described in the OECD TG 404 [2, 3]. Because systemic reactions play a minor role in modulating local skin toxicity potential of chemicals, skin irritation potential may be predicted by *in vitro* systems, provided they are sufficiently complex to mimic the skin barrier and the inherent reactivity of cells within the skin.

The method described here is based on a method initially developed and refined by L'Oréal for the EPISKIN™ model [4, 5]. The SOP was applied to the EpiDerm™ model with the goals of developing a common protocol for both systems that was able to predict skin irritation potential, according to the EU classification system, and of replacing the *in vivo* acute skin irritation test in rabbits [6, 7]. Upon review of existing information by the ECVAM Skin Irritation Task Force and an ECVAM Workshop, both the EPISKIN™ and EpiDerm™ skin irritation tests (SIT) were regarded as

H. Kandarova (✉)
MatTek In vitro Life Science Laboratories, Mlynske Nivy 73, 821 05 Bratislava, Slovakia
e-mail: hkandarova@mattek.com

M. Liebsch
Retired Head of Unit "Centre of Alternative Methods to Animal Experiments–ZEBET" at the BfR, Berlin, Germany

C. Eskes et al. (eds.), *Alternatives for Dermal Toxicity Testing*,
DOI 10.1007/978-3-319-50353-0_3

sufficiently promising predictors of skin irritancy potential and were ready to enter the formal validation study. Due to the under-prediction of several chemicals in the second Phase of the ECVAM validation study [8], ESAC recommended increasing the sensitivity of the EpiDerm™ SIT to better match *in vivo* rabbit data [9].

Following the recommendation of ESAC [9], the EpiDerm™ skin irritation test was further optimized by MatTek Corporation during 2006 and 2007. Use of an extended exposure time (60 min) and minor modification of exposure conditions improved the sensitivity of the assay. The applicability domain, prediction model (50% viability cutoff for identification of irritants) and the endpoint (MTT cytotoxicity assay) did not change. Thus the concept of a common protocol was maintained [10].

The predictive capacity of the modified EpiDerm™ SIT was initially assessed by MatTek Corporation, USA in an intra-laboratory study [10]. Transferability of the method was evaluated in 2007 in an external international validation study between four laboratories: ZEBET at the BfR, Berlin, Germany; BASF, Ludwigshafen, Germany; IIVS, Gaithersburg, MD and Zet-LSL, Linz, Austria [11, 12]. The validation trial was in accordance with the principles and criteria documented in OECD Guidance Document No. 34 on the Validation and International Acceptance of New or Updated Test Methods for Hazard Assessment [13] and ECVAM (2007) Performance Standards for applying human skin models to *in vitro* skin irritation [14].

In 2008, ESAC concluded that the Modified EpiDerm™ SIT had sufficient accuracy and reliability for prediction of R38 skin irritating and no-label (non-skin irritating) test chemicals [15]. The Modified EpiDerm™ SIT is an *in vitro* procedure that, depending on information requirements, allows determining the skin irritancy of chemicals as a stand-alone replacement test, as a screen, or within a testing strategy in combination with, if appropriate, a weight of evidence approach [16].

The test consists of a topical exposure of the neat test chemical to a reconstructed human epidermis (RhE) model followed by a cell viability test. Cell viability is measured by dehydrogenase conversion in cell mitochondria of MTT [(3–4,5-dimethyl thiazole 2-yl) 2,5-diphenyltetrazoliumbromide], into a blue formazan salt that is quantitatively measured after extraction from tissues [17]. The reduction of the viability of tissues exposed to chemicals in comparison to negative controls (treated with water) is used to predict the skin irritation potential. Comparative studies in RhE models employing various endpoints to predict skin irritancy of topical formulations have shown that the MTT endpoint has clear advantages, even over mechanistically based endpoints like the release of IL-1α [18, 19].

3.2 Current Validation Status

The ECVAM Scientific Advisory Committee (ESAC) formally endorsed the scientific validity of the Modified EpiDerm™ Skin Irritation Test (SIT) at its November, 2008 meeting [15]. ESAC concluded that the Modified EpiDerm™ SIT has sufficient sensitivity, specificity and accuracy for the prediction of skin irritating and non-irritating test chemicals. This assay is considered to be a validated, stand-alone *in vitro* replacement test for animal skin irritation testing and was adopted as EU.B46 an OECD TG 439 [16, 20].

3.3 Performance and Applicability of the Test Method

The EpiDerm™ SIT was developed and designed to predict skin irritation potential of neat test chemicals in the context of identification and classification of skin irritation hazard according to the EU classification system. Since the EU and GHS systems were harmonized in 2008, the EpiDerm™ SIT also allows for hazard identification of irritant substances in accordance with UN GHS [21]. The Modified EpiDerm™ SIT allows discrimination between irritants of category 2 and non-irritants. The test does not discriminate between non-mandatory subcategories of the UN GHS, i.e. it does not distinguish between GHS category 2 and category 3 irritants.

3.3.1 Reproducibility

The between and within laboratory reproducibility of the EpiDerm™ SIT test has been assessed twice, first in the original validation study during 2004–2006 [8] and later on in the follow-up validation trial performed by four independent laboratories in 2007 [11, 12].

The within-laboratory variability of the final version of the EpiDerm™ SIT was assessed for each laboratory using

1. Assessment of the frequency of non-qualified experimental runs as defined by the SOP (SD > 20%),
2. One-way ANOVA statistics,
3. Analysis of the within-laboratory standard deviation,
4. Box plot analysis for identification of outliers.

Amongst the 240 independent experiments, only ten experiments provided standard deviation above 20%. The frequency of the non-qualified experiments was very low (less than 5%), and the pre-defined 95% confidence interval of acceptable tests was confirmed. The most rigorous statistical measure applied in the validation study was a 1-way ANOVA. Significance levels of 5% and 1% were chosen to assess the variability between three independent runs for each of the tested chemicals. Representative data obtained in the Follow-up validation study with 20 reference chemicals are shown in Figs. 3.1 and 3.2.

A box plot analysis was performed for each chemical tested in the four laboratories. Amongst the 720 test results (20 chemicals (n = 3), three runs, four laboratories), only one significantly outlying value was identified and excluded from the data-set. Overall the protocol assessed in four laboratories was found to be reliable and robust and of similar variability as the EPISKIN™ method, which had gained the full regulatory acceptance in 2007.

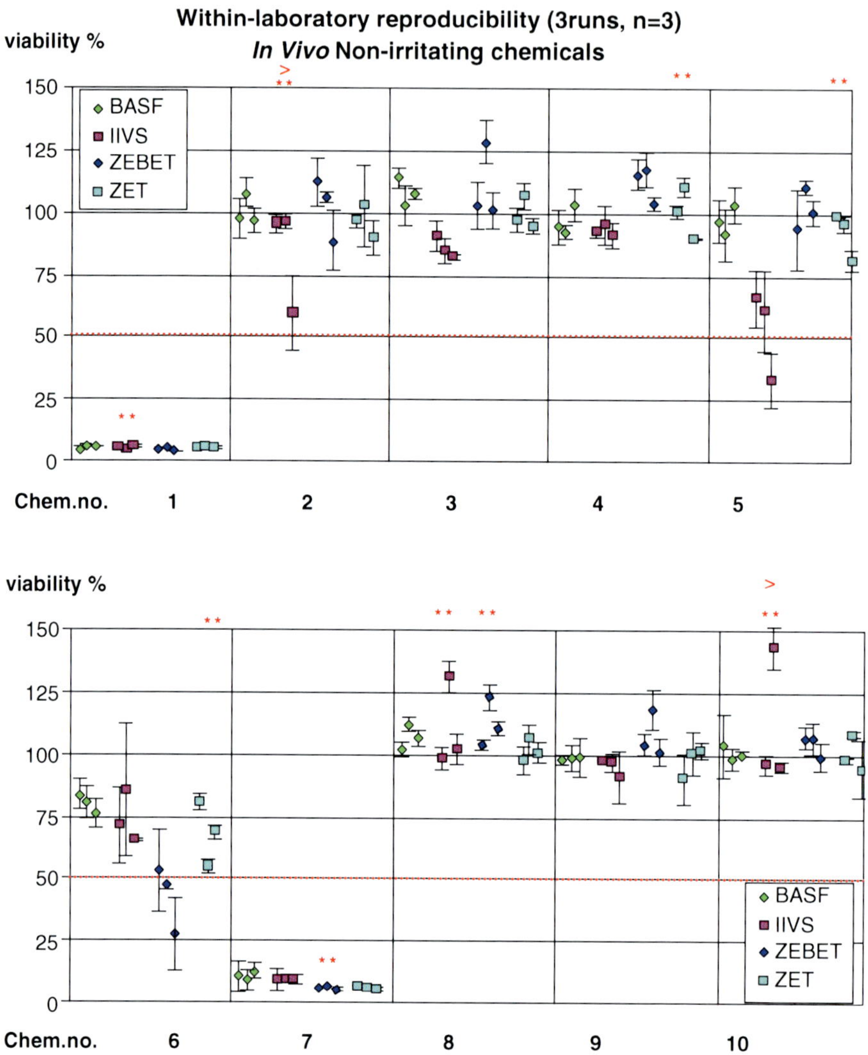

Fig. 3.1 Within laboratory reproducibility for *in vivo* non-irritating chemicals. **: 1-way ANOVA high differences between runs; >: SD over 20% between runs

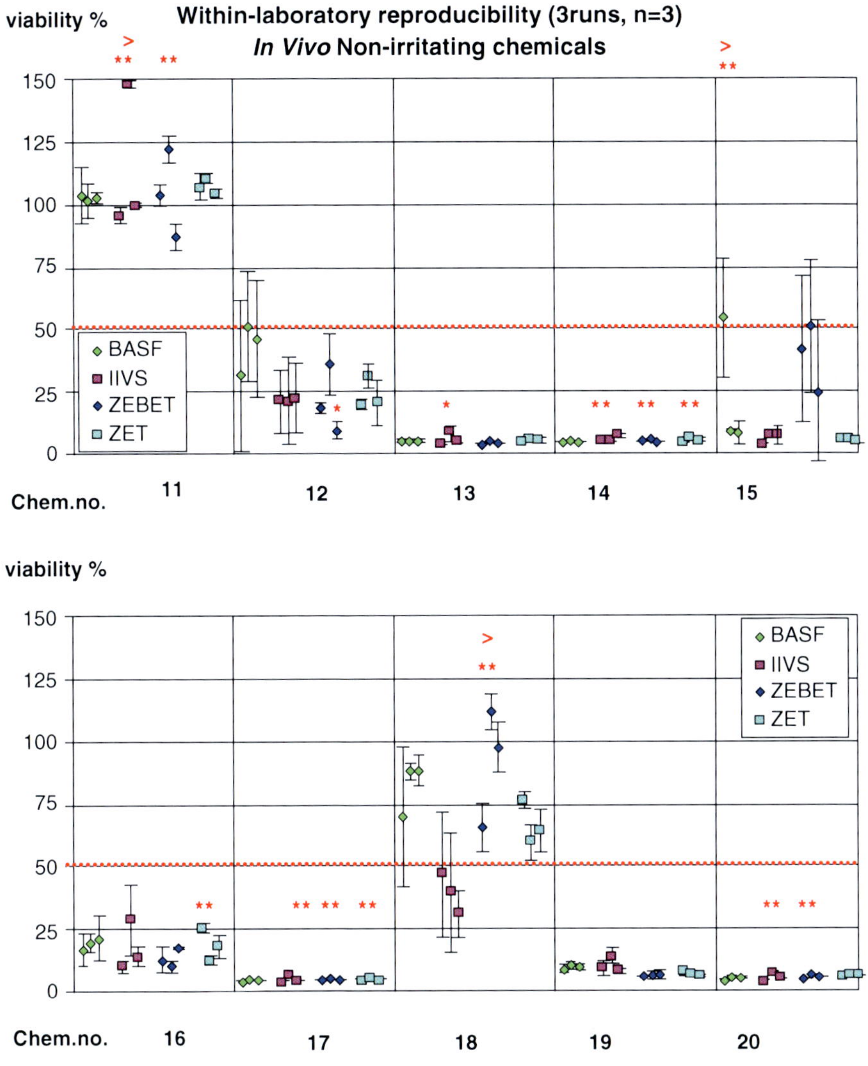

Fig. 3.2 Within laboratory reproducibility for *in vivo* irritating chemicals. **: 1-way ANOVA high differences between runs; >: SD over 20% between runs

3.3.2 Predictive Capacity

The predictive capacity of the final version of the EpiDerm™ SIT, that has been adopted and regulatory accepted as full replacement method, has been established using 55 chemicals and published by Kandarova et al. in 2007 [10]. The 2 × 2 Contingency table (Table 3.1) provides the statistical values demonstrating the high level of sensitivity and specificity of the method.

The follow-up validation study between four laboratories was conducted using 20 reference compounds listed in Table 3.2. These chemicals were selected by ECVAM as reference substances for future validation studies. Overall sensitivity and specificity and accuracy of almost 80% has been achieved for all three parameters (Table 3.3).

Table 3.1 2 × 2 contingency table for sensitivity, specificity and accuracy obtained with the EpiDerm™ SIT when testing 55 chemicals [10]

Modified SIT	EU DSD 2 classes Cut-off = 2.0 *N* = 55 experiments	UN GHS / EU CLP 2 classes Cut-off = 2.3 *N* = 55 experiments
Sensitivity	83.3% (20/24)	94.1% (16/17)
Specificity	77.4 (24/31)	71.1% (27/38)
Positive predictive value	74.1%	59.3%
Negative predictive value	85.7%	96.4%
Accuracy	80%	78.2%

Table 3.2 Twenty reference compounds tested in the follow-up validation study by four laboratories

No	Chemical name	CAS number	GHS	*In vivo* score (DIS)
1	1-bromo-4-chlorobutane	6940-78-9	No cat	0
2	Diethyl phthalate	84-66-2	No cat	0
3	Di-propylene glycol	25265-71-8	No cat	0
4	Naphthalene acetic acid	86-87-3	No cat	0
5	Allyl phenoxy-acetate	7493-74-5	No cat	0.3
6	Isopropanol	67-63-0	No cat	0.3
7	4-methyl-thio-benzaldehyde	3446-89-7	No cat	1
8	Methyl stearate	112-61-8	No cat	1
9	Allyl heptanoate	142-19-8	Cat 3	1.7
10	Heptyl butyrate	5870-93-9	Cat 3	1.7
11	Hexyl salicylate	6259-76-3	Cat 3	2
12	Terpinyl acetate	80-26-2	Cat 3	2
13	Tri-isobutyl phosphate	126-71-6	Cat 3	2
14	1-decanol	112-30-1	Cat 2	2.3
15	Cyclamen aldehyde	103-95-7	Cat 2	2.3
16	1-bromohexane	111-25-1	Cat 2	2.7
17	a-terpineol	98-55-5	Cat 2	2.7
18	Di-n-propyl disulphide	629-19-6	Cat 2	3
19	Butyl methacrylate	97-88-1	Cat 2	3
20	Heptanal	111-71-7	Cat 2	3,3

DIS Dominating irritating score

Table 3.3 2 × 2 contingency statistics for four testing laboratories and 20 substances

Laboratory	Sensitivity [%]	Specificity [%]	Accuracy [%]
BASF	73.3	80.0	76.7
ZEBET	76.7	73.3	75.0
IIVS	90.0	76.7	83.3
ZET	80.0	80.0	80.0
Overall	*80.0*	*77.5*	*78.8*

3.3.3 Applications and Limitations

The EpiDerm™ SIT was developed and designed to predict skin irritation potential of neat test chemicals in the context of identification and classification of skin irritation hazard according to the EU and GHS classification system. No clear applicability domain restrictions could be defined for EpiDerm™ SIT (similarly as for EPISKIN™ assay), except for testing gases, vapours and aerosols. Testing of these types of chemicals require special conditions not covered by the current test design.

The method is also not suitable for prediction of Cat 3 chemicals. Although IL-1α may provide some additional indication about possibly mild irritation effects of chemicals predicted as "no category", no prediction model has so far been validated and officially accepted by regulators.

One limitation of this assay method is a possible interference of the test chemicals with the MTT endpoint. A coloured test chemical, or one that directly reduces MTT (and thereby mimics dehydrogenase activity of the cellular mitochondria), may interfere with the MTT endpoint. However, these test chemical are a problem only if at the time of the MTT test (i.e. 42 h after test chemical exposure) sufficient amounts of the test chemical are still present on (or in) the tissues. In case of this unlikely event, the (true) metabolic MTT reduction and the contribution by a coloured test material or (false) direct MTT reduction by the test material can be quantified by a special procedure described in details in the SOP provided by MatTek [22, 23].

3.4 Comparison to Human Data

A set of limited human data from controlled 4 h–huma patch testing exists [24]. These data have been generated with scientific interest to evaluate whether the false negative results from the original EpiDerm™ SIT were obtained due to the insufficient exposure time *in vitro*, or whether the rabbit test provides oversensitive outcomes.

Of the 16 chemicals classified as irritants in the rabbit, only five substances were found to be significantly irritating to human skin. Concordance of the rabbit test with the 4-h. HPT was only 56%, whereas concordance of human epidermis models with human data was 76% (original EpiDerm™ SIT) and 70% (EPISKIN™). These results confirm observations that rabbits over-predicts skin effects seen in humans. Therefore, when validating *in vitro* methods, all available information, including human data, should be taken into account before making conclusions about their predictive capacity.

3.5 Brief Description of the Protocol: Experimental Procedure

3.5.1 Reconstituted Human Skin Model

The reconstructed tissue model EpiDerm™ (MatTek, Ashland, USA and MatTek IVLSL, Bratislava, Slovakia—ISO 9001:2008 certified) consists of normal, human-derived epidermal keratinocytes (taken from healthy volunteers negative to HIV, and Hepatitis) which have been cultured to form a multilayered, highly differentiated model of the human epidermis. It consists of organized basal, spinous and granular layers, and a multilayered stratum corneum containing intercellular lamellar lipid layers arranged in patterns analogous to those found *in vivo* [25].

The EpiDerm™ tissues (surface 0.63 cm^2) are cultured on specially prepared cell culture inserts and shipped to customers as kits, containing 24 tissues on shipping agarose together with necessary amount of culture media and 6-well plates. In addition the MTT kit (containing MTT concentrate, diluent, extractant, PBS and 24-well plate) can be provided by MatTek.

Quality controls of the test system: The EpiDerm™ System is manufactured according to defined quality assurance procedures compliant to GMP process. All biological components of the epidermis and the culture medium are tested by manufacturer for viral, bacterial, fungal and mycoplasma contamination. Barrier properties of each manufactured tissue lot are controlled by manufacturer. Per request, MatTek provides detailed information about ET50 experiment with Triton X-100 (1%) (chemical recommended as penetration marker by the OECD TG 439), information of tissue viability (MTT test), together with historical database of results.

Handling procedures for biological materials should be followed. It is recommended to wear gloves during handling with the skin and kit components (Table 3.4). After use, the epidermis, the material and all media in contact with it should be decontaminated prior to disposal (e.g. using 10% bleach or special containers).

3.5.2 Materials

3.5.2.1 Protocol Steps: Day 0–day Prior to Dosing

Note: Before any testing on the viable reconstructed human tissues is performed, it is recommended to perform the evaluation of the test chemical for interference with the measured endpoint (MTT assay). This procedure is described in details in the SOP that is provided together with the testing kit by MatTek.

1. EPI-200-SIT kits are shipped from MatTek facilities in USA and Slovakia (EU) every Monday.
2. Upon receipt of the shipment, examine all kit components for integrity. If there is a concern call MatTek immediately.
3. Record all information about supplied material into the MDS.
4. Place the DPBS into the refrigerator (5 ± 3 °C) and the vial containing the MTT concentrate in the freezer (−20 ± 5 °C).

Table 3.4 Material required for the conductance of the EpiDerm™ SIT

#	Description	Detail
(A) Material provided by MatTek Corporation with standard EPI-200-SIT Kit		
1	One sealed 24-well plate containing 24 inserts of EpiDerm™ embedded in transporting agar	EPI-200, 0.6 cm^2
2	Sterile 6-well plates used during the assay	8 pieces
3	Sterile 24-well plates used for MTT assay	2 pieces
4	One Bottle of DPBS	100 ml
5	One bottle Assay Medium, EPI-100-NMM	100 ml
6	One vial, containing the positive control chemical—5% SDS	2 ml
7	Sterile Nylon Mesh for application of liquid test materials	25 pieces
8	Protocol for Skin Irritation test according to the OECD TG 439	
(B) MTT-100 Assay Kit Components (ordered separately)		
1	One vial containing MTT concentrate (5 mg/ml)	2 ml
2	One vial MTT diluent (DMEM based culture medium)	8 ml
3	One bottle containing extracting solution isopropanol	60 ml
(C) Additional material and equipment needed		
1	Sterile Dulbecco's PBS (DPBS) without Ca^{2+} and Mg^{2+} (e.g. PAN or Biochrom)	2 L
2	Sterile, sharp blunt-edged forceps	
3	Positive displacement pipette for application of semi-solid test materials	30 μl
4	Mortar and pestle for grinding of granular solids	
5	Sterile disposable pipettes, pipette tips	20–200 μl 200–1000 μl
6	Sharp spoon—for application of solids. Aesculap, Purchase Number.: FK 623	NaCl weight: 25 mg
7	Bulb headed Pasteur pipettes—for spreading of test chemicals	
8	Parafilm	
9	Sterile cotton tip swabs	
10	Laminar flow hood—for work under sterile conditions	
11	Humidified incubator	37 °C, 5% CO_2, 95% relative humidity
12	96-well plate photometer equipped with filter 570 nm	
13	Laboratory balance	
14	Plate shaker	
15	Stop-watches	
16	Wash bottle	500 ml
17	Beakers—for washing and collecting DPBS	200 ml

Tissue conditioning:

1. Let the assay medium reach room temperature (20–25 °C). Do not pre-heat to 37 °C.
2. Pipette 0.9 ml of the assay medium into each well of sterile 6-well plates (For 24 inserts prepare eight 6-well plates. Use one 6-well plate for pre-incubation of three inserts).
3. Under sterile conditions, open the plastic bag containing the 24-well plate with epidermal tissues. Under a sterile airflow, remove the sterile gauze and

carefully (using sterile forceps) take out each insert containing the epidermal tissue. Remove any remaining agarose that adheres to the outer sides of the insert by gentle blotting on the sterile filter paper or gauze, and place the tissues in the empty, sterile 24-well plate.

4. Perform visual inspection of the inserts within the next 5 min. Record any tissue defects and excess moisture on the surface. Do not use tissues with defects or tissues with excessive moisture on the surface.
5. Dry the surface of the tissues with a sterile cotton tip swab and transfer tissues to a 6-well plate pre-filled with 0.9 ml medium. Place the plates for 60 ± 5 min into the incubator (37 ± 1 °C, 5 ± 1% CO2, 95% RH).
6. At the end of the first (60 min) pre-incubation period, transfer the inserts from upper wells into the lower wells of the 6-well plate. Further, pre-incubate the tissues (37 ± 1 °C, 5 ± 1% CO2, 95% RH) overnight for 18 ± 3 h.
7. Place the plates back into incubator for overnight pre-incubation.
8. Place the rest of the assay medium into the refrigerator (5 ± 3 °C) and the vial containing the MTT concentrate in the freezer (−20 ± 5 °C).
9. If necessary, prepare sufficient amount of rinsing DPBS for the next day (approximately 1.5 L per 24 inserts).
10. Prepare and sterilize all devices which will be used in the assay.

Day 1: Chemical exposure.
Note: Do not dose more than 18 tissues (=6 test articles including PC and NC in a block (SET), in order to be able to perform all steps as required by this protocol.

1. Place all devices, solution and chemicals necessary for the test into the sterile hood.
2. Prepare a sufficient number of 6-well plates pre-filled with 0.9 ml of assay medium in the upper row (1 plate = 1 chemical).
3. Remove the pre-equilibrated, 6-well plates from the incubator approximately 5 min before exposure to chemicals will begin.
4. Evaluate the surface of tissues and exclude completely wet tissues or tissues with any visible defects.
5. Remove any moisture using sterile cotton tip.
6. Before test chemical exposure, label all 6-well plate lids with the test material codes or names.
7. Apply 30 μl (liquid) or 25 mg (solid) of the undiluted test chemical, NC or PC to three single tissues each. Dose tissues at the time intervals needed later for rising off the test chemicals (optimal and highly recommended is 1 min interval).
8. Keep the plates with dosed tissues in the laminar flow hood, until the last tissue is dosed.
9. After dosing the last tissue, transfer *all plates for 35 ± 1* min to the humidified incubator (37 ± 1 °C, 5 ± 1% CO2, 95% RH).
10. After 35 min, remove *all plates* from the incubator, place them into the sterile hood and wait until the period of 60 min is completed for the first dosed tissue.

11. After the *60 ± 1 min* test chemical exposure, rinse the tissues with sterile DPBS, filling and emptying the tissue insert *15 times* to remove any residual test material. Use constant stream of DPBS applied from 1.5 cm distance from the tissue surface. (The stream of DPBS should not be too soft, otherwise, the test article will not be removed)
12. After the 15th rinse from washing bottle, completely submerge the insert 3 times in 150 ml DPBS (shake to remove all rests of test material).
13. Finally, rinse the tissue once from inside and once from outside with sterile DPBS. Remove excess of DPBS by gentle shaking the insert, blot insert on sterile blotting paper.
14. Transfer the blotted tissue inserts to new 6-well plates pre-filled with 0.9 ml of fresh assay medium.
15. After all inserts are washed, *DO NOT FORGET to carefully dry* the surface of each tissue with a sterile cotton tipped swab. In case that traces of the chemical are still present on the surface, try to remove it with the sterile wetted cotton swab. Record this procedure in the MDS. You may evaluate visually tissue surface under a dissecting stereoscope.
16. Incubate tissues in the incubator for next 24 ± 2 h. Record start time of incubation in the MDS.

Day 2–Change medium (mandatory—steps 1–2) and collect media for cytokine analysis (optional—steps 3–7)

1. At the end of the 24 ± 2 h. incubation period, pre-fill the lower row of the 6-well plates with 0.9 ml of fresh assay medium.
2. Transfer the inserts from the upper row of the 6-well plates into the lower row and place the 6-well plates back into the incubator for an additional 18 ± 2 h. post-incubation.
3. If the medium from the 24 h incubation will be analyzed for cytokine or chemokine release, prepare a sufficient number of sterile vials (e.g. cryotubes, volume 1.5 ml). Alternatively, the media can be stored in a labeled 24-well plate.
4. Mark the cryotubes or a 24-well plate with names or codes of the test chemicals and replicate code (e.g. a, b, c). Include the tissue lot number and date of the experiment. Use a water resistant marker.
5. Place the 6-well plates containing inserts on a plate shaker (500 rpm/min) for 5 min.
6. Transfer the medium (approximately 0.9 ml) from the 24-h incubation plates into the cryotubes or 24-well plate. Use fresh pipette tips between samples.
7. Close the vials properly. If used, the 24-well plate should be sealed with parafilm. Store the samples at −20 ± 5 °C (for up to 12 months) until analysis.

Day 3–MTT viability test

1. Prior to the MTT assay, label a sufficient number of 24-well plates.
2. Prepare MTT medium from frozen concentrate and pipette 300 μl of MTT medium in each well.

3. Remove inserts from the 6-well plates, blot the bottom of the inserts, and transfer them into the 24-well plates, pre-filled with 0.3 ml of MTT (1 mg/ml). Place the plates in the incubator (37 ± 1 °C, 5 ± 1% CO2, 95% RH), record the start time of MTT incubation in the MDS and incubate for *3 h ± 5 min*.
4. After MTT incubation is complete, gently blot the tissue on the absorbing paper and transfer inserts into new 24-well plates.
5. Immerse the inserts by gently pipetting 2 ml of isopropanol (extractant solution) into each insert. The level will rise above the upper edges of the insert, thus completely covering the tissues from both sides.
6. Seal the 24-well plates (e.g. with Parafilm or place into a sealable plastic bag) to inhibit extractant evaporation. Record start time of extraction in the MDS and extract formazan for at least 2 h at room temperature with gentle shaking on a plate shaker (120–200 rpm).
7. As an alternative, overnight extraction is also possible. Seal plates as described above and extract at room temperature in the dark, without shaking. Before using the extracts, shake for at least 15 min on plate shaker. After the extraction period is complete, pierce the inserts with an injection needle (~gauge 20, ~ 0.9 mm diameter) and allow the extract to run into the well from which the insert was taken. Afterwards the insert can be discarded. Before transferring the extract to 96 well plates pipette up and down 3x until the extractant solution is homogenous.
8. For each tissue, transfer 2 × 200 μl aliquots of the blue formazan solution into a 96-well flat bottom microtiter plate according to the fixed plate design given in spreadsheet. Use isopropanol as blanks.
9. Read OD in a 96-well plate spectrophotometer using a wavelength between 540 and 595 nm, preferably at *570 nm, without using a reference filter.*

3.5.2.2 Test Data

A blank, password protected MS EXCEL workbook *EpiDerm™-SIT-SPREAD.XLS* can be provided by MatTek. A copy should be made before the first data entry. The workbook consists of two single spreadsheets named: IMPORT and SPREAD. Data files of optical densities (ODs) generated by the microplate reader (without blank subtraction) are copied from the reader software to the Windows Clipboard and then pasted into the first spreadsheet of the EXCEL workbook. The blank corrections, calculation of results and statistical parameters are done automatically in the second part of the workbook. Use the fixed 96-well plate design as specified in the SOP provided by MatTek.

After data entry, the spreadsheet performs the following calculations:

1. Blank correction
2. For each individual tissue treated with a test chemical (TS), the positive control (PC) and the negative control (NC) the individual relative tissue viability is calculated according to the following formulas
   Relative viability TS (%) = [ODTS/Mean of ODNC] × 100.
   Relative viability NC (%) = [ODNC/mean of ODNC] × 100.
   Relative viability PC (%) = [ODPC/mean of ODNC] × 100.

Table 3.5 Prediction model

In vitro result	*In vivo* prediction
Mean tissue viability ≤50%	Irritant (I) (EU DSD R38 or UN GHS / EU CLP category 2)
Mean tissue viability >50%	Non-irritant (no classification)

3. For each test chemical, negative control, and the positive control, the mean relative viability of the three individual tissues is calculated and used for classification according to the Prediction Model.
4. The spreadsheet shows a graph of the results (% of relative viability ±SD)

Data interpretation procedure (Prediction Model)

According to the EU and GHS classification (R38/Category 2 or no class), an irritant is predicted if the mean relative tissue viability of three individual tissues exposed to the test chemical is reduced below 50% of the mean viability of the negative controls (Table 3.5).

3.5.3 Assay Quality Controls (OECD TG 439)

Assay acceptance criterion 1: negative control.

The *absolute OD* of the negative control (NC) tissues (treated with sterile DPBS) in the MTT-test is an indicator of tissue viability obtained in the testing laboratory after shipping and storing procedures and under specific conditions of use.

The assay meets the acceptance criterion if the mean OD570 of the NC tissues is >0.8 and ≤ 2.8.

Assay acceptance criterion 2: positive control.

A *5% SDS* (in H2O) solution is used as positive control (PC) and tested concurrently with the test chemicals. Concurrent means here the PC has to be tested in each assay, but not more than one PC is required per testing day. Viability of positive control should be within 95 ± 1% confidence interval of the historical data.

The assay meets the acceptance criterion if the mean viability of PC tissues expressed as % of the negative control tissues is ≤20%.

Assay acceptance criterion 3: standard deviation (SD).

Since in each test skin irritancy potential is predicted from the mean viability determined on three single tissues, the variability of tissue replicates should be acceptably low.

The assay meets the acceptance criterion if the SD calculated from individual % tissue viabilities of the three identically treated replicates is <18%.

Note: Chemicals that provide tissue viabilities in a range of 30–70% may provide high SD. If the high SD (above acceptance limits) is typical for the chemical and the classification of the chemical is consistent in all independent runs, it is recommended to accept this result, although the Assay Acceptance Criterion 3 is not met.

3.6 Role in a Testing Strategy

According to the UN GHS /EU CLP classification systems, the EpiDerm™ SIT method is able to identify Cat. 2 and No Cat. chemicals and can thus serve as a stand-alone skin irritation method for test chemicals identified as non-corrosives and in countries where the optional UN GHS Cat. 3 is not required. For authorities adopting the optional UN GHS Cat. 3, additional testing in an *in vitro* skin irritation test method not adopted by the OECD or in the *in vivo* test method may be required to resolve Cat. 3 from No Cat. If the EpiDerm™ SIT method results in Cat. 2, an *in vitro* skin corrosion test, if not previously performed, is required to determine the final classification (Cat. 2 (irritant) or Cat. 1(A, B or C) (corrosive) [26].

3.7 Perspectives from the Test Developer

3.7.1 Critical Steps in the Protocol

The EpiDerm™ SIT is a robust and easy to perform method that utilizes readily available laboratory equipment. The test can be performed by most laboratory personnel, provided that care is taken during the critical steps of exposure and washing.

Unequal spreading of a test material may lead to the false negative outcomes or high variability. Insufficient washing may lead to the over-prediction due to the additional 42 h post-exposure time. Omission of aseptic techniques and use of non-sterile tools may lead to contamination.

Special care should be taken when testing colorant materials (blue, deep red, violet) and MTT reducing materials since they will interfere with the MTT endpoint. MatTek and other tissue model suppliers have developed procedures to deal with such materials. The procedures are described in details in the SOPs provided with the testing kits/tissue products.

3.7.2 Possible Protocol Adaptations

If a test material sticks to the surface, it is possible to improve the washing technique and removal of the material by submerging the tissue completely into the DPBS for about 10 min. This step may also help achieve physiological pH levels after the tissues were exposed to materials with either low or high pH.

The EpiDerm™ SIT protocol has been developed to cover the needs of the REACH (Chemical regulations) regarding the classification and labeling purposes. Also, it partially covers the transport regulations. However, other protocols address the needs of the cosmetic industry for the assessment of mildness and skin tolerance of the cosmetic products, and of the medical devices industry with regard to the biocompatibility testing of the materials used (e.g. as implants). These protocols use long exposure times (typically 18–24 h) to better reflect the in-use conditions.

A protocol for testing final cosmetic products has been described by Faller et al. 2002 [18] In this study, 22 formulations, covering the full range of irritation responses *in vivo* and representing different cosmetic product classes, were tested in humans and *in vitro* with three different reconstructed human epidermis equivalents. The human data for the 22 coded products correlated well to the *in vitro* data. A coefficients of correlation of R = 0.94 for EpiDerm™, and of R = 0.90 and 0.84 for the other two models were obtained [18]. This study clearly demonstrated the usefulness and relevance of RhE equivalents for the *in vitro* assessment of the irritation potential of a series of cosmetic products. Moreover, it demonstrated the high correlation of the EpiDerm™ ET-50 approach (time-to-toxicity protocol) with results obtained in humans.

A protocol for medical devices testing (i.e. testing of irritation potential of extracts obtained from medical devices) has been published by Casas et al. in 2013 [27]. MatTek together with RIVM conducted follow up testing and optimization of the protocol [28] and the validation study of this protocol is on-going. Detailed SOP has been released to the validation laboratories and will be published in due course.

3.7.3 Challenges and Opportunities

The current method is not suitable for the prediction of the UN GHS optional Cat 3 chemicals (mild irritants). Although IL-1α may provide some additional indication about mild irritation effects of chemicals it is questionable, whether the chemicals classified as Cat 3 in rabbits would present any hazard to man. To develop a protocol that is capable of correctly predicting Cat 3 chemicals (mild irritants) it will be necessary to take into account human data.

3.8 Conclusions

The EpiDerm™ Skin Irritation Test (OECD TG 439), in its current form is a useful tool in the hands of modern toxicologists. Since the implementation of this assay into the EU and OECD test guidelines, the number of animals required for predicting skin irritation has decreased to a minimum. Further studies should focus on predicting UN GHS optional Cat. 3 chemicals and the applicability of the test method to correctly predict the irritation potential of complex mixtures.

References

1. United Nations (UN). Globally harmonized system of classification and labeling of chemicals (GHS), 2nd revised edn, UN New York and Geneva, 2007. Available: http://www.unece.org/trans/danger/publi/ghs/ghs_rev02/02files_e.html.
2. OECD. OECD guideline for testing of chemicals, No. 404: acute dermal irritation/corrosion. Organisation for economic cooperation and development, Paris; 2015.
3. Draize JH, Woodand G, Calvery HO. Methods for the study of irritation and toxicity of substances applied topically to the skin and mucous membranes. J Pharmacol Exp Ther. 1944;82:337–90.

4. Portes P, Grandidier M-H, Cohen C, Roguet R. Refinement of the Episkin protocol for the assessment of acute skin irritation of chemicals: follow-up to the ECVAM prevalidation study. Toxicol In Vitro. 2002;16:765–70.
5. Cotovio J, Grandidier M-H, Portes P, Roguet R, Rubinstenn G. The *In Vitro* acute skin irritation of chemicals: optimisation of the EPISKIN prediction model within the framework of the ECVAM validation process. ATLA. 2005;33:329–49.
6. Kandárová H, Liebsch M, Genschow E, Gerner I, Traue D, Slawik B, Spielmann H. Optimization of the EpiDerm™ test protocol for the upcoming ECVAM validation study on *in vitro* skin irritation tests. ALTEX. 2004;21(3/04):107–14.
7. Kandárová H, Liebsch M, Gerner I, Schmidt E, Genschow E, Traue D, Spielmann H. EpiDerm™ skin irritation test protocol–assessment of the performance of the optimized test. ATLA. 2005;33(4):351–67.
8. Spielmann H, Hoffmann S, Liebsch M, Botham P, Fentem J, Eskes C, Roguet R, Cotovio J, Cole T, Worth A, Heylings J, Jones P, Robles C, Kandárová H, Gamer A, Remmele M, Curren R, Raabe H, Cockshott A, Gerner I, Zuang V. The ECVAM international validation study on *in vitro* tests for acute skin irritation: report on the validity of the EPISKIN and EpiDerm™ assays and on the skin integrity function test. ATLA. 2007;35:559–601.
9. ESAC. Statement on the scientific validity of in-vitro tests for skin irritation testing, April 2007. 2007. Downloadable from http://ecvam.jrc.it/.
10. Kandárová H, Liebsch M, Gerner I, Schmidt E, Genschow E, Traue D, Spielmann H. The EpiDerm test protocol for the upcoming ECVAM validation study on skin irritation tests—an assessment of the performance of the optimized test. ATLA. 2005;33:351–67.
11. Liebsch M, Gamer A, Curren R, Frank J, Genschow E, Tharmann J, Remmele M, Bauer B, Raabe H, Barnes N, Hilberer A, Wilt N, Lorngjad-Schäfer MR, Schäfer C, Spiller E, Hayden P, Kanďárová H. Follow-up validation of the modified EpiDerm™ skin irritation test (SIT): results of a multicentre study of twenty reference test substances. 15th congress on alternatives to animal experimentation, September 19-21 Linz, Austria. ALTEX. 2008;25(Suppl 1):43.
12. Liebsch M, Kandarova H. Submission of the follow-up validation study of the modified EpiDerm™ skin irritation test for hazard identification and labelling of chemicals according to EU classification system. 23 April 2008; 2008. p. 58, 10 Annexes.
13. OECD. Guidance document no. 34 on the validation and international acceptance of new or updated test methods for hazard assessment. Paris: OECD; 2005.
14. ECVAM. Performance standards for applying human skin models to in vitro skin irritation; 2007. Available under download study document, at http://ecvam.jrc.ec.europa.eu.
15. ESAC. Statement on the scientific validity of in-vitro tests for skin irritation testing, November 2008; 2008. Downloadable from http://ecvam.jrc.it/
16. EU. B.46 Guideline. *In Vitro* skin irritation: reconstructed human epidermis (RHE) model test; 2009.
17. Mossman T. Rapid colorimetric assay for cell growth and survival application to proliferation and cytotoxicity assays. J Immunol Methods. 1983;65:55–63.
18. Faller C, Bracher N, Dami N, Roguet R. Predictive ability of reconstructed human epidermis equivalents for assessment of skin irritation of cosmetics. Toxicol In Vitro. 2002;16(5):557–72.
19. Faller C, Bracher M. Reconstructed skin kits: reproducibility of cutaneous irritating testing. Skin Pharmacol Appl Ski Physiol. 2002;15(Suppl 1):74–91.
20. OECD. Test no. 439: in vitro skin irritation reconstructed human epidermis test method. Organisation for economic cooperation and development, 23 July 2010. 2010.
21. ESAC. Esac statement on the performance standards (ps) for in vitro 2 skin irritation testing using reconstructed human epidermis; 2009. Downloadable from: https://eurl-ecvam.jrc.ec.europa.eu/validation-regulatory-acceptance/docs-skin-irritation-1/DOC9-ESAC_statement_updated-skin-irritation-PS_200909-22.pdf.
22. Kandárová H, Hayden P, Klausner M, Kubilus J, Sheasgreen J. An *In Vitro* skin irritation test (SIT) using the EpiDerm™ reconstructed human epidermal (RHE) model. JoVE. 2009;29 doi:10.3791/1366.

23. MatTek SOP In Vitro EpiDerm™ Skin Irritation Test (EPI-200-SIT) For use with MatTek Corporation's Reconstructed Human Epidermal Model EpiDerm™ (EPI-200-SIT). Doc. MK-24-007-0023, Version 07/11/2014; 2014. p. 35.
24. Jírová D, Basketter D, Liebsch M, Bendová H, Kejlová K, Marriott M, Kandárová H. Comparison of human skin irritation patch test data with *in vitro* skin irritation assays and animal data. Contact Dermatitis. 2010;62(2):109–16. doi:10.1111/j.1600-0536. 2009.01640.x.
25. Cannon CL, Neal PJ, Southee JA, Kubilus J, Klausner M. New epidermal model for dermal irritancy testing. Toxicol In Vitro. 1994;8:889–91.
26. OECD GD No. 203. Guidance document on an integrated approach on testing and assessment (IATA) for skin corrosion and irritation. Series on testing and assessment; 2014. http://www.oecd.org/officialdocuments/publicdisplaydocumentpdf/?cote=env/jm/mono(2014)19&doclanguage=en.
27. Casas JW, Lewerenz GM, Rankin EA, Willoughby JA Sr, Blakeman LC, McKim JM Jr, Coleman KP. *In vitro* human skin irritation test for evaluation of medical device extracts. Toxicol In Vitro. 2013;27(2013):2175–83.
28. Kandarova H, Willoughby JA, de Jong WH, Bachelor MA, Letasiova S, Milasova T, Breyfogle B, de la Fonteyne L, Coleman KP. Development, optimization, and standardization of an in vitro skin irritation test for medical devices using the reconstructed human tissue model EpiDerm. SOT 2015. The Toxicologist, 2015. p. 433. https://www.toxicology.org/pubs/docs/Tox/2015Tox.pdf.

An *In Vitro* Skin Irritation Test Using the SkinEthic™ Reconstructed Human Epidermal (RHE) Model

4

Nathalie Alépée, Marie Hélène Grandidier, Carine Tornier, and José Cotovio

4.1 Principle of the Test Method and Scientific Basis

Acute irritation is characterised by the non-immunological inflammatory response of living skin following injury caused by a single contact with an irritant substance. This response is local and reversible (unlike that produced by corrosion, which is irreversible). The *in vivo* evaluation of skin irritation is mainly based on semi-quantitative visual scoring (erythema and oedema). Besides morphological changes, irritation also involves more-complex, subjective and subtle phenomena, such as itching and burning sensations, which are not easily measurable [1]. Since cytotoxicity is also known (among other factors) to trigger irritation, it can be viewed as a first event likely to be shared by the effects of many irritants. Following mechanical or chemical assault, homeostatic mechanisms may be deregulated, leading to non-specific inflammation processes triggered by inflammatory mediators originating mainly from keratinocytes [2]. Cell and tissue damage lead to the release of inflammatory mediators, nerve stimulation, axonal reflexes, pain and itching [3–5]. The inflammatory response ultimately leads to observable phenomena such as localised skin swelling (oedema) and redness (erythema). Overall, clinical signs of irritation include the development of a rash, inflammation, swelling, scaling, and abnormal tissue growth in the affected area (Fig. 4.1).

Initially, to conduct the skin irritation assessment, most regulatory authorities required a standardized *in vivo* test in which—having first excluded skin corrosion potential—the chemical was applied to the skin of a maximum of three rabbits [6].

N. Alépée (✉) • M.H. Grandidier • J. Cotovio
L'Oréal Research and Innovation, 1, Avenue Eugène Schueller, 93600 Aulnay-sous-Bois Cedex, France
e-mail: nalepee@rd.loreal.com

C. Tornier
EPISKIN, 4 rue Alexander Fleming, 69366 Lyon, France

C. Eskes et al. (eds.), *Alternatives for Dermal Toxicity Testing*,
DOI 10.1007/978-3-319-50353-0_4

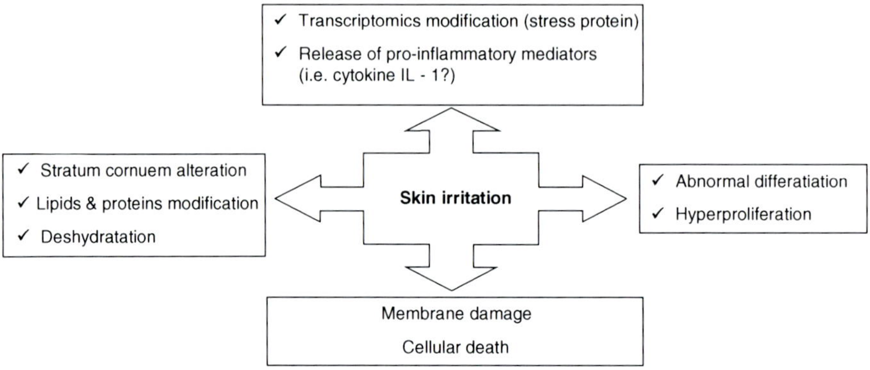

Fig. 4.1 Schematic of skin irritation effects

The ability of the chemical to induce erythema and/or oedema was scored per animal. A score of between 0 and 4 on the Draize scale, increasing with severity, was subjectively assigned on the basis of erythemal and oedemal effects, usually at 24, 48 and 72 h after application of the substance [7]. However, scientific concerns about the variability [8, 9] and predictive capacities of this animal test in terms of human health effects [10–12] were raised.

Animal welfare and, more recently, political pressure in Europe in areas such as legislation relating to chemicals and cosmetics have required the development of appropriate and validated alternative, *in vitro* test methods [13]. In the last 20 years, considerable scientific effort has gone into developing valid *in vitro* skin models to replace animal testing. Initial progress was made through the availability of bioengineered non-invasive methods applicable to the skin *in vivo*, such as trans-epithelial water loss and electrical resistance. These methods permitted the quantification of physiological changes and opened up new possibilities for *in vitro*/*in vivo* comparison [14, 15]. Based on these observations, various *in vitro* models such as primary human keratinocytes [16] and human skin equivalent models [17–19] were evaluated for their ability to assess cutaneous toxicity or irritation. Due to the increasing need for non-animal tests to predict human skin irritation, the European and Japanese Centers for the Validation of Alternative Methods (EURL-ECVAM/JACVAM) have focused their evaluation on four suitable *in vitro* reconstructed human epidermis test methods: these now-validated methods have similarly defined characteristics (Fig. 4.2) and include the SkinEthic™ RHE test method [20–22].

The three-dimensional SkinEthic™ RHE tissue, based on a pioneering concept by Dr. Prunerias, was first released by Martin Rosdy in 1989 [23, 24]. The SkinEthic™ RHE model consists of normal human keratinocytes cultured using a chemically defined growth medium at the air-liquid interface. It produces a highly differentiated and stratified epidermis model comprising main basal, supra basal, spinous and granular layers and a functional stratum corneum with a histological morphology comparable to *in vivo* human tissue [25, 26]. The validated SkinEthic™ RHE skin irritation test method involves a topical application of chemicals for 42 min followed by rinsing and post-incubation for 42 h. Irritant chemicals are

	Defined points	Flexible points
Cells	↳Non-transformed human keratinocytes	↳Source ↳Growth conditions
Test system (model)	↳Human epidermis-like structure ↳Stratum Corneum ↳Lipid profile (main *in vivo* classes) ↳Presence of a barrier function ↳Cytotoxic biomarker: IC-$_{50}$ or ET-$_{50}$ ↳Standard acceptability ranges ↳Containment: no leakage between compartments	↳Qualitative and model specific ↳Model specific ↳SDS, Triton X-100 or other ↳Model specific ↳Size and shape of models ↳Age of the tissues ↳Support membranes
Endpoint	- Viability assay, MTT test - normalized (%) versus controls	
Prediction Model	-Cut-off values (%)	↳ Test-method specific

Fig. 4.2 Specific key points: defined and flexible points

identified by their ability to decrease tissue viability (MTT reduction) below the defined threshold of 50% viability.

4.2 Current Validation Status

The reliability and relevance of the SkinEthic™ RHE skin irritation test method has been established through a rigorous, inter-laboratory validation study. Based on its scientific validity, this test method has been recommended for the testing of all classes of chemicals and for inclusion in tiered testing strategies [27]. The SkinEthic™ RHE test method was originally validated on the basis of the Performance Standards using the 20 defined reference chemicals (ESAC statement from November 2008; [28]). The SkinEthic™ RHE test method has been found scientifically valid in reliably predicting no-label and R38 (irritant) chemicals with respect to the previous EU classification scheme [29]. Re-evaluation based on recalculating the predictive values of the test method under the United Nations (UN) Globally Harmonized System of Classification and Labelling of Chemicals (GHS) was performed in 2008 and confirmed in April 2009 by ESAC for use under the UN GHS system as "applicable to all authorities" [29–31]. As a result, since 2010, the SkinEthic™ RHE test method has been accepted in the official Organisation for Economic Co-operation and Development (OECD) Test Guideline 439 (OECD TG439), allowing the identification of non-irritant and irritant substances and mixtures in accordance with UN GHS and the EU test method B.46 [32–34]. The SkinEthic™ RHE test method was also recently included as part of the Integrated Approach to Testing and Assessment (IATA) for Skin Irritation/Corrosion in OECD Guidance Document 203 [27, 35].

4.3 Performance and Applicability of the Test Method

4.3.1 Reproducibility

Two types of reproducibility were evaluated for the SkinEthic™ RHE test method: one by testing the same chemicals over time in a single laboratory (within-laboratory reproducibility, WLR) and the other by testing the same chemicals in different laboratories (between-laboratory reproducibility, BLR). WLR was calculated as the percentage of chemicals for which identical classifications were obtained in the three valid runs performed. BLR was calculated as the percentage of chemicals for which identical classifications were obtained between laboratories.

The reproducibility study involved evaluating the ten non-irritant and ten irritant reference test chemicals selected in accordance with the Performance Standard document [36]. The 20 chemicals were coded by Vitroscreen and subjected to blind tests in three laboratories: L'Oréal, Coty and Oroxcell. The same concordant classification was observed for 59 out of 60 items (98.3%) for the three laboratories when considering irritants versus non-irritants [20]. Therefore, none of those test substances showed a standard deviation (SD) > 18% in two laboratories. Only the allyl phenoxy-acetate gave a SD > 18% as unacceptable in the third laboratory, demonstrating the reproducibility of the test method. The proportion of identically classified test substances derived from the prediction model was 100% for two laboratories and 95% for the third laboratory, when considering all experiments [20]. In conclusion, regardless the analyses, low intra-and inter-run variability for all laboratories was observed with the negative and positive controls, and the 20 reference test substances indicated high intra- and inter-laboratory reproducibility.

4.3.2 Predictive Capacity

The study conducted by industry was submitted to EURL-ECVAM for evaluation and peer review. The SkinEthic™ test method was regarded by EURL-ECVAM as sufficiently similar to the validated EpiSkin™ method according to the European Classification System based on the Dangerous Substance Directive (DSD) [28]. Sensitivity and specificity for the 20 reference chemicals were 90% and 80%, respectively [20]. The results obtained in the three laboratories with an overall accuracy of 85% met EURL-ECVAM specificity (>80%) and sensitivity (>70%) requirements [36]. EURL-ECVAM also evaluated the test method in its in-house laboratory (called 'Correlate') with regard to transferability. Based on 19 of 20 test chemicals, a sensitivity of 90% and a specificity of 77.8% were reached (data available in

Annexe 5 of the OECD Explanatory Background Document; [33]). The same three test substances (1-bromo-4-chlorobutane, 4-methyl-thio-benzaldehyde and hexyl salicylate) were misclassified, as in other epidermis test methods [20, 22, 33]. No clear difference in the physicochemical properties between the correctly and incorrectly classified test substances was identified to explain this outcome [37]. Increasing the number of tests to 39 chemicals lead to similar predictive capacity with a sensitivity of 90%, a specificity of 80% and an overall accuracy of 85%, with 33 out of 39 test substances correctly classified [38].

In December 2008, the EU adopted and implemented the UN GHS [29] through the Classification, Labelling and Packaging (CLP) Regulation [39]. This regulation replaced the previous EU DSD legislation [40] on the classification of substances and mixtures. The CLP system continues to use two categories to distinguish non-classified (No Category) from irritant (Category 2) substances. However, according to the new rules for skin irritation classification and labelling (C&L) [29, 39], the cut-off score to distinguish between No Category and Category 2 substances was raised to 2.3 (UN GHS or CLP) from 2.0 (EU DSD). Consequently, substances with an *in vivo* score of between 2.0 and 2.3 that were considered irritant under EU DSD are now non-classified under UN GHS. This naturally led to a change in the specificity and sensitivity values. Since UN GHS defines irritants as substances with a score of 2.3 or more, the sensitivity of the SkinEthic™ test system was increased to 100% and the specificity decreased to 69.2% using the 20 reference chemicals. Overall accuracy was 80%, resulting in the test method being endorsed by the EURL-ECVAM and OECD Committees as a stand-alone replacement test method for the *in vivo* Draize rabbit test [41].

4.3.3 Applications and Limitations

This test is designed for mono- and multi-component test chemicals and mixtures. The protocol was established for liquid, viscous, semi-solid and solid chemicals. Topical application to the epidermis makes the method suitable for evaluating chemicals that are soluble or insoluble in water, volatile, creamy, sticky, fatty, powdered, etc." The inclusion of HPLC/UPLC-spectrophotometry to measure formazan in the procedures for the *in vitro* SkinEthic™ RHE test method also extends its applicability to strongly coloured chemicals [42]. The test method is not appropriate for testing gases and aerosols.

4.3.4 Comparison to Human Data

The *in vivo* Draize rabbit skin irritation test is an accepted regulatory method of classifying and labelling chemicals. As such, the classification and labelling results of this test were taken as the "gold standard" in the context of the validation study for the reconstructed human epidemis models. Several large-scale studies on human volunteers conducted in the 1990s concluded that the *in vivo* rabbit test often

Table 4.1 Summary table of *in vivo* and *in vitro* results

Chemical name	CAS number	EU CLP/UN GHS class	Human 4-h patch test	*In vitro* class
Isopropyl myristate	110-27-0	NC	NC	NC
Benzyl salicylate	118-58-1	NC	NC	NC
Isopropyl palmitate	142-91-6	NC	NC	NC
Dodecanoic acid	143-07-7	NC	NC	NC
Dipropylene glycol	25265-71-8	NC	NC	NC
Heptyl butyrate	5870-93-9	NC	NC	NC
Hexyl salicylate	6259-76-3	NC	NC	NC
Isopropanol	67-63-0	NC	NC	NC
Water	7732-18-5	NC	NC	NC
Naphthalene acetic acid	86-87-3	NC	NC	NC
10-Undecenoic acid	112-38-9	NC	NC	Cat 2
1-bromo-4-chlorobutane	6940-78-9	NC	NC	Cat 2
Eugenol	97-53-0	NC	NC	Cat 2
Methyl palmitate	112-39-0	Cat 2	NC	NC
Butyl methacrylate	97-88-1	Cat 2	NC	Cat 2
Alpha terpineol	98-55-5	Cat 2	NC	Cat 2
1-decanol	112-30-1	Cat 2	NC	Cat 2
1-Bromohexane	111-25-1	Cat 2	I	Cat 2
Heptanal	111-71-7	Cat 2	I	Cat 2
Nonanoic acid	112-05-0	Cat 2	I	Cat 2
Octanoic acid	124-07-2	Cat 2	I	Cat 2
Sodium lauryl sulphate (20% aq.)	151-21-3	Cat 2	I	Cat 2
Decanoic acid	334-48-5	Cat 2	I	Cat 2

over-predicts the severity of skin reactions and damage produced by chemicals, although there was also occasionally under-prediction [43–46]. Therefore, as defined by Jirova et al. [47], while concordance between the rabbit test and the results of the 4-h. HPT was rather poor (56%), the reconstructed human epithelium methods provided more convincing results. The results presented in Table 4.1 confirm observations that rabbit tests over-predict skin effects in humans. Given that the SkinEthic™ RHE test method was validated against the over-predicted rabbit test, prediction errs on the side of caution for the safety of consumers, which is essential in the context of risk assessment (Table 4.1).

4.4 Brief Description of the Protocol

Each test chemical (test material, negative and positive controls) is topically applied to three tissue replicates concurrently for 42 min at room temperature (RT), between 18 °C and 24 °C. Exposure to the test chemical is followed by rinsing with phosphate buffer saline (PBS) and mechanically dried. The epidermis is then transferred to a fresh medium and incubated at 37 °C for another 42 h. Cell viability is measured by enzymatic conversion of the vital dye MTT

Table 4.2 Prediction model of the SkinEthic™ RHE skin irritation test method

In vitro results	*In vivo* classification
Mean tissue cell viability ≤50%	Category 2 (Cat. 2)
Mean tissue cell viability >50%	Not classified (NC)

[3-(4,5-Dimethylthiazol-2-yl)-2,5-diphenyltetrazolium bromide, Thiazolyl blue; CAS number 298–93-1] into a blue formazan salt that is quantitatively measured after extraction from tissues [48]. Cell viability is assessed by incubating the tissues for 3 h with 0.3 mL MTT solution (1 mg/mL). The formazan crystals are extracted using 1.5 mL isopropanol for 2 h at RT and quantified by spectrophotometry at 570 nm wavelength. Sodium Dodecyl Sulphate (SDS 5%) and PBS treated epidermis are used as positive and negative controls, respectively. For each treated tissue, the cell viability is expressed as a percentage of the mean negative control tissues. The mean relative tissue cell viability above 50% predicts its non-irritancy potential. Irritant chemicals are identified by their ability to decrease cell viability below the defined threshold level (i.e. ≤50%, for UN GHS Category 2). The prediction model is defined as described below in Table 4.2. Details are provided in the SOP [49] and described in [20]. Key components of the protocol are also available at http://www.episkin.com.

4.5 Role in a Testing Strategy

The evaluation of the skin irritancy and corrosivity potential of a test chemical is a vital part of safety assessment. Alternatives to the rabbit Draize test for skin corrosivity have already received official approval, including human skin model tests using reconstructed human epidermal equivalents such as the SkinEthic™ RHE skin corrosion test method (see Chap. 10). For skin irritation, the SkinEthic™ RHE skin irritation test method was validated as a stand-alone test replacement for the rabbit Draize test (see above). In light of the full evaluation of local skin effects after a single dermal exposure using *in vitro* test methods, the OECD Guidance Document No. 203 on an Integrated Approach to Testing and Assessment (IATA) was established [27]. This IATA approach includes *in vitro* tests for skin corrosion (as described in OECD TG 431) and skin irritation (OECD TG 439) before considering testing on living animals [50].

The top-down approach (an *in vitro* skin corrosion test followed by an *in vitro* skin irritation test if the chemical is identified as non-corrosive in the first test) should be used when all available collected information and the weight-of-evidence (WoE) assessment result in a high a priori probability of the chemical being an irritant or a corrosive. The bottom-up approach (an *in vitro* skin irritation test followed by an *in vitro* skin corrosion test if the chemical is identified as an irritant in the first test) should be used only when all available collected information and the WoE assessment result in a high a priori probability of the chemical not being a skin irritant.

To demonstrate the application and relevance of both approaches using the SkinEthic™ RHE test methods, SkinEthic™ RHE irritation and corrosion data on

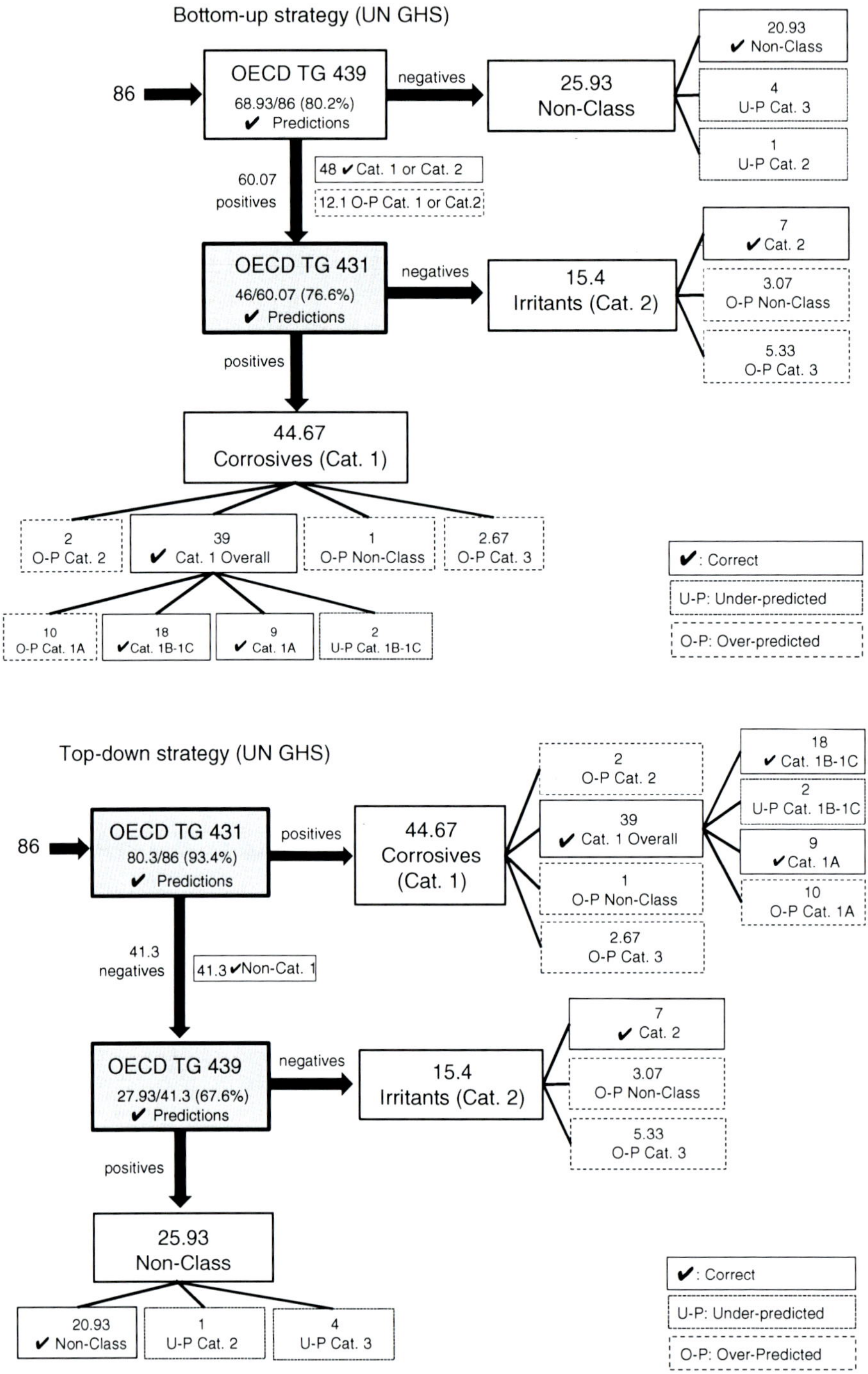

Fig. 4.3 *In vitro* classifications for the 86 test chemicals in the bottom-up testing and the top-down strategies based on the UN GHS classification system. ✔ correct (sub)-category classifications, *U-P* under-predicted, *O-P* over-predicted

86 substances were integrated in a bottom-up and top-down testing strategy to assess their capacity for hazard and safety assessment under UN GHS classifications ([35, 42, 51]). The results showed that the SkinEthic™ RHE model was applicable to a wide range of chemical classes and physical states. The bottom-up and top-down testing strategies showed an identical number of correct and incorrect classifications for the different (sub)-categories (Fig. 4.3). Overall strategies showed an accuracy of 89.5% in distinguishing between non-classified and classified substances, and 93.4% in distinguishing between corrosive and non-corrosive substances (Fig. 4.3). Furthermore, excellent sensitivities were obtained in predicting UN GHS category 1 chemicals (100%), followed by the category 2 irritant substances (70%), irrespective of the strategy and classification system used. Interestingly, none of the skin corrosive category 1B–and-1C and 1A chemicals were under-predicted as a skin irritant (Category 2) or non-classified, irrespective of the strategy and classification system used, suggesting that the SkinEthic™ RHE model ensures consumer safety when used in the context of the OECD recommended IATA. Only a single non-classified substance (2,4-Xylidine) was over-predicted as category 1B–and-1C and none as category 1A, suggesting that the SkinEthic™ RHE model also helps to avoid unnecessary over-labelling.

4.6 Perspectives from the Test Developer

4.6.1 Critical Steps in the Protocol

The critical steps of the standardized operating procedure could be listed as follows:

- Verify the absence of air bubble under the epidermis at each step;
- Test all test chemicals alone in separate plate;
- For liquids (16 μL ± 2 μL), dispense the substance onto the epidermis with a positive displacement pipette and apply a nylon mesh to gently spread the substance, taking care to cover the entire surface;
- For solids (10 ± 2 μL H_2O and 16 ± 2 mg test item), the substance should be crushed to a fine powder, ensuring good contact with the epidermis;
- For viscous and sticky chemicals, use a curved flat spatula or weigh directly on the nylon mesh;
- Apply the chemical-coated side of the nylon mesh to the epidermal surface;
- Carefully remove the nylon mesh before rinsing;
- Rinse the tissue thoroughly;
- Thoroughly protect the plate by stretching three parafilm layers over the plate to prevent the evaporation of the formazan during the extraction step.

4.6.2 Possible Protocol Adaptations

In all reconstructed epidermis test methods, the skin irritation potential of a chemical is determined by measuring tissue viability in treated tissues after topical application to the tissue surface. Tissue viability is determined by enzymatic reduction of MTT tetrazolium salt to purple reduced MTT (formazan) [48]. A known limitation of the photometric MTT-reduction assay is the possible interference of coloured test chemicals with the absorbance measurement of formazan. Analytical methods such as High/Ultra High Performance Liquid Chromatography (HPLC/UPLC) might be more appropriate to detect formazan in the *in vitro* assay. Cosmetics Europe undertook a study to establish and evaluate the use of this analytical method [42]. Based on the outcome of this project, it was concluded that this analytical endpoint detection system is relevant to all test methods, irrespective of the test system and test method used (e.g. SkinEthic™ RHE skin irritation assay). It was therefore recommended that the OECD Test Guideline 439 be revised to incorporate HPLC/UPLC-spectrophotometry as an additional endpoint detection system in the technical procedures for the *in vitro* SkinEthic™ RHE skin irritation test method [32].

4.6.3 Challenges and Opportunities

Challenges and opportunities might be seen in the context of the assessment of specific categories of ingredients (e.g. mixtures) as well as for UN GHS categorization using the SkinEthic™ RHE test method.

The SkinEthic™ RHE test method distinguishes between skin irritants (Cat. 2) and chemicals not classified for skin irritation (No Cat.). However, the test method is not designed to classify chemicals in the optional GHS Cat. 3 (mild irritants). Development of a test method exploiting quantitative analysis of expression profiles of relevant genes might be considered as such an approach was established and defined using the EpiSkin™ RhE-based test system [52].

Mixtures are defined as "a mixture or a solution composed of two or more substances in which they do not react" [34]. Since mixtures cover a wide spectrum of categories and compositions, the type of regulatory testing required may depend on the type of mixture. For example, cosmetic formulations can no longer be tested using animal studies in some parts of the world [53]. In contrast, biocides including mixtures may be subject to specific testing requirements [54]. As such, depending on the field and/or sector, the use of validated *in vitro* assays to assess mixtures is of relevance. Cases in which *in vitro* testing of preparations and mixtures could be useful and relevant include cosmetics, cleaning products, biocides and plant protection products might be very useful [55]. Although these mixtures had high-quality *in vivo* data, not all of them are publically available, allowing only limited comparisons between the *in vivo* and *in vitro* observed effects. Access to *in vivo* data will permit a better definition of the applicability domain of the test method for mixtures with complex physical properties such as hydrophobicity, sticky/buttery-like texture and waxy/creamy foam characteristics. Further investigation would also be beneficial for agrochemicals due to the limited-and-contradictory nature of information available and the difficulty in interpreting the data when the composition of the mixtures has not been identified—as reported for another RhE-based test method [56, 57].

4.7 Conclusions

The SkinEthic™ RHE skin irritation test method has gained international regulatory acceptance and has been adopted for the regulatory assessment of skin irritation to distinguish between EU CLP-UN GHS category 2 (irritant) and non-classified (No Category, non-irritant) chemicals (OECD TG 439). Intra- and inter-reproducibility findings indicate that the SkinEthic™ RHE model has high robustness in terms of its performance with an enlarged dataset of diverse chemicals and mixtures. Furthermore, the relevance of the integration of SkinEthic™ RHE skin irritation data in a bottom-up or top-down strategy has been demonstrated with a similar high accuracy for the determination of the potential hazard of chemicals.

References

1. Emmett EA. Toxic responses of the skin. In: Klaassen CD, Amdur MO, Doull J, editors. Casarett and Dull's toxicology. New York, NY, USA: Macmillan; 1986. p. 412–31.
2. Nickoloff BJ, Turka LA. Keratinocytes: key immunocytes of the integument. Am J Pathol. 1993;143:325–31.
3. Fluhr JW, Darlenski R, Angelova-Fischer I, Tsankov N, Basketter D. Skin irritation and sensitization: mechanisms and new approaches for risk assessment. 1. Skin irritation. Skin Pharmacol Physiol. 2008;21(3):124–35.
4. Kindt TJ, Osborne BA, Goldsby RA. Immunology. 6th ed. New York: Freeman and Company; 2006.
5. Welss T, Basketter DA, Schröder KR. In vitro skin irritation: facts and future. State of the art review of mechanisms and models. Toxicol In Vitro. 2004;18(3):231–43.
6. OECD. OECD Guideline for the testing of chemicals, no. 404: acute dermal irritation/corrosion. Organisation for economic cooperation and development, Paris, France; 2002.
7. Draize JH, Woodard G, Calvery HO. Methods for the study of irritation and toxicity of substances applied directly on the skin and mucous membranes. J Pharmacol Exp Ther. 1944;82:377–90.
8. Weil CS, Scala RA. Study of intra-and interlaboratory variability in the results of rabbit eye and skin irritation tests. Toxicol Appl Pharmacol. 1971;19:276–360.
9. Worth AP, Cronin MT. The use of bootstrap resampling to assess the variability of Draize tissue scores. ATLA. 2001;29(5):557–73.
10. Campbell RL, Bruce RD. Comparative dermatotoxicology. I. Direct comparison of rabbit and human primary skin irritation responses to isopropylmyristate. Toxicol Appl Pharmacol. 1981;59:555–63.
11. Robinson MK, Perkins MA, Basketter DA. Application of a four hour human patch test method for comparative and investigative assessment of skin irritation. Contact Dermatitis. 1998;38:194–202.
12. Robinson MK, McFadden JP, Basketter DA. Validity and ethics of the human 4 hour patch test as an alternative method to assess acute skin irritation potential. Contact Dermatitis. 2001;45:1–12.
13. Hartung T, Bremer S, Casati S, Coecke S, Corvi R, Fortaner S, Gribaldo L, Halder M, Janusch Roi A, Prieto P, Sabbioni E, Worth A, Zuang V. ECVAM's response to the changing political environment for alternatives: consequences of the European union chemicals and cosmetics policies. ATLA. 2003;31:473–81.
14. Van De Sandt J, Roguet R, Cohen C, Esdaile D, Ponec M, Corsini E, Barker C, Fusenig N, Liebsch M, Benford D, De Brugerolle De Fraissinette A, Fartasch M. The use of human

keratinocytes and human skin models for predicting skin irritation. ATLA. 1999;27: 723–43.
15. Zuang V, Rona C, Archer G, Berardesca E. Detection of skin irritation potential of cosmetics by non-invasive measurements. Skin Pharmacol Appl Ski Physiol. 2000;13:358–71.
16. Cohen C, Selvi-Bignon C, Barbuer A, Rougier A, Lacheretz F, Roguet R. Measurement of pro inflammatory mediators production by cultured keratinocytes: a predictive assessment of cutaneous irritation. In: Rougier A, Goldberg AM, Maibach HI, editors. Alternative methods in toxicology, In vitro skin toxicology, vol. Vol. 10. New York, NY, USA: Mary Ann Liebert; 1994. p. 83–96.
17. Faller C, Bracher M. Reconstructed skin kits: reproducibility of cutaneous irritancy testing. Skin Pharmacol Appl Ski Physiol. 2002;15:74–91.
18. Perkins MA, Osborne R, Rana FR, Ghassemi A, Robinson MK. Comparison of in vitro and in vivo human skin responses to consumer products and ingredients with a range of irritancy potential. Toxicol Sci. 1999;48:218–29.
19. Roguet R. Use of skin cells culture for in vitro assessment of corrosion and cutaneous irritancy. Cell Biol Toxicol. 1999;15:63–75.
20. Alépée N, Tornier C, Robert C, Amsellem C, Roux MH, Doucet O, Pachot J, Meloni M, de Brugerolle de Fraissinette A. A catch-up validation study on reconstructed human epidermis (SkinEthic™ RHE) for full replacement of the Draize skin irritation test. Toxicol In Vitro. 2010;24:257–66.
21. Kojima H, Ando Y, Idehara K, Katoh M, Kosaka T, Miyaoka E, Shinoda S, Suzuki T, Yamaguchi Y, Yoshimura I, Yuasa A, Watanabe Y, Omori T. Validation study of the in vitro skin irritation test with the LabCyte EPI-MODEL24. Altern Lab Anim. 2012;40:33–50.
22. Spielmann H, Hoffmann S, Liebsch M, Botham P, Fentem J, Eskes C, Roguet R, Cotovio J, Cole T, Worth A, Heylings J, Jones P, Robles C, Kandárová H, Gamer A, Remmele M, Curren R, Raabe H, Cockshott A, Gerner I, Zuang V. The ECVAM international validation study on in vitro tests for acute skin irritation: report on the validity of the EPISKIN and EpiDerm assays and on the skin integrity function test. ATLA. 2007;35:559–601.
23. Prunerias M, Régnier M, Woodley D. Methods for cultivation of keratinocytes with an air-liquid interface. J Investig Dermatol. 1983;81(sup n°1):28s–33s.
24. Rosdy M, Clauss LC. Terminal epidermal differentiation of human keratinocytes grown in chemically defined medium on inert filter substrates at the air–liquid interface. J Invest Dermatol. 1990;95:409–14.
25. Doucet O, Robert C, Zastrow L. Use of a serum-free reconstituted epidermis as a skin pharmacological model. Toxicol In Vitro. 1996;10:305–13.
26. Kandárová H, Liebsch M, Spielmann H, Genschow E, Schmidt E, Guest R, Whittingham A, Warren N, Gamer A, Remmele M, Kaufmann T, Wittmer E, De Wever B, Rosdy M. Assessment of the SkinEthic reconstituted human epidermis for skin corrosion testing according to OECD guideline 431. Toxicol In Vitro. 2006;20:547–59.
27. OECD. Guidance document on integrated approaches to testing and assessment of skin irritation/corrosion, series on testing and assessment, No. 203, OECD, Paris. 2014. Available at: http://www.oecd.org/env/ehs/testing/seriesontestingandassessmenttestingforhumanhealth.htm.
28. ECVAM. Statement of the ECVAM Scientific Advisory Committee (ESAC) on the scientific validity of in vitro tests for skin irritation testing. 2008. Available at: http://ecvam.jrc.it.
29. United Nations Economic Commission for Europe (UNECE). UN globally harmonised system of classification and labelling of chemicals. 2nd revised ed; 2008. http://www.unece.org/trans/danger/publi/ghs/ghs_rev02/02files_e.html. (as of April 2009)
30. ECVAM. Statement on the "Performance under UN GHS of three in vitro assays for skin irritation testing and the adaptation of the Reference Chemicals and Defined Accuracy Values of the ECVAM skin irritation Performance Standards", issued by the ECVAM Scientific Advisory Committee (ESAC30), 9 April 2009; 2009. Available at: http://ihcp.jrc.ec.europa.eu/our_activities/alt-animal-testing.
31. Griesinger C., Barroso J., and Zuang V.. ECVAM background document on the recent adaptations of the ECVAM performance standards for in vitro skin irritation testing in the context of

the drafting process of an EU Test Method and an OECD draft Test Guideline. Ispra, November 13, 2008; 2008. Available at: http://ecvam.jrc.ec.europa.eu/.
32. OECD. In vitro skin irritation: Reconstructed Human Epidermis Test Method. OECD Guideline for the Testing of Chemicals No. 439, OECD, Paris; 2015. Available at: http://www.oecd.org/env/testguidelines.
33. OECD. Explanatory background document to the OECD draft Test Guideline on in vitro skin irritation testing. Published in OECD Series on Testing and Assessment, No. 137, OECD, Paris; 2010. Available at: http://www.oecd.org/officialdocuments/displaydocumentpdf/?cote=env/jm/mono(2010)36&doclanguage=en.
34. United Nations (UN). Globally harmonized system of classification and labelling of chemicals (GHS), 2nd revised edn, UN New York and Geneva, 2013; 2013. Available at: http://www.unece.org/trans/danger/publi/ghs/ghs_rev05/05files_e.html.
35. Alépée N, Grandidier MH, Tornier C, Cotovio J. An integrated testing strategy for in vitro skin corrosion and irritation assessment using SkinEthic™ reconstructed human epidermis. Toxicol In Vitro. 2015;29:1779–92.
36. ECVAM. Statement on the validity of in vitro tests for skin irritation, issued by the ECVAM Scientific Advisory Committee (ESAC26), 27 April 2007; 2007. Available at: http://ihcp.jrc.ec.europa.eu/our_activities/alt-animal-testing.
37. Zuang V, Eskes C, Worth A, Cole T, Hoffmann S, Saliner AG, Netzeva T, Patlewicz G, Cockshott A, Gerner I. Report from the chemical selection sub-committee to the management team on potential reasons for the misclassification of chemicals in the EpiSkin and EpiDerm assays. Available under "Validation Study Documents"; 2007. Available at: http://ecvam.jrc.ec.europa.eu/.
38. Tornier C, Amsellem C, Fraissinette Ade Brugerolle A, Alépée N. Assessment of the optimized SkinEthic™ reconstructed human epidermis (RHE) 42 bis skin irritation protocol over 39 test substances. Toxicol In Vitro. 2010;24(1):245–56.
39. EC. Regulation (EC) no 1272/2008 of the European parliament and of the Council of 16 December 2008 on classification, labelling and packaging of substances and mixtures, amending and repealing directives 67/548/EEC and 1999/45/EC, and amending regulation (EC) no 1907/2006. Off J Eur Union. 2008;L353:1–1355.
40. EC. Commission Directive 2001/59/EC of 6 august 2001 adapting to technical progress for the 28th time Council Directive 67/548/EEC on the approximation of laws, regulations and administrative provisions relating to the classification, packaging and labelling of dangerous substances. Off J Eur Union. 2001;L225:1–333.
41. OECD. In vitro skin irritation: reconstructed human epidermis test method. Revised OECD Guideline TG 439 version originally adopted in 2010, OECD, Paris; 2013. Available at: http://www.oecd-ilibrary.org/environment/oecd-guidelines-for-the-testing-of-chemicals-section-4-health-effects_20745788.
42. Alépée N, Barroso J, De Smedt A, De Wever B, Hibatallah J, Klaric M, Mewes KR, Millet M, Pfannenbecker U, Tailhardat M, Templier M, McNamee P. Use of HPLC/UPLC-spectrophotometry for detection of formazan in in vitro reconstructed human tissue (RhT)-based test methods employing the MTT-reduction assay to expand their applicability to strongly coloured test chemicals. Toxicol In Vitro. 2015;29:741–61.
43. Basketter DA, Chamberlain M, Griffiths HA, York M. The classification of skin irritants by human patch test. Food Chem Toxicol. 1997;35:845–52.
44. Basketter DA, Jirova J, Kandarova H. Review of skin irritation/corrosion hazards on the basis of human data: a regulatory perspective. Interdiscip Toxicol. 2012;5(2):98–104.
45. Basketter DA, York M, McFadden JP, Robinson MK. Determination of skin irritation potential in the human 4-h patch test. Contact Dermatitis. 2004;51:1–4.
46. Robinson MK, Whittle E, Basketter DA. A two center study of the development of acute irritation responses to fatty acids. Am J Contact Dermatol. 1999;10:136–45.
47. Jirova D, Basketter D, Liebsch M, Bendova H, Keilova K, Marriott M, Kandárová H. Comparison of human skin irritation patch test data with in vitro skin irritation assays and animal data. Contact Dermatitis. 2010;6:109–16.

48. Mosmann T. Rapid colorimetric assay for cellular growth and survival: application to proliferation and cytotoxicity assays. J Immunol Methods. 1983;65:55–63.
49. SkinEthic™ RHE SOP, Version 2.0. SkinEthic skin irritation test-42bis test method for the prediction of acute skin irritation of chemicals: 42 minutes application +42 hours post-incubation; 2009. Available at: http://ihcp.jrc.ec.europa.eu/our_activities/alt-animal-testing.
50. OECD. In vitro skin corrosion: reconstructed human epidermis (RHE) test method. OECD Guideline for the Testing of Chemicals. Revised version originally adopted in 2004, OECD, Paris; 2014. Available at: http://www.oecd-ilibrary.org/environment/oecd-guidelines-for-the-testing-of-chemicals-section-4-health-effects_20745788.
51. Alépée N, Robert C, Tornier C, Cotovio J. The usefulness of the validated SkinEthic™ RHE test method to identify skin corrosive UN GHS subcategories. Toxicol In Vitro. 2014;28:616–25.
52. Groux, H., Lelievre, D., Cottrez, F., Auriault, C., Alépée, N., Meunier, J.R., and Cotovio, J.. Evaluation of IRR-IS®, an EpiSkin™ based model for quantifying chemical irritation potency. Oral Communication N°1.5, ESTIV2012, 17th October 2012, Lisbon, Portugal. Poster presented at 51st Annual Meeting of the Society of Toxicology, March 11–15, 2012, San Francisco, USA. The Toxicologist; 2012. p. 286. Available at: http://www.toxicology.org/AI/Pub/Tox/2012Tox.pdf.
53. EC. Regulation (EC) no 1223/2009 of the European Parliament and of the Council of 30 November 2009 on cosmetic products. Off J Eur Union. 2009;L342:59–209. Available at: http://eur-lex.europa.eu/legal-content/EN/TXT/?uri=CELEX:32009R1223
54. EC. Regulation (EC) No 528/2012 of the European Parliament and of the Council of 22 May 2012 concerning the making available on the market and use of biocidal products Text with EEA relevance. Off J Eur Union. 2012;L167:1–123
55. Molinari J, Eskes C, Andres E, Remoué N, Sá-Rocha VM, Hurtado SP, Barrichello C. Improved procedures for in vitro skin irritation testing of sticky and greasy natural botanicals. Toxicol In Vitro. 2013;27(1):441–50.
56. Eskes C, Detappe V, Koëter H, Kreysa J, Liebsch M, Zuang V, Amcoff P, Barroso J, Cotovio J, Guest R, Hermann M, Hoffmann S, Masson P, Alépée N, Arce LA, Brüschweiler B, Catone T, Cihak R, Clouzeau J, D'Abrosca F, Delveaux C, Derouette JP, Engelking O, Facchini D, Fröhlicher M, Hofmann M, Hopf N, Molinari J, Oberli A, Ott M, Peter R, Sá-Rocha VM, Schenk D, Tomicic C, Vanparys P, Verdon B, Wallenhorst T, Winkler GC, Depallens O. Regulatory assessment of in vitro skin corrosion and irritation data within the European framework: workshop recommendations. Regul Toxicol Pharmacol. 2012;62:393–403.
57. Kolle SN, Sullivan KM, Mehling A, van Ravenzwaay B, Landsiedel R. Applicability of in vitro tests for skin irritation and corrosion to regulatory classification schemes: substantiating test strategies with data from routine studies. Regul Toxicol Pharmacol. 2012;64(3):402–14.

In Vitro Skin Irritation Assay with the LabCyte EPI-MODEL

5

Hajime Kojima and Masakazu Katoh

5.1 Description of LabCyte EPI-MODEL24 [1]

5.1.1 The LabCyte EPI-MODEL24 Reconstructed Human Epidermis

LabCyte EPI-MODEL24 is a commercially available reconstructed human cultured epidermal model produced by Japan Tissue Engineering Co. Ltd., Japan. It consists of normal human epidermal keratinocytes whose biological origin is neonatal foreskin. In order to expand the human keratinocytes while maintaining their phenotype, they are cultured with 3T3-J2 cells as a feeder layer [2]. Reconstruction of human cultured epidermis is achieved by cultivating proliferating keratinocytes on an inert filter substrate with a surface area of 0.3 cm^2 at the air-liquid interface for 13 days using an optimized medium containing 5% fetal bovine serum. The result is a multilayer structure consisting of a fully differentiated epithelium with features of the normal human epidermis, including a stratum corneum (SC). The LabCyte EPI-MODEL24 is embedded in an agarose gel containing a nutrient solution and shipped in 24-well plates at around 18°C.

It is possible to observe in the ultrastructure a fully developed basement membrane zone, consisting of a highly developed lamina densa, lamina lucida, and anchoring filaments. Extrusion of lamellar bodies is observed at the interface between the stratum granular (SG) and SC layer. Lipid lamellae showing a characteristic electron-dense and electron-lucent pattern are also present. Keratohyalin

H. Kojima (✉)
National Institute of Health Sciences (NIHS), Setagaya-ku, Tokyo 158-8501, Japan
e-mail: h-kojima@nihs.go.jp

M. Katoh
Japan Tissue Engineering Co. Ltd., Gamagori City, Aichi 443-0022, Japan
e-mail: masakazu_katoh@jpte.co.jp

C. Eskes et al. (eds.), *Alternatives for Dermal Toxicity Testing*,
DOI 10.1007/978-3-319-50353-0_5

granules are ubiquitously present in the granular cells at the SG layer. The synthesis of specific lipids, including ceramides, which are known to be responsible for the water barrier of the SC, is detected in LabCyte EPI-MODEL24.

5.1.2 Quality Control for LabCyte EPI-MODEL24 [3]

LabCyte EPI-MODEL24 is manufactured according to a well-defined standard operating procedure (SOP). All production batches of LabCyte EPI-MODEL24 are inspected for viability, barrier function, and morphology.

The product is released only in accordance with stringent quality control procedures.

5.1.2.1 Tissue Viability: MTT Assay

Three replicate LabCyte EPI-MODEL24 tissues are subjected to an MTT assay as follows. Tissues are placed in the wells of 24-well plates containing 0.5 mL of 0.5-mg/mL MTT medium (Dojindo Co., Kumamoto, Japan) and were incubated for 3 h at 37°C in a 5% CO_2 humidified atmosphere. Formazan produced in the tissues is extracted with 300 μL of isopropanol and 200 μL of extract is measured at 570 nm and at 650 nm as a reference absorbance, with isopropanol as a blank. The mean of the OD values indicates tissue viability for each LabCyte EPI-MODEL24 batch. QC acceptance criteria is OD ≥ 0.8.

5.1.2.2 Barrier Function: 50% Inhibitory Concentration (IC50) Assay

To evaluate whether the stratum corneum in the LabCyte EPI-MODEL24 tissue resists the rapid penetration of the cytotoxic marker chemical sodium lauryl sulfate (SLS), the viability of the epidermis tissue was estimated in terms of the half maximal inhibitory concentration (IC50). 25 μL of SLS at concentrations of 0.1, 0.2, 0.3, and 0.4% (w/v) are applied to the LabCyte EPI-MODEL24, and cell viability is measured after 18 hours using an MTT assay. All experiments are performed in triplicate. The acceptable range for the LabCyte EPI-MODEL24 is shown in Table 5.1.

5.1.2.3 Morphology

A LabCyte EPI-MODEL24 tissue is fixed with 4% parafolmaldehyde and 2% sucrose in 0.1 M phosphate-buffered saline (PBS: pH 7.4) for more than 3 h and then embedded in paraffin. Five-micrometer vertical sections are cut and stained with hematoxylin and eosin for light-microscopic examination (Fig. 5.1).

Table 5.1 Acceptability ranges for negative control OD values and QC batch release criteria of the LabCyte EPI-MODEL24 according to the OECD TG 439 [4]

	Lower acceptance limit	Upper acceptance limit
Negative control OD values	≧0.7	≦2.5
QC batch release criteria (18-h treatment with SLS)	IC50 = 1.4 mg/mL	IC50 = 4.0 mg/mL

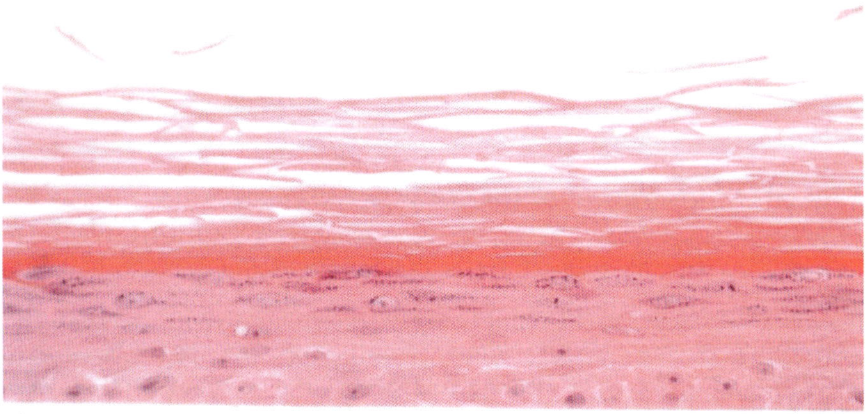

Fig. 5.1 Pathological slide of the LabCyte EPI-MODEL

QC Acceptance Criteria

Confirmation of the formation of multilayered epidermis-like tissue containing a stratum corneum.

5.2 Validation of the *In Vitro* Skin Irritation Test Method Using the LabCyte EPI-MODEL24

5.2.1 Study Plan

The LabCyte EPI-MODEL24 *in vitro* skin irritation test method based on a reconstructed human epidermis model (RhE) has undergone protocol optimization, and a multi-laboratory validation based on the European Center for the Validation of Alternative Methods (ECVAM) performance standards [5–7]. The validation report contained results that are in accordance with the revised reference chemicals described by the new ESAC statement 2009.

The objective of this study was to conduct a validation study of the *in vitro* skin irritation test method using LabCyte EPI-MODEL24 (LabCyte EPI-MODEL24 SIT) to assess both intra- and interlaboratory reproducibility as well as predictive capacity of this test method using set of 25 coded test chemicals for which high-quality *in vivo* data were available. The study was coordinated by the Japanese Society for Alternatives to Animal Experiment (JSAAE) and the Japanese Center for the Validation of Alternative Methods (JaCVAM). An additional objective was to conform more accurately to the classifications for skin irritation under the United Nations Globally Harmonised System of Classification and Labelling of Chemicals (UN GHS). The validation study was undertaken in accordance with the principles and criteria documented in the Organisation for Economic Co-operation and Development (OECD) *Guidance Document on the Validation and International Acceptance of New or Updated Test Methods for Hazard Assessment* (No. 34: [8]) and in accordance with the Modular Approach to validation [9].

5.2.2 Brief Description of the Protocol

According to SOP ver. 8.3 (refer to [10]), the LabCyte EPI-MODEL24 tissues were shipped from the supplier on Mondays and delivered to the recipients on Tuesdays. Upon receipt, the tissues were aseptically removed from the transport agarose medium, transferred to 24-well plates (BD Biosciences, CA, USA) with 0.5 mL of assay medium, and incubated overnight at 37°C in a 5% CO_2 humidified atmosphere. On Wednesday, the tissues were topically exposed to the test chemicals. For liquids test chemicals, 25 μL are applied with a micropipette. For solid test chemicals, 25 mg are applied from microtubes after application of 25 μL of sterile water. When necessary, the mixture is gently spread over the surface of the epidermis with a micro-spatula. Viscous liquids are applied using a micropipette with a cell-saver type tip. Each test chemical is applied to three tissues. In addition, three tissues serving as negative controls are treated with 25 μL distilled water, and three tissues serving as positive controls are exposed to 5% SLS. After a 15-min exposure, each tissue is carefully washed with PBS (Invitrogen, CA, USA) 15 times using a washing bottle to remove any remaining test chemical from the surface. The blotted tissues are then transferred to new 24-well plates containing 1 mL of fresh assay medium.

The treated and control tissues are incubated for 42 h at 37°C in a 5% CO_2 humidified atmosphere. After a post-incubation 42-h waiting period, the blotted tissues are transferred to new 24-well plates containing 0.5 mL of freshly prepared 0.5-mg/mL MTT medium (Dojindo Co., Kumamoto, Japan) for the MTT assay. Tissues are incubated for 3 h at 37°C in a 5% CO_2 humidified atmosphere and then transferred to and completely immersed in microtubes containing 300 μL of isopropanol. Formazan extraction is performed at room temperature, and the tissues are allowed to stand overnight. Subsequently, 200-μL extracts are transferred to a 96-well plate. The optical density is measured at 570 nm and 650 nm as a reference absorbance, with isopropanol as a blank.

The tissue viability is calculated as a percentage relative to the viability of the negative control using the following equation:

$$\text{Tissue Viability}(\%) = \text{Mean Measured OD sample} / \text{Mean Measured ODNC} \times 100,$$

where measured OD = (570 nm OD sample−570 nm OD blank)−(650 nm OD sample−650 nm OD blank).

Table 5.2 Required predictive values for sensitivity, specificity, and accuracy for any similar or modified test method to be considered valid in the performance standard for TG439 [13] and validation results on the LabCyte EPI-MODEL24 [14]

	Sensitivity (%)	Specificity (%)	Accuracy (%)
Performance standards	≥80	≥70	≥75
LabCyte EPI-MODEL24	82.3	72.6	77.5

In the second phase study, a prediction model for skin irritation potential in LabCyte EPI- MODEL24 was established per conditions for EPISKIN™ that are described in the ECVAM draft performance standards [11].

Acceptance criteria:

OD for the negative control of greater than 0.7 (Table 5.1).

Viability for the positive control of less than 40%.

The median of three replicate values for viability from tests with mean of tissue viability is used to classify a chemical according to the following prediction model.

Irritant if the median tissue viability ≦50% (if the test chemical is also predicted as noncorrosive).

Nonirritant if the median tissue viability >50% (in countries not adopting the UN GHS optional category 3 on mild irritants).

5.2.3 Performance and Applicability

In the second and third phase studies [12], 12 irritants and 13 nonirritants from the ECVAM performance standards based on the GHS-EU classification (GHS, 2013, Regulation (EC) No 1272/2008, [5, 7]) were tested by seven laboratories using LabCyte EPI-MODEL24 SIT. The assay demonstrated high reliability both within and between laboratories, and reliability of the positive control was 100%, with acceptable levels of accuracy: 77.5% overall accuracy, 82.3% overall sensitivity, and 72.6% overall specificity. These results indicate that the MTT assay is suitable for use as a stand-alone assay to distinguish between skin irritants and nonirritants. The prediction model and the results are described in Table 5.2.

5.2.4 Independent Peer Review

An independent peer review was conducted by volunteer OECD skin irritation experts. After the third phase study, the panel recommended that future work should focus on the following matters [3]: (1) the issue of misclassifying 1-bromohexane needed to be resolved, (2) an extensive analysis of the intra- and interlaboratory reproducibility needed to be documented appropriately with reference to the OECD performance standards, and (3) a need to assess variability between replicate tissues as well as to define acceptance criterion for each tissue. In order to comply better with performance standards, final classification for a complete run sequence at any given laboratory was required to be done using the mathematical mean rather than the median. Finally, appropriate documentation describing and demonstrating the adherence to GLP principles was requested [15]. These issues were each resolved in the catch-up validation and supplementary studies [14].

5.3 Test Guideline

The LabCyte EPI-MODEL24 was subsequently included in OECD TG 439 *in vitro* skin irritation: reconstructed human epidermis test method, which also provides information on the type of validation study used to validate the respective test methods [4]. The LabCyte EPI-MODEL24 SIT uses very similar protocols to the other adopted RhE models, and it is worth noting that all use a post-incubation period of 42 h. Variations are found primarily in four parameters that relate to the different barrier functions of the test methods: (a) preincubation time and volume, (b) application of test chemicals, (c) post-incubation volume, and (d) maximum acceptable variation, as shown in Table 5.3.

Table 5.3 Protocol parameters specific to each of the test methods included in TG439 [4]

	EpiSkin™ (SM)	EpiDerm™ SIT (EPI-200)	SkinEthic™ RHE™	LabCyte EPI-MODEL24 SIT
(a) Preincubation				
Incubation time	18–24 h	18–24 h	<2 h	15–30 h
Medium volume	2 mL	0.9 mL	0.3 mL	0.5 mL
(b) Chemical application				
For liquids	10 μL (26 μL/cm^2)	30 μL (47 μL/cm^2)	16 μL (32 μL/cm^2)	25 μL (83 μL/cm^2)
For solids	10 mg (26 mg/cm^2) + DW (5 μL)	25 mg (39 mg/cm^2) + DPBS (25 μL)	16 mg (32 mg/cm^2) + DW (10 μL)	25 mg (83 mg/cm^2) + DW (25 μL)
Use of nylon mesh	Not used	If necessary	Applied	Not used
Total application time	15 min	60 min	42 min	15 min
Application temperature	RT	a) at RT for 25 min b) at 37°C for 35 min	RT	RT
(c) Post-incubation volume				
Medium volume	2 mL	0.9 mL × 2	2 mL	1 mL
(d) Maximum acceptable variability				
Standard deviation between tissue replicates	SD ≤ 18	SD ≤ 18	SD ≤ 18	SD ≤ 18

RT room temperature, *DW* distilled water, *DPBS* Dulbecco's phosphate-buffered saline

5.4 Conclusion

Three validation studies were conducted by JSAAE in order to assess the performance of LabCyte EPI-MODEL24 SIT developed by J-TEC, and the results of these studies were submitted to OECD for inclusion into the OECD TG 439.

In the summary review report from the OECD, the peer review panel indicated the need to resolve an issue regarding the misclassification of 1-bromohexane. To this end, a rinsing operation intended to remove exposed chemicals was reviewed and SOP revised by J-TEC. Thereafter, in order to confirm general versatility of the revised SOP, a new validation management team was organized by JaCVAM to undertake a catch-up validation study that would compare the revised assay with similar *in vitro* skin irritation assays, per OECD TG 439 (2010). The catch-up validation and supplementary studies for LabCyte EPI-MODEL24 SIT using the revised SOPs were conducted at three laboratories. These results showed that the revised SOP of LabCyte EPI-MODEL24 SIT conformed more accurately to the classifications for skin irritation under UN GHS, thereby highlighting the importance of an optimized rinsing operation for the removal of exposed chemicals in obtaining consistent results from *in vitro* skin irritation assays [14].

References

1. Katoh M, Hamajima F, Ogasawara T, Hata KI. Assessment of the human epidermal model LabCyte EPI-MODEL for in vitro skin irritation testing according to the ECVAM-validated protocol. J Toxicol Sci. 2009;34(3):327–34.
2. Rheinwald JG, Green H. Serial cultivation of strains of human epidermal keratinocytes: the formation of keratinizing colonies from single cells. Cell. 1975;6:331–44.
3. OECD. Peer review report of the validation of the skin irritation testing using LabCyte EPI-MODEL 24, OECD Environmental Health and Safety Publications, Series on Testing and Assessment, No. 155. 2011.
4. OECD. In vitro skin irritation-reconstructed human epidermis test method, test guideline, No. 439. 2015.
5. ESAC. Statement on the validity of in-vitro tests for skin irritation. 2007.
6. ESAC. Statement on the scientific validity of in-vitro tests for skin irritation testing. 2008.
7. ESAC. Statement on the performance under UN GHS of three in-vitro assays for skin irritation testing and the adaptation of the reference chemicals and defined accuracy values of the ECVAM skin irritation performance standards. 2009.
8. OECD. Guidance document on the validation and international acceptance of new or up-dated test methods for hazard assessment, OECD Environmental Health and Safety Publications, Series on Testing and Assessment, No. 34. 2005.
9. Hartung T, Bremer S, Casati S, Coecke S, Corvi R, Fortaner S, Gribaldo L, Halder M, Hoffmann S, Roi AJ, Prieto P, Sabbioni E, Scott L, Worth A, Zuang V. A modular approach to the ECVAM principles on test validity. ATLA. 2004;32:467–72.
10. JaCVAM. 2015. Retrieved from: http://www.jacvam.jp/en/index.html
11. ECVAM. Performance standards for applying human skin models to in vitro skin irritation testing. 2007.
12. Kojima H, Ando Y, Idehara K, Katoh M, Kosaka T, Miyaoka E, Shinoda S, Suzuki T, Yamaguchi Y, Yoshimura I, Yuasa A, Watanabe Y, Omori T. Validation study of the in vitro skin irritation test with the LabCyte EPI-MODEL24. Altern Lab Anim. 2012;40(1):33–50.

13. OECD. Performance standards for the assessment of proposed similar or modified in vitro reconstructed human epidermis (RHE) test methods for skin irritation testing as described in TG 439. OECD Environmental Health and Safety Publications, Series on Testing and Assessment, No. 220. 2015.
14. Kojima H, Katoh M, Shinoda S, Hagiwara S, Suzuki T, Izumi R, Yamaguchi Y, Nakamura M, Kasahawa T, Shibai A. A catch-up validation study of an in vitro skin irritation test method using reconstructed human epidermis LabCyte EPI-MODEL24. J Appl Toxicol. 2014;34(7):766–74.
15. OECD. Compliance of laboratory suppliers with GLP principles. OECD Environmental Health and Safety Publications, Series on Testing and Assessment, No.5, 1999.

The epiCS® Skin Irritation Test (SIT) Method

6

Oliver Engelking, Horst W. Fuchs, and Dirk Weisensee

6.1 Principle of the Test Method and Scientific Basis

In vivo, chemical induced skin irritation is a non-immunogenic reaction, which appears shortly after exposure to the chemical. It is manifested by erythema and oedema, as the result of a cascade of events beginning with penetration of the stratum corneum and damage to the underlying layers of keratinocytes. Stressed, damaged or dying keratinocytes release mediators that initiate an inflammatory reaction, which acts on the cells in the dermis, particularly the stromal and endothelial cells. It is the dilation and increased permeability of the endothelial cells that produce the observed erythema and oedema *in vivo* [1]. In the regulatory context the United Nations (UN) Globally Harmonized System of Classification and Labelling of Chemicals (GHS) defines skin irritation as the production of reversible damage to the skin following the application of a test chemical for up to 4 h [2].

The epiCS® SIT method uses the reconstructed human epidermis epiCS® consisting of normal human-derived keratinocytes. Under specific culture conditions, they form a multilayered, highly differentiated model of the human epidermis. It is organised in basal, spinous and granular layers and a multilayered stratum corneum containing intercellular lamellar lipid layers representing main lipid classes analogous to those found *in vivo* [3]. In the epiCS® SIT method, the RhE represents the target organ *in vitro*. Due to the absence of any vascularisation in epiCS® tissues, it is used to measure the initiating events of the inflammatory cascade: cell and tissue damage.

To account for the reversibility of effects, a 'recovery' incubation time of 42 h after exposure to the chemicals is part of the protocol for the epiCS® SIT method. It is followed by an MTT assay to determine cell viability in epiCS® tissues: MTT

O. Engelking (✉) • H.W. Fuchs • D. Weisensee
CellSystems GmbH, Langeler Ring 5, 53842 Troisdorf, Germany
e-mail: oliver.engelking@cellsystems.de

C. Eskes et al. (eds.), *Alternatives for Dermal Toxicity Testing*,
DOI 10.1007/978-3-319-50353-0_6

[3-(4,5-dimethylthiazol-2-yl)-2,5-diphenyltetrazolium bromide] is reduced by mitochondrial activity into a blue formazan salt that is quantitatively measured after extraction from tissues [4]. Irritant chemicals are identified by their ability to decrease cell viability below defined threshold levels (i.e. ≤50%, for UN GHS category 2). Depending on the regulatory framework and applicability of the test guideline, chemicals that produce cell viabilities above the defined threshold level, may be considered non-irritants (i.e. > 50%, no category) [5].

6.2 Current Validation Status

The epiCS® skin irritation test (epiCS® SIT) method went through a multi-laboratory catch-up validation study based on the Performance Standards originally published in 2009 [6]. Three laboratories participated in the study: two contract research organisations (one in Europe, one in the USA) and one naive laboratory in Europe. The study was conducted in concordance with the guidance document on the validation and international acceptance of new or updated test methods for hazard assessment [7]. The European Union Reference Laboratory for alternatives to animal testing (EURL ECVAM) was involved in this catch-up validation study from planning to data analysis. Data had been submitted to EURL ECVAM in 2015 for independent scientific peer review by the EURL ECVAM Scientific Advisory Committee (ESAC) resulting in an ESAC opinion about the test method readiness for screening and regulatory use of epiCS® SIT method. As a consequence, the epiCS® SIT method will be submitted for integration into the OECD TG 439 in 2017.

6.3 Performance and Applicability of the Test Method

6.3.1 Reproducibility

Data on reproducibility for the epiCS® SIT method are derived from the catch-up validation study (see validation status). They were submitted to EURL ECVAM and were used to calculate the reproducibility within each laboratory over time in the three participating laboratories. The reference substances gave 90%, 95% and 100% concordant results in the three laboratories, which meets the acceptance criteria of ≥90% as defined in the Performance Standards [8].

The analysis of reproducibility between laboratories resulted in 84.2% concordance of the final classifications between the three laboratories, which meets the acceptance criterion of ≥80%.

6.3.2 Predictive Capacity

The predictive capacity of the epiCS® SIT method was determined in a catch-up validation study (see chapter on validation status) based on the comparison of the *in vitro* classification with the *in vivo* classification according to the UN GHS classification system.

The overall sensitivity in the three laboratories was found to be 86.7% which meets the acceptance criterion of ≥80% as defined in the Performance Standards (PS; [8]). The overall specificity was 80%, also meeting the PS criterion of ≥70%. Finally, the accuracy PS acceptance criterion of ≥75% was also met with an overall accuracy of 83.3% in the three laboratories.

6.3.3 Applications and Limitation

The epiCS® SIT method is applicable to solids, liquids, semi-solids and waxes. Liquids may be aqueous or nonaqueous; solids may be soluble or insoluble in water. The method is not applicable to gases and aerosols [5].

The epiCS® SIT method can be used for hazard identification of irritant chemicals (substances and mixtures) in accordance with UN GHS category 2.

None of the validated and regulatory accepted RhE based SIT methods can be used for classification of chemicals to the optional UN GHS category 3 (mild irritants). Therefore, depending on the regulatory framework in the country, it can be used to identify non-classified chemicals [5].

Coloured test chemicals or chemicals that become coloured during tissue treatment need the use of adapted controls. An interference of absorbance of up to 50% of the negative control can be corrected by calculation [5]. Stronger interference might need the use of different MTT detection methods (e.g. HPLC/UPLC) as indicated in the OECD Test Guideline 439 [5].

Reducing test substances may interfere with the MTT assay, resulting in reduction of MTT to blue formazan salt. This applies to substances, which stick to the tissue or penetrate the tissue and cannot completely be rinsed off the epidermis by the rinsing procedure before post-incubation. For correction, freeze-killed epiCS® tissues can be used as negative control tissues. MTT reduction detected with these tissues is caused by the chemical itself as the reducing mitochondrial activity of non-viable tissue can be excluded. A pre-test, in order to identify reducing chemicals before the actual tissue test, is recommended.

6.3.4 Comparison to Human Data

Only very limited human data are available. During the catch-up validation of the epiCS® SIT method, the 20 reference chemicals listed in the OECD Test Guideline 439 were used. Two out of these 20 chemicals (1-decanol and di-n-propyl-disulphide) are known to be irritant in the rabbit but for which there is reliable evidence that they are non-irritant in humans [9–11]. These substances are classified by the epiCS® model as skin irritants (UN GHS category 2) in two laboratories and no category in one laboratory.

6.4 Brief Description of the Protocol

The epiCS® SIT method is carried out with three replicates (three tissues per chemical). Negative and positive controls are run in parallel.

After reception, epiCS® tissues are incubated at 37°C in a humidified incubator to recover from transport stress for at least 4 h. Thereafter, 30 µl of the liquid or 30 mg of the solid test substance are applied topically to the tissue, which is exposed to the chemical for 20 min at room temperature. A nylon mesh is added on top of the chemical to allow for better repartition of the test chemicals on the tissue surface. By rinsing with DPBS, chemicals are removed from the epiCS® surface. The tissues are transferred to six well plates with new culture medium and post-incubated for 24 h (37°C, 5% CO_2). Used medium is exchanged and followed by a second post-incubation of 18 h (37°C, 5% CO_2). epiCS® tissues are transferred into MTT assay medium containing 1 mg/ml MTT and incubated for 3 h (37°C, 5% CO_2). The reduced formazan salt is extracted with isopropanol from the tissues and OD is determined at 550–570 nm. Relative tissue viability is calculated by division of OD values of the chemical treated tissue by OD value from negative control. A reduction of viability to ≤50% predicts the chemical as UN GHS category 2, values above 50% can be used to predict test chemicals as UN GHS no category.

6.5 Role in a Testing Strategy

The RhE-based test methods are able to identify category 2 and no category chemicals and can thus serve as stand-alone skin irritation methods for non-corrosives in countries where optional category 3 is not implemented. For authorities adopting category 3, additional testing in an *in vitro* skin irritation test method not adopted by the OECD or in the *in vivo* test method may be required to resolve category 3 from no category. In case RhE-based test methods result in category 2 prediction, an *in vitro* skin corrosion test, if not performed upfront, is required to determine the final classification (category 2 (irritant) or category 1 (A, B or C) (corrosive)).

6.6 Perspectives from the Test Developer

6.6.1 Critical Steps in the Protocol

The protocol for the epiCS® SIT method is available online [12].

The test is straightforward and no critical steps are involved.

6.6.2 Possible Protocol Adaptations

The protocol as part of the epiCS® SIT method is scientifically validated; therefore adaptations are not possible.

6.6.3 Challenges and Opportunities

The epiCS® tissues used in the epiCS® SIT method can be used for different toxicology applications like skin corrosion, skin sensitisation, phototoxicity and genotoxicity testing. Furthermore, the epidermis can be used for efficacy testing and as a tool for skin permeation assay. Protocols for these assays as well as further information on reconstructed human epidermis are available online at www.skin-invitro.com.

6.7 Conclusions

A validation study of the epiCS® SIT method was carried out, involving EURL ECVAM and the independent scientific peer reviewing by the EURL ECVAM Scientific Advisory Committee. The method was found to be highly reliable based on the reconstructed human epidermis epiCS®. The method will be proposed for incorporation into the OECD Test Guideline 439. The epiCS® tissues, however, can be used in a vast range of different applications in *in vitro* toxicology, research and efficacy testing.

References

1. OECD. New guidance document on an integrated approach on testing and assessment (IATA) for skin corrosion and irritation, Series on Testing and Assessment, No. 203. 2014.
2. UN. United Nations globally harmonized system of classification and labelling of chemicals (GHS), UN New York and Geneva. 2015. Available at: http://www.unece.org/trans/danger/publi/ghs/ghs_rev03/03files_e.html
3. Hoffmann J, Heisler E, Karpinski S, Losse J, Thomas D, Siefken W, Ahr H-J, Vohr HW, Fuchs HW. Epidermal-skin-test 1000 (EST-1000)—a new reconstructed epidermis for in vitro skin corrosivity testing. Toxicol In Vitro. 2005;19:925–9.
4. Welss T, Basketter DA, Schröder KR. In vitro skin irritation: fact and future. State of the art review of mechanisms and models. Toxicol In Vitro. 2004;18:231–43.
5. OECD. OECD guidelines for the testing of chemicals. In Vitro skin irritation: reconstructed human epidermis test method. 2015.
6. EC-ECVAM. Performance standards for in vitro skin irritation test methods based on Reconstructed human Epidermis (RhE). 2009. available online at: https://eurl-ecvam.jrc.ec.europa.eu/validation-regulatory-acceptance/docs-skin-irritation-1/DOC8-updated_ECVAM_PS_2009.pdf
7. OECD. Guidance document on the validation and international acceptance of new or updated test methods for hazard assessment, OECD series on testing and assessment No. 34, OECD, Paris. 2005. Available at: http://www.oecd.org/env/testguidelines
8. OECD. Document No. 220: Performance Standards for the Assessment of Proposed Similar or Modified In Vitro Reconstructed Human Epidermis (RHE) Test Methods for Skin Irritation Testing as Described in TG 439. OECD Environmental Health and Safety Publications, Series on Testing and Assessment. 2015. Available at: http://www.oecd.org/env/ehs/testing/series-testing-assessment-publications-number.htm.
9. Basketter DA, York M, McFadden JP, Robinson MK. Determination of skin irritation potential in the human 4-h patch test. Contact Dermatitis. 2004;51:1–4.

10. Jírová D, Basketter D, Liebsch M, Bendová H, Kejlová K, Marriott M, Kandarova H. Comparison of human skin irritation patch test data with in vitro skin irritation assays and animal data. Contact Dermatitis. 2010;62:109–16.
11. Jirova D, Liebsch M, Basketter D, Spiller E, Kejlova K, Bendova H, Marriott M, Kandarova H. Comparison of human skin irritation and photo-irritation patch test data with cellular in vitro assays and animal *in vivo* data. ALTEX. 2007;14:359–65.
12. CellSystems. epiCS® SKIN IRRITATION TEST, standard operating procedure (SOP epiCS SIT) version 5.1, 2017. Available online at: http://reconstructed-human-epidermis.com/user-data/downloads/manuels-sops/SOP-epiCS-skin-irritation.pdf

Part II

Skin Corrosion

Overview on Current Status of Alternative Methods and Testing Approaches for Skin Corrosion Testing

7

Chantra Eskes and Markus Hofmann

7.1 Background

Depending upon regulatory requirements, some geographical regions already allow the use of alternative methods for skin irritation and corrosion testing as full replacement of the animal testing, as it is the case in the European Union (EU), where a number of legislations require the use of alternative methods to animal toxicological testing. The EU Cosmetics Regulation, for example, prohibits animal testing of finished products since 2004 and of cosmetic ingredients since 2009. The animal testing ban was reinforced by a marketing ban of cosmetics tested on animals that entered into force since 2004 for the finished products and since 2013 for cosmetics containing ingredients tested on animals [1]. Furthermore, the EU regulation on the Registration, Evaluation, Authorisation and Restriction of Chemicals (REACH) 1907/2006 requires, depending on the tonnage level, that only *in vitro* testing is conducted for skin corrosion and irritation, or in case of need that the *in vivo* testing is conducted only if the test chemical falls outside of the applicability domain of the available *in vitro* methods or the results obtained from such methods do not allow a conclusive decision on (non-)classification and risk assessment ([2, 3]; see also Sect. 7.3 below). Finally, the EU regulation on classification, labelling and packaging of substances and mixtures (CLP), which implements the UN GHS classification and labelling of substances and mixtures in the EU, encourages the use of tiered weight-of-evidence evaluations and makes use of information from *in vitro*

C. Eskes (✉)
SeCAM Services and Consultation on Alternative Methods,
Via Campagnora 1, 6983 Magliaso, Switzerland
e-mail: chantra.eskes@secam-ce.eu

M. Hofmann
Swiss Federal Office of Public Health, Bern, Switzerland

C. Eskes et al. (eds.), *Alternatives for Dermal Toxicity Testing*,
DOI 10.1007/978-3-319-50353-0_7

testing in a tiered classification approach for skin corrosion and irritation (EU CLP; [4, 5]). Finally, the EU directive on the protection of animals used for scientific purposes states in article 13(1) that "Member States shall ensure that a procedure is not carried out if another method or testing strategy for obtaining the result sought, not entailing the use of a live animal, is recognised under the legislation of the Union" [6].

The area of skin corrosion represents one of the pioneering areas for the validation of alternative test methods, in which replacement alternatives have been validated and adopted in the regulation as early as 2000 in the European Union and in 2004 at the OECD level. If the animal *in vivo* study has been traditionally used to classify for potential skin corrosion and skin irritation hazard effects (such as the OECD TG 404 [7] originally adopted in 1981), current internationally agreed approaches recommend the use of Integrated Approach for Testing and Assessment (IATA) that allows to replace or minimize to the extent possible the use of *in vivo* animal testing while ensuring human safety [8]. The IATA is divided in three major parts including as Part 1 the use of existing information, physico-chemical properties and non-testing methods, as Part 2 a weight-of-evidence evaluation of all available information and as Part 3 the conduct of prospective testing first on *in vitro* methods for both skin corrosion and skin irritation testing before considering the use of *in vivo* animal testing as a last resort (see Sect. 7.3 for details). The IATA is considered applicable to both substances and mixtures, although it is acknowledged that there is a different amount of information available on the applicability of the different information sources of the IATA to mixtures and that such applicability may depend on the information available in each specific case to be assessed (see Sect. 7.3 below).

7.2 Classification for Skin Corrosion Hazard

The UN has published in 2003 the UN GHS classification system to favour harmonized classification of hazards across the world [9]. This classification system still uses the traditional *in vivo* animal test adopted by the OECD [7] originally developed by Draize et al. [10] as a reference. However, it also recommends the use of a tiered approach for skin corrosion and irritation which includes the use of *in vitro* data as well as other information sources such as existing human and animal data, physico-chemical properties and weight-of-evidence evaluation [9].

According to the UN GHS classification system, a substance is considered to be corrosive to skin in the *in vivo* animal test when it produces destruction of skin tissue (i.e. visible necrosis through the epidermis and into the dermis) in at least one tested animal after exposure for up to 4 h. According to the UN GHS classification system, corrosive substances are to be classified in category 1 where subcategorization is not required by a competent authority or where data are not sufficient for subcategorization. When data are sufficient and where required by a competent authority, substances may be classified in one of the three subcategories (see Table 7.1): subcategory 1A,

Table 7.1 Corrosion classification based on *in vivo* animal test data according to the UN GHS, UN transport packaging group and EU CLP classification systems

UN GHS[a] and EU CLP[b]		UN transport class 8 packaging group[c]	*In vivo* corrosive in at least one tested animal	
			Exposure	Observation
Category 1	Subcategory 1A	I	≤3 min	≤1 h
	Subcategory 1B	II	>3 min, ≤1 h	≤14 days
	Subcategory 1C	III[d]	>1 h, ≤4 h	≤14 days

[a]Recommends considering the use of a tiered approach for the evaluation of initial information and classification where applicable that includes existing human or animal data, existing *in vitro* data, pH-based assessment, SAR methods and a weight-of-evidence approach [9]

[b]Recommends the use of a testing and assessment strategy for classification of substances based on physico-chemical properties (organic (hydro)peroxide, pH), existing human and animal data, read-across or (Q)SAR methods and results from OECD adopted, validated or suitable *in vitro* test methods [11]

[c]In the UN model regulations for the transport of dangerous goods, allocation of substances listed in the Dangerous Goods List to the packaging groups in class 8 has been made on the basis of experience taking into account such additional factors as inhalation risk and reactivity with water (including the formation of dangerous decomposition products). In the absence of an entry in the dangerous goods list, it is recommended to take into account human experience from accidental exposure. In the absence of human experience, grouping shall be based on data obtained from experiments in accordance with OECD TG 404 or 435. A substance determined not to be corrosive according to OECD TG 430 or 431 may be considered not to be corrosive to skin for the purpose of the UN transport regulation without further testing [12]

[d]An additional alternative criterion for PG III is metal corrosion: corrosion rate on either steel or aluminium surfaces exceeding 6.25 mm a year at a test temperature of 55 °C when tested on both materials [12]

where responses are noted following up to 3 min exposure and up to 1 h observation; subcategory 1B, where responses are described following exposure between 3 min and 1 h and observations up to 14 days; and subcategory 1C, where responses occur after exposures between 1 h and 4 h and observations up to 14 days.

The European Union has implemented the UN GHS classification system by means of the EU CLP regulation (1272/2008; [4, 5]) replacing since December 2010 the former EU Dangerous Substances Directive 67/548/EEC establishing criteria for classification of substances (EU DSD; [13]) and since 2015 the EU Dangerous Preparation Directive 199/45/EC establishing criteria for classification of mixtures (EU DPD; [14]). The EU CLP is equivalent to the UN GHS and requires the use of the three corrosive subcategories, i.e. 1A, 1B and 1C. When data is not sufficient for subcategorization, a EU CLP category 1 classification only should be assigned [5]. In addition to the UN GHS and EU CLP classification systems, skin corrosives can also be classified for transport purposes according to the UN model regulations for the transport of dangerous goods [12], which is based on three packaging groups (PG I, II, III). Table 7.1 provides an overview of the UN GHS, UN transport packaging group as well as the EU CLP classification systems as defined based on the traditional *in vivo* animal test method [5, 9, 12].

7.3 *In Vitro* Prospective Testing

In cases where the evaluation of existing information in a weight-of-evidence approach as recommended within the IATA described in OECD GD 203 [8], indicates the need for prospective testing, all available existing information should be used to formulate a hypothesis of the most likely skin irritation/corrosion potential of the chemical. This hypothesis and the regulatory context under which a decision must be taken shall then guide the choice of the test methods to be used and the sequence of the prospective *in vitro* testing to be conducted in either a top-down or a bottom-up approach (for details see Sect. 7.4) [8, 15].

When the available collected information and the weight-of-evidence assessment result in a high a priori probability of the test chemical to be an irritant or a corrosive, a *top-down approach* should be used, starting with an *in vitro* method for the identification of skin corrosion hazard followed, in case the test chemical is identified as not being corrosive, by an *in vitro* method for the identification of skin irritation hazard. Conversely, when all available collected information and the weight-of-evidence assessment result in a high a priori probability of the test chemical not requiring classification for skin irritation and corrosion, a *bottom-up approach* should be used, starting with an *in vitro* method for identification of skin irritation followed, in case the test chemical is identified as being irritant, by an *in vitro* method for identification of skin corrosion. An example on the use of such approaches has been described using the SkinEthic™ RhE model, in which high accuracy values were reported using either a bottom-up or a top-down approach [16].

A number of validated and regulatory adopted *in vitro* test methods are recommended to be used within the bottom-up and top-down approaches for skin corrosion and irritation hazard classification (for details see Sect. 7.4). These methods have been validated according to internationally agreed principles [17] and adopted by the OECD so that they fall under the OECD international Mutual Acceptance of Data (MAD). According to MAD, test data generated in any OECD member country in accordance with these OECD test guidelines and following the principles of Good Laboratory Practice (GLP) should be accepted in other OECD member countries for assessment purposes and other uses relating to the protection of human health and the environment.

When limitations and domain of the validated and adopted *in vitro* tests are adequately considered, these tests can provide sufficient information for the decision on potential of the substance to cause skin irritation and/or corrosion. In case of *in vitro* skin corrosion testing, the most appropriate OECD TG for the test chemical and the specific purpose should be chosen. In particular, the applicability domain and the ability of the test methods to provide information on subcategorization may play an important role in the choice of the test method to be used.

In the EU, only *in vitro* testing should be conducted for substances manufactured or imported in quantities between 1 tonne and 10 tonnes per year [2, 3, 5]. Furthermore, for substances manufactured or imported in quantities of ≥10 tonnes per year, "an *in vivo* study for skin corrosion/irritation shall be considered only if the *in vitro* studies (…) are not applicable, or the results of these studies are not adequate for classification and risk assessment" [3, 5]. As a consequence, no *in vivo* testing should be conducted in cases where the substance falls under the scope of the adopted *in vitro* test methods, result in appropriate classification and labelling and there are no substance-specific limitations to using those tests [15].

According to the OECD IATA [8], animal testing for skin corrosion should be used only as a last resort when, e.g., discrimination between optional subcategories 1B and 1C for chemicals outside of the applicability domain of OECD TG 435 is required, or the test chemical cannot be tested with the *in vitro* test methods currently adopted by the OECD due to limitations or non-applicability. Furthermore, if additional testing is still needed, the OECD IATA suggests that other *in vitro* skin corrosion test methods not yet adopted by the OECD be used that may resolve specific optional or subcategorization issues [8]. This is the case for example of the EpiSkin™ test method for which the original prediction model (not adopted due to the limited data set of well-known *in vivo* corrosive subcategory 1C chemicals) may be considered in a weight-of-evidence approach to distinguish between the subcategories 1B from 1C [18].

7.4 *In Vitro* Alternative Methods for Skin Corrosion

Several *in vitro* assays for skin corrosion have undergone prevalidation [19] and validation studies in the 1990s [20, 21]. Such efforts led to the formal endorsement of the scientific validity of three *in vitro* alternatives which were adopted and included in the EU test guidelines in 2000 and in the OECD testing guidelines in 2004 and 2006 [22–26]. These assays are:

- The reconstructed human epidermal (RhE) models (OECD TG 431), including:
  - The EpiSkin™ Standard Model (SM) validated in 1998 following a formal prospective validation study [27]
  - The EpiDerm™ Skin Corrosion Test (SCT) validated in 2000 following formal prevalidation and catch-up validation studies [28]
  - The SkinEthic™ Reconstituted Human Epidermis (RhE) validated in 2006 for having met the performance standards as required in the OECD TG 431 [29–31]
  - The epiCS® (previously named EST-1000) validated in 2009 for having met the performance standards as required in the OECD TG 431 [31, 32]

    A follow-up study was further undertaken in the framework of the OECD to investigate the capability of these four reconstructed human epidermis models to correctly identify the UN GHS corrosive subcategories 1A, 1B and 1C, and the TG 431 has been updated accordingly [26, 33, 34].
- The *in vitro* skin corrosion rat skin transcutaneous electrical resistance (TER) test [35], which uses excised rat skin as a test system and its electrical resistance as an endpoint (OECD TG 430).
- The Corrositex® test [36, 37], based on the penetration of test chemicals through a hydrogenated collagen matrix (biobarrier) and supporting filter membrane and which was considered to be useful for acids, bases and their derivates (OECD TG 435).

For a full evaluation of local skin effects after a single dermal exposure, it is recommended that these assays are used within testing approaches such as the IATA recommended in the OECD GD 203 [8] or the testing and assessment strategy recommended by ECHA [15].

7.4.1 Reconstructed Human Epidermis (RhE) Test Methods

The three-dimensional RhE models are comprised of normal, human-derived epidermal keratinocytes, which have been cultured to form a multilayered, highly differentiated model of the human epidermis. It consists of organized basal, spinous and granular layers and a multilayered *stratum corneum* containing intercellular lamellar lipid layers arranged in patterns representing main lipid classes analogous to those found *in vivo*. The *in vitro* RhE models represent therefore the target organ of the species of interest.

The principles of the RhE test method is based on the premise that corrosive chemicals are able to penetrate the *stratum corneum* by diffusion or erosion and are cytotoxic to the cells in the underlying layers. Corrosive test chemicals are identified by their ability to produce a decrease in cell viability below defined threshold levels. The test chemical is applied topically to reconstructed human epidermis. Cell viability is measured by dehydrogenase conversion of the vital dye MTT into a blue formazan salt that is quantitatively measured after extraction from tissues.

The test methods described in the OECD TG 431 [26] allow the identification of corrosive and non-corrosive substances and mixtures. Furthermore, it allows identification of the subcategorization of corrosive substances and mixtures into the UN GHS subcategory 1A as well as into a combination of subcategories 1B and 1C [26, 33, 34, 38, 39]. All four commercially available models falling within OECD TG 431 are therefore able to subcategorize 1A versus 1B-and-1C versus no category [26, 33]. The standard operating procedures (SOPs) of these RhE models should be consulted when implementing and using one of these four models in a test laboratory:

- EpiSkin™ SM [18]
- EpiDerm™ SCT [40]
- SkinEthic™ RhE [41]
- epiCS® [42]

Based on the overall data set available (mainly composed of individual substances) that supported the test methods inclusion in the OECD TG 431 [20, 21, 34], the test guideline is considered applicable to a wide range of chemical classes and physical states including liquids (aqueous or nonaqueous), semi-solids, solids (soluble or insoluble in water) and waxes. In addition, OECD TG 431 is assumed to be applicable to mixtures as an extension of its applicability to substances. However, if additional information is available, this should be taken into account, in combination with the existing evidence, to evaluate the usefulness of a test method to assess mixtures [8]. In cases where evidence can be demonstrated on the non-applicability of the test guideline to a specific category of mixtures (e.g. following a strategy as proposed by Eskes et al. [43]), the test guideline should not be used for that specific category of mixtures [8, 24]. Similarly, in cases where evidence can be demonstrated on the non-applicability of test methods included in OECD TG 431 to a specific category of test chemicals, these test methods should not be used for that specific category of test chemicals [24]. Finally, the OECD TG 431 does not allow testing of gases and aerosols.

The OECD TG 431 does not allow discriminating skin corrosive subcategory 1B from subcategory 1C, despite the fact that the EpiSkin™ proposes a prediction model able to distinguish between these two corrosive subcategories [18]. This is due to the limited set of well-known *in vivo* corrosive subcategory 1C chemicals available to allow a formal assessment of the performances of the assays falling within the OECD TG 431. Furthermore, while the OECD TG 431 does not provide adequate information on skin irritation, it should be noted that OECD TG 439 [44] specifically addresses the health effect skin irritation *in vitro* and is based on the same RhE test system, though using another protocol. For a full evaluation of local skin effects after a single dermal exposure, the OECD Guidance Document No. 203 on an IATA for skin corrosion and irritation should be considered, which includes the conduct of *in vitro* tests for skin corrosion and skin irritation before considering testing in living animals [8].

Test chemicals absorbing light in the same range as MTT formazan and test chemicals able to directly reduce the vital dye MTT (to MTT formazan) may interfere with the tissue viability measurements and require the use of adapted controls for corrections. The type of adapted controls that may be required will vary depending on the type of interference produced by the test chemical and the procedure used to measure MTT formazan. Results for test chemicals producing non-specific interactions with MTT $\geq 50\%$ of the negative control should be taken with caution when OD is used as means of measurement. However, the use of HPLC/UPLC spectrophotometry as an alternative means of measuring the MTT formazan offers the possibility of evaluating the skin corrosion potential of strongly coloured test chemicals that could interfere with the standard OD measurements [45, 46]. Finally, fatty amine derivatives (characterized as cationic surfactants) were shown to have a tendency to be under-predicted with the test methods falling within the OECD TG 431 [47]. For these chemistries, an extended exposure period of 4 h is recommended for the EpiDerm™ SCT (see also Chap. 9 of this book).

7.4.2 Transcutaneous Electrical Resistance Test (TER)

In the transcutaneous electrical resistance test, corrosive chemicals are identified by their ability to produce a loss of normal *stratum corneum* integrity and barrier function, which is measured as a reduction in the TER below a certain threshold level [24, 48]. The transcutaneous electrical resistance is a measure of the electrical impedance of the skin, as a resistance value in kilo-ohms. The test chemical is applied for up to 24 h to the epidermal surfaces of skin discs (obtained from humanely killed rats aged 28–30 days), in a two-compartment test system in which the skin discs function as the separation between the compartments. For rat TER, a cut-off value of 5 kΩ has been selected based on extensive data for a wide range of substances where the vast majority of values were either clearly well above (often >10 kΩ) or well below (often <3 kΩ) this value. Generally, test chemicals that are non-corrosive in animals but are irritating or non-irritating do not reduce the TER below this cut-off value.

Furthermore, a dye-binding step is incorporated into the test procedure for confirming positive results specially for TER values around 5 kΩ [24, 48]. The dye-binding step is based on the dye sulforhodamine B and determines if the increase in ionic permeability is due to physical destruction of the *stratum corneum*. Indeed, exposure of certain non-corrosive test chemicals can result in a reduction of resistance below the cut-off of 5 kΩ allowing the passage of ions through the *stratum corneum*, thereby reducing the electrical resistance. For example, neutral organics and chemicals that have surface-active properties (including detergents, emulsifiers and other surfactants) can remove skin lipids making the barrier more permeable to ions. In case of skin corrosive effects where the *stratum corneum* is disrupted, the dye sulforhodamine B, when applied to the skin surface, rapidly penetrates and stains the underlying tissue. This particular dye is stable to a wide range of substance and is not affected by the extraction procedure. As a consequence, obtaining high dye contents may indicate a corrosive effect.

The test method described in OECD TG 430 [24] and following the recommended protocol [48] allows the identification of corrosive chemical and non-corrosive substances and mixtures. However, a limitation of the OECD TG 430, as demonstrated by the validation study, is that it does not allow the subcategorization of corrosive substances and mixtures in accordance with the UN GHS subcategories, which may have importance regarding transportation rules such as the UN [9, 12].

Based on the data set used in the validation (mainly substances) underlying the OECD TG 430 [20], the TG is considered applicable to a wide range of chemical classes and physical states including liquids (aqueous or nonaqueous), semi-solids, solids (soluble or insoluble in water) and waxes. However, since for specific physical states test items with suitable reference data are not always readily available, a comparably small number of waxes and corrosive solids were assessed during validation. The OECD TG 430 does not allow testing of gases and aerosols. Furthermore, in cases where evidence can be demonstrated on the non-applicability of the OECD TG 430 to a specific category of substances, the TG should not be used for that specific category of substances.

The TG 430 is assumed to be applicable to mixtures as an extension of its applicability to substances. However, if additional information is available, this should be taken into account, in combination with the existing evidence, to evaluate the usefulness of the test method to assess mixtures [8]. In cases where evidence can be demonstrated on the non-applicability of the OECD TG 430 to a specific category of mixtures (e.g. following a strategy as proposed by Eskes et al. [43]), the test guideline should not be used for that specific category of mixtures [8, 24].

While the OECD TG 430 does not provide adequate information on skin irritation, it should be noted that OECD TG 439 [44] specifically addresses the health effect skin irritation *in vitro*. For a full evaluation of local skin effects after a single dermal exposure, the OECD Guidance Document No. 203 on an IATA for skin corrosion and irritation should be considered, which includes the conduct of *in vitro* tests for skin corrosion and skin irritation before considering testing in living animals [8].

Finally the TER assay still makes use of animals from which the dorsal and flank hair are carefully removed at day 22 using small clippers. The clipped area is submerged in an antibiotic solution the first, third and fourth day, and the animals are humanely killed when 28–30 days old. Because of such procedures, depending on national regulations, this assay may or may not be considered as an animal experimentation. Indeed, the EU directive on the protection of animals used for scientific purposes (2010/63/EU; [6]) states that a "'procedure' means any use, invasive or non-invasive, of an animal for experimental or other scientific purposes, with known or unknown outcome, or educational purposes, which may cause the animal a level of pain, suffering, distress or lasting harm equivalent to, or higher than, that caused by the introduction of a needle in accordance with good veterinary practice".

7.4.3 *In Vitro* Membrane Barrier Test Method for Skin Corrosion (Corrositex®)

The *in vitro* membrane barrier test method for skin corrosion is described in detail in the OECD TG 435 [25]. It is also recommended by the UN model regulations for the transport of dangerous goods [12]. The only *in vitro* membrane barrier method currently endorsed as valid falling within the OECD TG 435 is the commercially available Corrositex® test method [49]. In Europe, although it was endorsed by the ECVAM's Scientific Advisory Committee (ESAC) as being scientifically valid, it has not been taken up in the EU legislation due to the fact that the Corrositex® test method was considered valid only for limited applicability domain of acids, bases and their derivates [36, 37].

The *in vitro* membrane barrier test method utilizes an artificial membrane designed to respond to corrosive chemicals in a manner similar to animal skin *in situ*. The system is comprised of two components: a synthetic macromolecular biobarrier and a chemical detection system (CDS) which allows detecting the membrane barrier damage caused by corrosive test chemicals after the application of the test chemical to the surface of the membrane barrier.

The classification assigned is based on the time (in minutes) it takes for a test chemical to penetrate through the membrane barrier and its supporting filter to the indicator solution. Penetration of the membrane barrier (or breakthrough) is measured by a change in the colour of a pH indicator dye or changes in other properties of the indicator solution such as physical appearance (flaking, precipitation, etc.). The time required for this change to occur (the breakthrough time) is reported to be inversely proportional to the degree of corrosivity of the test chemical, i.e. the longer it takes to detect a change, the less corrosive is the substance.

The OECD TG 435 allows the identification of corrosive test chemicals as well as the subcategorization of corrosive test chemicals according to the three UN GHS subcategories of corrosivity, i.e. 1A, 1B and 1C and to the UN transport packaging groups I, II and III for corrosivity hazard [9, 12]. In addition, the test method may be used to make decisions on the corrosivity and non-corrosivity of specific classes of chemicals including organic and inorganic acids, acid derivatives[1] and bases [36, 37].

In contrast to the OECD TG 430 and 431 which were validated mainly using individual substances, the validation data set of the Corrositex® test method falling within OECD TG 435 comprised both substances and mixtures. As a consequence, the method is considered applicable to both substances and mixtures. Furthermore, the *in vitro* membrane barrier test method may be used to test solids (soluble or insoluble in water), liquids (aqueous or nonaqueous) and emulsions.

A limitation of the Corrositex® test method is that many non-corrosive and some corrosive test chemicals may not qualify for testing, based on the compatibility test. Indeed, test chemicals that do not cause a detectable change in the compatibility test (i.e. induce a colour change of the CDS) cannot be tested with the membrane barrier test method and should be tested using other test methods. For instance, aqueous test chemicals with a pH in the range of 4.5–8.5 often do not qualify for testing, even though 85% of chemicals tested in this pH range were found to be non-corrosive in animal tests [37].

Finally, while the OECD TG 435 does not provide adequate information on skin irritation, it should be noted that OECD TG 439 [44] specifically addresses the health effects of skin irritation *in vitro*. For a full evaluation of local skin effects after a single dermal exposure, the OECD Guidance Document No. 203 on an IATA for skin corrosion and irritation should be considered, which includes the conduct of *in vitro* tests for skin corrosion and skin irritation before considering testing in living animals [8].

7.5 Comparison to the *In Vivo* Test Method

A summary of the major components of the regulatory *in vivo* and *in vitro* tests for skin corrosion is shown in Table 7.2.

The adopted *in vitro* reconstructed human epidermis methods are based on cultured tissues of the species of interest, i.e., humans. Although these models do not present all functional complexity that exist *in vivo*, i.e. the dermis and its features such as hair follicles, sebaceous glands and nerve and immune cells, such features seem to play a less important role in the mechanisms of skin corrosion than in the

[1] "Acid derivative" is a non-specific class designation and is broadly defined as an acid produced from a chemical either directly or by modification or partial substitution. This class includes anhydrides, halo acids, salts and other types of chemicals.

Table 7.2 Comparison of the principal method components of the regulatory accepted *in vivo* and *in vitro* test methods for skin corrosion

	In vivo skin corrosion test (OECD TG 404)	*In vitro* human skin model (OECD TG 431)	*In vitro* TER (OECD TG 430)	*In vitro* membrane barrier test—Corrositex® (OECD TG 435)
Model used	Albino rabbit	Three-dimensional reconstructed human epidermis, consisting of organized basal, spinous and granular layers, and a multilayered *stratum corneum* (0.38–0.63 cm^2 surface depending on the model)	Skin discs prepared from young rats, where 10–15 skin discs can be obtained per rat skin (0.79 cm^2 surface)	A synthetic macromolecular biobarrier and a chemical detection system (CDS) which indicates the presence of the test chemical
Number of replicates	1–3 animals based on severity of effects	At least two replicates for each exposure time	At least three skin discs	Two repeats in two batches
Dose and application of test chemical	0.5 mL (liquids) or 0.5 g (solids) applied to ~6 cm^2 of skin and covered with a gauze patch (~83.3 μL or mg/ cm^2). Solids might be moistened to ensure good skin contact	Liquids: 40–50 μL (79.4–131.6 μL/cm^2 depending on model). Solids: 20–25 mg (39.7–52.6 mg/cm^2 depending on model)	Liquids: 150 μL (~189.9 μL/cm^2). Solids: sufficient amount to cover surface and 150 μL of deionized water added on top of the solid	0.5 mL (liquids) or 0.5 g (solids) applied on membrane
Controls	The potential influence of the vehicle on irritation of the skin by the test chemical should be minimal, if any	Negative control: 0.9% NaCl or water Positive control: 8 N KOH or glacial acetic acid	Negative control: distilled water Positive control: 10 M hydrochloric acid	Negative control: e.g. 10% citric acid or 6% propionic acid Positive control: e.g. sodium hydroxide Vehicles or solvents should not alter integrity of the membrane barrier system and should not alter the corrosivity of the test chemical

(continued)

Table 7.2 (continued)

	In vivo skin corrosion test (OECD TG 404)	*In vitro* human skin model (OECD TG 431)	*In vitro* TER (OECD TG 430)	*In vitro* membrane barrier test—Corrositex® (OECD TG 435)
Exposure time	3 min, 1 h, 4 h applied in a sequential way, so that if corrosive effects are observed, the test is terminated. If no corrosive effects seen after 4 h exposure, the animal is observed up to 14 days	3 min at Room Temperature (RT) 1 h at RT or at 37 °C depending on the model In the EpiSkin™ model, also 4 h at RT	24 h at RT	The time needed for a test chemical to penetrate the membrane barrier is used to predict corrosivity. It is reported to be inversely proportional to the degree of corrosivity, i.e. the longer it takes to penetrate, the less corrosive is the substance
Washing	At the end of exposure time to remove test chemical	At the end of exposure time to remove test chemical	At the end of exposure time to remove test chemical	Not necessary
Endpoint(s) assessed	Grading of skin reactions Other reactions such as defatting of skin, clinical signs of toxicity and body weight, persistence of alopecia, hyperkeratosis, hyperplasia and scaling Histopathology may be carried out in case of equivocal responses	Cell viability: based on the principle that corrosive chemicals are able to penetrate the *stratum corneum* and are cytotoxic to the underlying layers	Transcutaneous electrical resistance: based on the principle that corrosive test chemicals can produce loss of *stratum corneum* integrity and barrier function, measured by the TER Dye binding: to determine if TER values below the cut-off but in absence of visual damage are due to increase in permeability or to skin corrosion	The time it takes a substance to penetrate through the membrane barrier Penetration is measured by colour or physical change of the chemical detection system

Interpretation of results	Results can be used to classify according to all UN GHS categories, optional categories and subcategories for skin corrosion and irritation	Able to distinguish between corrosive and non-corrosive, and discrimination between UN GHS subcategory 1A from combined subcategories 1B-and-1C. Not able to distinguish GHS subcategory 1B from 1C due to the limited set of well-known *in vivo* corrosive subcategory 1C chemicals	Able to distinguish between corrosives (UN GHS cat. 1) and non-corrosives	Allows identification of corrosives (GHS cat. 1) and subcategorization into the three GHS subcategories (1A, 1B and 1C)
Limitations	May overpredict human responses May be variable between laboratories Does not assess repetitive low-dose exposure Has the potential to cause considerable discomfort or pain to laboratory animals	Does not allow discrimination between subcategory 1B and 1C Not designed to provide information on skin irritation Not applicable to gases and aerosols Results obtained with test chemicals having non-specific interactions with MTT ≥50% should be taken with caution when OD is used as measurement for cell viability. This may be circumvented for coloured interfering test chemicals with the use of HPLC/UPLC as an alternative measurement	Not able to distinguish the three GHS subcategories (1A, 1B and 1C) Not designed to provide information on skin irritation Not applicable to gases and aerosols	Not designed to give information on skin irritation Not applicable to gases and aerosols Test chemicals not causing detectable changes in the chemical detection system cannot be tested In EU, considered valid only for acids, bases and their derivatives

inflammatory reactions that could lead to skin irritation. On the other hand, the adopted *in vitro* TER method makes use of excised rat skin which does include the dermis but no blood circulation. Finally, the adopted *in vitro* membrane barrier assay does only mimic the morphological features of the *in vivo* skin.

The various adopted *in vitro* models for regulatory purposes also mimic the mechanisms of skin corrosion occurring in the *in vivo* test. These encompass:

- Cell viability (reconstructed human epidermis models) based on the principle that corrosive chemicals are able to penetrate the *stratum corneum* and are cytotoxic to the underlying layers
- Loss of barrier function and integrity (TER assay), based on the principle that corrosive test chemicals can produce loss of *stratum corneum* integrity and barrier function
- Membrane barrier damage (membrane barrier test) presumably by the same mechanism(s) of corrosion that operate on living skin

With the exception of TER, the exposure times used with the adopted *in vitro* assays are comparable to those used *in vivo* (3 min, 1 h and 4 h for the RhE test methods and cut-offs of 3 min, 1 h and 4 h for the *in vitro* membrane barrier test), and the doses applied *in vitro* are similar or greater than those applied *in vivo* (for details see Table 7.2).

Unlike the *in vivo* test, the *in vitro* test methods make systematically use of positive and negative controls to check for the functionality of the test method. In addition, the *in vitro* test methods require ensuring technical proficiency by the laboratory prior to the routine use of the *in vitro* methods, by testing a list of recommended proficiency chemicals.

Overall, the adopted *in vitro* assays for skin corrosion are all able to distinguish between corrosive (UN GHS cat. 1) and non-corrosive test chemicals according to the UN GHS classification system. However, regarding the possibility of the assays to identify the UN GHS corrosive subcategories, the following currently applies:

- The membrane barrier test falling within the OECD TG 435 is considered valid to distinguish the three UN GHS subcategories 1A, 1B and 1C (as well as the three UN transport packaging groups I, II and III), even though its applicability is limited to test chemicals that are compatible with the chemical detection system of the assay and, in the EU, to acids, bases and their derivates.
- The RhE test methods falling within the OECD TG 431 are accepted to distinguish between the UN GHS subcategory 1A from a combination of subcategories 1B-and-1C but not to distinguish between the UN GHS subcategory 1B from the subcategory 1C due to the limited set of well-known *in vivo* corrosive subcategory 1C chemicals.
- The TER assay falling within the OECD TG 430 does not allow the subcategorization of corrosive test chemicals in accordance with the UN GHS subcategories.

Finally, none of the *in vitro* assays adopted for skin corrosion testing provide with adequate information on skin irritation. For that purpose, the *in vitro* method falling within the OECD TG 439 [44] should be used as it specifically addresses the health effects of skin irritation *in vitro*. For a full evaluation of local skin effects after a single dermal exposure, the OECD Guidance Document No. 203 on an IATA for skin corrosion and irritation should be considered [8].

References

1. EC. Regulation No 1223/2009 of the European Parliament and of the Council of 30 November 2009 on cosmetic products. Off J Eur Union. 2009; L342:59–209.
2. EC. Regulation No 1907/2006 of the European Parliament and of the Council of 18 December 2006 concerning the Registration, Evaluation, Authorisation and Restriction of Chemicals (REACH), establishing a European Chemicals Agency, amending Directive 1999/45/EC and repealing Council Regulation (EEC) No 793/93 and Commission Regulation (EC) No 1488/94 as well as Council Directive 76/769/EEC and Commission Directives 91/155/EEC, 93/67/EEC, 93/105/EC and 2000/21/EC. Off J Eur Union. 2006; L396:1–849.
3. EC. Regulation No 2016/863 of 31 May 2016 amending Annexes VII and VIII to Regulation (EC) No 1907/2006 of the European Parliament and of the Council on the Registration, Evaluation, Authorisation and Restriction of Chemicals (REACH) as regards skin corrosion/irritation, serious eye damage/eye irritation and acute toxicity. Off J Eur Union. 2016; L144:27–31.
4. EC. Regulation No 1272/2008 of the European Parliament and of the Council of 16 December 2008 on classification, labelling and packaging of substances and mixtures, amending and repealing Directives 67/548/EEC and 1999/45/EC, and amending Regulation (EC) No 1907/2006. Off J Eur Union. 2008; L353:1–1355.
5. EC. Regulation 2016/918 of 19 May 2016 amending, for the purposes of its adaptation to technical and scientific progress, Regulation (EC) No 1272/2008 of the European Parliament and of the Council on classification, labelling and packaging of substances and mixtures. Off J Eur Union. 2016; L156:1–103.
6. EC. Directive No 2010/63/EU of the European Parliament and of the council of 22 September 2010 on the protection of animals used for scientific purposes. Off J Eur Union. 2010; L276:33.
7. OECD guidelines for the testing of chemicals No. 404. Acute dermal irritation/corrosion. Paris, France: Organisation for Economic Cooperation and Development. Originally adopted in 1981. 2015. Available at: http://www.oecd-ilibrary.org/environment/oecd-guidelines-for-the-testing-of-chemicals-section-4-health-effects_20745788. Accessed 10 May 2016.
8. OECD Guidance Document No. 203: integrated approached to testing and assessment for skin irritation/corrosion. Environment, health and safety publications, series on testing and assessment. Organisation for Economic Cooperation and Development, Paris. 2014. Available at: http://www.oecd.org/env/ehs/testing/series-testing-assessment-publications-number.htm. Accessed 25 Apr 2016.
9. United Nations (UN). Globally harmonized system of classification and labelling of chemicals (GHS). Sixth revised edition. Part 3: Health hazards – chapter 3.2 skin corrosion/irritation. New York, USA, and Geneva, Switzerland: United Nations. p. 125–136. 2015. Available at: http://www.unece.org/fileadmin/DAM/trans/danger/publi/ghs/ghs_rev06/English/ST-SG-AC10-30-Rev6e.pdf. Accessed 25 Apr 2016.
10. Draize JH, Woodard G, Calvery HO. Methods for the study of irritation and toxicity of substances applied topically to the skin and mucous membranes. J Pharmacol Exp Ther. 1944;82:377–90.
11. ECHA. Guidance on the application of the CLP criteria. Guidance to regulation EC N. 1272/2008 on classification, labelling and packaging (CLP) of substances and mixtures.

Version 4.1. Chapter 3.2. Skin corrosion/irritation. 2015. Available at: https://echa.europa.eu/documents/10162/13562/clp_en.pdf. Accessed 26 Apr 2016.
12. UN. Volume I – Part 2. Classification. Recommendations on the transport of dangerous goods – model regulations. Nineteenth revised edition. 2015. Available at: http://www.unece.org/trans/danger/publi/unrec/rev19/19files_e.html. Accessed 26 Apr 2015.
13. EC. Directive No 2001/59/EC of 6 August 2001 adapting to technical progress for the 28th time Council Directive 67/548/EEC on the approximation of laws, regulations and administrative provisions relating to the classification, packaging and labelling of dangerous substances. Off J Eur Communities. 2001;L225:1–333.
14. EC. Directive No 1999/45/EC of the European Parliament and of the Council of 31 May 1999 concerning the approximation of laws, regulations and administrative provisions of the Member States relating to the classification, packaging and labelling of dangerous preparations. Off J Eur Communities. 1999;L200:1–68.
15. ECHA. Guidance on information requirements and chemical safety assessment. Chapter R.7a: endpoint specific guidance. Version 4.0. Subchapter R.7.2. Skin corrosion/irritation, serious eye damage/eye irritation and respiratory tract corrosion/irritation, p. 168–199. 2015. Available at: http://echa.europa.eu/documents/10162/13632/information_requirements_r7a_en.pdf. Accessed 25 Apr 2016.
16. Alépée N, Grandidier MH, Tornier C, Cotovio J. An integrated testing strategy for *in vitro* skin corrosion and irritation assessment using SkinEthic™ Reconstructed Human Epidermis. Toxicol In Vitro. 2015;29:179–1792.
17. OECD. Guidance Document No. 34: validation and international acceptance of new or updated test methods for hazard assessment. In: Environmental health and safety monograph series on testing and assessment. 2005. Available at: http://www.oecd.org/env/ehs/testing/series-testing-assessment-publications-number.htm. Accessed 10 May 2016.
18. EpiSkin™ SM (2011). INVITTOX protocol No. 118. EpiSkin™ skin corrosivity test. Available at: https://eurl-ecvam.jrc.ec.europa.eu/validation-regulatory-acceptance/skin-corrosion-docs/EpiSkin_INVITOX%20protocol%20corrosivity%20test_OECD.pdf. Accessed 10 May 2016.
19. Botham PA, Chamberlain M, Barratt MD, Curren RD, Esdaile DJ, Gardiner JR, Gordon VC, Hildebrand B, Lewis RW, Liebsch M, Logemann P, Osborne R, Ponec M, Régnier J-F, Steiling W, Walker AP, Balls M. A prevalidation study on *in vitro* skin corrosivity testing. The report and recommendations of ECVAM workshop 6. ATLA. 1995;23:219–55.
20. Fentem JH, Archer GEB, Balls M, Botham PA, Curren RD, Earl LK, Esdaile DJ, Holzhütter HG, Liebsch M. The ECVAM international validation study on *in vitro* tests for skin corrosivity. 2. Results and evaluation by the Management Team. Toxicol In Vitro. 1998;12:483–524.
21. Liebsch M, Traue D, Barrabas C, Spielmann H, Uphill P, Wilkins S, Wiemann C, Kaufmann T, Remmele M, Holzhütter HG. The ECVAM prevalidation study on the use of EpiDerm for skin corrosivity testing. ATLA. 2000;28:371–401.
22. EC. Annex I to Directive No 2000/33/EC adapting to technical progress for the 27th time Council Directive 67/548/EEC on the approximation of laws, regulations and administrative provisions relating to the classification, packaging and labelling of dangerous substances. Off J Eur Communities. 2000; L136:91–97.
23. EC. Regulation No 440/2008 of 30 May 2008 laying down test methods pursuant to Regulation (EC) No 1907/2006 of the European Parliament and of the Council on the Registration, Evaluation, Authorisation and Restriction of Chemicals (REACH). Off J Eur Union. 2008; L142:1–739.
24. OECD guidelines for the testing of chemicals No. 430. *In vitro* skin corrosion: transcutaneous electrical resistance test (TER). Paris, France: Organisation for Economic Cooperation and Development. 2015 (originally adopted in 2004). Available at: http://www.oecd-ilibrary.org/environment/oecd-guidelines-for-the-testing-of-chemicals-section-4-health-effects_20745788. Accessed 10 May 2016.
25. OECD guidelines for the testing of chemicals No. 435. *In vitro* membrane barrier test method for skin corrosion. Paris, France: Organisation for Economic Cooperation and Development. 2015 (originally adopted in 2006). Available at: http://www.oecd-ilibrary.org/environment/

oecd-guidelines-for-the-testing-of-chemicals-section-4-health-effects_20745788. Accessed 10 May 2016.
26. OECD guidelines for the testing of chemicals No. 431. *In vitro* skin corrosion: reconstructed human epidermis (RHE) test method. Paris, France: Organisation for Economic Cooperation and Development. 2016 (originally adopted in 2004). Available at: http://www.oecd-ilibrary.org/environment/oecd-guidelines-for-the-testing-of-chemicals-section-4-health-effects_20745788. Accessed 10 May 2016.
27. ESAC. Statement on the scientific validity of the EPISKIN™ test (an *in vitro* test for skin corrosivity). ATLA. 1998;26:277–80.
28. ESAC. Statement on the application of the Epiderm™ human skin model for skin corrosivity testing. ATLA. 2000;28:365–6.
29. ESAC. Statement on the application of the SkinEthic™ human skin model for skin corrosivity testing. 2006. Available at: http://ecvam.jrc.ec.europa.eu/ under "Publications", "ESAC statements". Accessed 18 Sep 2009.
30. Kandárová H, Liebsch M, Spielmann H, Genschow E, Schmidt E, Guest R, Whittingham A, Warren N, Gamer A, Remmele M, Kaufmann T, Wittmer E, De Wever B, Rosdy M. Assessment of the SkinEthic reconstituted human epidermis for skin corrosion testing according to OECD guideline 431. Toxicol In Vitro. 2006;20:547–59.
31. OECD Guidance Document No. 219: performance standards for the assessment of proposed similar or modified in vitro reconstructed human epidermis (RHE) test methods for skin corrosion testing as described in TG 431. OECD environmental health and safety publications, series on testing and assessment. 2015. Available at: http://www.oecd.org/env/ehs/testing/series-testing-assessment-publications-number.htm. Accessed 10 May 2016.
32. ESAC. ESAC statement on the scientific validity of an in vitro test method for skin corrosivity testing. 2009. Available at: http://ecvam.jrc.ec.europa.eu/ under "Publications", "ESAC statements". Accessed 18 Sep 2009.
33. Desprez B, Barroso J, Griesinger C, Kandárová H, Alépée N, Fuchs HW. Two novel prediction models improve predictions of skin corrosive sub-categories by test methods of OECD Test Guideline No. 431. Toxicol In Vitro. 2015;29:2055–80.
34. OECD Guidance Document No. 190: summary document on the statistical performance of methods in OECD test guideline 431 for sub-categorisation. Environment, health, and safety publications, series on testing and assessment. Organisation for Economic Cooperation and Development, Paris. 2013. Available at: http://www.oecd.org/env/ehs/testing/series-testing-assessment-publications-number.htm. Accessed 10 May 2016.
35. ESAC. Statement on the scientific validity of the rat skin transcutaneous electrical resistance (TER) test (an *in vitro* test for skin corrosivity). ATLA. 1998;26:275–7.
36. ESAC. Statement on the application of the CORROSITEX® assay for skin corrosivity testing. ATLA. 2001;29:96–7.
37. NIH. Corrositex®: an in vitro test method for assessing dermal corrosivity potential of chemicals. NIH Publication No. 99-4495. Research Triangle Park, NC: NIEHS; 1999.
38. Alépée N, Grandidier MH, Cotovio J. Sub-categorisation of skin corrosive chemicals by the EpiSkin™ reconstructed human epidermis skin corrosion test method according to UN GHS: revision of OECD test guideline 431. Toxicol In Vitro. 2014;28:131–45.
39. Alépée N, Rober C, Tornier C, Cotovio J. The usefulness of the validated SkinEthic™ RHE test method to identify skin corrosive UN GHS subcategories. Toxicol In Vitro. 2014;28:616–25.
40. EpiDerm™ SCT. SOP version MK-24-007-0024. Protocol for: in vitro EpiDerm™ skin corrosion test (EPI-200-SCT), for use with MatTek corporation's reconstructed human epidermal model EpiDerm. 2012. Available at: https://eurl-ecvam.jrc.ec.europa.eu/validation-regulatory-acceptance/skin-corrosion-docs/EpiDerm%20EPI-200-SCT%20%20Protocol%20MK-24-007-0024.pdf. Accessed 10 May 2016.
41. SkinEthic™ RHE. INVITTOX protocol SkinEthic™ skin corrosivity test. 2012. Available at: https://eurl-ecvam.jrc.ec.europa.eu/validation-regulatory-acceptance/eurl-ecvam/validation-regulatory-acceptance/skin-corrosion-docs/SkinEthic_INVITTOX_protocol_corrosivity_test_OECD_Jan2012.pdf. Accessed 10 May 2016.

42. epiCS® 4.1. SOP version in vitro skin corrosion: human skin model test epidermal skin test epiCS® CellSystems. 2012. Available at: http://cellsystems.de/files/epiCS/SOP-epiCS-skin-corrosion.pdf. Accessed 10 May 2016.
43. Eskes C, Detappe V, Koeter H, Kreysa J, Liebsch M, Zuang V, Amcoff P, Barroso J, Cotovio J, Guest R, Hermann M, Hoffmann S, Masson P, Alepee N, Arce LA, Bruschweiler B, Catone T, Cihak R, Clouzeau J, D'Abrosca F, Delveaux C, Derouette JP, Engelking O, Facchini D, Frohlicher M, Hofmann M, Hopf N, Molinari J, Oberli A, Ott M, Peter R, Sa-Rocha VM, Schenk D, Tomicic C, Vanparys P, Verdon B, Wallenhorst T, Winkler GC, Depallens O. Regulatory assessment of *in vitro* skin corrosion & irritation data within the European framework: workshop recommendations. Regul Toxicol Pharmacol. 2012;62:393–403.
44. OECD guidelines for the testing of chemicals No. 439. In vitro skin irritation: reconstructed human epidermis test method. Paris, France: Organisation for Economic Cooperation and Development. 215 (originally adopted in 2010). Available at: http://www.oecd-ilibrary.org/environment/oecd-guidelines-for-the-testing-of-chemicals-section-4-health-effects_20745788. Accessed 10 May 2016.
45. Alépée N, Barroso J, De Smedt A, De Wever B, Hibatallah J, Klaric M, Mewes KR, Millet M, Pfannenbecker U, Tailhardat M, Templier M, McNamee P. Use of HPLC/UPLC-spectrophotometry for detection of formazan in *in vitro* Reconstructed Human Tissue (RhT)-based test methods employing the MTT-reduction assay to expand their applicability to strongly coloured test chemicals. Toxicol In Vitro. 2015;29:741–61.
46. Alépée N, Hibatallah J, Klaric M, Mewes KR, Pfannenbecker U, McNamee P. Assessment of cosmetic ingredients in the *in vitro* reconstructed human epidermis test method EpiSkin™ using HPLC/UPLC-spectrophotometry in the MTT-reduction assay. Toxicol In Vitro. 2016;33:105–17.
47. Houthoff E, Rugen P, Hart D. Predictability of in vitro dermal assays when evaluating fatty amine derivatives. Toxicol In Vitro. 2015;29:1263–7.
48. TER. DB-ALM Protocol No. 115. SOP on the rat skin transcutaneous electrical resistance (TER) test. 2008. Available at: https://eurl-ecvam.jrc.ec.europa.eu/validation-regulatory-acceptance/skin-corrosion-docs/115.pdf. Accessed 11 May 2016.
49. Corrositex. DB-ALM protocol No. 116. SOP on the Corrositex® continuous time monitor assay. 2008. Available at: https://eurl-ecvam.jrc.ec.europa.eu/validation-regulatory-acceptance/skin-corrosion-docs/116.pdf. Accessed 11 May 2016.

The EpiSkin™ Human Epidermis Model for *In Vitro* Skin Corrosion of Test Chemicals

8

Nathalie Alépée, Marie Hélène Grandidier, and José Cotovio

8.1 Principle of Test Method and Scientific Basis

Common corrosives usually include strong acids, strong bases, or concentrated solutions of certain weak acids or weak bases. Their effect on living tissue such as skin is mainly based on acid-base reactions of amide hydrolysis and ester hydrolysis. Final clinical effects mostly depend on the substance. For example, hydrogen peroxide removes a bleached layer of skin, while nitric acid causes a characteristic yellow colour change in the skin. These reactions lead to chemical burns and define the usual mechanism of tissue damage caused by corrosives. Chemical burns follow the standard burn classification and may cause extensive and variable tissue damage. The exact symptoms of a chemical burn depend on the chemical involved. Symptoms and clinical signs include itching, bleaching or darkening of the skin, a burning sensation and tissue necrosis. Corrosive substances irreversibly damage the skin through the epidermis and into the dermis, beyond repair. The skin may be exposed to a wide range of chemicals through occupational exposure or consumer products such as solvents and detergents. Chemical production and related professional fields are examples of occupations in which chemical burns can occur.

Because chemicals may even lead to severe risks, there is a strong need for data on skin corrosion potential. These data would ensure a high level of protection in terms of human health and occupational safety as well as the safe transportation of chemicals (in line with UN Recommendations on the Transport of Dangerous Goods). It would also be necessary to achieve sustainable development while enhancing competitiveness and innovation. These requirements are reflected in legislation such as the Classification, Labelling, and Packaging (CLP) Regulation (1272/2008), the EU regulation on cosmetic products (EC 1223/2009) [1] and the

N. Alépée (✉) • M.H. Grandidier • J. Cotovio
L'Oréal Research and Innovation,
1, Avenue Eugène Schueller, 93600 Aulnay-sous-Bois Cedex, France
e-mail: nalepee@rd.loreal.com

C. Eskes et al. (eds.), *Alternatives for Dermal Toxicity Testing*,
DOI 10.1007/978-3-319-50353-0_8

REACH Regulation (1907/2006) [2]. Corrosion is also an endpoint addressed by the Environmental Protection Agency in its assessments of pesticide formulations and ingredients.

Currently, internationally accepted skin corrosion test methods are based on the fact that corrosive substances show cytotoxic effects following topical, short-term exposure of the epidermis. They include the traditional Draize rabbit test [3] as well as *in vitro* test methods based on reconstructed human epidermis technology, such as the EpiSkin™method. The test described herein is based on tissue viability evaluation after topical, short-term exposure to substances.

EpiSkin™is an *in vitro* three-dimensional reconstructed human epidermis model closely mimicking the biochemical and physiological properties of the upper parts of the human skin, i.e. the epidermis. In this model, keratinocytes are cultured at the air–liquid interface. The tissue is organised in basal, spinous and granular cell layers, but also presents a multi-layered functional *stratum corneum* containing intercellular lamellar lipid layers close to the main lipid classes found *in vivo* [4]. The test method involves topical application of the tested substance to the EpiSkin™ tissue, followed by tissue cell viability assessment. Corrosive chemicals are identified by their ability to decrease cell viability below defined threshold levels [5].

8.2 Current Validation Status

The reliability and relevance of the EpiSkin™ test method has been established through a controlled, inter-laboratory validation study overseen by the European Centre for the Validation of Alternative Methods (EURL-ECVAM). Based on its scientific validity, this test method has been recommended for all classes of chemical testing [6, 7] and for inclusion in tiered testing strategies as part of a tiered or weight-of-evidence evaluation [7]. Validation studies have reported that the EpiSkin™ test method was able to discriminate known skin corrosives and non-corrosives with an overall sensitivity of 82% (23/28) and specificity of 84% (27/32) in a database of 60 substances [8–11]. The test protocol provides promising indication of allowing distinction between severe and less-severe skin corrosives [11].

The EpiSkin™ test method was first accepted under the official Organisation for Economic Co-operation and Development (OECD) Test Guideline 431 [12] allowing the identification of non-corrosive and corrosive substances and mixtures in accordance with the United Nations (UN) Globally Harmonised System of Classification and Labelling of Chemicals (GHS) [12, 13]. In 2014, OECD Test Guideline 431 further supported the sub-categorisation of corrosive substances and mixtures into optional Category 1A, in accordance with the UN GHS, as well as a combination of Categories 1B and 1C using the EpiSkin™ test method [5, 14, 15]. Therefore, in the context of the revision of OECD TG 431, knowing that EpiSkin™'s Prediction Model already resulted in sufficiently accurate predictions, it has been used for comparison purposes of investigative analyses for others Reconstructed human Tissues models [15, 16]. EpiSkin™ was also recently included as part of the Integrated Approach to Testing and Assessment (IATA) in OECD Guidance Document 203 [17]for Skin Irritation/Corrosion.

8.3 Performance and Applicability of the Test Method

8.3.1 Reproducibility

Reproducibility was evaluated using two calculation methods: one by testing the same chemicals over time in a single laboratory (within-laboratory reproducibility, WLR) and the other by testing the same chemicals in different laboratories (between-laboratory reproducibility, BLR). WLR was calculated as the percentage of chemicals with identical classifications in three runs. BLR was the percentage of chemicals with identical classifications between laboratories.

8.3.1.1 Within-Laboratory Reproducibility (WLR)

In 1996–1997, an international skin corrosion validation study (SCVS) was conducted under the supervision of the EURL-ECVAM. Sixty coded chemicals were tested in three different laboratories (AFSSAPS, INRS, and Rhone Poulenc). Concordant classification was observed for 167 out of 180 items (92.7%) for the three laboratories when considering corrosives versus non-corrosives [11] (Table 8.1). In particular, the same prediction was observed for 42 out of 60 chemicals (70%), each in triplicate experiments and in all laboratories when considering corrosives versus non-corrosives [11]. In three cases (5%, 3/60), one laboratory had consistently different classification categories to those from the other laboratories. When considering UN GHS subcategory 1A versus 1B-and-1C versus non-corrosives, the WLR was still high, with 88.3% (159/180) reproducibility (Table 8.1). None of the UN GHS subcategory 1A substances were under-predicted as non-corrosives.

More recently, the OECD expanded its work plan to include a project to update OECD TG 431 on skin corrosion to allow sub-categorisation of corrosive chemicals [15]. Thus, in 2012, the OECD recommended testing 82 chemicals, including the chemical set initially used to validate all *in vitro* skin corrosive test methods, along with adapted controls for direct MTT reducers or colour interference. Eighty-five

Table 8.1 Within-laboratory reproducibility of the EpiSkin™ skin corrosion test method

	ECVAM SCVS laboratories			
Reproducibility	Rhone Poulenc	AFSSAPS	INRS	L'Oréal R&I
60 SCVS chemicals (corrosives vs. NC)	100% (60/60)	93.3% (56/60)	85.0% (51/60)	–
60 SCVS chemicals (1A vs. 1B–and-1C vs. NC)	95.0% (57/60)	93.3% (56/60)	76.7% (46/60)	–
55 SCVS chemicals (corrosives vs. NC)	100% (55/55)	92.7% (51/55)	85.5% (47/55)	92.7% (51/55)
55 SCVS chemicals (1A vs. 1B–and-1C vs. NC)	96.4% (53/55)	90.9% (50/55)	78.2% (43/55)	87.3% (48/55)
85 chemicals (corrosives vs. NC)	–	–	–	94.1% (80/85)
85 chemicals (1A vs. 1B-and-1C vs. NC)	–	–	–	88.2% (75/85)

chemicals were tested by L'Oréal using the EpiSkin™ test method in three independent runs [5]. Different combinations were used to calculate performance for the UN-GHS categories (1A, 1B, 1C, grouped category 1B-and-1C, and non-corrosive), with WLR estimated for the 85 chemicals altogether. To be able to compare results within the three laboratories that participated in the EURL ECVAM SCVS, WLR was also calculated for 55 EURL-ECVAM SCVS commercially available chemicals that were part of the 85 tested in this study [11].

When considering three different subcategories (1A, 1B-and-1C, and non-corrosive), 88.2% of the 85 chemicals (75/85) and 87.3% of the 55 SCVS chemicals (48/55) showed 100% concordance of classifications for the three independent runs performed at L'Oréal laboratories [5] (Table 8.1). In addition, for the same set of 55 SCVS chemicals, the other three participating laboratories found 100% concordance of classifications between three independent runs for 96.4% (53/55), 90.9% (50/55), and 78.2% (43/55) of the chemical set.

Overall, data obtained by L'Oréal laboratories can be compared with EURL ECVAM SCVS data from 20 years ago. In conclusion, the EpiSkin™ test method showed reproducible predictions based on cell viability values within-laboratories and over a long period of time. As such, the test method can reliably predict chemicals into the three different UN GHS skin corrosion subcategories.

8.3.1.2 Between-Laboratory Reproducibility (BLR)

The final classification of chemicals was based on the arithmetic mean value ($n = 3$) of cell viability at each exposure time (3 min, 1 and 4 h). For comparison, BLR was calculated considering (1) the three laboratories that participated in the EURL-ECVAM SCVS (using the data reported in [11]), (2) the three laboratories participating in the EURL-ECVAM SCVS plus L'Oréal R&I (using the data reported in [5]), and (3) three out of four laboratories including L'Oréal R&I (L'Oréal plus two of the laboratories that participated in the EURL ECVAM SCVS).

BLR was assessed through a skin corrosion validation study (SCVS) under the auspices of the EURL-ECVAM. Concordant classification was observed for 53 out of 60 chemicals (88.3%) in the three laboratories when considering corrosives versus non-corrosives [11] (Table 8.2). In five cases, results crossed classification boundaries in more than one laboratory related to UN GHS NC versus

Table 8.2 Between-laboratory reproducibility (BLR) of the EpiSkin™ skin corrosion test method

Reproducibility	3 ECVAM SCVS labs	L'Oréal R&I + 3 ECVAM SCVS labs	L'Oréal R&I + 2 ECVAM SCVS labs
60 SCVS chemicals (corrosives vs. NC)	88.3% (53/60)	–	–
60 SCVS chemicals (1A vs. 1B-and-1C vs. NC)	80.0% (48/60)	–	–
55 SCVS chemicals (corrosives vs. NC)	85.50% (47/55)	83.60% (46/55)	92.70% (51/55)
55 SCVS chemicals (1A vs. 1B-and-1C vs. NC)	81.80% (45/55)	74.50% (41/55)	87.30% (48/55)

subcategories 1B-and-1C, or UN GHS subcategories 1A versus 1B-and-1C. No chemical in UN GHS subcategory 1A was under-classified as non-corrosive.

Even after distinguishing between UN GHS subcategory 1A, 1B-and-1C, and non-corrosives, BLR was still high, with an accuracy of 80.0% (48/60) (Table 8.2).

When data acquired by L'Oréal laboratories in 2012–2013 are compared with the EURL-ECVAM SCVS data from almost 20 years ago, 74.5% of the chemicals (41/55) show 100% concordance on classifications for the four participating laboratories [5] (Table 8.2). When considering the three subcategories, 81.8% (45/55) to 87.3% (48/55) of the chemicals evaluated showed 100% concordance on classifications between three participating laboratories. When considering only corrosives versus non-corrosives, BLR was even higher, with an accuracy of between 85.5% (47/55) and 92.7% (51/55) (Table 8.2).

8.3.2 Predictive Capacity

The Management Team of the EURL-ECVAM SCVS defined several predictive capacity criteria to judge the scientific validity of the evaluated test methods. EpiSkin™ was the only one found suitable to distinguish between R35 (equivalent to UN GHS Category 1A) and R34 (equivalent to a combination of UN GHS Categories 1B and 1C) for all chemical types tested [11, 18]. The EpiSkin™ test method met the 'best possible' expected outcome for all over- and under-prediction criteria defined for non-corrosives and R34, while for R35, the under-predictions were acceptable but only slightly better than the 'unacceptable' outcome [11, 18].

More recently, the EU Classification, Labelling, and Packaging Regulation (EU CLP) system required the sub-categorisation of corrosive chemicals into the three UN GHS optional subcategories 1A, 1B, and 1C [19]. L'Oreal R&I was tasked with investigating the usefulness of the validated EpiSkin™ test method. The goal was to identify skin corrosive UN GHS Categories 1A, 1B, and 1C using the original validated prediction prediction model and adapted controls for direct MTT reduction. In total, 113 chemicals were tested, including 82 chemicals selected by the OECD expert group on skin corrosion. Predictive capacity was calculated on the basis of all individual predictions obtained for each chemical (Table 8.3). Calculations were performed for three categories (Category 1A, Category 1B-and-1C, and non-corrosive). Moreover, the predictive capacity of EpiSkin™ was calculated considering the data obtained by L'Oréal alone and in combination with the data obtained in the EURL-ECVAM SCVS. Thus, different chemicals ended up with a different number of independent classifications used to calculate predictive capacity, i.e. ranging from 12 classifications (40 chemicals from EURL-ECVAM SCVS that were not identified as MTT reducers in this study) to at least three classifications (chemicals that were not part of EURL-ECVAM SCVS or were identified as MTT reducers in this study). To prevent some chemicals from weighing differently in the calculation of predictive capacity, a weighted calculation was used to reflect the real

Table 8.3 EpiSkin™ predicted UN GHS/EU CLP classifications for the tested chemicals based on at least three runs

No.	Chemical name	CAS no.	Chemical category	Physical state	UN GHS *in vivo* class	Mean *in vitro* class
1	3,3-Dithiopropionic acid	1119-62-6	Organic acid	S	NC	NC
2	10-Undecenoic acid	112-38-9	Organic acid	S	NC	NC
3	L-(+)-glutamic acid hydrochloride	138-15-8	Organic acid	S	NC	NC
4	Dodecanoic acid (lauric acid)	143-07-7	Organic acid	S	NC	NC
5	alpha-Ketoglutaric acid	328-50-7	Organic acid	S	NC	1B/1C
6	Isostearic acid	54680-48-7	Organic acid	L	NC	NC
7	2-Hydroxyisobutyiric acid	594-61-6	Organic acid	S	NC	1B/1C
8	Oxalic acid dihydrate	6153-56-6	Organic acid	S	NC	1B/1C
9	4-Amino-5-methoxy-2--methylbenzensulphonic acid	6471-78-9	Organic acid	S	NC	NC
10	Naphthalene acetic acid	86-87-3	Organic acid	L	NC	NC
11	Sulphamic acid	5329-14-6	Inorganic acid	S	NC	NC
12	4-Amino-4H-1,2,4-triazole	584-13-4	Organic base	S	NC	NC
13	Hydrogenated tallow amine	61788-45-2	Organic base	S	NC	NC
14	1-(*o*-Tolyl)biguanide	93-69-6	Organic base	S	NC	NC
15	2,4-Xylidine (2,4-dimethylaniline)	95-68-1	Organic base	L	NC	NC
16	Potassium hydroxide (5%)	1310-58-3	Inorganic base	L	NC	1B/1C
17	Sodium carbonate (50%)	497-19-8	Inorganic base	L	NC	NC
18	Methyl laurate	111-82-0	Neutral organic	L	NC	NC
19	1-Decanol	112-30-1	Neutral organic	L	NC	NC
20	Methyl palmitate	112-39-0	Neutral organic	S	NC	NC
21	Methyl stearate	112-61-8	Neutral organic	L	NC	NC
22	Linalyl acetate	115-95-7	Neutral organic	L	NC	NC
23	Benzyl benzoate	120-51-4	Neutral organic	L	NC	NC
24	Tri-isobutyl phosphate	126-71-6	Neutral organic	L	NC	NC
25	Tetrachloroethylene	127-18-4	Neutral organic	L	NC	NC
26	Allyl heptanoate	142-19-8	Neutral organic	S	NC	NC

27	Isopropyl palmitate	142-91-6	Neutral organic	L	NC	NC
28	1,9-Decadiene	1647-16-1	Neutral organic	L	NC	NC
29	Benzylacetone (4-phenyl-2-butanone)	2550-26-7	Neutral organic	L	NC	NC
30	Dipropylene glycol monobutyl ether	29911-28-2	Neutral organic	L	NC	NC
31	Heptyl butyrate	5870-93-9	Neutral organic	L	NC	NC
32	*n*-Butyl propionate	590-01-2	Neutral organic	L	NC	NC
33	Butyl carbamate	592-35-8	Neutral organic	S	NC	NC
34	Methyl trimethylacetate	598-98-1	Neutral organic	L	NC	NC
35	2-Phenethyl alcohol (2-Phenylethanol)	60-12-8	Neutral organic	L	NC	NC
36	Di-*n*-propyl disulphide	629-19-6	Neutral organic	L	NC	NC
37	Isopropanol	67-63-0	Neutral organic	L	NC	NC
38	Butyl glycolate (polysolvan)	7397-62-8	Neutral organic	L	NC	NC
39	Allyl phenoxyacetate	7493-74-5	Neutral organic	L	NC	NC
40	Terpinyl acetate	80-26-2	Neutral organic	L	NC	NC
41	Diethyl phthalate	84-66-2	Neutral organic	L	NC	NC
42	Butyl methacrylate	97-88-1	Neutral organic	L	NC	NC
43	*o*-Methoxyphenol (guaiacol)	90-05-1	Neutral organic (Phenol)	L	NC	1B/1C
44	Eugenol	97-53-0	Neutral organic/Organic acid (Phenol)	L	NC	NC
45	4,4′-Methylene-bis-(2,6-ditert-butylphenol)	118-82-1	Neutral organic/Organic acid (Phenol)	S	NC	NC
46	Benzyl salicylate	118-58-1	Neutral organic/Organic acid (Phenol)	L	NC	NC
47	Hexyl salicylate	6259-76-3	Neutral organic/Organic acid (Phenol)	L	NC	NC
48	Sodium bicarbonate	144-55-8	Inorganic salt	S	NC	NC
49	Sodium bisulphite	7631-90-5	Inorganic salt	S	NC	NC
50	Phenethyl bromide (2-bromoethy benzene)	103-63-9	Neutral organic (Electrophile)	L	NC	NC
51	Cinnamaldehyde	104-55-2	Neutral organic (Electrophile)	L	NC	NC
52	Hydroxycitronellal	107-75-5	Neutral organic (Electrophile)	L	NC	NC

(continued)

Table 8.3 (continued)

No.	Chemical name	CAS no.	Chemical category	Physical state	UN GHS *in vivo* class	Mean *in vitro* class
53	1-Bromohexane	111-25-1	Neutral organic (Electrophile)	L	NC	NC
54	Triethylene glycol	112-27-6	Neutral organic (Electrophile)	L	NC	NC
55	Erucamide	112-84-5	Neutral organic (Electrophile)	S	NC	NC
56	*trans*-Cinnamaldehyde	14371-10-9	Neutral organic (Electrophile)	L	NC	NC
57	2-Ethoxyethyl methacrylate	2370-63-0	Neutral organic (Electrophile)	L	NC	NC
58	Dipropylene glycol	25265-71-8	Neutral organic (Electrophile)	L	NC	NC
59	Silane A-1430	2530-87-2	Neutral organic (Electrophile)	L	NC	NC
60	4-(Methylthio)-benzaldehyde	3446-89-7	Neutral organic (Electrophile)	L	NC	NC
61	3-Chloro-4-fluoronitrobenzene	350-30-1	Neutral organic (Electrophile)	S	NC	NC
62	1,6-Dibromohexane	629-03-8	Neutral organic (Electrophile)	L	NC	NC
63	2-Bromobutane	78-76-2	Neutral organic (Electrophile)	L	NC	NC
64	Isopropyl myristate	110-27-0	Soap/Surfactant	L	NC	NC
65	Sodium lauryl sulphate (20% aq.)	151-21-3	Soap/Surfactant	L	NC	NC
66	Sodium lauryl sulphate (50% aq.)	151-21-3	Soap/Surfactant	L	NC	NC
67	Sodium undecylenate (33%)	3398-33-2	Soap/Surfactant	L	NC	1B/1C
68	Fluoboric acid (hydrogentetrafluoroborate) (48%)	16872-11-0	Inorganic acid	L	1C	1A
69	*N,N*-Dimethylbenzylamine	103-83-3	Organic base	L	1C	1B/1C
70	Maleic anhydride	108-31-6	Neutral organic (Electrophile)	S	1C	1B/1C
71	Propionic acid	79-09-4	Organic acid	L	1B	1A
72	Butyric acid	107-92-6	Organic acid	L	1B	1A
73	Hydrobromic acid (48%)	10035-10-6	Inorganic acid	L	1B	1A
74	Phosphoric acid	7664-38-2	Inorganic acid	S	1B	1B/1C
75	Boron trifluoride-acetic acid complex	373-61-5	Inorganic acid/Organic acid	L	1B	1A
76	HCl + Sulphuric acid + Citric acid (5, 5, 5% wt)	–	Inorganic acid/Organic acid	L	1B	1B/1C
77	1-(2-Aminoethyl)piperazine	140-31-8	Organic base	L	1B	1B/1C

78	Ethanolamine	141-43-5	Organic base	Viscous	1B	1B/1C
79	Iron(III) chloride	7705-08-0	Inorganic salt	S	1B	1B/1C/1A
80	Octanoic acid (caprylic acid)	124-07-2	Organic acid	L	1B/1C	1B/1C
81	Glyoxylic acid monohydrate	563-96-2	Organic acid	S	1B/1C	1B/1C
82	Lactic acid	598-82-3	Organic acid	L	1B/1C	1B/1C
83	2-Methylbutyric acid	600-07-7	Organic acid	L	1B/1C	1B/1C
84	65/35 Octanoic/decanoic acid	68937-75-7	Organic acid	L	1B/1C	1B/1C
85	60/40 Octanoic/decanoic acid	68937-75-7	Organic acid	L	1B/1C	1B/1C
86	55/45 Octanoic/decanoic acid	68937-75-7	Organic acid	L	1B/1C	1B/1C
87	Carvacrol	499-75-2	Organic acid (Phenol)	L	1B/1C	1B/1C
88	2-*tert*-Butylphenol	88-18-6	Organic acid (Phenol)	L	1B/1C	1B/1C
89	Hydrochloric acid (14.4%)	7647-01-0	Inorganic acid	L	1B/1C	1B/1C
90	Cyclohexylamine	108-91-8	Organic base	L	1B/1C	1B/1C
91	*n*-Heptylamine	111-68-2	Organic base	L	1B/1C	1A
92	3-Methoxypropylamine	5332-73-0	Organic base	L	1B/1C	1B/1C
93	*N*,*N*-Dimethylisopropylamine	996-35-0	Organic base	L	1B/1C	1B/1C
94	Sodium bisulphate monohydrate	10034-88-5	Inorganic salt	S	1B/1C	1B/1C
95	Sodium bisulphate	7681-38-1	Inorganic salt	S	1B/1C	1B/1C
96	Allyl bromide	106-95-6	Neutral organic (Electrophile)	L	1B/1C	1B/1C
97	Glycol bromoacetate (85%)	3785-34-0	Neutral organic (Electrophile)	L	1B/1C	1B/1C
98	Methacrolein	78-85-3	Neutral organic (Electrophile)	L	1B/1C	1B/1C
99	Formic acid	64-18-6	Organic acid	L	1A	1A
100	Acetic acid	64-19-7	Organic acid	L	1A	1A
101	Bromoacetic acid	79-08-3	Organic acid	S	1A	1A
102	Acrylic acid	79-10-7	Organic acid	L	1A	1A
103	Boron trifluoride dihydrate	13319-75-0	Inorganic acid	L	1A	1A
104	Sulphuric acid (98%)	7664-93-9	Inorganic acid	L	1A	1A
105	Phosphorus tribromide	7789-60-8	Inorganic acid	L	1A	1A

(continued)

Table 8.3 (continued)

No.	Chemical name	CAS no.	Chemical category	Physical state	UN GHS *in vivo* class	Mean *in vitro* class
106	*N,N*-Dimethyldipropylenetriamine	10563-29-8	Organic base	L	1A	1B/1C
107	1,2-Diaminopropane	78-90-0	Organic base	L	1A	1A
108	Silver nitrate	7761-88-8	Inorganic salt	S	1A	1B/1C
109	Phenol	108-95-2	Organic acid	S	1A	1A
111	Dichloroacetyl chloride	79-36-7	Neutral organic (Electrophile)	L	1A	1A
110	Hexanoic acid	142-62-1	Organic acid	L	1	1A
112	Sulphuric acid (10%)	7664-93-9	Inorganic acid	L	1	1B/1C
113	Potassium hydroxide (10%)	1310-58-3	Inorganic base	L	1	1B/1C

NC Non-corrosive

predictive capacity of the test method. The ensuing predictive capacity of the EpiSkin™ skin corrosion test method, to classify chemicals according to three sub-categories (UN GHS Categories 1A, 1B-and-1C, and non-corrosive), is shown in Table 8.4.

The EpiSkin™ test method showed a specificity of 88.3% considering the 67 tested *in vivo* non-corrosive chemicals. Of the 7.86 false-positive classifications obtained, 11.2% were classified as UN GHS Category 1B-and-1C and 0.5% as UN GHS Category 1A (Table 8.4). The EpiSkin™ test method showed a sensitivity of 98.4%. There were two chemicals having false-negative classifications, i.e. Methacrolein and Iron (III) chloride, both UN GHS Category 1B chemicals under-predicted as non-corrosive in the EURL-ECVAM SCVS. None of the UN GHS Category 1A chemicals were under-predicted as non-corrosive. Of the classifications obtained for the *in vivo* UN GHS Category 1A chemicals, 83.3% were correctly predicted, 16.7% were under-predicted as UN GHS Category 1B-and-1C. Of the classifications obtained for the *in vivo* UN GHS Category 1B-and-1C chemicals, 79.8% were correctly predicted, 17.7% were over-predicted as UN GHS Category 1A, and 2.4% were under-predicted as non-corrosive.

The overall accuracy of the EpiSkin™ test method in distinguishing between corrosive and non-corrosive chemicals was 92.4% [5]. The accuracy in distinguishing between UN GHS Category 1A, Category 1B-and-1Cand non-corrosive chemicals was 83.1% when considering the L'Oréal data combined with the EURL-ECVAM SCVS data.

Table 8.4 Predictive capacity of the EpiSkin™ skin corrosion test method considering L'Oréal data for the complete set of 113 chemicals combined with the data for non-MTT reducers[a] obtained in the EURL ECVAM SCVS (weighted data)

Predictive capacity	Predictions/total	%
Specificity (non-corrosive correct predictions)	59.14/67	88.3
False positives (NC → C)	7.86/67	11.7
NC → Category 1B/1C	7.53/67	11.2
NC → Category 1A	0.33/67	0.5
Sensitivity (Corrosive correct predictions)	45.25/46	98.4
False negatives (C → NC)	0.75/46	1.6
Category 1A correct predictions	10.00/12	83.3
Category 1A → Category 1B/1C	2.00/12	16.7
Category 1A → NC	0.00/12	0.0
Category 1B/1C correct predictions	24.75/31	79.8
Category 1B/1C → Category 1A	5.50/31	17.7
Category 1B/1C → NC	0.75/31	2.4
Accuracy (Corrosive vs Non-corrosive)	104.39/113	92.4
Accuracy (Category 1A vs. Category 1B/1C vs. Non-corrosive)	93.89/113	83.1

[a]Since in the EULR ECVAM SCVS all data were acquired without the use of adapted controls for direct MTT reducers (water-killed tissues), the calculation of predictive capacity was performed using only the data obtained for non-MTT reducers obtained

8.3.3 Applications and Limitations

The test method allows the hazard identification of mono- and multi-component test substances (solids, liquids, and semi-solids). The liquids can be aqueous or non-aqueous; solids can be soluble or insoluble in water. Samples may be pure chemicals or dilutions. Where appropriate, solids should be ground to a powder before application; no other prior treatment of the sample is required. The test method is not applicable to the testing of gases and aerosols (although this is true for almost all reconstructed human epidermis test methods, included in OECD TG 431). A wide range of chemicals representing mainly individual substances was tested in the validation study. The empirical database of the validation and OECD studies amounted to 60 and 85 chemicals, respectively, covering a wide range of chemical classes [5, 8, 11]. The 113 chemicals presented in this paper include 27organic acids, 11 inorganic acids, 13 organic bases, three inorganic bases, 49 neutral organics, six inorganic salts, and four soap/surfactants.

Evaluation of coloured chemicals and chemicals that act directly on MTT (e.g. MTT reducer) could be performed. However, use of adapted controls, as described in the test method SOP, is needed to define the non-specific colour inherent to the chemical as well as the non-specific MTT reduction. Using standard photometry as the endpoint detection system, test results for chemicals inducing high non-specific MTT reduction and/or high non-specific colour should be taken with caution. Such coloured chemicals can be tested nevertheless by using an HPLC/UPLC spectrophotometry procedure (see Sect. 6.2), which may also be used with all types of test chemicals (coloured, non-coloured, MTT-reducers, and non-MTT reducers).

8.4 Brief Description of the Protocol

The assay is performed as described in Alépée et al. [5] and according to OECD Test Guideline 431 on *in vitro* skin corrosion, using the validated Standard Operating Procedure (SOP) available for download at: https://eurl-ecvam.jrc.ec.europa.eu/validation-regulatory-acceptance/topical-toxicity/skin-corrosion. The main difference between this approach and the original protocol used in Fentem et al. [11] is the inclusion of controls to correct direct MTT reduction by the test substances (using water-killed tissues), as described in the validated SOP.

Test substances are applied topically to the epidermal model (two epidermis units per test substance) at three different exposure periods: 3, 60, and 240 min at room temperature. Liquids (50 μL) are added using a positive displacement pipette. For solids, the chemical is crushed to a powder, if necessary, and 20 mg (in 100 μL of NaCl 9 g/L) is applied to the EpiSkin™ tissues. NaCl 9 g/L (50 μL) and glacial acetic acid (50 μL) are used as negative and positive controls, respectively. At the end of the exposure period, samples are rinsed with phosphate-buffered saline. Cell viability, used as the endpoint, is determined on the basis of cellular mitochondrial dehydrogenase activity, measured by tetrazolium salt MTT reduction ((3-4,5-dimethyl triazole-yl) 2,5-diphenyltetrazoliumbromide) and conversion into

Table 8.5 Prediction model of the EpiSkin™ skin corrosion test method

Exposure time	Viability (% of negative control)	UN GHS category
3 min	<35% after 3 min exposure	Skin corrosive Category 1A
3 min and 1 h	≥35% after 3 min exposure; and <35% after 1 h exposure	Skin corrosive Category 1B[a]
1 h and 4 h	≥35% after 1 h exposure; and <35% after 4 h exposure	Skin corrosive Category 1C[a]
4 h	≥35% after 4 h exposure	Non-corrosive

[a]If sub-categorisation into Categories 1B and 1C is not used as described, e.g. in the OECD TG 431, Category 1B-and-1Ccan be determined if the tissue viability is ≥35% after 3 min exposure and <35% after 4 h exposure

blue formazan salt that is quantitatively measured after extraction from tissues [20]. The tissues are incubated for 3 h in a MTT solution (0.3 mg/mL; 2 mL per well) at 37 °C, 5% CO_2, and >95% humidified atmosphere. MTT is reduced from a yellow-coloured tetrazolium soluble salt into blue formazan precipitate by succinate dehydrogenase in the mitochondria of living cells. The precipitated formazan is extracted overnight with acidified isopropanol (0.5 mL, 0.04 N HCl in isopropanol) at room temperature, protected from light. Each tube should be mixed thoroughly before reading to verify that all the formazan is correctly extracted. The precipitated extract is then quantified by spectrophotometry at a wavelength of 570 nm. The extracted MTT formazan may be quantified using either a standard absorbance (OD_{570}) measurement or an HPLC/UPLC-spectrophotometry procedure (see Sect. 6.2). For each treated tissue, the viability is expressed as a % relative to negative control tissues (mean). The final viability, for the three measurement times, allows the classification of the substance according to the prediction model (Table 8.5).

8.5 Role in Testing Strategy

The OECD has adopted an Integrated Approach to Testing and Assessment (IATA) for skin corrosion and irritation (OECD GD 203) that provides guidance—through a modular approach to classification and labelling—on how to integrate and use existing and new information on the corrosive and irritant hazard potential of chemicals. The goal is to keep the use of animals to a minimum while ensuring human safety [17]. IATA comprises well-described and characterised modules. Each of them contains one or more individual information sources of a similar type. One of the modules addresses the use of validated and accepted *in vitro* or *ex vivo* tests for skin irritation and skin corrosion when no existing human and/or animal data are available. A sequential bottom-up or top-down testing strategy is proposed to assess the potential need for skin irritation and corrosion classification, as described in Fig. 8.1.

In vitro EpiSkin™ skin irritation and skin corrosion test methods have been formally validated and adopted for the regulatory assessment of both skin irritation (OECD TG 439) and skin corrosion (OECD TG 431) of substances, respectively

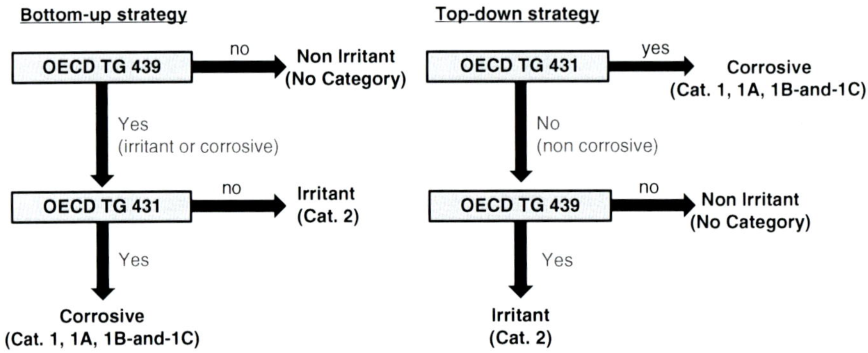

Fig. 8.1 Overview of the sequential bottom-up and top-down testing strategies based on OECD IATA Guidance Document 203 for the classification of chemicals according to their skin corrosion and irritancy potential

[15, 21, 22]. EpiSkin™ skin irritation and corrosion data were integrated into either a sequential bottom-up or a top-down testing strategy (based on a set of 87 chemicals) to investigate the usefulness and limitations of the two approaches using statistical data analysis, according to the EU CLP/UN GHS classification system [13, 19].

The skin irritation and skin corrosion test methods showed high predictive capacities in both bottom-up and top-down testing sequences. The overall classification of the 87 chemicals, according to the EU CLP/UN GHS classification system, remained very similar. However, the respective predictive capacities of the test methods did differ slightly, depending on whether they were used as a first or a second step in the testing sequence (Fig. 8.2).

When used as a first step, in a top-down approach, the skin corrosion method (OECD TG 431) correctly identified 100% (35/35) of the Category 1 chemicals. It should be noted that seven category 1B-and-1Cchemicals were over-predicted as Category 1A and one Cat. 1A chemical was under-predicted as Category 1B-and-1C (Fig. 8.2). Similarly, this approach correctly identified 76.9% (10/13) of Category 2 and 84.6% (33/39) of non-classified chemicals. However, methyl palmitate (UN GHS Cat. 2, consistently false negative in *in vitro* tests) was misclassified as non-classified.

When the skin irritation method (OECD TG 439) was used as a first step in a bottom-up strategy, 84.8% (33/39) of non-classified chemicals were correctly identified, whereas six non-classified chemicals were over-predicted as either irritants (five chemicals) or a corrosive (one chemical). Using the skin corrosion test method as the second step led to correct identification of 76.9% (10/13) of Category 2 irritants and 100% (35/35) of Category 1 corrosive chemicals. Overall, few false negatives were generated (e.g. di-*n*-propyl disulphide and methyl palmitate, both UN GHS Cat.2).

Both testing approaches (Bottom-Up and Top-Down) resulted in comparable and high overall performances. EU CLP classified (UN GHS Cat. 1 and Cat. 2)

EU CLP – Top Down		EpiSkin				Sum
		Non Class.	Cat. 2	Cat. 1B-1C	Cat. 1A	
TG 431	Cat. 1A	0	0	1	10	11
↓	Cat. 1B-1C	0	0	17	7	24
TG 439	Cat. 2	1	10	2	0	13
	Non Class.	33	5	1	0	39
Sum		34	15	21	17	87

EU CLP – Bottom Up		EpiSkin				Sum
		Non Class.	Cat. 2	Cat. 1B-1C	Cat. 1A	
TG 439	Non Class.	33	5	1	0	39
↓	Cat. 2	3	10	0	0	13
TG 431	Cat. 1B-1C	0	0	17	7	24
	Cat.1A	0	0	1	10	11
Sum		36	15	19	17	87

Fig. 8.2 Sequential testing approaches using both EpiSkin™ skin irritation and skin corrosion test methods. Contingency table for 87 tested chemicals according to EU CLP classification

chemicals were identified with a very high sensitivity (94.2–95.7%), the EU CLP non-classified (UN GHS optional Cat. 3 and No Cat.) chemicals with an appropriate specificity (70.4–70.7%), so that high overall accuracy values were obtained for the identification of EU CLP classified versus non-classified chemicals by both approaches (82.6–83.2%). Furthermore, very high sensitivities were obtained for the identification of UN GHS/EU CLP Cat. 1 chemicals (97.9–99.8%), very high specificities for non-Cat. 1 chemicals (92.6–93.0%), and very high accuracies for the identification of skin corrosives vs. non-corrosives by both approaches (94.9–95.5%). Overall accuracies of 71.7–72.2% were found for predicting the single (sub)categories (EU CLP non-classified, EU CLP/UN GHS Cat. 2, Subcat. 1B/1C and Subcat. 1A). Results also indicated the testing strategies to be more predictive than the individual assays and to tend to err towards safety. Finally, no *in vivo* UN GHS/EU CLP Subcat. 1A was under-predicted as non-Cat. 1, and no EU CLP non-classified chemical was over-predicted as Cat. 1A, showing that no extreme mis-classifications occur when using either the bottom-up or top-down testing strategies.

8.6 Perspectives From the Test Developer

8.6.1 Critical Steps in the Protocol

The critical steps of the standardised operating procedure include the following:

- The absence of air bubble under the epidermis should be verified at each step;
- All test substances should be tested alone in separate plate;

- For viscous and sticky chemicals, a curved flat spatula should be used or the test chemical shall be weighed directly on the nylon mesh (pre-testing compatibility of the test chemical with the nylon mesh should be considered);
- Tissues should be rinsed thoroughly;
- Always make use of a freshly prepared MTT solution (preparation to be used within 3 h);
- The epidermis should be gently detached from the matrix and turned with the epidermis topical side against the matrix before formazan extraction is conducted with isopropanol acid.

8.6.2 Possible Protocol Adaptations

Test chemicals may interfere with the MTT assay, either by direct reduction of the MTT into blue formazan and/or by colour interference if the test chemical absorbs—naturally or due to treatment procedures—in the same OD range as formazan (570 ± 30 nm). Additional controls should be used to detect and correct potential interference such as the non-specific MTT reduction (NSMTT) control and/or the non-specific colour (NSC_{living} or NSC_{killed}) controls [5]. This is especially important when a chemical is difficult to remove from the tissue by rinsing or when it penetrates the epidermis and is therefore present in the tissues when the MTT viability test is performed. A detailed description of how to correct direct MTT reduction and interference from colouring agents is available in the SOP (www.episkin.com).

For coloured test chemicals that are not compatible with the standard absorbance (OD) measurement (strong interference with the MTT assay), the alternative HPLC/UPLC-spectrophotometry procedure to measure MTT formazan may be used. The HPLC/UPLC-spectrophotometry allows the separation of the MTT formazan from the test chemical before its quantification. As a result, NSC_{living} or NSC_{killed} controls are not required when using HPLC/UPLC. NSMTT controls should nevertheless be used if the chemical is suspected of directly reducing MTT or has a colour that impedes the assessment of the capacity to directly reduce MTT. When using HPLC/UPLC-spectrophotometry to measure MTT formazan, viability is calculated as the percentage of the MTT formazan peak area obtained with living tissues exposed to the test chemical relative to the MTT formazan peak obtained with the concurrent negative control. As such, HPLC/UPLC-spectrophotometry may be used with all types of test chemicals (coloured, non-coloured, MTT reducers, and non-MTT reducers) for the measurement of MTT formazan [23]. Due to the diversity of HPLC/UPLC-spectrophotometry systems, qualification of the device must be demonstrated before use, ensuring compliance with acceptance criteria based on a set of standard qualification parameters similar to those described in the U.S. Food and Drug Administration guidance for industry on bio-analytical method validation [23, 24]. Cosmetics Europe has undertaken a study that focuses on evaluation of skin corrosive chemicals. The results obtained in this study provide further support for Within Laboratory Reproducibility of HPLC-UPLC-spectrophotometry for measurement of formazan on methacrolein and boron trifluoride dihydrate [25]. Taking into account the low variability between the three tissue replicates and between the three independent experiments, the results demonstrated that the test method protocol is robust irrespective of whether OD or HPLC/UPLC-spectrophotometry is used as the endpoint detection system for measurement of formazan. Finally, use of HPLC/UPLC-spectrophotometry for measurement of formazan is incorporated into the OECD TGs 431 (skin corrosion) and in the Scientific Committee on Consumer Safety Note of Guidance [15, 26].

8.6.3 Challenges and Opportunities

Because chemicals may lead to severe risks, there is a strong need for data on skin corrosion potential. These data would ensure a high level of protection in terms of human health and occupational safety as well as the safe transportation of chemicals and mixtures (in line with UN Recommendations on the Transport of Dangerous Goods). It would also be necessary to achieve sustainable development while enhancing competitiveness and innovation.

The EpiSkin™ test method may be used as a stand-alone test method for the detection or exclusion of corrosive effects of chemicals. Negative results with this test method will require an additional *in vitro* skin irritation test—if not performed beforehand—to determine whether the chemical should be classified Category 2 (irritant) or if it does not require classification (No Category). This *in vitro* strategy replaces the *in vivo* test according to OECD TG 404 for decision-making. As a result, the integration of all existing and all newly generated information on the corrosion and irritation hazard potential of chemicals for final decisions on classification and labelling is needed [17]. For a final product, the skin corrosion potential will actually depend on a number of factors such as the final concentration of the ingredient in the finished product, the presence of 'neutralising' substances, the excipients used, the exposure route, the use conditions, and applied concentrations [27].

The OECD TG 431 further allows the sub-categorisation of corrosive chemicals into UN GHS skin corrosive sub-categories Category 1A or Category 1B-and-1C but does not allow the distinction between Category 1B and 1C. The original prediction model from the EpiSkin™ test method allows sub-categorisation of corrosive chemicals into the three Categories 1A, 1B, and 1C. However, it should be noted that its ability to discriminate between Categories 1B and 1C could not be formally evaluated/validated due to the lack of high-quality *in vivo* reference data that could be benchmark *in vitro* results [5, 8, 11, 16].

8.7 Conclusions

Pre-validation and validation studies have been completed for the *in vitro* human epidermis model of commercially available EpiSkin™. Based on its scientific validity, the test method has been recommended for the testing of all classes of chemicals and for inclusion in tiered testing strategies as part of integrated approaches for testing and assessment.

References

1. EC. Regulation (EC) no 1223/2009 of the European Parliament and of the council of 30 November 2009 on cosmetic products. Off J Eur Union. 2009;L342:59–209. Available at: http://eur-lex.europa.eu/legal-content/EN/TXT/?uri=CELEX:32009R1223
2. EC. Regulation (EC) no 1907/2006 of the European Parliament and of the council of 18 December 2006 concerning the registration, evaluation, authorisation and restriction of chemicals (REACH), establishing a European Chemicals Agency, amending directive 1999/45/EC and repealing council regulation (EEC) No. 793/93 and commission regulation (EC) No. 1488/94 as well as council directive 76/769/EEC and commission directives 91/155/EEC, 93/67/EEC, 93/105/EC and 2000/21/EC. Off J Eur Union. 2006;L396:1–1355.
3. Draize JH, Woodard G, Clavery HO. Methods for the study of irritation and toxicity of substances applied topically to the skin and mucous membranes. J Pharmacol Exp Ther. 1944;82:377–90.
4. Tinois E, Gaetani Q, Gavraud B, Dupond D, Rougier A, Pouradier-Dupteil X. The EPISKIN model: successful reconstruction of human epidermis *in vitro*. In: Rougier A, Goldberg M, Maibach HI, editors. *In vitro* skin toxicology. New York: Mary Ann Liebert; 1994. p. 133–40.
5. Alépée N, Grandidier MH, Cotovio J. Sub-categorisation of skin corrosive chemicals by the EpiSkin™ reconstructed human epidermis skin corrosion test method according to UN GHS: revision of OECD test guideline 431. Toxicol *In Vitro*. 2014;28:131–45.
6. Balls M, Corcelle G. Statement on the scientific validity of the EPISKIN™ test (an *in vitro* test for skin corrosivity). ATLA. 1998;26:278–80.
7. ICCVAM(Interagency Coordinating Committee on the Validation of Alternative Methods). ICCVAM evaluation of EPISKIN, EpiDerm (EPI-200), and the rat skin transcutaneous electrical resistance (TER) assay: *in vitro* test methods for assessing dermalcorrosivity potential of chemicals. NIH Publication No. 02-4502. Research Triangle Park, NC: National Institute of Environmental Health Sciences; 2002. Available at: http://iccvam.niehs.nih.gov/methods/epiddocs/epis_brd.pdf
8. Barratt MD, Brantom PG, Fentem JH, Gerner I, Walker AP, Worth AP. The ECVAM international validation study on *in vitro* test for skin corrosivity. 1. Selection and distribution of the test chemicals. Toxicol *In Vitro*. 1998;12:471–82.
9. Botham PA, Hall TJ, Dennett R, McCall JC, Basketter DA, Whittle E, et al. The skin corrosivity test *in vitro*. Results of an inter-laboratory trial. Toxicol *In Vitro*. 1992;6:191–4.
10. Botham PA, Chamberlain M, Barratt MD, Curren RD, Esdaile DJ, Gardner JR, et al. A prevalidation study on *in vitro* skin corrosivity testing: the report and recommendations of ECVAM workshop 6. ATLA. 1995;23:219–55.
11. Fentem JH, Archer GEB, Balls M, Botham PA, Curren RD, Earl LK, Esdaile DJ, Holzhutter H-G, Liebsch M. The ECVAM international validation study on *in vitro* tests for skin corrosivity. 2. Results and evaluation by the management team. Toxicol *In Vitro*. 1998;12:483–524.
12. OECD. *In vitro* skin model. OECD guideline for the testing of chemicals No. 431, OECD, Paris. 2004. Available at: http://www.oecd-ilibrary.org/environment/test-no-431-invitro-skin-corrosion-human-skin-model-test_9789264071148-en.
13. UN. United Nations globally harmonized system of classification and labelling of chemicals (GHS). 6th revised ed. New York: UN; 2015.
14. OECD. Summary document on the statistical performance of methods in OECD test guideline 431 for sub-categorisation. In: OECD series on testing and assessment no. 190. Paris: OECD; 2013.
15. OECD. *In vitro* skin corrosion: reconstructed human epidermis test method. In: OECD guideline for the testing of chemicals no. 431. Paris: OECD; 2015. Originally adopted in 2004, OECD, Paris. Available at http://www.oecd.org/env/testguidelines.
16. Desprez B, Barroso J, Griesinger C, Kandárová H, Alépée N, Fuchs HW. Two novel prediction models improve predictions of skin corrosive sub-categories by test methods of OECD test guideline no. 431. Toxicol *In Vitro*. 2015;29:2055–80.

17. OECD. Guidance document on integrated approaches to testing and assessment of skin irritation/corrosion. In: Series on testing and assessment, no. 203. Paris: OECD; 2014. Available at: http://www.oecd.org/env/ehs/testing/seriesontestingandassessmenttestingforhumanhealth.htm.
18. EC-ECVAM. Statement on the scientific validity of the EpiSkin™ test (an *in vitro* test for skin corrosivity), issued by the ECVAM Scientific Advisory Committee (ESAC10), 3 April 1998. 1998. Available at: http://ecvam.jrc.ec.europa.eu.
19. EU CLP. Regulation (EC) no 1272/2008 of the European Parliament and of the council of 16 December 2008 on classification, labelling and packaging of substances and mixtures, amending and repealing directives 67/548/EEC and 1999/45/EC, and amending regulation (EC) no 1907/2006. Off J Eur Union. 2008;L353:1–1355.
20. Mosmann T. Rapid colorimetric assay for cellular growth and survival: application to proliferation and cytotoxicity assays. J Immunol Methods. 1983;65:55–63.
21. OECD. *In vitro* skin irritation: reconstructed human epidermis test method. In: OECD guideline for the testing of chemicals no. 439. Paris: OECD; 2010. Available at:http://www.oecd-ilibrary.org/environment/test-no-439-in-vitro-skinirritation_9789264090958-en.
22. OECD. *In vitro* skin irritation: reconstructed human epidermis test method. Originally adopted in 2010. Paris: OECD; 2015. Available at: http://www.oecd-ilibrary.org/environment/oecd-guidelines-for-the-testing-of-chemicals-section-4-health-effects_20745788
23. Alépée N, Barroso J, De Smedt A, De Wever B, Hibatallah J, Klaric M, Mewes KR, Millet M, Pfannenbecker U, Tailhardat M, Templier M, McNamee P. Use of HPLC/UPLC-spectrophotometry for detection of formazan in *in vitro* reconstructed human tissue (RhT)-based test methods employing the MTT-reduction assay to expand their applicability to strongly coloured test chemicals. Toxicol *In Vitro*. 2015;29:741–61.
24. US FDA. 2001. Guidance for industry: bioanalytical method validation. U.S. Department of Health and Human Services, Food and Drug Administration. Available at: http://www.fda.gov/downloads/Drugs/Guidances/ucm070107.pdf.
25. Alépée N, Hibatallah J, Klaric M, Mewes KR, Pfannenbecker U, McNamee P. Assessment of cosmetic ingredients in the *in vitro* reconstructed human epidermis test method EpiSkin™ using HPLC/UPLC-spectrophotometry in the MTT-reduction assay. Toxicol *In Vitro*. 2016;33:105–17.
26. Scientific Committee on Consumer Safety (SCCS). The SCCS notes of guidance for the testing of cosmetic ingredients and their safety evaluation 9th revision. SCCS/1564/15 revised version of 25 April 2016. 2016. Available at URL: http://ec.europa.eu/health/scientific_committees/consumer_safety/docs/sccs_o_190.pdf.
27. Zuang V, Alonso MA, Botham PA, Eskes C, Fentem J, Liebsch M, van de Sandt JJ. Skin irritation and corrosion. Altern Lab Anim. 2005;33(Suppl 1):35–46.
28. EpiSkin™ Protocol No. 118. EpiSkin™ Skin Corrosivity Test. 2017. Available at: http://www.ecvam-dbalm.jrc.ec.europa.eu/.

The EpiDerm™ Skin Corrosion Test (EpiDerm™ SCT)

9

Helena Kandarova and Manfred Liebsch

9.1 Principle of the Test Method and Scientific Basis

The potential for chemical induced skin corrosion is an important consideration in establishing procedures for the safe handling, packing and transport of chemicals. Various systems for classification of corrosive potential are included in international regulatory requirements.

The EpiDerm™ Skin Corrosion Test (SCT) is based on the experience that corrosive chemicals are cytotoxic after a short term exposure to the stratum corneum of the epidermis, if cytotoxicity is immediately determined after chemical exposure. It is designed to predict and classify skin corrosion potential of chemicals by using a three dimensional human epidermis model EpiDerm™. The EpiDerm™ SCT is able to reliably discriminate chemicals that are corrosive to skin from non-corrosive chemicals, and can therefore be used for the classification of skin corrosion hazard according to the UN Globally Harmonised System (GHS) for classification [1]. The EpiDerm™ SCT with its modification of the Prediction model can be also used to distinguish between sub-categories of corrosivity [2–4].

The EpiDerm™ SCT has been scientifically described for the first time by Perkins et al. [5], later on evaluated, optimized and in its final form validated by ZEBET at the BfR in the framework of ECVAM activities. In order to replace the skin corrosion test in animals, ECVAM has in the 1990s supported twice a formal validation study for predicting skin corrosion by alternative methods. The first validation study [6], conducted between 1995 and 1997, was performed with two commercially available

H. Kandarova (✉)
MatTek *In Vitro* Life Science Laboratories, Mlynské Nivy 73, 821 05 Bratislava, Slovakia
e-mail: hkandarova@mattek.com

M. Liebsch
Retired Head of Unit "Centre of Alternative Methods to Animal Experiments – ZEBET" at the BfR, Berlin, Germany

C. Eskes et al. (eds.), *Alternatives for Dermal Toxicity Testing*,
DOI 10.1007/978-3-319-50353-0_9

in vitro test systems, the Skin2 ZK1350 and EPISKIN™ and it also included two other methods CORROSITEX (analytical assay) and TER test (*ex vivo* method). The TER test [7] and both skin model assays have shown usefulness for corrosion testing, however of the skin model assays only EPISKIN™ was considered as assay fully meeting the acceptance criteria set by the Management Team [8]. Unfortunately, the Skin2 model was withdrawn from the market very shortly after the ECVAM validation study without possibility of method optimization. Also, the commercial availability of EPISKIN™ model (at that time belonging to SADUC-Biomatériaux Imedex, France) was restricted following the completion of the validation study. It was due to the transfer of the ownership of the EPISKIN™ technology from Imedex to L'Oréal and construction of new production laboratories in Lyon-Gerlan.

Due to the lack of commercial availability of the validated methods for the skin corrosion endpoint, ECVAM supported a catch-up validation study of the EpiDerm™ SCT in 1998. The study was performed according to the ECVAM pre-validation scheme, to allow for refinement of the test protocol and the prediction model, as well as for independent assessment of the performance of the refined methodology in a final blind trial in three laboratories [9].

The validation study revealed that the EpiDerm™ SCT protocol provided a highly balanced prediction of 88% sensitivity and 86% specificity, which was regarded as the best predictivity that an *in vitro* skin corrosivity test could be expected to achieve [9] taking into account the inherent variability of the *in vivo* reference test [6].

Based on the successful validation studies with EPISKIN™ and EpiDerm™ models [8, 10], in 2002 the National Coordinators of OECD Test Guideline Programme (WNT) endorsed New Draft Test Guidelines TG 431 (Human Skin Model) for *In Vitro* Skin Corrosion Testing which was originally adopted by the OECD in 2004 [4]. The EpiDerm™ SOP has been successfully transferred to and validated with the SkinEthic™ RHE [11, 12] and EST-1000 [13, 14]. The two models, that underwent catch-up validation studies were also endorsed for skin corrosion testing and later on implemented into the OECD TG 431 [4].

The OECD TG 431 initially allowed only discrimination between corrosive and non-corrosive chemicals, however in light of the REACH and transport regulations, and after thorough analysis of all available data from EPISKIN™, EpiDerm™, SkinEthic™ and EST-1000 validation studies and new data generated by the skin models suppliers [3] the OECD updated the OECD TG 431 concerning sub-categorisation information of corrosives [4].

9.2 Current Validation Status

The ECVAM Scientific Advisory Committee (ESAC) formally endorsed the scientific validity of the EpiDerm™ Skin Corrosion Test (SCT) at its 14th Meeting [10]. ESAC concluded that the EpiDerm™ SCT has sufficient sensitivity, specificity and accuracy for the prediction of skin corrosive and non-corrosive effect of test substances. EpiDerm™ SCT is considered to be a validated, stand-alone *in vitro* replacement test for animal skin corrosion testing and is adopted as EU.B41 since in 2000 and OECD TG 431 since 2004.

Regarding sub-categorization (prediction of corrosive potency) the EpiDerm™ SCT is able under the current prediction model (described in the latest version of OECD TG 431 [4]) to classify the severely corrosive chemicals (cat 1A) with a sensitivity of >90% based on 3 min exposure time and 50% viability cut off. Furthermore, a modified Prediction Model to assess the skin corrosion potential in two steps (to minimize the over-prediction of corrosive effects of cat 1A), has been adopted in the latest version of the OECD TG 431 [4], in which a cut off of 25% at 3 min is used to discriminate the optional sub-category 1A from a combination of optional Sub-categories 1B-and-1C.

9.3 Performance and Applicability of the Test Method

The EpiDerm™ SCT was initially developed and designed to predict skin corrosion potential of neat test chemicals in the context of identification and classification of skin corrosion hazard according to the EU Dangerous Substance Directive (DSD) classification system (i.e. R34/R35 and no class), later replaced by the UN GHS by means of the EU Classification Labelling and Packaging (CLP) classification system [15]. The EpiDerm™ SCT also allows for hazard identification of corrosive substances in accordance with UN GHS (corrosive vs. non-corrosive), and is also able to partially discriminate between the corrosive subcategories of the UN GHS. Discrimination between categories have importance mainly for the transport of chemicals. The EpiDerm™ SCT is able to discriminate between cat 1A (severely corrosive chemicals) and the merged cat 1B-and-1C matching exactly the classes of the former EU DSD system.

9.3.1 Reproducibility

The between and within laboratory reproducibility of the EpiDerm™ SCT was assessed in the original validation study [9]. In phases I and II, simple biostatistical methods were applied. Contingency tables were used for the assessment of the predictive value of the test in phase I. For the assessment of intra-laboratory and inter-laboratory variability in phase II, the percentage concordances between assays and laboratories were determined. Data obtained in phase III of the validation study (24 chemicals tested in three laboratories) were analysed by applying a two-factor ANOVA, regarding the factor "chemical" as non-random and the factor "laboratory" as random sources for data. The possible presence of systematic inter-laboratory deviation was checked: (a) visually, by plotting the viability values of two laboratories against each other in bivariate scattergrams, as shown in Fig. 9.1; and (b) by applying the non-parametric sign test. This test is robust against deviations of the residual distribution from the normal Gaussian distribution.

Later on, as a part of the follow-up work performed at the request of the OECD expert group, additional data on 80 chemicals were produced in three independent runs by MatTek [2]. Most of these chemicals have also been tested in the Phase I of the ECVAM's validation study, however only with the 3 min time-point. In the

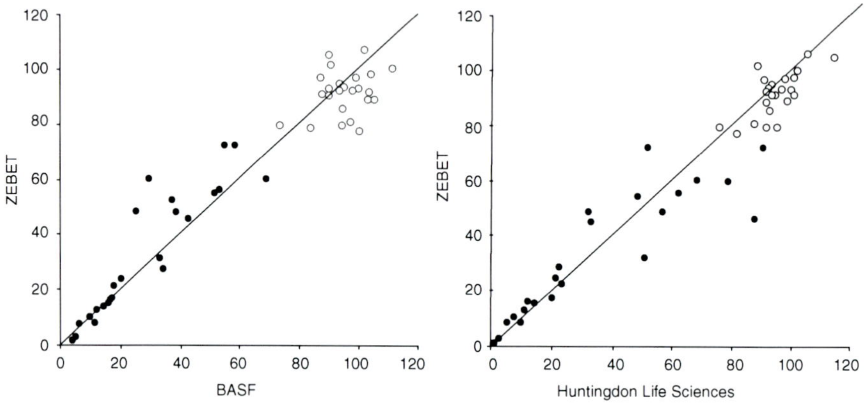

Fig. 9.1 Inter-laboratory reproducibility obtained in Phase III of the validation study (datasets of BASF, ZEBET and Huntington Life Science). The two graphs displayed in figure show 48 mean tissue viability values obtained from testing 24 chemicals in two independent tests with a 3 min exposure in the blind trial. Each graph compares the data of two laboratories which, in the case of ideal concordance, would be placed along the 45° line. *Dot:* corrosive *in vivo*, *open dot:* non-corrosive *in vivo*. The graphs display a sufficient inter-laboratory concordance, characterised by an excellent concordance at the lower and upper ends of the scale and some more variability in the middle

"OECD" follow up study, data for both 3 min and 60 min exposure times were generated and MTT correction step was applied for the MTT reducing chemicals. The overall results of this testing and proposal for a new prediction model for sub-categorization were published by Desprez et al. [3]. Furthermore, data obtained for the 3 min exposure during the original validation [9], and data that were produced almost 15 years afterwards in the OECD follow up study demonstrated excellent long-term reproducibility of the model and the assay.

9.3.2 Predictive Capacity

The EpiDerm™ SCT allows discrimination between corrosives and non-corrosives with a sensitivity of 88% and a specificity of 86% [9]. Sensitivity achieved in the original validation study could have been further improved to more than 95% if an MTT correction step would have been performed for the MTT reducing chemicals tested in the Phase III of the validation. Unfortunately, this very useful procedure developed by IIVS and described in Liebsch et al. [9], has been implemented into the SOP only after the validation.

More recently, the EpiDerm™ SCT test has been evaluated for its ability to discriminate the corrosive subcategories UN GHS Cat. 1A vs. the merged Cat. 1B-and-1C. The statistical parameters for the OECD TG 431 implemented prediction model, which is based on the originally validated viability cut-offs, and the new PM, proposed by Desprez et al., are shown in Table 9.1.

The new PM consists of a two-step approach. In the step 1, corrosive and non-corrosive chemicals are classified based on the originally validated PM [9]. In the second step, the group of corrosive chemicals undergoes sub-categorization based on the 25% cell viability cut-off obtained at 3 min. of exposure.

Table 9.1 Contingency tables for EpiDerm™ SCT [3]

In vivo	*In vitro* predictions		
(a) EpiDerm™ SCT using original PM (50% cut off at 3 min exposure time), 240 predictions (80 chemicals tested 3 times independently)			
	1A	1 BC	NC
1A	**91.7%**	8.33%	0
1 BC	41.9%	**58.1%**	0
NC	2.7%	23.4%	**73.9%**
Overall accuracy	70.4%		
(b) EpiDerm™ SCT using the new PM (25% cut off at 3 min exposure time, 2 steps procedure), 240 predictions (80 chemicals tested 3 times independently)[a]			
	1A	1 BC	NC
1A	**83.3%**	16.67%	0
1 BC	29.0%	**71.0%**	0
NC	2.7%	23.4%	**73.9%**
Overall accuracy	74.2%		

[a]The Episkin™ sensitivity for Cat. 1A, merged Cat. 1B-and-1C and specificity for the same dataset is 83.3%, 76.3%, 79.2% resulting in overall accuracy of 78.8%

The new PM significantly improves the over-prediction rate for cat 1B-and-1C chemicals, whilst maintaining high sensitivity as requested by regulators. With this new PM for EpiDerm™, the sensitivity and specificity for sub-categorization purposes of EpiDerm™ SCT and EPISKIN™ SCT is almost the same.

9.3.3 Applications and Limitations

The EpiDerm™ SCT was developed and designed to predict skin corrosion potential of neat test chemicals in the context of identification and classification of skin corrosion hazard according to the EU and GHS classification system. No clear applicability domain restrictions could be defined for EpiDerm™ SCT, except for testing gases, vapours and aerosols. Testing of these types of chemicals require special conditions not covered by the current test design. Possible toxic interference of volatile chemicals across plate wells can be avoided by sealing the wells with an adhesive cover sheet, or testing the volatile chemicals on separate plates.

The test is able to partially discriminate between the corrosive subcategories of the UN GHS, i.e. Cat. 1A vs. the merged Cat. 1B-and-1C. It does not distinguish between GHS category 1B and category 1C. No prediction model for this type of prediction using RhE models has been so far adopted by regulators.

One limitation of this test method is a possible interference of the test substances with the MTT endpoint. A coloured test substance or one that directly reduces MTT (and thereby mimics dehydrogenase activity of the cellular mitochondria) may interfere with the MTT endpoint. However, these test chemicals are a problem only if at the time of the MTT test sufficient amounts of the test chemical are still present on (or in) the tissues.

In case of this event, the (true) metabolic MTT reduction and the contribution by a coloured test material or (false) direct MTT reduction by the test material can be quantified by a special procedure described in details in the SOP provided by [16].

9.3.4 Comparison to Human Data

Limited human data that exist on skin irritation and corrosion endpoint suggest, that reconstructed human tissue models and the validated *in vitro* assays for assessment of skin corrosion err on the side of safety. Some over-predictions have been reported for materials with low pH by cleaning industry [17].

9.4 Brief Description of the Protocol

9.4.1 Reconstituted Human Skin Model

The reconstructed tissue model EpiDerm™ (MatTek, Ashland, USA and MatTek IVLSL, Bratislava, Slovakia) consists of normal, human-derived epidermal keratinocytes which have been cultured to form a multilayered, highly differentiated model of the human epidermis. It consists of organized basal, spinous and granular layers, and a multilayered stratum corneum containing intercellular lamellar lipid layers arranged in patterns analogous to those found *in vivo* [18].

The epidermal cells are taken from healthy volunteers negative to HIV, and Hepatitis. The EpiDerm™ tissues (surface 0.63 cm^2) are cultured on specially prepared cell culture inserts and shipped to customers as kits, containing 24 tissues on shipping agarose together with necessary amount of culture media and 6-well plates. In addition the MTT kit (containing MTT concentrate, diluent, extractant, PBS and 24-well plate) can be provided by MatTek.

The EpiDerm™ kit is manufactured according to defined quality assurance procedures compliant to GMP process and ISO 9001:2008. All biological components of the epidermis and the culture medium are tested by manufacturer for viral, bacterial, fungal and mycoplasma contamination. Barrier properties of each manufactured tissue lot are controlled by the manufacturer. Per request, MatTek provides detailed information about ET50 experiment with Triton X-100 (1%) (chemical recommended as penetration marker by the OECD TG 431), information of tissue viability (MTT test), together with historical database of results.

Handling procedures for biological materials should be followed. It is recommended to wear gloves during handling with the skin and kit components. After use, the epidermis, the material and all media in contact with it should be decontaminated prior to disposal (e.g. using 10% bleach or special containers).

9.4.2 Materials

All material that is needed to conduct the EpiDerm™ skin corrosion test is summarised in Table 9.2. The basic EpiDerm™ SCT kit (EPI-200-SCT) contains 24 units of the standard EpiDerm™ model (EPI-200) embeded in transporting agar, bottle of assay medium, sterile 6 and 24-well plates, small amount of DPBS and one vial of the contol material that can be used to test barrier properies of the EpiDerm™ model (Triton X-100). In addition, MatTek also offeres kit for conducting MTT-assay that contains MTT-concentrate, MTT-diluent and extracting solution. Further details on the material and reagents required for the EpiDerm™ SCT can be found in Table 9.2.

Table 9.2 Material required for the conductance of the EpiDerm™ SCT

#	Description	Detail
(A) Material provided by MatTek Corporation with standard EPI-200-SCT Kit:		
1	One sealed 24-well plate containing 24 inserts of EpiDerm™ embedded in transporting agar	EPI-200, 0.6 cm^2
2	Sterile 6-well plates used during the assay	4 pieces
3	Sterile 24-well plates used for MTT assay	2 pieces
4	Two bottles of DPBS	100 mL
5	One bottle Assay Medium, EPI-100-ASY	100 mL
6	One vial, containing the positive control chemical—1% Triton	10 mL
7	Sterile Nylon Mesh for application of liquid test materials	25 pieces
8	Protocol for Skin Corrosion test according to the OECD TG 431	
(B) MTT-100 Assay Kit Components (ordered separately):		
1	One vial containing MTT concentrate (5 mg/mL)	2 mL
2	One vial MTT diluent (DMEM based culture medium)	8 mL
3	One bottle containing extracting solution isopropanol	60 mL
(C) Additional material and equipment needed		
1	Sterile Dulbecco's PBS (DPBS) without Ca2+ and Mg2+ (e.g. PAN or Biochrom)	2 L
2	Sterile, sharp blunt-edged forceps	
3	Positive displacement pipette for application of semi-solid test materials	50 µL
4	Mortar and pestle for grinding of granular solids	
5	Sterile disposable pipettes, pipette tips	20-200 µL 200–1000 µL
6	Sharp spoon—for application of solids. Aesculap, Purchase Number.: FK 623	NaCl weight: 25 mg
7	Bulb headed Pasteur pipettes—for spreading of test substances	
8	Parafilm	
9	Sterile cotton tip swabs	
10	Laminar flow hood—for work under sterile conditions	
11	Humidified incubator	37°C, 5% CO_2, 95% relative humidity
12	96-well plate photometer equipped with filter 570 nm	
13	Laboratory balance	
14	Plate shaker	
15	Stop-watches	
16	Wash bottle	500 mL
17	Beakers—for washing and collecting DPBS	200 mL
18	Potassium Hydroxide, 8 N, to be used as positive control in the SCT assay	Sigma P4494

9.4.3 Protocol steps: DAY 0–Day prior to dosing

Note: Before any testing on the viable reconstructed human tissues is performed, it is recommended to perform the evaluation of the test substance for interference with the measured endpoint (MTT assay). This procedure is described in details in the SOP [16].

1. EPI-200-SCT kits are shipped from MatTek facilities in USA and Slovakia (EU) every Monday.
2. Upon receipt of the shipment, examine all kit components for integrity. If there is a concern call MatTek immediately.
3. Record all information about supplied material into the MDS.
4. Place the DPBS into the refrigerator (5 ± 3°C) and the vial containing the MTT concentrate in the freezer (−20 ± 5°C).
5. If the test is not performed on the day of receipt, store the EpiDerm™ tissues in the refrigerator at 4°C until next day. If you plan to determine any additional endpoints to MTT viability measurements, place the tissues immediately upon arrival into the EpiDerm™ Maintenance Medium and pre-incubate 1 h (37°C, 5%. CO_2, humidified atmosphere). Afterwards replace the medium and continue with overnight pre-incubation.

9.4.4 DAY 1: Tissue Conditioning (Pre-Incubation) and Chemical Exposure

1. Prepare *two* 6-well plates for *four* chemicals and the negative (NC) and positive control (PC) for the *3 min application*. Pipette 0.9 mL Maintenance Medium in each well.
2. Prepare *two* 6-well plates for *four* chemicals and the negative (NC) and positive control (PC) for the *1 h application*. Pipette 0.9 mL Maintenance Medium in each well.
3. Remove the shipped multiwell plate from the plastic bag. Open the 24-well plate under a sterile airflow and remove the sterile gauze. Carefully take out each insert containing the epidermal tissue, rapidly remove any remaining agarose that adheres to the outer sides of the insert by gentle blotting on the sterile filter paper, and immediately place it in a well of the prepared 6-well plate. Act quickly as the epidermal cultures dry out rapidly when not in contact with medium. Make sure that no air bubbles are formed underneath the insert!
4. Mark the 6-well plates as shown in Fig. 9.2.
5. Place the 6-well plates containing the tissues into a humidified (37°C, 5% CO_2) incubator for 1 h.

3 minutes exposure

NC tissue 1	Chemical 1 tissue 1	Chemical 2 tissue 1
NC tissue 2	Chemical 1 tissue 2	Chemical 2 tissue 2

1 hour exposure

NC tissue 1	Chemical 1 tissue 1	Chemical 2 tissue 1
NC tissue 2	Chemical 1 tissue 2	Chemical 2 tissue 2

Chemical 3 tissue 1	Chemical 4 tissue 1	PC 1 tissue 1
Chemical 3 tissue 2	Chemical 4 tissue 2	PC 2 tissue 2

3 minutes exposure

Chemical 3 tissue 1	Chemical 4 tissue 1	PC 1 tissue 1
Chemical 3 tissue 2	Chemical 4 tissue 2	PC 2 tissue 2

1 hour exposure

Fig. 9.2 Plate design for the skin corrosion testing

9.4.5 Preparations for the Main Test (Performed During the Pre-Incubation Time on a Day of Experiment)

1. Prepare the MTT medium according to the procedure below.
   (a) Thaw the MTT concentrate (MTT-100-CON) and dilute with the MTT diluent (MTT-100-DIL). Store the remaining MTT solution in the dark at 4°C for later use on the same day (do not store until next day since MTT will degrade with time).
   (b) If you are preparing your own MTT, after preparation of stock solution of MTT (5 mg/mL in DPBS) and filtration using a sterile 0.45 μm filter, add 2 mL of the stock-solution to 8 mL Maintenance Medium (final concentration: 1 mg MTT/mL medium). MTT Stock solution can be stored frozen (−20 °C) up to 1 month. Since MTT is toxic wear protective gloves during manipulation with MTT solution!
2. Prepare two 24-well plates to be used as "holding and MTT plates" one for the 3 min experiment, the other for the 1 h experiment
3. Pipette 300 μL of either maintenance medium or MTT medium in each well and place the plates in the incubator

9.4.6 1-Hour Application

Note: Dosing time interval is dictated by rinsing procedure. If the technician has performed the test already, 45 s intervals is sufficient for both application and washing procedures. However, if the test is performed for the first time, 1 min dosing interval is recommended.

1. After 1 h of pre-incubation, transfer each insert to new 6-well plates with fresh medium (0.9 mL per well). Alternatively, aspirate the pre-incubation medium from the 6-well plates and pipette 0.9 mL of fresh medium into each well.
2. Set the timer to 1 h and start it. Add 50 μL of H_2O (negative control) into the first insert atop the tissue. After 45 s repeat the procedure with the second tissue. Proceed with test material 1–4 (50 μL: liquids, 25 mg + 25 μL H_2O: solids) and the positive control in the same manner until all tissues are dosed.
   Dosing interval scheme for the 60 min experiment:

0:00–0:45—tissue 1 (NC)	4:30–5:15—tissue 7 (C3)
0:45–1:30—tissue 2 (NC)	5:15–6:00—tissue 8 (C3)
1:30–2:15—tissue 3 (C1)	6:00–6:45—tissue 9 (C4)
2:15–3:00—tissue 4 (C1)	6:45–7:30—tissue 10 (C4)
3:00–3:45—tissue 5 (C2)	7:30–8:15—tissue 11 (PC)
3:45–4:30—tissue 6 (C2)	8:15–9:00—tissue 12 (PC)

3. After the 60 min period of exposure for the first tissue is complete, using forceps remove the first insert from the 6-well plate. Using a wash bottle gently rinse the tissue with DPBS (= fill and empty insert 20 times in a constant stream of DPBS) to remove any residual test material. Remove excess DPBS by gently shaking the insert and blot bottom on blotting paper. Place insert in the prepared holding plate
4. Once all tissues have been rinsed and are in the holding plate, dry the surface with cotton swab, remove inserts from the holding plate, blot bottom and transfer into the 24-well plate, prepared for the MTT assay.
5. Place plate in the incubator, record start time of MTT incubation and incubate for 3 h (37°C, 5% CO_2).

9.4.7 3 Min Application

1. Start the timer for 3 min. Add 50 μL H_2O (negative control) into the first insert atop the tissue. After 45 s repeat the procedure with the second tissue. Following 45 s intervals enable to dose four tissues. Afterwards, washing of the tissue 1 has to start.
2. After the 3 min period of exposure for the first four tissues is complete, start the timer for 3 min and with forceps remove the first insert from the 6-well plate. Using a wash bottle gently rinse the tissue with DPBS (= fill and empty insert 20 times in a constant stream of DPBS) to remove any residual test material. Remove excess DPBS by gently shaking the insert and blot bottom on blotting paper. Place insert in the prepared holding plate. After 45 s repeat the procedure with the second insert, after 1:30 min with the third…etc.

Dosing interval scheme for the 3 min experiment:

SET 1 = NC and chemical 1 (6 min)
Dosing:
0:00–0:45—tissue 1 (NC)
0:45–1:30—tissue 2 (NC)
1:30–2:15—tissue 3 (C1)
2:15–3:00—tissue 4 (C1)
Rinsing
3:00–3:45—tissue 1 (NC)
3:45–4:30—tissue 2 (NC)
4:30–5:15—tissue 3 (C1)
5:15–6:00—tissue 4 (C1)
SET 2 = chemical 2, 3 (6 min)
SET 3 = chemical 4, PC (6 min)

3. Proceed with all test materials (50 μL: liquids, 25 mg + 25 μL H_2O: solids) and the positive control in the same manner until all tissues are dosed and rinsed.
4. Once all tissues have been rinsed and are in the holding plate, carefully dry the surface of the tissue with a cotton swab. Afterwards remove inserts from the holding plate, blot bottom and transfer into the 24-well plate, prepared for the MTT assay. Place plates in the incubator, record start time of MTT and incubate for 3 h (37°C, 5% CO_2).

9.4.8 MTT Test and Reading

1. After the 3 h MTT incubation period is complete, gently blot the tissue on the absorbing paper and transfer inserts into new 24-well plates.
2. Immerse the inserts by gently pipetting 2 mL of isopropanol (extractant solution) into each insert. The level will rise above the upper edges of the insert, thus completely covering the tissues from both sides.
3. Seal the 24-well plates (e.g. with Parafilm or place into a sealable plastic bag) to inhibit extractant evaporation. Record start time of extraction in the Methods Documentation Sheet (MDS) and extract formazan for at least 2 h at room temperature with gentle shaking on a plate shaker (120–200 rpm).
4. As an alternative, overnight extraction is also possible. Seal plates as described above and extract at room temperature in the dark, without shaking. Before using the extracts, shake for at least 15 min on plate shaker. After the extraction period is complete, pierce the inserts with an injection needle (~gauge 20, ~0.9 mm diameter) and allow the extract to run into the well from which the insert was taken. Afterwards the insert can be discarded. Before transferring the extract to 96 well plates pipette up and down 3× until the extractant solution is homogenous.

5. For each tissue, transfer 2 × 200 μL aliquots of the blue formazan solution into a 96-well flat bottom microtiter plate according to the fixed plate design given in spreadsheet. Use isopropanol as blanks.
6. Read OD in a 96-well plate spectrophotometer using a wavelength between 540 and 595 nm, preferably at 570 nm, without using a reference filter.

9.4.9 Test Data

A blank MS EXCEL workbook EpiDerm-SCT-SPREAD.XLS can be provided by MatTek for the data analysis. A copy should be made before the first data entry. The workbook consists of two single spreadsheets named: IMPORT and SPREAD. Data files of optical densities (ODs) generated by the microplate reader (with blank subtraction) are copied from the reader software to the Windows Clipboard and then pasted into the first spreadsheet of the EXCEL workbook. The blank corrections, calculation of results and statistical parameters are done automatically in the second part of the workbook. Use the fixed 96-well plate design as specified in the SOP provided by MatTek.

After data entry, the spreadsheet performs the following calculations:

1. For each individual tissue treated with a test chemical (TC), the positive control (PC) and the negative control (NC) the individual relative tissue viability is calculated according to the following formulas

$$\text{Relative viability TC}(\%) = [\text{ODTC / Mean of ODNC}] \times 100$$

$$\text{Relative viability NC}(\%) = [\text{ODNC / mean of ODNC}] \times 100$$

$$\text{Relative viability PC}(\%) = [\text{ODPC / mean of ODNC}] \times 100$$

2. For each test chemical, negative control, and the positive control, the mean relative viability of the two individual tissues is calculated and used for classification according to the Prediction Model.
3. The spreadsheet shows a graph of the results (% of relative viability ± Difference)

9.4.10 Data Interpretation Procedure (Prediction Model)

A single testing run composed of at least two tissue replicates should be sufficient for a test chemical when the resulting classification is unequivocal. However, in cases of borderline results, such as non-concordant replicate measurements, a second run may be considered, as well as a third one in case of discordant results between the first two runs.

The prediction model for the EpiDerm™ skin corrosion test method associated with the UN GHS classification system and currently implemented into the OECD TG 431 [4] is shown in Table 9.3:

Table 9.3 Prediction model of the EpiDerm™ Skin Corrosion Test

Viability measured after exposure time points (t = 3 and 60 min)	Prediction to be considered
Step 1 (all chemicals)	
<50% after 3 min exposure	Corrosive
≥50% after 3 min exposure AND <15% after 60 min exposure	Corrosive
≥50% after 3 min exposure AND ≥15% after 60 min exposure	Non-corrosive
Step 2 (corrosive chemicals)	
<25% after 3 min exposure	Sub-category 1A
≥25 after 3 min exposure	A combination of optional Sub-categories 1B-and-1C

9.4.11 Assay Quality Controls (OECD TG 431)

9.4.11.1 Assay Acceptance Criterion 1: Negative Control (NC)

The absolute OD of the H_2O treated NC tissues in the MTT-test is an indicator of tissue viability obtained in the testing laboratory after the shipping and storing procedure and under specific conditions of the assay. Tissue viability is meeting the acceptance criterion if the mean OD of the mean of NC is ≥0.8 and ≤2.8.

9.4.11.2 Assay Acceptance Criterion 2: Positive Control (PC)

8 N KOH (Sigma P4494) is used as PC and has to be tested once on each testing day. 1 h exposure of the PC will reveal a mean tissue viability less than 15%.

9.4.11.3 Assay Acceptance Criterion 3: Variability

Based on the test design the experiments can be performed on two or three tissue replicates per exposure time. The SD (if N = 3) or Difference (if N = 2) should not exceed 20%. In the range between 20% and 100% viability the coefficient of variation (CV) is an additional acceptance criterion. Based on the results from the original validation study, it should not exceed 0.3.

9.5 Role in a Testing Strategy

The EpiDerm™ SCT method may be used as a stand-alone test method for the detection or exclusion of corrosive effects of test chemicals. A negative result in these test methods will require an additional *in vitro* skin irritation test, if not performed upfront, to determine if the chemical should be classified Cat. 2 (irritant) or if it does not require classification (No Cat.), and thus replace the *in vivo* test according to OECD TG 404.

The EpiDerm™ SCT also allows for the sub-categorisation of corrosive chemicals into Cat. 1A or Cat. 1B-and-1C but does not permit the distinction of the latter into Cat. 1B and Cat. 1C. [19].

9.6 Perspectives from the Test Developer

9.6.1 Critical Steps in the Protocol

The EpiDerm™ SCT is an easy to perform method that utilizes readily available laboratory equipment. The test can be performed by most laboratory personnel. Critical steps of this procedure are exact timing and sufficient removal of the test articles from the tissue surface. Unequal spreading of a test material may lead to the false negative outcomes or high variability. Insufficient washing may lead to the over-prediction of the corrosive effect.

Special care should be taken when testing colorant materials (blue, deep red, violet) and MTT reducing materials since they will interfere with the MTT end-point. MatTek and other tissue model suppliers have developed procedures to deal with such materials. The procedures are described in detail in the SOPs provided with the tissue products.

9.6.2 Possible Protocol Adaptations

Some amines, when tested in EPISKIN™ and EpiDerm™ Skin Corrosion Tests have shown tendency for under-prediction of corrosive effects [20]. Based on the non-published experimental data with this group of chemicals, it is recommended to extend exposure time in the EpiDerm™ SCT to 4 h for these materials. MTT correction step will be necessary, since amines, besides causing delayed corrosion effects will also reduce MTT chemically and could cause false negative predictions. The cut-off 20% cell viability at 4 h should be applied to improve the prediction of these groups of chemicals.

9.6.3 Challenges and Opportunities

Test methods based on reconstructed human epidermis tissues are playing a prominent role in topical toxicology and for risk assessment purposes. History has shown that test methods can be improved and further modified even after their validation to serve better purposes of the modern toxicology. For complete replacement of the animal tests, it will be important to understand better the applicability domains, the strengths and weaknesses of the particular assays described in the current guidelines. This will be very important with regard to their correct and justified use in the testing strategies and assessments *in vitro* such as described in the OECD GD 203 on IATA [19].

One of the remaining challenges in the development of *in vitro* test methods for skin corrosion is the need to correlate the *in vitro* predictions to *in vivo* rabbit data. This has been very challenging to achieve for materials that do not cause corrosion in human but only in rabbit. It seems to be almost impossible to develop a reliable method for further distinguishing cat. 1B from cat. 1C chemicals, since the rabbit test itself does not provide reliable and reproducible reference data for method development and validation purposes.

9.7 Conclusions

The EpiDerm™ Skin Corrosion Test (OECD TG 431), was one of the first test methods accepted for regulatory use by the EU and OECD Competent Authorities after scientific validation and peer review. The protocol represents a robust, reliable and reproducible *in vitro* method. Furthermore, it is easily transferable to other tissue models as shown in the catch up validation studies with SkinEthic™ and EpiCS® (former EST-1000). Since the adoption of this test method within EU and OECD test guidelines, the number of animals required for predicting skin corrosion has significantly decreased. In combination with the *in vitro* skin irritation test and QSARs it may completely replace the need of *in vivo* topical toxicity testing in animals for most of the chemicals under the EU chemicals regulation. Adoption of the new prediction model for sub-categorization allows to further enhance this method and enable a better prediction of corrosive potency as required by certain specific regulations.

References

1. UN GHS. Harmonised integrated classification system for human health and environmental hazards of chemical substances and mixtures. Sixth revision. 2015. 521 p. http://www.unece.org/fileadmin/DAM/trans/danger/publi/ghs/ghs_rev06/English/ST-SG-AC10-30-Rev6e.pdf.
2. Kandárová H, Letasiova S, Milasova T, Hayden P, Klausner M. Analysis of the Validated Epiderm Skin Corrosion Test (EpiDerm SCT) and a Prediction Model for Sub-Categorization According to the UN GHS and EU CLP. ALTEX Proceedings 2, LINZ 2013, p. 47. 2013. Available at: http://www.altex.ch/resources/linz_2013_entire_issue.pdf.
3. Desprez B, Barroso J, Griesinger C, Kandárová H, Alépée N, Fuchs HW. Two novel prediction models improve predictions of skin corrosive sub-categories by test methods of OECD Test Guideline No. 431. Toxicol In Vitro. 2015;29(8):2055–80.
4. OECD. Guidelines for the testing of chemicals No. 431: in vitro skin corrosion: human skin model test. Paris: OECD, . 8 p. Originally adopted in 2004; 2016.
5. Perkins MA, Osborne R, Johnson GR. Development of an in vitro method for skin corrosion testing. Fundam Appl Toxicol. 1996;31:9.18.
6. Fentem JH, Archer GEB, Balls M, Botham PA, Curren RD, Earl LK, Esdaile DJ, Holzhutter H-G, Liebsch M. The ECVAM international validation study on in vitro tests for skin corrosivity. 2. Results and evaluation by the Management Team. Toxicol In Vitro. 1998;12:483–524.
7. ECVAM. ESAC statement on the scientific validity of the rat skin transcutaneous electrical resistance (TER) test (an in vitro test for skin corrosivity) dated 3rd April 1998. ATLA. 1998;26(3):275–80.
8. ECVAM. ESAC statement on the scientific validity of the EPISKIN test (an in vitro test for skin corrosivity), dated 3rd April 1998. ATLA. 1998;26(3):275–80.
9. Liebsch M, Traue D, Barrabas C, Spielmann H, Uphill P, Wilkins S, McPherson JP, Wiemann C, Kaufmann T, Remmele M, Holzhütter H-G. The ECVAM prevalidation study on the use of EpiDerm for skin corrosivity testing. ATLA. 2000;28:371–401.
10. ECVAM. ESAC statement on the application of the EPIDERM human skin model for skin corrosivity testing, dated 21st March 2000. ATLA. 2000;28(3):365–7.
11. Kandárová H, Liebsch M, Spielmann H, Genschow E, Schmidt E, Guest R, Whittingham A, Warren N, Gamer A, Remmele M, Kaufmann T, Wittmer E, De Wever B, Rosdy M. Assessment of the SkinEthic reconstituted human epidermis for skin corrosion testing according to OECD guideline 431. Toxicol In Vitro. 2006;20:547–59.

12. ECVAM. Statement on the application of the SkinEthicTM human skin model for skin corrosivity testing, issued by the ECVAM Scientific Advisory Committee (ESAC25), 17 November 2006. 2006. Available at: http://ecvam.jrc.ec.europa.eu. (25) EC-ECVAM (2009).
13. Hoffmann J, et al. Epidermal-skin-test 1000 (EST-1000) – a new reconstructed epidermis for in vitro skin corrosivity testing. Toxicol In Vitro. 2005;19:925–9.
14. ECVAM. ESAC statement on the scientific validity of an in-vitro test method for skin corrosivity testing: the EST-1000, issued by the ECVAM Scientific Advisory Committee (ESAC30), 12 June 2009. 2009. Available at: http://ecvam.jrc.ec.europa.eu.
15. EU. Guidance on labelling and packaging in accordance with Regulation (EC) No 1272/2008 and its updates (2016). 2008. https://echa.europa.eu/documents/10162/13562/clp_labelling_en.pdf.
16. MatTek. EpiDerm™ SOP, Version MK-24-007-0024 (November 2014). Protocol for: in vitro EpiDerm™ skin corrosion test (EPI-200-SCT), For use with MatTek Corporation's reconstructed human epidermal model EpiDerm. 2014.
17. Scheel J, Heppenheimer A, Lehringer E, Kreutz J, Poth A, Ammanna H, Reisinger K, Banduhn N. Classification and labeling of industrial products with extreme pH by making use of in vitro methods for the assessment of skin and eye irritation and corrosion in a weight of evidence approach. Toxicol In Vitro. 2011;25(2011):1435–47.
18. Cannon CL, Neal PJ, Southee JA, Kubilus J, Klausner M. New epidermal model for dermal irritancy testing. Toxicol In Vitro. 1994;8:889–91.
19. OECD GD No. 203. Guidance document on an integrated approach on testing and assessment (IATA) for skin corrosion and irritation. In: Series on testing and assessment. 11-Jul-2014. 2014. http://www.oecd.org/officialdocuments/publicdisplaydocumentpdf/?cote=env/jm/mono(2014)19&doclanguage=en.
20. Houthoff E, Rugen P, Hart D. Predictability of in vitro dermal assays when evaluating fatty amine derivatives. Toxicol In Vitro. 2015;29(2015):1263–7.

10 Assessment of the Human Epidermis SkinEthic™ RHE Model for *In Vitro* Skin Corrosion Testing of Chemicals

Nathalie Alépée, Marie-Hélène Grandidier, and José Cotovio

10.1 Principle of the Test Method and Scientific Basis

Corrosive agents are chemicals that cause irreversible damage and destruction of the skin, often burning through several layers of tissue. Reactions are characterized by specific serious signs such as ulcers, bleeding, bloody scabs, and discoloration. Corrosivity represents the most extreme form of skin hazard, in which the skin can be destroyed beyond the body's ability to heal easily. Chemical burns can occur when skin is exposed to a corrosive substance such as a strong acid or base able to damage proteins (amide hydrolysis) or modify lipids (ester bonds hydrolysis).

According to current international regulatory requirements, assessment of skin corrosion is mandatory for all chemicals placed on the market. Data on skin corrosion potency are required by several pieces of legislation, notably the Classification, Labelling and Packaging (CLP) Regulation (1272/2008) [1], the Cosmetics Directive (76/768/EEC), which was repealed from July 2013 by the EU regulation on cosmetic products (EC 1223/2009) [2], and the REACH Regulation (1907/2006) [3]. Corrosivity data are also collected by regulatory agencies concerned with the transportation of hazardous chemicals, in the event of accident and chemical spill (Fig. 10.1).

To replace the Draize rabbit skin test [6], significant efforts have been undertaken to develop alternative *in vitro* test methods to replace *in vivo* testing. The SkinEthic™ Reconstructed Human Epidermis (RHE) test method has been developed on the premise that corrosive chemicals are able to penetrate the *stratum corneum* by diffusion or erosion and are cytotoxic to the cells in the underlying layers.

N. Alépée (✉) • M.-H. Grandidier • J. Cotovio
L'Oréal Research and Innovation, 1, Avenue Eugène Schueller, 93600 Aulnay-sous-Bois Cedex, France
e-mail: nalepee@rd.loreal.com

C. Eskes et al. (eds.), *Alternatives for Dermal Toxicity Testing*,
DOI 10.1007/978-3-319-50353-0_10

EU DSD	No label		R38	Cause burns (R34)	Cause severe burns (R35)	
UN GHS	No Category		Cat.2	Cat.1C	Cat.1B	Cat.1A
UN GHS	No Category	Cat. 3	Cat.2	Cat.1C	Cat.1B	Cat.1A
Model Regulations Good (TDG)				Class 8 PG III	Class 8 PG II	Class 8 PG I
	Not classified		Irritant	Corrosive		

Fig. 10.1 EU DSD, EU CLP, UN GHS, and TDG classifications. *EU DSD* Former European Classification System based on the Dangerous Substance Directive prior to adoption of the UN GHS system. *EU CLP* Current European Union system of classification, labelling, and packaging of chemical after adoption of the UN GHS system. Indeed, in the EU, the UN GHS system is implemented through the regulation on classification, labelling, and packaging of substances and mixtures (CLP) ([1, 4]) *UN GHS* UN Globally Harmonized System of Classification and Labelling of Chemicals (GHS) [4], *TDG* Transportation of Dangerous Goods. Regarding transportation rules, class 8 includes the following packaging groups: class 8 packaging groups—PG I, very dangerous; PG II, medium danger; PG III, minor danger. Cat: category 1A, 1B, 1C [5]

The SkinEthic™ RHE model consists in fully differentiated three-dimensional epidermal tissue grown from normal human keratinocytes in a chemically defined medium at the air-liquid interface [7, 8]. The model is histologically similar to *in vivo* human epidermis and features a functional permeability barrier, which is one of the main functions of viable skin. The SkinEthic™ RHE corrosion test method has been developed and internationally accepted using the decrease in cell viability after exposure to a chemical [9, 10]. Briefly, cell viability determination is based on cellular mitochondrial dehydrogenase activity, measured by MTT reduction and conversion into blue formazan salt that is quantified after extraction from tissues [11]. The decrease in cell viability in treated tissues is compared to negative controls and expressed as a percentage. The percentage reduction in viability is used to predict the corrosion potential.

More recently, the EU CLP classification system—based on the United Nations Globally Harmonized System of Classification and Labelling of Chemicals [4]—required subcategorization of corrosive chemicals into the three UN GHS optional subcategories: skin corrosion subcategories 1A, 1B, and 1C (Fig. 10.1). In parallel, during an international workshop organized by the Swiss regulatory authorities, it was recommended that further investigations be carried out on the reconstructed human epidermis models originally validated to discriminate between corrosive and non-corrosive chemicals. This was in order to evaluate their usefulness in identifying the three optional subcategories of corrosive chemicals under the UN GHS classification system (1A, 1B, and 1C) or at least to distinguish subcategory 1A from subcategory 1B and 1C skin corrosive chemicals [12]. Investigation of the usefulness of the validated SkinEthic™ RHE test method to discriminate between UN GHS skin corrosive subcategory 1A and subcategory 1B and 1C substances was then further explored [13, 14]. More than 80 substances were tested based on the OECD Expert Panel on Skin Irritation and Corrosion recommendation since the original validation studies of the SkinEthic™ RHE protocol underwent refinements. Those improvements include the possibility to correct MTT interacting substances as implemented in the skin irritation SkinEthic™ RHE test method [15].

10.2 Current Validation Status

The SkinEthic™ Reconstructed Human Epidermis (RHE) test method has been adopted to distinguish between corrosive and non-corrosive chemicals as well to identify UN GHS subcategories 1A and 1B and 1C [16] within the context of Organization for Economic Co-operation and Development Test Guideline 431 [17].

Historically, the SkinEthic™ RHE test method has been formally validated by the European Centre for the Validation of Alternative Methods (ECVAM, recently renamed EURL ECVAM Scientific Advisory Committee (ESAC)) to distinguish between corrosive and non-corrosive chemicals [9]. The method was scientifically validated as sufficiently similar in regard to its structural and functional characteristics and its performance in reference to the performance standards as required by OECD TG 431. The SkinEthic™ RHE test method underwent an external validation study conducted by ZEBET. A blind trial was conducted in three laboratories (ZEBET, Safepharm, and BASF), where several chemicals from the EURL ECVAM validation trials were tested using the SkinEthic™ RHE test method. Performance and reproducibility standards were met as defined by OECD TG 431 ([17, 18]; ESAC, 2006 revised in [19]).

Until 2005, SkinEthic™ RHE tissues were cultivated on 0.63 cm^2 inserts. Due to manufacturing constraints, the size of the inserts was harmonized and the method adapted accordingly by applying a reduced volume of test chemical and a reduced isopropanol extraction volume during the MTT reduction assay. The test method was assessed with 25 representative test chemicals from different chemical classes including 12 OECD reference test chemicals similarly classified as compared to *in vivo* and fully compliant with OECD performance and reproducibility requirements [15, 17].

Changing classification criteria in the European Union from the EU Dangerous Substances Directive [20] to the EU Classification, Labelling, and Packaging Regulation [1] had substantial consequences on regulatory classification for skin corrosion. The EU CLP classification system, based on UN GHS classification [4], required that corrosive chemicals be subcategorized into the three UN GHS optional subcategories. Tests involving 84 test chemicals (including chemicals from EURL ECVAM Skin Corrosion Validation Studies (SCVS) and new chemicals identified by the OECD Expert Panel on Skin Corrosion (OECD TG 431)) showed that the SkinEthic™ RHE test method was highly sensitive and allowed reliable identification of UN GHS subcategories 1A and 1B and 1C [13, 16]. The initial prediction model (cutoff value of 50% cell viability at 3 min) for SkinEthic™ RHE resulted in quite high over-prediction rates of category 1B and 1C chemicals that were over-predicted as category 1A, whilst providing high correct predictions rates of category 1A and non-corrosive chemicals. Switching from the original PM to the novel PMs (PMvar1 or PMvar2) allows to obtain higher correct predictions for categories 1B and 1C in a range of 60–70% [14]. The SkinEthic™ RHE test method has been adopted within the context of OECD TG 431 to distinguish subcategories of corrosive and non-corrosive chemicals [17].

10.3 Performance and Applicability of the Test Method

10.3.1 Reproducibility

To allow adequate determination of inter-laboratory reproducibility, a blind trial was conducted in three laboratories (ZEBET, Safepharm, and BASF) in which the 12 endorsed OECD reference chemicals were tested. Results obtained with the SkinEthic™ epidermal model were reproducible both within and between laboratories and over time [18]. Performance of the SkinEthic™ RHE test method with regard to its intra-laboratory reproducibility was evaluated in the three laboratories. Performance of 88.8% (32/36) was found for the 12 reference chemicals tested altogether in the three laboratories when distinguishing between corrosives and non-corrosives. Assessment of classification concordance was further evaluated on 25 chemicals by Tornier et al. [15]. Only 2 out of 25 tested chemicals were differently classified between three independent experiments, presenting a within-laboratory reproducibility of 92.0%.

In the context of the OECD TG 431 [21] revision, a total of 84 chemicals identified by the OECD Expert Panel on Skin Corrosion were tested in three independent experiments. Independent statistical data analysis was performed to evaluate between-run variability and classification concordance of the SkinEthic™ RHE test method for subcategorization of skin corrosive chemicals, according to the EU CLP classification system [16]. Only 6 out of the 80 OECD chemicals were differently classified between three independent experiments when distinguishing between corrosive and non-corrosive chemicals, in accordance with the prediction model. Concordance of 92.5% in classifications was then obtained when discriminating between corrosive and non-corrosive chemicals (Table 10.1). Of all the 84 tested chemicals (including the four chemicals from ECVAM SCVS that were not part of the OECD list), 92.9% (78/84) showed 100% concordance in classifications between the three independent runs [13]. Furthermore, 87.5% concordance in classifications was found when distinguishing between UN GHS subcategories 1A and 1B and 1C and non-corrosives (70 concordant predictions in the three classes out of the 80 tested chemicals).

Table 10.1 Within-laboratory reproducibility of the SkinEthic™ RHE skin corrosion test method. Assessment of concordant classifications

Data set/analysis	Within-laboratory reproducibility
NC vs. C (80 OECD chemicals)	92.5% (74/80)
Cat. 1A vs. cat. 1B and 1C vs. NC (80 OECD chemicals)	87.5% (70/80)
NC vs. C (80 OECD chemicals + 4 subst. SCVS)	92.9% (78/84)

SCVS skin corrosion validation study, *NC* non-corrosives, *C* corrosives

Table 10.2 Predictive capacity of the SkinEthic™ RHE skin corrosion test method

	Reference chemicals (30)		All OECD chemicals (80)		All chemicals (84)
Predictive capacity	Total nb.	%	Total nb.	%	%
Subcategory 1A correct predictions	27/30	90.0	29/36	80.6	NA
Subcategory 1A → non-corrosive	0/30	0.0	0/36	0.0	
Subcategories 1B and 1C correct predictions	24/30	80.0	59/93	63.4	
Subcategories 1B and 1C → subcategory 1A	6/30	20.0	27/93	29.0	
Sensitivity (corrosive correct predictions)	60/60	100.0	122/129	94.6	94.9
Specificity (non-corrosive correct predictions)	24/30	80.0	82/111	73.9	74.6
Accuracy (non-corrosive vs. corrosive)	84/90	93.3	204/240	85.0	85.7
Accuracy (NC vs. subcategories 1B and 1C vs. subcategory 1A)	75/90	83.3	170/240	7.8	NA

NA: Not applicable. Four chemicals from the ECVAM SCVS that were not part of the OECD list were also tested: one non-corrosive and three corrosives. Three corrosive chemicals are relevant in assessing the overall sensitivity of the test method but were excluded from the OECD list because available *in vivo* data do not allow their corrosive classification to be subcategorized

10.3.2 Predictive Capacity

The predictive capacity of the SkinEthic™ RHE test method for 84 chemicals, tested in three independent experiments, is described in Table 10.2. Detailed specifications on tested chemicals, cell viabilities, and predictive capacities are given by Alépée et al. [13].

Among the 30 OECD reference chemicals, 8 non-corrosive chemicals were correctly predicted as such and 2 were false positives, corresponding to a specificity of 80%. For corrosive chemicals, the sensitivity was 100% since all were correctly classified. Overall accuracy over 25 test chemicals was 93.3%. Furthermore, discrimination involving UN GHS subcategory 1A versus 1B and 1C versus non-corrosives leads to 83.3% concordance of classifications (75 concordant predictions in the three classes out of the 90 tests).

Of the 84 chemicals identified by the OECD Expert Panel on Skin Corrosion and the four additional chemicals from the SCVS, the SkinEthic™ RHE test method showed 74.6% specificity (85/114). None of the 38 non-corrosive tested chemicals was overclassified as UN GHS category 1A. Overall, the SkinEthic™ RHE test method showed 94.9% sensitivity (131/138). The accuracy of SkinEthic™ RHE in distinguishing between corrosive and non-corrosive chemicals was 85.7% (216/252).

122 out of 129 corrosive classifications (36 category 1A and 93 categories 1B and 1C from all OECD chemicals) were correctly classified and 7 were false negative. This leads to 94.6% sensitivity. All false-negative classifications were *in vivo* GHS subcategory 1B and 1C chemicals, showing that none of UN GHS subcategory 1A chemicals was under-classified as non-corrosive. Furthermore, of the 111 non-corrosive classifications, 82 were correctly predicted as such, which corresponds to 73.9% specificity. However, 29 false positives (26.1%) were classified as skin corrosion UN GHS subcategories 1B and 1C with no over-prediction as subcategory 1A.

Concerning the 93 chemicals in UN GHS subcategory 1B and 1C classifications, 59 chemicals were correctly predicted as subcategories 1B and 1C (63.4%), 27 were overclassified as UN GHS subcategory 1A (29.0%), and 7 were under-classified as non-corrosive (7.5%). Regarding the identification of UN GHS subcategory 1A for the 36 subcategory 1A classifications, 29 were correctly predicted (80.6%), with seven under-classifications as UN GHS subcategories 1B and 1C (19.4%). No subcategory 1A was under-predicted as non-corrosive. Overall accuracy in distinguishing between UN GHS subcategory 1A, UN GHS subcategories 1B and 1C, and non-corrosive chemicals was 70.8% (170/240).

10.3.3 Applications and Limitations

The SkinEthic™ RHE skin corrosion test method is applicable to both chemicals and mixtures, although only limited information on the testing of mixtures is available, in contrast to chemicals for which information is available for a wide range of chemical classes. The 84 tested chemicals include 26 organic acids, 9 inorganic acids, 2 acid mixtures, 12 organic bases, 3 inorganic bases, 24 neutral organics, 6 inorganic salts, and 2 soap/surfactants. Physical states include liquids, solids, and semi-solids. Liquids may be aqueous or nonaqueous; solids may be soluble or insoluble in water. However, the method is not appropriate for testing gases and aerosols (even though this is true for all reconstructed human epidermis test methods, including OECD TG 431).

Evaluation of chemicals that act directly on MTT, such as MTT reducers (those that are naturally coloured or become coloured during tissue treatment), could be performed. Nevertheless, adapted controls—as described in the test method Standard Operating Procedure (SOP)—are needed to define non-specific MTT reduction and non-specific colour inherent to the chemical or developed once the chemical has been applied to the tissue. Using standard spectrophotometry as the read-out results obtained for chemicals inducing high non-specific MTT reduction and/or non-specific colour should be taken with caution. However, the use of the alternative HPLC/UPLC spectrophotometry as an endpoint measurement instead of the optical density (OD) measurement allows the determination of the cell viability for strongly coloured chemicals enhancing the applicability domain to this chemical type (see Sect. 10.6.2).

10.3.4 Comparison to Human/*In Vivo* Data

Skin corrosion produces irreversible damage. However, human data are usually not sufficient to subcategorize chemicals according to their corrosion potential (e.g. UN GHS subcategories 1A, 1B, and 1C) as required in some regulatory frameworks and legislation. A clear case for subcategory 1A classification (corresponding to 3 min contact in *in vivo* protocol involving rabbits) would be an accidental splash that gives rise to skin necrosis. In cases where prolonged exposure is needed before necrosis occurs (not to be confounded with delayed effects), subcategories 1B and 1C seem more reasonable. The distinction between category 1B and category 1C (corresponding to 1 and 4 h exposure in rabbits, respectively) may not be so obvious in practice. Simple classification as Cat.1 (without subcategorization) should be used, if distinction between category 1A and categories 1B and 1C is not clearly apparent. Different comparisons were performed and performance was determined (see paragraph on Predictive Capacity, above).

Despite the fact that information might be gained from evaluation of single parameters within a tiered approach (e.g. caustic alkalies with extreme pH should be considered dermal corrosives), it is worth taking into account all available information and making a hazard assessment based on the overall weight of evidence. Particular care must be taken in classifying certain types of chemicals such as acids and bases, inorganic salts, aldehydes, phenols, and surfactants. For mixtures containing strong acids or bases, the pH should be used as a classification criterion since it is a better indicator of corrosion. Such correlation has been proven using reconstructed human epidermis test methods [22].

10.4 Brief Description of the Protocol

The test chemical is applied topically to a three-dimensional SkinEthic™ model, consisting in fully differentiated epidermal tissue grown from normal human keratinocytes in a chemically defined medium at the air-liquid interface. The tissue is organized in basal, spinous and granular cell layers, but also presents a multi-layered functional *stratum corneum* containing intercellular lamellar lipid layers close to

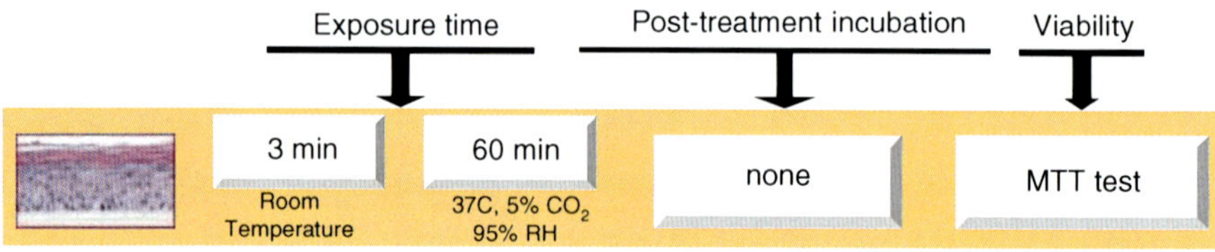

Fig. 10.2 *In vitro* skin corrosion SkinEthic™ RHE testing method

the main lipid classes found *in vivo* [8]. The SkinEthic™ RHE skin corrosion test method is performed using a SOP in accordance with the relevant OECD TG 431 [16, 24]. Key details and components of the protocol to perform the skin corrosion test are available at http://www.episkin.com.

Briefly, for each exposure time (3 min and 60 ± 5 min), liquid/viscous (80 μL/cm²) or solid/sticky chemicals (40 mg/cm²) are applied to RHE tissue samples (Fig. 10.2). Negative (sterile H_2O treated tissue sample) and positive (8 N KOH treated tissue sample) controls are evaluated concurrently with chemicals. At the end of each exposure time (3 min at room temperature and 60 min at 37 °C, with 5% CO_2 and 95% saturated humidity), the RHE tissue samples are rinsed with PBS without Ca^{++} and Mg^{++}. Tissue viability is then assessed via MTT reduction by incubating SkinEthic™ RHE tissue samples in a MTT solution (1 mg/mL) for 180 ± 15 min at 37°c, with 5% CO_2 and a >95% humidified atmosphere. After 2 h or overnight extraction at room temperature, the formazan crystals are dissolved in isopropanol. Quantification of cell viability after 3 min and 1 h treatment of the tissue is then obtained by comparing the optical density of the extracts measured at 570 nm and expressed as a percentage relative to the negative control (treated with H_2O).

The experiment is qualified when the acceptance criteria described in the standard operating procedure and in OECD TG 431 are met, i.e.: (1) negative control (sterile H_2O treated tissue sample) is qualified if mean OD is ≥0.8 and ≤3.0; (2) positive control (8 N KOH treated tissue sample) is qualified if it is classified as corrosive at 1 h; and (3) the difference in viability between the two tissue replicates should not exceed 30% in the range of 20–100% viability, and for optical density ≥0.3.

The prediction model used to classify a test chemical as non-corrosive, UN GHS skin corrosive subcategory 1A or skin corrosive subcategories 1B and 1C is shown in Table 10.3. A chemical is classified as "skin corrosive subcategory 1A" if relative tissue viability has decreased below 15% after 3 min of treatment irrespective of the result obtained at 1 h of treatment. If tissue viability is ≥15% after 3 min Treatment, but decreases below 15% after 1 h of treatment, the chemical is classified as "skin corrosive subcategories 1B and 1C". Finally, a chemical is classified as "non-corrosive" if the relative tissue viability is ≥15% after 3 min and 1 h treatment.

Table 10.3 Predictions models (PM) applied to the cell viability values obtained using the SkinEthic™ RHE test method to classify test chemicals into the EU CLP/UN GHS optional subcategories

Exposure Time	Viability (% of negative control)		UN GHS / EU CLP Subcategory
3 min	<15 % (independent of result obtained at 1 hour)	→	Skin corrosive Category 1A
3 min & 1 hour	≥15% at 3 min. and <15% at 1 hour	→	Skin corrosive Category 1B-and-1C
3 min & 1 hour	≥15% at 3 min. and ≥15% at 1 hour	→	Non Corrosive

From Desprez et al., 2015

Step1: Corr. vs. Non-Corr.	{v3<50} OR {v3≥50 AND v60<15} → Corr.		v3≥50 AND v60≥15 → Non-Corr.
Step2: Cat. 1A vs. Cat. 1B/1C	v3<x → Cat. 1A	v3≥x → Cat. 1B/1C	—

Establishment of composite indicator 'vfin'		**Theoretical range of values for 'vfin'**
v3<50	→ vfin = v3	[0; 50]
v3≥50 AND v60<15	→ vfin = v3 + v60	[50; 115]
v3≥50 AND v60≥15	→ vfin = (2 × v3) + v60	[115; 300]

Criteria	**Prediction**
vfin<y	Cat. 1A
y≤vfin<z	Cat. 1B-and1C
vfin≥z	Non-Corrosive

10.5 Role in a Testing Strategy

In 2014, the OECD adopted the first Guidance Document (GD) on an Integrated Approach to Testing and Assessment (IATA) for Skin Corrosion and Skin Irritation [21]. The GD proposes an IATA to identify chemicals' potential to cause skin corrosion or skin irritation based on classification and labelling according to UN GHS [4]. The IATA comprises modules consisting of non-testing information (e.g. existing human/*in vivo*/*in vitro* skin irritation and corrosion data, QSARs, and read-across), weigh-of-evidence analyses, and prospective *in vitro* testing. With regard to *in vitro* testing, OECD IATA introduces the possibility to test the *in vitro* skin irritation/corrosion potential of a chemical using either a top-down approach (an *in vitro* skin corrosion test followed by an *in vitro* skin irritation test) or a bottom-up approach (an *in vitro* skin irritation followed by an *in vitro* skin corrosion test).

In the context of testing strategy, the SkinEthic™ RHE skin corrosion test method is not a stand-alone method. This method must be coupled with the irritation test method for a top-down or bottom-up approach. In the top-down approach, if the result concludes that the substance is corrosive, then the chemical belongs to category 1. If the conclusion is non-corrosive, an *in vitro* irritation test must be conducted. Based on the outcome of this test, the chemical is then either classified as category 2 or is nonclassified. In contrast, in the bottom-up approach, if the chemical is identified as non-irritant, it is considered as nonclassified; otherwise, a corrosion test is necessary. If the result of this test is corrosive, then the chemical is considered category 1; if non-corrosive, it is considered a category 2 (Fig. 10.3).

The *in vitro* SkinEthic™ RHE methods have been formally validated and adopted for the regulatory assessment of skin irritation (OECD TG 439) and corrosion (OECD TG 431) of chemicals [17, 21, 23]. A study based on the SkinEthic™ RHE skin irritation and corrosion data on 83 chemicals, used either in a sequential bottom-up or top-down testing strategy, assessed their predictive capacity (Fig. 10.3). For the EU CLP system, both strategies showed an accuracy of 90.0% to distinguish nonclassified from classified chemicals. Furthermore, a 76% accuracy was found in distinguishing between UN GHS subcategory 1A, subcategories 1B and 1C, irritants (Cat.2), and nonclassified (NC) chemicals out of the 83 tested chemicals [25]. 7 out of 43 chemicals were over-predicted as either irritant ($n = 4$) or a corrosive ($n = 3$). Only one false negative result (methyl palmitate, UN GHS Cat.2) was obtained, which is a chemical consistently identified as false negative (No category) in all reconstructed human epidermis test methods included in the OECD *in vitro* Test Guidelines.

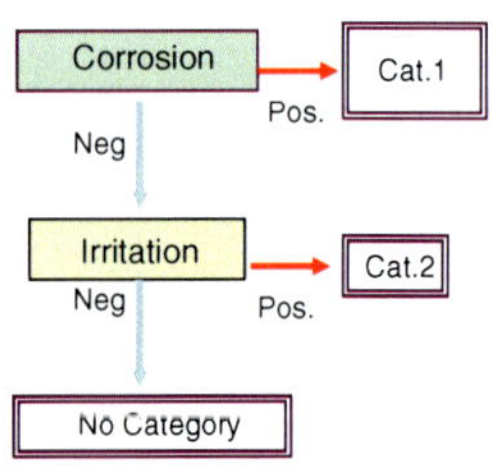

EU CLP – Top Down		TG 404 Non Class.	Cat. 2	Cat. 1B-1C	Cat. 1A	Sum
TG 431	Cat. 1A	0	0	2	9	11
↓	Corrosive	0	0	1	0	1
TG 439	Cat. 1B-1C	0	0	18	10	28
	Cat. 2	1	10	2	0	13
	Non Class.	25	4	1	0	30
Sum		26	14	24	19	83

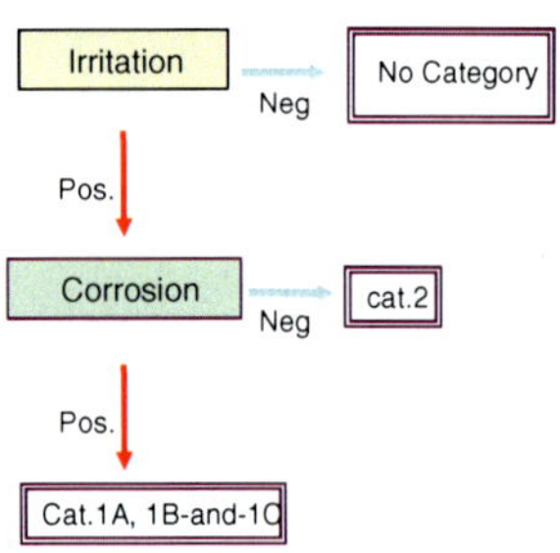

EU CLP – Bottom Up		TG 404 Non Class.	Cat. 2	Cat. 1B-1C	Cat. 1A	Sum
TG 439	Non Class.	25	4	1	0	30
↓	Cat. 2	1	10	2	0	13
TG 431	Cat. 1B-1C	0	0	18	10	28
	Corrosive	0	0	1	0	1
	Cat. 1A	0	0	2	9	11
Sum		26	14	24	19	83

Fig. 10.3 Main steps of the IATA recommended in OECD GD 203 and contingency table for 83 tested chemicals based on the EU-CLP classification system (median classification applied)

The integration of SkinEthic™ RHE skin irritation and corrosion data in a bottom-up or top-down strategy showed a similar high accuracy for the determination of the potential hazard of chemicals. The sequence of testing should be guided by a weight-of-evidence approach as recommended in the OECD IATA.

10.6 Perspectives from the Test Developer

10.6.1 Critical Steps in the Protocol

The critical steps of the standardized operating procedure include the following:

- The absence of air bubble under the epidermis should be verified at each step.
- For liquids (40 ± 3 μL), the test chemical is dispensed onto the epidermis with a positive displacement pipette, and a nylon mesh is applied to gently spread the substance; be sure to cover all the surface.
- For solids (20 ± 2 μL H_2O and 20 ± 3 mg test item), the test chemical should be crushed to a fine powder. Ensure good contact with the epidermis.
- Viscous and sticky chemicals are applied using a curved flat spatula or weighed directly on the nylon mesh; apply the chemical-coated side of the nylon mesh to the epidermal surface.
- The nylon mesh should be carefully removed before rinsing.
- Tissues should be rinsed thoroughly.
- Plates should be thoroughly protected by stretching 3 parafilm layers over the plate to prevent the evaporation of the formazan during the extraction step.

10.6.2 Possible Protocol Adaptations

In the *in vitro* SkinEthic™ RHE test method, the skin corrosion potential of a chemical is determined by measuring cell viability, which is determined using the MTT assay by enzymatic reduction of MTT tetrazolium salt to reduced MTT (formazan). Formazan is quantified photometrically with the results being expressed as the percentage viability of the chemical-treated tissues relative to the negative control. A known limitation of the MTT photometric assay is possible interference by, for example, coloured test chemicals and chemicals that have the ability to directly reduce the MTT with optical density absorbance (OD) measurement of formazan. To address this limitation, Cosmetics Europe undertook a project that evaluated the use of HPLC/UPLC as an endpoint detection system to extend the applicability of *in vitro* RhT test methods to include strongly coloured chemicals [26]. Twenty six chemicals chosen to create a balanced chemical set of coloured and noncoloured chemicals were tested using the SkinEthic™ RHE skin corrosion test method. The formazan solvent extracts were measured by OD within L'Oréal R&I that performed the *in vitro* test method and by HPLC/UPLC in three participating laboratories. The standard deviation data, well below 18%, demonstrated a very high level of reproducibility between the participating laboratories in measuring the formazan extract samples from the *in vitro* skin corrosion test method using HPLC/UPLC. For 92% (24 out of 26 test chemicals), high concordance for the measurement of cell viability using the OD or HPLC/UPLC detection systems was observed (difference < 5%). The two remaining chemicals (CAS# 74578-10-2 and #176429-22-4), which were incompatible with OD since the non-specific colour was ≥50% at 1 h, could be correctly classified using HPLC/UPLC, bypassing the known limitation of the MTT photometric assay. A comparison of these classifications for all test chemicals (24) for which measurements could be made in both endpoint detection systems (considered compatible with OD measurement) identified 100% concordance. On the basis of the results obtained, it was concluded that this HPLC/UPLC analytical endpoint detection system is relevant to the *in vitro* SkinEthic™ RHE test method but also to all reconstructed human tissue test methods irrespective of the model used within the OECD Test Guideline 431 [17].

10.6.3 Challenges and Opportunities

Corrosivity is not a risk factor that usually occurs with cosmetics. But corrosivity could occasionally arise after a manufacturing error, misuse of chemicals by the consumer, or the unrestricted transport of chemicals. As a result, additional data to support negative predictions might be requested depending on national requirements and legislation. In this case, in the context of the acute topical irritation effect, a chemical that is not predicted to cause corrosive effects by the SkinEthic™ RHE test method would require additional testing to establish a definitive classification as either irritant or non-corrosive (also relevant to all RHE test methods).

OECD TG 431 allows for the subcategorization of corrosive chemicals into category 1A or categories 1B and 1C but does not permit the distinction of the latter into category 1B and category 1C due to the lack of high-quality *in vivo* reference data against which to benchmark the *in vitro* results. Scientific evaluation of the capacity of all RHE test methods is therefore hampered by the lack of relevant (human and animal) reference data.

Opportunities might also be explored by considering the applicability of the method to hazard identification for lower-concentrated mixtures without further testing when the evaluation of the mixture has been already performed using the bridging principle. Briefly, a mixture will be classified using the criteria for substances, taking into account the testing and evaluation strategies to develop data for the skin corrosion endpoint. If a tested mixture classified in the highest subcategory for corrosion is concentrated, a more concentrated mixture should be classified in the highest corrosion subcategory without additional testing. Moreover, if a tested mixture classified in the highest category for skin/eye irritation is concentrated and does not contain corrosive ingredients, a more concentrated mixture should be classified in the highest irritation category without additional testing. If a mixture is diluted with a diluent which has an equivalent or lower corrosivity/irritancy classification than the least corrosive/irritant original ingredient and which is not expected to affect the corrosivity/irritancy of other ingredients, then the new mixture may be classified as equivalent to the original mixture. Further testing might be needed to confirm this, preferably using an appropriate validated *in vitro* test before any regulatory acceptance.

10.7 Conclusions

Currently, internationally accepted test methods for skin corrosion testing include the traditional animal test (Draize rabbit test) as well as *in vitro* test methods, including validated test methods based on reconstructed human epidermis technology. The SkinEthic™ RHE skin corrosion test method has gained international regulatory acceptance and has been adopted for the regulatory assessment of skin corrosion to distinguish between UN GHS subcategories 1A, 1B and 1C, and non-corrosives (OECD TG 431). Intra- and inter-reproducibility findings indicate that the SkinEthic™ RHE model has high robustness in terms of its performance with an enlarged dataset of diverse chemicals. Furthermore, the relevance of the integration of SkinEthic™ RHE skin corrosion data in either a bottom-up or a top-down strategy (as recommended in [21]) has been demonstrated to have a similar high accuracy for the determination of the potential hazard of chemicals. The sequence of testing of a chemical should be guided by a weight-of-evidence approach as recommended in the OECD IATA.

References

1. EC Regulation (EC) No 1272/2008 of the European Parliament and of the Council of 16 December 2008 on classification, labelling and packaging of substances and mixtures, amending and repealing Directives 67/548/EEC and 1999/45/EC, and amending Regulation (EC) No 1907/2006. Off J Eur Union. 2008;L353:1–1355.
2. EC. Regulation (EC) No. 1223/2009 of the European Parliament and of the Council of 30 November 2009 on cosmetic products. Off J Eur Union. 2009;L342:59–209. Available at: http://eur-lex.europa.eu/legal-content/EN/TXT/?uri=CELEX:32009R1223
3. EC. REGULATION (EC) No. 1907/2006 of the European Parliament and of the Council of 18 December 2006 concerning the Registration, Evaluation, Authorisation and Restriction of chemicals (REACH), establishing a European Chemicals Agency, amending Directive 1999/45/EC and repealing Council Regulation (EEC) No. 793/93 and Commission Regulation (EC) No. 1488/94 as well as Council Directive 76/769/EEC and Commission Directives 91/155/EEC, 93/67/EEC, 93/105/EC and 2000/21/EC. Off J Eur Union. 2006;L396:1–1355.
4. UN. United Nations globally harmonized system of classification and labelling of chemicals (GHS), Sixth revised edition. New York: UN; 2015a.
5. United Nations, UN. Part 2. Classification. In: Recommendations on the transport of dangerous goods – model regulations – Volume 1. Nineteenth revised edition. 2015b.
6. Draize JH, Woodard G, Clavery HO. Methods for the study of irritation and toxicity of substances applied topically to the skin and mucous membranes. J Pharmacol Exp Ther. 1944;82:377–90.
7. Rosdy M, Clauss L-C. Terminal epidermal differentiation of human keratinocytes grown in chemically defined medium on inert filter substrates at the air–liquid interface. J Invest Dermatol. 1990;95:409–14.
8. Rosdy M, Pisani A, Ortonne J-P. Production of basement membrane components by a reconstructed epidermis cultured in the absence of serum and dermal factors. Br J Dermatol. 1993;129:227–34.

9. EC ECVAM. Statement on the application of the SkinEthic™ human skin model for skin corrosivity testing, issued by the ECVAM Scientific Advisory Committee (ESAC25), 17 November 2006. 2006. Available at: http://ecvam.jrc.ec.europa.eu.
10. OECD. *In vitro* skin irritation: reconstructed human epidermis test method. In: OECD guideline for the testing of chemicals No. 439. Paris: OECD; 2010. http://www.oecd-ilibrary.org/environment/test-no-439-in-vitro-skinirritation_9789264090958-en.
11. Mosmann T. Rapid colorimetric assay for cellular growth and survival: application to proliferation and cytotoxicity assays. J Immunol Methods. 1983;65:55–63.
12. Eskes C, Detappe V, Koëter H, Kreysa J, Liebsch M, Zuang V, Amcoff P, Barroso J, Cotovio J, Guest R, Hermann M, Hoffmann S, Masson P, Alépée N, Arce LA, Brüschweiler B, Catone T, Cihak R, Clouzeau J, D'Abrosca F, Delveaux C, Derouette JP, Engelking O, Facchini D, Fröhlicher M, Hofmann M, Hopf N, Molinari J, Oberli A, Ott M, Peter R, Sá-Rocha VM, Schenk D, Tomicic C, Vanparys P, Verdon B, Wallenhorst T, Winkler GC, Depallens O. Regulatory assessment of *in vitro* skin corrosion and irritation data within the European framework: workshop recommendations. Regul Toxicol Pharmacol. 2012;62(2):393–403.
13. Alépée N, Robert C, Tornier C, Cotovio J. The usefulness of the validated SkinEthic™ RHE test method to identify skin corrosive UN GHS subcategories. Toxicol *In Vitro*. 2014;28:616–25.
14. Desprez B, Barroso J, Griesinger C, Kandárová H, Alépée N, Fuchs HW. Two novel prediction models improve predictions of skin corrosive sub-categories by test methods of OECD Test Guideline No. 431. Toxicol *In Vitro*. 2015;29:2055–80.
15. Tornier C, Roquet M, Fraissinette AB. Adaptation of the validated SkinEthic™ Reconstructed Human Epidermis (RHE) skin corrosion test method to 0.5 cm^2 tissue sample. Toxicol *In Vitro*. 2010;24:1379–85.
16. OECD. Summary document on the statistical performance of methods in OECD test guideline 431 for sub-categorisation. In: OECD series on testing and assessment No. 190. Paris: OECD; 2013.
17. OECD. *In vitro* skin corrosion: reconstructed human epidermis test method. In: OECD guideline for the testing of chemicals No. 431. Paris: OECD; 2016. Originally adopted in 2004. Available at http://www.oecd.org/env/testguidelines.
18. Kandárová H, Liebsch M, Spielmann H, Genschow E, Schmidt E, Guest R, Whittingham A, Warren N, Gamer A, Remmele M, Kaufmann T, Wittmer E, De Wever B, Rosdy M. Assessment of the SkinEthic Reconstituted Human Epidermis for skin corrosion testing according to OECD guideline 431. Toxicol *In Vitro*. 2006;20:547–59.
19. ESAC. Statement on the "Performance under UN GHS of three *in vitro* assays for Skin Irritation testing and the Adaptation of the Reference chemicals and defined accuracy values of the ECVAM Skin Irritation Performance Standards". 2009. Available at: http://ihcp.jrc.ec.europa.eu/our_labs/eurl-ecvam/validation-regulatoryacceptance/docs-skin-irritation-1/DOC6-ESAC2009.pdf. (Accessed 15 July 13).
20. EC. Commission Directive 2001/59/EC of 6 August 2001 adapting to technical progress for the 28th time Council Directive 67/548/EEC on the approximation of laws, regulations and administrative provisions relating to the classification, packaging and labelling of dangerous substances. Off J Eur Union. 2001;L225:1–333.
21. OECD. Guidance document on integrated approaches to testing and assessment of skin irritation/corrosion. In: Series on testing and assessment, No. 203. Paris: OECD; 2014. Available at: http://www.oecd.org/env/ehs/testing/seriesontestingandassessmenttestingforhumanhealth.htm.
22. Basketter D, Jírová D, Kandárová H. Review of skin irritation/corrosion hazards on the basis of human data: a regulatory perspective. Interdiscip Toxicol. 2012;5:98–104.
23. OECD. *In vitro* skin irritation: reconstructed human epidermis test method. Originally adopted in 2010. Paris: OECD; 2015. Available at: http://www.oecd-ilibrary.org/environment/oecd-guidelines-for-the-testing-of-chemicals-section-4-health-effects_20745788.

24. SkinEthic™ RHE Skin Corrosion Test. 2017. Available at: [http://www.ecvam-dbalm.jrc.ec.europa.eu/].
25. Alépée N, Grandidier MH, Tornier C, Cotovio J. An integrated testing strategy for *in vitro* skin corrosion and irritation assessment using SkinEthic™ Reconstructed Human Epidermis. Toxicol *In Vitro*. 2015b;29:1779–92.
26. Alépée N, Barroso J, De Smedt A, De Wever B, Hibatallah J, Klaric M, Mewes KR, Millet M, Pfannenbecker U, Tailhardat M, Templier M, McNamee P. Use of HPLC/UPLC-spectrophotometry for detection of formazan in *in vitro* reconstructed human tissue (RhT)-based test methods employing the MTT-reduction assay to expand their applicability to strongly coloured test chemicals. Toxicol *In Vitro*. 2015a;29:741–61.

The epiCS® Skin Corrosion Test Method 11

Oliver Engelking, Dirk Weisensee, and Horst W. Fuchs

11.1 Introduction and Principle of the Test Method

In vitro assessment of the human health endpoint skin corrosion makes use of reconstructed human epidermis (RhE), which is obtained from normal human epidermal keratinocytes. During the *in vitro* culture of the keratinocytes, there are several steps that lead to differentiation from monolayer cell cultures into a 3D model which closely mimics the histological, morphological, biochemical and physiological properties of the upper parts of the human skin, i.e. the epidermis. At the air-liquid interface, the RhE models develop a highly differentiated structure comprising viable cell layers forming the stratum basale, stratum spinosum and stratum granulosum in the lower regions and several layers of finally differentiated keratinocytes forming the stratum corneum at the apical site (5) [2].

As recommended by the OECD guideline TG 431, epiCS® meet the general conditions of a multilayered tissue with a functional stratum corneum and the functional conditions of a stable and sufficiently high cell viability, skin barrier and reproducibility within or between laboratories. Due to this reconstructed skin barrier which resembles the barrier *in vivo* very closely, this test method was established to replace the assessment of skin corrosivity using laboratory animals.

The principle of the RhE *in vitro* test method for skin corrosion is based on the hypothesis that corrosive chemicals are able to penetrate the stratum corneum by diffusion or erosion and are cytotoxic to the underlying cell layers. Skin corrosion refers to the production of irreversible damage to the skin manifested as visible necrosis through the epidermis (and into the dermis) following the application of a test chemical. This was defined by the UN GHS (United Nations Globally Harmonized System of Classification and Labelling of Chemicals) [3] and provided

O. Engelking (✉) • D. Weisensee • H.W. Fuchs
CellSystems GmbH, Langeler Ring 5, 53842 Troisdorf, Germany
e-mail: oliver.engelking@cellsystems.de

C. Eskes et al. (eds.), *Alternatives for Dermal Toxicity Testing*,
DOI 10.1007/978-3-319-50353-0_11

the basis for an *in vitro* procedure that allows the identification of non-corrosive (NC) and corrosive (Cat. 1) substances and mixtures as described in the test guideline OECD TG 431 (2016). After refinement of the method using a larger set of chemicals, it became possible to discriminate also between the corrosive subcategory 1A and a combination of subcategories1B and 1C by several RhE test methods including epiCS®. The epiCS® test method has shown to have a high reliability and relevance for specific testing purposes like skin corrosion and skin irritation testing (see Chap. 6 on epiCS® skin irritation test (SIT) method) as well as *in vitro* skin sensitization testing (see Chap. 20 on epidermal equivalent (EE) potency assay). The functional conditions of the epiCS® test method are very well characterized and include viability, barrier function, morphology, reproducibility and quality control, demonstrating high performance and predictive capacity of the epiCS® test system.

The degree of tissue necrosis is determined by the MTT assay [4]. This assay measures cell viability of the tissues after topical exposure with the test chemicals for 3 min or 60 min. The tetrazolium salt is reduced by intracellular dehydrogenases and converted into coloured formazan. Two or three tissues are used per treatment together with the negative control and positive control. After exposure, tissues are rinsed and blotted, and assay medium is replaced by MTT assay medium. After 3 h MTT incubation, tissues are washed with PBS and blotted, and the blue formazan salt is extracted with isopropanol. The optical density of the formazan extract is determined spectrophotometrically at 540–570 nm, and the cell viability is calculated for each tissue as percent of the mean of the negative control tissues. The corrosivity potential of unknown chemicals can be predicted from the mean tissue viabilities obtained after 3 and 60 min exposure in comparison to the negative control tissues treated with H_2O. A chemical or mixture is classified as corrosive if the relative tissue viability is decreased below 50% after a 3 min exposure period (UN GHS Cat. 1). Furthermore, a test chemical classified as non-corrosive with a value ≥50% viability after 3 min will be classified corrosive (UN GHS Cat. 1) if the relative tissue viability decreases below 15% after exposure for 60 min. Finally for those test chemicals identified as corrosives, if cell viability at 3 min is <15%, the test chemical is predicted to be within the corrosive subcategory 1A, whereas if cell viability at 3 min exposure is ≥15%, the test chemical is predicted to be within the corrosive combined subcategories 1B and 1C [1].

In some cases enough information may be available from structurally related compounds to make classification decisions. Likewise, pH extremes like ≤2 and ≥11.5 may indicate dermal corrosive effects. However, the use of extreme pH alone as a marker to classify an acid-containing product as corrosive to skin might result in over-prediction. Consequently, it is recommended to test substances or mixtures having extreme pH, in a validated *in vitro* test method to investigate the corrosive capacity or to confirm a non-corrosive classification.

11.2 Current Validation Status

The *in vitro* skin corrosion test makes use of reconstructed human epidermis (RhE) test methods and is described by the OECD guidelines for the testing of chemicals (OECD TG 431, 2016).

Table 11.1 Predictive capacity (specificity, sensitivity and overall accuracy) of the epiCS® (formerly EST1000) skin corrosion test method (corrosive vs. non-corrosive)

Specificity (%)	84.7 (61/72)
Sensitivity (%)	100 (72/72)
Overall accuracy (%)	92.4 (133/144)

An inter-laboratory study with the epiCS® human reconstituted epidermis (RhE) model (formerly EST-1000) was conducted and reviewed by an independent ECVAM Scientific Advisory Committee (ESAC) Peer Review [5]. Based on the study results, the non-commission members of the ECVAM Scientific Advisory Committee endorsed on 12 June 2009 the statement that the epiCS® test method for skin corrosion testing can be used for reliably predicting the corrosive potential of chemical substances. The epiCS® test method was therefore considered to meet the Performance Standards as determined in the OECD test guideline TG 431 on *in vitro* skin corrosion testing using human skin model tests [5].

In this context, epiCS® was validated successfully with 12 reference chemicals and obtained regulatory approval for skin corrosion testing in accordance with the OECD test guideline 431 [1, 5]. On the basis of the individual predictions of the four participating testing laboratories for the 12 reference chemicals (three tests per reference chemical per laboratory), the predictive capacity shown in Table 11.1 was obtained.

In this study each of the 12 chemicals were tested three times in four laboratories. The total number of test results was therefore 144, with 72 results concerning non-corrosives ($n = 6$ chemicals) and 72 results concerning corrosives ($n = 6$ chemicals). In this validation study, 11 out of 12 reference chemicals of the OECD TG 431 were correctly predicted using epiCS®. Only tetrachloroethylene was incorrectly predicted by three laboratories as a skin corrosive (false-positive prediction), with the forth laboratory making a correct prediction as non-corrosive.

Recently, the EU Classification, Labelling and Packaging Regulation (EU CLP) system required the subcategorization of corrosive chemicals into the UN GHS optional subcategories 1A and 1B and 1C. This is mainly relevant for regulation on transportation of goods, whereas protection measures for human health are not affected by subcategorization, i.e. the protection measures for human health remain the same independent of the corrosive subcategory.

A further study assessed whether the epiCS® skin models can reproducibly discriminate the corrosive subcategory 1A (strong corrosive) from combined subcategories 1B and 1C (weak corrosive). Eighty chemicals including solids, semi-solids and liquids of different chemical classes (e.g. electrophiles, organic bases and acids, neutral organics, surfactants, inorganic salts and acids, phenols), selected by the OECD expert group on skin corrosion, were tested in two independent runs. Freeze-inactivated tissues were used to correct for direct MTT reduction and interference by colouring agents. The results are shown in Sect. 15.3.2, demonstrating high predictive capacity values.

The classification is based on the following two-step prediction model:

Step 1: Non-corrosive (NC) classification is achieved if the viability is ≥50% after 3 min and ≥15% after 1 h of chemical exposure. A chemical is classified corrosive if the viability is <50% after 3 min and/or <15% after 1 h of exposure.

Step 2: For those test chemicals identified as corrosives in Step 1, if cell viability at 3 min is <15%, the test chemical is predicted to be within the optional corrosive subcategory 1A, whereas if cell viability at 3 min exposure is ≥15%, the test chemical is predicted to be within the optional corrosive combined subcategories 1B and 1C [1].

The possibility for subcategorization was adopted within OECD TG 431 which included the epiCS® test method for the subcategorization of corrosive substances and mixtures into optional category 1A, in accordance with the UN GHS [3], as well as a combination of categories 1B and 1C [1, 6].

11.3 Performance and Applicability of the Test Method

11.3.1 Predictive Capacity

With regard to the latest testing guideline, the results with the epiCS® test method obtained after testing 80 chemicals in two independent runs demonstrate correctly classified corrosive and non-corrosive chemicals with a high sensitivity (87, 50%) and specificity (71, 62%). The overall accuracy regarding subcategorization into UN GHS subcategories 1A and 1B and 1C is 69, 81%. According to the current prediction model using the two-step approach (see above), which is described in the latest testing guideline [1], predictions for category 1B and 1C chemicals could be further improved (Tables 11.2 and 11.3).

The results demonstrate that the epiCS® skin corrosion test method is able to correctly identify corrosive and non-corrosive chemicals according to the UN GHS classification system and also distinguish between UN GHS category 1A and subcategory 1B and 1C chemicals with the current PM. Modifications of the original prediction model were adopted within the OECD TG 431 that resulted in improved prediction capacity especially for the identification of subcategory 1B and 1C chemicals [1, 7].

11.3.2 Applications and Limitations

The epiCS® *in vitro* skin corrosion test method is applicable to solids, liquids, semisolids and waxes. Liquids may be aqueous or nonaqueous; solids may be soluble or insoluble in water. The method is not applicable to gases and aerosols [1].

The epiCS® *in vitro* skin corrosion test method can be used for hazard identification of corrosive and non-corrosive chemicals (substances and mixtures) and further distinguish UN GHS category 1A from categories 1B and 1C.

A possible limitation is the interference of a test chemical with the endpoint MTT. To identify colour interference, spectral analysis of a coloured chemical in water (environment during exposure) and/or isopropanol (extracting solution) should be performed to evaluate if the test chemical requires additional controls. If the test chemical in water and/or isopropanol absorbs light in the range of 570 ± 30 nm, colourant controls should be performed. A test chemical may also directly reduce MTT, thus mimicking cellular dehydrogenase activity. This property

Table 11.2 Predictive capacity of *in vitro* subcategorization with the epiCS® skin corrosion test method (non-corrosive vs. 1A or 1B and 1C corrosivity) using the new prediction model

	epiCS® *In vitro* sub-categorization results		
In vivo results	1A	1B-and-1C	NC
1A	87.50% Correct prediction for 1A	12.50% 1A Under-predicted as 1B-and-1C	0.00% 1A Under-predicted as NC
1B-and-1C	32.79% 1B-and-1C Over-predicted as 1A	60.66% Correct prediction for 1B-and-1C	6.56% 1 B-and-1 C Under-predicted as NC
NC	0.00% NC Over-predicted as 1A	28.38% NC Over-predicted as 1B-and-1C	71.62% Correct prediction for NC
	Accuracy = 69.81%		

In this approach the corrosivity categories can be discriminated in a second step: a viability value <15% after 3 min exposure is assigned to category 1A; a viability value ≥15% after 3 min exposure is assigned to categories 1B and 1C

Table 11.3 Predictive capacity of *in vitro* subcategorization with the epiCS® skin corrosion test method (non-corrosive vs. 1A or 1B and 1C corrosivity) in comparison with other RhE methods

Statistics on entire set of chemicals (n = 80 chemicals tested over 2 or 3 runs, i.e. 159–240 classifications)		
	Other RhE models	epiCS®
Overclassifications		
1B and 1C overclassified 1A	21.5–31.2%	32.8%
NC overclassified 1B and 1C	20.7–27.0%	28.4%
NC overclassified 1A	0.0–2.7%	0.0%
NC overclassified corrosive	20.7–27.0%	28.4%
Global overclassification rate (all categories)	17.9–24.5%	25.8%
Underclassifications		
1A underclassified 1B and 1C	16.7%	12.5%
1A underclassified NC	0.0%	0.0%
1B and 1C underclassified NC	0.0–7.5%	6.6%
Global underclassification rate (all categories)	3.3–5.4%	4.4%
Correct classifications		
1A correctly classified	83.3%	87.5%
1B and 1C correctly classified	61.3–76.3%	60.7%
NC correctly classified	73.0–79.3%	71.6%
Accuracy (predictive capacity)	70.0–78.8%	69.8%

NC non-corrosive

of the test chemical is only a problem, if at the time of the MTT test (after the test chemical has been rinsed off), there is still sufficient amount of the test chemical present on (or in) the tissues. In this case the (true) metabolic MTT reduction and the (false) direct MTT reduction can be differentiated and quantified by a procedure described in the SOP epiCS® skin corrosion test ([8]—Sect. 6.2). In brief, the test chemical should be incubated for 3 h (37 ± 1 °C, 5 ± 1% CO_2, 95% RH) with the

MTT assay medium without the skin model to evaluate direct MTT-reducing activity of the test chemical. In case of direct interaction, the MTT-reducing chemicals are applied to freeze-killed tissues that possess no metabolic activity but absorb and bind the test substance similar to viable tissues. Each MTT-reducing test chemical is applied to two freeze-killed tissues. In addition, two freeze-killed tissues are left untreated, and the entire protocol is performed in parallel to the assay performed with viable epiCS® tissues. The freeze-killed epiCS® tissues should undergo 2–3 freeze-thaw cycles beforehand to ensure complete inactivation of intracellular reducing enzymes.

Other limitations include the fact that the surface of epiCS® being hydrophobic, so that even spreading of aqueous substances is sometimes not possible. When a nylon mesh is used as a spreading support for liquid test chemicals, it has to be considered that some test chemicals may react with the mesh itself. If an interaction between test substance and the mesh is noticed microscopically, the test substance has to be applied without using a mesh as a spreading aid. Finally, the method is not designed to be compatible with highly volatile test substances. However, possible toxic interference across plate wells can be avoided by sealing the wells with an adhesive cover sheet or testing volatile chemicals on separate plates.

11.4 Brief Description of the Protocol

Upon reception, tissues should be conditioned by pre-incubation for release of transport stress-related compounds and debris. After overnight pre-incubation tissues are transferred to fresh epiCS® culture medium and topically exposed with the test chemicals for 3 min and 60 min, respectively. Two or three tissues each are used per treatment, negative control and positive control. After exposure tissues are rinsed and assay medium is replaced by 300 μl MTT assay medium per tissue in a 24-well plate. Freshly prepared MTT is used at a concentration of 1 mg/ml. After 3 h incubation, tissues are washed with PBS and the blue formazan salt is extracted with isopropanol. The optical density of the formazan extract is determined spectrophotometrically at 540–570 nm, and cell viability is calculated for each tissue as % of the mean of the negative control tissues. Skin corrosivity potential of the test materials is classified according to the remaining cell viability obtained after 3 min or 60 min exposure with the test chemical.

11.4.1 60 min and 3 min Exposure in Detail

Liquids (50μl) are applied apically. In case of solids, the test material is grinded in a mortar with a pestle. 25 mg of test substance is applied apically using a sharp application spoon with fine ground test material. To increase the contact between tissue surface and the solid chemical, 50 μl H_2O should be applied first for wetting the tissue surface. To avoid possible toxic interference across wells, it is recommended to use one plate per chemical, in particular if volatile substances are tested.

8 N KOH serves as a positive control (PC) and should be tested once per testing day. One hour exposure of the PC should result in a mean tissue viability less than 20%.

Furthermore, it is recommended to start with the 60 min exposure first. For this purpose, 50 μl H_2O (negative control) is added onto the first epiCS® surface and if needed a mesh is applied above. After 1 min the procedure is repeated with the second tissue. After 2 min the third tissue is dosed and so on. After dosage of all epiCS® tissues (50 μl for liquids, 25 mg + 50 μl H_2O for solids), the 6-well plates are stored in the incubator (37 ± 1 °C, 5 ± 1% CO_2, 95% RH) for the rest of the exposure time until 1 h exposure is reached for the first tissue dosed. The mesh is removed and the tissue gently rinsed with a wash bottle with PBS (the insert is filled and emptied 20 times in a constant soft stream of PBS). All tissues are rinsed in an interval of 1 min. The excess PBS is removed by gently shaking the insert and blot bottom with blotting paper, and the insert is placed in the prepared holding plate containing culture medium. The inserts are transferred into the 24-well plate, which is prepared for the MTT assay (1 mg/ml MTT in assay medium, 300 μl per insert). The plate is then placed in the incubator for 3 hours (37 ± 1 °C, 5 ± 1% CO_2, 95% RH). A similar procedure is to be conducted for the 3 min exposure time, with the exception that all is done at room temperature.

11.4.2 MTT Procedure

After completion of the 3 h MTT incubation (1 mg/ml MTT in assay medium, 300 μl per insert), the inserts are removed, dried by blotting the bottom on blotting paper and transferred into a new 24-well plate containing 2 ml of the isopropanol extraction solution. The extraction plate is sealed and stored over night without shaking at 4–8 °C or with shaking (100 rpm) at room temperature for 2 h. Inserts are pierced and discarded, and 2 × 200 μl of the blue formazan extract is transferred into 2 wells of a 96-well plate per tissue. The read out of optical density (OD) is done in a spectrophotometer at 540–570 nm without reference filter, since the "classical" reference filter often used in the MTT test (630 nm) is still within the absorption curve of formazan.

Cell viability is calculated for each tissue as the percentage of the mean negative control cell viability. Skin corrosivity potential of the test materials is classified according to the cell viability obtained after 3 min or 60 min exposure with the test chemical according to the prediction model mentioned above (Sect. 15.2).

11.5 Perspectives from the Test Developer

The protocol for the epiCS® *in vitro* skin corrosion test method is robust and does not need further adaptations. Detailed instructions of how to perform the test are described in the SOP for epiCS® [8], and it is important to stick to the recommendations mentioned there. For example, the method is not designed to be compatible with highly volatile test substances, and therefore the use of one six-well plate per chemical is important to avoid cross contamination into neighbouring cavities. Moreover, the exact timeline for chemical incubation, washing and subsequent MTT incubation should be followed. Therefore, the intermediate step with a holding plate was introduced, before entering all treated tissues into the MTT medium at once.

In a further study, the data of different RhE methods were compared and the different prediction models (PM) for each test system evaluated. Optimization of the PM yielded an improved predictive capacity for each of the *in vitro* test methods [7].

11.6 Conclusions

Results demonstrate that the epiCS® skin corrosion test method is able to correctly identify corrosive and non-corrosive test chemicals and to distinguish between UN GHS category 1A and UN GHS category 1B and 1C test chemicals. The epiCS® skin corrosion test method with the most recent recommended prediction model, which led to higher accuracy of the test method and improved prediction capacity especially with respect to subcategorization into categories 1B and 1C, is mentioned in the latest revised version of the OECD TG 431 (2016).

References

1. OECD. Test Guideline (Nr. 431) for the testing of chemicals – *In vitro* skin Corrosion: reconstructed human epidermis (RHE) test method. 2016.
2. Hoffmann J, Heisler E, Karpinski S, Losse J, Thomas D, Siefken W, Ahr HJ, Vohr HW, Fuchs HW. Epidermal-skin-test 1000 (EST-1000)—a new reconstructed epidermis for *in vitro* skin corrosivity testing. Toxicol In Vitro. 2005;19:925–9.
3. UN. United Nations globally harmonized system of classification and Labelling of chemicals (GHS), 6h revised ed. New York and Geneva: UN; 2015. Available at: http://www.unece.org/trans/danger/publi/ghs/ghs_rev06/06files_e.html.
4. Mosmann T. Rapid colorimetric assay for cellular growth and survival: application to proliferation and cytotoxicity assays. J Immunol Methods. 1983;65:55–63.
5. ESAC. Peer review panel consensus report on the EST1000 in vitro test method for assessing skin corrosion in vitro. 2009.
6. OECD. Summary document on the statistical performance of methods in OECD test guideline 431 for sub-categorisation, OECD Environment, Health and safety Publications (EHS), series on testing and assessment, no. 190. Paris: OECD Publications; 2013.
7. Desprez B, Barroso J, Griesinger C, Kandárová H, Alépée N, Fuchs HW. Two novel prediction models improve predictions of skin corrosive sub-categories by test methods of OECD test guideline no. 431. Toxicol In Vitro. 2015;29:2055–80.
8. epiCS® SOP. Version 7.0. *In Vitro* Skin corrosion: human skin model test epidermal skin test 1000 (epiCS®) including sub-categorization of corrosive chemicals with epiCS. CellSystems, 2017. Available at: http://www.skininvitro.com.

The *In Vitro* Membrane Barrier Test Corrositex® for Skin Corrosion

12

Rich Ulmer and Amy Wang

12.1 Principle of the Test Method and Scientific Basis

Corrositex is a standardized, quantitative, *in vitro* test for skin corrosivity based upon the determination of the time required for a test material to pass through a biobarrier membrane. The test system is comprised of two components, a synthetic macromolecular biobarrier and a chemical detection system (CDS). The basis of this test method is that it detects membrane barrier damage caused by corrosive test substances [1], presumably by the same mechanism(s) of corrosion that operate on living skin. Membrane barrier damage caused by a corrosive test material after the application of the test material to the surface of the artificial membrane barrier is detected by the CDS, i.e., by either color change of the indicator solution or consistency change at sample/testing fluid interface below the barrier. The assay system is depicted in Fig. 12.1.

The scientific basis of the Corrositex method is the relationship between the ability of a chemical or formulation to destroy the integrity of the biobarrier and its corrosive potential *in vivo*. The description of major components and the mechanism of action of the Corrositex test are briefly shown in Figs. 12.2 and 12.3.

When a test material diffuses through the biobarrier, it mixes with and alters the pH of the CDS. A color change occurs when the pH of the CDS falls below 4.5 or rises

---

Rich Ulmer has been President, Chief Executive Officer, and Director of InVitro International since 1994. Amy Wang has been a Director of R&D of InVitro International since 2009.

R. Ulmer (✉) • A. Wang
InVitro International, Placentia, USA
e-mail: invitro@invitrointl.com

C. Eskes et al. (eds.), *Alternatives for Dermal Toxicity Testing*,
DOI 10.1007/978-3-319-50353-0_12

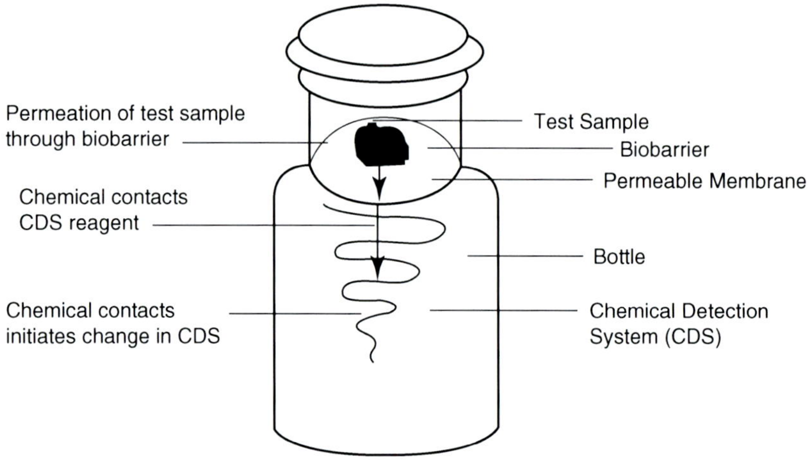

Fig. 12.1 Schematic diagram depicting the biobarrier and chemical detection system of the Corrositex® test method

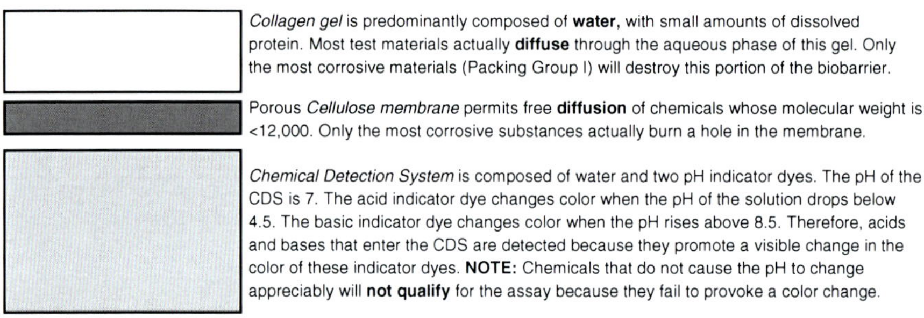

Fig. 12.2 Description of the components of the Corrositex® test

above 8.5. The *time* that is required to cause this change in pH is governed by three factors: (1) the *strength* of the acid or base, (2) the *rate of diffusion* of the test material, and (3) for very corrosive substances, the *rate of destruction* of the biobarrier.

These mechanisms are depicted in the graph shown in Fig. 12.3. In this example, concentrated and dilute hydrochloric acid (HCL) were analyzed with the Corrositex® test. Because the rate of diffusion is proportional to concentration, the concentrated acid diffused through the biobarrier more rapidly than the dilute acid solution. Additionally, the concentrated HCL reacted chemically (hydrolyzed) with collagen and cellulose to cause destruction of the biobarrier. As a result, the concentrated acid entered the CDS and caused a color change in less than 2 min. By contrast, the dilute acid required 25 minutes to diffuse through the biobarrier and induce a visible color change. Based on the prediction model of the test method, the concentrated acid was considered to be a Packing Group I material and the dilute acid was a Packing Group II substance.

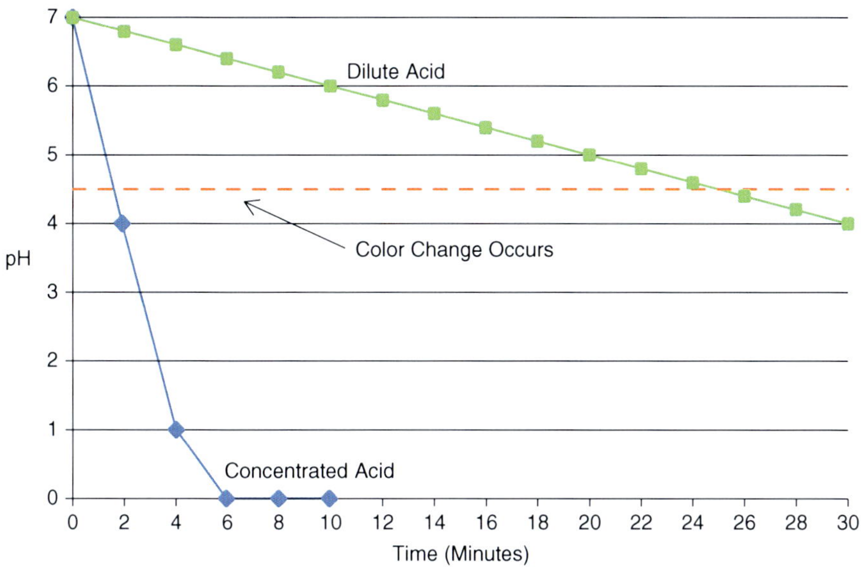

Fig. 12.3 An example of diffusion time and acid strength govern "breakthrough" times

12.2 Current Validation Status

Corrositex® has been granted regulatory approval by the US DOT since 1993 (Exemption DOT-E 10904 Revision 8) [2]. The US DOT limits the use of Corrositex® to specific classes of chemicals, including acids, acid derivatives, acyl halides, alkylamines and polyalkylamines, bases, chlorosilanes, and metal halides and oxyhalides.

Corrositex® has also been accepted by the Transport Canada—Permit for Equivalent Level of Safety SU 5807 (Ren.1) as an alternative test method to determine the UN packing group for not fully specified products or substances that have a primary or subsidiary classification of Class 8 [3].

In 2000, the National Institute of Environmental Health Sciences (NIEHS) and the major US federal regulatory agencies including the Environmental Protection Agency (EPA) (http://www.epa.gov), the Occupational Safety and Health Administration (OSHA) (http://www.osha.gov), and the Consumer Product Safety Commission (CPSC) (http://www.cpsc.gov) accepted the use of the Corrositex® skin model test as a replacement for the animal test for skin corrosivity. The Food and Drug Administration (FDA) (http://www.fda.gov) also endorsed the acceptability of the method but stated that corrosivity testing for the types of products it regulates is likely to be limited. The Department of Transportation (http://www.dot.gov) had already accepted the method for certain chemicals since 1993 (see above). Then in 2013, OSHA (29CFR1910.1200 Appendix A) stated *in vitro* alternatives (such as Corrositex) that have been scientifically validated shall be used to make classification decisions.

In 2000, the European Union ECVAM Scientific Advisory Committee (ESAC) unanimously endorsed the statement that the Corrositex® assay is a scientifically valid test method. The applicability domain is stated to identify noncorrosives and skin corrosive subcategories 1A, 1B, and 1C. Corrositex® was considered applicable to specific classes of chemicals, i.e., organic and inorganic acids, acid derivatives, and bases [4].

Since 2006, the *In Vitro* Membrane Barrier Test Method for Skin Corrosion was adopted as the OECD Test Guideline No 435, which is applicable to Corrositex® [5]. Furthermore, in 2014, the OECD published a guidance document on an integrated approach on testing and assessment (IATA) for skin corrosion and irritation. OECD Test Guideline 435 was the third adopted *in vitro* test method for skin corrosion (ENV/JM/MONO (2014)19). Within its approved applicability domain, OECD accepted Corrositex to identify noncorrosives and skin corrosive subcategories 1A, 1B, and 1C according to the Globally Harmonized System (GHS) for classification (equivalent to UN Packing Group I, II, and III).

12.3 Performance and Applicability of the Test Method

The Corrositex test method allows the identification of corrosive chemical substances and mixtures and allows the subcategorization of corrosive substances according to the GHS classification system (Table 12.1). In addition, such a test method may be used to make decisions on the corrosivity and noncorrosivity of specific classes of chemicals, e.g., organic and inorganic acids, acid derivatives, and bases for certain transport testing purposes. The *in vitro* membrane barrier test method may be used to test solids (soluble or insoluble in water), liquids (aqueous or nonaqueous), and emulsions. The samples may be pure chemicals, dilutions, formulations, or waste. No prior treatment of the sample is required.

However, test chemicals (substances and mixtures) not causing a detectable change in the compatibility test (i.e., color change in the chemical detection system) cannot be tested with the membrane barrier test method [1]. In particular aqueous substances with a pH in the range of 4.5–8.5 often do not qualify for testing; however, 85% of chemicals tested in this pH range were noncorrosive in animal tests [1].

Table 12.1 Designation of UN packing groups/GHS skin corrosion categories [6]

	Corrositex time (minutes)			
Corrositex category 1	0–3 min	>3–60 min	>60–240 min	>240 min
Corrositex category 2	0–3 min	>3–30 min	>30–60 min	>60 min
UN packing group	PG I	PG II	PG III	Noncorrosive
GHS skin corrosion category	GHS skin corrosion category 1			
GHS skin corrosion subcategories	Subcategory 1A	Subcategory 1B	Subcategory 1C	not Catergory 1

Another potential limitation of this method is misinterpretation of category when the test material has an intense color. The misinterpretation can be easy to reduce by measuring the pH value as well as observing the color change.

12.4 Brief Description of the Protocol

The Corrositex test method is performed in three steps: qualify, categorize, and classify. Before following the Corrositex testing protocol, the biobarrier discs must be prepared and refrigerated at 2–8°C for at least 2 h. When making the biobarrier discs, the entire content of the Corrositex diluent vial is slowly added and mixed with the biobarrier matrix powder vial kept in a water bath at 68–70°C for about 20 minutes to ensure complete and uniform solubilization. After solubilization of the matrix, the hot solution (200 μL) is pipetted into each disc placed in 24-well plate. The plate is then sealed with plastic wrap and stored at 2–8°C.

The first step, a qualification test, is done to insure that the test sample and the CDS reagent are compatible. Test material (150 μL or 100 mg) is added to the "qualify" test tube. If a color change or consistency change at sample/testing fluid interface is observed, the sample is judged to be compatible with the detection system and the remainder of the test is performed.

The second step of the Corrositex test utilizes appropriate indicator solutions to permit categorization of the test sample as either a Corrositex Category 1 or Corrositex Category 2 material. Corrositex Category 1 materials are typically strong acids/bases, while Corrositex Category 2 materials are typically weak acids/bases. A "categorization screen" is conducted to enable the test material to be assessed against the appropriate scoring scale. Test material (150 μL or 100 mg) is added to Tube A and Tube B. The tubes are then mixed and the resulting colors observed. If a color observed in either tube matches to a color on the chart provided by InVitro International, record that category. If no color change is seen, add two drops of the "confirm" reagent to tube B. The tube is then mixed and the resulting colors are used to confirm the proper category.

The third step of the test is performed by applying the test sample to the biobarriers. When the chemical permeates through or destroys the full thickness of the biobarriers, it comes into contact with the CDS which then undergoes a color change. This color change is visually observed and the time required for the color change to occur is recorded. Four replicate vials are used for each test material. For each assay, at least one vial for the positive control and one vial for the negative control are employed. In details, a biobarrier disc is placed into a vial containing CDS, and approximately 500 μL or 500 mg of the test material is added to the biobarrier disc. The vial is then observed continuously, against a solid white background, for the first 10 min, and then at approximately 10 min intervals for 1 h (Category 2 test materials) or 4 h (Category 1 test materials), or until the breakthrough of the chemical occurs. Breakthrough time is recorded in minutes and seconds. The assay results are considered to be acceptable if the positive control and negative control time fall in predefined time ranges. The UN packing group and

GHS skin corrosion category are predicted using the mean breakthrough times shown in Table 12.1.

12.5 Role in a Testing Strategy

Corrositex serves several purposes in a responsible commercial corporation's corrosive testing strategy. Probably its primary purpose is to meet corporate ethical safety responsibilities and accountability. Next, Corrositex's role is to meet regulatory standards as much as possible globally. Thirdly, Corrositex plays a major role in replacing the use of live animals in what is generally acknowledged as a painful, even cruel use of rabbits or rats. Finally, Corrositex plays an important role in reducing both the cost and time spent to make a proper corporate decision determining noncorrosive or corrosive/hazardous status of a material or commercial good.

Furthermore, in 2014, the OECD published a guidance document on an IATA for skin corrosion and irritation in which the OECD Test Guideline 435 is part of the recommended information sources. IATA describes several modules which group information sources and analysis tools and also provide guidance on how to (1) integrate and use existing test and non-test data for the assessment of skin irritation and skin corrosion potentials of chemicals and (2) propose an approach when further testing is needed, including when negative results are found [5]. This IATA approach includes the conduct of *in vitro* tests for skin corrosion (such as described in the OECD TG 435) and skin irritation before considering testing in living animals. Positive results from *in vitro* test methods can be used to classify a chemical as corrosive or Non-corrosive without the need for animal testing, thus reducing and refining the use of animals and avoiding the pain and distress that might occur if animals were used for this purpose.

12.6 Perspectives from the Test Developer

The preparation of the biobarrier membrane discs is probably the most critical step in the assay since the classification assigned is based on the time it takes a substance to penetrate through the membrane barrier to the indicator solution. When mixing the biobarrier matrix powder and diluent, it is important to make sure that the powder is completely dissolved before applying the solution to the discs. It is also important to ensure that no air bubbles are formed either when filling the pipette with solution or when applying into each membrane disc evenly. In general, the biobarrier is stable for 7 days if it is wrapped and stored at 2–8°C. For each assay, the positive and negative control should be used to ensure the quality of the biobarrier discs.

Furthermore, when performing the qualification step, it is important to focus attention immediately on the vial in case a change in color or consistency is a short-lived change. Finally, when performing the categorization step, besides the color change from Tube A and Tube B, it is also recommended to measure the pH value in Tube A and Tube B to ensure the final category of the test material.

12.7 Conclusions

Based on current available information, the Corrositex test method is best suited for the determination of corrosivity of acidic and alkaline compounds. Consequently, manufacturers and shippers of these types of products, typically referred to as possible Class 8 corrosives, would be expected to benefit most from the use of this *in vitro* alternative to the *in vivo* skin corrosivity test method.

References

1. ICCVAM. Corrositex®. An in vitro test method for assessing dermal corrosivity potential of chemicals. The results of an independent peer review evaluation coordinated by ICCVAM, NTP and NICEATM. NIEHS, NIH Publication No. 99–4495; 1999.
2. U.S. DOT (2013). Exemption DOT-E-10904, 8th Rev. Washington, DC: U.S. DOT.
3. Transport Canada. Permit for equivalent level of safety SU 5807 (Ren.1). Ottawa, ON: Transport Canada; 2002.
4. ECVAM. Statement on the application of the corrositex® assay for skin corrosivity testing. In: 15th meeting of ECVAM scientific advisory committee (ESAC), Ispra, Italy. ATLA 29(2), 96–97: No. 203, Paris: OECD; 2001.
5. OECD. Guidance document on integrated approaches to testing and assessment of skin irritation/corrosion, series on testing and assessment, No. 203. Paris: OECD; 2014. Available at: www.oecd.org/officialdocuments/publicdisplaydocumentpdf/?cote=ENV/JM/MONO(2014)19&doclanguage=en
6. United Nations (UN). Globally harmonized system of classification and labeling of chemicals (GHS). 6th revised en. New York and Geneva: UN; 2015. Chap. 3.2.

The Rat Skin Transcutaneous Electrical Resistance (TER) Test

13

Robert Guest

13.1 Principle of the Test Method and Scientific Basis

The rat skin transcutaneous electrical resistance (TER) test (hereafter referred to as the TER test) was initially developed and used by Oliver and co-workers in the 1980s to screen chemicals for their skin corrosion potential and to guide humane *in vivo* skin testing. After protocol refinement, followed by intra-laboratory, inter-laboratory, pre-validation and validation studies [1–6], the test was adopted as OECD Test Guideline 430 in 2004, being revised in 2013 and again in 2015 [7]. The test can be used to determine the skin corrosion potential of chemicals or chemical mixtures without the use of living animals, thereby avoiding the pain or discomfort, which can accompany serious skin damage. Whilst described as an *in vitro* procedure, the TER test can be more correctly defined as an *ex vivo* method as it is conducted on skin discs obtained from a humanely killed young rat. *In vivo*, skin corrosion can be identified as irreversible damage to the skin manifested as visible necrosis through the epidermis and into the dermis [8]. The skin effects can be seen as areas of necrotic skin or ulceration, which may sometimes be preceded or accompanied by severe inflammation (erythema and oedema), and may lead to scab (crust) and scar tissue formation with or without bleeding and alopecia. Depending upon the degree of skin corrosivity of the chemical, some skin effects (e.g. visible necrosis) may occur within minutes of exposure, whilst others may take longer to develop. Scar tissue may even take weeks to form.

The scientific basis of the TER test is that corrosive chemicals or chemical mixtures have the potential to produce a loss of integrity of the stratum corneum (the surface layer of the epidermis) and a reduction in the normal barrier function of the skin. These effects can be measured by recording the passage of ions through skin

R. Guest
Envigo, Derby, UK
e-mail: robert.guest@envigo.com

C. Eskes et al. (eds.), *Alternatives for Dermal Toxicity Testing*,
DOI 10.1007/978-3-319-50353-0_13

discs using a 'Wheatstone bridge' apparatus, which provides a measure of the electrical impedance of the skin as a resistance value in kilo ohms (kΩ). The TER of the skin is dependent upon a number of factors, including the species and strain of the animal, region of the body, age and condition, and so these have to be standardized in order to ensure reproducibility of measurements. The skin of a young rat is used because the sensitivity in testing of chemicals for skin corrosion potential has been previously demonstrated, and rat skin is the only source that has been formally validated [6]. The threshold for corrosion in rat skin, of the age and strain specified in the validated TER test method, using the prescribed apparatus and procedures, is a value of 5 kΩ. This value has been selected on the basis of extensive data for a large number of substances and mixtures [4–6]. The value is considered to be robust because almost all corrosive substances produce a mean TER value of less than 3 kΩ in this system, whilst almost all non-corrosive substances produce a TER value of greater than 10 kΩ. However, in addition to corrosive substances, some substances that are not corrosive to skin, for example, some surfactants and neutral organic chemicals, are able to cause a reduction of the barrier properties of the skin without producing irreversible tissue damage. For this reason a 'dye-binding' procedure is incorporated into the TER test to allow identification of false-positive results.

In outline, the skin corrosivity of each test chemical or mixture is determined using three skin discs freshly prepared from the dorsal pelt of a humanely killed young rat. Each skin disc is secured tightly over the end of a PTFE (polytetrafluoroethylene) tube, with the epidermis facing inside the tube. The end of the PTFE tube with skin disc in place is submersed in electrolyte solution in order to form a two-compartment test system. The test chemical is then applied to the epidermal surface of the skin. After a contact time of up to 24 h, the test item is washed from the epidermal surface, and following hydration of the skin using electrolyte, the TER is measured. Negative and positive controls are incorporated into the test, each consisting of exposure of three skin discs to a negative control or positive control substance. If the mean TER of the skin discs treated with the test item is below, or is close to 5 KΩ, but the possibility of this being not due to true corrosivity is suspected, as might be the case for some surfactants and neutral organic chemicals, then a dye-binding step is conducted.

13.2 Current Validation Status

The TER test was originally used by industry in the early 1980s, primarily as a screening test to predict skin corrosion potential rather than the degree of corrosivity and to guide humane *in vivo* skin testing [1, 2]. During the next 10 years, a number of protocol refinements resulted in optimization of the skin contact time (24 h) and reduction of the rate of false-positive results [3]. An inter-laboratory trial was undertaken [4], and a formal pre-validation study took place between 1993 and 1994 in two laboratories using 25 corrosive and 25 non-corrosive chemicals [5]. A formal international validation study was conducted in three test facilities between

1996 and 1997, under the auspices of ECVAM, using 60 chemicals [6]. The results met the predefined criteria for acceptable under- and over-prediction rates, and on 3 April 1998 the ECVAM Scientific and Advisory Committee (ESAC) unanimously endorsed the TER test as scientifically validated for use as a replacement for the animal skin corrosion test for distinguishing between corrosive and non-corrosive chemicals [9].

The rat skin TER test was also accepted for its intended use by the European Scientific Committee for Cosmetic Products and Non-food Products (SCCNFP) [10]. On 8 June 2000, the test was adopted as Method B.40 of Annex V to Council Directive 67/548/EEC under the 27th adaptation to technical progress, superseded in 2008 by the EU 'Test Methods Regulation' to bring the test methods in line with the EU REACH chemicals directive [11]. In 2001, a summary report focusing on the performance of the TER test was prepared on behalf of the US National Toxicology Program Interagency Center for the Evaluation of Alternative Toxicological Methods (NICEATM) [12]. The authors concurred with the ECVAM conclusion that the test is both reliable and reproducible, but that for some chemical or product classes (e.g. cleaners and detergents), the small number of chemicals and/or the unbalanced distribution of corrosive and non-corrosive chemicals did not allow accurate conclusions to be made on the performance of this assay for these chemical classes. The summary report was incorporated into a background review document (BRD) published by ICCVAM in August 2001 [13] which provided a number of recommendations following review of available validation studies. A draft OECD Test Guideline 430 was issued for the test method on 27 March 2002, and OECD formally adopted the finalized test guideline on 13 April 2004. This was further revised on 26 July 2013 to include a set of performance standards as Annex 1, for the assessment of new or modified TER test methods having similarity to the validated reference method (VRM) in accordance with the principles of OECD Guidance Document No. 34 [14]. OECD TG 430 was revised again on 28 July 2015 [7], to include reference to the guidance document on an integrated approach to testing and assessment (IATA) for skin corrosion and irritation [15]. The revised test guideline no longer included the performance standards set out as Annex 1 in the previous version, but in August 2015, these were published as a separate OECD Guidance Document, No. 218 of the OECD Series on Testing and Assessment [16]. This stand-alone document is intended for use by developers of new or similar test methods for generating the validation data that is required by the OECD for inclusion of a new or modified test method into the Test Guideline 430 and before use of the test for regulatory purposes.

13.3 Performance and Applicability of the Test Method

13.3.1 Reproducibility

The repeatability and reproducibility of the TER test in three different studies have been summarized [12].

In the inter-laboratory trial reported by Botham et al. in 1992 [4], no statistically significant level of inter-laboratory variability was found for corrosives (6 chemicals), non-corrosives (14 chemicals) or for all test materials (20 chemicals), but an intra-laboratory analysis was not possible. In the pre-validation study reported by Botham et al. in 1995 [5], the agreement for the classifications obtained by both of the participating laboratories was 92% (23 of 25 corrosive and 23 of 25 non-corrosive chemicals).

In the ECVAM validation study reported by Fentem et al. in 1998 [6], 60 chemicals were each tested twice by each of three laboratories. Based on their analyses, ECVAM concluded that inter- and intra-laboratory variability was approximately equivalent, with no evidence of systematic differences between experiments within a laboratory. Of the 60 chemicals tested, 37 gave the same skin corrosion classification in both experiments in all three laboratories. For 10 of the remaining 23 chemicals, only one experiment resulted in a classification differing from the other 5 predictions. Although there were differences for some chemicals between experiments within, and between laboratories, ECVAM concluded that the rat skin TER assay was reliable and reproducible. Due to the lack of quantitative data for individual chemicals in the published studies, no independent evaluation of repeatability or reproducibility for the rat skin TER assay could be conducted. However, after reviewing the intra- and inter-laboratory evaluations conducted by ECVAM, it was concluded that the analyses were appropriate and that the conclusions were accurate.

Based on the performances obtained with the TER test for the 24 Reference Chemicals listed in the OECD Performance Standards Document No. 218 [16], any future similar or modified TER test method must achieve a within-laboratory reproducibility (WLR) equal to or higher than 90% and a between-laboratory reproducibility (BLR) equal to or higher than 80%, when tested with those 24 chemicals.

13.3.2 Predictive Capacity

The result of the TER test can be used to determine whether the test chemical will require classification as a Category 1 corrosive substance or if it does not require classification as corrosive to skin, according to the UN GHS [8]. The results cannot be used to assign skin-corrosive subcategories 1A, 1B or 1C.

Using data obtained from the Botham et al. (1992 and 1995) and ECVAM validation studies [4–6] for 122 chemicals and chemical mixtures, the rat skin TER test method was able to discriminate between known skin corrosives and non-corrosives with an overall accuracy of 81% (99/122), a sensitivity of 94% (51/54) and specificity of 71% (48/68). Based on the ECVAM validation study alone, which consisted of 355 trials on 60 chemicals [6], the overall accuracy was 79% (282/355), with a sensitivity of 88% (140/159) and specificity of 72% (142/196). So when the data from the ECVAM validation study [6] were evaluated independently from other studies Botham et al. [4, 5], the overall performance was very similar.

Table 13.1 Sensitivity, specificity and accuracy required for proposed or similar test methods to the VRM

Sensitivity	Specificity	Accuracy
≥90% (actual for rat skin TER[a], 93%)	≥75% (actual for rat skin TER[a], 75%)	≥82.5% (actual for rat skin TER[a], 84%)

[a]Values based on the results from the Validated Reference TER test method (VRM) for the 24 Reference Chemicals

The performance characteristics for the TER test remained consistent when evaluated against various chemicals classes, including organic and inorganic bases and base mixtures, organic and inorganic acids and acid mixtures.

Based on the predictive capacity obtained with the TER test for the 24 Reference Chemicals listed in the OECD Performance Standards Document No. 218 [16], any future similar or modified TER test method must achieve a sensitivity, specificity and accuracy values as shown in Table 13.1 to be considered valid to discriminate corrosive from non-corrosive chemicals.

13.3.3 Applications and Limitations

On the basis of the data generated in the ECVAM validation study for 60 chemicals over a wide range of chemical classes and physical states including liquids, semi-solids, solids and waxes, and taking into account the relative simplicity of the mechanism of action of corrosives, it was concluded that the test method would be generally applicable across all chemical classes [6]. However, a comparably small number of waxes and corrosive solids were assessed during validation.

In practical terms, the rat skin TER test can be conducted on any chemical, substance, mixture or formulation that can be applied uniformly over the surface of the skin disc and can be removed from the skin disc at the end of the exposure period to allow measurement of the skin TER. If it is demonstrated that the TER test is not applicable to a specific category of substance, then the test should no longer be used for that category. The test may not be appropriate for substances that cannot be removed from the skin surface, such as some adhesives or paints since this would likely result in elevated TER values. It is not possible to test gases or aerosols using the validated protocol, and currently, OECD TG 430 does not allow testing of these forms of substance. It may be conceivable that these forms of material could be tested using a modified TER test method.

Some non-corrosive surfactants and neutral organics may disrupt the stratum corneum in the TER test causing a lowering of the TER below 5 kΩ, in which case the additional dye-binding step should be conducted. The TER test can be used to assess coloured substances, but if it is necessary to conduct the dye-binding step, the possibility of the coloration interfering with measurements in the dye-binding step should be considered.

The TER test is assumed to be applicable to testing of mixtures as an extension of its applicability to substances. However, since mixtures cover a wide spectrum of categories and composition, in cases where it can be shown that the test is not applicable to a specific category of mixtures, e.g. following a strategy as proposed by Eskes et al. [17], the OECD TG 430 should not be used for that specific category of mixtures.

Before use of the Test Guideline on a mixture for generating data for an intended regulatory purpose, it should be considered whether, and if so why, it may provide adequate results for that purpose. Such considerations are not needed, when there is a regulatory requirement for testing of the mixture.

13.3.4 Comparison to Human Data

No published comparisons of rat skin TER skin corrosion and human skin corrosion data have been identified.

13.4 Description of the Protocol

OECD TG 430 describes the experimental methodology for the validated reference model of the TER test, and also its protocol is described in the ECVAM DB-ALM Protocol Number 115 [18]. Whilst the test is relatively inexpensive and simple to conduct, it is essential that the test facility is able to house, handle and maintain laboratory rats in accordance with national regulations and recognized standards of animal care and welfare. In some countries, national law may regulate the preparation of rats for provision of the skin to be used in the TER test, whilst in some other countries, preparation and use of the rat may not be regarded as a 'regulated' procedure. Prior to the routine use of the TER test, and for adherence to OECD TG 430, laboratories should demonstrate technical proficiency by correctly classifying the skin corrosivity of the proficiency substances listed in the Test Guideline. It is also essential that the test facility can generate acceptable results using the recommended positive and negative control materials and can demonstrate reproducibility of the test method.

13.4.1 Apparatus Required to Conduct the Test

- A low-voltage alternating current Wheatstone bridge apparatus, e.g. LCR 6401 Databridge (H Tinsley & Co., Croydon, Surrey, UK).
- Electrodes and PTFE tubes with rubber 'O'-rings. Specifications can be found in Fig. 13.1.
- Resistance box for calibration of the Wheatstone bridge apparatus, as shown in Fig. 13.2.
- Bench centrifuge
- Spectrophotometer (for measuring optical density of extracted sulforhodamine dye solution).

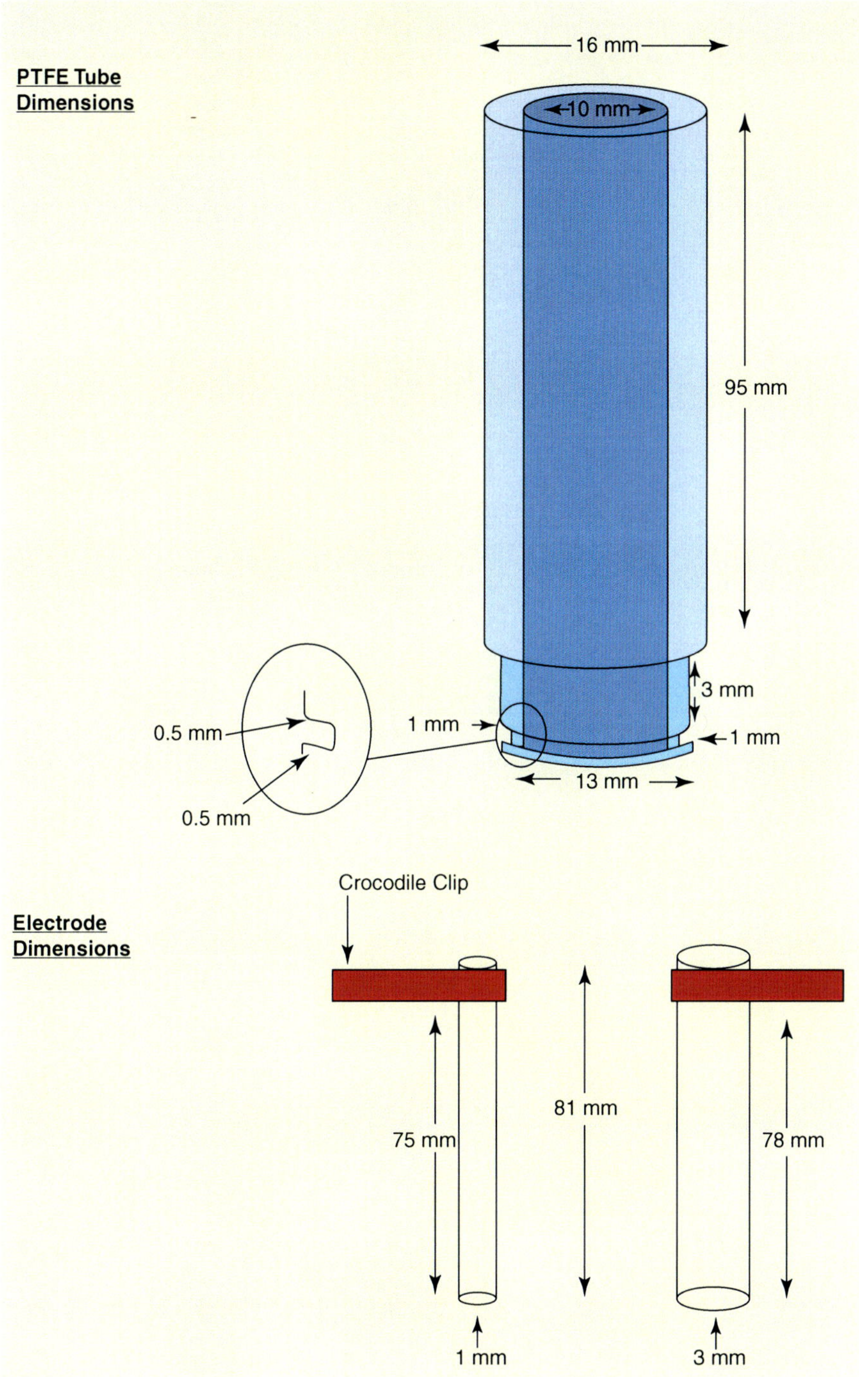

Fig. 13.1 PTFE tube and electrode dimensions (From DB-ALM Protocol n° 115)

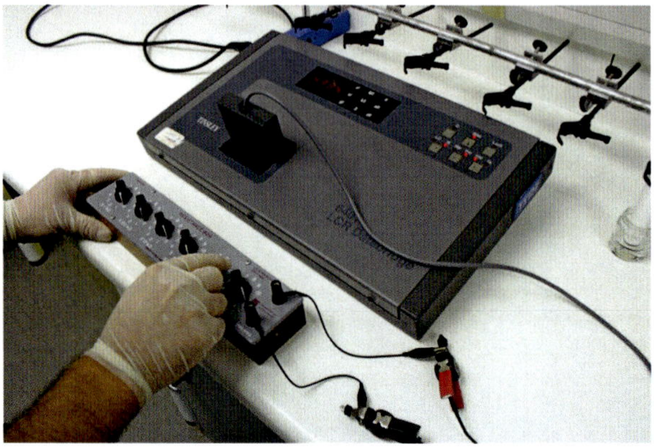

Fig. 13.2 Calibration of the Wheatstone bridge apparatus

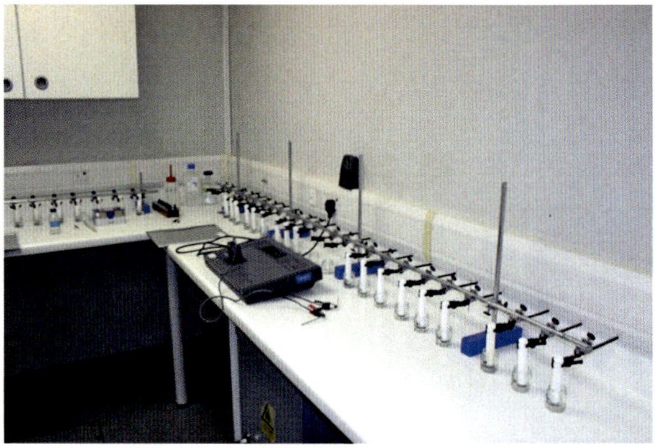

Fig. 13.3 TER apparatus

- Other materials: crocodile clips, spring ('Terry') clips (optional), disposable receptacles to serve as receptor chambers, scalpel, disposable plastic syringes, disposable vials for extraction of dye from skin discs, petroleum jelly.

  Figure 13.3 shows a typical arrangement of the TER apparatus.

13.4.2 Materials Required for the TER Step

- Antibiotic solution (e.g. a mixture of 8000 μg/mL streptomycin, 800 μg/mL penicillin, 10 μg/mL amphotericin, 10 μg/mL chloramphenicol). It is also acceptable to use mixtures of antibiotics containing glutamine, which are commercially available.
- Electrolyte solution: Magnesium sulphate heptahydrate 154 mM.
- Tap water for rinsing test material from the skin surface.
- De-ionised water for moistening of solid test materials and to also serve as negative control material.

- Test chemical: A sufficient amount is required to allow the application of 150 μL of a
- liquid or a uniform layer of a solid to each of three skin discs.
- 10 M hydrochloric acid to serve as positive control material.
- 70% ethanol (for lowering surface tension at the skin surface).

13.4.3 Materials for Dye-Binding Step (if Required)

- 10% solution of sulforhodamine B dye (Acid Red 52; C.I. 45,100; CAS No. 3520–42-1) in distilled water
- De-ionised water for washing skin discs
- 30% (v/v) aqueous solution of sodium dodecyl sulphate for extraction of sulforhodamine dye from skin discs

The procedure section of OECD TG 430 covers all of the technical and practical aspects of the test. The first part of this section discusses the source and specifications of the animals used for the test, the key features being the use of a standard laboratory strain of rat (Wistar derived is recommended) of either sex but of a specific age, 22 days at the time of hair removal and 28–30 days at the time of preparation of skin discs. The age of the rat is critical to ensure that the hair follicles are in the dormant (telogen) phase. The use of older animals can result in lower than expected TER values.

Careful preparation of the skin and skin discs is critical to the successful conduct of the test. During removal of hair from the dorsum and flanks, it is important to avoid abrasion of the skin. The clipped skin should be washed with antibiotic solution after clipping and again on the third or fourth day after the first wash. This procedure is conducted to inhibit bacterial growth on the skin and therefore minimize the possibility of changes in the barrier properties. Animals should then be used within 3 days of the second wash.

A sufficient number of skin discs can be obtained from the pelt of one animal to conduct one study or possibly two studies when sharing the negative and positive controls. The skin used must be stripped of excess subcutaneous fat by carefully peeling it away. A total of eleven discs are usually required, two for the initial quality control procedure and three for each of the test, positive control and negative control materials. Any skin discs that have abrasions or other defects should be rejected. Ideally, the skin discs should be used straight away, but if storage is necessary, data should be generated to show that the specific storage conditions do not result in unusual results for the positive and negative controls. The Test Guideline describes the removal of skin discs of 20 mm in diameter before placement onto the end of the PTFE tubes, but an alternative, and possibly more convenient, procedure is to stretch the complete strip of prepared skin over the end of the PTFE tube and then press fit a rubber 'O'-ring over the end of the tube to hold the skin in place, whilst the excess tissue is removed using a scalpel blade. The rubber 'O'-ring must be sealed to the end of the tube using a thin film of petroleum jelly, but care must be taken not to contaminate the skin disc as this could affect the skin impedance (Fig. 13.4a–e). A spring clip can be used to support the tube in the receptor chamber, but alternatively, the tube can be held in place by a clamp attached to a retort stand. It is important that the skin disc is fully submerged in the electrolyte.

To ensure the integrity of the skin, a quality control check must be conducted before commencement of the test. Two discs are used for this purpose, and a resistance value of greater than 10 kΩ in each indicates suitability of the skin.

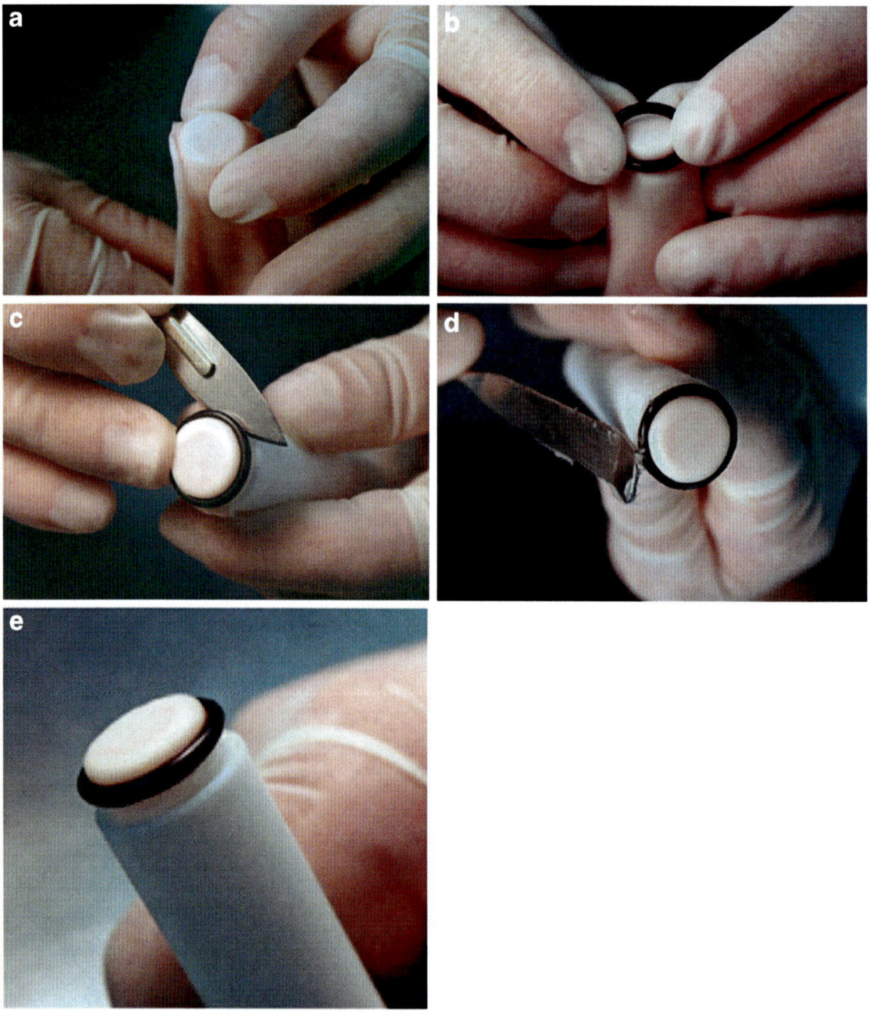

Fig. 13.4 (**a**–**e**) Skin disc preparation. (**a**) The prepared skin is place over one end of a PTFE tube, epidermal surface in contact with the tube. (**b**) A rubber 'O'-ring is press fitted over the end of the PTFE tube. (**c**) Excess skin is trimmed away. (**d**) 'O'- ring is sealed using petroleum jelly. (**e**) Skin disc *in situ*

For each testing run (experiment), skin discs from a single animal should be used. Three skin discs are exposed to the test chemical, three to a positive control chemical and three to a negative control chemical. The concurrent positive and negative controls are included to demonstrate adequate performance of the experimental model. For the VRM, the positive and negative control test chemicals used are 10 M hydrochloric acid and distilled water, respectively. 150 μL of a liquid test chemical, or a sufficient quantity of solid test chemical to cover the skin disc, must be applied as evenly as possible to the epidermal surface. Application of liquids can be achieved using a syringe, and solids can be applied using a small spatula or sharp spoon. Moistening of solids is recommended and is achieved by addition of 150 μL of de-ionised water followed by gentle agitation of the tube. For some solids (e.g. waxes) moistening may not be appropriate, but instead, assurance of good skin contact might be possible only after prior softening or melting (a temperature of 30 °C is recommended). The use of higher temperatures must be documented and an assessment made of whether this could affect the outcome of the study. The test and control chemicals are allowed to remain on the skin surface for 24 h at a temperature of 20–23 °C. At the end of the exposure period, the test chemical and control materials are removed from the skin discs by washing using a jet of tap water (at, or below, room temperature) until no further material can be removed.

The skin impedance is measured as TER by using a low-voltage, alternating current Wheatstone bridge. General specifications of the bridge are 1–3 volt operating voltage, a sinus or rectangular shaped alternating current of 50–1000 Hz and a measuring range of at least 0.1–30 kΩ. The Databridge used in the validation study measured inductance, capacitance and resistance up to values of 2000H, 2000 μF and 2 MΩ, respectively, at frequencies of 100 Hz or 1 kHz, using series or parallel values. For the VRM, measurements are recorded in resistance, at a frequency of 100 Hz and using series values.

Prior to measuring the TER, the surface tension of the skin must be reduced by addition of a volume of 70% aqueous ethanol sufficient to cover the epidermis, which is then discarded before addition of 3 mL of electrolyte solution (154 mM magnesium sulphate solution). The Databridge electrodes are placed on either side of the skin disc to measure the resistance in kΩ/skin disc as shown in Fig. 13.5. It is important that the position of the electrodes is as described in the Test Guideline when conducting the TER measurement. Electrode dimensions and the length of the electrode exposed below the crocodile clips are shown in Fig. 13.1. The clip attached to the thicker (inner) electrode is rested on the top of the PTFE tube during resistance measurement to ensure that a consistent length of electrode is submerged in the MgSO4 solution. The thinner (outer) electrode is positioned inside the receptor chamber so that it rests on the bottom of the chamber. The distance between the crocodile clip and the bottom of the PTFE tube is maintained as a constant, because this distance affects the resistance value obtained. Consequently, the distance between the inner electrode and the skin disc should be constant and minimal (e.g. 1–2 mm in the VRM).

If the measured resistance value is greater than 20 kΩ, this may be due to the remains of the test chemical coating the epidermal surface of the skin disc. Further removal of this coating can be attempted, for example, by sealing the PTFE tube with a gloved thumb and shaking it for approximately 10 s. The electrolyte solution

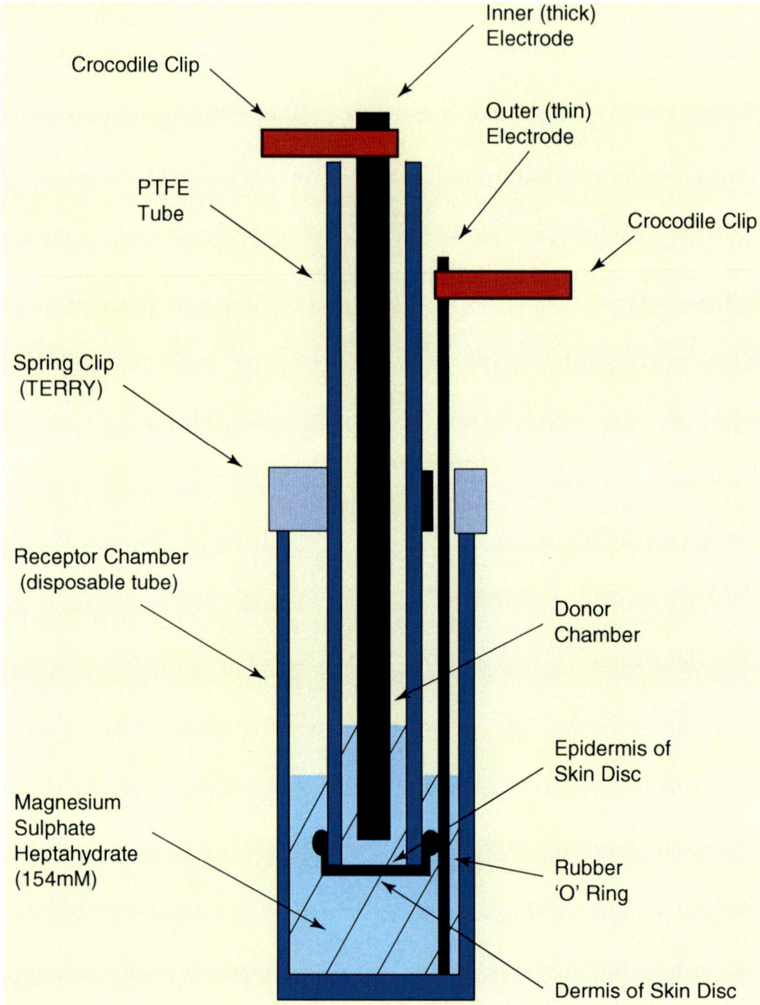

Fig. 13.5 Transcutaneous Electrical Resistance Test 2-compartment System (from DB-ALM Protocol n° 115)

is then discarded, and the resistance measurement is repeated with fresh electrolyte solution.

When the skin discs are removed from the PTFE tubes after measurement of the TER, the epidermal surface and the whole skin disc should be examined, and any skin changes should be recorded, e.g. perforation, blanching and thickening.

If it is suspected that the test material might have the potential to reduce the resistance of the skin without causing dermal corrosion, for instance, in the case of some neutral organic chemicals or surface-active agents, or if a TER value of less than 5 kΩ is obtained and there is no visual evidence of damage to the skin disc (e.g. erosion, perforation or obvious blanching), then a dye-binding step, using sulforhodamine B, must be conducted on all of the skin discs. The procedure is adequately described in the Test Guideline and results in production of solutions of the dye extracted from skin discs using a 30% (w/v) SDS solution in distilled water. The optical density (OD) of the extracted dye solutions is measured at 565 nm in a spectrophotometer.

13.4.4 Calculation of Sulforhodamine B Dye Content

The sulforhodamine B dye content in µg/disc can be calculated from the optical density values. Whilst OECD Test Guideline 430 recommends that the dye content is determined for each skin disc by the use of an appropriate calibration curve, a calculation method was used in the protocol used in the 1996–1997 ECVAM validation study [16]:

Using a sulforhodamine B dye molar extinction coefficient of 8.7×10^4 and a molecular weight of 580 (with no correction for the purity of the dye being made) and the example of a measured optical density of 0.973:

$$\frac{0.973}{8.7} \times 10^{-4} = 11.2 \times 10^{-6} \text{ equivalent to } 11.2\,\mu\text{M solution or } 11.2\,\mu\text{mol}/\text{L}$$

$$11.2 \times 580 \times 10^{-6} = 6496 \times 10^{-6}\,\text{g}/\text{L} = 6.496 \times 10^{-3}\,\text{g}/\text{L}$$

The dye is extracted in 5 mL of solvent, therefore:

$$\frac{6.496 \times 10^{-3}}{200} = 0.325 \times 10^{-4}\,\text{g}/\text{L} = 32.5 \times 10^{-6}\,\text{g}/\text{L}$$

The solution is diluted 1 in 5 (v/v) prior to measurement of OD, therefore:

$$32.5 \times 10^{-6} \times 5 = 162.5 \times 10^{-6} \text{ equivalent to } 162.5\,\mu\text{g}/\text{disc}$$

The values for each of the three discs can then be used to calculate the mean dye content.

An adaptation of this calculation method is that if an optical density of 1.000 is assumed, the dye content of the skin disc is calculated to be 167 µg (providing the dye is extracted in 5 mL of solvent and the solution is diluted 1 in 5 (v/v) prior to measurement of the OD). Hence, to allow simple routine calculation of the dye content of each treated skin disc, the optical density of the extracted dye measured at 565 nm (OD_{565}) can be multiplied by a constant value of 167 µg to give the dye content in µg/disc. To demonstrate this using the above example of a measured optical density of 0.973:

$$0.973 \times 167 = 162.5\,\mu\text{g}/\text{disc}$$

Table 13.2 Acceptable mean TER values for positive and negative controls

Control	Substance	Resistance range (kΩ)
Positive	10 M hydrochloric acid	0.5–1.0
Negative	Sterile distilled water	10–25

Table 13.3 Acceptable dye-binding results for positive and negative controls

Control	Substance	Dye content range (μg/disc)
Positive	10 M hydrochloric acid	40–100
Negative	Sterile distilled water	15–35

13.4.5 Acceptability Criteria

The mean TER values and mean dye content for the concurrent positive and negative controls must fall within acceptable ranges; otherwise, the test is considered invalid. The acceptable ranges are given in Tables 13.2 and 13.3, respectively.

13.4.6 Interpretation of Results

Interpretation of the results of the test must take into account the mean TER values, and where appropriate the mean dye content, for the test material treated discs.

The prediction model for the rat skin TER test, associated with UN GHS [8], and given in OECD TG 430 (9), is as follows:

The test chemical is considered to be non-corrosive to skin:

1. If the mean TER value obtained for the test chemical is greater than 5 kΩ.
2. The mean TER value obtained for the test chemical is less than or equal to 5 kΩ:
   a. The skin discs show no obvious damage.
   b. The mean disc dye content is less than the mean disc dye content of the 10 M HCl positive control obtained concurrently.

The test chemical is considered to be corrosive to skin:

1. If the mean TER value obtained for the test chemical is less than or equal to 5 kΩ, and the skin discs are obviously damaged, for example, they are perforated (Fig. 13.6) or are blanched (Fig. 13.7).
2. The mean TER value obtained for the test chemical is less than or equal to 5 kΩ:
   a. The skin discs show no obvious damage.
   b. The mean disc dye content is greater than or equal to the mean disc dye content of the 10 M HCl positive control obtained concurrently.

A testing run (experiment) composed of at least three replicate skin discs should be sufficient for a test chemical when the classification is unequivocal. However, in cases of borderline results, such as non-concordant replicate measurements and/or mean TER equal to 5 ± 0.5 kΩ, a second independent testing run (experiment) should be considered, as well as a third one in case of discordant results between the first two testing runs (experiments).

Fig. 13.6 Example of perforation of skin disc following exposure to test chemical

Fig. 13.7 Example of a blanched skin disc (*right*) as compared to normal skin disc (*left*)

13.5 Role in a Testing Strategy

The TER test may be used as part of an integrated approach to testing and assessment (IATA) for skin corrosion and irritation [15]. It can be used to determine or confirm the potential of a test chemical to cause skin corrosion and so can be used as a stand-alone test for classification purposes, for example, classification to UN GHS or EU CLP [8, 19], providing that subcategorization of skin corrosivity is not required. Alternatively, it can be used to screen for skin corrosion potential as part of wider programme of tests to identify human health hazards. In this situation, the skin corrosion test should be conducted before any other tests because if the chemical is shown to be corrosive to skin, it is possible to waive testing for skin irritation and eye irritation. Furthermore, testing for oral, dermal or inhalation toxicity using the neat chemical would not be recommended due to the potential to cause corrosive effects (although skin corrosion potential may not always correlate precisely with corrosivity via the oral and inhalation routes due to factors such as acid/alkaline reserve and the dosage delivered). If a test chemical is shown to be non-corrosive in the TER test, and an assessment of skin irritation is required, this should be conducted using a validated *in vitro* skin irritation test to determine if the chemical requires classification for skin irritancy potential. One use of the TER test can be to test a range of dilutions of a chemical or formulations containing different concentrations of active ingredients in order to determine the thresholds for skin corrosion. This can provide useful information for selection of suitable non-corrosive concentrations for use in other toxicological studies or for product development purposes.

13.6 Perspectives from a Test User

The rat skin TER test was originally used by industry to screen chemicals and formulations for skin corrosion without the use of living animals. The use of the test extended to the assessment of skin corrosion to satisfy regulatory requirements, and it has undoubtedly resulted in reduced usage of live animals for assessment of skin corrosion potential. The test allows an *ex vivo* assessment of skin corrosion potential using full-thickness rat skin in which a visual assessment of skin damage can be made in addition to changes in the TER. Whilst it is a simple, rapid and relatively inexpensive method, the use of the test has declined in recent years due to the availability of methods that do not require the use of an animal and which can be used to generate data for subcategorization of corrosivity of substances and mixtures in accordance with classification schemes such as UN GHS or EU CLP, e.g. OECD 431 *In Vitro* Skin Corrosion: Reconstructed Human Epidermis (RHE) Test Method [20] and OECD 435 *In Vitro* Membrane Barrier Test Method [21]. The TER test is still used by a small number of companies in situations when a simple determination of skin corrosion potential is required without subcategorization, and/or for comparison of TER test data with historical data for similar test chemicals, or where a test chemical is outside the applicability domain of other validated skin corrosion test methods.

When conducting proficiency testing according to OECD TG 430, it has been noted that TER values much higher than 5 kΩ (even as high as the values obtained for negative controls) can be obtained following a 24-h exposure of skin discs to the chemical 2-tert-butylphenol (CAS Number 88–18-6), which is classified as a corrosive chemical (UN GHS Category 1B–and-1C) according to *in vivo* skin corrosivity data. Test user experience has shown that blanching of the skin discs can also be noted at the end of the exposure period (unpublished data). In the ECVAM validation study conducted using exposure times of 24 h and 2 h [6], two of the six runs of testing failed to identify the chemical as corrosive. After the 24-h exposure period, all TER values were greater than 5 kΩ, and after the 2-h exposure period, TER values ranging from below 2 kΩ to above 10 kΩ were recorded. It is thought that coagulative necrosis of the skin discs may occur following exposure to 2-tert-butylphenol and that after a 24-h exposure, the skin changes can actually result in an increase in the TER. It is therefore recommended that when blanching of one or more of the skin discs is noted, further investigations be undertaken in order to establish if the test chemical is or is not corrosive (e.g. testing of the chemical using additional exposure times, histopathological examination, use of a different *in vitro* skin corrosion test method).

13.6.1 Strengths of the TER Test Method

The rat skin TER test method:

- Is a scientifically validated test method for discrimination between skin corrosives and non-corrosives.
- Does not require the use of live animals.
- Uses fresh, full-thickness skin so it is a scientifically relevant model.
- Is relatively simple to conduct. No special skills are required.
- Does not require highly specialized equipment.
- Is robust.
- Is relatively inexpensive (about the same as a rabbit skin irritation study).
- Is rapid and can be conducted within a small laboratory area.
- Requires only a small amount of test sample (<0.5 g in total).
- Is applicable to most classes of substance (including most coloured materials).
- Has an objective endpoint (only minimal subjective assessment of skin effects is required).
- The 24-h exposure may increase the likelihood of identification of corrosive substances that have a delayed mechanism of corrosivity.

13.6.2 Weaknesses of the TER Test Method

The rat skin TER test method:

- Requires the use of an animal and of a specific age.
- Requires a facility that is able to accommodate laboratory animals.
- Requires a time period before testing whilst the animal is obtained, acclimatized and subjected to skin preparation.
- May be (or may become) a regulated animal procedure in some countries.
- May not be appropriate for testing of materials with adhesive properties (e.g. paints).
- Uses skin that is derived from a non-human source.

- Does not permit subcategorization of corrosives as required by a number of classification schemes.
- There is limited potential for further refinement, e.g. to assess skin irritancy.

13.6.3 Critical Steps in the Protocol

The main aspects of the protocol that are critical to satisfactory performance of the TER test and reproducibility of measurements are:

- The species, strain, age and condition of the animal from which the skin pelts are obtained. These are specified in the test method and great care must be taken in preparation of the pelts and skin discs. It is essential to avoid abrasion of the skin during removal of hair and to ensure removal of the subcutaneous fat.
- Uniform application of the test chemical over the epidermal surface of the skin discs.
- Satisfactory removal of the test chemical from the epidermal surface prior to measurement of the TER. The washing procedure is described in the test method, including the procedure for removing chemicals that adhere to the epidermal surface, but in some cases it may not be possible to remove all traces of the test chemical.
- The properties and dimensions of the test apparatus. These may influence the TER values obtained. The 5 kΩ corrosive threshold was developed from data obtained with the specific apparatus and procedures described in the ECVAM validation study and OECD Test Guideline. Different threshold and control values may apply if the test conditions are altered or a different apparatus is used.
- Correct positioning of the electrodes during measurement of the TER. Details are provided in the test protocol.

13.6.4 Possible Protocol Adaptations

In principal, it would be possible to modify the TER test protocol to use the skin obtained from other animals, including the skin from humans. The use of the human skin would eliminate the factor of species differences in prediction of human skin corrosivity but would introduce the complication of obtaining supplies of the human skin of acceptable quality and consistency of TER to provide an acceptable level of reproducibility. It is considered that this adaption is unlikely to be pursued given that the predictive capacity of the rat skin TER test is already high, as demonstrated in validations studies against *in vivo* animal data.

If a false-negative result is suspected following the 24-h exposure period (e.g. blanching of the skin accompanied by a TER of greater than 5 kΩ), the possibility of coagulative necrosis of the skin having occurred could be investigated by measurement of thickness of the skin discs and comparison to negative controls (to detect increases in skin thickness) and histopathological examination. Assessment of an exposure time of less than 24 h could also be investigated to determine if a shorter exposure time would result in a TER of less than 5 kΩ.

As described in Sect. 13.4, if the dye-binding step is conducted, an adaptation of the calculation method can be used for simple, routine determination of the dye content of each treated skin disc (providing that the dye extraction and dilution of the resultant solution is conducted according to the protocol). The optical density of the extracted dye measured at 565 nm (OD_{565}) can simply be multiplied by a constant value of 167 μg to give the dye content in μg/disc. The values for each of the three discs can then be used to calculate the mean dye content.

13.6.5 Challenges and Opportunities

The main challenge to conduct of the TER test is the pre-test requirement to obtain, prepare and care for a laboratory rat in accordance with animal welfare regulations. On the basis that the test requires the use of the skin obtained from a laboratory rat, the future use of the TER test is likely to be limited to testing of substances that fall outside the applicability domain of alternative validated skin corrosion tests that do not require the use of skin from animals, i.e. OECD TG 431 and OECD TG 435. When testing for skin corrosion potential is required, it should be noted that the results of a TER test will not permit subcategorization of skin corrosivity according to the UN GHS or EU CLP schemes. A substance shown to be corrosive in the TER test would require classification as GHS Category 1, without subcategorization.

13.7 Conclusions

The rat skin transcutaneous electrical resistance test is a validated test for assessment of skin corrosion potential of chemicals including substances and mixtures that does not require the use of use live animals. It has been adopted as the OECD TG 430 "*In Vitro* Skin Corrosion: Transcutaneous Electrical Resistance Test Method (TER)" and is a fairly rapid, relatively simple and inexpensive test to conduct. However, it does require the use of a humanely killed rat, and in some countries this may be regulated by national law relating to animal use. Even before adoption as an OECD Test Guideline, the TER test had been used by industry as a screening test for assessment of skin corrosion. It is still used by industry, but not extensively due to the more common use of validated skin corrosion test methods that do not require the housing of animals and use of the animal skin, and which, unlike the TER test, can also be used to generate data for subcategorization of corrosivity of substances and mixtures in accordance with classification schemes such as UN GHS or EU CLP. With the increasing desire to minimize the use of animals and animal tissues for testing of chemicals, it is feasible that the use of the rat skin TER test will further decline, whilst the use of methods not requiring the use of the animal skin will increase.

References

1. Oliver GJA, Pemberton MA, Rhodes C. An in vitro skin corrosivity test - modifications and validation. Fd Chem Toxicol. 1986;24:507–12.
2. Oliver GJA, Pemberton MA, Rhodes C. An in vitro model for identifying skin-corrosive chemicals I Initial Validation. Toxicol In Vitro. 1988;2:7–17.
3. Barlow A, Hirst RA, Pemberton MA, Rigden A, Hall TJ, Botham PA, Oliver GJA. Refinement of an in vitro test for the identification of skin corrosive chemicals. Toxicol Methods. 1991;1:106–15.
4. Botham PA, Hall TJ, Dennett R, McCall JC, Basketter DA, Whittle E, Cheeseman M, Esdaile DJ, Gardner J. The skin corrosivity test in vitro: results of an interlaboratory trial. Toxic In Vitro. 1992;6:191–4.
5. Botham PA, Chamberlain M, Barratt MD, Curren RD, Esdaile DJ, Gardner JR, Gordon VC, Hildebrand B, Lewis RW, Liebsch M, Logemann P, Osborne R, Ponec M, Regnier JF, Steiling W, Walker AP, Balls M. A prevalidation study on *in vitro* skin corrosivity testing. The report and recommendations of ECVAM workshop 6. ATLA. 1995;23:219–55.
6. Fentem JH, Archer GEB, Balls M, Botham PA, Curren RD, Earl LK, Esdaile DJ, Holzhütter H-G, Liebsch M. The ECVAM international validation study on in vitro tests for skin corrosivity 2. Results and evaluation by the Management Team. Toxic In Vitro. 1998;12:483–524.
7. OECD. Guideline for testing of chemicals (no. 430). In Vitro skin corrosion: transcutaneous electrical resistance test (TER). Originally adopted in 2004. Paris: Organisation for Economic Cooperation and Development; 2015.
8. United Nations (UN). Globally Harmonised System of Classification and Labelling of Chemicals (GHS). Sixth Revised Edition. Chapter 3.2 Skin Corrosion/Irritation. New York, Geneva: United Nations; 2015.
9. European Commission (1998). ECVAM statement on the scientific validity of the rat skin Transcutaneous Electrical Resistance (TER) test (an in vitro test for skin corrosivity). European commission joint research centre, environment institute, Ispra (VA), Italy presenting the results of the 10th ECVAM Scientific Advisory Committee (ESAC) meeting on 31 March 1998.
10. SCCNFP. 1998. Opinion on in vitro methods to assess skin Corrosivity in the safety evaluation of cosmetic ingredients or mixtures of ingredients adopted by the plenary session of the SCCNFP on 25 November 1998. Information obtained online at http://ec.europa.eu/health/scientific_committees/consumer_safety/opinions/sccnfp_opinions_97_04/sccp_out47_en.htm.
11. European Commission. B.40. *In Vitro* skin corrosion: transcutaneous electrical resistance test (TER). Annex, part B of Commission regulation (EC) no 440/2008 of 30 may 2008 laying down test methods pursuant to regulation (EC) no 1907/2006 of the European Parliament and of the Council on the registration, evaluation, authorisation and restriction of chemicals (REACH). Off J Eur Commun Legis. 2008;142(1):384–93.
12. Tice R., Haneke K.. Summary report of the rat skin transcutaneous electrical resistance (TER) In Vitro assay for assessing dermal Corrosivity. Drafted may 13, 1999, revised July 24, 2001. (tab 5.2 in BRD). National Institute of Environmental Health Sciences (NIEHS). Durham, NC: Research Triangle Park; 2001.
13. The National Toxicology Program (NTP) Interagency Center for the Evaluation of Alternative Toxicological Methods (NICEATM). Episkin™, Epiderm™, and rat skin Transcutaneous Electrical Resistance (TER). In vitro test methods for assessing the dermal corrosivity potential of chemicals. Background review document. National Institute of Environmental Health Sciences (NIEHS). Durham, NC: Research Triangle Park; 2001.
14. OECD. OECD series on testing and assessment no. 34. Guidance document on the validation and international acceptance of new or updated test methods for hazard assessment. Paris: Organisation for Economic Cooperation and Development; 2005.

15. OECD. OECD series on testing and Assessment no. 203. New guidance document on an integrated approach for testing and assessment on skin corrosion and irritation. Paris: Organisation for Economic Cooperation and Development; 2014.
16. OECD. OECD series on testing and Assessment no. 218. Performance standards for the Assessment of proposed similar or modified test methods for skin corrosion testing as described in TG 430. Paris: Organisation for Economic Cooperation and Development; 2015.
17. Eskes C, Detappe V, Koëter H, Kreysa J, Liebsch M, Zuang V, Amcoff P, Barroso J, Cotovio J, Guest R, Hermann M, Hoffmann S, Masson P, Alépée N, Arce LA, Brüschweiler B, Catone T, Cihak R, Clouzeau J, D'Abrosca F, Delveaux C, Derouette JP, Engelking O, Facchini D, Fröhlicher M, Hofmann M, Hopf N, Molinari J, Oberli A, Ott M, Peter R, Sá-Rocha VM, Schenk D, Tomicic C, Vanparys P, Verdon B, Wallenhorst T, Winkler GC, Depallens O. Regulatory Assessment of *In Vitro* skin corrosion and irritation data within the European framework: workshop recommendations. Regul Toxicol Pharmacol. 2012;62:393–403.
18. ECVAM DB-ALM. INVITTOX Protocol No. 115. Rat Skin Transcutaneous Electrical Resistance (TER) Test. Available at: http://ecvam-dbalm.jrc.ec.europa.eu/view_doc.cfm?iddoc=703&tdoc=prot.
19. European Commission. Regulation (EC) no 1272/2008 of the european parliament and of the council of 16 December 2008 on classification, labelling and packaging of substances and mixtures, amending and repealing directives 67/548/EEC and 1999/45/EC, and amending regulation (EC) no 1907/2006. Off J Eur Commun Legis. 2009;L353:87–91.
20. OECD. Guideline for testing of chemicals (no. 431). In Vitro skin corrosion: reconstructed human epidermis (RHE) test method. Paris: Organisation for Economic Cooperation and Development; 2016.
21. OECD. Guideline for testing of chemicals (no. 435). In vitro membrane barrier test method. Paris: Organisation for Economic Cooperation and Development; 2015.

Part III

Skin Sensitization

Overview on Current Status and Combination of Test Methods

14

Erwin L. Roggen

14.1 Introduction

The *mode of action (MOA)* concept, the *Tox21* strategy, the concept of *pathways of toxicity* and the *adverse outcome pathway (AOP)* framework aim at toxicity testing and assessment based on in-depth understanding of the *in vivo* physiological and toxicological processes in humans and on their relation to specific toxicological endpoints [1–5].

New technologies and paradigms are currently transforming these concepts into applicable animal-free toxicity testing systems by implementation of libraries of generic profiles of genes (genomics), proteins (proteomics) and metabolites (metabonomics) describing molecular initiators, pathways and key events of toxicity within tissues, organisms and biological systems [6].

14.2 The Mouse Local Lymph Node Assay (LLNA): A 'Surrogate' Gold Standard for Human Toxicity

The LLNA is the only animal-based test that underwent formal validation, which makes it the preferred gold standard for development of animal-free test methods for sensitization, when sufficient human data are not available. Compared with other *in vivo* methods (e.g. in guinea pig), the LLNA offers important advantages with respect to animal welfare, including a requirement for reduced animal numbers as well as reduced pain and trauma. In addition to hazard identification, the LLNA is used for determining the relative skin sensitizing potency of contact allergens as a pivotal contribution to the risk assessment process [7].

E.L. Roggen
3Rs Management and Consulting ApS, Kongens Lyngby, Denmark
e-mail: 3rsmc.eu@gmail.com

C. Eskes et al. (eds.), *Alternatives for Dermal Toxicity Testing*,
DOI 10.1007/978-3-319-50353-0_14

In the light of its importance for the development of useful animal-free test methods, the Cosmetic Directive prohibiting the use of animals, and the European Directive making (available) animal-free test methods the first choice, it is relevant to review the learnings provided by the LLNA with focus especially on misleading results, weight of evidence arguments for classification, labelling and risk assessments and the available body of information for the development of animal-free methods targeting defined chemical classes (see Chap. 15).

14.3 The Current Toolbox for Animal-Free Assessment of Skin Sensitization

Decades of intensive research have provided mechanistic understanding, toxicity pathways and components of these pathways, involved in *in vivo* sensitization as well as elicitation responses to xenobiotic allergens [8–10]. Ankley et al. [1] used the available information to suggest an MOA pathway that comprises distinct but interrelated events. This MOA pathway formed the basis for the development of the recently published OECD flow diagram of the AOP for skin sensitization induction [11].

14.3.1 Bioavailability: The Chemical Has the Capability to Reach the Viable Cell Layers

Before any of the suggested key events can occur, the compounds have to acquire access to the immune system. Useful *in vitro* and *in silico* methodologies for assessing skin absorption and systemic availability of chemicals exist [12, 13]. With the methodologies available, efforts should now go to the development and implementation of methods for quantification of compound disposition in, e.g. the skin, to obtain information on kinetics, potential tissue bioaccumulation and actual exposure at cellular level.

14.3.2 Haptenation: The Chemical Reacts Covalently with a 'Carrier Protein'

Sensitizing chemicals are generally reactive, electrophilic chemicals that form covalent bonds with nucleophilic nuclei on proteins (haptens), although some chemicals require activation by host enzymes (pro-haptens) or by oxidative derivatization (pre-haptens) to acquire sufficient electrophilicity. Occasionally, non-electrophilic binding occurs through disulphide exchange or co-ordination bonds [14]. The reaction rate and mechanism by which the chemical is reacting with the nucleophilic groups on the protein influence its allergenic potency [15, 16].

Tools for animal-free assessment of haptenation are well established and are extensively described in Chap. 16. The direct peptide reactivity assay (DPRA) is based on measurement of the reactivity of the hapten with two different peptides and allows for the detection of the majority of reactive chemicals [17]. Adding an incubation step with horseradish peroxidase and hydrogen peroxide (PPRA) provides a straightforward approach for detecting the peptide reactivity of pro-haptens [18]. The available data indicate that chemical reactivity with lysine appears to drive a type I sensitization response, while reactivity with cysteine results in a type IV response [19].

In vitro studies suggest that the specificity of the covalent modification is time and dose dependent and that the target proteins *in vitro* become more general and less discriminative over time and with increasing concentrations of the chemical [20]. A better understanding is required of the features (if any) on both chemical and targeted protein that make the resulting hapten–protein complex a sensitizer, determine potency and drive T helper cell type 1 (Th1)–Th2 skewing. This would render the current test methods into *in vitro* protein haptenation assays that provide a more complete data set on the tested chemicals.

14.3.3 Inflammation: Innate Recognition Followed by Activation of Innate Immunity

There is increasing evidence underpinning the central role of innate immune responses and inflammation in skin sensitization. Reactive oxygen species (ROS) play a central role in allergen-induced sensitization induction [21–23]. ROS production results in degradation of endogenous hyaluronic acid (HA) and TLR2 and TLR4 activation. The available evidence strongly points at TLR2 and TLR4 as crucial for activation of the Th1 responses in skin sensitization and allergic contact dermatitis [24, 25]. ROS also signal the 'nucleotide-binding domain and leucine-rich repeat containing family' protein 3 (NLRP3) inflammasome resulting in IL-1b, IL-18 and IL-33 activation. Especially, IL-18 is consistently found to play a role in skin sensitization induction, but not irritant contact dermatitis [26].

The acquired knowledge has resulted in several assays performed with human primary keratinocytes (KCs) or KC cell lines either as a submerged culture or as reconstructed human epidermis (RhE). The most advanced test methods are described in subsequent chapters and assess oxidative stress, IL-18 release or cytotoxicity.

14.3.3.1 Assessing Chemical Reactivity and Oxidative Stress

The relevance of 'nuclear factor erythroid 2-related factor 2' (Nrf2) – 'Kelch-like ECH-associated protein 1' (Keap1) pathway to skin sensitization is explained by the direct reactivity of most sensitizing materials to key cysteine residues of Keap1, an Nrf2 repressor protein. Test chemicals that exclusively react with lysine should therefore be considered outside the chemical applicability domain [27, 28]. The KC cell line-based methods (e.g. KeratinoSens™, LuSens) are typically cell-based reporter gene assay for screening substances with a full dose–response assessment. The induction of a luciferase gene under the control of the antioxidant response element (ARE) is determined (Chaps. 17 and 18). Other assays are based on RhE (e.g. SENS-IS, SenCeeTox). The SENS-IS method is based on the quantitative analysis of specific biomarkers for irritation and ARE gene expression in combination with a proprietary gene set expressed in EPISKIN™ (SkinEthic™) upon exposure to chemicals (Chap. 25). Similarly, the SensCeeTox approach combines markers for cell viability, Nrf2/ARE gene expression and direct reactivity over concentration and time in a proprietary algorithm (Chap. 26).

14.3.3.2 Assessing Inflammation by Measuring IL-18 Levels

IL-18 plays a proximal role in skin sensitization induction, but not in irritant contact dermatitis or asthma, by enhancing the secretion of pro-inflammatory mediators and by favouring Th-1-type immune response [21, 22]. The NCTC 2544 IL-18 test was shown to be potentially useful for identification of skin sensitizers (Chap. 19). Respiratory sensitizers and non-sensitizing irritants were consistently negative in this assay [21, 22]. In combination with a RhE-based irritation test (Chap. 20), a good potency categorization was obtained [29]. Gibbs et al. [30] integrated both tests for identification and classification of skin sensitizing chemicals, including chemicals of low water solubility or stability. This human *in vitro* assay appears is currently being subjected to additional testing of a larger chemical set to fully evaluate the utility of this assay and to establish a definitive prediction model.

14.3.3.3 Assessing Chemical Sensitizers Using a Specific Gene Signature

The analysis of chemical-induced changes in gene expression by the HaCaT human keratinocyte cell revealed ten genes that with high accuracy could discriminate sensitizers and non-sensitizers, including irritants. An algorithm was developed to compare changes in gene regulation of chemicals of unknown class to that induced by chemicals of known class. A chemical was assigned the most predominant class indicated by these algorithms [31].

While the involvement of TLR2, TLR4 and ROS signalling, and NLRP3 inflammasome assembly in skin sensitization is well established, there is growing evidence that these mechanisms also play a role in lung sensitization [32]. Thus, a better understanding is required about the subtle balance between danger signals or intracellular interactions promoting distinct immune phenotypes. Furthermore, it is imperative to understand how reactivity rate and mechanisms of haptenation affect this balance.

14.3.4 Dendritic Cell Activation: From Innate Responses to Dendritic Cell (DC) Maturation

It is generally accepted that activation of DCs results in prominent phenotypic and functional changes including enhanced levels of MHC class I and co-stimulatory molecules (e.g. cluster of differentiation (CD54, CD80 and CD86) and receptors that are essential for migration) and antigen-presenting capacity [33].

Extensive genomic analysis of monocyte-derived dendritic cells (Mo-DCs), human monocytic leukaemia cell line (THP-1) and MUTZ-3 cells exposed to skin sensitizers exerting cysteine and cysteine/lysine reactivity has identified genes describing the primary pathways of skin sensitization, i.e. signalling through transcription factors Nrf2 and aryl hydrocarbon receptor (AHR), and protein ubiquitination [28, 34]. Lysine-reactive chemicals appeared to be less efficient [28].

By stratifying the sensitizing chemicals into chemical reactivity groups, a number of canonical pathways known to be involved in the biology of sensitization were confirmed, while novel pathways were identified. Sensitizers with different reactivity mechanisms or potency were further shown to engage different pathways, indicating that the biological endpoint of T-cell priming is achieved through different upstream mechanisms [15].

The available tools for assessing DC activation and maturation can be grouped based on the read-out for sensitization, i.e. specific CD marker, gene profile or protein signature.

14.3.4.1 Assessing Phenotypic Changes

Peripheral blood mononuclear cells (PBMCs) or CD34+ hematopoietic progenitor cells are applied to identify chemicals as sensitizers based on their capacity to trigger expression of CD86/CD54, IL-1b release or internalization of MHC class II molecules (Chap. 23). However, generation of DC like cells often is considered time-consuming and complicated, with donor-to-donor variability introducing uncertainty about negative results. Therefore, cell lines with some characteristics of DC have gained much attention. The most advanced DC maturation test is the human cell line activation test (h-CLAT), which uses THP-1 cell and measures changes in CD86 and CD54 expression levels [35]. There are indications that the h-CLAT correlates with the LLNA and may have the potential to provide information about the potency of the test chemical (Chap. 21). The MUSST (recently renamed 'U-Sens') and the modified MUSST assays are possible alternatives to the h-CLAT assay (Chap. 22). These protocols are based on the U937 cell line and predict a chemical to be a sensitizer if it induces a dose-dependent increase in CD86 expression [36].

A foetal skin-derived dendritic cell (FSDC) line was described to discriminate sensitizers and irritants through differential effects on CD40 and CXCR4 protein expression (Chap. 29). The assessed sensitizers increased CD40 and CXCR4 levels, while irritants decreased the expression of both proteins. Since these observations were similar to those obtained with monocyte-derived dendritic cells, the FSDC test method was suggested as a potential tool for *in vitro* assessment of chemicals for their sensitizing potential [37].

14.3.4.2 Assessing Changes in Gene Expression

Transcriptomic analyses revealed a discriminating gene expression profile in human CD34+ progenitor-derived DC after exposure to skin sensitizers versus non-sensitizers. By comparing their responsiveness towards a non-sensitizing danger signal and a sensitizer, VITOSENS® gene markers CREM and CCR2 appeared to display a specific response [38] (29). Functional and transcriptional analysis of various myeloid cell lines has clearly demonstrated the significance of the MUTZ-3 cell line as a model for functional studies of inflammatory responses [39]. The genomic allergen rapid detection (GARD) test is a MUTZ-3-based assay for assessing chemical sensitizers utilizing genomic biomarker prediction signatures to generate prediction calls of unknown chemicals as skin sensitizers, respiratory sensitizers or non-sensitizers, including irritants [34]. The predictive performance of the GARD for skin sensitization was assessed in an in-house validation study (Chap. 27).

14.3.4.3 Assessing Changes in Protein Expression

More than 200 proprietary skin and lung markers emerged from the EU-funded FP6 project Sens-it-iv using the mass spectroscopy-based proteomic biomarker discovery platform of Proteome Sciences [40]. Specific assays were developed using its tandem mass tagging technology combined with selected reaction monitoring mass spec. SensiDerm™ applies a biomarker panel comprising ten proteins which were shown to be differentially expressed in MUTZ-3 cells in response to sensitizers compared to non-sensitizers (Chap. 28).

A better understanding of how the early gene changes contribute (or not) to the expression of maturation markers may help to understand and resolve the reasons behind the low specificity of the available test methods. Furthermore, understanding better the association between pathway activity and chemical class will help the development of assays for subcategorization.

14.3.5 Dendritic Cell Migration: Translating the Message into Specific Actions

The molecular mechanisms driving migration of DC to and from peripheral tissues were reviewed [41]. Fibroblasts play a key role both as advisors helping the KCs and Langerhans cells (LCs) to discriminate irritants from sensitizers, which in many cases are irritants themselves, and as guides helping the LCs out of the epidermis into the dermis and further towards lymphatic vessels [42]. Using a full-thickness tissue-engineered skin model containing fully functional MUTZ-3-derived LCs (MUTZ-LC), the MUTZ-LCs were demonstrated to mature and to acquire the ability to migrate towards C-X-C motif ligand (CXCL)12 and C-C motif ligand (CCL)19/21 in a comparable manner with primary LCs in skin explants [43].

The acquired knowledge has resulted in a DC migration assay, which is based on carboxyfluorescein succinimidyl ester (CFSE)-labelled MUTZ-3 cells. The discriminating feature of the assay is that irritant-induced migration is CCL5 dependent, while sensitizer-induced migration is CXCL12 dependent. The read-out of the test is the ratio between migration towards CXCL12 or to CCL5 [44].

While the preliminary data on 12 chemicals are promising (no misclassification), further evaluation performed with more chemicals is required. The test is also expensive and rather complicated which may hamper its application by industry. More work is required to refine the test to make it more attractive for industrial use.

14.3.6 T-Cell Priming and Proliferation: The Turning Point

Primary T-cell responses in lymph nodes require contact-dependent information exchange between T cells and DCs. The available evidence indicates that T-cell priming by DCs occurs in three successive stages. Transient serial encounters during the first activation phase (T-cell activation) are followed by a second phase of stable contacts culminating in cytokine production antigen-driven T-cell proliferation, which triggers a transition into a third phase of high motility and rapid proliferation (antigen-independent and IL-12-driven proliferation) [45, 46]. Studies in mice have exposed the induction of two functionally polarized populations of T cells, distinguished by patterns of cytokine production [47]. Th17 cells were shown to play a crucial role in allergen-specific cellular and humoral immune responses through the activation of both contact hypersensitivity and airway hyperresponsiveness. It has been suggested that IL-17 has activities similar to the pro-inflammatory cytokines IL-1 and TNF-a, which are known to have crucial roles in the induction of other cytokines, chemokines and adhesion molecules. It is also known that IL-17 itself is a potent inducer of IL-1 and TNF-α acting on macrophages and keratinocytes [48, 49].

The human T-cell priming assay (hTCPA) described in Chap. 31 is based on a coculture system that measures the effect derived from the contact of freshly isolated T cells with autologous DC cells previously activated and chemically modified by the test substance. The test is usually repeated on T cells derived from five different donors to minimize donor-to-donor variability. T cells are isolated again and restimulated with autologous DC and the same control chemicals to assess antigen specificity [50]. Comparison of the proliferation during stimulation and restimulation is an important parameter. The second, more reliable and robust read-out is when the expression of IFN-γ and TNF-α is measured simultaneously. Preliminary results demonstrated the capability of this assay to effectively predict antigenicity of chemicals, including drugs.

Our understanding of the T-cell populations that are activated by xenobiotics is increasing. It is, however, not clear yet how Th1–Th2 skewing and the balance between regulatory and effector T cells is controlled. T-cell stimulation is a pivotal event, being part of the sensitization induction phase as well as the clinical phase. Potency assessment performed with T-cell-based assays needs therefore to build on an in-depth understanding of mechanisms behind potency of sensitization induction on one hand and severity of clinical symptoms on the other hand.

14.4 Skin-Based Methods for Assessing the Clinical Outcome of T-Cell Stimulation

From the previous it should be clear that there are a number qualified test methods available for assessing the sensitizing potential of chemical substances. However, these methods rely on responses by either KCs or isolated immune cells (e.g. DCs) upon exposure to a chemical challenge, while *in vivo* responses to a challenge are the result of alterations in cell-cell interactions and micro-environments [42, 51]. Furthermore, the available tools identify hazard but do not provide information about the clinical consequence of this hazard.

These aspects are to some extent covered by Skimune™ (Chap. 30). This approach involves taking blood and skin tissue samples from human donors and culturing, or growing, them in the lab. The first step of the approach, like the hTCPA discussed earlier, involves treatment of monocyte-derived DCs with a chemical substance and culturing with autologous T cells. Subsequently, the activated cells are cocultures with autologous skin. T-cell proliferation and interferon-γ secretion are measured in separate assays, while the skin explants are analysed histologically. By observing how the skin sample reacts to its own immune cells, potentially allergenic substances can be identified.

While such approaches provide information about the clinical relevance of the chemical-specific T cells, no information is acquired with respect to skin penetration, bioavailability and early inflammatory responses leading to sensitization.

14.5 Integrated Approaches for Testing and Assessment

Each of the test methods described in this chapter represents one of the key events currently believed to be essential for the development of skin sensitization [52]. It is anticipated that full replacement of animal-based testing and assessment of chemical substances for their potential of being a sensitizer requires testing strategies combining several of the test methods described. However, it is not clear yet how many different key events should be addressed.

While the implementation of testing strategies for safety assessment is not new (OECD TG 404, 2002; OECD TG 405, 2002), the definition and set-up of such strategies differ considerably among the various industry sectors, while variations on the theme exist within a specific sector or even company [6, 53]. In an effort to harmonize reporting, the OECD has been coordinating efforts aiming at the establishment of a guidance document on the reporting of integrated approaches to testing and assessment (IATA). This has recently resulted in the endorsement by OECD of two guidance documents on the reporting of IATAs [54, 55].

In the area of skin sensitization, several strategies have been assessed, while more are coming (Table 14.1). Bauch and co-workers (2012) developed a strategy based on protein reactivity, activation of the Keap-1/Nrf2 signalling pathway and DC activation. The accuracy of the emerging prediction model was 94% as compared to human data and 83% as compared to the LLNA for the 54 chemicals tested. Nukada et al. [59] suggested that a tiered testing strategy using the h-CLAT and DPRA had a practical utility in skin sensitization screening of 101 chemicals, on the basis of a 96% sensitivity. In an effort to bridge the gap between qualitative and quantitative approaches, Jaworska et al. [60–62] developed a methodology to design a testing strategy in the form of a Bayesian network. For the development of the most recent ITS-3 strategy, 207 chemicals were used for which *in vitro* (h-CLAT, KeratinoSens™), *in chemico* (DPRA) and *in silico* test data were available. The

Table 14.1 Overview of integrated approaches for testing and assessment

Strategy/IATA	Input data	Chemicals (*n*)	Purpose	Accuracy
Bauch et al. [56]	DPRA, LuSens, mMUSST or DPRA, KeratinoSens, mMUSST	54	Hazard ID	WoE is positive if 2 of 3 *in vitro* tests are positive. Human hazard, 94% LLNA hazard: 83%
Gomes et al. [57] Stacking metamodel	PhysChem properties, various (Q)SAR models, DPRA, KeratinoSens, U-Sens	165	Hazard ID	Performance for validation set ($n = 40$), LLNA hazard: 92.5%
Natsch et al. [58]	MUSST, KeratinoSens™ and DPRA	145	Hazard ID	Positive if 2 of 3 *in vitro* tests are positive. LLNA hazard, 81%
Nukada et al. [59]	h-CLAT and DPRA	101	Hazard ID and potency	The sum of individual scores: LLNA hazard, 85% and LLNA potency (3 classes), 71% The straightforward tiered system: LLNA hazard, 86%, and LLNA potency (3 classes), 73%
Jaworska et al. [60–62] Bayesian network ITS-1, 2 and 3	DPRA, KeratinoSens, U937 (sub for U-SENS), Bioavailability (C-free, AUC120, log Kow), TIMES-SS Predictions	207 (ITS-3)	Hazard ID and potency	ITS-3 results with external test set ($n = 60$) LLNA hazard, 100% GHS potency (3 classes), 96% LLAN potency (4 classes), 89%
van der Veen et al. [63]	QSARs, DPRA, HaCaT gene signature, h-CLAT	41	Hazard ID	Tiered testing strategy. LLNA hazard, 100%
Patlewicz et al. [64]	PhysChem properties, (Q)SARs, DPRA, glutathione depletion assay, KeratinoSens, h-CLAT, U937, LLNA, GPMT, other relevant data (e.g. skin irritation/corrosion, Ames)	100	Hazard ID	*In silico* and *in chemico* data alone showed an LLNA hazard accuracy of 73.85%. Info from other endpoints (e.g. Ames mutagenicity) improved accuracy to 87.6%
Natsch et al. [65]	KeratinoSens, Cor1C420-assay, TIMES-SS and PhysChem properties	312	Hazard ID and potency	KeratinoSens™ Positive and/or adduct forming, LLNA hazard accuracy of 82.8% ($n = 312$) Global analysis yielded a regression of *in vitro* data to LLNA potency with an R2 of 60% predicting LLNA EC3 with a mean error of 3.5-fold ($n = 244$) Correlation of chemicals tested positive *in vitro* with human data showed an R2 of 49% ($n = 63$) (similar correlation between LLNA and human)

(continued)

Table 14.1 (continued)

Strategy/IATA	Input data	Chemicals (n)	Purpose	Accuracy
Urbish et al. [66] '2 out of 3' prediction model	DPRA, h-CLAT, KeratinoSens, U-SENS	213	Hazard ID	'2 out of 3' positive tests, rates a positive prediction LLNA hazard, 79% Human hazard, 90%
Hirota et al. [67] Artificial neural network	(1), h-CLAT/DPRA, (2) DPRA/ARE assay, (3) SH test/ARE assay, (4) h-CLAT/DPRA/ARE assay, (5) h-CLAT/SH test/ARE assay	139	Hazard ID and potency	The best performing models included: h-CLAT/DPRA model LLNA hazard, 69.1% (n = 139) h-CLAT/DPRA/ARE model LLNA hazard, 69.6% (n = 69) h-CLAT/SH test/ARE model LLNA hazard, 75.3% (n = 73)
Takenouchi et al. [68] Integrated testing strategy (ITS) and sequential testing strategy (STS)	h-CLAT, DPRA and DEREK	139 (chemicals with log Kow >3.5 were excluded)	Hazard ID and potency	LLNA hazard data ITS, 89% (n = 114 of 128) LLNA hazard data STS, 85% (n = 111 of 129) LLNA potency data 3 classes, 74% (ITS) LLNA potency data 3 classes, 73% (STS)
Strickland et al. [69] Machine learning approaches incl. logistic regression and support vector machine	DPRA, h-CLAT, KeratinoSens™, the OECD QSAR Toolbox 6 PhysChem properties	LLNA hazard, 94 training set and 26 external test set Human hazard, 72 training set and 24 external test set	Hazard ID	LLNA hazard: 7 models with the highest accuracy 89–96% (test set) and 96–99% (training set) used a support vector machine (SVM) approach with different combinations of predictor variables Human hazard: 6 models with the highest accuracy 92% used: (1) DPRA, h-CLAT and read-across; (2) DPRA, h-CLAT, read-across and KeratinoSens; or (3) DPRA, h-CLAT, read-across, KeratinoSens and log P
Asturiol D et al. [70]	DPRA, KeratinoSens™, h-CLAT, complemented with predictions from several software packages	269	Hazard ID	Consensus of two classification trees based on descriptors for protein reactivity and structural features LLNA hazard, 93%

accuracy of the strategy for predicting LLNA hazard (two classes) was 100%, GHS potency classification (three classes) 96% and LLNA potency (four classes) 89% using an external test set of 60 chemicals. Natsch et al. [58] tested a database of 145 chemicals in the MUSST, KeratinoSens™ and the DPRA. The tests and combination of tests were evaluated for predictivity. Furthermore, analysis of the dose–response parameters of the individual tests indicated a correlation to sensitization potency (Natsch et al.) [65]. van der Veen et al. [63] designed a tiered strategy based on the complementary characteristics of QSARs, DPRA, KeratinoSens™, HaCaT gene signature, IL-18 release and h-CLAT that was able to correctly identify all 41 chemicals tested. In line with the previous reports, Urbisch et al. [66] and Hirota et al. [67] showed that various combinations of descriptors from several *in vitro* sensitization test models were more predictive than the individual descriptors. In general, the nonanimal test methods exhibited good predictivities when compared to local lymph node assay (LLNA) data and even better predictivities when compared to human data. However, better performance was obtained with the '2 out of 3' prediction model, which achieved accuracies of 90% or 79% when compared to human or LLNA data, respectively (Urbisch et al.) [66]. Hirota et al. [67] observed that some combinations (e.g. h-CLAT + DPRA) performed better than others (e.g. DPRA + antioxidant response element-based assay). Takenouchi et al. [68] evaluated a testing strategy incorporating the h-CLAT, DPRA and DEREK, by comparison to LLNA data using a data set of 139 chemicals. The combinations of the methods were based on integrated testing strategy (ITS) concept and a sequential testing strategy (STS). After exclusion of the negative results for chemicals with log Kow > 3.5, the sensitivity and accuracy of ITS and STS improved to 97% and 89% and 98% and 85%, respectively. Moreover, both strategies showed good correlation with LLNA on three potency classifications, yielding accuracies of 74% (ITS) and 73% (STS). Most recently, Strickland et al. [69] evaluated an IATA including machine learning approaches to predict human skin sensitization hazard. Data from three *in chemico* or *in vitro* assays (DPRA, h-CLAT and KeratinoSens™), six physicochemical properties and an *in silico* read-across prediction of skin sensitization hazard were combined into 12 variable groups, which were evaluated for their predictive capacity using two machine learning approaches, logistic regression and support vector machine. Models were trained on 72 substances and tested on an external set of 24 substances. The 6 models with the highest accuracy of 92% for human hazard used: (1) DPRA, h-CLAT and read-across; (2) DPRA, h-CLAT, read-across and KeratinoSens; or (3) DPRA, h-CLAT, read-across, KeratinoSens and log P. The models were better predictors of human skin sensitization hazard than the LLNA (88%), any of the alternative methods alone (63–79%) or test batteries combining data from the individual methods (75%). The seven models with the highest accuracy for predicting LLNA hazard (89–96% test set and 96–99% for training set) used a support vector machine (SVM) approach with different combinations of predictor variables. The performance statistics of the SVM models were higher than any of the nonanimal tests alone and higher than simple test battery approaches using these methods [71]. Finally, Asturiol et al. [70] combined DPRA, KeratinoSens™, h-CLAT and predictions from several software packages for hazard identification. As compared to the LLNA, consensus of two classification trees based on descriptors for protein reactivity and structural features revealed an accuracy of 93%.

14.6 Summary

During the last decade, several methods for assessing skin and respiratory sensitization have emerged. Some have entered the pre-validation process, and other less advanced assays generated interesting contributions to the molecular understanding of sensitization mechanisms. There are also promising animal-free strategies emerging, which in contrast to the *in vivo* studies distinguish between skin and respiratory sensitizers. Evaluation of the potency of chemical sensitizers with *in vitro* methods may become reality in the near future.

Our understanding of the molecular mechanisms driving skin sensitization and contact dermatitis is substantial. While the most advanced tools can be used for classification, our understanding of the relation between reactivity rate, mechanism of haptenation, protein target selection, pathway activation and T-cell skewing is still not sufficient to fully describe chemicals using animal-free testing methods. More efforts should be addressed to refine existing methods and to further develop new methods that lead to an improved awareness of the real mechanisms of a chemical in triggering a sensitization reaction in exposed human beings.

The currently available methods have proven complementary. Most important however is the observation that detailed analysis of chemicals producing misleading results may help to define limitations of the respective tests as well as of the database derived from animal studies.

Several testing strategies have emerged, and more will emerge in the near future. Currently, the usefulness of these strategies in the context of an IATA for skin sensitization is being evaluated by several groups in the context of the OECD initiatives.

References

1. Ankley GT, Bennett RS, Erickson RJ, Hoff DJ, Hornung MW, Johnson RD, Mount DR, Nichols JW, Russom CL, Schmieder PK, Serrano JA, Tietge JE, Villeneuve DL. Adverse outcome pathways: a conceptual framework to support ecotoxicology research and risk assessment. Environ Toxicol Chem. 2010;29:730–41.
2. OECD. Proposal for a template and guidance on developing and assessing the completeness of adverse outcome pathways. 2012.
3. OECD. Guidance document on developing and assessing adverse outcome pathways. 2013.
4. US-EPA. Guidelines for carcinogen risk assessment. Washington, DC: U.S. Environmental Protection Agency; 2005.
5. US National Research Council. Toxicity testing in the 21st century:a vision and a strategy. Washington, DC: The National Academies Press; 2007.
6. Berg N, DeWever B, Fuchs HW, Gaca M, Krul C, Roggen EL. Toxicology in the 21st century – working our way towards a visionary reality. Toxicol In Vitro. 2011;25:874–81.

7. Rovida C, Ryan C, Cinelli S, Basketter D, Dearman R, Kimber I. The local lymph node assay (LLNA). Curr Protoc Toxicol. 2012;20:1–14.
8. Basketter DA, Kimber I. Contact hypersensitivity. In: McQueen CA, editor. Comprehensive toxicology, vol. 5. 2nd ed. Kidlington: Elsevier; 2010. p. 397–411.
9. Corsini E, Roggen EL. Immunotoxicology: opportunities for non-animal test development. Altern Lab Anim. 2009;37:387–97.
10. Roggen EL. Application of the acquired knowledge and implementation of the Sens-it-iv toolbox for identification and classification of skin and respiratory sensitizers. Toxicol In Vitro. 2012;27:1122–6. doi:10.1016/j.tiv.2012-09.019.
11. OECD. The adverse outcome pathway for skin sensitization initiated by covalent binding to proteins. Part 1: scientific evidence, series on testing and assessment. 2013;168.
12. Basketter D, Pease C, Kasting G, Kimber I, Casati S, Cronin M, et al. Skin sensitization and epidermal disposition: the relevance of assessment. Altern Lab Anim. 2007;35:137–54.
13. Smith CK, Hotchkiss SAM. Allergic contact dermatitis: chemical and metabolic mechanisms. London: Taylor & Francis Ltd; 2001. doi:10.1034/j.1600-0536.2001.450224-2.x.
14. Chipinda, I, Hettick JM, Siegel PD. Haptenation: chemical reactivity and protein binding. J Allergy. 2011; 839682. doi: 10.1155/2011/839682
15. Albrekt AS, Johansson H, Börje A, Borrebaeck CAK, Lindstedt M. Differentially regulated signalling pathways in MUTZ-3 cells stimulated with skin sensitizers. BMC Pharmacol Toxicol. 2014.; http://www.biomedcentral.com/2050-6511/15/5.
16. Chipinda I, Ajibola RO, Morakinyo MK, Ruwona TB, Simoyi RH, Siegel PD. Rapid and simple kinetics screening assay for electrophilic dermal sensitizers using nitrobenzenethiol. Chem Res Toxicol. 2010;23:918–25.
17. Gerberick F, Aleksic M, Basketter D, Casati S, Karlberg AT, Kern P, et al. Chemical reactivity measurement and the predictive identification of skin sensitisers. The report and recommendations of ECVAM workshop 64. Altern Lab Anim. 2008;36:215–42.
18. Gerberick F, Troutman J, Foertsch L, Vasallo JD, Quijano M, Dobson RLM, et al. Investigation of peptide reactivity of pro-hapten skin sensitizers using a peroxidase-peroxide oxidation system. Toxicol Sci. 2009;112:164–74.
19. Lalko JF, Dearman RJ, Gerberick GF, Troutman JA, Api AM, Kimber I. Reactivity of chemical respiratory allergens in the peroxidase peptide reactivity assay. Toxicol In Vitro. 2012;27:651–61.
20. Manchanda T, Hess D, Dale L, Ferguson SG, Rieder MJ. Haptenation of sulfonamide reactive metabolites to cellular proteins. Mol Pharmacol. 2002;62:1011–26.
21. Corsini E, Galbiati V, Mitjans M, Galli CL, Marinovich M. NCTC 2544 and IL-18 production: a tool for the identification of contact allergens. Toxicol In Vitro. 2013;27:1127–34.
22. Corsini E, Galbiati V, Nikitovic D, Tsatsakis AFM. Role of oxidative stress in chemical allergens induced skin cells activation. Food Chem Toxicol. 2013;61:74–81.
23. Ray PD, Huang BW, Tsuji Y. Reactive oxygen species (ROS) homeostasis and redox regulation in cellular signalling. Cell Signal. 2012;24:981–90.
24. Esser PR, Wölfle U, Dürr C, von Loewenich FD, Schempp CM, Freudenberg MA, et al. Contact sensitizers induce skin inflammation via ROS production and hyaluronic acid degradation. PLoS One. 2012;7:e41340.
25. Martin SF, Dudda JC, Bachtanian E, Lembo A, Liller S, Durr C, et al. Toll-like receptor and IL-12 signaling control susceptibility to contact hypersensitivity. J Exp Med. 2008;205:2151–62.
26. Antonopoulos C, Cumberbatch M, Mee JB, Dearman RJ, Wei X, Liew FY, et al. IL-18 is a key proximal mediator of contact hypersensitivity and allergen-induced Langerhans cell migration in murine epidermis. J Leukoc Biol. 2008;83:361–7.
27. Emter R, van der Veen JW, Adamson G, Ezendam J, van Loveren H, Natsch A. Gene expression changes induced by skin sensitizers in the KeratinosensTM™ cell line: discriminating Nrf2-dependent and Nrf2-independent events. Toxicol In Vitro. 2013;27:2225–32.
28. Migdal C, Botton J, El Ali Z, Azoury ME, Guldemann J, Gimenez-Arnau E, et al. Reactivity of chemical sensitizers toward amino acids in cellulo plays a role in the activation of the Nrf2-ARE pathway in human monocyte dendritic cells and the THP-1 cell line. Toxicol Sci. 2013;133:259–74.

29. Dos Santos GG, Spiekstra SW, Sampat-Sardjoepersad SC, Reinders J, Scheper RJ, Gibbs S. A potential in vitro epidermal equivalent assay to determine sensitizer potency. Toxicol In Vitro. 2011;25:347–57.
30. Gibbs S, Corsini E, Spiekstra SW, Galbiati V, Fuchs HW, Degeorge G, Troese M, Hayden P, Deng W, Roggen EL. An epidermal equivalent assay for identification and ranking potency of contact sensitizers. Toxicol Appl Pharmacol. 2013;15:529–41.
31. van der Veen JW, Pronk TE, van Loveren H, Ezendam J. Applicability of a keratinocyte gene signature to predict skin sensitizing potential. Toxicol In Vitro. 2013;27:314–22.
32. Roggen EL. In vitro approaches for detection of chemical sensitization. Basic Clin Pharmacol Toxicol. 2014;115:32–40.
33. Tan JKH, O'Neil HC. Maturation requirements for dendritic cells in T cell stimulation leading to tolerance versus immunity. J Leukoc Biol. 2005;78:319–24.
34. Johansson H, Lindstedt M, Albrekt AS, Borrebaeck CAK. A genomic biomarker signature can predict skin sensitizers using a cell-based in vitro alternative to animal tests. BMC Genomics. 2011;12:399–417.
35. Ashikaga T, Yoshida Y, Hirota M, Yoneyama K, Itagaki H, Sakaguchi H, et al. Development of an in vitro skin sensitization test using human cell lines; human cell line activation test (h-CLAT). I. Optimization of the h-CLAT protocol. Toxicol In Vitro. 2006;20:767–73.
36. Python F, Goebel C, Aeby P. Assessment of the U937 cell line for the detection of contact allergens. Toxicol Appl Pharmacol. 2007;220:113–24.
37. Neves BM, Cruz MT, Francisco V, Gonçalo M, Figueiredo A, Duarte CB, Lopes MC. Differential modulation of CXCR4 and CD40 protein levels by skin sensitizers and irritants in the FSDC cell line. Toxicol Lett. 2008;177:74–82.
38. Lambrechts N, Nelissen I, Van Tendeloo V, Witters H, Van Den Heuvel R, Hooyberghs J, Schoeters G. Functionality and specificity of gene markers for skin sensitization in dendritic cells. Toxicol Lett. 2011;203:106–10.
39. Lundberg K, Albrekt A-S, Nelissen I, Santegoets S, de Gruijl TD, Gibbs S, et al. Transcriptional profiling of human dendritic cell populations and models - unique profiles of in vitro dendritic cells and implications on functionality and applicability. PLoS One. 2013;8:e52875.
40. Thierse HJ, Budde P, Dietz L, Ohnesorge S, Eikelmeier S, Conde M, Zucht HD, Schulz-Knappe P. Proteomic identification of allergen-regulated proteins and allergen-protein interaction networks in assisting biomarker and assay development. In: Roggen EL, Weltzien H-U, Hermans H, editors. Progress towards novel testing strategies for in vitro assessment of allergens. Kerala: Transworld Research Network; 2011. p. 145–66.
41. Villablanca EJ, Russo V, Mora JR. Dendritic cell migration and lymphocyte homing imprinting. Histol Histopathol. 2008;23:897–910.
42. Ouwehand K, Scheper RJ, de Gruijl TD, Gibbs S. Epidermis-to dermis migration of immature Langerhans cells upon topical irritant exposure is dependent on CCL2 and CCL5. Eur J Immunol. 2012;40:2026–34.
43. Ouwehand K, Spiekstra SW, Waaijman T, Scheper RJ, de Gruijl TD, Gibbs S. Langerhans cells derived from a human cell line in a full-thickness skin equivalent undergo allergen-induced maturation and migration. Tech Adv. 2011;90:1028–33.
44. Rees B, Spiekstra SW, Carfi M, Ouwehand K, Williams CA, Corsini E, et al. Inter-laboratory study of the in vitro DC migration assay for identification of contact allergens. Toxicol In Vitro. 2011;25:2124–3445.
45. Jelley-Gibbs DM, Strutt TM, McKinstry KK, Swain SL. Influencing the fates of CD4 T cells on the path to memory: lessons from influenza. Immunol Cell Biol. 2008;86:343–52.
46. Mempel TR, Henricksen SE, von Andrian UH. T-cell priming by dendritic cells in lymph nodes occurs in three distinct phases. Nature. 2004;8:154–9.
47. Xu H, Dilulio NA, Fairchild RL. T cell populations primed by hapten sensitization in contact sensitivity are distinguished by polarized patterns of cytokine production: interferon gamma-producing(Tc1) effector CD8+ T cells and interleukin (IL) 4/IL-10 negative regulatory CD4+ T cells. J Exp Med. 1996;183:1001–12.

48. Albanesi C, Scarponi C, Cavani A, Federici M, Nasorri F, Girolomoni G. Interleukin-17 is produced by both Th1 and Th2 lymphocytes, and modulates interferon-gamma and interleukin-4-induced activation of human keratinocytes. J Invest Dermatol. 2000;115:81–7.
49. Nakae S, Komiyama Y, Nambu A, Sudo K, Iwase M, Homma I, et al. Antigen-specific T cell sensitization is impaired in IL-17-deficient mice, causing suppression of allergic cellular and humoral responses. Immunity. 2002;17:375–87.
50. Richter A, Schmucker SS, Esser PR, Traska V, Weber V, Dietz L, et al. Human T cell priming assay (hTCPA) for the identification of contact allergens based on naive T cells and DC–IFN-c and TNF-a readout. Toxicol In Vitro. 2013;27:1180–5.
51. Roggen EL, Lindstedt M, Borrebaeck C, Verheyen GR. Interactions between dendritic cells and epithelial cells in allergic disease. Toxicol Lett. 2006;162:71–82.
52. Adler S, Basketter D, Creton S, Pelkonen O, van Benthem J, Zuang V, Andersen KE, Angers-Loustau A, Aptula A, Bal-Price A, Benfenati E, Bernauer U, Bessems J, Bois FY, Boobis A, Brandon E, Bremer S, Broschard T, Casati S, Coecke S, Corvi R, Cronin M, Daston G, Dekant W, Felter S, Grignard E, Gundert-Remy U, Heinonen T, Kimber I, Kleinjans J, Komulainen H, Kreiling R, Kreysa J, Leite SB, Loizou G, Maxwell G, Mazzatorta P, Munn S, Pfuhler S, Phrakonkham P, Piersma A, Poth A, Prieto P, Repetto G, Rogiers V, Schoeters G, Schwarz M, Serafimova R, Tähti H, Testai E, van Delft J, van Loveren H, Vinken M, Worth A, Zaldivar JM. Alternative (non-animal) methods for cosmetics testing: current status and future prospects-2010. Arch Toxicol. 2011;85:367–485.
53. Kinsner-Ovaskainen A, Rzepka R, Rudowski R, Coecke S, Cole T, Prieto P. Acutoxbase, an innovative database for in vitro acute toxicity studies. Toxicol In Vitro. 2009;23:476–85.
54. OECD. Guidance document on the reporting of defined approaches to be used within integrated approaches for testing and assessment - ENV/JM/HA(2016)10. 2016.
55. OECD. Guidance document on the reporting of defined approaches and individual information sources to be used within integrated approaches to testing and assessment (IATA) for skin sensitization - ENV/JM/HA(2016)11. 2016.
56. Bauch C, Kolle SN, Ramirez T, Eltze T, Fabian E, Mehling A, Teubner W, van Ravenzwaay B, Landsiedel R. Putting the parts together: combining in vitro methods to test for skin sensitizing potential. Regulat Toxicol Pharmacol. 2012;63:489–504.
57. Gomes C, Hicham N, Thomas M, Ibanez F, Collin JF, Saporta G. Stacking prediction for a binary outcome. In: Compstat, Aug 2012, Limassol, Cyprus. 2012; pp.271–282.
58. Natsch A, Ryan CA, Foertsch L, Emter R, Jaworska J, Gerberick F, Kern P. A dataset on 145 chemicals tested in alternative assays for skin sensitization undergoing prevalidation. J Appl Toxicol. 2013;33(11):1337–52. doi:10.1002/jat.2868.
59. Nukada Y, Miyazawa M, Kazutoshi S, Sakaguchi H, Nishiyama N. Data integration of non-animal tests for the development of a test battery to predict the skin sensitizing potential and potency of chemicals. Toxicol In Vitro. 2013;27(2):609–18.
60. Jaworska J, Dancik Y, Kern P, Gerberick F, Natsch A. Bayesian integrated testing strategy to assess skin sensitization potency: from theory to practice. J Appl Toxicol. 2013;33(11):1353–64. doi:10.1002/jat.2869.
61. Jaworska J, Harol A, Kern PS, Gerberick GF. Integrating non-animal test information into an adaptive testing strategy - skin sensitization proof of concept case. ALTEX. 2011;28(3):211–25.
62. Jaworska J, Natsch A, Ryan C, Strickland J, Ashikaga T, Miyazawa M. Bayesian integrated testing strategy (ITS) for skin sensitization potency assessment: a decision support system for quantitative weight of evidence and adaptive testing strategy. Arch Toxicol. 2015;89(12):2355–83.
63. van der Veen JW, Rorije E, Emter R, Natsch A, van Loveren H, Ezendam J. Evaluating the performance of integrated approaches for hazard identification of skin sensitizing chemicals. Regul Toxicol Pharmacol. 2014;69:371–9.
64. Patlewicz G, Casati S, Basketter DA, Asturiol D, Roberts DW, Lepoittevin JP, Worth AP, Aschberger K. Can currently available non-animal methods detect pre and pro-haptens relevant for skin sensitization? Regul Toxicol Pharmacol. 2016;82:147–55. Aug 26 S0273-2300(16)30228-8

65. Natsch A, Emter R, Gfeller H, Haupt T, Ellis G. Predicting skin sensitizer potency based on in vitro data from KeratinoSens and kinetic peptide binding: global versus domain-based assessment. Toxicol Sci. 2015;143(2):319–32. doi:10.1093/toxsci/kfu229.
66. Urbisch D, Mehling A, Guth K, Ramirez T, Honarvar N, Kolle S, Landsiedel R, Jaworska J, Kern PS, Gerberick F, Natsch A, Emter R, Ashikaga T, Miyazawa M, Sakaguchi H. Assessing skin sensitization hazard in mice and men using non-animal test methods. Regul Toxicol Pharmacol. 2015;71:337–51.
67. Hirota M, Fukui S, Okamoto K, Kurotani S, Imai N, Fujishiro M, Kyotani D, Kato Y, Kasahara T, Fujita M, Toyoda A, Sekiya D, Watanabe S, Seto H, Takenouchi O, Ashikaga T, Miyazawa M. Evaluation of combinations of in vitro sensitization test descriptors for the artificial neural network-based risk assessment model of skin sensitization. J Appl Toxicol. 2015;35:1333–47.
68. Takenouchi O, Fukui S, Okamoto K, Kurotani S, Imai N, Fujishiro M, Kyotani D, Kato Y, Kasahara T, Fujita M, Toyoda A, Sekiya D, Watanabe S, Seto H, Hirota M, Ashikaga T, Miyazawa M. (2015) test battery with the human cell line activation test, direct peptide reactivity assay and DEREK based on a 139 chemical data set for predicting skin sensitizing potential and potency of chemicals. J Appl Toxicol. 2015;35(11):1318–32.
69. Strickland J, Zang Q, Paris M, Lehmann DM, Allen D, Choksi N, Matheson J, Jacobs A, Casey W, Kleinstreuer N.Multivariate models for prediction of human skin sensitization hazard. J Appl Toxicol. 2016; doi: 10.1002/jat.3366.
70. Asturiol D, Casati S, Worth A. Consensus of classification trees for skin sensitisation hazard prediction. Toxicol In Vitro. 2016;36:197–209. doi:10.1016/j.tiv.2016.07.014.
71. Strickland J, Zang Q, Kleinstreuer N, Paris M, Lehmann DM, Choksi N, Matheson J, Jacobs A, Lowit A, Allen D, Casey W. Integrated decision strategies for skin sensitization hazard. J Appl Toxicol. 2016;36(9):1150–62. doi:10.1002/jat.3281.

The Local Lymph Node Assay 15

David A. Basketter, Ian Kimber, and G. Frank Gerberick

15.1 The Early History of the LLNA

The early history of the local lymph node assay (LLNA) evolved from an appetite to develop alternative predictive methods for the identification of contact allergens that avoided some of the pitfalls of the earlier guinea pig techniques [1] and that would also deliver animal welfare benefits. The approach was based on harnessing the then understanding of the immunological events that drive the acquisition of skin sensitisation. In particular the aim was to employ mice rather than guinea pigs and to engineer an approach that would permit the accurate identification of contact allergens based on events associated with the acquisition of sensitisation, rather than the elicitation of an allergic reaction (as employed by the standard guinea pig test methods) [2]. The plan also was to develop a method that would support screening of larger numbers of chemicals in a shorter timeframe than required for guinea pig assays and where decisions would be based upon objective and quantitative read-outs [3].

These goals were achieved by the development of what is now called the local lymph node assay (LLNA). The efficacy of the method was established in a collaborative research programme involving four UK laboratories, assisted by funding from the UK government [4]. However, those first few years of design and development were coincident with the wider scientific community identifying the

D.A. Basketter (✉)
DABMEB Consultancy Ltd, Sharnbrook MK44 1PR, UK
e-mail: dabmebconsultancyltd@me.com

I. Kimber
Faculty of Life Sciences, University of Manchester,
Michael Smith Building, Oxford Road, Manchester M13 9PT, UK

G.F. Gerberick
The Procter and Gamble Company, Central Product Safety, Mason, OH 45040, USA

C. Eskes et al. (eds.), *Alternatives for Dermal Toxicity Testing*,
DOI 10.1007/978-3-319-50353-0_15

refinement, reduction and replacement of animals (3Rs) as being an important objective. This gave work on development and refinement an added impetus because it was clear that the LLNA offered important 3R advantages with respect to both reduction and refinement. As a consequence, work continued with the LLNA, comparing its ability to identify skin sensitisers with the experience in the guinea pig as well is what was understood from human predictive testing. Ring trials were also conducted under blinded conditions as well as under what were regarded as 'field conditions', i.e. the normal manner in which the LLNA would be used in practice [5, 6]. Importantly it was demonstrated that the assay generated reliable and reproducible results in multiple laboratories with the newly publicised OECD positive control sensitisers [7]. There flowed from these investigations the initiation of interlaboratory trials, which involved a number of facilities in the USA in addition to some of those from the UK [8–10]. Finally, analyses were undertaken to demonstrate the temporal stability of local lymph node assay predictions [11].

The outputs of this body of work encouraged the view that the LLNA was an alternative method for hazard identification, which provided both refinement and reduction animal welfare benefits, as well as technical and scientific advantages, and was ready for formal validation [12, 13]. In parallel, agencies in Europe and in the USA, in collaboration with the OECD, had begun to develop draft guidelines concerning the formal validation of alternative methods, recognising that standards had to be set that would ensure alternative methods, including *in vitro* techniques, and would produce results which were reliable and fit for purpose and which could also be accepted globally. Although Europe could lay claim to the institution of the first official body focused on alternatives, the European Centre for the Validation of Alternative Methods (ECVAM), it was actually the USA in the form of its multiagency representative body, the Interagency Coordinating Committee on the Validation of Alternative Methods (ICCVAM), that produced the first official guidance on the validation of such alternatives. Because of this, the decision was taken by the originators of the LLNA to submit their body of data following the ICCVAM guidelines to that agency for their formal consideration. The submission was made in 1998, the independent peer review took place in 1999, and the official publications arising from the outcome appeared a little later [14, 15]. A complementary paper from the LLNA originators detailing the validation information was also published [16]. As is well known, the LLNA was thereby regarded as a valid alternative to the older guinea pig methods and proceeded to the OECD, where it became an official test guideline [17], updated a number of years later [18].

15.2 Lessons on Hazard Identification and Validation

In a sense, that should be the end of the story, at least for skin sensitisation hazard identification. However the reality was somewhat different. It must be borne in mind that both for the developers of the assay and for the host of regulators and reviewers involved in the formal validation process, the experience was completely novel. In effect, the validation of the LLNA represented the first practical learning exercise in validation [15]. The originators of the assay subsequently published a paper reflecting that experience, noting in particular some of the key areas that were made by the independent validation review process [19]. Perhaps the most important of these were that the review process insisted on making changes to both the protocol and to the prediction model. It is to be hoped, indeed expected, that validation in the future has learned that this is anathema: the protocol and the associated prediction model submitted must be sacrosanct; those undertaking the independent review (a very necessary and important step) have to decide only whether they are relevant and reliable in respect of their stated purpose ([19, 20].

Another lesson that was learned in the LLNA validation process concerned the vexed question of applicability domains. In the LLNA validation, 200 substances had been evaluated, probably still the record for any validation submission. However questions still arose during the validation process, with concerns that certain classes of material had not been adequately evaluated [14, 15]. How this impacted the LLNA is not crucial here, rather what is important is that, for the future, the lesson to be learned is that applicability domain questions must only be asked in relation to the mechanism of action and/or mechanistic chemistry relevant to the toxicological endpoints under examination. For example, a statement such as 'pesticides were not assessed in the validation' is not a legitimate question. The assay does not know the use that humans will make of the substance. What the assay should do is reflect the necessary mechanistic elements of the toxicological endpoints such that questions of the type mentioned are always irrelevant.

What else does the LLNA teach us about the validation of alternatives? Perhaps one of the most significant things is that once an assay is submitted for validation much of the ownership (and certainly a lot of the control) is taken away from the assay originators. This is right and proper as it is essential for global acceptance of data that there is no question of bias. From the test developers' perspective, it means that, at the latest, at the point where the protocol and prediction model are fixed, it is in their very best interests to be entirely open and transparent about all the results, including the data that does not fit the expected pattern. In the case of the LLNA, the crucial examples were the false-negative results normally obtained with nickel, the most common human skin sensitiser, and the positive results obtained with sodium lauryl sulphate, the most widely used standard skin irritant substance, very widely regarded as entirely non-sensitising [16, 21]. Openness with this information and a willingness to discuss it contributed significantly to the ultimate consensus that the LLNA actually worked for the large majority of substances and could therefore be regarded as a valid alternative to the guinea pig methods. The essential lesson is the openness and transparency are central to building a consensus concerning the merits of an alternative assay.

Another lesson from the LLNA validation process turned out to be this: a successful validation outcome is the beginning of a new journey with an assay, one where many groups start to use the method, groups with no previous experience, and therefore some of whom find difficulties in conduct, interpretation or both. The LLNA experience was that this generated a considerable need for advice and training, much of this directed towards the test developers. Other difficulties that arose regarding the wider use of the assay included an unwillingness to handle the radioactive thymidine necessary to generate the endpoint data [22]. This spawned a whole series of efforts by others to examine alternative endpoints for the assay (many of which had already been explored by the test developers) and from which, ultimately, two options received the benefit of catch-up validation approval and the publication, some 10 years after the original guideline, of their own modified OECD guidelines [23].

A variety of other issues have arisen post-validation. Perhaps the most significant amongst these has been the concern surrounding false-positive and false-negative results, with the former dominating criticism from some industrial toxicologists. An extensive body of publications has appeared in the last few years on this topic, although it is not the purpose of this chapter to examine this question in any detail [21, 24, 25]. Rather it is to provide an alert on the type of problem which can occur with any validated new method. Probably the most important learning, not just for industrial toxicologists, but especially for regulators, is that no assay is or ever has been perfect. A paper on the specific topic was published by the developers of the LLNA to emphasise that there will be substances which give misleading results and times when the assay does not work [26]. That is true of *every* current toxicological test and will be true of *all* of those in the future, even those based entirely on chemistry and computer predictions! All this is very wisely written in guidance for chemical legislation in Europe, where any decisions, for example, on classification and labelling, must always be taken based on the weight of the evidence, not on the basis of the results from a single assay [27]. It is worth reflecting that the authors of this chapter are unable to think of a single example where the only information on skin sensitisation potential comes from only a single assay; there is always other information to consider, whether that be chemistry, read-across data, structure activity relationships, results of genetic toxicology screening, data on irritancy, skin penetration, acute toxicity and so on. For many substances, for this endpoint at least, there may also be human data, which can be taken into account, with detailed guidance on this aspect also having recently been published [27].

15.3 From Identification to Characterisation of Hazards

One of the interesting aspects of the development of alternative assay is the potential for its purpose to evolve from screening assay to a regulatory hazard identification method and, from there, to progress with further development as the data can contribute to hazard characterisation. In the case of the LLNA, the change from screening test to hazard identification method occurred early and led to the validation process. However, even before formal validation had occurred, quantitative data from the method had been used in a QSAR study, and the possibility of deploying a dose response data to measure relative potency had been understood and published [28]. By the time of validation, the details of this process were published [29–31]. Shortly after validation, more substantial comparisons of potency predictions with human and other data were made available [19, 32, 33]. This led to the potential for use of the assay in risk assessment (see later for details). Then, with several further years of analysis, the ability of the LLNA to give useful insights on potency became embedded in regulatory toxicology, so that more potent skin sensitisers could be assessed and risk managed more appropriately. The underlying point is that the initial intention of the developers of the LLNA in the late 1980s changed, evolved, over the subsequent years, and the same sort of thing is likely to happen with newer *in vitro* alternatives.

Another development which is likely to occur with a validated assay is the misuse of the method. For the LLNA, the most obvious misuse, as it was with the guinea pig methods, was its adoption for the testing of formulations. The assay was neither developed nor validated for this purpose, although the same could be said for the guinea pig methods which were also developed for the testing of chemical substances but became misused for the testing of formulations. Of particular concern in this respect is the misuse of these assays as a replacement for risk assessment. The same misuse occurs with other sensitisation tests including the human repeated insult patch test, none of them being a substitute for a proper risk assessment process [34].

15.4 The LLNA in Sensitisation Risk Assessment

It was long recognised that guinea pig sensitisation tests, whilst capable of hazard identification, were of limited utility in terms of the assessment of the relative potency of a sensitiser [31, 35]. This meant that the data they produced made risk assessment difficult. The post-validation development of the LLNA as a tool for potency measurement facilitated a step change in risk assessment for skin sensitisation, effectively bringing it into line with the mainstream approaches used for sub-acute/chronic toxicology endpoints.

The measurement of the relative potency of a skin sensitiser in the LLNA is termed the EC3 value and is the concentration of a chemical necessary to cause a threefold stimulation in draining lymph nodes compared to the concurrent vehicle-treated control [29, 36, 37]. Substantial sets of EC3 data have been published [38]. Their correlation with what is understood of potency in humans has also been examined in detail on several occasions [39–43]. The particular focus has been on how well EC3 data predict the induction threshold (i.e. the no effect level) in a human repeated insult patch test (HRIPT), as that figure represents the point of departure for quantitative risk assessment (QRA). Complete details of the QRA technique are published and need not be repeated here [44, 45]. However, the impact of the replacement of the LLNA by *in vitro* alternatives is likely to be felt most keenly in the area of risk assessment, simply because the target for validation of alternative is essentially only hazard identification, not hazard characterisation [46]. Although it is an area of much active research, it remains unclear to what extent *in vitro* methods alone can provide information on sensitisation potency. A range of strategies is being investigated, including both simpler data combinations (e.g. [47]) and more sophisticated statistical approaches (e.g. [48]) or neural network modelling (e.g. [49]).

15.5 The Role of the LLNA in *In Vitro* Validation

In this section of the chapter, the focus has been placed on how the LLNA has influenced the development and validation of *in vitro* alternatives for skin sensitisation. Perhaps the most obvious aspect of this has been the way in which the plethora of publications on the *in vivo* assays has made available large volumes of experimental data against which new *in vitro* alternatives can be compared. In addition, the objective nature of the LLNA has undoubtedly lent a level of credibility to its datasets which makes them appealing to *in vitro* test developers, as well as to those whose responsibility it is to provide independent peer review on such assays. Of particular note has been the publication of a combined list of over 300 defined chemical substances evaluated in the LLNA [38]. This database, in combination with information from guinea pig assays and human experience, ultimately provides a very good foundation for validation of alternatives that seek to replace the *in vivo* assays currently used for skin sensitisation hazard identification. In this respect, it is pertinent to note that the human evidence can predominate over the *in vivo* data, good examples being the false positive results from the LLNA obtained with sodium lauryl sulphate, copper chloride and xylene, plus the false negative data obtained with nickel salts. It is very important that in the evaluation of *in vitro* alternatives, including how they are assessed in integrated testing strategies, the attention remains on predicting human skin sensitisation hazard rather than replicating the 85–90% accurate predictions from *in vivo* assays. This means it is necessary to pay particular attention to the clinical data [50].

The second (and arguably at least as important) way in which the LLNA has influenced the development of skin sensitisation is in the context of hazard characterisation, i.e. potency estimation. Accordingly, the LLNA database already mentioned is being used as a key tool in the development of *in vitro* methods, which deliver information that assists in characterisation of the relative potency of a skin sensitiser (e.g. [47]). As already mentioned above, the LLNA, through its dose response, can provide well-differentiated insights concerning the relative potency of an identified skin sensitising chemical. Indeed it is recognised that there may be at least five orders of magnitude of difference in the relative potency of sensitisers assessed in the LLNA [38]. This information increasingly is being used in regulatory toxicology for refined classification and labelling and is central to our efforts to make further improvements in quantitative risk assessment for this endpoint [51]. Again, it is important to mention that the ultimate benchmark for authentication of potency predictions must be the actual relative potency in humans. An approach and a dataset regarding this human information have been published recently [39, 52].

15.6 Quo Vadis?

The development, validation and evolution of the LLNA are often taken as a practical example of relevance for *in vitro* alternatives in skin sensitisation, as well as a wider paradigm for nonanimal alternatives in toxicology. In many senses, this has to be a good thing, since, as illustrated in this chapter, several important lessons can be taken from the experience. However, it is also worth reflecting that the LLNA arose in a rather different era, one in which there was not the prospect of a testing ban in Europe, applied independently of the development and acceptance of viable *in vitro* methods [53]. The authors of this review chapter sense that in the non-cosmetic areas of toxicological endeavour, there will be a heightened resistance to change, to the adoption of *in vitro* methods. In part this is due to the arbitrary deadline applied to cosmetics but also due to the advantages in skin sensitisation assessment that have been conferred by the LLNA and which make its complete replacement much more challenging. Thus, although OECD Guidelines for a trio of *in vitro* skin sensitisation alternatives have just been published [54] (and several others are under active consideration), the time when the current *in vivo* test guidelines are withdrawn seems still very distant.

References

1. Botham PA, Basketter DA, Maurer T, Mueller D, Potokar M, Bontinck WJ. Skin sensitization - a critical review of predictive test methods in animal and man. Food Chem Toxicol. 1991;29:275–86.
2. Kimber I, Weisenberger C. A murine local lymph node assay for the identification of contact allergens. Arch Toxicol. 1989;63:274–82.
3. Kimber I, Basketter DA. The murine local lymph node assay; collaborative studies and new directions: a commentary. Food Chem Toxicol. 1992;30:165–9.
4. Kimber I, Hilton J, Botham PA, Basketter DA, Scholes EW, Miller K, Robbins MC, Gray TJB, Waite SJ. The murine local lymph node assay. Results of an interlaboratory trial. Toxicol Lett. 1991;55:203–13.
5. Basketter DA, Scholes EW, Kimber I, Botham PA, Hilton J, Miller K, Robbins MC, Harrison PTC, Waite SJ. Interlaboratory evaluation of the local lymph node assay with 25 chemicals and comparison with guinea pig tests. Toxicol Methods. 1991;1:30–43.
6. Scholes EW, Basketter DA, Sarll AE, Kimber I, Evans CD, Miller K, Robbins MC, Harrison PTC, Waite SJ. The local lymph node assay: results of a final interlaboratory validation under field conditions. J Appl Toxicol. 1992;12:217–22.
7. Basketter DA, Selbie E, Scholes EW, Lees D, Kimber I, Botham PA. Results with OECD recommended positive control sensitizers in the maximisation, Buehler and local lymph node assays. Food Chem Toxic. 1993;31:63–7.
8. Kimber I, Hilton J, Dearman RJ, Gerberick GF, Ryan CA, Basketter DA, Scholes EW, Ladics GS, Loveless SE, House RV, Guy A. An international evaluation of the local lymph node assay and comparison of modified procedures. Toxicology. 1995;103:63–73.
9. Kimber I, Hilton J, Dearman RJ, Gerberick GF, Ryan CA, Basketter DA, Lea L, House RV, Ladics GS, Loveless SE, Hastings KL. Assessment of the skin sensitization potential of topical medicaments using the local lymph node assay: an interlaboratory exercise. J Toxicol Environ Health. 1998;53:563–79.
10. Loveless SE, Ladics GS, Gerberick GF, Ryan CA, Basketter DA, Scholes EW, House RV, Hilton J, Dearman RJ, Kimber I. Further evaluation of the local lymph node assay in the final phase of an international collaborative trial. Toxicology. 1996;108:141–52.
11. Dearman RJ, Hilton J, Evans P, Harvey P, Basketter DA, Kimber I. Temporal stability of local lymph node assay responses to hexyl cinnamic aldehyde. J Appl Toxicol. 1998;18:281–4.
12. Basketter DA, Gerberick GF, Kimber I, Loveless SE. The local lymph node assay - a viable alternative to currently accepted skin sensitisation tests. Food Chem Toxicol. 1996;34:985–97.
13. Chamberlain M, Basketter DA. The local lymph node assay: status of validation. Food Chemical Toxicol. 1996;34:999–1002.
14. Dean JH, Twerdok LE, Tice RR, Sailstad DM, Hattan DG, Stokes WS. ICCVAM evaluation of the murine local lymph node assay. II conclusions and recommendations of an independent scientific peer review panel. Reg Toxicol Pharmacol. 2001;34:258–73.
15. Sailstad DM, Hattan D, Hill RN, Stokes WS. ICCVAM evaluation of the murine local lymph node assay. I. The ICCVAM review process. Reg Toxicol Pharmacol. 2001;34:249–57.

16. Gerberick GF, Ryan CA, Kimber I, Dearman RJ, Lea LJ, Basketter DA. Local lymph node assay validation assessment for regulatory purposes. Am J Cont Derm. 2000;11:3–18.
17. OECD. Local lymph node assay test guideline no 429. Paris: Organisation for Economic Cooperation and Development; 2002.
18. OECD. Test guideline 429: The local lymph node assay. Paris: Organisation for Economic Cooperation and Development; 2010a.
19. Basketter DA, Blaikie L, Dearman RJ, Kimber I, Ryan CA, Gerberick GF, Harvey P, Evans P, White IR, Rycroft RJG. Use of the local lymph node assay for the estimation of relative contact allergenic potency. Contact Dermatitis. 2000;42:344–8.
20. Balls M, Amcoff P, Bremer S, Casati S, Coecke S, Clothier R, et al. Zuang V; ECVAM. The principles of weight of evidence validation of test methods and testing strategies. The report and recommendations of ECVAM workshop 58. Altern Lab Anim. 2006;34(6):603–20.
21. Basketter DA, Gerberick GF, Kimber I. Strategies for identifying false positive responses in predictive sensitisation tests. Food Chem Toxicol. 1998;36:327–33.
22. Basketter DA, Cockshott A, Corsini E, Gerberick GF, Idehara K, Kimber I, et al. An evaluation of performance standards and non-radioactive endpoints for the local lymph node assay. ATLA. 2008;36:243–57.
23. Basketter DA, Ball N, Cagen S, Carrillo J-C, Certa H, Eigler D, et al. Application of a weight of evidence approach to analysing discordant sensitization datasets: implication for REACH. Regul Toxicol Pharmacol. 2009a;55:90–6.
24. Kreiling R, Hollnagel HM, Hareng L, Eigler D, Lee MS, Griem P, et al. Comparison of the skin sensitizing potential of unsaturated compounds as assessed by the murine local lymph node assay (LLNA) and the guinea pig maximization test (GPMT). Food Chem Toxicol. 2008;46(6):1896–904.
25. Basketter DA, McFadden JF, Gerberick GF, Cockshott A, Kimber I. Nothing is perfect, not even the local lymph node assay. A commentary and the implications for REACH. Contact Dermatitis. 2009b;60:65–9.
26. OECD. Guidelines for test of chemicals nos 442a and 442b. Paris: Organisation for Economic Cooperation and Development; 2010b.
27. ECHA. Guidance on the Application of the CLP Criteria. 2013. Version 4.0 November 2013. www.echa.europa.eu/documents/10162/13562/clp_en.pdf. Last accessed 20 oct 2014.
28. Basketter DA, Roberts DW, Cronin M, Scholes EW. The value of the local lymph node assay in quantitative structure activity investigations. Contact Dermatitis. 1992;27:137–42.
29. Basketter DA, Lea L, Cooper K, Dickens A, Briggs D, Pate I, Dearman RJ, Kimber I. A comparison of statistical approaches to derivation of EC3 values from local lymph node assay dose responses. J Appl Toxicol. 1999;19:261–6.
30. Kimber I, Basketter DA. Contact sensitization: a new approach to risk assessment. Hum Ecol Risk Assess. 1997;3:385–95.
31. Kimber I, Basketter DA, Berthold K, Butler M, Garrigue JL, Lea LJ, Newsome C, Roggeband R, Steiling W, Stropp G, Waterman S, Wiemann C. Skin sensitisation testing in potency and risk assessment. Toxicol Sci. 2001;59:198–208.
32. Gerberick GF, Robinson MK, Ryan CA, Dearman RJ, Kimber I, Basketter DA, Wright Z, Marks JG. Contact allergenic potency: correlation of human and local lymph node assay data. Am J Cont Derm. 2001;12:156–61.
33. Ryan CA, Gerberick GF, Cruse LW, Basketter DA, Lea LJ, Blaikie L, Dearman RJ, Warbrick EV, Kimber I. Activity of human contact allergens in the murine local lymph node assay. Contact Dermatitis. 2000;43:95–102.
34. Basketter DA. The human repeated insult patch test in the 21st century: a commentary on ethics and validity. Cutan Ocul Toxicol. 2009;28:49–53.
35. Basketter DA, Andersen KE, Lidén C, van Loveren H, Boman A, Kimber I, Alanko K, Berggren E. Evaluation of the skin sensitising potency of chemicals using existing methods and considerations of relevance for elicitation. Contact Dermatitis. 2005a;52:39–43.
36. Basketter DA, Gerberick GF, Kimber I. The local lymph node assay EC3 value: status of validation. Contact Dermatitis. 2007;57:70–5.

37. Gerberick GF, Ryan CA, Kern PS, Schlatter H, Dearman RJ, Kimber I, Patlewicz G, Basketter DA. Compilation of historical local lymph node assay data for the evaluation of skin sensitization alternatives. Dermatitis. 2005;16:157–202
38. Kern PS, Gerberick GF, Ryan CA, Kimber I, Aptula A, Basketter DA. Historical local lymph node data for the evaluation of skin sensitization alternatives: a second compilation. Dermatitis. 2010;21:8–32.
39. Api AM, Basketter DA, Lalko J. Correlation between experimental human and murine skin sensitization induction thresholds. Cutan Ocul Toxicol. 2014;28:1–5.
40. Basketter DA, Clapp C, Jefferies D, Safford RJ, Ryan CA, Gerberick GF, Dearman RJ, Kimber I. Predictive identification of human skin sensitisation thresholds. Contact Dermatitis. 2005b;53:260–7.
41. Basketter DA, McFadden JP. Cutaneous allergies. In: Dietert RR, Luebke RW, editors. Immunotoxicity, immune dysfunction and chronic disease. New York: Humana Press; 2012. p. 103–26.
42. Griem P, Goebel C, Scheffler H. Proposal for a risk assessment methodology for skin sensitization based on sensitization potency data. Regul Toxico Pharmaol. 2003;38:269–90.
43. Schneider K, Akkan Z. Quantitative relationship between the local lymph node assay and human skin sensitization assays. Regul Toxicol Pharmacol. 2004;39:245–55.
44. Api AM, Basketter DA, Cadby PA, Cano M-F, Ellis G, Gerberick GF, Griem P, McNamee PM, Ryan CA, Safford B. Dermal sensitization quantitative risk assessment (QRA) for fragrance ingredients. Regul Toxicol Pharmacol. 2008;52:3–23.
45. Robinson MK, Gerberick GF, Ryan CA, McNamee P, White I, Basketter DA. The importance of exposure assessment of skin sensitization risk. Contact Dermatitis. 2000;42:251–9.
46. Vocanson M, Nicolas J-F, Basketter DA. In vitro approaches to the identification and characterization of skin sensitisers. Exp Rev Dermatol. 2013;8:395–405.
47. Natsch A, Emter R, Gfeller H, Haupt T, Ellis G. Predicting skin sensitizer potency based on in vitro data from KeratinoSens and kinetic peptide binding: global versus domain-based assessment. Toxicol Sci. 2015;143:319–32.
48. Jaworska JS, Natsch A, Ryan C, Strickland J, Ashikaga T, Miyazawa M. Bayesian integrated testing strategy (ITS) for skin sensitization potency assessment: a decision support system for quantitative weight of evidence and adaptive testing strategy. Arch Toxicol. 2015;89(12):2355–83.
49. Tsujita-Inoue K, Hirota M, Ashikaga T, Atobe T, Kouzuki H, Aiba S. Skin sensitization risk assessment model using artificial neural network analysis of data from multiple in vitro assays. Toxicol In Vitro. 2014;28:626–39.
50. Basketter DA, White IR, McFadden JP, Kimber I. Skin sensitization: integration of clinical data into hazard identification and risk assessment. Human Exp Toxicol. 2015;34(12):1222–30.
51. IFRA. International dialogue for the evaluation of allergens. 2014. http://www.ideaproject.info/uploads/Modules/Documents/idea-annual-review-2014---summary-distributed-to-the-participants.pdf. Last accessed 10 February 2015.
52. Basketter DA, Alepee N, Ashikaga T, Barroso J, Gilmour N, Goebel C, Hibatallah J, Hoffmann S, Kern P, Martinozzi-Teissier S, Maxwell G, Millet M, Reisinger K, Sakaguchi H, Schepky A, Tialhardat M, Templier M. Categorisation of chemicals according to their relative human skin sensitizing potency. Dermatitis. 2014;25:11–21.
53. EU. Directive 2003/15/EC of the European parliament and the council of 27 February 2003 amending council directive 76/768/ EEC on the approximation of the laws of the members states relating to cosmetic products. Off J Eur Union. 2003;L66:26–35.
54. OECD. Guidelines for test of chemicals nos 442c and 442d. Paris: Organisation for Economic Cooperation and Development; 2015.

Utilization of Peptide Reactivity Assays for the Prediction of Skin Sensitization

16

G. Frank Gerberick, John A. Troutman, Leslie M. Foertsch, and Petra S. Kern

16.1 Principle of the Test Method and Scientific Basis

Allergic contact dermatitis (ACD) is the clinical manifestation of skin sensitization and an important toxicological endpoint for occupational and consumer health risk assessments. *In vivo* models for evaluating the potential of novel materials to act as skin allergens are used routinely and globally accepted for use in identifying hazard. Based on the murine local lymph node assay (LLNA), both hazard and potency information are used for quantitative risk assessments via determination of the concentration that produces a threshold positive response and classification of the relative skin-sensitizing potency [1–5]. The current ban on animal testing of cosmetic and toiletry ingredients in Western Europe makes *in vivo* models obsolete and require the cosmetic industry to rely solely on alternatives for assessing the risk of contact allergy from new chemicals. One focus area has been the development of *in chemico* assays based on reactivity.

The ability of a chemical to react with skin proteins is thought to play a key role in the development of skin sensitization [6–9]. Contact allergens are generally low molecular weight molecules that have the ability to penetrate the stratum corneum and covalently modify skin proteins through a variety of reactivity mechanisms. Chemical covalent binding to nucleophilic centers in skin proteins is regarded to be the molecular initiating event in the skin sensitization AOP [10] without which skin sensitization would not occur. Thus if a chemical is capable

G.F. Gerberick (✉) • J.A. Troutman • L.M. Foertsch
The Procter and Gamble Company, Mason Business Center, Cincinnati, OH 45040, USA
e-mail: gerberick.gf@pg.com

P.S. Kern
The Procter and Gamble Company, Beijing Innovation Center, Beijing, China
e-mail: Kern.ps@pg.com

C. Eskes et al. (eds.), *Alternatives for Dermal Toxicity Testing*,
DOI 10.1007/978-3-319-50353-0_16

of reacting with proteins either directly or after appropriate biotransformation, then it may have the potential to act as a contact allergen. Reaction products are immunogenic and stimulate the migration and maturation of Langerhans cells from the epidermis to local skin-draining lymph nodes for antigen presentation to naive T cells, resulting in the clonal expansion of antigen-specific memory T cells [11].

Skin sensitizers (or their metabolites) are generally electrophilic and reactive toward nucleophilic sites on proteins. This feature has been exploited by investigators that are interested in developing non-animal approaches for evaluating the skin sensitization potential of novel materials based on reactivity with nucleophile-containing peptides. Although there are other nucleophilic amino acids, the e-NH2 group of lysine and the –SH group of cysteine are relatively strong nucleophiles and often cited in the literature for assessing hapten reactivity [6, 9, 12–14]. Because protein reactivity is a key step in the induction of ACD, investigators have reported the use of a number of model nucleophiles in an effort to develop *in chemico* approaches for predicting the skin sensitization potential of new chemicals based on reactivity. Published reports have included the use of synthetic peptides containing single or multiple nucleophilic amino acids, glutathione, human serum albumin, and 4-nitrobenzenethiol [7, 15–28]. Our laboratory has been on two approaches: the Direct Peptide Reactivity Assay (DPRA) and Peroxidase Peptide Reactivity Assay (PPRA). The development and use of these assays for assessing skin sensitization potential of chemicals have been published previously [21, 22, 29, 30].

16.2 Current (Pre)Validation Status

DPRA – Through the development of DPRA, three informal interlaboratory studies were conducted. These studies were critical to the development of a robust SOP for the assay. In 2009, the DPRA and the experience gained through conducting these interlaboratory studies were shared with EURL ECVAM along with all of the intralab data generated to that point. EURL ECVAM coordinated a validation study of the DPRA using a modular approach [31]. The process involved a thorough evaluation of the DPRA protocol and its transferability and within- and between-laboratory reproducibility among other things including its ability to contribute to the subcategorization of skin-sensitizing chemicals for regulatory purposes. Overall, the recommendation was that the DPRA is a robust and reliable test method that should be further evaluated in the context of its use as part of an integrated approach [32]. Finally, EURL ECVAM supports the development of an OECD test guideline for the DPRA. A project proposal has already been submitted to the OECD, and finalization of the guideline is expected early in 2015.

PPRA – EURL ECVAM has not formally evaluated the PPRA. However, a robust SOP has been developed, and one interlaboratory study is currently being conducted with 3 laboratories and 24 test chemicals (see below).

16.3 Protocol Description

DPRA - The assay utilizes to synthetic peptides containing either a lysine or cysteine residue, Ac-RFAAKAA-COOH (lysine peptide) and Ac-RFAACAA-COOH (cysteine peptide). Peptide stock solutions are prepared to a final concentration of 1.25 mM in either 100 mM ammonium acetate buffer, pH 10.2 (lysine peptide), or 100 mM phosphate buffer, pH 7.5 (cysteine peptides). Test chemical solutions at a concentration of 100 mM are prepared in acetonitrile or other water-miscible solvent. Samples are prepared to contain 0.5 mM peptide and either 5 mM or 25 mM test chemical for a peptide, test chemical ratio of 1:10 (cysteine reaction) or 1:50 (lysine reaction). In the case of the 1:10 ratio reactivity samples, 750 μL of peptide solution, 200 μL of acetonitrile, and 50 μL of test chemical solution are added to each autosampler vial. Similarly for the 1:50 ratio reactivity samples, 750 μL of peptide solution and 250 μL of test chemical solution are added to each autosampler vial. All reaction mixtures are prepared in triplicate. Samples without the test chemicals are also prepared in triplicate to serve as controls. The reaction vials are incubated in the dark for 24 h at room temperature prior to HPLC analysis.

In addition to reaction samples and controls, calibration standards are also prepared from the peptide stock solution by diluting them with appropriate buffer and acetonitrile to the following concentrations: 0.0156, 0.0313, 0.0625, 0.125, 0.25, and 0.50 mM.

Sample analysis is carried out using an HPLC coupled with a UV detector (220 nm) as the chromatographic system which utilizes a C18 reverse phase column (Zorbax SB-C18 (2.1 × 100 mm) or similar). A 6–10 μL injection of the reactivity samples is made onto the column. The two mobile phases consist of 0.1% TFA in water (A) and 0.085% TFA in acetonitrile (B), and a gradient of 90% (A) to 60% (A) over 20 min at a flow rate of 0.3 ml/min is used for the separation. Peptide reactivity with the test chemicals is reported as percent peptide depletion. Peptide depletion is determined as the reduction of the peptide concentration in the samples relative to the average peptide concentration of the controls.

For each test chemical, the mean peptide depletion of the triplicates is calculated for both the cysteine and lysine peptides. Using the means of cysteine and lysine depletion, the overall average peptide depletion is calculated and a determination of a sensitizer or non-sensitizer based on a decision tree model [22]. Chemicals with an average cysteine and lysine depletion <6.37% are considered to have minimal reactivity. Meanwhile, chemicals with average depletion between 6.37 and 22.62% are considered to have low reactivity, chemicals between 22.62 and 42.27% have moderate reactivity, and chemicals above 42.47% are assigned to the high reactivity class. Generally, chemicals with moderate to high reactivity are associated with moderate to strong skin sensitization potency, whereas those categorized as having minimal to low reactivity include weak and non-sensitizers. For hazard identification, chemicals with high, moderate, and low reactivity are grouped as sensitizers, whereas chemicals with minimal reactivity are considered non-sensitizers.

PPRA - Reactivity assessments in the PPRA are performed in discrete reactions with a cysteine- or lysine-containing synthetic peptide. Chemical reactivity is determined by measurement of peptide depletion following 24 h of incubation under ambient conditions in 0.1 M potassium phosphate buffer (pH 7.4). The final reaction volume is 0.3 mL. Reactions with cysteine peptide (20 mM) are performed in the presence and absence of a horseradish peroxidase (3 U/mL) and hydrogen peroxide (100 μM) oxidation system (HRP/P) for identifying potential hapten and pro-/pre-hapten sensitizers [29]. Reactivity of test chemicals to cysteine-based synthetic peptide is determined in reactions containing 10 μM desferroxamine and test article at concentrations of 0.039, 0.078, 0.16, 0.31, 0.63, 1.3, 2.5, and 5 mM. The final organic content is $\leq$1%. Reactivity in the absence of enzyme (direct reactivity) is assessed in test chemical incubations devoid of HRP/P. Incubations with cysteine peptide and no test chemical, in the absence and presence of HRP/P, serve as zero-depletion reference controls for comparisons to direct and enzyme-mediated test chemical reactivity determinations, respectively. Post-incubation processing of cysteine-containing samples included a 30-min incubation step with dithiothreitol (DTT) to reverse apparent dimerization of thiols in the cysteine peptide nucleophile.

Reactions with lysine peptide (5 μM) are performed under direct (no HRP/P) test conditions [30]. The final test chemical concentrations targeted for lysine-containing reactions are 0.011, 0.034, 0.10, 0.31, 0.93, 2.8, 8.3, and 25 mM in $\leq$25% organic solvent. Sample incubates with cysteine and lysine are analyzed by high-performance liquid chromatography with tandem mass spectrometry (LC/MS/MS) detection. Reactions are prepared and analyzed in batches of up to 12 chemicals that include 2 controls (eugenol and glutaraldehyde) and reference controls.

A provisional prediction model based on peptide depletion and estimation of reactivity potency is used to classify test chemicals for hazard (yes/no) and binning reactivity into one of three categories (i.e., minimally reactive, reactive, and highly reactive). Concentration-response data for chemicals that are positive for hazard are analyzed to estimate the effective concentration of test chemical that depletes peptide by 25% (i.e., EC25) by fitting a three-parameter log-logistic model to peptide depletion data or by linear interpolation. Chemicals with an EC25 $\geq$ 0.1 mM are considered "reactive," and chemicals with an EC25 < 0.1 mM are considered "highly reactive." Both "reactive" and "highly reactive" chemicals are classified as "sensitizers" for hazard prediction. Chemicals that are classified as "minimally reactive" are considered "non-sensitizers" for hazard prediction.

16.4 Test Method Performance and Applicability

DPRA – The DPRA has been easily transferred to many laboratories that have experience in HPLC methods. The data obtained from the assay has also been shown to be reproducible both within and between laboratories. During the assay's development phase, transferability between laboratories was assessed in three interlaboratory studies (data not published) and later during the validation study coordinated by EURL ECVAM, which assessed reproducibility both within and between laboratories [33].

Even from its early development, the DPRA has shown a significant correlation between peptide depletion and skin sensitization potential [21]. In an initial dataset of 82 chemicals, when depletion data was compared to local lymph node assay (LLNA) EC3 values, the DPRA data resulted in an 89% prediction accuracy (sensitizer versus non-sensitizer) [22]. Upon expansion of the chemical dataset to 145 chemicals, the assay demonstrated an accuracy of 80%. This decrease in accuracy can be attributed to both limitations in the assay's applicability domain and limitations in the LLNA predictions [25, 26].

The DPRA has been shown to be applicable to chemicals covering a wide variety of organic functional groups, reaction mechanisms, sensitization potencies, and physicochemical properties. However, some limitations do exist. The most notable limitation is the lack of a metabolic component. Although some pre-/pro-haptens have tested positive in the DPRA, this is not the case with all such test chemicals. Pre-haptens, molecules that must undergo oxidation to become reactive, will often test positive in the DPRA because exposure to air during the reaction setup offers sufficient exposure to oxygen. This limitation to the DPRA has been addressed in the PPRA, which incorporates a metabolic component.

Additional limitations include solubility of the test material, incompatibility with metal-containing compounds, and no straightforward way to handle complex mixtures. Test chemicals must be soluble in a water-miscible solvent to prepare a 100 mM solution. If the test material is not completely soluble at this concentration, a more dilute test chemical solution can be used. Additionally if the test chemical has very low water solubility, it may fall out of solution when the test chemical solution is mixed with the peptide solution. In either of these situations, results exhibiting reactivity may be used to support a positive prediction (potential skin sensitizer), but no firm conclusion on the lack of reactivity should be drawn from a negative result. The assay is based upon covalent binding between test chemical and peptide, so it is not designed to predict the skin sensitization of metal compounds. Metal-containing compounds have been shown to react with protein via non-covalent binding mechanisms. The DPRA prediction model is based upon molar ratios between test chemical and peptide. Therefore, it has not been validated for complex mixtures. However, it may be useful for testing simple, multi-constituent substances. If the composition is known (molecular weight and percentage of each constituent), the molecular weight of test chemical needed to prepare the 100 mM solution can be calculated using a weighted average of the molecular weights of the mixture's constituents.

PPRA – The PPRA is applicable to test chemicals that are soluble in acetonitrile or other nonreactive, water-miscible solvents. The reliability and performance of the PPRA and its prediction model have not been fully evaluated against a large set of chemicals. However, preliminary test results that were determined during method development and evaluation have indicated that the assay could be used in a tiered screening approach to determine skin sensitization potential in conjunction with additional information. These preliminary results for 70 chemicals that included 19 pre-/pro-haptens and 26 haptens resulted in a prediction accuracy of 83% [30]. Subsequent refinements of the method have resulted in better dose-response characterization, hazard prediction, and enhanced lysine sensitivity. To evaluate the utility and transferability of the assay, pre-validation studies with three laboratories are currently underway. Twenty-four chemicals tested under blinded conditions in three independent runs within each laboratory will be used to assess both intra- and interlaboratory reproducibility for qualitative hazard prediction and quantitative reactivity classification.

16.5 Perspectives from the Test Developer

DPRA – One particular watch out when conducting the DPRA is precipitate formation in the reaction mixture. Precipitation frequently occurs in the reaction mixture over time, and this is not a major concern. When precipitate occurs shortly after preparing the reaction mixture, care should be taken when analyzing the data.

Co-elution between peptide and test chemical is an additional issue that may be encountered when running the assay. Depending on the degree of overlap between the test chemical and peptide, peptide depletion may or may not be reasonably estimated. The data must be interpreted carefully so as not to overstate the results.

The assay was initially developed using an HPLC with a UV detector for analyzing the samples. This analytical system was chosen because it is commonly available to many laboratories. Over the years, mass spectrometry has become more common and the equipment has become more accessible. With minor adjustments to the protocol, it is possible to utilize mass spectrometry instruments for sample analysis.

PPRA – The PPRA was developed as a refinement to the DPRA in which loss of peptide is determined following 24-h co-incubation with test chemical under ambient lab conditions in the presence and absence of horseradish peroxidase-hydrogen peroxide (HRP/P) oxidation system. Concentration-response data for chemicals that are positive for hazard are analyzed to estimate the effectiveness of test chemical that depletes peptide by 25% (EC25). Recent refinements of the PPRA are significant improvements and will help to meet the critical need of finding reliable non-animal methods for predictions of skin sensitization potential in the future.

16.6 Role in a Testing Strategy/Conclusions

Considering the Adverse Outcome Pathway (AOP) for skin sensitization, there is general agreement within the scientific community that it is unlikely that one single alternative method will be able to provide sufficient information to replace the use of animals for skin sensitization [34]. Due to the complexity of allergic reactions, the chemical diversity of sensitizing chemicals, and the varying applicability domains of the different *in vitro* tests, the development of predictive integrated testing strategies remains the most realistic approach to reduce or replace animal testing for the assessment of skin sensitization potential [35, 36].

The initial event of haptenation is the major determinant of the skin sensitization process, and thus the protein-binding properties of a chemical should be intrinsically linked to its sensitization potential [37]. Peptide reactivity assays, as outlined here for the DPRA and PPRA, therefore play a key role in dermal sensitization toxicity testing but should always be considered in combination with other information in the context of integrated approaches such as weight of evidence (WoE) or integrated testing strategies (ITS).

When determining the skin sensitization potential of a chemical, one needs to distinguish between the capability of the alternative method to identify the sensitization hazard (allergen or non-allergen) and estimation of the correct potency of a sensitizing molecule. The accuracy of the DPRA or PPRA for identifying the sensitization hazard and distinguishing sensitizers from non-sensitizers is high [22, 26, 38]. In addition to supporting identification of sensitizers/non-sensitizers, the DPRA and PPRA may also be able to contribute to the assessment of sensitizing potency, e.g., by supporting, within an integrated approach, the subcategorization of sensitizers according to the United Nations Globally Harmonized System of Classification and Labeling of Chemicals [39] or to determine the sensitization potency for the purpose of risk assessment. More work however would be required to determine to which extent exactly the DPRA or PPRA results relate to potency categories. As the DPRA is an *in chemico* test method lacking metabolic capacity, substances that require metabolic (pro-haptens) or abiotic activation (pre-haptens) may not be detected by the DPRA and hence would provide limited information in a testing strategy. These restrictions could be overcome by using PPRA data, which within its restrictions allow also for the detection of pre- and pro-haptens. In addition, DPRA and PPRA specifically detect peptide reactivity associated with lysine and cysteine amino acids only. All complementary information to address the missing complexity may be derived from test methods addressing other key events involved in skin sensitization as well as non-testing methods including read-across information.

Peptide reactivity assays, i.e., DPRA, have been an integral part and will always remain one of the necessary key methods in testing strategies, including the "2 out of 3" weight of evidence (WoE) approach [26, 38, 40], an ITS based on Bayesian networks [36], and an ITS which includes an assessment of potency [41, 42].

References

1. Basketter DA, Evans P, Fielder RJ, Gerberick GF, Dearman RJ, Kimber I. Local lymph node assay—validation and use in practice. Food Chem Toxicol. 2002;40:593–8.
2. Dearman RJ, Basketter DA, Kimber I. Local lymph node assay—use in hazard and risk assessment. J Appl Toxicol. 1999;19:299–306.
3. Gerberick GF, Ryan CA, Kimber I, Dearman RJ, Lea LJ, Basketter DA. Local lymph node assay: validation assessment for regulatory purposes. Am J Dermat. 2000;11:3–18.
4. Kimber I, Dearman RJ, Scholes EW, Basketter DA. The local lymph node assay: development and applications. Toxicology. 1994;93:13–31.
5. Kimber I, Dearman RJ, Basketter DA, Ryan CA, Gerberick GF. The local lymph node assay: past, present and future. Contact Dermatitis. 2002;47:315–28.

6. Dupuis G, Benezra C, editors. Allergic contact dermatitis to simple chemicals: a molecular approach. New York: Marcel Dekker Inc.; 1982.
7. Gerberick F, Aleksic M, Basketter D, Casati S, Karlberg A-T, Kern P, Kimber I, Lepoittevin J-P, Natsch A, Ovigne JM, et al. Chemical reactivity measurement and the predictive identification of skin sensitizers: the report and recommendations of ECVAM workshop 64. ATLA. 2008;36:215–42.
8. Landsteiner K, Jacobs J. Studies on the sensitization of animals with simple chemical compounds. J Exp Med. 1936;64:625–39.
9. Lepoittevin J-P, Basketter DA, Goossens A, Karlberg A-T, editors. Allergic contact dermatitis: the molecular basis. Berlin: Springer; 1998.
10. OECD. The adverse outcome pathway for skin sensitisation initiated by covalent binding to proteins. Part 1: scientific evidence. Series on testing and assessment no. 168. 2012. Online at: http://www.oecd.org/officialdocuments/publicdisplaydocumentpdf/?cote=ENV/JM/MONO(2012)10/PART1&docLanguage=En
11. Ryan CA, Hulette BC, Gerberick GF. Review: approaches for the development of cell based in vitro methods for contact sensitization. Toxicol In Vitro. 2001;15:43–55.
12. Ahlfors SR, Sterner O, Hansson C. Reactivity of contact allergenic haptens to amino acid residues in a model carrier peptide, and characterization of formed peptide-hapten adducts. Skin Pharmacol Appl Ski Physiol. 2003;16:59–68.
13. Alvarez-Sanchez R, Basketter DA, Pease C, Lepoittevin J-P. Studies of chemical selectivity of hapten, reactivity, and skin sensitization potency. 3. Synthesis and studies on the reactivity toward model nucleophiles of the 13C-labeled skin sensitizers, 5-chloro-2-methylisothia- zol-3-one (MCI) and 2-methylisothiazol-3-one (MI). Chem Res Toxicol. 2003;16:627–37.
14. Meschkat E, Barratt MD, Lepoittevin J-P. Studies of the chemical selectivity of hapten, reactivity, and skin sensitization potency. 1. Synthesis and studies on the reactivity toward model nucleophiles of the (13)C-labeled skin sensitizers hex-1-ene- and hexane-1,3-sultones. Chem Res Toxicol. 2001;14:110–7.
15. Aleksic M, Thain E, Roger D, Saib O, Davies M, Li J, Aptula A, Zazzeroni R. Reactivity profiling: covalent modification of single nucleophile peptides for skin sensitization risk assessment. Toxicol Sci. 2009;108:401–11.
16. Alvarez-Sanchez R, Basketter DA, Pease C, Lepoittevin J-P. Covalent binding of the 13C-labeled sensitizers 5-chloro-2- methylisothiazol-3-one (MCI) and 2-methylisothiazol-3-one (MI) to a model peptide and glutathione. Bioorg Med Chem Lett. 2004a;14:365–8.
17. Alvarez-Sanchez R, Divkovic M, Basketter DA, Pease C, Panico M, Dell A, Morris H, Lepoittevin J-P. Effect of glutathione on the covalent binding of the 13C-labeled skin sensitizer 5-chloro-2- methylisothiazol-3-one (MCI) to human serum albumin: identification of adducts by NMR, MALDI-MS and nano-ES MS/MS. Chem Res Toxicol. 2004b;17:1280–8.
18. Aptula AO, Patlewicz G, Roberts DW, Schultz TW. Non- enzymatic glutathione reactivity and in vitro toxicity: a non-animal approach to skin sensitization. Toxicol In Vitro. 2006;20:239–47.
19. Chipinda I, Ajibola RO, Morakinyo MK, Ruwona TB, Simoyi RH, Siegel PD. Rapid and simple kinetic screening assay for electrophilic dermal sensitizers using nitrobenzenethiol. Chem Res Toxicol. 2010;23:918–25.
20. Enoch SJ, Roberts DW, Cronin MTD. Electrophilic reaction chemistry of low molecular weight respiratory sensitizers. Chem Res Toxicol. 2009;22:1447–53.
21. Gerberick GF, Vassallo JD, Bailey RE, Chaney JG, Morrall SW, Lepoittevin J-P. Development of a peptide reactivity assay for screening contact allergens. Toxicol Sci. 2004;81:332–43.
22. Gerberick GF, Vassallo JD, Foertsch LM, Price BB, Chaney JG, Lepoittevin J-P. Quantification of chemical peptide reactivity for screening contact allergens: a classification tree model approach. Toxicol Sci. 2007;97:417–27.
23. Katon H, Okamoto M, Yamashita K, Nakamura Y, Fukumori Y, Nakai K, Kaneko H. Peptide-binding assessment using mass spectrometry as a new screening method for skin sensitization. J Toxicol Sci. 2003;28:19–24.
24. Natsch A, Gfeller H. LC-MS-based characterization of peptide reactivity of chemicals to improve the in vitro prediction of skin sensitization potential. Toxicol Sci. 2008;106:464–78.

25. Natsch A, Gfeller H, Rothaupt M, Ellis G. Utility and limitations of a peptide reactivity assay to predict fragrance allergens in vitro. Toxicol In Vitro. 2007a;21:1220–6.
26. Natsch A, Ryan CA, Foertsch L, Emter R, Jaworska J, Gerberick F, Kern P. A dataset on 145 chemicals tested in alternative assays for skin sensitization undergoing prevalidation. J Appl Toxicol. 2007b;33:1337–52.
27. Roberts DW, Patlewicz G, Kern PS, Gerberick GF, Kimber I, Dearman R, Ryan C, Basketter D, Aptula AO. Mechanistic applicability domain: classification of a local lymph node assay dataset for skin sensitization. Chem Res Toxicol. 2007;20:1019–30.
28. Schultz TW, Yarbrough JW, Johnson EL. Structure-activity relationships for reactivity of carbonyl-containing compounds with glutathione. SAR QSAR Environ Res. 2005;16:313–22.
29. Gerberick GF, Troutman JA, Foertsch LM, Vassallo JD, Quijano M, Dobson RL, Goebel C, Lepoittevin JP. Investigation of peptide reactivity of pro-hapten skin sensitizers using a peroxidase-peroxide oxidation system. Toxicol Sci. 2009;112:164–74.
30. Troutman JA, Foertsch LM, Kern PS, Dai HJ, Quijano M, Dobson RL, Lalko JF, Lepoittevin JP, Gerberick GF. The incorporation of lysine into the peroxidase peptide reactivity assay for skin sensitization assessments. Toxicol Sci. 2011;122:422–36.
31. Hartung T, Bremer S, Casati S, Coecke S, corvi R, Fortaner S, Gribaldo L, Halder M, Hoffman S, Roi AJ, Prieto P, Sabbioni E, Scott L, Worth A, Zuang V. A modular approach to the ECVAM principles on test validity. Altern Lab Anim. 2004;32:467–72.
32. EURL ECVAM. Recommendation on the direct peptide reactivity assay (DPRA) for skin sensitization testing. European Commission Joint Research Center Institute for Health and Consumer Protection. 2013. http://ihcp.jrc.ec.europa.eu/.
33. EURL ECVAM. Direct peptide reactivity assay (DPRA) validation study report. European Comission Joint Research Center Institute for Health and Consumer Protection. 2012. Accessible online at: https://eurl-ecvam.jrc.ec.europa.eu/eurl-ecvam-recommendations.
34. Adler S, Basketter D, Creton S, Pelkonen O, van Benthem J, Zuang V, Andersen KE, Angers-Loustau A, Aptula A, Bal-Price A. Alternative (non-animal) methods for cosmetics testing: current status and future prospects-2010. Arch Toxicol. 2011;85:367–485.
35. Jaworska J, Harol A, Kern PS, Gerberick GF. Integrating non-animal test information into an adaptiveTesting strategy – skin sensitization proof of concept case. ALTEX. 2011;28:211–25.
36. Jaworska J, Dancik Y, Kern P, Gerberick F, Natsch A. Bayesian integrated testing strategy to assess skin sensitization potency: from theory to practice. J Appl Toxicol. 2013;33(11):1353–64.
37. Roberts DW, Aptula AO. Determinants of skin sensitisation potential. J Appl Toxicol. 2008;28(3):377–87.
38. Bauch C, Kolle SN, Ramirez T, Eltze T, Fabian E, Mehling A, et al. Putting the parts together: combining in vitro methods to test for skin sensitizing potentials. Regul Toxicol Pharmacol. 2012;63:489–504.
39. UN GHS. Globally harmonized system of classification and labeling of chemicals, fourth revised condition. New York, Geneva: United Nations; 2011. ST/SG/AC. 10/30 Rev4
40. van der Veen JW, Rorije E, Emter R, Natsch A, van LH, Ezendam J. Evaluating the performance of integrated approaches for hazard identification of skin sensitizing chemicals. Regul Toxicol Pharmacol. 2014;69(3):371–9.
41. Nukada Y, Miyazawa M, Kazutoshi S, Sakaguchi H, Nishiyama N. Data integration of non-animal tests for the development of a test battery to predict the skin sensitizing potential and potency of chemicals. Toxicol In Vitro. 2013;27:609–18.
42. Tsujita-Inoue K, Hirota M, Ashikaga T, Atobe T, Kouzuki H, Aiba S. Skin sensitization risk assessment model using artificial neural network analysis of data from multiple in vitro assays. Toxicol In Vitro. 2014;28:626–39.

17 The KeratinoSens™ Assay for Skin Sensitization Screening

Andreas Natch

17.1 Principle of the Test Method and Scientific Basis

The mechanism of skin sensitization by low-molecular-weight chemicals has been described in detail [1], and it was summarized in an adverse outcome pathway by the OECD [2]. Chemical modification of skin proteins by the sensitizing agent is widely accepted as the molecular initiating event (MIE) in skin sensitization [3]. Therefore, from a molecular point of view, the key feature all skin sensitizers have in common is their intrinsic electrophilicity or their potential to be metabolically transformed to electrophilic, protein-reactive chemicals. This is exploited by *in chemico* methods, directly quantifying reactivity of chemicals [4]. Recently, several studies have tried to find molecular markers at the level of the transcriptome, which are activated by sensitizers, and thus could serve as *in vitro* predictive markers to detect sensitizers in cell-based assays. In theory, such a molecular marker would need to respond to the electrophilic features of skin sensitizers, as it is difficult to conceive another common molecular descriptor of skin sensitizers which would lead to canonical receptor activation by sensitizers but not by non-sensitizers. Indeed, a signaling pathway which responds specifically to electrophilic chemicals had been identified: The repressor protein Keap-1 (Kelch-like ECH-associated protein 1) binds to the transcription factor Nrf2 (nuclear factor (erythroid-derived 2)-like 2) and targets this transcription factor to ubiquitination and proteasomal degradation in the absence of electrophilic/oxidative stress. In the presence of electrophiles, on the other hand, Keap-1 is being modified at reactive cysteine interfaces on the surface [5], which leads to liberation of Nrf2. Free Nrf2 can then migrate to the nucleus and bind to the antioxidant/electrophile response element (ARE/EpRE) and activate a battery of mainly cytoprotective genes [6] (see Fig. 17.1). We could show that this pathway is indeed a valuable cellular endpoint

A. Natch
Givaudan Suisse SA, Ueberlandstrasse 138, CH-8600 Duebendorf, Switzerland
e-mail: natsch@givaudan.com

C. Eskes et al. (eds.), *Alternatives for Dermal Toxicity Testing*,
DOI 10.1007/978-3-319-50353-0_17

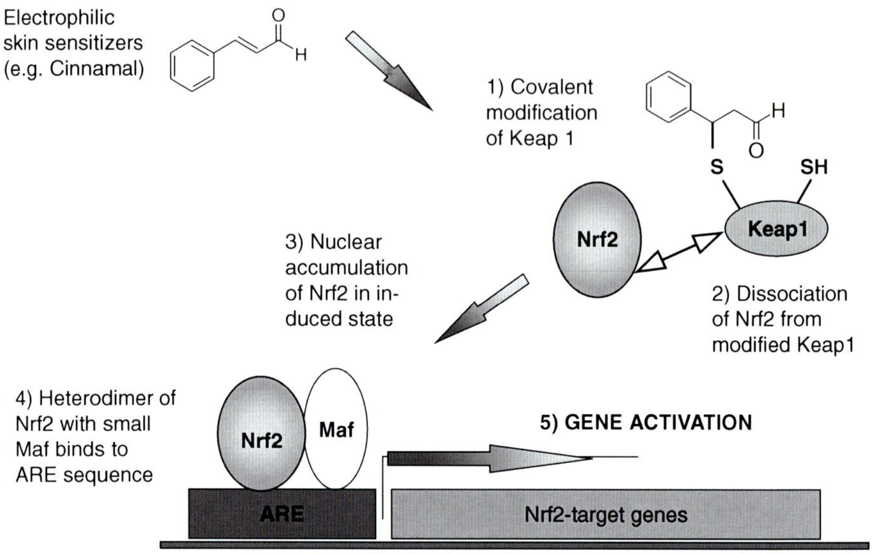

Fig. 17.1 The key molecular mechanism of Nrf2 activation

to detect skin sensitizers *in vitro* [7, 8]. This result was confirmed by several independent laboratories [9–13], and the accumulated transcriptome data indicate that it is the molecular pathway most widely and reproducibly upregulated by skin sensitizers [14]. The involvement of the Keap1-Nrf2-ARE regulatory pathway in skin sensitization has lately been demonstrated in a number of *in vivo* studies on Nrf2-knockout mice ([12, 15, 16]. Based on the most recent studies, the Nrf2 pathway appears to mainly represent a pathway of defense (PoD) activated by skin sensitizers, since overall it appears that knockout of Nrf2 leads to an enhanced sensitization response, although the first knockout study had reached an opposite conclusion [15].

The KeratinoSens™ cell line is an immortalized adherent cell line derived from HaCaT human keratinocytes transfected with a selectable plasmid. This plasmid contains the luciferase gene under the transcriptional control of the SV40 promoter fused with the ARE sequence from the human AKR1C2 gene. The AKR1C2 gene was identified as one of the key genes upregulated by contact sensitizers in dendritic cells [17–18]. A clone with a stable chromosomal integration of this plasmid was selected and expanded as the KeratinoSens™ cell line. It could be shown that activation of luciferase in the KeratinoSens™ cells closely reflects activation of the endogenous AKR1C2 gene. In addition, activation of the luciferase gene in the KeratinoSens™ cell line was shown to be dependent on the presence of Nrf2 [19], and the luciferase induction also closely mimics the induction of other endogenous, Nrf2-regulated genes. Therefore measuring luciferase activation in KeratinoSens™ is a very valid surrogate for a direct screening of the Nrf2 activation. Induction of cytoprotective pathways in keratinocytes in response to electrophiles and oxidative stress is addressing the second key event of the skin sensitization adverse outcome pathway (AOP; [2]), one of the steps considered crucial in the AOP concept for skin sensitization.

17.2 Current Validation Status

The KeratinoSens™ assay has been tested against a broad range of low-molecular-weight chemicals with known skin sensitization potential [20–25], and it was found to respond to skin sensitizers from a broad range of so-called applicability domains, i.e., chemicals reacting with proteins by different mechanisms.

The assay underwent a validation study organized by the test developer lab and performed in the lead lab and four naïve laboratories [26]. This study on 28 chemicals tested in three repetitions in five laboratories was the basis for a submission to the European Union Reference Laboratory for Alternatives to Animal Testing (EURL ECVAM) [27]. EURL ECVAM analyzed the data, identified data gaps, and requested additional studies. Based on the complete dataset, EURL ECVAM then proceeded to a peer review and finally issued a recommendation [27]. Given the limitations discussed below and in alignment with the recommendations for other assay targeting the skin sensitization endpoint, EURL ECVAM recommended using the assay as part of an integrated approach for testing and assessment (IATA). The recommendation then served as a basis for the drafting of an OECD guideline, which was recently adopted as guideline 442d.

17.3 Performance and Applicability of the Test Method

17.3.1 Reproducibility

For the set of coded chemicals tested during the validation studies, the KeratinoSens™ protocol yielded concordant predictions within the Givaudan laboratory (86%; $N = 14$) and between the five laboratories participating in the ring trial (86%, $N = 21$). In addition, dose-response curves measured in independent experiments yield highly reproducible results; thus, as an example in Fig. 17.2, the EC1.5 values (concentration inducing luciferase 1.5-fold) and the IC50 values for cytotoxicity measured in independent experiments (separated by a time gap of 3 years) in our laboratory are shown. This example indicates that not only the yes/no predictions are reproducible but that also the dose-response information, which can feed into potency assessment, is highly reliable over time.

17.3.2 Predictive Capacity

The EURL ECVAM peer review evaluated the predictivity found in the different studies. The recommendation reported that the accuracy of the test method in

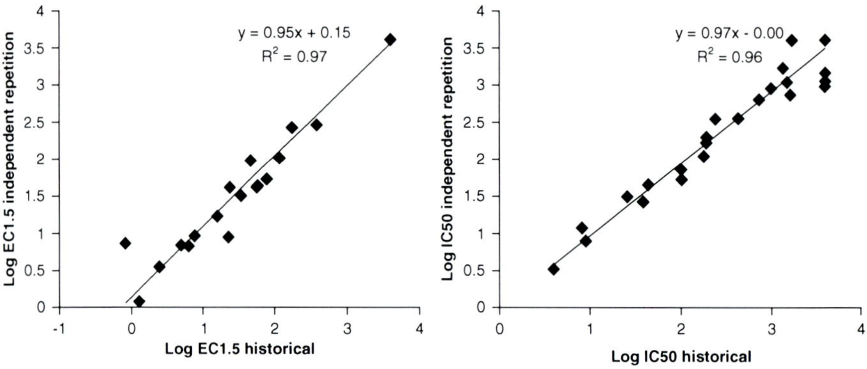

Fig. 17.2 Intralaboratory reproducibility of dose-response information obtained with the KeratinoSens™ assay. Shown EC1.5 values (*left*) and IC50 values for cytotoxicity (*right*)

predicting the *in vivo* classification (sensitizer/non-sensitizer) determined on the basis of existing evidence from LLNA, GPMT, Buehler test, and human data for the 21 chemicals evaluated blindly in the validation study was 90% (sensitivity 87%, specificity 100%). However since the chemicals selected by the lead lab to be used in the validation study have already been used to develop and optimize the KeratinoSens™ prediction model, it was considered likely that these values reflected a best-case scenario. When calculating the predictive capacity on the basis of a larger set of data generated in-house by Givaudan, sensitivity and specificity was about 75% ($n = 77$ sensitizers and 104 non-sensitizers). A recently published study correlating KeratinoSens™ data with classifications in the LLNA reported an accuracy of 77% (sensitivity 79% and specificity 72%) for a set of 145 chemicals [24]. Thus it is plausible that these figures might reflect the actual performance of the test in discriminating between sensitizers and non-sensitizers (analysis taken from [27]).

On a larger database of 244 molecules, accuracy was at 79.9% compared to *in vivo* data mainly from the LLNA [23], and on a recent data compilation on 173 chemicals accuracy was at 73% vs. LLNA data only. In that same compilation, for 103 chemicals, accuracy vs. LLNA and vs. human data could be compared; in that comparison, accuracy was at 82% for human data and at 74% vs. LLNA data [25]. Differences in the values between studies come on the one hand from differences in the datasets evaluated; on the other hand, some studies just compared to LLNA only, while in our analysis clear human evidence overruling LLNA was taken into consideration if available (i.e., by correcting known false-positives such as surfactants and false-negatives in the LLNA when classifying the *in vivo* reference data).

17.3.3 Applications and Limitations

During the validation by EURL ECVAM and through the testing of large databases, applicability and limitations were explored. This led to a detailed assessment of the limitations of the assay, and these conclusions are summarized below (mainly taken from the EURL ECVAM statement [27]).

17.3.3.1 Solubility of Test Substances

Chemicals that are not soluble in either water or DMSO, being these the two solvents prescribed by the SOP, cannot be tested in the KeratinoSens™. Chemicals with a calculated octanol/water partition coefficient (cLogP) up to 5 were successfully tested with the method.

17.3.3.2 Solvent Effects

As with many *in vitro/in chemico* assays, chemicals which are not stable in the prescribed solvents because of hydrolysis or direct interaction cannot be reliably tested.

17.3.3.3 Mechanistic Limitations

As the key mechanism leading to the activation of the Keap1-Nrf2-ARE pathway appears to be the electrophilic reaction of stressors with nucleophilic thiols (e.g., cysteine sulfhydryl groups) of Keap-1, it is possible that skin sensitizing chemicals with selective reactivity toward other nucleophiles may not be reliably identified by the KeratinoSens™, thereby leading to false-negative results. Thus it is obvious from the screening results that anhydrides, which are known for a preferential reactivity with lysine residues, are negative in the assay. However, there is scientific evidence that the Nrf2 pathway can be activated by other types of modification of Keap-1 cysteines such as oxidation or glutathionylation and that, moreover, the Nrf2 transcription factor may be controlled by other signaling pathways. It is therefore plausible that some sensitizing chemicals not covalently modifying Keap-1 cysteines (e.g., amine-reactive chemicals) can nevertheless activate the Nrf2 pathway, leading to true-positive responses in the KeratinoSens™ assay. Complementary information from peptide reactivity assays may help addressing this uncertainty, in particular assays able to distinguish between cysteine and lysine reactivity.

17.3.3.4 Metabolic Limitations

While a number of pro-haptens requiring enzymatic oxidation or deamination are reported to be correctly classified by the KeratinoSens™, pro-haptens requiring P450 activation are reported not to be identified by the assay. A possibility to incorporate a metabolic system in the KeratinoSens™ assay is to use Alachlor-induced rat liver S9 fractions, as is practiced widely in genotoxicity studies [28], but this does not form part of the validated protocol. In addition, it was found that the group of chemicals becoming Nrf2 activators in KeratinoSens™ only after metabolic activation by S9 is rather small and mainly includes some phenolic or methoxylated compounds [28].

A variety of pre-haptens have been reported as correctly predicted by the assay (e.g., 1,4-phenylenediamine, hydroquinone, and isoeugenol). However, some pre-haptens reported to have a slower rate of spontaneous oxidation (e.g., limonene) may require an oxidation step before the actual experiment. However, this is also the case in the animal test—significant response in the LLNA for chemicals like limonene and linalool is only obtained after an intentional strong exposure to oxygen, which must not be relevant to the use of the chemical in real life [29].

17.3.3.5 Potency Classes

Most of the misclassifications generated by the KeratinoSens™ concern chemicals that are moderate and weak sensitizers *in vivo*, while the false-negative rate for strong sensitizers is clearly lower. This should be kept in mind when interpreting negative results.

17.3.4 Comparison to Human Data

As indicated above, comparison with human data was made recently for a set of 102 chemicals for which both human and LLNA data are available [25]. Part of this evaluation is reproduced in Table 17.1. Interestingly, in this dataset, the *in vitro* methods including KeratinoSens™ do have a clearly better predictivity for human than for LLNA data. Prediction of human data by LLNA is included in the table, and actually the KeratinoSens™ prediction of human data (82% accuracy) is similar to the prediction of human data by the LLNA (also 82% accuracy).

Recently we also compared KeratinoSens™ dose-response and peptide reactivity data to human potency information. Especially the EC3 value derived from KeratinoSens™ (i.e., concentration for threefold luciferase induction) gave a significant correlation to human potency [23]. In combination with reactivity data, the correlation is in a similar range as reported for the correlation between LLNA EC3 and human data. The correlation (logarithmic R^2 around 0.45) however is not very strong for both the LLNA and the *in vitro* data, but this may be partly due to limitations in the human potency dataset.

Table 17.1 Predictivity of KeratinoSens™ and other *in vitro* assays for LLNA and human data (reproduced from Urbisch et al., Elsevier, 2014 under CC license http://dx.doi.org/10.1016/j.yrtph.2014.12.008)

	Human data				LLNA data			
Cooper statistics	Se [%][a]	Sp [%]	Acc [%]	*N*	Se [%]	Sp [%]	Acc [%]	*N*
KeratinoSens™	82	84	82	102	74	73	74	103
DPRA	84	84	84	102	77	85	79	105
h-CLAT	89	64	82	98	86	68	81	101
"2 out of 3" approach[b]	90	90	90	101	81	83	82	103
LLNA	91	64	82	111	–	–	–	–

[a]*Se* sensitivity, *Sp* specificity, *Acc* accuracy
[b]For explanation of "2 out of 3" approach, see below under 22.5

17.4 Brief Description of the Protocol

17.4.1 Experimental Steps

KeratinoSens™ cells are grown for 24 h in 96-well plates with a start inoculum of 10,000 cells/well. The medium is then replaced with medium containing the test substance and a final level of 1% of the solvent, DMSO. Each test substance is tested at 12 twofold dilutions ranging from 0.98 to 2000 μM. Each 96-well test plate may contain seven serially diluted test substances, and it always contains six wells with the solvent control, one well with no cells for background value and five wells with the positive control, cinnamic aldehyde, in five different concentrations. In each repetition, three parallel replicate plates are run with this same setup, and a fourth parallel plate is prepared from the same stock solution and the same cell suspension for cytotoxicity determination. Plates are then sealed with an adhesive foil to avoid any cross-contamination by volatile substances and incubated for 48 h with the test substances. At the end of the incubation period, cells are washed and lysed, and luciferase activity in the cell lysate is determined and in the parallel plates cytotoxicity (with the MTT assay) is measured. For an example of dose-response curves obtained, see Fig. 17.3. This full procedure needs to be repeated at least two times, and in case of non-congruent or borderline results, a third repetition is needed. Cytotoxicity can also be assessed with an alternative assay using the PrestoBlue® reagent, which allows measuring cell viability in the same cells prior to luciferase determination (manuscript submitted).

17.4.2 Data Processing

Data evaluation is automatically performed by a standardized Excel template which forms part of the SOP. The test plates are read by a plate reader, and the generated

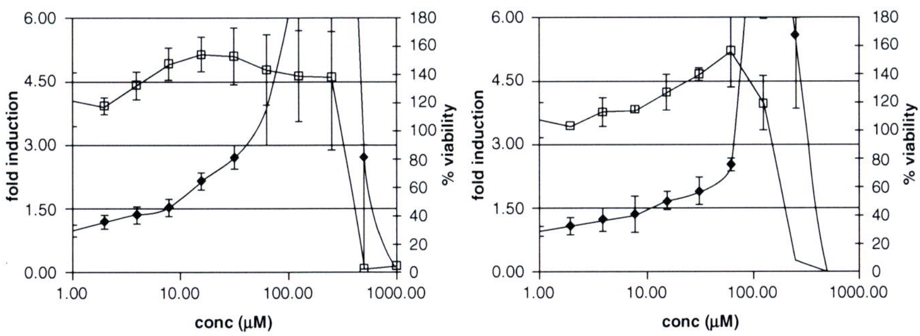

Fig. 17.3 Typical dose-response curves measured in KeratinoSens™. *Left*: cinnamic aldehyde. *Right*: paraphenylenediamine. Shown is fold-luciferase induction (*closed diamonds*) and % viability (*open squares*)

raw data are directly pasted into this template, and all data processing is performed automatically by a publically available Excel sheet. For both the MTT and the luciferase data, first the background value recorded in an empty well without added cells is subtracted. For the MTT data the % viability is then calculated for each well in the test plate in relation to average of the six solvent control wells. For the luciferase data the average value of the six solvent control wells is set to 1, and for each well in the test plate the fold induction is calculated in relation to this value.

The following parameters are then calculated from these processed raw data:

I_{max}	Maximal fold gene induction of the luciferase gene over the full dose-response up to 1000 µM
EC 1.5	Concentration in µM for 1.5-fold gene induction
EC 2	Concentration in µM for twofold gene induction
EC 3	Concentration in µM for threefold gene induction
Pos/Neg	Rating of chemical according to prediction model
reps. Positive	number of independent repetitions positive/number of repetitions done
IC50	Concentration in µM for 50% reduction of cell viability
IC30	Concentration in µM for 30% reduction of cell viability

17.4.3 Prediction Model

Chemicals are rated positive if the following conditions are met (see also Scheme 17.1):

- The I_{max} indicates >1.5-fold gene induction, and this induction is statistically significant above the solvent control in a particular repetition as determined by Student's t-test. The EC1.5 value is below 1000 µM in all three repetitions or in at least two repetitions. (If the I_{max} is exactly equal to 1.5, the chemical is still rated negative and no EC1.5 value is calculated by the evaluation sheet.)
- At the lowest concentration with a gene induction above 1.5-fold (i.e., at the EC 1.5 determining value), the cellular viability is above 70%.
- There is an apparent overall dose-response for luciferase induction, which is similar between the repetitions.

17.5 Role in a Testing Strategy

17.5.1 Hazard Assessment

Already in the original publication on KeratinoSens™, we proposed to combine KeratinoSens™ data with at least a peptide reactivity assessment, and this proposal is also supported by the recent ECVAM statement. In a very simple approach, we combined KeratinoSens™ data with an LC-MS-based peptide reactivity assay

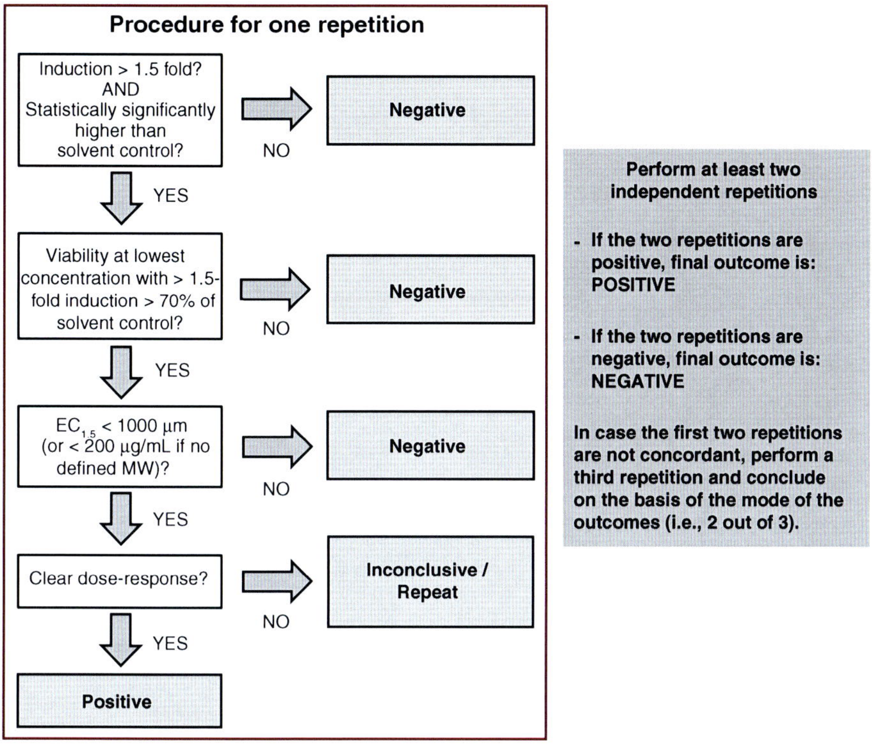

Scheme 17.1 Prediction model of the KeratinoSens™ assay

to predict hazard. A conservative approach is to use only evidence for adduct formation (i.e., observable covalent peptide modification) in the LC-MS assay for positive classification, as this is the actual MIE in the AOP. This yields high specificity, as it avoids false-positives due to peptide oxidation, but at the cost of limited sensitivity (some sensitizers do oxidize the peptide but do not form observable adducts; these are correctly predicted if peptide depletion is used for a positive rating). However, positive evidence can then be combined, rating each chemical positive which is either positive in KeratinoSens™ and/or forms direct peptide adducts [22]. This simple combination enhances sensitivity as compared to KeratinoSens™ alone with only minor loss in specificity, the main reason being the positive identification of exclusively amine-reactive chemicals missed in KeratinoSens™[23].

Another straightforward approach to combine KeratinoSens™ data into a hazard assessment strategy is the "2 out of 3" approach, whereby KeratinoSens™ data are combined with data from the DPRA and dendritic cell activation [25, 30]. Any two congruent results drive the final assessment in this approach, which clearly enhances accuracy, esp. for prediction of human data (see Table 17.1). In this approach it is sufficient to run two tests for most chemicals and perform the third test only in case of discordant results. The most cost-efficient way is to perform the rapid and cheap

KeratinoSens™ and DPRA assays first and then decide on the necessity of performing an h-CLAT which involves technically more demanding evaluation of cell surface markers.

17.5.2 Potency Prediction

The dose-response parameters for both gene induction and, to a lesser extent, cytotoxicity correlate to potency in the LLNA [24] but are as such clearly not sufficient for potency prediction. Therefore different ways to use them in an integrated testing strategy were explored. In the Bayesian net approach, KeratinoSens™ data are used along with DPRA, dendritic cell activation, *in silico* prediction based on TIME SS, and bioavailability calculations. These information sources are grouped in a mechanistically based network, whereby the individual outputs affect the predicted probability distribution of four LLNA potency classes [31]. The big advantage of this approach is the fact that there is a probabilistic information on the robustness of the prediction on the one hand, and on the other hand this approach can work with partial evidence, offering the possibility of minimizing redundant testing. The drawback is the necessity of data discretization (leading to information loss) and the prediction of LLNA classes only (and not a continuous potency scale), as this mathematical approach works with classes and is not applicable to continuous scales.

We have recently explored the simpler approach of multiple regression analysis. In this approach LLNA or human potency is predicted by a linear regression equation of logarithmically transformed peptide reactivity, luciferase induction, cytotoxicity, and evaporation data. This analysis was further refined, by creating specific models for mechanistic domains of chemicals in case there was sufficient *in vivo* data for similarly acting chemicals [23]. This refinement led to a clearly improved potency prediction, and a strategy to combine the two approaches was presented.

Besides these more formalized approaches, KeratinoSens™ and the other *in vitro/in chemico* data may in the future also be used for "*in vitro*-based read-across." While in classical read-across, a structural relationship is used to directly relate to the animal data of related compounds, in *in vitro*-based read-across, we should know the *in vitro* data for both the test chemical and the related chemicals used for read-across, while the *in vivo* data are only available on the latter. *In vitro* data may then be used for a refined read-across prediction ("is the new chemical rated more or less potent *in vitro* as compared to the related chemicals with known *in vivo* outcome?"). Such an approach may be preferable in chemical domains with some but only limited *in vivo* data (i.e., insufficient to build a robust statistical domain-based model). This approach will be facilitated by the large accumulating databases of *in vitro* and *in vivo* data.

17.6 Perspectives from the Test Developer

17.6.1 Critical Steps in the Protocol

Technically the assay is straightforward, and it is based on established assay setups in 96-well plates with adherent cell lines. Therefore little hurdles were encountered when transferring the assay to other laboratories, and stability of assay results over time is very good (see also Fig. 17.1). The key technical limitation, as for all assays in submerged cultures, is the water solubility of the test chemicals. This may especially affect the confidence in negative predictions for highly insoluble chemicals, which are negative at the maximal dose of the prediction model (1000 μM) but are not really dissolved at this concentration. In general, testing of chemicals with cLogP >7 should be avoided, while in the range of cLogP 5–7, results should be treated with care taking other evidence into account.

17.6.2 Possible Protocol Adaptations

A counter screen to retest chemicals initially tested negative in the presence of S9 mix was discussed in Sect. 22.3.3 [28]. This was found useful for a structurally relatively narrow group of chemicals, but it may also be further explored if completely novel types of chemicals are being tested in KeratinoSens™. Another simple adaptation of the protocol is to use the PrestoBlue® viability assay which is performed on the same cells which are subsequently used for luciferase determinations (manuscript submitted).

17.6.3 Challenges and Opportunities

A more challenging approach would be to further develop the model from the 2D culture into a 3D tissue culture model. This would facilitate application of poorly soluble chemicals and would help to bring a more realistic *in vivo*-like exposure scenario (i.e., topical application to the *stratum corneum* of a 3D tissue).

This approach is taken by, e.g., the emerging SENSIS® model, whereby Nrf2 endogenous genes are measured in treated primary 3D tissues using RT-PCR technique. However, this is a more resource-intensive detection method as compared to luciferase technology.

We have therefore tried to transform the KeratinoSens™ model into a 3D culture. Since the HaCaT (and thus KeratinoSens™) cells have lost the ability for terminal differentiation when cultured at the air-liquid interface, a co-culture was established whereby a fraction of KeratinoSens™ cells are seeded along with primary keratinocytes. Using adapted media, these cell mixtures differentiated into a fully stratified epidermis when cultured at the air-liquid interface, still expressing stable levels of the Nrf2-inducible luciferase. A second, constitutively expressed Renilla luciferase was expressed in the KeratinoSens™ cells used in this approach to facilitate signal normalization. Topical application of sensitizers in the acetone/olive oil vehicle used in the LLNA led to induced luciferase expression. However, repeated testing revealed a batch-to-batch variability which may pose difficulties in a formal validation process.

Another approach, which might be explored, is to add another reporter gene regulated by a non-redundant pathway offering additional information, especially for chemicals currently predicted as false-negatives. Technically we have shown that dual luciferase expression is possible and that double transgenic cell lines keep the predictivity of the original KeratinoSens™ cell line. However, to date we have investigated a limited set of additional markers [19]. In most cases (based on a limited dataset however), these markers gave redundant information or, in the case of the marker Fos, would increase sensitivity by detecting amine-reactive chemicals but at the cost of reduced specificity by responding to the irritating surfactant SDS, and more work is needed to identify a robust nonredundant additional genetic marker which increases sensitivity without major loss in specificity. However, as indicated above, this additional information (e.g., on amine-reactive chemicals) may also come from an *in chemico* test used in an integrated testing strategy and must not come from a cellular assay. Finally, while the approaches discussed above would all improve sensitivity, we also observe a number of false-positives, and a significant effort was made to understand those and potentially be able to discriminate false-positives and true-positives. Often these false-positive responses occur only at high concentrations, and these chemicals are thus predicted as very weak sensitizers in an integrated assessment [23].

Oxidative stress may in some cases induce the Nrf2 pathway. While for some chemicals this occurs only at cytotoxic concentrations, which is addressed in the prediction model, some chemicals also can trigger oxidative stress at non-cytotoxic concentrations. Indeed, many skin sensitizers are able to trigger oxidative stress, and this is even considered to constitute part of the danger signal essential for induction of skin sensitization. However, it is of relevance to understand this signal in those chemicals which trigger oxidative stress at non-cytotoxic doses, but are not sensitizers, thereby triggering the false-positive response. This is especially critical for some phenolic compounds from plants, for which a high interest for cosmetic use exists, but which can act as inducers of the Nrf2 pathway.

17.7 Conclusions

KeratinoSens™ is an established rapid screening test with a high reproducibility and a technical simplicity which facilitated adaptation by several CROs to date. The analysis of the use of KeratinoSens™ data in integrated testing strategies for both hazard and potency predictions has advanced rapidly in the recent 2 years along with the progression through ECVAM validation and OECD adoption. Limitations in the predictive capacity and opportunities for developing improved or complementary assays were identified based on the screening of large sets of reference chemicals. These databases containing both yes/no predictions according to the prediction model but also detailed dose-response information have all been made publicly available which facilitates creation of novel integrated testing strategies implementing KeratinoSens™ data and which can be used for *in vitro*-based read-across.

References

1. Karlberg AT, Bergstrom MA, Borje A, Luthman K, Nilsson JL. Allergic contact dermatitis--formation, structural requirements, and reactivity of skin sensitizers. Chem Res Toxicol. 2008;21:53–69.
2. OECD. The adverse outcome pathway for skin sensitisation initiated by covalent binding to proteins, part 1: scientific evidence. Oecd Environment, Health and Safety Publications, Series On Testing and Assessment no. 168. 2012.
3. Ahlfors SR, Sterner O, Hansson C. Reactivity of contact allergenic haptens to amino acid residues in a model carrier peptide, and characterization of formed peptide-hapten adducts. Skin Pharmacol Appl Ski Physiol. 2003;16:59–68.
4. Gerberick F, Aleksic M, Basketter D, Casati S, Karlberg AT, Kern P, Kimber I, Lepoittevin J-P, Natsch A, Ovigne JM, Rovida C, Sakaguchi H, Schultz T. Chemical reactivity measurement and the predictive identification of skin sensitisers. ATLA. 2008;36:215–42.
5. Wakabayashi N, Dinkova-Kostova AT, Holtzclaw WD, Kang MI, Kobayashi A, Yamamoto M, Kensler TW, Talalay P. Protection against electrophile and oxidant stress by induction of the phase 2 response: fate of cysteines of the Keap1 sensor modified by inducers. Proc Natl Acad Sci U S A. 2004;101:2040–5.
6. Motohashi H, Yamamoto M. Nrf2-Keap1 defines a physiologically important stress response mechanism. Trends Mol Med. 2004;10:549–57.
7. Natsch A. The Nrf2-Keap1-ARE toxicity pathway as a cellular sensor for skin sensitizers--functional relevance and a hypothesis on innate reactions to skin sensitizers. Toxicol Sci. 2010;113:284–92.
8. Natsch A, Emter R. Skin sensitizers induce antioxidant response element dependent genes: application to the in vitro testing of the sensitization potential of chemicals. Toxicol Sci. 2008;102:110–9.
9. Ade N, Leon F, Pallardy M, Peiffer JL, Kerdine-Romer S, Tissier MH, Bonnet PA, Fabre I, Ourlin JC. HMOX1 and NQO1 genes are upregulated in response to contact sensitizers in dendritic cells and THP-1 cell line: role of the Keap1/Nrf2 pathway. Toxicol Sci. 2009;107:451–60.
10. Johansson H, Lindstedt M, Albrekt AS, Borrebaeck CA. A genomic biomarker signature can predict skin sensitizers using a cell-based in vitro alternative to animal tests. BMC Genomics. 2011;12:399.
11. Miyazawa M, Takashima A. Development and validation of a new in vitro assay designed to measure contact allergen-triggered oxidative stress in dendritic cells. J Dermatol Sci. 2012;68:73–81.
12. van der Veen JW, Pronk TE, van Loveren H, Ezendam J. Applicability of a keratinocyte gene signature to predict skin sensitizing potential. Toxicol In Vitro. 2013;27(1):314–22. doi: 10.1016/j.tiv.2012.08.023.
13. Vandebriel RJ, Pennings JL, Baken KA, Pronk TE, Boorsma A, Gottschalk R, Van Loveren H. Keratinocyte gene expression profiles discriminate sensitizing and irritating compounds. Toxicol Sci. 2010;117:81–9.
14. Neves BM, Goncalo M, Figueiredo A, Duarte CB, Lopes MC, Cruz MT. Signal transduction profile of chemical sensitisers in dendritic cells: an endpoint to be included in a cell-based in vitro alternative approach to hazard identification? Toxicol Appl Pharmacol. 2011;250:87–95.

15. Kim HJ, Barajas B, Wang M, Nel AE. Nrf2 activation by sulforaphane restores the age-related decrease of T(H)1 immunity: role of dendritic cells. J Allergy Clin Immunol. 2008;121:1255–1261.e7.
16. El Ali Z, Gerbeix C, Hemon P, Esser PR, Martin SF, Pallardy M, Kerdine-Römer S. Allergic skin inflammation induced by chemical sensitizers is controlled by the transcription factor Nrf2. Toxicol Sci. 2013;134(1):39–48. doi:10.1093/toxsci/kft084.
17. Ryan CA, Gildea LA, Hulette BC, Dearman RJ, Kimber I, Gerberick GF. Gene expression changes in peripheral blood-derived dendritic cells following exposure to a contact allergen. Toxicol Lett. 2004;150(3):301–316.
18. Gildea LA, Ryan CA, Foertsch LM, Kennedy JM, Dearman RJ, Kimber I, Gerberick GF. Identification of gene expression changes induced by chemical allergens in dendritic cells: opportunities for skin sensitization testing. J Invest Dermatol. 2006;126:1813–1822.
19. Emter R, van der Veen JW, Adamson G, Ezendam J, van Loveren H, Natsch A. Gene expression changes induced by skin sensitizers in the KeratinoSens cell line: discriminating Nrf2-dependent and Nrf2-independent events. Toxicol In Vitro. 2013;27:2225–32.
20. Ball N, Cagen S, Carrillo JC, Certa H, Eigler D, Emter R, Faulhammer F, Garcia C, Graham C, Haux C, Kolle SN, Kreiling R, Natsch A, Mehling A. Evaluating the sensitization potential of surfactants: integrating data from the local lymph node assay, guinea pig maximization test, and in vitro methods in a weight-of-evidence approach. Regul Toxicol Pharmacol. 2011;60:389–400.
21. Delaine T, Niklasson IB, Emter R, Luthman K, Karlberg AT, Natsch A. Structure-activity relationship between the in vivo skin sensitizing potency of analogues of phenyl Glycidyl ether and the induction of Nrf2-dependent luciferase activity in the KeratinoSens in vitro assay. Chem Res Toxicol. 2011;24(8):1312.
22. Emter R, Ellis G, Natsch A. Performance of a novel keratinocyte-based reporter cell line to screen skin sensitizers in vitro. Toxicol Appl Pharmacol. 2010;245:281–90.
23. Natsch A, Emter R, Gfeller H, Haupt T, Ellis G. Predicting skin sensitizer potency based on in vitro data from keratinosens and kinetic peptide binding: global versus domain-based assessment. Toxicol Sci. 2014;143:319–32.
24. Natsch A, Ryan CA, Foertsch L, Emter R, Jaworska J, Gerberick F, Kern P. A dataset on 145 chemicals tested in alternative assays for skin sensitization undergoing prevalidation. J Appl Toxicol. 2013;33:1337–52.
25. Urbisch D, Mehling A, Guth K, Ramirez T, Honarvar N, Kolle S, Landsiedel R, Jaworska J, Kern PS, Gerberick F, Natsch A, Emter R, Ashikaga T, Miyazawa M, Sakaguchi H. Assessing skin sensitization hazard in mice and men using non-animal test methods. Regul Toxicol Pharmacol. 2015;71(2):337–51. doi:10.1016/j.yrtph.2014.12.008.
26. Natsch A, Bauch C, Foertsch L, Gerberick F, Normann K, Hilberer A, Inglis H, Landsiedel R, Onken S, Reuter H, Schepky A, Emter R. The intra- and inter-laboratory reproducibility and predictivity of the KeratinoSens assay to predict skin sensitizers in vitro: results of a ring-study in five laboratories. Toxicol In Vitro. 2011;25:733–44.
27. ECVAM. EURL ECVAM recommendation on the KeratinoSens™ assay for skin sensitisation testing. 2014 http://ihcp.jrc.ec.europa.eu/our_labs/EURL-ECVAM/eurl-ecvam-recommendations/recommendation-keratinosens-skin-sensitisation. Last accessed 26 March 2014.
28. Natsch A, Haupt T. Utility of rat liver S9 fractions to study skin-sensitizing Prohaptens in a modified KeratinoSens assay. Toxicol Sci. 2013;135:356–68.
29. Kern S, Dkhil H, Hendarsa P, Ellis G, Natsch A. Detection of potentially skin sensitizing hydroperoxides of linalool in fragranced products. Anal Bioanal Chem. 2014;406:6165–78.
30. Bauch C, Kolle SN, Ramirez T, Eltze T, Fabian E, Mehling A, Teubner W, van Ravenzwaay B, Landsiedel R. Putting the parts together: combining in vitro methods to test for skin sensitizing potentials. Regul Toxicol Pharmacol. 2012;63:489–504.
31. Jaworska J, Dancik Y, Kern P, Gerberick F, Natsch A. Bayesian integrated testing strategy to assess skin sensitization potency: from theory to practice. J Appl Toxicol. 2013;33:1353–64.

18 LuSens: Shedding Light on Skin Sensitization

Tzutzuy Ramirez, Annette Mehling, and Robert Landsiedel

18.1 Introduction

Allergic contact dermatitis (ACD), the clinically relevant outcome of skin sensitization, is one of the most prevalent skin diseases. It is estimated that 15–20% of the general population will be sensitized at some point during the course of their lives [1, 2] with the prevalence possibly increasing [3, 4]. ACD can be associated with substantial morbidity; affect quality of life and the sensitization state—the prerequisite for the development of ACD—is usually a lifelong effect. Exposure to contact allergens frequently results in ACD, not only in the consumer setting but also in the work environment, and is a major cause of occupational disease. This results in high costs for healthcare systems and the economy as well as in an impairment of the quality of life for the patients [5]. Testing the skin sensitization potential of a substance is therefore necessary for the hazard assessment of any substance which may come into contact with human skin; this test is indeed demanded by substance regulations such as REACH.

18.2 Brief Overview of Current Regulations

The high prevalence, and the social and economic impact resulting from sensitization, is reflected by the requirement for the evaluation of the sensitization potential of a substance placed on the market. Most of the substance legislations worldwide require information on skin sensitization, amongst others, the European Chemicals Regulation concerning the Registration, Evaluation, Authorisation and Restriction of Chemicals

T. Ramirez (✉) • R. Landsiedel
BASF SE, Experimental Toxicology and Ecology, Ludwigshafen, Germany
e-mail: tzutzuy.ramirez-hernandez@basf.com; robert.landsiedel@basf.com

A. Mehling
BASF Personal Care and Nutrition GmbH, Düsseldorf, Germany
e-mail: Annette.mehling@basf.com

C. Eskes et al. (eds.), *Alternatives for Dermal Toxicity Testing*,
DOI 10.1007/978-3-319-50353-0_18

(REACH; Regulation (EC) No 1907/2006). With the implementation of REACH, an assessment of the intrinsic potential of a substance to cause sensitization (hazard) is part of the mandatory base set of toxicological endpoints to be evaluated for all substances to be registered. According to the UN Globally Harmonized System (GHS) for classification and labelling of chemicals, a skin-sensitizing substance can be assigned the label of a strong sensitizer (Cat 1A) or "another" sensitizer (Cat 1B) if sufficient data is available to make such a distinction. The two GHS subcategories affect concentration limits in formulations which are regarded to be non-hazardous.

In the regulatory setting, the sensitizing potential of substances is generally evaluated using animal tests, such as the murine local lymph node assay (LLNA; [6–8]), which gives a measure of the induction phase of sensitization, or the guinea pig tests, which provide information on both induction and elicitation phases [9]. Europe is currently probably a trailblazer in the area of new chemical and cosmetic legislations but also in promoting the use of 3R (refinement, reduction, replacement) methods for toxicological assessments. The European Chemicals Legislation (REACH, [10]) specifically mentions that animal testing should only be performed as a last resort [11]. Annex XI of REACH gives an indication of how non-standard procedures, which currently applies to non-animal testing methods, can be used and with an explicit mention of the weight of evidence (WoE) approach, defined as the conclusion derived "from several independent sources of information leading to the assumption/conclusion that a substance has or has not a particular dangerous property, while the information from each single source alone is regarded insufficient to support this notion". As of March 11, 2013, the European Union imposed an animal testing ban on both cosmetic products and their ingredients, which is accompanied by a marketing ban, if animal tests were conducted after this date for the purpose of the cosmetics regulation [12]. Within the Tox21 and ToxCast projects, the United States Environmental Protection Agency has implemented a different approach in which high-throughput methods are used to screen chemical libraries and address different toxicity pathways and modes of action (MoA) in humans with goal of defining tools that can be used by risk assessors in their decision processes (http://www.epa.gov/ncct/Tox21/). The increasing awareness and interest in animal welfare aspects on a global basis is resulting in various countries making strides in moving away from animal testing, in particular of cosmetic products, and resulting in similar legislations banning animal testing for this market segment. Yet, currently, there is no non-animal alternative test method for the endpoint of skin sensitization available that has gained full regulatory acceptance as a stand-alone method. During the last decades, extensive work has been conducted to develop non-animal test methods able to replace current animal test methods for the predictive identification of skin sensitizers (reviewed in, e.g. [13, 14]). Given the complexity of the sensitization pathway, a combination of tests will be needed to achieve reliable predictions of the skin sensitization potential of a substance.

18.3 The Adverse Outcome Pathway for Skin Sensitization

The steps leading to skin sensitization are relatively well described, and in contrast to many other toxicological endpoints, human data is available to substantiate predictions on sensitization potentials. The sequence of events leading to skin sensitization and ultimately allergic contact dermatitis has recently been described by the OECD in the document titled "The Adverse Outcome Pathway (AOP) for Skin Sensitisation Initiated by Covalent Binding to Proteins" [15, 16].

Contact allergies develop in two stages: (1) the sensitization phase in which antigen-/allergen-specific T cells are generated and (2) the elicitation phase in which renewed contact with the allergen leads to the allergic response (the adverse outcome). The AOP described by the OECD identifies eleven events involved, whereby four are considered to be key events in the AOP. The skin, and in particular the stratum corneum, provides an effective barrier against the entry of substances into the body. Once the substance has penetrated into the skin, the initiating event of the sensitization process is the molecular interaction of the substance, typically a low molecular weight (LMW) electrophilic substance termed hapten, with the skin proteins creating a complete antigen (key event 1). The electrophilic hapten may be generated from a pro- or pre-hapten via metabolic or abiotic transformation [17]. Key event 2 is the activation of keratinocytes. During contact with the hapten, keratinocytes must generate "danger signals" to generate the proper context for an allergic response to develop (reviewed in [18, 19]). Amongst these responses, the oxidative and electrophilic stress-driven expression of genes under the control of the antioxidant response element (ARE) as part of the Keap1/Nrf2 pathway is well described [20]. The third key event is the activation of dendritic cells. Dendritic cells take up and process antigens and present fragments, migrate to the lymph nodes and present the antigen to naïve T cells. Only mature dendritic cells can activate naïve T cells and are characterized by the upregulation of cell surface markers such as CD54 and CD86. Key event 4 is characterized by the proliferation of the antigen-specific T cells and the generation of antigen-specific memory T cells (reviewed in [21, 22]). Upon renewed contact with the same hapten, this will be presented to the preformed allergen-specific T cells which will ultimately lead to the inflammatory response typical for allergic contact dermatitis ([14, 23], Fig. 18.1).

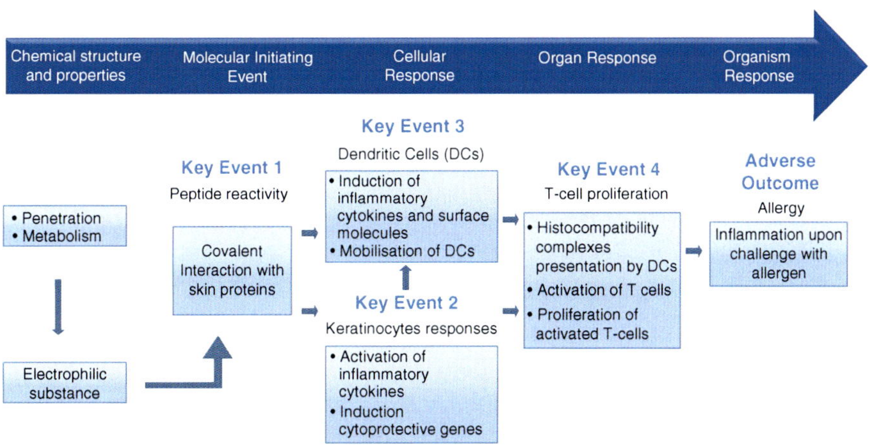

Fig. 18.1 Overview of the skin sensitization adverse outcome pathway as described by the OECD. The AOP consists of a number of primarily sequential steps. The four key events, namely, (*1*) peptide reactivity, (*2*) keratinocyte activation, (*3*) dendritic cell activation and (*4*) T-cell proliferation, are highlighted (adapted from [15])

18.4 The Keap1/Nrf2 Pathway

A substantial amount of data is now available demonstrating that a wide range of skin sensitizers is able to activate the Keap1/Nrf2 pathway [24, 25]. Under physiological conditions, the Kelch-like ECH-associated protein 1 (Keap1) sequesters the transcriptional regulator nuclear factor-erythroid 2-related factor 2 (Nrf2) in the cytoplasm provoking its proteasomal degradation [26]. The Keap1 protein contains highly reactive cysteine (Cys) residues which, if covalently modified, result in the dissociation of the Keap1/Nrf2 complex, and Nrf2 is released. Nrf2 can then translocate into the nucleus, where it can form a complex with other molecules. This complex then binds to the antioxidant response element (ARE) in the promoter region of several genes, thereby activating the downstream transcription of the ARE-dependent genes (Fig. 18.2). The majority of skin sensitizers are electrophilic and are therefore able to react with the cysteine residues of Keap1, thereby enhancing Nrf2 release [25, 27, 28]. These findings led to the development of reporter cell lines designed to monitor the Nrf2 antioxidant response pathway. To date, a number of cell lines have been generated, among those, it can be mentioned AREc32 (a human breast cancer-based cell line [29]), KeratinoSens™ (a human keratinocyte-based cell line; [27]) and LuSens (a human keratinocyte-based cell line; [30]). These cell lines have been established and further developed to identify potential skin sensitizers with the human keratinocyte-based assay having been validated with a large set of test substances for its use to detect skin sensitizers. A more recent study describes the activation of the Nrf2 pathway in the human monocytic cell line (THP-1) after exposure to skin sensitizers [31].

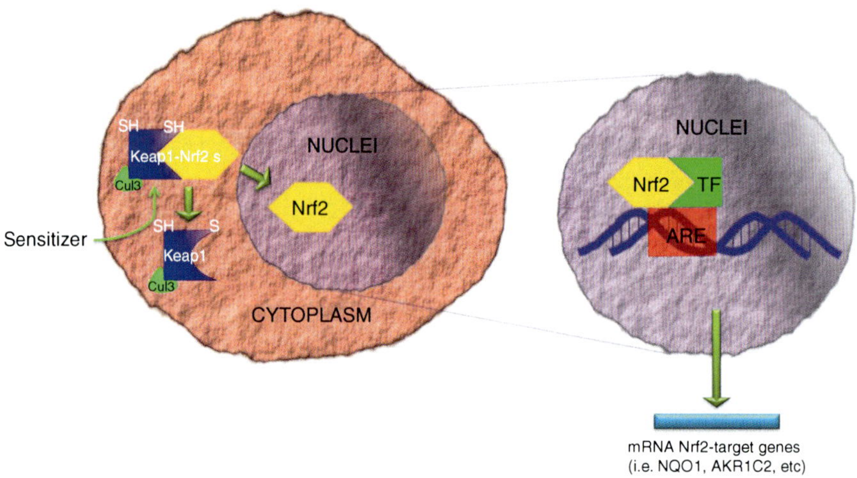

Fig. 18.2 Activation of the Keap1/Nrf2 pathway by skin sensitizers. Covalent binding of the Cys residues (SH) leads to dissociation of the Keap1/Nrf2 complex. Nrf2 is released and translocated into the nucleus where it binds to other transcription factors (TF) and activates genes containing an ARE sequence in their promoter region

Keratinocytes represent the predominant cell type in the skin and play a pivotal role in the activation of the dendritic cells of the skin, the Langerhans cells, following stimulation by pathogens and haptens [14, 32]. The KeratinoSens™ and LuSens reporter cell lines are derived from immortalized human keratinocytes and measure the activation of the Nrf2-Keap1 pathway via luminescent reaction. Due to the nature of both cell lines, the evidence obtained only reflects the final activation of the Nrf2 pathway and indirectly gives a measure of other steps, such as binding of the hapten to Keap1 and translocation of Nrf2 into the nucleus.

18.5 LuSens Assay

The LuSens assay is an *in vitro* method for the identification of skin sensitizers using a genetically modified human keratinocyte cell line, LuSens cells [30]. It uses a reporter gene for luciferase placed under the control of the antioxidant response element (ARE) and hence monitors Nrf-2 transcription factor activity. The measured endpoint is the upregulation of luciferase activity after 48 h of incubation with test substances. This upregulation is an indicator for the activation of the Keap1/Nrf2/ARE signalling pathway [20, 24, 25, 34].

To make a statement on the skin sensitization hazard of test substance, a LuSens assay, comprising at least two but a maximum of three independent repetitions, needs to be carried out. In a valid repetition (i.e. meeting all acceptance criteria, see section 18.8 of this chapter), sensitizing potential of the substance is indicated if the luciferase activity equals or exceeds a 1.5-fold induction compared to the vehicle control at concentrations that do not reduce cell viability to more than 70%. The third repetition is only required when the first two repetitions are not concordant (e.g. one repetition is positive and the other is negative). If both results of the first two repetitions of an experiment are either positive or negative, the experiment is completed.

18.6 LuSens Cells

The LuSens cells were developed by transfection of pGL4.20-ARE-Luc2 into an immortalized keratinocyte cell line. The ARE sequence originated from the NADPH:quinone oxidoreductase 1 gene from rat (ggtaccagtctagagtcacagtgacttggcaaaatcgctagc). Nevertheless, due to the high homology of the ARE sequences, the response element is also functional in humans and is upregulated by contact sensitizers. This allows a quantitative measurement of luciferase gene induction (via luminescence), as an indicator of the activity of the Nrf2 transcription factor in cells following exposure to potentially sensitizing substances. The luciferase signal is therefore a surrogate for the activation of endogenous Nrf2-dependent genes by sensitizers, and the strict dependence of the luciferase signal in the recombinant cell line on Nrf2 has been demonstrated in an in-house validation study [30] and in a multi-laboratory validation study [45].

The cell lines harbouring the luciferase reporter gene construct were selected via resistance to the antibiotic puromycin, which was the selection marker included in the transfected plasmid. From these experiments, a population expressing luciferase under the control of ARE was obtained and the clone with the most suitable characteristics selected. Briefly, following transfection with pGL4.20-ARE-Luc2, several clones were selected and separately cultured in order to expand single-cell colonies and select a clone with (a) the best signal to noise ratio of the luciferase expression and (b) good dynamic range to react in a dose-dependent fashion to a weak sensitizer (i.e. ethylene glycol dimethacrylate (EGDMA)). For this selection, 24 clones were tested, and 2 further selected to test a set of 25 test substances with known sensitization potential. From these experiments, clone 16 demonstrated the best proficiency to identify skin sensitizers and the best dynamic range of luciferase expression after exposure to sensitizers. This clone was further cultured to establish the LuSens cell line. In addition, this clone was also subjected to molecular characterization to confirm the stable integration of the reporter gene construct. For this purpose, a cytogenetic analysis and genomic sequencing for verification of the presence of the ARE sequence were performed. The cytogenetic analysis was consistent with a cell line of human origin, presenting a hypertriploid karyotype with an average number of 77 chromosomes (modal range from 74 to 80) and 6 marker chromosomes (M1–M6), of which 3 showed 2 copies each [30]. In addition, the genomic sequencing demonstrated that the sequence of the ARE of the reporter gene construct was present without any modification and that genomic integration had occurred (data not shown). When a Blast search of the ARE sequence identified in the LuSens cell line was compared to DNA sequences from the NCBI genome database (http://www.ncbi.nlm.nih.gov/genome), the sequence from LuSens clone 16 provided a 100% identity with the NAD(P)H dehydrogenase [quinone] 1 gene (Nqo1) sequence of *Rattus norvegicus* strain BN Sprague-Dawley, thus confirming that the sequence had not been modified during integration (Sequence ID: ref. AC_000087.1). It is important to highlight that ARE consensus sequence, responsible for ARE functionality amongst species (i.e. human, mouse, rat), TMAnnRTGAYnnnGCRwww (where M = A or C, R = A or G, Y = C or T, W = A or T, S = G or C) ([35], Fig. 18.3), is present in the rat ARE sequence found in LuSens cells and is therefore also functional in human cells. Further molecular

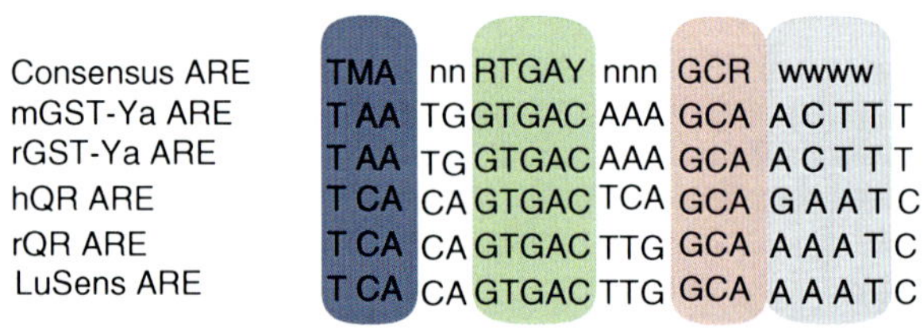

Fig. 18.3 The ARE consensus sequence. The known AREs from GST and QR genes are aligned. Nucleotides at essential positions are highlighted. The consensus for nucleotides at the essential and preference sites are indicated. The abbreviations follow standard IUPAC nomenclature (M: A or C, R: A or G, Y: C or T, w: A or T, S: G or C; modified from [35])

characterization indicated stable insertion of the pGL4.20-ARE-Luc2 sequence in proximity to DNA sequence of chromosome 14. In addition, recent experiments from our laboratory have demonstrated accurate responses in the LuSens assay even with cells that had been cultured without the selection antibiotic for over 30 passages in culture indicating stability and robustness of the cell line.

18.7 Assay

The LuSens assay consists of two phases: range finder experiment and main experiment. The range finder is important to select the adequate dose range that will be used in the main experiment; therefore, in this range finder, only viability by means of MTT is evaluated. From the range finder, the concentration in which cell viability corresponds to no less than 75% (CV75) is calculated. The highest tested concentration in the main experiment is then 1.2xCV75 (or 2000 μM if no cytotoxicity is observed). Briefly, for range finder experiments, LuSens cells are used for pre-culture at passage 4 and should present 80–90% confluent. Cells are trypsinized and resuspended in 9 mL DMEM containing 10% FBS Superior (Biochrome, Germany). From the cell suspension, 10,000 cells (in 120 μL) are seeded in each well of a clear flat-bottom 96-well plate (TPP, Germany). Cells are incubated for 24 h at 37 °C in a humidified atmosphere containing 5% CO_2. Prior to treatment, test substances are dissolved in DMSO to prepare stock solutions, 12 mM ethylene glycol dimethylacrylate (EGDMA as positive control, Sigma, Germany) and 500 mM DL-lactic acid (as negative control, Sigma, Germany). 200 mM stock solution (100x) of test substances are prepared. For treatment, 50 μL of the test substance dilutions are applied to have in triplicate following test concentrations: 0.976, 1.953, 3.906, 7.812, 15.625, 31.25, 62.5, 125, 250, 500, 1000 and 2000 μM and 120 μM EGDMA and 5000 μM DL-lactic acid to a final volume of 200 μL. The plates are sealed with a breathable tape (Nunc/Thermo, Germany) and incubated for 48 h at 37 °C in a humidified atmosphere containing 5% CO_2. After 48 h of treatment, viability is measured by means of MTT assay and using a photometer. Then the concentration at which the viability is reduced to no more than 75% (CV75) compared to vehicle control (VC) is calculated.

For the main experiment, the expression of luciferase at non-cytotoxic concentrations is evaluated. Briefly, cells were seeded into clear and white flat-bottom 96-well plates (TPP, Switzerland or Perkin Elmer, VWR, Germany; per well: 10 000 cells contained in 200 μL of cell suspension). The ones seeded in clear plates are used for the cytotoxicity assay, whereas, the white plates for the luminescent assay. Prior treatment, cells are incubated for 24 h at 37 °C in a humidified atmosphere containing 5% CO_2. For treatment, test substances are dissolved in DMSO (100x stock solution) at concentrations according to the preliminary cytotoxicity data. Substances are further diluted (1:25) in DMEM containing only 1% FBS superior (Biochrome, Germany) to obtain 4x stock solution. Final DMSO concentration on the cells should not exceed 1%. The highest tested concentration is 1.2x CV75.

Treatment is performed by applying 50 µL of the test substance dilution to each well (final volume, 200 µL) for 48 h. Each substance is tested at six concentrations in triplicate. Positive and negative controls are prepared as indicated in the range finder experiment and dose in a similar fashion. After treatment, ARE activation is measured by luminescent reaction using a luminometer. The fold induction (FI) of the luminescent signal is calculated by dividing the relative luminescence units (RLU) of the treated cells (TC) by the RLU of vehicle control (VC)-treated cells using following equation:

FI = (RLU TC)/(RLU VC). In parallel, the MTT assay is performed as indicated in the range finder experiment.

18.8 Acceptance Criteria and Prediction Model

For acceptance of a repetition, the average induction for the positive control (120 µM EGDMA) should be equal or above 2.5 luciferase fold induction and it should have a relative viability of at least 70%. The induction triggered by the negative control (5000 µM DL-Lactic acid) as well as the basal expression of the cells should be below 1.5 luciferase fold induction as compared to the induction of the solvent control. The coefficient of variability of at least 21 solvent control wells should be below 20%. At least 3 test concentrations must be within viability limits (i.e. relative viability of at least 70%). Moreover, in case a result is to be considered negative, at least one concentration must be cytotoxic (i.e. have a cell viability below 70%), or the maximum concentration of 2000 µM must have been tested. The mean basal expression of the blank (only cells) should be below 1.5 luciferase fold induction (relative to blank corrected solvent control). If any of these criteria is not met, the repetition is considered not valid and needs to be repeated. According to the prediction model, a test compound is considered to have sensitizing potential when the luciferase induction is above or equal to 1.5-fold compared to the vehicle control in 2 (or more) consecutive non-cytotoxic tested concentrations, whereby at least three tested concentrations must be non-cytotoxic. A test compound is considered to not to have sensitizing potential in this test if the above effects are not observed.

In order to come to a conclusion on the skin sensitization hazard of a substance, one complete experiment needs to be conducted. A complete experiment consists of two valid independent repetitions according to the above-described acceptance criteria. If the first two repetitions come to the same result (i.e. either being negative or being positive), no further testing is required. In case that the first two repetitions give discordant results (i.e. one is negative and the other is positive), a third independent repetition needs to be conducted to complete the experiment. The skin-sensitizing potential of a test substance is determined by the result of the majority of the repetitions of an experiment. If two of two or two of three repetitions are negative/positive, the substance is considered as negative/positive.

18.9 Predictivity

The predictive capacity of the LuSens assay has been evaluated in an in-house validation study using a set of 74 substances, which included the LLNA performance standards. The predictivity of the LuSens assay was evaluated using Cooper statistics [36], in which sensitivity, specificity and accuracy of the method were evaluated. For 69 or 72 tests substances for which human or LLNA data were available, it was obtained a sensitivity of 83 or 74%, a specificity of 81 or 74% and an accuracy of 83 and 74%, in comparison to human or LLNA data, respectively. These data indicate that the predictive capacity of LuSens to identify skin sensitizers was comparable to other non-animal methods, in particular to the KeratinoSens™ [30].

18.10 Advantages and Limitations of the LuSens Assay

On a set of 74 test substances with known sensitization potential, the LuSens assay has demonstrated good intra-laboratory reproducibility of 93% and a proficiency to identify skin sensitizers. The LuSens assay correctly predicted 57 of 69 or 53 of 74 substances when compared to human or LLNA data, respectively. From these data the following predictivity values were calculated: sensitivity of 83% or 73%, specificity of 81% or 74% and an overall accuracy of 83% or 74% when compared to human or LLNA data, respectively. As previously mentioned, LuSens assay, similar to KeratinoSens™, detects skin sensitizers through the activation of the Nrf2 pathway; however, different to KeratinoSens™, the luciferase gene in LuSens is under the control of ARE element from the rat NADPH:quinone oxidoreductase (nqo1). The introduction of cytotoxicity range finder in the assay provides the advantage of the selection of non-cytotoxic concentrations for the assessment of the luciferase expression in the presence of test substances. The inclusion of negative and positive control in the same plate where test substance is applied allows a better assessment of the validity of a given run.

As reported by Urbisch and coworkers [37], since the KeratinoSens™ and the LuSens assay address keratinocyte activation, both methods could be interchangeable. From a set of 69 test substances that have been tested in both assays, it was calculated an overall interchangeability of 88%. Moreover, their integration in the "2 out of 3 approach" (including the direct peptide reactivity assay and the human cell line activation test) resulted in similar accuracies towards prediction of sensitization when compared to LLNA or human data. In addition, it was described that ARE-based assays in general provide similar good predictions of the sensitization potential of Michael acceptors, quinone precursors, Schiff base formers, nucleophilic substitutions and nucleophilic substitutions in aromatic compounds. Contrastingly, the sensitization potential of acylating agents was not reliably predicted, since acylating agents most likely transfer their acyl group to lysine residues, effect that might not trigger the activation of the Keap1/Nrf2 pathway, as accomplished by reactivity on the cysteine residues.

The validation of a new method might contribute to the international acceptance and recognition under the regulatory framework; during this process, several aspects towards reproducibility, robustness and predictive capacity are evaluated [38, 39]. In this context, an inter-laboratory validation of LuSens assay in five different laboratories from the USA, Germany and Switzerland is under evaluation. The data provided from the testing of reference chemicals foreseen in the OECD TG to keratinocyte-based ARE-Nrf2 luciferase reporter gene test method [33, 40] are promising and suggest that LuSens could fulfil all the needed requirements to be considered a me-too assay to KeratinoSens™.

18.11 Use of the LuSens Assay in Testing Strategies for Hazard Assessment

The current consensus of the scientific and regulatory community is that due to the complexity of the sensitization process, no single test can sufficiently cover the sensitization process. This is also reflected in the OECD guidelines for the "*In Chemico* Skin Sensitisation: Direct Peptide Reactivity Assay (DPRA)" and the keratinocyte-based ARE-Nrf2 luciferase reporter gene test method (DPRA and ARE-Nrf2 luciferase test method, [33, 40]). In these documents it is stated that "It is however likely that combinations of non-animal methods (*in silico, in chemico, in vitro*) within Integrated Approaches to Testing and Assessment (IATA) will be needed to be able to substitute for the animal tests currently in use given the specific AOP mechanistic coverage of each of the currently available non-animal test methods". There are differing definitions of IATA, integrated testing strategies (ITS) and weight of evidence (WoE), and in the following testing strategies are approaches that cover different combinations of tests from different information sources which can be effectively combined in a quantifiable fashion to satisfy an information need, in this case regulatory hazard assessments. The ARE-based assays provide information on keratinocyte activation—key event 2 of the AOP.

An extensive database has now been published in which a number of non-animal test methods were evaluated, namely, the DPRA, the KeratinoSens™ assay, the LuSens assay, the (modified) myeloid U937 skin sensitization test ((m)MUSST) and the human cell line activation test (hCLAT) as well as a very simple testing strategy termed the "2 out of 3 weight of evidence approach" [37]. The test methods evaluated cover three of the four key events described by the OECD AOP: key event 1 (peptide reactivity), DPRA; key event 2 (keratinocyte activation), the KeratinoSens™ and LuSens assays; and key event 3 (dendritic cell activation), the (m)MUSST and hCLAT. The "2 out of 3 WoE approach" was first described by Bauch and coworkers [41]. The prediction model of this approach is very straightforward in that preferably three tests covering three different key events are conducted, and the results of two tests govern the prediction, i.e. if two are positive, the substance is considered to have a sensitizing potential, or if two are negative, the substance does not need to be classified as a sensitizer. In the [41] study, 54 substances were evaluated, and the results were compared to both human and LLNA data. When compared to human data, the "2 out of 3 WoE approach" consisting of the DPRA, LuSens and (m)MUSST exhibited a very high accuracy of 94%; when compared to LLNA data, an accuracy of 83% was achieved. The LLNA had an accuracy of 90% in this study when compared to the human data. In a study using 145 substances published by Natsch and coworkers [42], a "2 out of 3 WoE approach" using the DPRA, the KeratinoSens™ assay and the MUSST achieved an accuracy of 81% in comparison to the LLNA data; no comparison was made with human data. In a recently published paper by Urbisch et al. [37], in which over 200 substances were studied, the predictivity of the "2 out of 3 WoE approach" was also assessed. To this accord, the DPRA, the KeratinoSens™ assay and the hCLAT were used as these methods are in the advanced stages of acceptance as OECD guidelines, and the "2 out of 3 WoE approach" exhibited a sensitivity, specificity and accuracy of 90% when compared to human data (114 substances), whereas the LLNA had an accuracy of only 82% for this data set. When comparing the data set to available LLNA data (180 chemicals), the "2 out of 3 WoE approach" yielded an accuracy of 79%. In all studies the "2 out of 3 WoE approach" yielded a better accuracy than the individual methods and where human data was available a better accuracy than the LLNA (Fig. 18.4). In the Urbisch et al. [37] study, the interchangeability of the LuSens assay with the KeratinoSens™ assay was calculated to be 88%. Based on that test set, the KeratinoSens™ assay offered an accuracy of 85% or 91% and the LuSens assay a comparable accuracy of 83% or 93% when compared to LLNA or human data, respectively. These studies indicate that the LuSens assay can be effectively integrated into testing strategies for hazard assessments.

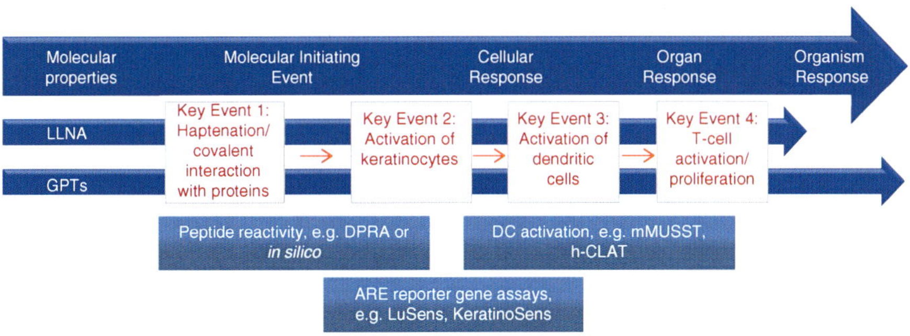

	Bauch et al., 2012 compared to LLNA/ human data (n=54)	Natsch et al., 2013 compared to LLNA data (n=145)	Urbisch et al., 2014 compared to LLNA (n=180)/ to human (n=101)
Accuracy	83/94%	81%	82/90%

Fig. 18.4 Predictivity of the combination of assays including essential steps of the skin sensitization AOP. In the context of the skin sensitization AOP, the DPRA evaluates the protein/peptide reactivity of a substance (key event 1); the KeratinoSens™ and LuSens assays provide information on the keratinocyte activation (key event 2), and, for example, the hCLAT and mMUSST provide information on the dendritic cell activation (key event 3). Thus, when these methods are combined on testing strategies, they cover the first three key events of the sensitization process, providing mechanistic relevance and supporting the scientific rationale for using a combination of these methods in an AOP-based ITS

18.12 Concluding Remarks

The LuSens assay is an *in vitro* keratinocyte-based method that uses the Nrf2 pathways to identify skin sensitizers. It uses a genetically modified keratinocyte cell line harbouring the luciferase gene under the control of the ARE element of the rat Nqo1 gene. Hence, it can be used to indirectly monitor binding of the Cys residues of Nrf2—therefore an intracellular indicator of protein reactivity—and the subsequent binding of Nrf2 to the ARE and the expression of the downstream genes, in this case luciferase of post-exposure to chemicals that could possess a sensitization potential. The assay identifies a potential sensitizer via luciferase activity equal or above 1.5-fold induction compared to the vehicle control under sufficiently non-cytotoxic conditions. The method has been validated for its predictivity and accuracy in identifying potential skin sensitizers by testing more than 60 test substances. The LuSens test method underwent a Performance Standard-based validation study based on which it was reviewed and received positive recommendation by the EURL ECVAM Scientific Advisory Committee to be use for supporting the discrimination between skin sensitisers and non-sensitisers in accordance with the UN GHS [43, 45]. Moreover, reports on the interchangeability of both methods suggest that LuSens can be used as an alternative to KeratinoSens™. Since this method addresses key event 2 of the skin sensitization AOP, it can be used as part of a testing strategy, and data is available indicating that use in the "2 out of 3 WoE" approach can result in predictivities comparable to that of the LLNA.

References

1. Bruckner AL, Weston WL, Morelli JG. Does sensitization to contact allergens begin in infancy? Pediatrics. 2000;105(1):e3.
2. Thyssen JP, Linneberg A, Menné T, Johansen JD. The epidemiology of contact allergy in the general population–prevalence and main findings. Contact Dermatitis. 2007;57(5):287–99.
3. Lunder T, Kansky A. Increase in contact allergy to fragrances: patch-test results 1989–1998. Contact Dermatitis. 2000;43(2):107–9.
4. Nguyen SH, Dang TP, MacPherson C, Maibach H, Maibach HI. Prevalence of patch test results from 1970 to 2002 in a multi-centre population in North America (NACDG). Contact Dermatitis. 2008;58(2):101–6.
5. Peiser M, Tralau T, Heidler J, Api AM, Arts JH, Basketter DA, English J, Diepgen TL, Fuhlbrigge RC, Gaspari AA, Johansen JD, Karlberg AT, Kimber I, Lepoittevin J-P, Liebsch M, Maibach HI, Martin SF, Merk HF, Platzek T, Rustemeyer T, Schnuch A, Vandebriel RJ, White IR, Luch A. Allergic contact dermatitis: epidemiology, molecular mechanisms, in vitro methods and regulatory aspects. Current knowledge assembled at an international workshop at BfR, Germany. Cell Mol Life Sci. 2012;69(5):763–81.
6. OECD. OECD TG 429. Skin sensitization, Local Lymph Node Assay. 2010.
7. OECD. OECD test guidelines (TG) 442A. Skin Sensitization: Local Lymph Node Assay: DA. 2010.
8. OECD. OECD test guidelines (TG) 442B. Skin Sensitization: Local Lymph Node Assay: BrdU-ELISA. 2010.
9. OECD. OECD TG 406. Skin Sensitization. 1992.
10. EC. Reach; Regulation No 1907/2006 of the European parliament and of the council of 18 December 2006 concerning the Registration, Evaluation, authorisation and restriction of

chemicals (REACH), establishing a European chemicals agency, amending directive 1999/45/EC and repealing Council regulation (EEC) no 793/93 and commission regulation (EC) no 1488/94 as well as Council directive 76/769/EEC and commission directives 91/155/EEC, 93/67/EEC, 93/105/EC and 2000/21/EC. Off J Eur Union Legis. 2006; 396:1–849.

11. ECHA-14-A-07-EN. The use of alternatives to testing on animals for the REACH regulation second report under article 117(3) of the REACH regulation. 2014. ED-04-14-483-EN-N - ISBN: 978–92–9244-593-5. DOI: 10.2823/22471. 2 June 2014.
12. EU Regulation 1223/2009 of the European Parliament and of the Council of 30 November 2009 on cosmetic products. Official Journal of the European Union L 342/59-209. 22.12.2009.
13. Ladics GS, Fry J, Goodman R, Herouet-Guicheney C, Hoffmann-Sommergruber K, Madsen CB, Penninks A, Pomés A, Roggen EL, Smit J, Wal JM. Allergic sensitization: screening methods. Clin Transl Allergy. 2014;4(1):13.
14. Mehling A, Eriksson T, Eltze T, Kolle S, Ramirez T, Teubner W, van Ravenzwaay B, Landsiedel R. Non-animal test methods for predicting skin sensitization potentials. Arch Toxicol. 2012;86:1273–95.
15. OECD. ENV/JM/MONO(2012)10/PART1. The adverse outcome pathway for skin Sensitisation Initiated by Covalent binding to proteins PART 1: scientific evidence. Series on testing and assessment no.168. 2012.
16. OECD. ENV/JM/MONO(2012)10/PART2. The adverse outcome pathway for skin Sensitisation Initiated by Covalent binding to proteins part 2: use of the AOP to develop chemical categories and integrated assessment and testing approaches. Series on testing and assessment no.168. 2012.
17. Gerberick F, Aleksic M, Basketter D, Casati S, Karlberg AT, Kern P, Kimber I, Lepoittevin J-P, Natsch A, Ovigne JM, Rovida C, Sakaguchi H, Schultz T. Chemical reactivity measurement and the predicitve identification of skin sensitisers. The report and recommendations of ECVAM workshop 64. Altern Lab Anim. 2008;36(2):215–42.
18. Gallucci S, Matzinger P. Danger signals: SOS to the immune system. Curr Opin Immunol. 2001;13(1):114–9.
19. McFadden JP, Basketter DA. Contact allergy, irritancy and danger. Contact Dermatitis. 2000;42(3):123–7.
20. Natsch A, Emter R. Skin sensitizers induce antioxidant response element dependent genes: application to the in vitro testing of the sensitization potential of chemicals. Toxicol Sci. 2008;102(1):110–9.
21. Steinman RM. The dendritic cell system and its role in immunogenicity. Annu Rev Immunol. 1991;9:271–96. Review
22. Steinman RM, Banchereau J. Taking dendritic cells into medicine. Nature. 2007;449(7161):419–26.
23. Banchereau J, Steinman RM. Dendritic cells and the control of immunity. Nature. 1998;392(6673):245–52.
24. Ade N, Leon F, Pallardy M, Peiffer JL, Kerdine-Romer S, Tissier MH, Bonnet PA, Fabre I, Ourlin JC. HMOX1 and NQO1 genes are upregulated in response to contact sensitizers in dendritic cells and THP-1 cell line: role of the Keap1/Nrf2 pathway. Toxicol Sci. 2009;107(2):451–60.
25. Natsch A. The Nrf2-Keap1-ARE toxicity pathway as a cellular sensor for skin sensitizers--functional relevance and a hypothesis on innate reactions to skin sensitizers. Toxicol Sci. 2010;113(2):284–92.
26. Kobayashi A, Kang MI, Okawa H, Ohtsuji M, Zenke Y, Chiba T, Igarashi K, Yamamoto M. Oxidative stress sensor Keap1 functions as an adaptor for Cul3-based E3 ligase to regulate proteasomal degradation of Nrf2. Mol Cell Biol. 2004;24(16):7130–9.
27. Emter R, Ellis G, Natsch A. Performance of a novel keratinocyte-based reporter cell line to screen skin sensitizers in vitro. Toxicol Appl Pharmacol. 2010;245(3):281–90.
28. Motohashi H, Yamamoto M. Nrf2-Keap1 defines a physiologically important stress response mechanism. Trends Mol Med. 2004;10(11):549–57.

29. Emter R, Natsch A. Skin sensitizers induce antioxidant response element dependent genes: application to the in vitro testing of the sensitization potential of chemicals. Toxicol Sci. 2008;102(1):110–9.
30. Ramirez T, Mehling A, Kolle SN, Wruck CJ, Teubner W, Eltze T, Aumann A, Urbisch D, van Ravenzwaay B, Landsiedel R. LuSens: a keratinocyte based ARE reporter gene assay for use in integrated testing strategies for skin sensitization hazard identification. Toxicol In Vitro. 2014;28(8):1482–97.
31. Migdal C, Botton J, El Ali Z, Azoury ME, Guldemann J, Giménez-Arnau E, Lepoittevin J-P, Kerdine-Römer S, Pallardy M. Reactivity of chemical sensitizers toward amino acids in cellulo plays a role in the activation of the Nrf2-ARE pathway in human monocyte dendritic cells and the THP-1 cell line. Toxicol Sci. 2013;133(2):259–74.
32. Mehling A, Beissert S. Dendritic cells under investigation in autoimmune disease. Crit Rev Biochem Mol Biol. 2003;38(1):1–21.
33. OECD. OECD TG 442D: in vitro skin Sensitisation: ARE-Nrf2 luciferase test method. 2015.
34. Vandebriel RJ, Pennings JL, Baken KA, Pronk TE, Boorsma A, Gottschalk R, Van Loveren H. Keratinocyte gene expression profiles discriminate sensitizing and irritating compounds. Toxicol Sci. 2010;117(1):81–9.
35. Wasserman WW, Fahl WE. Comprehensive analysis of proteins which interact with the antioxidant responsive element: correlation of ARE-BP-1 with the chemoprotective induction response. Arch Biochem Biophys. 1997;344(2):387–96.
36. Cooper JA, Saracci R, Cole P. Describing the validity of carcinogen screening-tests. Br J Cancer. 1979;39(1):87–9.
37. Urbisch D, Mehling A, Guth K, Ramirez T, Honarvar N, Kolle S, Landsiedel R, Jaworska J, Kern PS, Gerberick F, Natsch A, Emter R, Ashikaga T, Miyazawa M, Sakaguchi H. Assessing skin sensitization hazard in mice and men using non-animal test methods. Regul Toxicol Pharmacol. 2014; pii: S0273-2300(14)00309–2.
38. Hartung T, Bremer S, Casati S, Coecke S, Corvi R, Fortaner S, Gribaldo L, Halder M, Hoffmann S, Roi AJ, Prieto P, Sabbioni E, Scott L, Worth A, Zuang V. A modular approach to the ECVAM principles on test validity. Altern Lab Anim. 2004;32:467–72. http://www.epa.gov/ncct/Tox21/.
39. OECD. Guidance document on the validation and international acceptance of new or updated test methods for hazard assessment, Series on testing and assessment, no. 34. Paris: OECD; 2005.
40. OECD. OECD TG 442C: in Chemico skin Sensitisation: direct peptide reactivity assay (DPRA). 2015.
41. Bauch C, Kolle SN, Ramirez T, Eltze T, Fabian E, Mehling A, Teubner W, van Ravenzwaay B, Landsiedel R. Putting the parts together: combining in vitro methods to test for skin sensitizing potentials. Regul Toxicol Pharmacol. 2012;63(3):489–504.
42. Natsch A, Ryan CA, Foertsch L, Emter R, Jaworska J, Gerberick F, Kern P. A dataset on 145 chemicals tested in alternative assays for skin sensitization undergoing prevalidation. J Appl Toxicol. 2013;33(11):1337–52.
43. ESAC (2016). ESAC opinion on the BASF-coordinated Performance Standards-based validation of the LuSens test method for skin sensitisation testing. Available at: [http://publications.jrc.ec.europa.eu/repository/bitstream/JRC103706/esac_opinion_2016-04_lusens_final.pdf]
44. OECD. OECD TG 442D: in Chemico skin Sensitisation: ARE-Nrf2 Luciferase Test Method. 2015
45. Ramirez T, Stein N, Aumann A, Remus T, Edwards A, Norman KG, Ryan C, Bader JE, Fehr M, Burleson F, Foertsch L, Wang X, Gerberick F, Beilstein P, Hoffmann S, Mehling A, van Ravenzwaay B, Landsiedel R. Intra- and inter-laboratory reproducibility and accuracy of the LuSens assay: A reporter gene-cell line to detect keratinocyte activation by skin sensitizers. Toxicol in Vitro. 2016;32:278–286

NCTC 2544 and IL-18 Production: A Tool for the Identification of Contact Allergens

19

Emanuela Corsini and Valentina Galbiati

19.1 Principle of the Test Method and Scientific Basis

The NCTC 2544 assay is based on the selective induction of intracellular interleukin-18 (IL-18) by contact allergens in the human keratinocyte cell line NCTC 2544. This assay was developed within the SENS-IT-IV project sponsored by the European Union (LSHB-CT-2005-018681). The assay proved to be useful in the identification and discrimination of contact allergens from respiratory sensitizers and irritants [1–4]. NCTC 2544 is a commercially available skin epithelial-like cell line originating from normal human skin. Due to their anatomical location and critical role in skin inflammatory and immunological reactions, the use of keratinocytes and skin organotypic culture as a simplified *in vitro* model to evaluate the potential toxicity of chemicals destined for dermal application is amply justified [5, 6]. Keratinocytes sense haptens and in turn initiate a program of enhances or de novo expression of inflammatory molecules representing the starting point of primary inflammation [1, 2]. IL-18, formerly known as IFN-γ-inducing factor, is a potent inducer of IFN-γ by activated T cells [7]. We had focused our attention on IL-18 since this cytokine has been shown to play a key proximal role in the induction of allergic contact sensitization and to favor Th-1 type immune response [7–9]. Human keratinocytes constitutively express IL-18 mRNA and protein [10], and work published by [10, 11] showed the induction of IL-18 following exposure to contact sensitizers. In the NCTC 2544 assay, cells are typically exposed to four non-cytotoxic concentrations (cell viability higher of 80%, as assessed by MTT reduction assay) of the chemical under investigation for 24 h. Cell-associated IL-18 is then evaluated by ELISA. The majority of sensitizers so far tested, including pre- and pro-haptens (viz., p-phenylenediamine, eugenol, isoeugenol, and cinnamyl alcohol), induced a

E. Corsini (✉) • V. Galbiati
Laboratory of Toxicology, Department of Pharmacological and Biomolecular Sciences, Università degli Studi di Milano, Milan, Italy
e-mail: emanuela.corsini@icloud.com

C. Eskes et al. (eds.), *Alternatives for Dermal Toxicity Testing*,
DOI 10.1007/978-3-319-50353-0_19

dose-related increase in IL-18, whereas the majority of irritants and respiratory allergens failed. A total of 47 chemicals were tested (of which two should be excluded as not pure to a subsequent GC-MS, gas chromatography-mass spectrography, analysis), with an overall accuracy of 84.4%. In Table 19.1 the list of the chemicals tested in the NCTC 2544 assay and their *in vitro* and *in vivo* classification are reported.

Table 19.1 Chemicals tested in the NCTC 2544 assay and their *in vivo* classification

Chemical	CAS number	*In vitro* EC1.2	LLNA classification
Respiratory allergens			
Ammonium hexachloroplatinate	16919-58-7	Negative	Not available
Diphenylmethane diisocyanate (MDI)	101-68-8	Negative	Strong
Glutaraldehyde	111-30-8	Negative	Strong
Hexamethylenediisocyanate	822-06-0	*Negative/positive*[a]	Extreme
Maleic anhydride	108-31-6	Negative	Strong
Toluene diisocyanate	584-84-9	Negative	Strong
Trimellitic anhydride	552-30-7	Negative	Moderate
Contact allergens			
2,4-dinitrochlorobenzene	97-00-7	Positive	Extreme
4-nitrobenzylbromide	100-11-8	Positive	Extreme
Oxazolone	15646-46-5	Positive	Extreme
2-bromo-2-bromomethyl glutaronitrile	35691-65-7	*Positive/negative*[a]	Strong
Cobalt chloride	7646-79-9	Positive	Strong
Formaldehyde	50-00-0	Positive	Strong
Lauryl gallate	1166-52-5	Positive	Strong
p-phenylenediamine	106-50-3	Positive	Strong
2-mercaptobenzothiazole	149-30-4	*Positive/negative*[a]	Moderate
Cinnamaldehyde	104-55-2	Positive	Moderate
Diethylmaleate	141-05-9	Positive	Moderate
Glyoxal	107-22-2	Positive	Moderate
Isoeugenol	97-54-1	Positive	Moderate
Phenylacetaldehyde	122-78-1	*Negative*[b]	Moderate
Tetramethylthiuram	137-26-8	Positive	Moderate
Benzocaine	94-09-7	Positive	Weak
Cinnamyl alcohol	104-54-1	Positive	Weak
Citral	5392-40-5	Positive	Weak
Eugenol	97-53-0	Positive	Weak
Hexyl cinnamic aldehyde	101-86-0	*Negative*[b]	Weak
Phenyl benzoate	93-99-2	*Negative*	Weak
Resorcinol	108-46-3	Positive	Weak
Malachite green	569-64-2	Positive[c]	Not available
Nickel sulfate	7786-81-4	Positive	False negative
Non-sensitizers/irritants			
Ammonium persulfate	7727-54-0	Negative	Not available

Table 19.1 (continued)

Chemical	CAS number	*In vitro* EC1.2	LLNA classification
Benzaldehyde	100-52-7	Negative	Not classified
Chlorobenzene	108-90-7	Negative	Not classified
Diethyl phthalate	84-66-2	Negative	Not classified
Ethyl vanillin	121-32-4	*Positive*	Not classified
Glycerol	56-81-5	Negative	Not classified
Hydroxybenzoic acid	99-96-7	Negative	Not classified
Lactic acid	50-21-5	*Negative/positive*[a]	Not classified
Octanoic acid	124-07-2	*Negative/positive*[a]	Not classified
Phenol	108-95-2	Negative	Not classified
Propylene glycol	57-55-6	Negative	Not classified
Salicylic acid	69-72-7	Negative	Not classified
Sodium lauryl sulfate	151-21-3	Negative	Not classified
Sulfamic acid	5329-14-6	Positive	Not available
Tween 80	9005-65-6	Negative	Not classified
Zinc sulfate	7733-02-0	Negative	Not classified

The shading is related to the LLNA potency classification. In bold the chemicals wrongly classified

[a]Chemical was initially correctly identified, but in subsequent trials it was wrongly classified

[b]Subsequent GC-MS analysis to assess purity revealed a significative presence of impurities

19.2 Current (Pre) Validation Status

The assay underwent pre-validation in a two-tiered approach project together with the epidermal equivalent (EE) assay (pre-validation of a novel-tiered approach to determine the skin-sensitizing capacity and potency of chemicals) sponsored by the Dutch association ZonMw. The primary aim was to evaluate the reproducibility and transferability of the NCTC 2544 assay with 13 sensitizers and 16 non-sensitizers.

The study was structured and conducted in two sequential phases:

Phase A: test method transfer to the trained laboratories

Phase B: assessment of the protocol performance by testing chemicals, under blind conditions, in all the laboratories

The study failed to demonstrate the ability of the NCTC 2544 assay to discriminate contact sensitizers from non-contact sensitizers. The predictive capacity was very low, as all laboratories, with the exception of our laboratory, predicted all chemicals (except the sensitizer p-phenylenediamine) as non-sensitizers. As lead lab, a sensitivity of 90.9% and a specificity of 81% were achieved for the chemicals tested (excluding phenylacetaldehyde and hexyl cinnamic aldehyde, as these two chemicals were found to contain approximately 50% of impurities). The mistake in this study was to start Phase B before a successful transfer of the method. Looking carefully at the data obtained during the transfer phase published by [12], DNCB

[c]There is evidence in human to support contact allergy

was correctly identified as a sensitizer in four out of six laboratories, and resorcinol was correctly classified as a sensitizer by two out of six laboratories. Even if the different labs did not perform equally good in this phase of the study, with regard to the correct classification of the selected chemicals, the method was transferred, the optimal culture conditions were established in all six labs, and it was decided to enter the pre-validation phase testing 29 blinded chemicals. This carelessness effectively compromised the possible success of the study. The mistake was to not pay the proper attention to it and to go straight to the second phase. Furthermore, looking at the results obtained in the ZonMw pre-validation study, in some of the laboratories, the intracellular IL-18 content was very high (different FCS? not optimal cell density conditions?), which may decrease the dynamic range of the assay; the CV80 was not reached in the IL-18 assay, which is particularly important if no IL-18 induction is observed.

19.3 Performance and Applicability of the Test Method

Overall, the method was transferred to naïve laboratories [12], suggesting the potential use of the test in immunotoxicity testing strategies, with the foresight to use training chemicals to establish the assay in the laboratory to ensure that correct classification occurs. The failure of the pre-validation study pointed out the necessity of operating a number of training experiments in the naïve laboratories. As part of the training and transfer plan, laboratories should demonstrate technical proficiency, using the proficiency substances listed below, to establish the assay in their laboratories, prior to beginning any "official" use of the assay, to ensure that correct classification occurs.

19.3.1 Reproducibility

In the ZonMW study, the between-laboratory reproducibility in terms of the classification S versus NS was 79.3% for 23 of the 29 chemicals, the same prediction was obtained in the 6 laboratories.

19.3.2 Predictive Capacity

In Table 19.2 the contingency table compiling all data available is reported. Overall 45 chemicals were tested, and a sensitivity of 86.4% and a specificity of 82.6% with an overall accuracy of 84.4% were achieved.

19.3.3 Applications and Limitations

NCTC 2544 assay can be used as screening, replacement, and reduction test for hazard identification and general classification and labeling of skin sensitizers.

Table 19.2 Contingency table for the NCTC 2544 assay

		Predicted classification		
		Contact sensitizer	Non-contact sensitizer	Total
Chemical classification	Sensitizers	19	4	22
	Non-sensitizer	3	19	23
	Total	22	23	45
Sensitivity	86.4			
Specificity	82.6			
Accuracy	84.4			

Fisher's exact test. In the analysis of all data available and reported in Table 19.1, the two impure chemicals were excluded (phenylacetaldehyde and hexyl cinnamic aldehyde), while for chemicals reporting an inconsistent classification in different trials, they were considered as wrongly classified

Factors such as compound solubility, chemical stability in water, and metabolic activation, which may mask the potential allergenicity of some chemicals, must be considered in case of negative results. This may be the case of diphenylmethane diisocyanate (MDI) and anhydrides, highly unstable in water. Trimellitic anhydride (TMA) is readily hydrolyzed in water to trimellitic acid, and MDI produces inert, solid, insoluble polyurea. The hydrolysis of isocyanate in aqueous solution is rapid, with a half-life of less than 20 seconds. If submerged cell culture may be unfavorable for many of the respiratory sensitizers, due to chemical instability, we have successfully tested IL-18 production in reconstituted human epidermis, which allows application in organic solvent, i.e., acetone/olive oil, to overcome this problem. The use of reconstituted epidermis may indeed beat the limitation of chemical solubility and stability in traditional submerged cell culture, a major drawback of several *in vitro* assays. Some chemicals, designated "pro-haptens," require a metabolic transformation step to act as haptens capable of activating the immune system. NCTC 2544 possesses both phase I and II metabolic activation capacity, with a good expression of cytochrome P450-dependent enzymatic activities [13]. A comparison between the phase I enzyme activities expressed in normal human keratinocytes and in several human keratinocyte cell lines, namely, HaCaT, SVK14, and NCTC 2544, was established by [14]. From that study it appeared that in NCTC 2544 cell line, both basal and induced levels of 7-ethoxycoumarin O-deethylase (ECOD) activity are higher than those found in other cell lines. This is particularly relevant for an *in vitro* method, as lack or limited metabolic capacity is often one of the major drawbacks for many *in vitro* models. Finally, as the NCTC2544 assays use a threshold-based prediction model (SI $\geq$ 1.2), it may have an inherent limitation in the detection of some weak sensitizers: the cell activation induced by very weak sensitizers may remain just below the thresholds set for a positive result, and the molecules will not be correctly classified.

19.3.4 Comparison to Human Data

Not available for the NCTC2544 assay.

19.4 Brief Description of the Protocol

For intracellular IL-18 content, 1.25×10^5 NCTC2544 cells seeded in 24-well plate are typically used after overnight adherence. Cells are incubated with four increasing concentrations of the selected chemicals or their corresponding vehicle for 24 h. Final concentration of DMSO should not exceed 0.2%. Each concentration is tested in quadruplicate. PPD 60 μg/ml is used as positive control. Each chemical should be tested on three separate occasions (different days).

The relative amounts of IL-18 protein present in the cell lysates are given in pg/ml. The result is then expressed as pg/mg of total cell protein as assessed by the BCA protein determination method. The pg/mg is calculated by the following equation:

$$\text{Intracellular IL-18} = \frac{\text{IL-18 pg/ml in cell lysate}}{\text{Total protein content in cell lysate mg/ml}} = \text{IL-18 pg/mg}$$

19.4.1 Prediction Model

If the fold increase in intracellular IL-18 is ≥1.2 and the increase in IL-18 is statistically significant from vehicle-treated cells (Dunnett multiple comparisons test), the chemical is classified as contact sensitizer. If the fold increase in intracellular IL-18 is <1.2 and there is no statistical significance, the chemical is classified as non-contact sensitizer. The positive control should be included in each run. The positive control meets the acceptance if the fold increase in intracellular IL-18 is >1.5 compared to vehicle-treated cells. For a given chemical, the same classification must be obtained in two out of three independent experiments. The 1.2-fold increase is indented for at least one of the concentrations tested. The 1.2 induction is observed at CV ≥ 80%.

A negative response is obtained when no increase in intracellular IL-18 is observed following chemical treatment. As additional criteria, some cytotoxicity (CV 80 ± 10%) must be observed in at least two experiments (especially for potential negative compounds). When no toxic effect is observed up to the highest test concentration allowed in the assay, a careful examination of the solubility data should be performed to ensure that the test substance has indeed been solubilized. Negative test results obtained with poorly or insoluble substances should not be considered sufficient for classification decisions. For details see [12, 15].

19.5 Role in a Testing Strategy

We propose the NCTC 2544 assay as screening, replacement, reduction test for hazard identification and general classification and labeling of skin sensitizers. The test may also be complementary to the LLNA, allowing for a rapid and easy discrimination of contact sensitizers from respiratory allergens. Regarding to relationship with potency, there is a general trend for IL-18 induction at lower concentrations for extreme/strong sensitizers, whereas higher concentrations (>100 μg/ml) are needed in the case of weak sensitizers [1, 2], which may be useful for GHS classification.

19.6 Perspectives from the Test Developer

The NCTC 2544 assay holds promise. A 96-well format assay was also developed [4], and the method was also adapted for the assessment of photoallergens [16, 17]. We believe, however, that the release of IL-18 in reconstituted epidermis represents the future of the assay. Briefly, in the 96-well format, to speed up the assay and reduce cell manipulation (no cell-washing steps and no protein determination), after the treatment, cells are lysed in culture medium directly adding Triton X-100 (final concentration 0.5%). In this case the total IL-18 (intracellular plus released) is assessed. Also under this experimental condition, a selective increase in total IL-18 was observed only following treatment with contact allergens, whereas both irritants and respiratory allergens failed, indicating the possibility to use total IL-18 to identify contact allergens in a high-throughput manner. Regarding the photoallergy, in [16, 17] the possibility to use the NCTC2544 assay to identify photoallergens and discriminate from phototoxic chemicals was explored. The effect of UVA irradiation over NCTC2544 cells treated with increasing concentrations of 15 compounds including photoallergens (benzophenone, 4-ter-butyl-4-methoxy-dibenzoylmethane, 2-ethylexyl-p-methoxycinnamate, ketoprofen, 6-methylcumarin), photoirritant and photoallergen (4-aminobenzoic acid, chlorpromazine, promethazine), photoirritants (acridine, ibuprofen, 8-methoxypsoralen, retinoic acid), and negative compounds (lactic acid, SDS, and p-phenylenediamine) was investigated. At the maximal concentration assayed with non-cytotoxic effects (CV80 under irradiated condition), all tested photoallergens induced a significant and a dose-dependent increase of intracellular IL-18 following UVA irradiation, whereas photoirritants failed, indicating that the NCTC2544 assay may be useful for the *in vitro* evaluation of the photoallergic potential of chemicals.

19.6.1 Critical Steps in the Protocol

Critical points identified in the performance of the NCTC 2544 assay are the cell density [3, 4], and the time cells have been cultured before use in experiments [1, 2]. If cells reach confluence at the moment of treatment, the ability to identify contact allergens is lost; therefore, a careful check for the optimal cell density using PPD, as reference contact allergen, is critical. In our hands, a cell density of $1\text{–}2.5 \times 10^5$ cells/ml gave optimal stimulation. Due to the use of different fetal calf serum, a different cellular doubling time can be expected in the different laboratories. Each laboratory should check for the optimal cell density, using the positive control PPD as a reference compound. Regarding time after thawing, cells should be used between 3 weeks after thawing and up to 6 months after.

The measurements are to be performed within the described cytotoxic range (>CV80 relative to vehicle-treated cells) and cell density (<80% confluence when treatment starts), at a maximal test concentration 1000 μg/ml. Among experiments, differences in the absolute value of IL-18 were observed, also in vehicle-treated cells. In control cells, historically, the intracellular IL-18 content varied from 300 to 2500 pg/mg. This may be due to differences in the number of cells seeded and adhering to plate wells, to the use of a different cell batch as well as to differences in the IL-18 ELISA and protein assessment performance, which may result in over- or under-estimation of IL-18 content. For this reason, the stimulation index is calculated. As long as cells don't reach confluence and are used after 3 weeks from thawing, the discriminatory capacity is preserved independently from the absolute value of IL-18.

19.6.2 Possible Protocol Adaptations

As part of the training and transfer plan, it is wise that trainees then use training chemicals (proficiency substances) to establish the assay in their laboratories. Laboratory should perform three valid runs prior to beginning any "official" use of the assay, to ensure that correct classification occurs. The proficiency substances suggested to demonstrate technical proficiency of the laboratory are listed in Table 19.3.

A formal training phase should include:

- Phase Ia: training of the participating laboratories (theoretical and possibly procedural training) with the proficiency substances mentioned above
- Phase Ib: test method transfer to the trained laboratories and verification of the test method protocol to prove successful method transfer before being allowed to proceed to the blind testing phase with ten chemicals well characterized (meet the transfer acceptance criteria)

19.6.3 Challenges and Opportunities

As further opportunities of the NCTC2544 assay, the 96-well format assay [4] and the assessment of photoallergens [16, 17] can be mentioned.

Table 19.3 Proficiency substances suggested to demonstrate technical proficiency of laboratory

Chemical	Cas N°	*In vivo* potency
2,4-dinitrochlorobenzene	97-00-7	Extreme
p-phenylenediamine	106-50-3	Strong
Cinnamaldehyde	104-55-2	Moderate
Resorcinol	108-46-3	Weak
Glycerol	56-81-5	Negative

19.7 Conclusions

We propose the NCTC 2544 assay as screening, replacement, and reduction test for hazard identification and general classification and labeling of skin sensitizers. Our data suggest that IL-18 production by NCTC 2544 cells may represent an interesting *in vitro* model for the screening of potential contact allergens, able to discriminate contact allergens from respiratory allergens and irritants. The test may also be complementary to the LLNA, allowing for a rapid and easy discrimination of contact sensitizers from respiratory allergens. It is, however, necessary that any new laboratories perform three valid runs with suggested chemicals prior to beginning any "official" use of the assay, to ensure that correct classification occurs.

References

1. Corsini E, Galbiati V, Mitjans M, Galli CL, Marinovich M. NCTC 2544 and IL-18 production: a tool for the identification of contact allergens. Toxicol In Vitro. 2013a;27:1127–34.
2. Corsini E, Galbiati V, Nikitovic D, Tsatsakis AM. Role of oxidative stress in chemical allergens induced skin cells activation. Food Chem Toxicol. 2013b;61:74–81.
3. Corsini E, Mitjans M, Galbiati V, Lucchi L, Galli CL, Marinovich M. Use of IL-18 production in a human keratinocyte cell line to discriminate contact sensitizers from irritants and low molecular weight respiratory allergens. Toxicol In Vitro. 2009;23:789–96.
4. Galbiati V, Mitjans M, Lucchi L, Viviani B, Galli CL, Marinovich M, Corsini E. Further development of the NCTC 2544 IL-18 assay to identify in vitro contact allergens. Toxicol In Vitro. 2011;25:724–32.
5. Corsini E, Roggen EL. Immunotoxicology: opportunities for non-animal test development. Altern Lab Anim. 2009;37:387–97.
6. Gober MD, Gaspari AA. Allergic contact dermatitis. Curr Dir Autoimmun. 2008;10:1–26.
7. Okamura H, Tsutsui H, Komatsu T, Yutsudo M, Hakura A, Tanimoto T, Torigoe K, Okura T, Nukada Y, Hattori K, et al. Cloning of a new cytokine that induces IFN-γ production by T cells. Nature. 1995;378:88–91.
8. Antonopoulos C, Cumberbatch M, Mee JB, Dearman RJ, Wei X, Liew FY, Kimber I, Groves RW. IL-18 is a key proximal mediator of contact hypersensitivity and allergen-induced Langerhans cell migration in murine epidermis. J Leukoc Biol. 2008;83:361–7.
9. Cumberbatch M, Dearman RJ, Antopoulos C, Groves RW, Kimber I. Interleukin-18 induces Langerhans cell migration by a tumor necrosis factor-α and IL-1β-dependent mechanism. Immunology. 2001;102:323–30.
10. Naik SM, Cannon G, Burbach GJ, Singh SR, Swerlick RA, Wilcox JN, Ansel JC, Caughman SW. Human keratinocytes constitutively express interleukin-18 and secrete biologically active interleukin-18 after treatment with pro-inflammatory mediators and dinitrochlorobenzene. J Invest Dermatol. 1999;113:766–72.
11. Van Och FM, Van Loveren H, Van Wolfswinkel JC, Machielsen AJ, Vandebriel RJ. Assessment of potency of allergenic activity of low molecular weight compounds based on IL-1alpha and IL-18 production by a murine and human keratinocyte cell line. Toxicology. 2005;210:95–109.
12. Teunis M, Corsini E, Smits M, Madsen CB, Eltze T, Ezendam J, Galbiati V, Gremmer E, Krul C, Landin A, Landsiedel R, Pieters R, Rasmussen TF, Reinders J, Roggen E, Spiekstra S, Gibbs S. Transfer of a two-tiered keratinocyte assay: IL-18 production by NCTC2544 to determine the skin sensitizing capacity and epidermal equivalent assay to determine sensitizer potency. Toxicol In Vitro. 2013;27(3):1135–50.

13. Gelardi A, Morini F, Dusatti F, Penco S, Ferro M. Induction by xenobiotics of phase I and phase II enzyme activities in the human keratinocyte cell line NCTC 2544. Toxicol In Vitro. 2001;15:701–11.
14. Cotovio J, Leclaire J, Roguet R. Cytochrome P450-dependent enzyme activities in normal adult human keratinocytes and transformed human keratinocytes. In Vitro Toxicol. 1997;10:207–16.
15. Galbiati V, Corsini E. The NCTC 2544 IL-18 assay for the in vitro identification of contact allergens. Curr Protoc Toxicol. 2012. Chapter 20:Unit 20.8. doi:10.1002/0471140856.tx2008s54
16. Galbiati V, Martínez V, Bianchi S, Mitjans M, Corsini E. Establishment of an in vitro photoallergy test using NCTC2544 cells and IL-18 production. Toxicol In Vitro. 2013;27:103–10.
17. Galbiati V, Bianchi S, Martínez V, Mitjans M, Corsini E. NCTC 2544 and IL-18 production: a tool for the in vitro identification of photoallergens. Toxicol In Vitro. 2014a;28:13–7.

Epidermal Equivalent (EE) Potency Assay 20

Susan Gibbs and Sander W. Spiekstra

20.1 Principle of the Test Method and Scientific Basis

The epidermal equivalent (EE) Potency Assay is aimed at ranking sensitizing compounds according to their potency (extreme, strong, moderate and weak sensitizing potency) with the aid of a reconstructed human epidermal equivalent [1–3]. Correct chemical classification (potency assessment) as well as chemical labelling (sensitizer identification) is of importance when considering the need to totally replace *in vivo* animal testing for hazard and risk assessment of potentially sensitizing compounds. The EE potency assay is based on our understanding from clinical observations that sensitizer potency is directly related to the irritant potency of the chemical [4–7]. Therefore, by definition, the assay does not enable sensitizers to be distinguished from non-sensitizers (irritants). Local trauma induces in addition to cell death, also IL-1alpha release from epidermal keratinocytes. This IL-1alpha will then trigger the innate immune system resulting in a general inflammatory response [8–10]. In addition to IL-1alpha secretion, sensitizers in particular will result in an increase in IL-18 production [11–14].

The EE potency assay is a modification of the EURL ECVAM-validated EE assay, which assesses corrosive and irritant properties of a chemical by applying the undiluted chemical directly to the EE [15, 16]. In the EE potency assay, a dose response of the diluted chemical is performed by applying the chemical topically to the stratum corneum with the aid of a filter paper disc in a similar manner to patch

S. Gibbs (✉)
Department of Dermatology, VU University Medical Centre, Amsterdam, The Netherlands

Department of Oral Cell Biology, Academic Centre for Dentistry Amsterdam (ACTA), University of Amsterdam and VU University, Amsterdam, The Netherlands
e-mail: S.Gibbs@vumc.nl

S.W. Spiekstra
Department of Dermatology, VU University Medical Centre, Amsterdam, The Netherlands

C. Eskes et al. (eds.), *Alternatives for Dermal Toxicity Testing*,
DOI 10.1007/978-3-319-50353-0_20

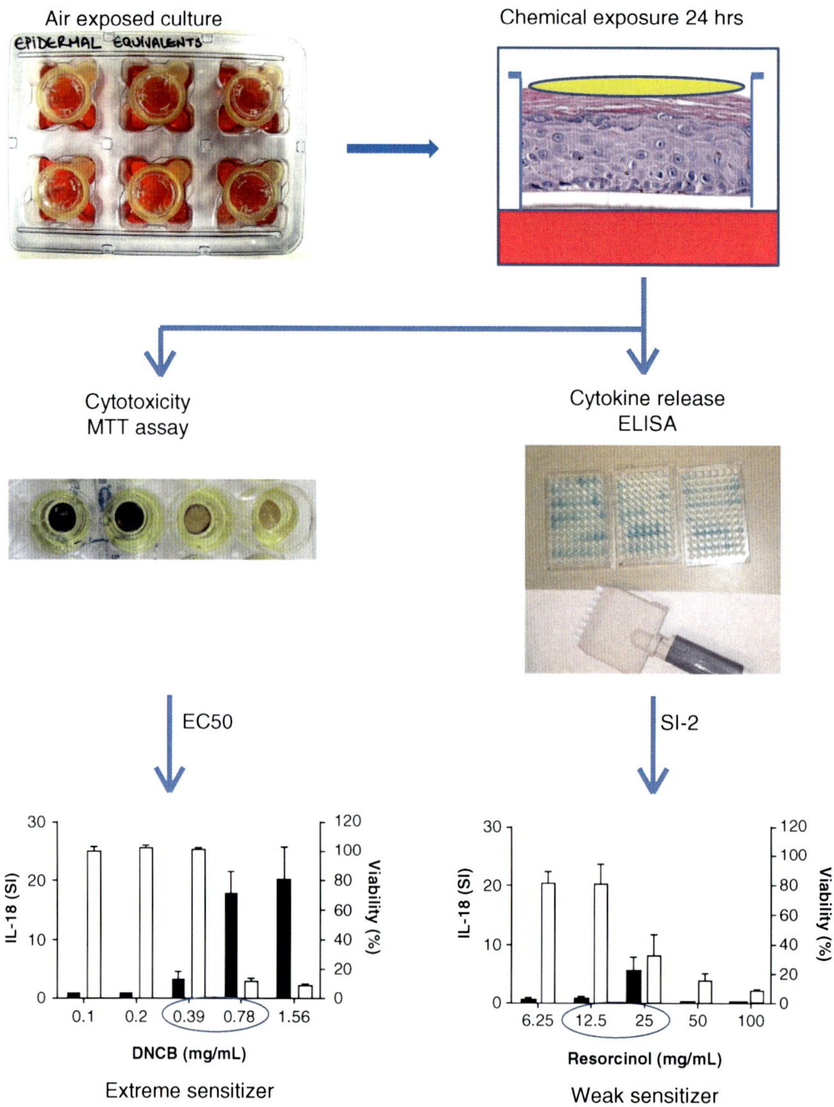

Fig. 20.1 Schematic diagram illustrating the EE potency assay

testing human individuals in an allergy clinic. The primary readout of the EE potency assay is the chemical concentration, which reduces cell viability (as assessed by MTT assay) by 50% (EC50 value). Additional readouts are the chemical concentration which results in a twofold increase in IL-1alpha or IL-18 release [3, 17, 18]. The lower the EC50 value, IL-1alpha (SI-2) or IL-18 (SI-2), the greater the sensitizer potency (Fig. 20.1).

This assay may be combined in a tiered approach (tier 2) with any other assay, which is used to identify a sensitizer (tier 1); for example, the keratinocyte IL-18 assay also described in this book [19–22]. Furthermore the EE potency assay has recently been combined with an IL-18 readout within a single EE in order to assess both chemical labelling and classification [17, 18].

Commercially available EE models, including epiCS® (Cell Systems, Biotechnology GmbH, Troisdorf, Germany) and RHE® (Reconstructed Human Epidermis) (SkinEthic™ Laboratories, Lyon, France), as well as in-house reconstructed EE made from primary human keratinocytes (VUMC-EE) can be used in the EE potency assay [3, 17, 18]. All epidermal models comprise a fully differentiated epidermis with viable and cornified cell layers. Differentiation, epithelial stratification and cornification are induced by culturing at the air-liquid interface using transwell inserts with a polycarbonate membrane (Fig. 20.1).

The major advantage of the assay is that the use of EE allows a chemical exposure that mimics human exposure (topical application) and therefore overcomes drawbacks of traditional submerged culture, including chemical solubility and stability in culture medium. In the EE assay, chemicals are applied topically to the stratum corneum of EE with the aid of a chemical-impregnated filter paper disc for 24 h. The standard operating procedure has been optimized especially for epiCS® and VUMC-EE. Minor modifications, e.g. exposure times, readout cut-off values, volumes used during chemical exposure, etc. may be required for other EE models.

20.2 Current (Pre)Validation Status

The EE potency assay underwent pre-validation in a project together with the NCTC2544 assay (pre-validation of a novel tiered approach to determine the skin sensitizing capacity and potency of chemicals) sponsored by the Dutch ZonMw programme Dierproeven Begrensd III (114011015) and the EU Frame Programme 6 Integrated Project Sens-it-iv (LSHB-CT-2005-018681). The primary aim of the EE potency assay was to evaluate the reproducibility and transferability of the EE potency assay with 13 sensitizers. The study was structured and conducted in two sequential phases using the epiCS® EE model: phase A, test method transfer from the lead laboratory to four naive laboratories within Europe [23] and phase B, assessment of the protocol performance by testing chemicals, under blind conditions, in four European laboratories [3]. Overall, the study succeeded in ranking sensitizers according to their potency and showed good correlation with human DSA05 and NOEL as well as animal LLNA data. In phase A, technology transfer went very well, and all five laboratories succeeded in obtaining similar and relevant EC50 values for the two test chemicals (DNCB, resorcinol) [23]. Four laboratories proceeded into the phase B pre-validation phase and tested 13 blinded chemicals [3]. Currently the EE potency assay is undergoing further technology transfer to naive partners and epiCS®, RHE and MatTek EpiDerm™ EE models in preparation for further validation in a ring study including America, Asia and Europe. The EE potency assay is being tested in combination with an IL-18 readout within a single EE in order to assess both chemical labelling and classification [17, 18].

20.3 Performance and Applicability of the Test Method

The method was easily transferred from the in-house VUMC-EE model to commercially available EE (EpiCS®, RHE,) and to naive laboratories [3, 17, 18, 23] strongly indicating the potential use of the test in immunotoxicity testing strategies. This can be attributed to the fact that no cell expansion and cell culture is necessary which often is the main contributor to experimental variation between laboratories. It is possible that some factors may vary between different EE such as the optimal duration and method of chemical exposure (filter paper vs direct), the vehicle used, the base line release of IL-1alpha and IL-18 and the barrier function of the EE, which in turn will influence the EC50 value. Therefore it is advised to optimize the procedure for each type of EE before progressing with potency testing. The assay is easy to implement in any laboratory with standard experience in chemical exposure followed, MTT assay and ELISA.

20.3.1 Reproducibility

Within-lab reproducibility: for each chemical, two independent runs are required. The percentage of chemicals having the same prediction in both runs (% concordance) ranged from 65 to 100% for the four participating laboratories in the pre-validation study [3].

Between-lab reproducibility: the percentage of chemicals having the same prediction in four laboratories was 35% concordance. However, the two best performing laboratories had 77% concordance [3].

20.3.2 Predictive Capacity

From the 13 blinded chemicals selected for pre-validation, 2 chemicals were identified as problematic chemicals which fell outside of the applicability domain of the assay as no EC50 value could be obtained (p-phenylenediamine and cobalt (II) chloride interfered with the colorimetric MTT assay) [3]. Therefore 11 of the 13 blinded chemicals were studied in detail. For these 11 chemicals, in addition to the linear correlation of EC50 and IL1 alpha values with animal and human data, a binary prediction model in which an EC50 $\geq$ 7 mg/ml = weak to moderate sensitizer and EC50 < 7 mg/ml = strong to extreme sensitizer was used. An average of 77% accuracy (range 65–82%), 69% sensitivity (range 50–83%) and 84% specificity (range 80–100%) was obtained for the four laboratories [3]. Notably, for the average linear correlation obtained by the four laboratories, EC50 and IL-1alpha$_{2x}$

values correlated best with human DSA_{05} (Spearman $r = 0.845$; $p = 0.006$ and $r = 0.929$; 0.002, respectively) followed by human NOEL and least with murine LLNA-EC3. However, it should be noted that no human data was available for oxazolone, which was the major outlier in the LLNA-EC3 correlation. Two of the laboratories further correlated IL-18_{2x} with DSA05 and obtained a Spearman correlation $r = 0.833$; $p = 0.015$ [3].

20.3.3 Applications and Limitations

The EE potency assay can be applied to any commercially available EE model which consists of a differentiated epidermis and stratum corneum and which is cultured at the air-liquid interface. It can be easily implemented in any laboratory with basic cell culture, ELISA and MTT assay experience. It can be used for chemical classification (potency assessment) of sensitizing compounds. Water-insoluble chemicals, of unknown molecular mass, can be tested although care should be taken that vehicles are not used at irritant concentrations. The vehicles that have been successfully tested on EE are water, DMSO (maximum 1% v/v) and acetone: olive oil (AOO) (4:1). The assay is suitable for testing chemical haptens as well as pro-haptens [1]. The major limitation of the EE potency assay is that it does not distinguish sensitizers from non-sensitizers. This limitation has recently been overcome by combining the EE potency assay with an IL-18 readout within a single EE construct [17, 18]. Alternatively, this limitation can be overcome by combining the EE potency assay with other assays capable of making this distinction (see other chapters in this book). Examples are the IL-18 keratinocyte NCTC assay [20]; dendritic cell maturation-based assays (e.g. hCLAT, MUSST) [1, 24] or dendritic cell migration assays (e.g. MUTZ-DC migration assay) [17, 18, 21]. Another limitation is that not all chemicals exhibit enough cytotoxicity to enable an EC50 or cytokine SI-2 value to be obtained (Table 20.1). If this is not due to chemical insolubility at 200 mg/ml, then it is possible that the chemical does not penetrate the stratum corneum and/or is an extremely weak sensitizer. The respiratory sensitizers ammonium hexachloroplatinate, diphenylmethane diisocyanate, trimellitic anhydride did not reduce metabolic activity nor increase IL-1alpha secretion at 200 mg/ml and therefore may be seen as a failure in this skin assay or classed as a very weak sensitizer [1, 24]. At the moment, the following chemicals have been identified as falling outside of the applicability domain of the assay. Cobalt II chloride, lauryl gallate and p-phenylenediamine interfered with the MTT colorimetric assays whereas for tetramethyl thiuram disulphide maximum solubility was reached in the vehicle at chemical concentration <200 mg/ml and before an EC50 value could be obtained according to the SOP [1, 3, 24]. Another relevant limitation of the assay is the expense of using multiple commercial EE in the dose response required to obtain an EC50 value for a chemical. However, this is compensated by the fact that very few labour hours are required to perform the assay as no expensive culture time and complicated analysis is required.

Table 20.1 Putative prediction model with EE potency assay EC50 value: extreme ≤1.5, strong >1.5 ≤ 10.0, moderate >10.0 ≤ 20.0, weak >20 ≤ 200, very weak >200

Chemical + CAS	Human cat.	Human NOEL (μg/cm²)	Human DSA₀₅ (μg/cm²)	LLNA-EC₃ (%)	*In vivo* references	EE potency assay score (data from references) [1–3, 17, 18, 23, 24]	
Extreme						*epiCS®*	*VUMC*
Oxazolone (15646-46-5)	ND	ND	ND	0.001–0.003	[25–27]	Extreme	Moderate
1-Chloro-2,4-dinitrobenzene (97-00-7)	1	8.8	2.1–5.5	0.006–0.131	[5, 25–35]	Extreme	Extreme
4-Nitrobenzylbromide (100-11-8)	ND	ND	ND	0.05	[25]	–	Extreme
Strong							
Formaldehyde (50-00-0)	2	37	89–411	0.27–0.99	[29, 30, 36–43]	Strong	–
Methyldibromo glutaronitrile (35691-65-7)	2	ND	ND	0.9	[30]	–	Strong
Glyoxal (107-22-2)	2	ND	ND	0.5–1.4	[25, 30, 44]	–	Weak
Cinnamaldehyde (104-55-2)	2	200, 400, 591	157–1111	0.2–3.1	[5, 28–30, 33, 36, 39, 40, 43, 45–53]	Strong	Moderate
Phenyl acetaldehyde (122-78-1)	ND	591	133–938	3–8.8	[25, 30, 36, 41, 42, 48, 54]	Strong	–
Isoeugenol (97-54-1)	2	69, 250	775–1333	0.5–5.0	[5, 28–30, 38, 43, 52, 55]	Moderate	Strong
2-Bromo-2-(bromomethyl) pentanedinitrile (90632-86-3)	ND	ND	ND	ND	–	–	Strong

Moderate							
Citral (5392-40-5)	3	200, 779, 1400	310–1691	1.2–13.0	[24, 25, 29, 48, 49, 56, 57]	Strong	–
Eugenol (97-53-0)	3	1938, 3200	5926	4.9–40.9	[26–30, 38, 43, 58, 59]	Strong	Weak
2-Mercaptobenzothiazole (149-30-4)	3	ND	1642–2269	1.0–6	[29, 30, 37, 39, 60, 61]	Moderate	Very weak
Weak							
Resorcinol (108-46-3)	4	ND	ND	5.5–6.3	[28, 30]	Weak	Weak
Benzocaine (94-09-7)	4	2000	3831–41,667	1.8–37	[30, 40, 43, 47, 62–64]	Moderate	–
Cinnamyl alcohol (104-54-1)	3	2000–3000	635–13,122	19.1–21	[25, 30, 49, 56, 65, 66]	–	Strong
α-Hexylcinnamaldehyde (101-86-0)	5	23,622	ND	1.2–17.6	[30, 36, 37, 67–69]	Weak	–
Very weak							
Phenyl benzoate (93-99-2)	3	ND	52,489	1.2–20	[25, 30, 37, 49, 67]	–	Very weak
diphenylmethane diisocyanate (26,447-40-5)	ND	ND	ND	ND		–	Very weak
Trimellitic anhydride (552-30-7)	ND	ND	ND	0.22	[64, 70]	–	Very weak
Ammonium hexachloroplatinate (16919-58-7)	ND	ND	ND	ND		–	Very weak

(continued)

Table 20.1 (continued)

Chemical + CAS	Human cat.	Human NOEL (μg/cm^2)	Human DSA$_{05}$ (μg/cm^2)	LLNA-EC$_3$ (%)	*In vivo* references	EE potency assay score (data from references) [1–3, 17, 18, 23, 24]	
Failure							
p-Phenylenediamine (106-50-3)	1	10	6.9–345	0.001–2.2	[5, 25, 28–30, 38, 40, 43, 50, 51, 64]	MTT disturbance	MTT disturbance
Lauryl gallate (1166-52-5)	2	ND	ND	0.3	[30]	–	MTT disturbance
Tetramethylthiuram disulphide (137-26-8)	3	ND	3832–5388	5.2–6.0	[30, 36, 39, 40, 45]	–	Solubility problem
Cobalt (II) chloride[b] (7791-13-1)	ND	ND	171, 453	0.4, 0.8	[29, 30, 39, 40, 67, 71]	MTT disturbance	–

The chemicals are listed according to their potency values obtained from a combined assessment of all data available from the human category scale, human NOEL, human DSA$_{05}$ and murine LLNA-EC$_3$ experiments. When human and murine data were conflicting or limited, the human data were prioritized in the ranking above murine data

Human category scale: 1 = extensive evidence of contact allergy in relation to degree of exposure and size of exposed population; 2 = a frequent cause of contact allergy but of less significance compared with induction of skin sensitization in a HRIPT category 1; 3 = a common cause of contact allergy, perhaps requiring higher exposure compared with category 2; 4 = infrequent cause of contact allergy in relation to level of exposure; 5 = a rare cause of contact allergy except perhaps in special circumstances [30]

Human NOEL (μg/cm^2) = no observed effect level; all available data for NOEL is shown

Human DSA$_{05}$ (μg/cm^2) = induction dose per skin area (DSA) that produces a positive response in 5% of the tested population

The LLNA-EC$_3$ values are expressed as % according to [67]: potency classification is based on the mathematical estimation of the concentration of chemical necessary to obtain a threshold positive response (SI = 3); this is termed the EC$_3$ value. Chemicals with an EC$_3$ value (%) >10 to <100 are classified as weak, >1 to <10 moderate, >0.1 to <1 strong and <0.1 extreme

ND: indicates no data available found by the authors; – = no data available

For LLNA data, a range of values is shown which was obtained from ICCVAM report Annex II-1 (see Comparative LLNA, Guinea Pig, and Human Data Used in the Performance Evaluation. Annex II-1. LLNA Data for 196 Substances Used for the Evaluation of Skin Sensitization Potency http://iccvam.niehs.nih.gov/docs/immunotox_docs/LLNA-pot/3b-AppC-BRD-AnnexII-1.pdf. Last assessed 1 January 2015). For NOEL and DSA$_{05}$ data, ranges were obtained from [30] and ICCVAM report Annex II-2 (see Human Data for LLNA Potency Evaluation. Annex II-2. http://iccvam.niehs.nih.gov/docs/immunotox_docs/LLNA-pot/3b-AppC-BRD-AnnexII-2.pdf. Last assessed 10 January 2015

20.3.4 Comparison to Human Data

The EE potency assay has been extensively correlated to both human and murine LLNA data [1, 3, 17, 18]. Table 20.1 shows all results obtained to date from the test developer lab using epiCS® and in-house VUMC-EE models. Notably, in the pre-validation study with four European laboratories, the EE potency assay correlated best with human DSA_{05} followed by human NOEL and least with murine LLNA-EC3 [3, 17, 18]. However, it should be noted that no human data was available for oxazolone which was the major outlier in the LLNA-EC3 correlation.

20.4 Brief Description of the Protocol

The assay is performed according to the SOP in order to determine the EC_{50} value of a chemical of unknown or known identity, solubility and molecular mass, and the release of IL-1alpha and IL-18 associated with the chemical exposure [23]. First the maximum solubility of the chemical is identified by dissolving the compound in the vehicles: acetone/olive oil (AOO) 4:1 and 1% DMSO in assay medium. The vehicle with the highest dissolving capacity is chosen. Dilutions are made starting at 200 mg/ml until a clear solution is reached. Then, in order to prepare the chemical for topical exposure of EE, twofold serial dilutions are made from the highest soluble concentration of the chemical (preferably 200 mg/ml) until 0.10 mg/ml is reached. Note this single dose response enables an extreme to a very weak sensitizer to be classified according to its EC50 value in a single dose response and is a simplification of the multiple dose findings (broad dose A, B and fine dose) published in the full SOP [23]. Two independent runs are performed for each chemical. Whereas chemical dose responses are tested in single-fold within each independent run, control conditions (unexposed, vehicle(s) and positive control) should preferably be tested in duplicate per independent run. Each chemical should be tested on a separate plate. EE cultures are topically exposed to chemicals, impregnated in 8 mm diameter paper filters (designed for the use with 8 mm Finn Chambers) for 24 h. Hereafter, cytotoxicity is measured by MTT assay and the release of IL-1α and IL-18 by ELISA. For sensitizer potency, the EC_{50} value and cytokine (SI-2) values are determined. The EC50 value and cytokine (SI-2) value of the unknown chemical are then correlated to the values obtained for a standard test panel of chemicals [17, 18].

Primary readout parameter: cytotoxicity (MTT assay) expressed as EC_{50} value (EC_{50} = effective chemical concentration required to reduce EE metabolic activity—corresponding to cell viability—to 50% of the maximum value of vehicle treated EE).

Secondary readout parameter: IL-18 or IL-1alpha expressed as SI-2 (chemical concentration resulting in: ≥twofold increase in cytokine release as assessed by ELISA compared to vehicle-treated EE (SI-2).

Negative control: vehicle-exposed EE should not result in more than 30% cytotoxicity compared to unexposed EE as this would indicate inferior barrier function in the EE batch.

Positive control (putative, still under investigation): each laboratory should identify the optimal concentration of DNCB resulting in an EC100 and EC50 value as well as a cytokine SI-2 value.

Data analysis: EC50 values are obtained by non-linear regression analysis using dose response formula's ($r > 0.9$) based on the chemical concentration required to reduce metabolic activity (corresponding to cell viability) to 50% of the value obtained by vehicle exposed EE. IL-1alpha or IL-18 SI-2 values are obtained by non-linear regression analysis based on the chemical concentration required to increase cytokine release by twofold compared to vehicle-exposed EE. A linear correlation with animal and human data is then made, or alternatively a binary prediction model can be used in which EC50 $\geq$ 7 mg/ml = weak to moderate sensitizer and EC50 < 7 mg/ml = strong to extreme sensitizer [3].

20.5 Role in a Testing Strategy

We propose that the EE potency assay is combined with the assay which assesses release of IL-18 (SI-5) to provide a single test for identification and GHS classification of skin sensitizing chemicals [17, 18]. Alternatively it can be used in combination with any assay, which can identify a sensitizer from a non-sensitizer [19, 21, 24, 72]. Many examples are provided in this book. In this way, it will function as a tier 2 assay which determines sensitizer potency (classification) of chemicals already labelled as a sensitizer in a tier 1 assay.

20.6 Perspectives from the Test Developer

The EE potency assay holds promise since technology transfer to naive laboratories and different commercially available EE has already proven to be successful [1, 23]. Since chemicals are applied topically to the stratum corneum in a similar manner to human exposure, a future perspective lies in the area of testing mixtures. Most consumer products consist of mixtures, which may be composed of chemicals with an irritant or sensitizing property. It is unknown how such combinations will influence final irritant or sensitizing potency of the combined product. Furthermore, the possibility exists to investigate similar ingredients dissolved in a wide range of vehicles (e.g. ointments, lotions, creams) in order to determine the influence of different vehicles on sensitizer potency. Already we have reported that for SkinEthic™ RHE, viability and IL-18 release are influenced when the chemicals DNCB, citral or eugenol are dissolved in ethanol, acetone, olive oil or dimethylformamide [17, 18]. Additionally photoallergens can be studied in combination with exposure of EE to solar simulators again reflecting a superior physiologically relevant test model compared to conventional models using submerged cell cultures [73].

20.6.1 Critical Steps in the Protocol

It is most important that EE are handled upon arrival in the laboratory according to the procedures of the supplier in order to maintain viability. Furthermore, it is important that the vehicle itself does not result in more than 30% cytotoxicity since the EC50 value of the test chemical is expressed relative to the vehicle control. If the vehicle does show >30% cytotoxicity, the run should be discarded. This could be due to an impaired barrier function of the batch of EE or alternatively that the vehicle itself has irritant properties. In order to identify very weak sensitizers, it is important that the maximum solubility of the chemical reaches 200 mg/ml. If this level of solubility is not reached, then it is possible that an EC50 is not obtained, and then the result will be inconclusive.

20.6.2 Possible Protocol Adaptations

It is strongly advised that during the technology transfer, new laboratories implement the EE potency assay into their laboratory with a set of three training chemicals (DNCB, extreme; cinnamaldehyde, strong/moderate; resorcinol, weak). Correct classification should be obtained with these training chemicals before proceeding to test chemicals. Furthermore, each laboratory should ensure that vehicle exposure does not result in >30% cytotoxicity, and also optimization of the chemical concentration of the positive control DNCB is required to ensure maximum cytotoxicity EC 0.

With regard to the prediction model, currently a linear correlation with animal and human data is made, or alternatively a binary prediction model can be used in which EC50 $\geq$ 7 mg/ml = weak to moderate sensitizer and EC50 < 7 mg/ml = strong to extreme sensitizer [3]. Reviewing all data available from the test developer laboratory (VUmc) (see Table 20.1) [1, 3, 17, 18], it may be possible to extend the prediction model for the EC50 value to the following: extreme $\leq$1.5, strong >1.5 $\leq$ 10.0, moderate >10.0 $\leq$ 20.0, weak >20 $\leq$ 200 and very weak >200. As more data becomes available from different laboratories and different EE models, it will be possible to test this putative prediction model.

20.6.3 Challenges and Opportunities

Further challenges are to combine the EE potency assay with an IL-18 readout within a single EE in order to assess both chemical labelling and classification [17, 18]. Since water-insoluble substances can be tested with the EE potency assay, an excellent opportunity now exists for testing mixtures and photoallergens.

20.7 Conclusions

We propose the EE potency assay as a screening, replacement, reduction test for hazard labelling of skin sensitizers. The test allows for a rapid and easy labelling of contact sensitizers. Continual addition of data obtained from known chemicals will provide a golden standard classification table for correlation and prediction of unknown chemicals.

References

1. dos Santos GG, Spiekstra SW, Sampat-Sardjoepersad SC, Reinders J, Scheper RJ, Gibbs S. A potential in vitro epidermal equivalent assay to determine sensitizer potency. Toxicol In Vitro. 2011;25:347–57.
2. Spiekstra SW, dos Santos GG, Scheper RJ, Gibbs S. Potential method to determine irritant potency in vitro - comparison of two reconstructed epidermal culture models with different barrier competency. Toxicol In Vitro. 2009;23:349–55.
3. Teunis MA, Spiekstra SW, Smits M, Adriaens E, Eltze T, Galbiati V, Krul C, Landsiedel R, Pieters R, Reinders J, Roggen E, Corsini E, Gibbs S. International ring trial of the epidermal equivalent sensitizer potency assay: reproducibility and predictive-capacity. ALTEX. 2014;31:251–68.
4. Agner T, Johansen JD, Overgaard L, Volund A, Basketter D, Menne T. Combined effects of irritants and allergens. Synergistic effects of nickel and sodium lauryl sulfate in nickel- sensitized individuals. Contact Dermatitis. 2002;47:21–6.
5. Basketter DA, Kan-King-Yu D, Dierkes P, Jowsey IR. Does irritation potency contribute to the skin sensitization potency of contact allergens? Cutan Ocul Toxicol. 2007;26:279–86.
6. Bonneville M, Chavagnac C, Vocanson M, Rozieres A, Benetiere J, Pernet I, Denis A, Nicolas JF, Hennino A. Skin contact irritation conditions the development and severity of allergic contact dermatitis. J Invest Dermatol. 2007;127:1430–5.
7. McLelland J, Shuster S, Matthews JN. 'Irritants' increase the response to an allergen in allergic contact dermatitis. Arch Dermatol. 1991;127:1016–101.
8. Edele F, Esser PR, Lass C, Laszczyk MN, Oswald E, Struh CM, Rensing-Ehl A, Martin SF. Innate and adaptive immune responses in allergic contact dermatitis and autoimmune skin diseases. Inflamm Allergy Drug Targets. 2007;6:236–44.
9. Martin SF, Jakob T. From innate to adaptive immune responses in contact hypersensitivity. Curr Opin Allergy Clin Immunol. 2008;8:289–93.
10. Martin SF, Esser PR, Weber FC, Jakob T, Freudenberg MA, Schmidt M, Goebeler M. Mechanisms of chemical-induced innate immunity in allergic contact dermatitis. Allergy. 2011;66:1152–63.
11. Antonopoulos C, Cumberbatch M, Mee JB, Dearman RJ, Wei XQ, Liew FY, Kimber I, Groves RW. IL-18 is a key proximal mediator of contact hypersensitivity and allergen-induced Langerhans cell migration in murine epidermis. J Leukoc Biol. 2008;83:361–7.
12. Corsini E, Galbiati V, Mitjans M, Galli CL, Marinovich M. NCTC 2544 and IL-18 production: a tool for the identification of contact allergens. Toxicol In Vitro. 2013;27:1127–34.
13. Cumberbatch M, Dearman RJ, Antonopoulos C, Groves RW, Kimber I. Interleukin (IL)-18 induces Langerhans cell migration by a tumour necrosis factor-alpha- and IL-1beta-dependent mechanism. Immunology. 2001;102:323–30.
14. Okamura H, Tsutsi H, Komatsu T, Yutsudo M, Hakura A, Tanimoto T, Torigoe K, Okura T, Nukada Y, Hattori K. Cloning of a new cytokine that induces IFN-gamma production by T cells. Nature. 1995;378:88–9.
15. Fentem JH, Archer GE, Balls M, Botham PA, Curren RD, Earl LK, Esdaile DJ, Holzhutter HG, Liebsch M. The ECVAM international validation study on in vitro tests for skin corrosivity. 2. Results and evaluation by the management team. Toxicol In Vitro. 1998;12:483–524.

16. Spielmann H, Hoffmann S, Liebsch M, Botham P, Fentem JH, Eskes C, Roguet R, Cotovio J, Cole T, Worth A, Heylings J, Jones P, Robles C, Kandarova H, Gamer A, Remmele M, Curren R, Raabe H, Cockshott A, Gerner I, Zuang V. The ECVAM international validation study on in vitro tests for acute skin irritation: report on the validity of the EPISKIN and EpiDerm assays and on the skin integrity function test. Altern Lab Anim. 2007;35:559–601.
17. Gibbs S, Corsini E, Spiekstra SW, Galbiati V, Fuchs HW, Degeorge G, Troese M, Hayden P, Deng W, Roggen E. An epidermal equivalent assay for identification and ranking potency of contact sensitizers. Toxicol Appl Pharmacol. 2013;272:529–41.
18. Gibbs S, Spiekstra S, Corsini E, McLeod J, Reinders J. Dendritic cell migration assay: a potential prediction model for identification of contact allergens. Toxicol In Vitro. 2013;27:1170–9.
19. Corsini E, Roggen EL. Immunotoxicology: opportunities for non-animal test development. Altern Lab Anim. 2009;37:387–97.
20. Galbiati V, Mitjans M, Lucchi L, Viviani B, Galli CL, Marinovich M, Corsini E. Further development of the NCTC 2544 IL-18 assay to identify in vitro contact allergens. Toxicol In Vitro. 2011;25:724–32.
21. Rovida C, Martin SF, Vivier M, Weltzien HU, Roggen E. Advanced tests for skin and respiratory sensitization assessment. ALTEX. 2013;30:231–52.
22. van Och FM, van LH, Van Wolfswinkel JC, Machielsen AJ, Vandebriel RJ. Assessment of potency of allergenic activity of low molecular weight compounds based on IL-1alpha and IL-18 production by a murine and human keratinocyte cell line. Toxicology. 2005;210:95–109.
23. Teunis M, Corsini E, Smits M, Madsen CB, Eltze T, Ezendam J, Galbiati V, Gremmer E, Krul C, Landin A, Landsiedel R, Pieters R, Rasmussen TF, Reinders J, Roggen E, Spiekstra S, Gibbs S. Transfer of a two-tiered keratinocyte assay: IL-18 production by NCTC2544 to determine the skin sensitizing capacity and epidermal equivalent assay to determine sensitizer potency. Toxicol In Vitro. 2013;27:1135–50.
24. Reisinger K, Hoffmann S, Alepee N, Ashikaga T, Barroso J, Elcombe C, Gellatly N, Galbiati V, Gibbs S, Groux H, Hibatallah J, Keller D, Kern P, Klaric M, Kolle S, Kuehnl J, Lambrechts N, Lindstedt M, Millet M, Martinozzi-Teissier S, Natsch A, Petersohn D, Pike I, Sakaguchi H, Schepky A, Tailhardat M, Templier M, van Vliet E, Maxwell G. Systematic evaluation of non-animal test methods for skin sensitisation safety assessment. Toxicol In Vitro. 2014;29:259–70.
25. Gerberick GF, Ryan CA, Kern PS, Schlatter H, Dearman RJ, Kimber I, Patlewicz GY, Basketter DA. Compilation of historical local lymph node data for evaluation of skin sensitization alternative methods. Dermatitis. 2005;16:157–202.
26. Gerberick GF, House RV, Fletcher ER, Ryan CA. Examination of the local lymph node assay for use in contact sensitization risk assessment. Fundam Appl Toxicol. 1992;19:438–45.
27. Loveless SE, Ladics GS, Gerberick GF, Ryan CA, Basketter DA, Scholes EW, House RV, Hilton J, Dearman RJ, Kimber I. Further evaluation of the local lymph node assay in the final phase of an international collaborative trial. Toxicology. 1996;108:141–52.
28. Basketter DA, Gerberick F, Kimber I. The local lymph node assay and the assessment of relative potency: status of validation. Contact Dermatitis. 2007;57:70–5.
29. Basketter DA, Scholes EW. Comparison of the local lymph node assay with the guinea-pig maximization test for the detection of a range of contact allergens. Food Chem Toxicol. 1992;30:65–9.
30. Basketter DA, Alepee N, Ashikaga T, Barroso J, Gilmour N, Goebel C, Hibatallah J, Hoffmann S, Kern P, Martinozzi-Teissier S, Maxwell G, Reisinger K, Sakaguchi H, Schepky A, Tailhardat M, Templier M. Categorization of chemicals according to their relative human skin sensitizing potency. Dermatitis. 2014;25:11–21.
31. Betts CJ, Dearman RJ, Heylings JR, Kimber I, Basketter DA. Skin sensitization potency of methyl methacrylate in the local lymph node assay: comparisons with guinea-pig data and human experience. Contact Dermatitis. 2006;55:140–7.
32. Friedmann PS, Moss C, Shuster S, Simpson JM. Quantitation of sensitization and responsiveness to dinitrochlorobenzene in normal subjects. Br J Dermatol. 1983;109(Suppl 25):86–8.
33. NTP. Assessment of contact hypersensitivity to Dinitrochlorobenzene, potassium dichromate and methyl salicylate in BALB16 female mice. Research Triangle Park, NC: National Institute of Environmental Health Sciences Report; 1997.

34. Rees JL, Friedmann PS, Matthews JN. Sex differences in susceptibility to development of contact hypersensitivity to dinitrochlorobenzene (DNCB). Br J Dermatol. 1989;120:371–4.
35. Kimber I, Hilton J, Dearman RJ, Gerberick GF, Ryan CA, Basketter DA, Scholes EW, Ladics GS, Loveless SE, House RV. An international evaluation of the murine local lymph node assay and comparison of modified procedures. Toxicology. 1995;103:63–73.
36. Basketter DA, Kimber I. Predictive testing in contact allergy: facts and future. Allergy. 2001;56:937–43.
37. Basketter DA, Clapp C, Jefferies D, Safford B, Ryan CA, Gerberick F, Dearman RJ, Kimber I. Predictive identification of human skin sensitization thresholds. Contact Dermatitis. 2005;53:260–7.
38. Kimber I, Hilton J, Botham PA, Basketter DA, Scholes EW, Miller K, Robbins MC, Harrison PT, Gray TJ, Waite SJ. The murine local lymph node assay: results of an inter-laboratory trial. Toxicol Lett. 1991;55:203–13.
39. Kligman AM. The identification of contact allergens by human assay. II. Factors influencing the induction and measurement of allergic contact dermatitis. J Invest Dermatol. 1966;47:375–92.
40. Kligman AM. The identification of contact allergens by human assay. 3. The maximization test: a procedure for screening and rating contact sensitizers. J. Invest Dermatol. 1966;47:393–409.
41. Ryan CA, Gerberick GF, Cruse LW, Basketter DA, Lea L, Blaikie L, Dearman RJ, Warbrick EV, Kimber I. Activity of human contact allergens in the murine local lymph node assay. Contact Dermatitis. 2000;43:95–102.
42. Ryan CA, Cruse LW, Skinner RA, Dearman RJ, Kimber I, Gerberick GF. Examination of a vehicle for use with water soluble materials in the murine local lymph node assay. Food Chem Toxicol. 2002;40:1719–25.
43. Marzulli FN, Maibach HI. The use of graded concentrations in studying skin sensitizers: experimental contact sensitization in man. Food Cosmet Toxicol. 1974;12:219–27.
44. Basketter DA, Scholes EW, Kimber I. The performance of the local lymph node assay with chemicals identified as contact allergens in the human maximization test. Food Chem Toxicol. 1994;32:543–7.
45. Basketter DA, Gerberick GF, Kimber I, Loveless SE. The local lymph node assay: a viable alternative to currently accepted skin sensitization tests. Food Chem Toxicol. 1996;34:985–97.
46. Danneman PJ, Booman KA, Dorsky J, Kohrman KA, Rothenstein AS, Sedlak RI, Steltenkamp RJ, Thompson GR. Cinnamic aldehyde: a survey of consumer patch-test sensitization. Food Chem Toxicol. 1983;21:721–5.
47. Kimber I, Hilton J, Weisenberger C. The murine local lymph node assay for identification of contact allergens: a preliminary evaluation of in situ measurement of lymphocyte proliferation. Contact Dermatitis. 1989;21:215–20.
48. Opdyke DL. Monographs on fragrance raw materials. Food Cosmet Toxicol. 1979;17:241–75.
49. RIFM. In: NICEATM, editor. Data submission to NICEATM. Woodcliff Lake, NJ: RIFM; 2007.
50. Warbrick EV, Dearman RJ, Lea LJ, Basketter DA, Kimber I. Local lymph node assay responses to paraphenylenediamine: intra- and inter-laboratory evaluations. J Appl Toxicol. 1999;19:255–60.
51. White JM, Kullavanijaya P, Duangdeeden I, Zazzeroni R, Gilmour NJ, Basketter DA, McFadden JP. p-Phenylenediamine allergy: the role of Bandrowski's base. Clin Exp Allergy. 2006;36:1289–93.
52. Wright ZM, Basketter PA, Blaikie L, Cooper KJ, Warbrick EV, Dearman RJ, Kimber I. Vehicle effects on skin sensitizing potency of four chemicals: assessment using the local lymph node assay. Int J Cosmet Sci. 2001;23:75–83.
53. Basketter DA, Kimber I. Assessing the potency of respiratory allergens: uncertainties and challenges. Regul Toxicol Pharmacol. 2011;61(3):365–72. doi: 10.1016/j.yrtph.2011.10.001.
54. Basketter DA, Evans P, Gerberick GF, Kimber I. Factors affecting thresholds in allergic contact dermatitis: safety and regulatory considerations. Contact Dermatitis. 2002;47:1–6.
55. Basketter DA, Cadby P. Reproducible prediction of contact allergenic potency using the local lymph node assay. Contact Dermatitis. 2004;50:15–7.

56. Steltenkamp RJ, Booman KA, Dorsky J, King TO, Rothenstein AS, Schwoepp EA, Sedlak RI, Smith TH, Thompson GR. Cinnamic alcohol: a survey of consumer patch-test sensitization. Food Cosmet Toxicol. 1980;18:419–24.
57. Ashby J, Basketter DA, Paton D, Kimber I. Structure activity relationships in skin sensitization using the murine local lymph node assay. Toxicology. 1995;103:177–94.
58. Lalko J, Isola D, Api AM. Ethanol and diethyl phthalate: vehicle effects in the local lymph node assay. Int J Toxicol. 2004;23:171–7.
59. Kimber I, Weisenberger C. Anamnestic responses to contact allergens: application in the murine local lymph node assay. J Appl Toxicol. 1991;11:129–33.
60. Basketter DA, Selbie E, Scholes EW, Lees D, Kimber I, Botham PA. Results with OECD recommended positive control sensitizers in the maximization, Buehler and local lymph node assays. Food Chem Toxicol. 1993;31:63–7.
61. Scholes EW, Basketter DA, Sarll AE, Kimber I, Evans CD, Miller K, Robbins MC, Harrison PT, Waite SJ. The local lymph node assay: results of a final inter-laboratory validation under field conditions. J.Appl. Toxicology. 1992;12:217–22.
62. Basketter DA, Scholes EW, Wahlkvist H, Montelius J. An evaluation of the suitability of benzocaine as a positive control skin sensitizer. Contact Dermatitis. 1995;33:28–32.
63. Griem P, Goebel C, Scheffler H. Proposal for a risk assessment methodology for skin sensitization based on sensitization potency data. Regul Toxicol Pharmacol. 2003;38:269–90.
64. van Och FM, Slob W, de Jong WH, Vandebriel RJ, van LH. A quantitative method for assessing the sensitizing potency of low molecular weight chemicals using a local lymph node assay: employment of a regression method that includes determination of the uncertainty margins. Toxicology. 2000;146:49–59.
65. Jordan WP Jr, King SE. Delayed hypersensitivity in females. The development of allergic contact dermatitis in females during the comparison of two predictive patch tests. Contact Dermatitis. 1977;3:19–26.
66. Elahi EN, Wright Z, Hinselwood D, Hotchkiss SA, Basketter DA, Pease CK. Protein binding and metabolism influence the relative skin sensitization potential of cinnamic compounds. Chem Res Toxicol. 2004;17:301–10.
67. Basketter DA, Lea LJ, Dickens A, Briggs D, Pate I, Dearman RJ, Kimber I. A comparison of statistical approaches to the derivation of EC3 values from local lymph node assay dose responses. J Appl Toxicol. 1999;19:261–6.
68. Dearman RJ, Wright ZM, Basketter DA, Ryan CA, Gerberick GF, Kimber I. The suitability of hexyl cinnamic aldehyde as a calibrant for the murine local lymph node assay. Contact Dermatitis. 2001;44:357–61.
69. Gamer AO, Nies E, Vohr HW. Local lymph node assay (LLNA): comparison of different protocols by testing skin-sensitizing epoxy resin system components. Regul Toxicol Pharmacol. 2008;52:290–8.
70. Basketter DA, Smith Pease CK, Patlewicz GY. Contact allergy: the local lymph node assay for the prediction of hazard and risk. Clin Exp Dermatol. 2003;28:218–21.
71. Ikarashi Y, Ohno K, Tsuchiya T, Nakamura A. Differences of draining lymph node cell proliferation among mice, rats and guinea pigs following exposure to metal allergens. Toxicology. 1992;76:283–92.
72. Galbiati V, Mitjans M, Corsini E. Present and future of in vitro immunotoxicology in drug development. J Immunotoxicol. 2010;7:255–67.
73. Gibbs MS, De BG, Mulder A, Mommaas AM, Ponec M. Melanosome capping of keratinocytes in pigmented reconstructed epidermis--effect of ultraviolet radiation and 3-isobutyl-1-methyl-xanthine on melanogenesis. Pigment Cell Res. 2000;13:458–66.

The h-CLAT Method

21

Hitoshi Sakaguchi and Takao Ashikaga

21.1 Principle of the Test Method and Scientific Basis

The mechanism of skin sensitization consists of two steps: an induction phase and an elicitation phase. In the induction phase, a hapten that is a small molecular weight compound (generally less than 500 Da) must first penetrate the stratum corneum and be absorbed into the epidermis, where it can bind to skin protein, thereby forming an immunogen. These modified proteins may then be recognized by Langerhans cells (LCs). Subsequently, the LCs become activated and migrate from the epidermis to the draining lymph node where they present antigen to naïve T cells. This recognition then stimulates the generation and proliferation of a population of memory T cells. In the elicitation phase, the same hapten is absorbed and binds to protein. The immunogen is recognized by LCs and existing memory T cells. Inflammation (e.g., erythema and oedema) occurs as a result of antigen presentation. Since the mechanism of skin sensitization is very complex and it is extremely difficult to reproduce all mechanisms in a single *in vitro* system, the development of an *in vitro* method should consider or address various aspects of the induction phase of the sensitization process such as the potential dermal penetration of a chemical, the protein/peptide interaction, activation of keratinocytes, and the initiation of an antigen-specific immune response [2]. LCs play a critical role during the induction phase of skin sensitization because of their ability to initiate immune responses by processing and presenting antigens following exposure to chemical allergens. Upon antigen capture, LCs differentiate, mature, and migrate to the draining lymph nodes

H. Sakaguchi (✉)
Kao Corporation, Tochigi, Japan
e-mail: sakaguchi.hitoshi@kao.co.jp

T. Ashikaga
Shiseido, Co., Ltd., Yokohama, Japan
e-mail: takao.ashikaga@to.shiseido.co.jp

C. Eskes et al. (eds.), *Alternatives for Dermal Toxicity Testing*,
DOI 10.1007/978-3-319-50353-0_21

where they present the allergens to naïve T cells and trigger their proliferation [3, 4]. During this process, LC maturation is characterized by the up-regulation of class II major histocompatibility complex antigen (MHC class II) and the expression of co-stimulatory molecules such as CD86, CD54, CD80, and CD40 [5, 6]. It is difficult to obtain a sufficient number of LCs from the epidermis. To solve this limitation, some studies have used human peripheral blood mononuclear cells (PBMC) or CD34+ hematopoietic progenitor cells that were cultured in the presence of specific cytokines as antigen-presenting cells (APCs) instead of LCs for the development of an *in vitro* method. Although the use of dendritic cells (DCs) for predicting allergens has potential, there are still some technical problems with the routine use of these cells in skin sensitization tests. These problems include availability of human blood and donor-to-donor variability [7, 8].

However, the use of human cell lines instead of DCs has provided a technical solution to these latter problems. THP-1 cells, which were first derived from the peripheral blood of a one-year-old male with acute monocytic leukaemia [9, 10], are a human monocytic leukaemia cell line. Following phorbol 12-myristate 13-acetate (PMA) treatment, THP-1 cells differentiate into macrophages. Treatment of THP-1 cells with allergens has been reported to augment the cell surface markers CD40, CD54, CD80, CD83, CD86, and HLA-DR. Of these markers, there have been many reports that CD86 is up-regulated by treatment of THP-1 cells with some allergens [11–14]. In addition to CD86, CD54has also been reported as an activation marker following treatment with some allergens [14–17]. Other cell surface markers, including MHC class II molecule internalization [12], CD1a, CD40, and CD80 [18], were also reported to be augmented by some allergens. THP-1 cells can be bought from many major cell banks including the American Type Culture Collection (ATCC). Based on the above information, Kao Corporation and Shiseido Co., Ltd. (in Japan) developed an *in vitro* skin sensitization test termed the human Cell Line Activation Test (h-CLAT). The h-CLAT is based on phenotypic changes in THP-1 cells, which should simulate the function of LCs. Examination of the phenotypic and functional changes induced in APC, including LCs, by test agents is one of the most important approaches for the development of *in vitro* sensitization methods. The h-CLAT method is based on the activation process of APCs (augmentation of CD86 and/or CD54 expression in THP-1 cells) [15, 19]. Therefore, the endpoint of h-CLAT is an important mechanism in the induction phase and is a necessary part of a series of skin sensitization processes.

21.2 Current (Pre) Validation Status

In order to confirm the transferability and the between-laboratory reproducibility of h-CLAT, ring studies among both Japanese and European cosmetics industries have been performed. Both the studies among the seven Japanese laboratories and among the five European laboratories showed a high potential of this method regarding both transferability and between-laboratory reproducibility [20, 21]. Through these activities, the h-CLAT protocol has been modified to improve these performances.

In 2008, test developers proposed the conduction of a validation study of h-CLAT to the European Union Reference Laboratory for Alternatives to Animal Testing (EURL ECVAM). The study was designed to generate information according to a modular approach, with the primary objective of fully assessing the reliability of the h-CLAT method (i.e., its transferability and within and between laboratory reproducibility). The h-CLAT method has been formally evaluated in a EURL ECVAM coordinated validation study since 2009 in collaboration with the Japanese Center for the Validation of Alternative Methods (JaCVAM). This validation study demonstrated that the h-CLAT test method is transferable to laboratories experienced in cell culture techniques and flow cytometric analysis [22]. The within-laboratory and between-laboratory reproducibility ($n = 15$ and $n = 24$, respectively), as characterized on the basis of concordant classifications of the chemicals employed, were both in the order of 80%. Subsequently an independent peer review by the EURL ECVAM Scientific Advisory Committee (ESAC) has been carried out and it was considered scientifically valid by EURL ECVAM recommendation [22]. EURL ECVAM does not recommend h-CLAT as a stand-alone method for the determination of skin sensitization hazard. However, according to EURL ECVAM, based on the outcome of the validation study and reports from the scientific literature, data generated with the h-CLAT method should prove valuable as part of Integrated Approaches to Testing and Assessment (IATA) together with complementary information (e.g., *in chemico* or other *in vitro* data, QSAR (Quantitative Structure Activity Relationship) or read-across predictions). EURL ECVAM concluded that they fully support the development of an OECD (Organization for Cosmetic Co-operation and Development) Test Guideline (TG) for the h-CLAT method. The OECD has recently adopted a new TG No. 442E for the h-CLAT method [1] after 2 years of discussion at the OECD expert working group of the national coordinators of the test guidelines programme (WNT).

21.3 Performance and Applicability of the Test Method

21.3.1 Reproducibility

For assessment of within-laboratory reproducibility (WLR), the haptens DNCB (M.W. = 202.6, categorized as 'extreme' by the murine local lymph node assay (LLNA)) and Ni (M.W. = 262.9, categorized as 'moderate' by LLNA) were evaluated as sensitizers, with SLS (M.W. = 288.4, categorized as 'false positive' by LLNA) evaluated as a non-sensitizer. These compounds were used as positive (DNCB and Ni) and negative (SLS) controls for THP-1 cell and Foetal Bovine Serum (FBS)selection. The skin sensitization potential of DNCB, Ni, and SLS was evaluated correctly in all of the 20 trials that were evaluated on different days. These data indicate good WLR for h-CLAT [21]. These data are shown in Table 21.1.

In the EURL ECVAM pre-validation study, WLR was assessed using 15 coded chemicals. The results demonstrated an overall WLR of 80%. EURL ECVAM has published a recommendation regarding the h-CLAT method after the formal validation [22].

Table 21.1 Within-laboratory reproducibility

DNCB (4 μg/mL)				Ni (100 μg/mL)				SLS (54 μg/mL)			
No.	CD86 RFI	CD54 RFI	Viability (%)	No.	CD86 RFI	CD54 RFI	Viability (%)	No.	CD86 RFI	CD54 RFI	Viability (%)
1	255	342	88	1	195	1107	83	1	80	121	88
2	477	365	90	2	361	1555	85	2	66	129	96
3	268	210	92	3	217	978	83	3	59	100	86
4	502	358	90	4	225	1897	86	4	66	161	86
5	298	306	90	5	256	1629	85	5	50	91	88
6	358	309	87	6	248	1555	84	6	75	136	91
7	463	365	84	7	217	1169	84	7	62	91	80
8	306	249	88	8	252	1423	84	8	71	92	90
9	331	258	89	9	189	1052	85	9	57	136	79
10	264	496	89	10	199	903	81	10	70	114	93
11	246	476	83	11	213	895	80	11	85	92	96
12	269	525	89	12	202	785	84	12	62	122	93
13	341	506	88	13	195	951	81	13	66	71	96
14	224	341	85	14	187	1360	80	14	81	90	94
15	241	351	86	15	195	689	84	15	54	121	94
16	218	342	85	16	192	1030	84	16	50	80	91
17	217	400	87	17	179	872	88	17	54	121	94
18	217	319	87	18	158	591	86	18	36	119	85
19	228	399	87	19	153	484	85	19	57	69	94
20	272	284	92	20	191	1064	86	20	78	146	90
Mean ($N = 20$)	300	360	88	Mean	211	1099	84	Mean	64	110	90
SD ($N = 20$)	88	87	2	SD	45	369	2	SD	12	25	5

RFI relative fluorescence intensity, *SD* standard deviation

Table 21.2 Between-laboratory reproducibility

Chemical	Potency	Markers	Lab A		Lab B		Lab C		Lab D		Lab E	
Methylchloroisothiazolinone/ methylisothiazolinone (act. 1.5%)	Extreme	CD86	-		+	+ 2/3	+	+ 2/3	+	+ 3/3	+	+ 2/3
		CD54				- 0/3		- 0/3		- 0/3		- 0/3
DNCB	Extreme	CD86	+	+ 3/3	+	+ 3/3	+	+ 3/3	+	+ 3/3	+	+ 3/3
		CD54		+ 3/3		+ 3/3		+ 3/3		+ 3/3		+ 3/3
1,4-Dihydroquinone	Strong	CD86	+	+ 3/3	+	+ 3/3	+	+ 3/3	+	+ 2/3	+	+ 3/3
		CD54		- 1/3		- 1/3		- 1/3		- 0/3		+ 3/3
1,4-Phenylenediamine	Strong	CD86	-		+	+ 3/3	+	+ 2/3	+	+ 3/3	+	+ 3/3
		CD54				+ 2/3		- 0/3		+ 2/3		- 1/3
Methyldibromo glutaronitrile	Strong	CD86	-		+	+ 3/3	+	+ 2/3	+	+ 2/3	+	+ 3/3
		CD54				- 1/3		+ 3/3		+ 2/3		+ 2/3
2-Mercaptobenzothiazole	Moderate	CD86	-		+	- 0/3	+	- 0/3	+	+ 3/3	+	- 0/3
		CD54				+ 3/3		+ 3/3		+ 3/3		+ 3/3
Cinnamic aldehyde	Moderate	CD86	-		-	- 1/3	+	+ 3/3	+	+ 2/3	+	+ 3/3
		CD54				- 0/3		+ 2/3		+ 3/3		+ 2/3
Tetramethylthiuram disulfide	Moderate	CD86	-		+	+ 3/3	+	+ 3/3	+	+ 3/3	+	+ 3/3
		CD54				+ 3/3		+ 3/3		+ 3/3		+ 3/3
Hexyl cinnamic aldehyde	Weak	CD86	-	- 1/3	-	- 0/3	-	- 1/3	-	- 1/3	-	- 0/3
		CD54		- 0/3		- 1/3		- 0/3		- 0/3		- 1/3
Eugenol	Weak	CD86	+	+ 2/3	+	- 1/3	+	+ 2/3	+	+ 3/3	+	- 1/3
		CD54		+ 2/3		+ 3/3		+ 2/3		+ 2/3		+ 3/3
Benzalkonium chloride	NS	CD86	+	+ 2/3	-	- 0/3	-	- 0/3	-	- 0/3	-	- 0/3
		CD54		- 0/3		- 0/3		- 0/3		- 0/3		- 0/3
Sodium lauryl sulfate	NS	CD86	-	- 0/3	-	- 0/3	-	- 0/3	-	- 0/3	-	- 0/3
		CD54		- 0/3		- 0/3		- 0/3		- 0/3		- 0/3
Vanillin	NS	CD86	-	- 0/3	-	- 0/3	-	- 0/3	-	- 1/3	-	- 0/3
		CD54		- 0/3		- 0/3		- 0/3		- 0/3		- 0/3
Glycerol	NS	CD86	-		-	- 0/3	-	- 0/3	-	- 0/3	-	- 0/3
		CD54				- 0/3		- 0/3		- 0/3		- 0/3
Salicylic acid	NS	CD86	-		+	- 0/3	+	- 0/3	+	- 0/3	+	- 1/3
		CD54				+ 3/3		+ 3/3		+ 3/3		+ 3/3

The chemicals highlighted in grey are allergens and the chemicals in white are non-allergens. For each chemical, a data box indicates the results for CD86 expression (right upper data); CD54 expression (right bottom data); and final result ('+' or '-', at left) based on a prediction model using the CD86 and CD54 data on the right hand side. For the final outcome, a '+' for either CD86 or CD54 would result in that experiment being designated a '+'. Also, any overall positive outcome is highlighted in grey. The figure in parentheses shows how many experiments of the three conducted are over the positive criteria. NS non-sensitizer

For assessment of between-laboratory reproducibility (BLR), ten sensitizers and five non-sensitizers were evaluated by five (or four) laboratories, at the doses individually determined by each laboratory, in the COLIPA (the current name of this organization is Cosmetics Europe) Ring Trials [21]. The results of the COLIPA Ring Trials study are shown in Table 21.2. One sensitizer, hexyl cinnamic aldehyde, was evaluated as 'false-negative' between all laboratories. Moreover, salicylic acid was evaluated as 'false positive' between all laboratories. There was some variability regarding the outcome of CD86 and CD54 between each laboratory, but the final result of h-CLAT was the same between laboratories except for benzalkonium chloride. Therefore, 14 of 15 chemicals were judged as the same result by five (or four) laboratories. A Japanese inter-laboratory study was also conducted by seven laboratories, using eight chemicals [20]. The h-CLAT method showed approx. 96% of predicted performance in a total of 56 tests (seven laboratories, eight chemicals);

there were two false negatives in the evaluation. Ethylene diamine and eugenol each showed a false negative result in one laboratory. In conclusion, these data indicate good BLR for the h-CLAT. In the EURL ECVAM pre-validation study, BLR was assessed using 24 coded chemicals. The results demonstrated an overall BLR of 79.2%. EURL ECVAM has published a recommendation regarding the h-CLAT method after the formal validation [22].

21.3.2 Predictive Capacity

In total, 100 chemicals that were known to have sensitization potential based on LLNA have been evaluated using h-CLAT [19]. To cover a range of relative allergenic potencies, test chemicals were selected according to the results of LLNA. A total of 72 sensitizers were evaluated: eight extreme, 16 strong, 25 moderate, and 23 weak allergens as classified by LLNA. Twenty-eight non-sensitizers were also evaluated. The accuracy between h-CLAT and LLNA was 84% (sensitivity: 88%; Specificity:75%; positive predictivity: 90%; negative predictivity: 70%). Thus, h-CLAT is expected be a useful method for predicting the skin sensitization potential of chemicals. These data are shown in Table 21.3. When compared with *in vivo* data, there are both some false negatives and false positives; nine chemicals (including Isoeugenol and Phthalic anhydride) were classified as 'false negative' and seven chemicals (including 1-Bromobutane and Diethylphthalate) were classified as 'false positive' in h-CLAT.

The h-CLAT method has also been used to evaluate a total of 143 chemicals with log Kow values above and below 3.5, which included 37 chemicals with relatively low water solubility [23]. The sensitivity and overall accuracy were 83% and 80%, respectively. These data suggested that h-CLAT could successfully detect sensitizers with log Kow values up to 3.5. When chemicals with log Kow values greater than 3.5 that were deemed positive by h-CLAT were included with the 112 chemicals that had log Kow values up to 3.5, the sensitivity and accuracy in terms of the resulting applicable 128 chemicals out of the 143 chemicals reached 95% and 88%, respectively.

21.3.3 Applications and Limitations

On the basis of the current available data, the h-CLAT method shows broad applicability. However, h-CLAT can produce some false negatives. Possible causes of false negatives are listed below [1, 20, 23].

1. Lack of solubility of the test chemical in the medium. Test chemicals with a log Kow of greater than 3.5 may still be tested at lower concentrations at which they are soluble. In such a case, a positive result could still be used to support the identification of the test chemical as a skin sensitizer but a negative result should not be considered.

Table 21.3 Accuracy between LLNA and h-CLAT

Chemical	LLNA		h-CLAT						
	Potency category	EC3 (%)	Result	CD86	CD54	EC150 (CD86)	EC200 (CD54)	CV75 (μg/mL)	Vehicle
Oxazolone	Extreme	0.003	P	+	–	2.71	–	166.6	DMSO
Diphenylcyclopropenone		0.003	P	–	+	–	3.92	6.0	DMSO
MCI/MI (act. 1.5%)		0.005	P	+	–	2.21	–	3.2	Saline
p-Benzoquinone		0.0099	P	+	+	2.68	2.25	4.3	DMSO
1-Benzoylacetone		0.04	P	–	+	–	39.6	92.8	DMSO
2,4-Dinitrochlorobenzene: DNCB		0.05	P	+	+	2.3	2.66	5.0	DMSO
4-Nitrobenzyl bromide		0.05	P	+	+	0.95	0.91	3.6	DMSO
Potassium dichromate		0.08	P	+	+	2.09	1.06	3.2	Saline
Glutaraldehyde (act. 50%)	Strong	0.1	P	+	+	2.78	2.7	5.3	Saline
1,4-Dihydroquinone		0.11	P	+	–	2.13	–	5.0	Saline
1,4-Phenylenediamine		0.16	P	+	–	2.09	–	36.7	Saline
Phthalic anhydride		0.16	N	–	–	–	–	>400#	DMSO
Maleic anhydride		0.16	P	–	+	–	298.4	658.0	DMSO
Benzyl bromide		0.2	P	+	+	3.2	2.86	7.5	DMSO
Benzoyl peroxide		0.30	N	–	–	–	–	41.0	DMSO
Lauryl gallate		0.30	P	+	+	2.02	2.91	8.2	DMSO
Propyl gallate		0.32	P	–	+	–	32.5	125.0	DMSO
Cobalt chloride		0.38	P	–	+	–	35.5	208.3	Saline
2-aminophenol		0.4	P	+	–	0.89	–	6.0	DMSO
Chloramine T		0.4	P	+	+	259.2	291	314.7	Saline
2-Nitro-1,4-phenylenediamine		0.5	P	–	+	–	276.7	490.7	DMSO
Formaldehyde (act. 37%)		0.61	P	+	+	4.3	5.14	5.8	Saline
Iodopropynylbutylcarbamate		0.87	P	–	+	–	8.15	12.8	DMSO
Methyldibromoglutaronitrile		0.9	P	+	+	10.8	9.42	9.9	DMSO

(continued)

Table 21.3 (continued)

Chemical	LLNA Potency category	EC3 (%)	h-CLAT Result	CD86	CD54	EC150 (CD86)	EC200 (CD54)	CV75 (μg/mL)	Vehicle
Isoeugenol	Moderate	1.2	N	–	–	–	–	112.5	DMSO
1-Naphthol		1.3	P	+	+	18.3	12.7	57.2	DMSO
Glyoxal (act. 40%)		1.4	P	–	+	–	286.7	396.0	Saline
2-Hydroxyethyl acrylate		1.4	P	+	+	10.8	19.49	24.4	Saline
2-Mercaptobenzothiazole		1.7	P	–	+	–	57.5	169.0	DMSO
Methylisothiazolinone (act.9.7%)		1.9	P	+	+	9.23	7.89	24.7	Saline
3-Dimethylaminopropylamine		2.2	P	+	+	131	165.8	276.7	Saline
Ethylene diamine		2.2	P	+	–	265.7	–	271.7	Saline
l,2-Benzisothiazolin-3-one		2.3	P	–	+	–	0.54	1.83	DMSO
Methyl-2-nonynoate		2.5	P	+	–	91.8	–	191.7	DMSO
Cinnamic aldehyde		3.0	P	+	+	10.2	12.3	28.0	DMSO
Phenylacetaldehyde		3.0	P	+	+	11.5	13.4	27.0	DMSO
3-Aminophenol		3.2	P	–	+	–	117.1	243.2	DMSO
3-Propylidenephthalide		3.7	P	+	+	31.2	55.6	113.0	DMSO
Benzylideneacetone		3.7	P	+	+	25.8	34.6	35.3	DMSO
a-Methyl cinnamic aldehyde		4.5	P	+	+	40.3	70.16	139.0	DMSO
Nickel sulphate		4.8	P	+	+	42.2	45.3	150.0	Saline
Tetramethylthiuramdisulfide		5.2	P	+	+	9.11	2.46	10.0	DMSO
3,4-Dihydrocoumarin		5.6	P	–	+	–	531.7	810.0	DMSO
Resorcinol		5.7	P	–	+	–	200.9	613.4	Saline
Diethylenetriamine		5.8	N	–	–	–	–	1221.6	Saline
Diethyl maleate		5.8	P	+	–	64	–	120.0	DMSO
2-Methoxy-4-methyl-phenol		5.8	P	+	+	30.8	207.4	280.0	DMSO
4-Chloroaniline		6.5	P	+	–	87.7	–	200.0	DMSO
Trimellitic Anhydride		9.2	P	+	–	81.2	–	>250#	DMSO

1-Bromohexane	Weak	10	N	–	–	–	–	163.3	DMSO
Amyl cinnamic aldehyde		11	P	–	+	–	25.94	24.2	DMSO
Hexyl cinnamic aldehyde		11	N	–	–	–	–	37.0	DMSO
2,3-Butanedion		11	P	+	+	24.4	46.9	94.0	Saline
Citral		13	P	+	+	8.41	15	24.0	DMSO
Eugenol		13	P	+	+	64.4	137.2	143.2	DMSO
Abietic acid		15	N	–	–	–	–	89.6	DMSO
Oxalic acid		15	P	+	–	310.33	–	>1000	DMSO
4-Allylanisole		18	P	–	+	–	207.6	186.0	DMSO
Lillial		19	P	–	+	–	44.6	42.2	DMSO
Phenyl benzoate		20	P	+	+	166.2	236.4	>500#	DMSO
Cinnamic alcohol		21	P	+	+	147	196.6	350.0	DMSO
Cyclamen aldehyde		22	N	–	–	–	–	49.8	DMSO
Benzocaine		22	P	+	–	74.2	–	545.5	DMSO
Imidazolidinyl urea		24	P	+	+	42.3	39.3	40.1	Saline
Geraniol		26	P	+	–	123.3	–	140.0	DMSO
Ethyleneglycoldimethacrylate		28	P	–	+	–	535.43	563.6	DMSO
Linalool		30	P	–	+	–	68.3	290.0	DMSO
Penicillin G		30	P	–	+	–	3754.3	>5000	Saline
Butyl glycidyl ether		31	N	–	–	–	–	185.3	DMSO
Hydroxycitronellal		33	P	+	+	35.4	26.8	700.0	DMSO
Pyridine		72	P	+	–	2370	–	4166.7	Saline
Aniline		89	P	+	+	927.9	550.8	930.0	DMSO

(continued)

Table 21.3 (continued)

Chemical	LLNA		h-CLAT						
	Potency category	EC3 (%)	Result	CD86	CD54	EC150 (CD86)	EC200 (CD54)	CV75 (μg/mL)	Vehicle
Acetanisole	Non-sensitizer	NC	N	–	–	Not done		426.0	DMSO
Benzalkonium chloride			N	–	–			3.0	Saline
Benzoic acid			N	–	–			>1000	DMSO
1-Bromobutane			P	+	+			500.0	DMSO
1-Butanol			N	–	–			>1000	DMSO
Chlorobenzene			P	+	–			597.2	DMSO
Dextran			N	–	–			>5000	Saline
Diethyl phthalate			P	–	+			600.0	DMSO
Dimethyl formamide			N	–	–			>5000	Saline
Ethyl benzoylacetate			N	–	–			571.1	DMSO
Ethyl vanillin			N	–	–			569.5	DMSO
Glycerol			N	–	–			>5000	Saline
4-Hydroxybenzoic acid			N	–	–			>1000	DMSO
2-Hydroxypropyl methacrylate			N	–	–			>1000	DMSO
Isopropanol			N	–	–			>5000	Saline
Lactic acid			N	–	–			2800	Saline
6-Methyl coumarin			N	–	–			276.9	DMSO
Methyl salicylate			N	–	–			542.4	DMSO
Octanoic acid			P	–	+			359	DMSO
Propylene glycol			N	–	–			>1000	Saline
Propyl paraben			P	+	+			106.7	DMSO
Saccharin			N	–	–			>1000	DMSO
Salicylic acid			P	–	+			>1000	DMSO
Streptomycin sulphate			N	–	–			>1000	Saline
Tween-80			N	–	–			>5000	Saline
Vanillin			N	–	–			650.0	DMSO
Zinc sulphate			P	–	+			670.8	Saline
Sodium lauryl sulphate	False positive		N	–	–			60.0	Saline

P positive, *N* negative, *NC* not classified, *EC150* effective concentration 150, *EC200* effective concentration 200

2. Limited metabolic capability of the THP-1 cells. Pro-haptens (i.e., chemicals requiring enzymatic activation to exert their sensitization activity) and pre-haptens (i.e., chemicals activated by auto oxidation) may also provide negative results in h-CLAT.
3. Limited information is currently available regarding the applicability of this method for a mixture of chemicals. Therefore, when evaluating a mixture, it is important to consider this limitation of applicability.
4. Fluorescent test chemicals can be assessed with the h-CLAT, nevertheless, strong fluorescent test chemicals emitting at the same wavelength as FITC or PI will interfere with the flow cytometric detection and thus cannot be correctly evaluated using FITC-conjugated antibodies or PI.

It is therefore thought that h-CLAT can evaluate various chemicals except for those above-mentioned.

21.3.4 Comparison with Human Data

A total of 66 test chemicals with known human sensitizing potential were evaluated by h-CLAT to determine h-CLAT performance for predicting human sensitizing potential [24].Of the 51 tested sensitizers, 45 were positive in h-CLAT, indicating relatively high sensitivity. Also, 10 of 15 non-sensitizers were correctly detected as negative. The overall accuracy between the human data and h-CLAT outcome was 83%. Furthermore, h-CLAT could accurately predict the human sensitizing potential of 23 tested chemicals that were amines, heterocyclic compounds, or sulphur compounds. These data indicate the utility of h-CLAT for predicting the human skin sensitizing potential of a variety of chemicals.

21.4 Brief Description of the Protocol

In this page, only a brief description of the h-CLAT protocol is provided. If more detailed information is needed, please refer to the OECD TG 442E [1] and DB-ALM Protocol 158 [25].

21.4.1 Cells and Culture

THP-1 cells are purchased from a reliable cell bank (ATCC, Number: TIB-202). Cells are cultured in RPMI 1640 medium with 10% FBS (v/v), 0.05 mM 2-mercaptoethanol, and appropriate antibiotics (100 U/mL of penicillin and 100 µg/mL of streptomycin).

RPMI-1640: GIBCO, #22400-089 (containing 25 mM HEPES buffer and L-glutamine).

FBS: GIBCO, #10099-141.FBS is to be inactivated by heating to 56 °C for 30 min and tested by verifying the reactivity of THP-1 cells.

2-Mercaptoethanol: GIBCO, #21985-023.
Antibiotics (e.g., Penicillin-Streptomycin, GIBCO, #15140-122).

21.4.2 Preparation of Cells

For testing, THP-1 cells are seeded at a density of either 0.1 or 0.2 × 10^6 cells/mL, and are pre-cultured for 72 or 48 h in culture flasks, respectively. On the day of testing, cells harvested from culture flask are resuspended with fresh culture medium at a density of 2 × 10^6 cells/mL. Subsequently, cells are distributed with 500 μL media into a 24-well flat-bottom plate (1 × 10^6 cells/well) or with 80 μL media into a 96-well flat-bottom plate (1.6 × 10^5 cells/well).

21.4.3 Dose Finding Assay (Propidium Iodide (PI) Assay)

Test chemicals are dissolved in saline, medium, or dimethyl sulphoxide (DMSO, ≥99% purity) to a final concentration of 100 mg/mL or 500 mg/mL. Test chemicals that are not soluble in saline are dissolved in DMSO and diluted. Based on the 100 mg/mL (in saline) or 500 mg/mL (in DMSO) solutions of the test chemicals, two-fold serial dilutions are made using the corresponding solvent to obtain the stock solutions (eight doses) to be tested in the h-CLAT method. These stock solutions are then further diluted 50-fold (for saline) or 250-fold (for DMSO) into the culture medium (working solutions). These working solutions are finally used for treatment with a further two-fold dilution factor. The culture medium or working solutions are mixed 1:1 (v/v) with the cell suspensions prepared in the 24-well or 96-well flat-bottom plate. The treated plates are then incubated for 24 ± 0.5 h at 37 °C under 5% CO_2. After 24 ± 0.5 h of exposure, cells are transferred into sample tubes and collected by centrifugation. The supernatants are discarded and the remaining cells are resuspended with 600 μL of phosphate-buffered saline containing 0.1% bovine serum albumin (FACS buffer). The cell suspension (200 μL) is transferred into a 96-well round-bottom plate and washed twice with 200 μL of FACS buffer. Finally, the cells are resuspended in 200 μL of FACS buffer and 10 μL of PI solution is added (final concentration of PI is 0.625 μg/mL).

Cellular PI uptake is analysed using flow cytometry with the acquisition channel FL-3.A total of 10,000 living (PI negative) cells are acquired. Cell viability can be calculated by the cytometer analysis program using the following equation. When cell viability is low, up to 30,000 cells, including dead cells, should be acquired. Alternatively, the data acquisition can be finished 1 min after initiation.

$$\text{Cell Viability} = \frac{\text{Number of living cells}}{\text{Total Number of acquired cells}} \times 100$$

The CV75 value, i.e., the concentration at which 75% of the THP-1 cells survive (25% cytotoxicity), is calculated by log-linear interpolation using the following equation:

$$\log \mathrm{CV75} = \frac{(75-C)\times \log(B)-(75-A)\times \log(D)}{A-C}$$

where

A is the minimum value of cell viability over 75%.

C is the maximum value of cell viability below 75%.

B and *D* are the concentrations showing the value of cell viability *A* and *C*, respectively. The CV75 value is used to determine the concentration of test chemicals in CD86/CD54 expression measurement described below.

21.4.4 CD86/CD54 Expression Measurement

The appropriate solvent (basically, saline, or DMSO) is used to dissolve the test chemicals. The test chemicals are first diluted to the concentration corresponding to 100-fold (for saline) or 500-fold (for DMSO) of the 1.2 × CV75 value determined in the dose finding assay. If the CV75 is not determined (i.e., if sufficient cytotoxicity is not observed in the dose finding assay), the highest soluble concentration of test chemical prepared with each solvent should be used as the starting dose. Then, 1.2-fold serial dilutions are made using the corresponding solvent to obtain the stock solutions (eight doses ranging from 0.335 × CV75 to 1.2 × CV75) to be tested in the h-CLAT method. The stock solutions are then further diluted 50-fold (for saline) or 250-fold (for DMSO) into the culture medium (working solutions).These working solutions are finally used for treatment with a further two-fold dilution factor. Test chemicals and control substances prepared as working solutions are mixed with suspended cells at a 1:1 ratio, and the cells are incubated for 24 h. Alternative concentrations may be used upon justification (e.g., in case of poor solubility or cytotoxicity).

DNCB should be tested as the positive control in each assay, at a final concentration of 4.0 μg/mL, and should yield approximately 70–90% cell viability. Alternatively, the CV75 of DNCB, which is determined in each test facility, could be also used as the positive control dose. For each test chemical and control substance, one experiment is needed to derive a prediction. Each experiment consists of at least two independent runs (n = 2). After 24 h of exposure, cells are transferred into sample tubes and collected by centrifugation. The supernatants are discarded and the remaining cells are resuspended with 600 μL of FACS buffer. The cells are split into three aliquots of 180 μL into a 96-well round-bottom plate. After centrifugation, the cells are resuspended in 200 μL of blocking solution (FACS buffer containing 0.01% (w/v) globulin) and are incubated at 4 °C for 15 minutes. After centrifugation, the cells are stained with 50 μL of FITC-labelled anti-CD86, anti-CD54 or mouse IgG1 (isotype) antibodies at 4 °C for 30 min. The antibodies should be diluted in FACS buffer using a diluting of 3:25 (v/v, for CD86) or 3:50 (v/v, for CD54 and IgG1). After three washes with 200 μL of FACS buffer, the cells are resuspended in 200 μL of FACS buffer and 10 μL of PI solution is added (the final concentration of PI is 0.625 μg/mL). The expression levels of CD86 and CD54, and cell viability, are analysed using flow cytometry.

21.4.5 Data and Reporting

The expression of CD86 and CD54 is analysed with flow cytometry with the acquisition channel FL-1. Based on the geometric mean fluorescence intensity (MFI), the relative fluorescence intensity (RFI) of CD86 and CD54 for positive control cells and chemical-treated cells are calculated according to the following equation:

$$\mathrm{RFI} = \frac{\text{MFI of chemical treated cells} - \text{MFI of chemical treated isotype control cells}}{\text{MFI of solvent treated control cells} - \text{MFI of solvent treated isotype control cells}}$$

The cell viability of the isotype control cells is also calculated according to the equation described in the dose finding study (PI assay). Each chemical is tested in at least two independent runs to derive a single prediction (positive or negative). Each independent run is performed on a different day or on the same day provided that for each run: (a) independent fresh stock solutions and working solutions of the test chemicals and antibody solutions are prepared and (b) independently harvested cells are used (i.e., cells are collected from different culture flasks); however, cells may be derived from the same passage. If the RFI of CD86 is equal to or greater than 150% at any tested dose (≥50% of cell viability) in at least two independent runs and/or if the RFI of CD54 is equal to or greater than 200% at any tested dose (≥50% of cell viability) in at least two independent runs, the prediction is considered as positive. Otherwise, it is considered as negative. In case the first two independent runs are not concordant, a third run needs to be performed and the final prediction will be based on the mode of the conclusions from the three individual runs (i.e., 2 out of 3). Test chemicals with a log Kow of up to 3.5 have been successfully tested by this test method [23]. Test chemicals with a Log Kow of greater than 3.5 may still be tested at lower soluble concentrations. In such cases, a negative result should not be considered, whereas a positive result could still be used to support the identification of the test chemical as a skin sensitizer.

For the test chemicals considered to be sensitizers, two effective concentrations (EC) values, the EC150 for CD86 and EC200 for CD54, i.e. the concentration at which the test chemicals induced a RFI of 150 or 200, can be calculated by the following equations:

$$\mathrm{EC200}\ (\text{for CD54}) = B_{\text{dose}} + \left[\left(200 - B_{\text{RFI}}\right) / \left(A_{\text{RFI}} - B_{\text{RFI}}\right) \times \left(A_{\text{dose}} - B_{\text{dose}}\right)\right]$$

$$\mathrm{EC150}\ (\text{for CD86}) = B_{\text{dose}} + \left[\left(150 - B_{\text{RFI}}\right) / \left(A_{\text{RFI}} - B_{\text{RFI}}\right) \times \left(A_{\text{dose}} - B_{\text{dose}}\right)\right]$$

where

A_{dose} is the lowest concentration in μg/mL with RFI > 150 (CD86) or 200 (CD54).

B_{dose} is the highest concentration in μg/mL with RFI < 150 (CD86) or 200 (CD54).

A_{RFI} is the RFI at the lowest concentration with RFI > 150 (CD86) or 200 (CD54).

B_{RFI} is the RFI at the highest concentration with RFI < 150 (CD86) or 200 (CD54).

For the purpose of more precisely deriving the EC150 and EC200 values, three independent runs should be performed. The EC150 and EC200 values are the median value calculated from three independent runs. When only two of three independent runs meet the positive criteria, the higher EC150 or EC200 value is adopted. Whereas it is not always possible to derive the EC150 and/or EC200 value for positive chemicals, the value could potentially contribute to the assessment of sensitizing potency.

The following acceptance criteria should be met when using the h-CLAT method.

- The cell viabilities of medium and solvent control are more than 90%.
- For the positive control (DNCB), the RFI values of both CD86 and CD54 are over the positive criteria (CD86: RFI ≥ 150 and CD54: RFI ≥ 200) and cell viability is more than 50%.
- For the solvent control (DMSO), the RFI values of both CD86 and CD54 should not exceed the positive criteria (CD86: RFI ≥ 150 and CD54: RFI ≥ 200).
- For both medium and DMSO controls, the MFI ratio of both CD86 and CD54 to isotype control should be >105%.
- The cell viability of tested chemicals at more than four tested doses in each run should be ≥50%.

Negative results are acceptable only for test chemicals exhibiting cell viability at 1.2 × CV75of less than 90%. Negative results with cell viability of 90% or higher are discarded. The dose finding study should be redone to determine the CV75 determination. Positive results for test chemicals of any cell viability at 1.2 × CV75 are acceptable. It should be noted that when 5000 μg/mL in saline, 1000 μg/mL in DMSO, or the highest soluble concentration is used as the maximal test concentration of a test chemical, the results are acceptable.

21.5 Role in Testing Strategy

There is general agreement regarding the key biological events underlying skin sensitization. The current knowledge of the chemical and biological mechanisms associated with skin sensitization has been summarized in the form of an Adverse Outcome Pathway (AOP). The first key event is the covalent binding of electrophilic substances to nucleophilic centres in skin proteins. The second key event in this AOP takes place in keratinocytes and includes inflammatory responses as well as gene expression associated with specific cell signalling pathways such as the antioxidant/electrophile response element (ARE)-dependent pathways. The third key event is the activation of DC, typically assessed by expression of specific cell surface markers, chemokines, and cytokines. The fourth key event is T-cell proliferation, which is indirectly assessed in the murine LLNA. The h-CLAT method is proposed to address the third key event (dendritic cell activation) of the skin sensitization AOP by quantifying changes in the expression of cell surface markers associated with the process of DC activation (i.e., CD86 and CD54), in the human leukaemia cell line THP-1, following exposure to sensitizers. The measured expression levels of CD86 and CD54 cell surface markers are then used to support discrimination between skin sensitizers and non-sensitizers.

The current consensus among the scientific community is that one single non-animal test will not be sufficient as a stand-alone method to cover the endpoint of skin sensitization but that use of an integrated testing strategy (ITS) will be necessary. DC activation has a major role in the skin sensitization process. However, since DC activation represents only one key event in the skin sensitization AOP, information generated with test methods that measure markers of DC activation may not be sufficient on its own to conclude regarding the absence of the skin sensitization potential of chemicals. Therefore, data generated with the h-CLAT method should be considered in the context of integrated approaches such as IATA and should be combined with other complementary information, such as that derived from *in vitro* assays, that address other key events of the skin sensitization AOP as well as that from non-testing methods, including read-across from chemical analogues. Examples of the use of h-CLAT data in combination with other information have been reported in the literature [26–30].

As described, the h-CLAT method supports the discrimination of skin sensitizers from non-sensitizers when used in integrated approaches such as IATA. In the context of the IATA, the Direct Peptide Reactivity Assay(DPRA) evaluates the protein/peptide reactivity of a substance, the KeratinoSens™ and LuSens assays evaluate keratinocyte activation, and the h-CLAT, Myeloid U937 Skin Sensitization Test (MUSST) and modified MUSST (mMUSST) methods evaluate dendritic cell activation. Thus, together they cover the first three of the four key events of the sensitization process, supporting the scientific rationale for using a combination of these methods in an ITS. The '2-out-of-3' WoE approach was proposed for the ITS [29]. This '2-out-of-3' WoE approach provides slightly higher accuracies compared to the predictivities of the single assays. A tiered strategy was designed based on the complementary characteristics of the included methods and was compared to a majority voting approach [30]. This tiered testing strategy was able to correctly identify all 41 chemicals tested. In terms of the total number of experiments required, the tiered testing strategy requires less experiments compared to the majority voting approach.

The h-CLAT method may also potentially contribute to the assessment of sensitizing potency. Two EC values, the EC150 for CD86 and the EC200 for CD54, can be calculated from the h-CLAT data. It was reported that a statistically significant correlation was observed between the EC150 and the EC200 with the EC3 of LLNA [28]. These results indicated that EC150 and EC200 values can be used as indicators for estimation of the EC3 in LLNA. NICEATM (NTP Interagency Center for the Evaluation of Alternative Toxicological Methods) is working with other NTP (National Toxicology Progrum) scientists and industry experts to create an ITS to combine information from multiple testing methods for identification of potential skin sensitizers [26]. The sequential or tiered test batteries that require the results of the h-CLAT, DPRA and Derek were evaluated for predicting sensitizing potential and potency of chemicals in a predetermined way [27]. The testing strategy uses a Bayesian network to analyse data from non-animal tests and other information about a test substance, such as chemical structure and solubility, to identify potential skin sensitizers [31]. Another study shows that an artificial neural network analysis of data from multiple *in vitro* assays is a useful approach for prediction of the skin sensitization potency of chemicals [28].

21.6 Perspectives from the Test Developer

21.6.1 Critical Steps in the Protocol

Cell maintenance is the most important point of the method. THP-1 cells should be maintained at densities ranging from 0.1×10^6 to 0.8×10^6 cells/mL. Cells are routinely passaged every 2–3 days at a density of $0.1–0.2 \times 10^6$ cells/mL. Do not allow the cell density to exceed 1×10^6 cells/mL. In case of overgrowth, the reactivity of the THP-1 cells to sensitizers dramatically worsens. Thus, overgrowth conditions for THP-1 should be avoided during pre-culture in order to maintain the response of THP-1 cells to sensitizers and to more clearly distinguish sensitizers and non-sensitizers [32]. In addition, in order to obtain reliable results, it is essential that properly growing cell cultures are used. Therefore, facilities should accumulate their own historical data of the doubling time and set the acceptance range according to these data.

In all of our previous studies, we have used THP-1 cell lots obtained from the American-Type Culture Collection (ATCC). THP-1 cells are also commonly available in other cell banks. Even though THP-1 cells have been established as a cell line, the cells do not always have exactly the same properties. We compared three newly obtained THP-1 cell lots from American, European, and Japanese cell banks, against our reference THP-1 cells from the ATCC [33]. One THP-1 cell lot purchased from a Japanese cell bank did not induce CD86/CD54 augmentation following Ni treatment. The cell bank disclosed to us that the newly obtained THP-1 cell lot is different from the ATCC lot in terms of DNA pattern sequence. This information indicated that some unexpected events (e.g., contamination of other cells, mix-up of cells) occurred. Therefore, we strongly suggest that THP-1 cells are obtained from the ATCC.

In order to use appropriate cells, a reactivity check should be performed every time before method. Only THP-1 cells which pass this reactivity check can be used. The reactivity check should be performed according to the SOP (Standard Operating Procedure) using both positive controls (2, 4-dinitrochlorobenzene and nickel sulphate) and a negative control (lactic acid). There are some critical points regarding flow cytometer use. The basic calibration of the flow cytometer should be performed with appropriate calibration beads following the manufacturer's instructions. The flow cytometer must be set before testing. Attention should be paid to the maintenance of the cytometer in accordance with the manufacturer's instructions. In particular, the process of washing should be conducted very carefully because insoluble chemicals could flow into the flow line.

21.6.2 Possible Protocol Adaptations

It may be possible to use other fluorochromes such as APC, PE, and PE-Cy5 instead of FITC, if the overall comparability can be ensured. It would also be possible to use a serum-free medium if the compatibility can be proved. However, the test developer would like to point out that these attempts will be difficult.

Regarding endpoints, some indicators could be added to the h-CLAT method. IL-8 production was predominantly induced in THP-1 cells following allergen stimulation [34]. In addition, changes in the cell-surface thiols of THP-1 cells may be useful indicators for an *in vitro* sensitization assay [35].

21.6.3 Challenges and Opportunities

Several projects using h-CLAT such as Cosmetics Europe, The Research Institute for Fragrance Materials (RIFM), and The Interagency Coordinating Committee on the Validation of Alternative Methods (ICCVAM) will enable the conduction of safety assessment of chemicals regarding sensitization potential without the use of animals in the near future. OECD adaptation will promote the prevalence of this method. At this time, there are several contract research organizations that can conduct the h-CLAT method in Europe, Japan, and the USA. Anyone who wants to try to do the h-CLAT can go to the ECVAM database site [25].

Some new assays that have tried to expand the applicability domain of h-CLAT have been reported. An *in vitro* photosensitization assay was developed based on h-CLAT [36]. Many potential sensitizers are not directly immunogenic but require activation outside or inside the skin by non-enzymatic oxidation (pre-haptens) or by metabolic transformation (pro-haptens) prior to being able to induce an immune response. This necessary activation step has not yet been actively integrated into the h-CLAT method. However, co-culture of THP-1 cells with HaCaT keratinocytes, which allows cross talk between HaCaT and THP-1 cells, appears to be suitable for the detection of pro-haptens and has been shown to be useful for the prediction of sensitization potential of pro-haptens [37]. The h-CLAT method will be able to contribute to the evaluation of other toxicological endpoints in addition to skin sensitization. For example, biocompatibility biomaterials and dental restorative materials could be evaluated by using h-CLAT [38].

21.7 Conclusions

The h-CLAT method is based on the activation process of APCs (augmentation of CD86 and/or CD54 expression in THP-1 cells). The activation process by which DCs change from antigen processing to antigen presenting cells is considered to be a key event in the acquisition of skin sensitization. Therefore, the endpoint of the h-CLAT method is an important mechanism in the induction phase and is a necessary part of a series of skin sensitization processes. Many papers related to the h-CLAT method have been published, and these publications have shown the usefulness of this method for prediction of the skin sensitization potential of targeted chemicals. The formal validation study, conducted by EURL ECVAM, demonstrated that the h-CLAT test method is transferable to laboratories experienced in cell culture techniques and flow cytometry analysis and that the within-laboratory and between-laboratory reproducibility were both in the order of 80%. Moreover, the OECD has recently adopted a new TG No. 442E for the h-CLAT method [1]. Since h-CLAT will not be sufficient as a stand-alone method to cover the endpoint of skin sensitization, data generated with the h-CLAT method should be considered in the context of integrated approaches such as IATA and should be combined with other complementary information. Taking into consideration, the concentration-response information generated by the method, it is plausible that h-CLAT may potentially contribute within an IATA to the characterization of skin sensitization potency. Some approaches, to expanding the applicability domain of h-CLAT, including combination with other alternative methods, are being investigated. Test developers believe that the h-CLAT method will play an important role in *in vitro* skin sensitization testing.

References

1. OECD. OECD Guideline for the testing of chemicals, for a new test guideline, Test No. 442E: in vitro skin sensitization: human Cell Line Activation Test (h-CLAT). 2016. http://www.oecd-ilibrary.org/environment/test-no-442e-in-vitro-skin-sensitisation_9789264264359-en;jsessionid=2711x8ivdmkft.x-oecd-live-03. Accessed 29July 2016.
2. Basketter DA, Casati S, Gerberick GF, Griem P, Philips B, Worth A. Skin sensitisation. Altern Lab Anim. 2005;33:83–103.
3. Banchereau J, Steinman RM. Dendritic cells and the control of immunity. Nature. 1998;392:245–52.
4. Hart DNJ. Dendritic cells: unique leukocyte population which control the primary immune response. Blood. 1997;90:3245–87.
5. Aiba S, Katz SI. Phenotypic and functional characteristics of in vivo-activated Langerhans cells. J Immunol. 1990;145:2791–6.
6. Ozawa H, Nakagawa S, Tagami H, Aiba S. Interleukin-1 beta and granulocyte macrophage colony-stimulating factor mediate Langerhans cell maturation differently. J Invest Dermatol. 1996;106:441–5.
7. Aiba S, Terunuma A, Manome H, Tagami H. Dendritic cells differently responded to haptens and irritants by their production of cytokines and expression of co-stimulatory molecules. Eur J Immunol. 1997;27:3031–8.
8. Rougier N, Redziniak G, Mougin D, Schmitt D, Vincent C. In vitro evaluation of the sensitizaiton potential of weak contact allergens using Langerhans-like dendritic cells and autologous T cells. Toxicology. 2000;145:73–82.
9. Tsuchiya S, Yamabe M, Yamaguchi Y, Kobayashi Y, Konno T, Tada K. Establishment and characterization of a human acute monocytic leukemia cell line (THP-1). Int J Cancer. 1980;26:171–6.
10. Tsuchiya S, Kobayashi Y, Goto Y, Okumura H, Nakae S, Konno T, Tada K. Induction of maturation in cultured human monocytic leukemia cells by a phorbol diester. Cancer Res. 1982;42: 1530–6.
11. Ade N, Leon F, Pallardy M, Peiffer JL, Kerdine-Romer S, Tissier MH, Bonnet PA, Fabre I, Ourlin JC. HMOX1 and NQO1 genes are up-regulated in response to contact sensitizers in dendritic cells and THP-1 cell line: role of the Keap1/Nrf2 pathway. Toxicol Sci. 2009;107:451–60.

12. Ashikaga T, Hoya M, Itagaki H, Katumura Y, Aiba S. Evaluation of CD86 expression and MHC class II molecule internalization in THP-1 human monocyte cells as predictive endpoints for contact sensitizers. Toxicol In Vitro. 2002;16:711–6.
13. Tietze C, Blömeke B. Sensitization assays: monocyte-derived dendritic cells versus a monocytic cell line (THP-1). J Toxicol Environ Health A. 2008;71:965–8.
14. Yoshida Y, Sakaguchi H, Ito Y, Okuda M, Suzuki H. Evaluation of the skin sensitization potential of chemicals using expression of co-stimulatory molecules, CD54 and CD86, on the naïve THP-1 cell line. Toxicol In Vitro. 2003;17:221–2.
15. Ashikaga T, Yoshida Y, Hirota M, Yoneyama K, Itagaki H, Sakaguchi H, Miyazawa M, Ito Y, Suzuki H, Toyoda H. Development of an in vitro skin sensitization test using human cell lines; human Cell Line Activation Test (h-CLAT). I. Optimization of the h-CLAT protocol. Toxicol In Vitro. 2006;20:767–73.
16. Mitjans M, Viviani B, Lucchi L, Galli CL, Marinovich M, Corsini E. Role of p38 MAPK in the selective release of IL-8 induced by chemical allergen in naive THp-1 cells. Toxicol In Vitro. 2008;22:386–95.
17. Sakaguchi H, Ashikaga T, Miyazawa M, Yoshida Y, Ito Y, Yoneyama K, Hirota M, Itagaki H, Toyoda H, Suzuki H. Development of an in vitro skin sensitization test using human cell lines; human Cell Line Activation Test (h-CLAT). II. An inter-laboratory study of the h-CLAT. Toxicol In Vitro. 2006;20:774–84.
18. Miyazawa M, Ito Y, Yoshida Y, Sakaguchi H, Suzuki H. Phenotypic alterations and cytokine production inTHP-1 cells in response to allergens. Toxicol In Vitro. 2007;21:428–37.
19. Ashikaga T, Sakaguchi H, Sono S, Kosaka N, Ishikawa M, Nukada Y, Miyazawa M, Ito Y, Nishiyama N, Itagaki H. A comparative evaluation of in vitro skin sensitisation tests: the human cell-lineactivation test (h-CLAT) versus the local lymph node assay (LLNA). Altern Lab Anim. 2010;38:275–84.
20. Ashikaga T, Sakaguchi H, Okamoto K, Mizuno M, Sato J, Yamada T, Yoshida M, Ota N, Hasegawa S, Kodama T, Okamoto Y, Kuwahara H, Kosaka N, Sono S, Ohno Y. Assessment of the human Cell Line Activation Test (h-CLAT) for skin sensitization; results of the first Japanese inter-laboratory study. Altern Anim Test Exp. 2008;13:27–35.
21. Sakaguchi H, Ryan C, Ovigne JM, Schroeder KR, Ashikaga T. Predicting skin sensitization potential and inter-laboratory reproducibility of a human Cell Line Activation Test (h-CLAT) in the European Cosmetics Association (COLIPA) ring trials. Toxicol In Vitro. 2010;24:1810–20.
22. EC EURL-ECVAM. Recommendation on the human Cell Line Activation Test (h-CLAT) for skin sensitisation testing. 2013. Accessible at: https://eurl-ecvam.jrc.ec.europa.eu/eurl-ecvam-recommendations.
23. Takenouchi O, Miyazawa M, Saito K, Ashikaga T, Sakaguchi H. Predictive performance of the human Cell Line Activation Test (h-CLAT) for lipophilic chemicals with high octanol-water partition coefficients. J Toxcol Sci. 2013;38:599–609.
24. Nukada Y, Ashikaga T, Sakaguchi H, Sono S, Mugita N, Hirota M, Miyazawa M, Ito Y, Sasa H, Nishiyama N. Predictive performance for human skin sensitizing potential of the Human Cell Line Activation Test (h-CLAT). Contact Dermatitis. 2011;65:343–53.
25. DB-ALM (IVITTOX). Protocol 158: human Cell Line Activation Test (h-CLAT). 2014. http://ecvam-dbalm.jrc.ec.europa.eu/. Accessed 6 Jan 2015.
26. National Toxicology Program. Development of integrated testing strategies. 2014. http://ntp.niehs.nih.gov/pubhealth/evalatm/integrated-testing-strategies/index.html. Accessed 7 Jan 2015.
27. Takenouchi O, Fukui S, Okamoto K, Kurotani S, Imai N, Fujishiro M, Kyotani D, Kato Y, Kasahara T, Fujita M, Toyoda A, Sekiya D, Watanabe S, Seto H, Hirota M, Ashikaga T, Miyazawa M. Test battery with the human cell line activation test, direct peptide reactivity assay and DEREK based on a 139 chemical data set for predicting skin sensitizing potential and potency of chemicals. J Appl Toxicol. 2014;35:1318–32.
28. Tsujita-Inoue K, Hirota M, Ashikaga T, Atobe T, Kouzuki H, Aiba S. Skin sensitization risk assessment model using artificial neural network analysis of data from multiple in vitro assays. Toxicol In Vitro. 2014;28:626–39.

29. Urbisch D, Mehling A, Guth K, Ramirez T, Honarvar N, Kolle S, Landsiedel R, Jaworska J, Kern PS, Gerberick F, Natsch A, Emter R, Ashikaga T, Miyazawa M, Sakaguchi H. Assessing skin sensitization hazard in mice and men using non-animal test methods. Regul Toxicol Pharmacol. 2014; doi:10.1016/j.yrtph.2014.12.008.
30. Van der Veen JW, Rorije E, Emter R, Natch A, van Loveren H, Ezendam J. Evaluating the performance of integrated approaches for hazard identification of skin sensitizing chemicals. Regul Toxicol Pharmacol. 2014;69:371–9.
31. Jaworska J, Natsch A, Ryan C, Strickland J, Ashikaga T, Miyazawa M, Sakaguchi H. Bayesian integrated testing strategy (ITS) for skin sensitization potency assessment: a decision support system for quantitative weight of evidence and adaptive testing strategy. Arch Toxicol. 2015;89:2355–83.
32. Mizuno M, Yoshida M, Kodama T, Kosaka N, Okamoto K, Sono S, Yamada T, Hasegawa S, Ashikaga T, Kuwahara H, Sakaguchi H, Sato J, Ota N, Okamoto Y, Ohno Y. Effects of preculture conditions on the human Cell Line Activation Test (h-CLAT) results: results of the 4th Japanese inter-laboratory study. Altern Anim Test Exp. 2008;13:70–82.
33. Kosaka M, Okamoto K, Mizuno M, Yamada T, Yoshida M, Kodama T, Sono S, Ashikaga T, Sato J, Ota N, Hasegawa S, Okamoto Y, Kuwahara H, Sakaguchi H, Ohno Y. A study of the criteria for selection of THP-1 cells in the human Cell Line Activation Test (h-CLAT): results of 2nd Japanese inter-laboratory study. Altern Anim Test Exp. 2008;13:55–62.
34. Nukada Y, Miyazawa M, Kosaka N, Ito Y, Sakaguchi H, Nishiyama N. Production of IL-8 in THP-1 cells following contact allergen stimulation via mitogen-activated protein kinase activation or tumor necrosis factor-alpha production. J Toxicol Sci. 2008;33:175–85.
35. Suzuki M, Hirota M, Hagino S, Itagaki H, Aiba S. Evaluation of changes of cell-surface thiols as a new biomarker for in vitro sensitization test. Toxicol In Vitro. 2009;23:687–96.
36. Hoya M, Hirota M, Suzuki M, Hagino S, Itagaki H, Aiba S. Development of an in vitro photosensitization assay using human monocyte-derived cells. Toxicol In Vitro. 2009;23:911–8.
37. Hennen J, Aeby P, Goebel C, Schettgen T, Oberli A, Kalmes M, Blömeke B. Cross talk between keratinocytes and dendritic cells: impact on the prediction of sensitization. Toxicol Sci. 2011;123:501–10.
38. Tamura A, Fukumoto I, Yui N, Matsumura M, Miura H. Increasing the repeating units of ethylene glycol-based dimethacrylates directed toward reduced oxidative stress and co-stimulatory factors expression in human monocytic cells. J Biomed Mater Res. 2014; doi:10.1002/jbm.a.35251.

U-SENS™: A U937 Cell Line Activation Test for Skin Sensitization

22

Nathalie Alépée, Cécile Piroird, and Laurent Nardelli

22.1 Principle of the Test Method and Scientific Basis

A skin sensitizer refers to a chemical that will lead to an allergic response following a skin contact as defined by the United Nations Globally Harmonized System of Classification and Labelling of Chemicals (UN GHS) [1]. Predictive assessment of the contact allergenic potential of chemicals is performed using guinea pigs, mice, animal-free approaches and human beings [2]. With the exception of the mouse local lymph node assay (LLNA), *in vivo* models for skin sensitization assess sensitization induction by monitoring clinical reactions—such as erythema, oedema and ear swelling—elicited by challenging the exposed individuals with the test chemical. While the available animal models determine global organ and organism responses to assess the sensitization potential of chemicals, animal-free approaches only address the key events describing the Mode of Action (MOA) pathway for skin sensitization as proposed by Adler et al. [3] and others [4–6]. These key mechanistic events underpinning the skin sensitization process which lead to allergic contact dermatitis (ACD) in humans have been formally described in an adverse outcome pathway (AOP) for skin sensitization by the OECD [7]. This AOP captures the impact of skin exposure and describes key events starting from the molecular initiating event, i.e. covalent binding of a chemical to skin proteins (protein haptenation), via intermediate cellular events like keratinocyte and dendritic cell (DC) activation, to the final determining result which is the induction of hapten-specific T cells. The latter, after being challenged by the chemical, will be the key effector cells in the clinical expression of skin sensitization (ACD).

N. Alépée (✉) • C. Piroird • L. Nardelli
L'Oréal Research and Innovation, 1, Avenue Eugène Schueller, 93600 Aulnay-sous-Bois Cedex, France
e-mail: nalepee@rd.loreal.com

C. Eskes et al. (eds.), *Alternatives for Dermal Toxicity Testing*,
DOI 10.1007/978-3-319-50353-0_22

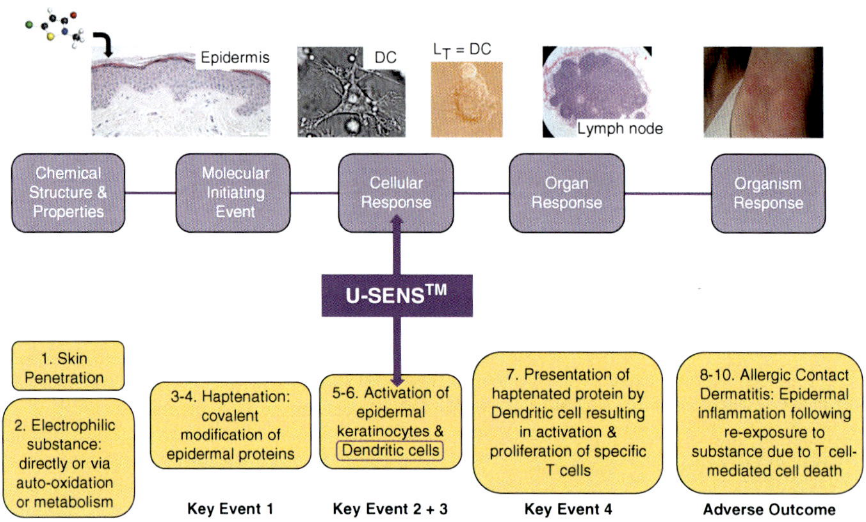

Fig. 22.1 The U-SENS™ assay among the key events in the Adverse Outcome Pathway for Skin Sensitization

Dendritic cell activation upon exposure to sensitizers leads to functional changes. For example, there are clear changes in cytokine secretion (e.g. TNF-α IL-1β) and in the expression of some chemokine receptors such as CCR7 and CXCR4 [8]. Additionally, during dendritic cell maturation, co-stimulatory and intercellular adhesion molecules such as HLA-DR, HLA-ABC, CD40, CD80, CD83, CD86 and ICAM-1/CD54 can be up-regulated [8, 9]. Most of the *in vitro* test methods measure the activation of the cell surface marker CD86, which has been established as mechanistically relevant and predictive [10, 11].

The U-SENS™ assay quantifies the induction of the CD86 protein marker expression, associated with DC maturation *in vivo*. The assay is performed on the human myeloid U937 cell line, closely related to monocytes and dendritic cells. The assay therefore addresses one of the biological mechanisms covered by key event 3 of the skin sensitization AOP [7, 12, 13] (see Fig. 22.1).

22.2 Current (pre)Validation Status

In order to allow for regulatory acceptance of the U-SENS™ assay, a validation study (combining two multicentric studies conducted in 2013 and 2014) including four laboratories and testing up to 38 chemicals designed to assess reliability was carried out according to internationally agreed principles [14]. As such, the design features for the validation exercise—such as the sample sizes for within-laboratory reproducibility (WLR) and between-laboratory reproducibility (BLR) assessments—were defined to allow for an assessment of the reliability and preliminary predictive capacity of the U-SENS™ assay. In terms of validation modules [15], the first five modules were addressed. The test method was defined in detail in a standard protocol (module 1). WLR (module 2) was successfully demonstrated. On average, over four laboratories, WLR was 93% (70/75) in terms of concordance of classifications. In an earlier study by Piroird et al. [10], the U-SENS™ test method was shown to be transferred to three laboratories (module 3) and was confirmed by the validation study [16]. For the chemicals tested three times, the more frequent (or median) prediction was considered for the between-laboratory (BLR, module 4) analysis. Overall 84.2% (32/38) chemicals were identically classified in all four laboratories independently of the studies. In addition to the validation study, the assessment of the predictive capacity (module 5) of U-SENS™ was defined as 90% sensitivity, 71% specificity and 86% accuracy based on the dataset of 175 chemicals.

The test method has been independently peer reviewed by the European Union Reference Laboratory for Alternatives to Animal Testing (EURL ECVAM) Scientific Advisory Committee (ESAC) (ESAC opinion No. 2016-03) [17] and been incorporated in OECD Test Guidelines 442E [18]. Therefore considering all available evidence and input from regulators and stakeholders, the U-SENS™ was recommended by EURL ECVAM to be used as part of an IATA to support the discrimination between sensitizers and non-sensitizers for the purpose of hazard classification and labelling [1]. Examples of the use of U-SENS™ data in combination with other information, including historical data are also reported elsewhere in the literature [13].

22.3 Performance and Applicability of the Test Method

22.3.1 Reproducibility

Two types of reproducibility were evaluated for the U-SENS™ test method: one obtained by testing the same chemicals over time in a single laboratory (within-laboratory reproducibility, WLR) and the other by testing the same chemicals in different laboratories (between-laboratory reproducibility, BLR). WLR was calculated as the percentage of chemicals for which 100% concordant classifications were obtained in the three runs performed. BLR was calculated as the percentage of chemicals for which 100% concordant classifications were obtained between laboratories. In total, 321 experiments, representing the complete data set, were analysed.

WLR was assessed in four participating laboratories (Bioassay, CiToxLab, L'Oréal and WIL Research). Each of them tested the same 15 chemicals in three independent experiments, for which three differently coded test samples were provided. The same concordant classification was observed for 54 out of 60 items (90.0%) for the four laboratories when considering positives versus negatives (Table 22.1). Note that the WLR of the lead laboratory in the first study was comparably high (20/21; 95.2%). WLR conducted in a more extensive exercise within L'Oréal demonstrated the same concordance for 34 out of 35 chemicals (97.1%)

Table 22.1 Final classifications (*N* Negative, *P* Positive) for all laboratories

CAS number	Name	Reference classification based on LLNA data	L'Oréal Classification			Bioassay Classification			CiToxLAB Classification			WIL Research Classification			L'Oréal automation Classification		
			E1	E2	E3	E1	E2	E3	E1	E2	E3	E1	E2	E3	E1	E2	E3
100-11-8	4-Nitrobenzyl bromide	S	P	P	P	P			P			P			P		
104-54-1	Cinnamic alcohol	S	P			P			P			P			P		
106-50-3	1,4-Phenylenediamine	S	P			P			P			P			P		
107-15-3	Ethylene diamine	S	P	P	P	P	P	P	P	P	P	P	P	P	P	P	P
108-45-2	m-Phenylenediamine	S	P	P	P	P			P			P			P		
108-46-3	Resorcinol	S	P	P	P	P	P	P	P	P	P	P	P	P	P	P	P
109-55-7	3-Dimethylaminopropylamine	S	P	P	P	P	P	P	P	P	P	P	P	P	P	P	P
1166-52-5	Lauryl gallate	S	P	P	P	P			P			P			P		
120-51-4	Benzyl benzoate	S	P	P	P	P			P			P			N		
137-26-8	Tetramethylthiuram disulfide	S	P	P	P	P			P			P			P		
140-67-0	4-Allylanisole	S	P	P	P	P			P			P			P		
141-05-9	Diethyl maleate	S	P	P	P	P			P			P			P		
2682-20-4	Methylisothiazolinone	S	P	P	P	P	P	P	P	P	P	P	P	P	P	P	P
3344-77-2	12-Bromo-1-dodecanol	S	P	P	P	P			P			P			P		
35,691-65-7	Methyldibromo glutaronitrile	S	P	P	P	P			P			P			P		
591-27-5	m-Aminophenol	S	P	P	P	P			P			P			P		
615-50-9	Toluene diamine sulphate	S	P	P	P	P	P	P	P	P	P	P	P	P	P	P	P
80-62-6	Methylmethacrylate	S	P	P	P	P	P	P	P	P	P	P	P	P	P	P	P
93-99-2	Phenyl benzoate	S	P	P	P	P			S			S			S		
100-51-6	Benzyl alcohol	NS	N	N	N	N	N	N	*N*	P	P	N	N	N	P	P	*N*
108-90-7	Chlorobenzene	NS	**N**	**N**	**P**	**N**			**P**			**N**			**N**		
109-65-9	1-Bromobutane	NS	N			N			N			N			N		
110-54-3	Hexane	NS	N	N	N	N	N	N	N	N	*P*	N	N	N	N	N	N
119-36-8	Methyl salicylate	NS	N			N			N			N			N		
121-33-5	Vanillin	NS	**P**			**N**			**P**			**N**			**P**		
121-57-3	Sulfanilic acid	NS	N			N			N			N			N		
1330-20-7	Xylene	NS[a]	N			N			N			N			N		

25,322-68-3	Polyethylene glycol	NS	**N**	**N**	**N**	**N**	**N**	**N**	*N*	**P**	**P**	**N**	**N**	**N**	N	N	N
3810-74-0	Streptomycin sulfate	NS	N	N	N	N	N	N	P	P	*N*	N	N	N	N	N	N
50-21-5	Lactic acid	NS	N	N	N	N	N	N	*P*	N	N	N	N	N	N	N	N
56-81-5	Glycerol	NS	N	N	N	N	N	N	N	N	N	N	N	N	N	N	N
65-85-0	Benzoic acid	NS	N	N	N	N	N	N	*P*	N	N	N	N	N	N	N	N
69-72-7	Salicylic acid	NS	N	N	N	N			N			N			N		
75-35-4	Vinylidene dichloride	NS	N	N	N	N	N	N	N	N	N	N	N	N	N	N	N
77-92-9	Citric acid	NS	N			N			N			N			N		
81-07-2	Saccharin	NS	N	N	N	N	N	N	N	N	N	N	N	N	N	N	N
92-48-8	6-Methylcoumarin	NS	**P**	**P**	**P**	**N**			**P**			**P**			**P**		
99-96-7	4-Hydroxybenzoic acid	NS	N			N			N			N			N		

NS non-sensitizer, *S* Sensitizer, *Italic* chemicals with discordant experiments within a laboratory, *Bold* chemicals with discordant experiments between laboratories, *E* experiment

[a]Classification based on human data (false positive in the LLNA compared to humans [19])

Table 22.2 Within- and between-laboratory reproducibility (WLR/BLR) of the U-SENS™ test method

	Study 2014		Ring trial 2013			
	nb Chem.	WLR 4 Lab	nb Chem.	WLR 1 lab	Overall	EC EURL ECVAM
WLR	15	90%	21	95.2%	91%	≥85%
BLR	24	83.3%	14	85.7%	84.2%	≥80%/3 labs

(Tables 22.1 and 22.2). When considering all evaluations, 91.3% of the items (74/81) showed 100% concordance of classifications between the three independent runs performed (Table 22.2).

To calculate BLR, the final classification for each chemical in each laboratory was obtained using the arithmetic median value (more frequent) over the three runs performed. Table 22.1 summarizes the classification for all laboratories. The proportion of chemicals concordantly classified between laboratories was 85.7 and 83.3% in 2013 and 2014, respectively. Overall, 84.2% (32/38) of chemicals were identically classified in all four laboratories (Table 22.2). Similar performances are obtained for 38 chemicals when combining the study results with those of an earlier multicentre study, as well as with an automated version of the U-SENS™ [16].

22.3.2 Predictive Capacity

The predictive capacity of the U-SENS™ test method was evaluated as part of a larger systematic evaluation of non-animal test methods [20], during the validation study merging the two multicentric studies conducted in 2013 [10] and 2014 [16], and after a larger evaluation of the test method involving 175 chemicals [10].

As shown in [20], the study was initiated and supervised by the skin tolerance task force of Cosmetics Europe with contributions from EURL ECVAM. Each test method was challenged with a common set of ten chemicals. Regarding the U-SENS™ test method, formerly known as "MUSST", an overall accuracy of 10/10 was obtained (100% sensitivity, 100% specificity) (Table 22.3).

During the validation study, the performance of the U-SENS™ test method was evaluated using 19 sensitizing and 19 non-sensitizing chemicals in four different laboratories. A sensitivity of 99%, specificity of 88% and accuracy of 94% were obtained, although assessing predictive capacity was not a primary objective of the study. Regarding specificity, a total of eight chemicals had at least one false positive classification (Table 22.3). Results generated in the validation study [16] and other published studies [10] overall indicate that, compared with LLNA results, the accuracy in distinguishing skin sensitizers (i.e. UN GHS Cat.1) from non-sensitizers is 86% ($N = 166$) with a sensitivity of 91%

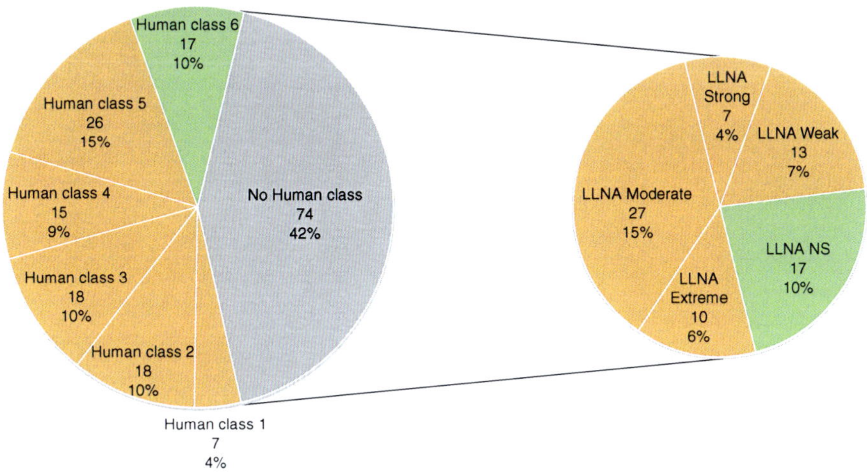

Fig. 22.2 Distribution of human and LLNA potency classes of 175 chemicals. The number below each class corresponds to the number of chemicals in the respective class. The percentage reflects the class representation within the overall data set of 175 chemicals

(118/129) and a specificity of 65% (24/37). False negative predictions compared to LLNA with the U-SENS™ are more likely to concern chemicals showing a weak to moderate skin sensitization potency (i.e. UN GHS subcategory 1B) than chemicals showing a strong skin sensitization potency (i.e. UN GHS subcategory 1A).

Moreover, in order to better define the predictivity of the U-SENS™ test method, a set of 175 chemicals was evaluated [10]. A primary eligibility criterion for the chemical set selection was the availability of robust *in vivo* data to allow a proper comparative evaluation of *in vitro* results. As such, availability of both human data (from Categorization of Chemicals According to Their Relative Human Skin Sensitizing Potency—[19]) and/or LLNA data (OECD Test Guideline 429) with *in vivo* skin sensitization classification were considered. However, human data was the main criterion considered for the prediction. Based on human, LLNA and EU-CLP classifications, all potency classes, from extreme sensitizer to non-sensitizer, were represented. The respective distributions are summarized in Fig. 22.2. The predictivity of the U-SENS™ test method, based on the global set of 175 chemicals, showed a high sensitivity of 90%, a specificity of 71% and an overall accuracy of 86% with a kappa value of 58% (i.e. good agreement despite the unbalanced dataset (81% sensitizers *in vivo* versus 19% non-sensitizers *in vivo*) (Table 22.3 and Fig. 22.3). This indicates the potential of U-SENS™ to contribute to the discrimination between sensitizers and non-sensitizers, although its use as a standalone method for this purpose is not recommended since DC activation addresses only one step in the skin sensitization pathway [10].

Table 22.3 Predictive capacity [10, 16, 20]

	Nb of test items	Laboratory	Specificity	Sensitivity	Accuracy
Cosmetics Europe set	10 (7 S + 3 NS)	L'Oréal	100%	100%	100%
Validation set	38 (19 S + 19 NS)	Bioassay	100%	95%	98%
		CiToxLAB	68%	100%	84%
		L'Oréal	89%	100%	95%
		WIL Research	95%	100%	98%
		Mean	*88%*	*99%*	*94%*
U-SENS database	175 (141 S + 34 NS) (NS includes human class 6 only)	L'Oréal	71%	90%	86%
	175 (115 S + 60 NS) (NS includes human classes 5 & 6)	L'Oréal	55%	96%	82%

S Sensitizer; *NS* Non-Sensitizer

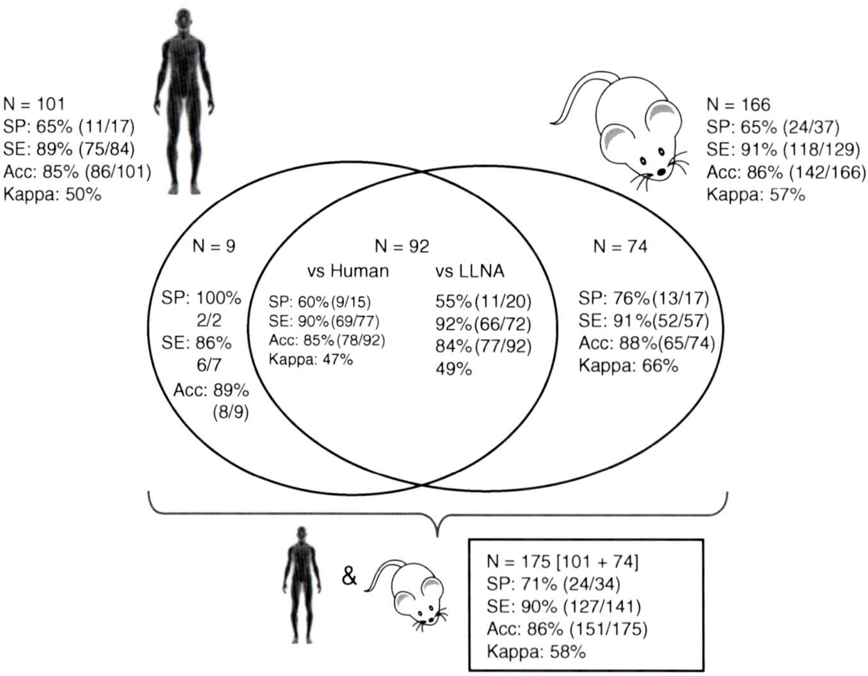

Fig. 22.3 Predictive performance of the U-SENS™ test method on 175 chemicals compared to human and/or *in vivo* LLNA data. Human categories were dichotomized by considering category 6 [19] as non-sensitizers and the categories 1–5 as sensitizers. N: number of chemicals. The values in parenthesis are the number on chemicals correctly predicted out of the evaluated chemicals. *SP* specificity, *SE* sensitivity, *Acc* Accuracy

22.3.3 Applications and Limitations

The U-SENS™ assay is applicable to all mono-chemicals or mixtures that are soluble in the aqueous testing conditions and compatible with flow cytometry analysis. It has been shown to be applicable to a broad range of chemicals covering relevant ranges of chemical classes (fragrances, dyes, preservatives, actives, UV filter, and non-cosmetics ingredients), reaction mechanisms, skin sensitization potency (as determined by *in vivo* studies) and physico-chemical properties. Based on the reaction mechanistic domains proposed by Aptula and Roberts [21], several protein reactivity classes (Michael acceptors, Schiff base formation, bi-molecular and aromatic nucleophilic substitutions and acyl transfer agents) were integrated into the Toxtree application (Fig. 22.4) [23].

Pre- or pro-haptens were therefore evaluated. The predictive capacity of the U-SENS™ assay was comparable regardless of the reactivity classes. Chemicals that have been reported to be pre- or pro-haptens were correctly predicted by the U-SENS™ assay (Table 22.4).

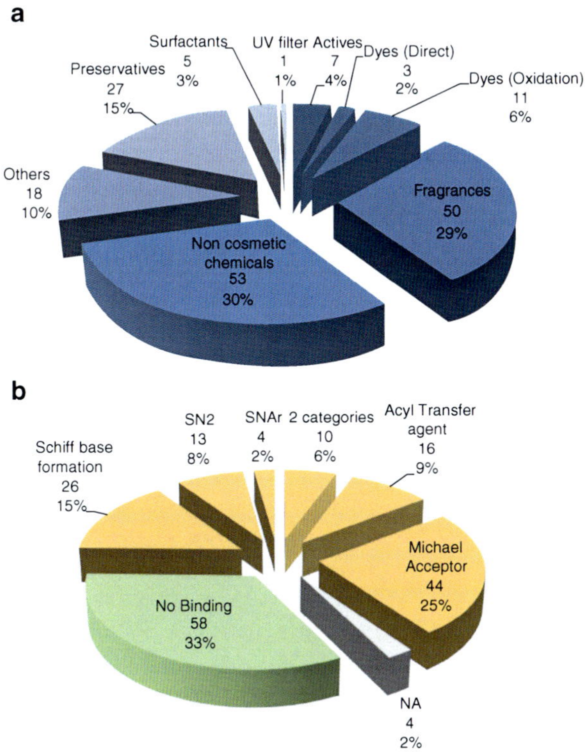

Fig. 22.4 Distribution of the evaluation set related to 175 chemicals grouped by (**a**) use categories and (**b**) the corresponding reactivity classes according to Toxtree v2.6.0 [22]. SN2 and SNAr correspond to the nucleophilic substitutions (bi-molecular and aromatic, respectively)

Table 22.4 Pre-haptens (chemicals activated by auto-oxidation) and Pro-haptens (chemicals requiring enzymatic activation) correctly predicted by the U-SENS™ test method

Chemical name	CAS number	Type	Human class	LLNA class	*In vivo* class	U-SENS™ prediction
1,4-Phenylenediamine	106-50-3	Pre-hapten	1	Strong	S	P
Isoeugenol	97-54-1		2	Moderate	S	P
1,4-Dihydroquinone	123-31-9		3	Strong	S	P
3-Methylcatechol	488-17-5		NA	Extreme	S	P
2-aminophenol	95-55-6	Pro-hapten	2	Strong	S	P
Eugenol	97-53-0		3	Weak	S	P
Cinnamic alcohol	104-54-1		3	Weak	S	P
Geraniol	106-24-1		4	Weak	S	P
Aniline	62-53-3		4	Weak	S	P
1-Naphthol	90-15-3		NA	Moderate	S	P
2-Methoxy-4-methyl-phenol	93-51-6		NA	Moderate	S	P

S Sensitizer, *P* Positive, *NA* Not applicable

Table 22.5 Lipophilic chemicals characterized by a high octanol-water partition correctly predicted by the U-SENS™ test method

Chemical name	CAS number	Log Kow (KOWWIN)	Human class	LLNA class	*In vivo* class	U-SENS™ prediction
Abietic acid	514-10-3	6.46	3	Weak	S	P
Hexyl cinnamic aldehyde	101-86-0	4.82	5	Moderate	S	P
Benzyl benzoate	120-51-4	3.54	5	Weak	S	P
Undec-10-enal	112-45-8	4.12	NA	Moderate	S	P
12-Bromo-1-dodecanol	3344-77-2	5.11	NA	Moderate	S	P
1-Bromohexane	111-25-1	3.63	NA	Weak	S	P
Benzyl cinnamate	103-41-3	4.06	NA	Weak	S	P
Cyclamen aldehyde	103-95-7	3.91	NA	Weak	S	P

S Sensitizer, *P* Positive *NA* Not applicable

Although questions have been raised regarding the pertinence of testing lipophilic chemicals with high octanol-water partition coefficients in a water based cell assay, suggesting criteria of exclusion from the applicability domain of another dendritic cell activation assay [23], the U-SENS™ assay also correctly predicted lipophilic tested chemicals (Table 22.5).

When focusing on the false negative and false positive chemicals, no specific reactivity class can be highlighted or excluded from the applicability domain, the misclassified chemicals being equally distributed among all classes. False negatives cannot be excluded when evaluating poorly water-soluble chemicals (e.g. some polymers). Colour interference of dyes (fluorescent dyes) and fluorescence used in flow cytometry may also potentially be a source of false negative results although the known *in vivo* sensitizers dyes being tested were correctly predicted

Table 22.6 Dyes chemicals correctly predicted by the U-SENS™ test method

Chemical name	CAS number	Colipa	Human class	LLNA class	*In vivo* class	U-SENS™ prediction
1,4-Phenylenediamine	106-50-3	A007	1	Strong	S	P
2-Nitro-1,4-phenylenediamine	5307-14-2		2	Strong	S	P
Toluene diamine sulphate	615-50-9		2	Strong	S	P
2-Aminophenol	95-55-6	A014	2	Strong	S	P
Metol	55-55-0	A022	3	Strong	S	P
Resorcinol	108-46-3	A011	4	Moderate	S	P
Aniline	62-53-3		4	Weak	S	P
p-Benzoquinone	106-51-4		NA	Extreme	S	P
Bandrowski's base	20048-27-5		NA	Extreme	S	P
p-Aminophenol	123-30-8	A016	NA	Strong	S	P
3-Phenylenediamine	108-45-2	A003	NA	Strong	S	P
1-Naphthol	90-15-3	A017	NA	Moderate	S	P
N,N-bis(2-hydroxyethyl)-p-phenylenediamine sulfate	54381-16-7	A050	NA	Moderate	S	P
3-Aminophenol	591-27-5	A015	NA	Moderate	S	P

S Sensitizer, *P* Positive *NA* Not applicable

in U-SENS™ test method (Table 22.6). Nevertheless, strong fluorescent test chemicals emitting at the same wavelength as fluorescein isothiocyanate (FITC) or as propidium iodide (PI), might interfere with the flow cytometric detection and thus cannot be correctly evaluated using FITC-conjugated antibodies (potential false negative) or PI (viability not measurable). In such a case, other fluorochrome-tagged antibodies or other cytotoxicity markers, respectively, can be used as long as it can be shown they provide similar results as the FITC-tagged antibodies or PI, e.g. by testing the proficiency chemicals identified in the OECD TG 442E [18].

Membrane disrupting chemicals can lead to false positive results due to a non-specific increase of CD86 expression, as 3 out of 7 false positives relative to the *in vivo* reference classification were surfactants [10]. As such positive results with surfactants should be considered with caution whereas negative results with surfactants could still be used to support the identification of the test chemical as a non-sensitizer.

In the light of the above, positive results with surfactants and negative results with strong fluorescent test chemicals should be interpreted in the context of the stated limitations and together with other information sources within the framework of Integrated Approaches to Testing and Assessment (IATA).

Limited information is currently available on the applicability of the U-SENS™ method to multi-constituent substances/mixtures [10]. The test method is nevertheless technically applicable to the testing of multi-constituent substances and mixtures. However, before use of this test method on a mixture for generating data for an intended regulatory purpose, it should be considered whether, and if so why, it may provide adequate results for that purpose. Moreover, when testing multi-constituent substances or mixtures, consideration should be given to possible interference of cytotoxic constituents with the observed responses. Finally, results should be interpreted with care for chemicals that may interfere with CD86 induction pathways due to their own biological activity (some vegetal extracts).

22.3.4 Comparison to Human Data

Ideally, the real gold standard in establishing an *in vitro* sensitization predictive test should be human data, but analysis of such data requires a considerable degree of expert judgment since, if Human Repeated Insult Patch Test (HRIPT) reports are not available, data compilations must take into account the incidence in the general population, or in occupational usage related to exposure. For this reason, such data most often concern sensitizing chemicals, seldom non-sensitizing ingredients. Among the different sources reporting human data compilations [19, 24, 25], the human classification proposed by Basketter et al. [19] was chosen for the comparison with 101 *in vitro* classifications. Compared with human results, the accuracy in distinguishing skin sensitizers (i.e. UN GHS Cat.1; human class 1–4) from non-sensitizers (human class 5–6) is 77% (N = 101) with a sensitivity of 100% (58/58) and a specificity of 47% (20/43). When confronting the 101 human data set with S ranging from class 1 to 5 and NS in class 6, the U-SENS™ test method showed a high specificity of 65% (11/17), a good sensitivity of 89% (75/84) and an overall accuracy of 85% (86/101) (Fig. 22.3). Taken together, this information indicates the usefulness of the U-SENS™ method to contribute to the identification of skin sensitization hazards. However, the accuracy values given here for the U-SENS™ as a stand-alone test method are only indicative, since the test method should be considered in combination with other sources of information in the context of an IATA. This is the most extensive set used to evaluate the predictive performance of an *in vitro* assay against human data, since, until now, this exercise was restricted to specific cases [26] or limited to small sets [27, 28]. Due to the good degree of correlation between EC3 values and intrinsic human potency [29], LLNA remains a strong reference and often constitutes the method of choice to evaluate the reliability of new *in vitro* tests for sensitization hazard identification.

22.4 Brief Description of the Protocol

The U-SENS™ DataBase service on ALternative Methods to animal experimentation (DB-ALM) protocol no. 183 should be employed when implementing the test method in any laboratory [30]. Briefly, in the U-SENS™ assay, the modulation of the CD86 membrane marker in U937 cells, a human histiocytic lymphoma cell line (clone CRL-1593.2 [31]) used as a DC surrogate is measured by flow cytometry following 45 ± 3 h of exposure to at least four concentrations of test chemical selected amongst usable concentrations pre-defined in the DB-ALM protocol N°183, as illustrated in Fig. 22.5. The test method is designed to discriminate between sensitizing and non-sensitizing chemicals whereby chemicals are classified as sensitizers if the CD86-IgG1 percent of positive cells exceeds a defined threshold (i.e. Stimulation Index ≥150) compared to the vehicle control, in at least two independent

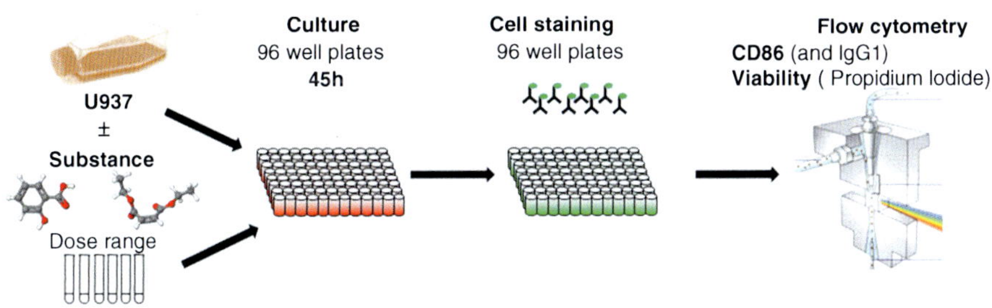

Fig. 22.5 The U-SENS™ DB-ALM protocol N°183

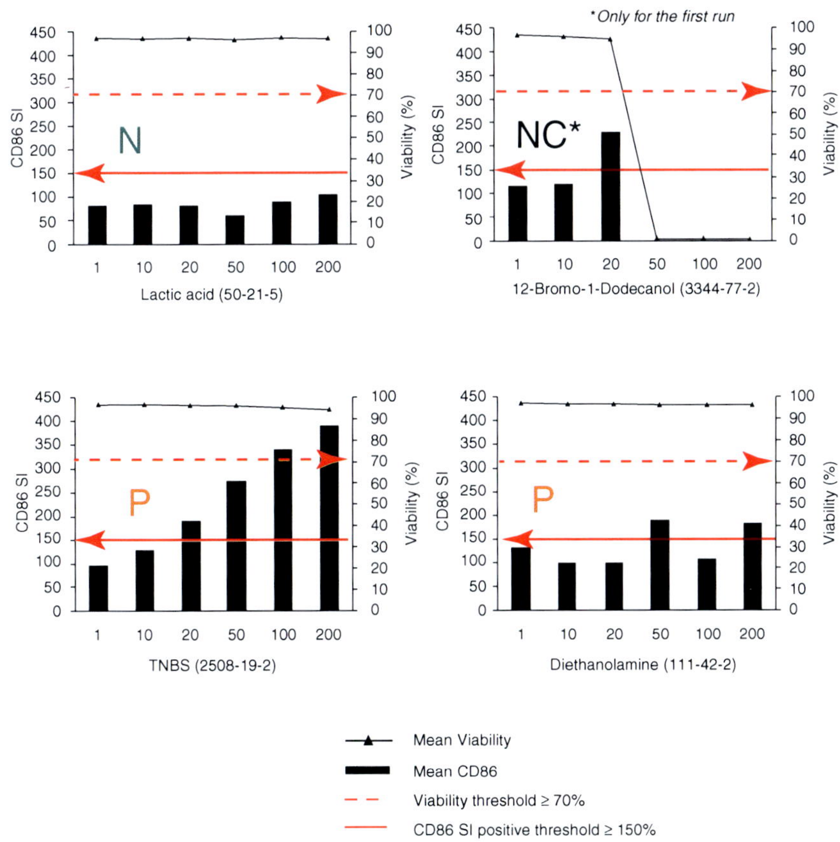

Fig. 22.6 Examples of CD86 S.I. and viability mean values obtained to derive an individual conclusion. *N* Negative, *P* Positive, *NC* Not conclusive

measurements (i.e. run repetitions). Cell viability is measured concurrently by Propidium Iodide staining and CD86 values are considered for the prediction only if cell viability is ≥70%. Examples of dose response curves obtained with a sensitizer and a non-sensitizer are shown in Fig. 22.6. The positive (picrylsulfonic acid, TNBS)

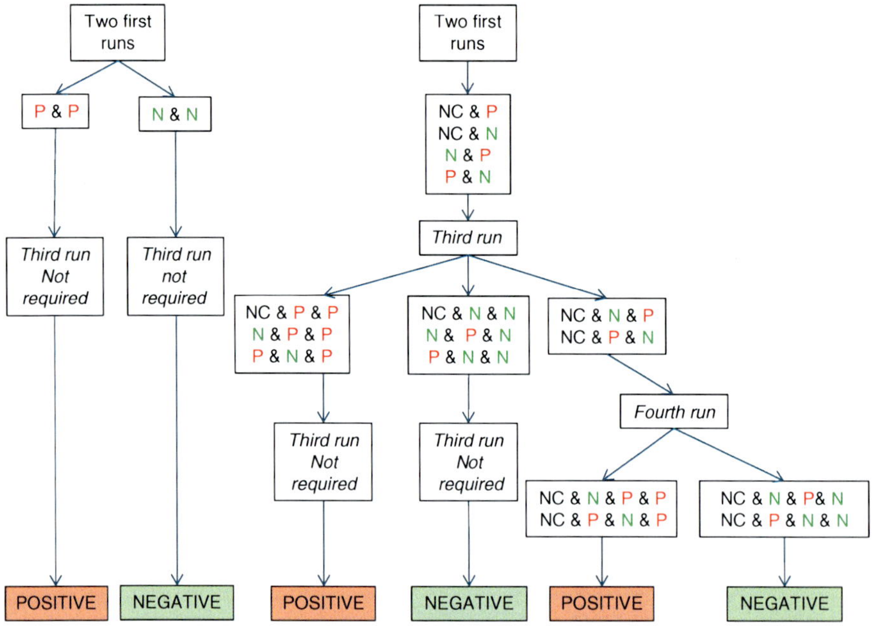

Fig. 22.7 Prediction model used in the U-SENS™ test method (based on the U-SENS™ DataBase service on ALternative Methods to animal experimentation protocol [30])

and the negative (lactic acid) controls are concurrently performed on the test chemicals and acceptance criteria are applied to discard results if the inductions from the controls are outside the acceptable range. Acceptance criteria are also applied to the test system (viability and CD86 of basal expression untreated U937 cells) and for test chemicals and vehicle controls (complete cell culture medium or dimethyl sulfoxide: DMSO).

Briefly, each test chemical is tested in at least four concentrations and in at least two independent runs (performed on a different day) to derive a single prediction (NEGATIVE or POSITIVE). As reported in Fig. 22.7, the prediction model (PM) for the U-SENS™ assay should be applied as following:

- The individual conclusion of an U-SENS™ run is considered Negative (hereinafter referred to as N) if the S.I. of CD86 is less than 150% at all non-cytotoxic concentrations (cell viability ≥70%) and if no interference is observed (cytotoxicity, solubility regardless of the non-cytotoxic concentrations at which the interference is detected). In all other cases: S.I. of CD86 higher or equal to 150% and/or interferences observed, the individual conclusion of an U-SENS™ run is considered Positive (hereinafter referred to as P).
- An U-SENS™ prediction is considered NEGATIVE if at least two independent runs are negative (N) (Fig. 22.7). If the first two runs are both negative (N), the U-SENS™ prediction is considered NEGATIVE and a third run does not need to be conducted.

- An U-SENS™ prediction is considered POSITIVE if at least two independent runs are positive (P) (Fig. 22.7). If the first two runs are both positive (P), the U-SENS™ prediction is considered POSITIVE and a third run does not need to be conducted.
- Because a dose finding assay is not conducted, there is an exception if, in the first run, the S.I. of CD86 is higher or equal to 150% at the highest non-cytotoxic concentration only. The run is then considered to be concluded NOT CONCLUSIVE (NC), and additional concentrations (between the highest noncytotoxicity concentration and the lowest cytotoxicity concentration) should be tested in additional runs. In case a run is identified as NC, at least two additional runs should be conducted, and a fourth run in case runs 2 and 3 are not concordant (N and/or P independently) (Fig. 22.7). Follow-up runs will be considered positive even if only one non-cytotoxic concentration gives a CD86 equal or above 150%, since the concentration setting has been adjusted for the specific test chemical. The final prediction will be based on the majority result of the three or four individual runs (i.e. 2 out of 3 or 2 out of 4) (Fig. 22.7).

22.5 Role in a Testing Strategy

No single *in vitro* or *in silico* approach is able to predict the sensitizing property or potency of a new chemical without animals. Quality information will rely on a wise combination of various approaches [6, 13, 32–34]. In such context, the U-SENS™ test method, addressing the key event 3 of the AOP, is part of The European Cosmetics Industry Trade Association current (Cosmetics Europe) programme. Following the evaluation of 16 non-animal test methods using ten chemicals [20], the U-SENS™ assay was selected for the testing strategy comprising more than 100 chemicals, for which both LLNA and human data are available. The potential contribution of read-out parameters (EC150 and CV70)—instead of currently applied prediction models—to the strategy is being explored. This will allow re-assessment of testing strategies already proposed using new data. Ultimately, the testing strategy—combined with bioavailability and skin metabolism data and exposure consideration—could provide a data integration approach for skin sensitization safety assessments for cosmetic ingredients. In addition, the U-SENS™ assay was part of a decision strategy for skin sensitization hazard identification based on *in silico*, *in chemico* and *in vitro* data analysed using a statistic "stacking" meta-model [33, 35, 36]. Considering all available evidence and input from regulators and stakeholders, the U-SENS™ was recommended by EURL ECVAM [17] to be used as part of an IATA to support the discrimination between sensitizers and non-sensitizers for the purpose of hazard classification and labelling. In its guidance document on the reporting of structured approaches to data integration and individual information sources used within IATA for skin sensitization, the OECD currently discusses a number of case studies describing different testing strategies and prediction models [12]. As such, the current approach has already been submitted and included to the OECD as a case study [13].

22.6 Perspectives from the Test Developer

22.6.1 Critical Steps in the Protocol

The U-SENS™ test method has been proven to be sufficiently stable and reproducible for routine testing, as demonstrated during the validation study. Prior to routine use of the test method, laboratories should demonstrate technical proficiency, using the ten Proficiency chemicals listed in Annex II of OECD TG 442E [18]. Moreover, test method users should maintain a historical database of data generated with the reactivity checks and with the positive and solvent/vehicle controls as described in the DB-ALM Protocol N°183 [30], and use these data to confirm the reproducibility of the test method in their laboratory is maintained over time.

Some key parameters optimized and standardized in order to ensure valid and reproducible data should be highlighted.

- The U-SENS™ assay is based on measurements performed using a flow cytometer, which is a sensitive piece of equipment. That is why the flow cytometer should regularly undergo maintenance and daily/weekly cleaning procedures. Study personnel should be skilled in cell cultures and flow cytometry procedures to avoid any issues related to an external impact factor during the experiment. The facility where the flow cytometer is located should keep a controlled temperature ~20 °C (air conditioned) whenever the flow cytometer is on, in order to avoid drift in CD86 measurement. This is especially important if using a plate sampler.
- A thorough quality control of the U937 cells must be performed. At the end of the 45 ± 3 h incubation treatment period, the mean viability of the triplicate untreated U937 cells must be >90% and the CD86 basal expression of untreated U937 cells within the range of ≥2% and ≤25%. This is important to ensure a sufficiently dynamic range of the response.
- The first run final chemical concentrations should be, by default, 1, 10, 20, 50, 100, and 200 μg/mL. The test concentrations should be selected from 27 predefined options in the range of 0.1–200 μg/mL based on the results of previous runs in order to show and/or confirm dose-dependency of CD86 increase at non-toxic doses or the absence of CD86 increase up to the maximum non-toxic dose [30].
- The chemical should be incubated with U937 cells in a 96-well plate, covered by sealing tape to prevent possible collateral effects of volatile chemicals.
- Some practice is recommended to master the U-SENS™ assay. It is important that the study personnel practice the U-SENS™ assay back in their laboratory, after training, by performing two runs per week for 3–6 weeks prior to routine testing and assessment. A trained experimenter can run at least eight test items in a run and can perform two runs in 1 week (up to 16 different tests items per week). An U-SENS experiment consisted of at least two independent runs in two different weeks to overall define a call for a chemical.

22.6.2 Possible Protocol Adaptations

As an add-on of the validation exercise, all coded samples were evaluated in a medium-throughput laboratory that had developed an automated U-SENS™ version. This internal laboratory conducted testing similar to that in the other four participating laboratories. In the multicentric study, 15 chemicals were evaluated in three independent experiments, for which three differently coded test samples were provided. The resulting classifications summarized in Table 22.1 demonstrated the same concordance for 14 out of 15 chemicals (93.3%) between the three independent runs performed. For the calculation of BLR, the final classification for each 38 test chemicals in this laboratory was obtained using the arithmetic median value of viability over the three runs performed. The proportion of chemicals concordantly classified between the automated approach used in one laboratory and the manual approaches used in the other four laboratories was 32/38 = 84.2% (Table 22.1). The performance was in all respects very similar to that of the other laboratories performing the U-SENS™ assay manually. Specificity was 84.2% (16/19) and sensitivity was 94.7% (18/19), resulting in a concordance of 89.5% (34/38) [16].

Overall, the U-SENS™ assay was sufficiently robust and standardized for automation. As the method is performed in a 96-well plate format, the throughput might be enhanced for testing purposes.

22.6.3 Challenges and Opportunities

In the context of the integrated approaches to testing and assessment and in particular the Integrated Testing Strategies, the question of the interchangeability of test methods addressing the same key event of the AOP was raised [37]. For their analysis, Urbisch and co-workers compiled the results of the U937-CD86 [26], the mMUSST [38] and the h-CLAT assays [39]. To obtain the most complete analysis, the U-SENS™ results were also compared to the results of the three assays used as surrogates for dermal dendritic cells. This produced respective interchangeabilities of about 73–78%. The differences observed can be explained by positive outcomes observed in the U-SENS™ assay while negative outcomes were obtained in at least one of the other three assays. U-SENS™ appeared to be more sensitive than the other methods as it allowed the correct classification of a number of known LLNA strong to extreme or human class 1 and 2 sensitizers, which were misclassified by at least one of the other assays (Urbisch et al. 2015; [10]). Overall, the U-SENS™ assay relies on a more conservative approach than other assays do.

With the promising U-SENS™ predictive performance to distinguish between LLNA S/NS classes, the extent to which the EC150 and CV70 values (alone or in combination) might correlate with EU-CLP categories and discriminate the more potent sensitizers from the weaker and non-sensitizers was initially considered [10].

Further investigation as to whether these EC150/CV70 parameters could be informative for human potency evaluation is ongoing through the Cosmetics Europe program.

22.7 Conclusions

The U-SENS™ method models dendritic cell activation upon exposure to test chemicals. Like dendritic cells, upon contact with sensitizers, U937 human histiocytic lymphoma cells are activated and increase the CD86 expression. The transferability, intra- and inter-reproducibility of the U-SENS™ assay was demonstrated in four laboratories. Moreover, an automated evaluation study conducted in parallel showed similar performances. In summary, with a sensitivity over 90% in predicting the skin sensitization of chemicals, the U-SENS™ test method could complement the battery of assays already validated and approved to assess skin sensitization. Considering all available evidence, the U-SENS™ method was recommended by EC EURL ECVAM to be used as part of an IATA for the purpose of hazard classification and labelling and has been adopted in OECD Test Guidelines 442E.

References

1. UN. United Nations Globally Harmonized System of Classification and Labelling of Chemicals (GHS). ST/SG/AC.10/30/Rev.6, Sixth Revised Edition, New York and Geneva: United Nations. 2015. Available at: http://www.unece.org/fileadmin/DAM/trans/danger/publi/ghs/ghs_rev06/English/ST-SG-AC10-30-Rev6e.pdf.
2. Thyssen JP, Gimenez-Arnau E, Lepoittevin JP, Menne T, Boman A, Schnuch A. The critical review of methodologies and approaches to assess the inherent skin sensitization potential (skin allergies) of chemicals part I. Contact Dermatitis. 2012;66:11–24.
3. Adler S, Basketter D, Creton S, Pelkonen O, van Benthem J, Zuang V, et al. Alternative (non-animal) methods for cosmetics testing: current status and future prospects. Arch Toxicol. 2010;85:367–485.
4. Aeby P, Ashikaga T, Bessou-Touya S, Schepky A, Gerberick F, Kern P, Marrec-Fairley M, Maxwell G, Ovigne JM, Sakaguchi H, Reisinger K, Tailhardat M, Martinozzi-Teissier S, Winkler P. Identifying and characterizing chemical skin sensitizers without animal testing: Colipa's research and method development program. Toxicol In Vitro. 2010;24:1465–73.
5. Basketter D, Casati S, Gerberick GF, Griem P, Philips B, Worth A. Skin sensitisation. Altern Lab Anim. 2005;33(Suppl 1):83–103.
6. Goebel C, Aeby P, Ade N, Alepee N, Aptula A, Araki D, Dufour E, Gilmour N, Hibatallah J, Keller D, Kern P, Kirst A, Marrec-Fairley M, Maxwell G, Rowland J, Safford B, Schellauf F, Schepky A, Seaman C, Teichert T, Tessier N, Teissier S, Weltzien HU, Winkler P, Scheel J. Guiding principles for the implementation of non-animal safety assessment approaches for cosmetics: skin sensitisation. Regul Toxicol Pharmacol. 2012;63:40–52.
7. OECD. The adverse outcome pathway for skin sensitization initiated by covalent binding to proteins. Part 1: scientific evidence. Series on Testing and Assessment No. 168. 2012. Online at: http://www.oecd.org/officialdocuments/publicdisplaydocumentpdf/?cote=ENV/JM/MONO(2012)10/PART1&docLanguage=En.

8. Dos Santo GG, Reinders J, Ouwhand K, Rustemeyer T, Scheper RJ, Gibbs S. Progress on the development of human in vitro dendritic cell based assays for assessment of skin sensitizing potential of compounds. Toxicol Appl Pharmacol. 2009;236:372–82.
9. Vandebriel RJ, van Loveren H. Non-animal sensitization testing: state-of-the-art. Crit Rev Toxicol. 2010;40:389–404.
10. Piroird C, Ovigne JM, Rousset F, Martinozzi Teissier S, Gomes C, Cotovio J, Alépée N The myeloid U937 skin sensitization test (U-SENS) addresses the activation of dendritic cell event in the adverse outcome pathway for skin sensitization. Toxicol In Vitro. 2015 doi: 10.1016/j.tiv.2015.03.009.
11. Sakaguchi H, Ashikaga T, Miyazawa M, Kosaka N, Ito Y, Yoneyama K, Sono S, Itagaki H, Toyoda H, Suzuki H. The relationship between CD86/CD54 expression and THP-1 cell viability in an in vitro skin sensitization test-human cell line activation test (h-CLAT). Cell Biol Toxicol. 2009;25:109–26.
12. OECD. Guidance document on the reporting of defined approaches and individual information sources to be used within integrated approaches to testing and assessment (IATA) for skin sensitisation- ENV/JM/HA(2016)11. Organisation for Economic Cooperation and Development, Paris. 2016. Available at: https://community.oecd.org/community/iatass.
13. OECD. Series on testing & assessment No. 256: guidance document on the reporting of defined approaches and individual information sources to be used within integrated approaches to testing and assessment (IATA) for skin sensitisation, annex 1 and annex 2. ENV/JM/HA(2016)29. Organisation for Economic Cooperation and Development, Paris. 2016. Available at: https://community.oecd.org/community/iatass.
14. OECD. OECD Series on Testing and Assessment No. 34: Guidance Document on the Validation and International Acceptance of New or Updated Test Methods for Hazard Assessment. Organisation for Economic Co-operation and Development, Paris. 2005. Available at: http://www.oecd.org/chemicalsafety/testing/seriesontestingandassessmentpublicationsbynumber.htm.
15. Hartung T, Bremer S, Casati S, Coecke S, Corvi R, Fortaner S, Gribaldo L, Halder M, Hoffmann S, Roi AJ, Prieto P, Sabbioni E, Scott L, Worth A, Zuang V. A modular approach to the ECVAM principles on test validity. Altern Lab Anim. 2004;32:467–72.
16. Alépée N, Piroird C, Aujoulat M, Dreyfuss S, Hoffmann S, Hohenstein A, Meloni M, Nardelli L, Gerbeix C, Cotovio J. Prospective multicentre study of the U-SENS test method for skin sensitization testing. Toxicol In Vitro. 2015;30:373–82.
17. EC EURL ECVAM. ESAC Opinion No. 2016-03 on the L'Oréal-coordinated study on the transferability and reliability of the U-SENS™ test method for skin sensitisation testing. EUR 28178 EN. 2016. doi 10.2787/815737. Available at: http://publications.jrc.ec.europa.eu/repository/handle/JRC103705.
18. OECD. Guideline for the testing of chemicals No. 442E: in vitro skin sensitisation: human cell line activation test (U-SENS™). Organisation for Economic Cooperation and Development, Paris. 2017. Available at: http://www.oecd.org/env/testguidelines.
19. Basketter D, Alépée N, Ashikaga T, Barroso J, Gilmour N, Goebel C, Hibatallah J, Hoffmann S, Kern P, Martinozzi-Tessier S, Maxwell G, Reisinger K, Sakaguchi H, Schepky A, Tailhardat M, Templier M. Categorization of chemicals according to their relative human skin sensitizing potency. Dermatitis. 2014;25(1):11–21.
20. Reisinger K, Hoffmann S, Alépée N, Ashikaga T, Barroso J, Elcombe C, Gellatly N, Galbiati V, Gibbs S, Groux H, Hibatallah J, Keller D, Kern P, Klaric M, Kolle S, Kuehnl J, Lambrechts N, Lindstedt M, Millet M, Martinozzi-Teissier S, Natsch A, Petersohn D, Pike I, Sakaguchi H, Schepky A, Tailhardat M, Templier M, van Vliet E, Maxwell G. Systematic evaluation of non-animal test methods for skin sensitisation safety assessment. Toxicol In Vitro. 2014;29:259–70.
21. Aptula AO, Roberts DW. Mechanistic applicability domains for non-animal-based prediction of toxicological end points: general principles and application to reactive toxicity. Chem Res Toxicol. 2006;19(8):1097–105.

22. Patlewicz G, Jeliazkova N, Safford RJ, Worth AP, Aleksiev B. An evaluation of the implementation of the Cramer classification scheme in the Toxtree software. SAR QSAR Environ Res. 2008;19(5–6):495–524.
23. Takenouchi O, Miyazawa M, Saito K, Ashikaga T, Sakaguchi H. Predictive performance of the human cell line activation test (h-CLAT) for lipophilic chemicals with high octanol-water partition coefficients. J Toxicol Sci. 2013;38(4):599–609.
24. De Groot AC, Weyland JW, Kathon CG. A review. J Am Acad Dermatol. 1988;18:350–8.
25. Schlede E, Aberer W, Fuchs T, Gerner I, Lessmann H, Maurer T, Rossbacher R, Stropp G, Wagner E, Kayser D. Chemical substances and contact allergy -244 substances ranked according to allergenic potency. Toxicology. 2003;193(3):219–59.
26. Natsch A, Ryan CA, Foertsch L, Emter R, Jaworska J, Gerberick F, Kern P. A dataset on 145 chemicals tested in alternative assays for skin sensitization undergoing prevalidation. J Appl Toxicol. 2013;33(11):1337–52.
27. Bauch C, Kolle SN, Ramirez T, Eltze T, Fabian E, Mehling A, Teubner W, van Ravenzwaay B, Lansiedel R. Putting the parts together: combining in vitro methods to test for skin sensitizing potentials. Regul Toxicol Pharmacol. 2012;63(3):489–504.
28. Nukada Y, Ashikaga T, Sakaguchi H, Sono S, Mugita N, Hirota M, Miyazawa M, Ito Y, Sasa H, Nishiyama N. Predictive performance for human skin sensitizing potential of the human cell line activation test (h-CLAT). Contact Dermatitis. 2011;65(6):343–53.
29. Basketter DA, Clapp C, Jefferies D, Safford B, Ryan CA, Gerberick F, Dearman RJ, Kimber I. Predictive identification of human skin sensitization thresholds. Contact Dermatitis. 2005; 53:260–7.
30. DB-ALM. Protocol no. 183: U937 Cell Line Activation Test for Skin Sensitization (U-SENS™). 2017. 30pp. Accessible at: http://ecvam-dbalm.jrc.ec.europa.eu/.
31. Sundström C, Nilsson K. Establishment and characterization of a human histiocytic lymphoma cell line (U-937). Int J Cancer. 1976;17:565–77.
32. Maxwell G, Aeby P, Ashikaga T, et al. Skin sensitisation: the colipa strategy for developing and evaluating non-animal test methods for risk assessment. ALTEX. 2011;28(1):50–5.
33. Nocairi H, Gomes C, Thomas M, Saporta G. Improving stacking methodology for combining classifiers: applications to cosmetic industry. Electron J Appl Stat Anal. 2016;9(2):340–61.
34. Rovida C, Alépée N, Api AM, Basketter DA, Bois FY, Caloni F, Corsini E, Daneshian M, Eskes C, Ezendam J, Fuchs H, Hayden P, Hegele-Hartung C, Hoffmann S, Hubesch B, Jacobs MN, Jaworska J, Kleensang A, Kleinstreuer N, Lalko J, Landsiedel R, Lebreux F, Luechtefeld T, Locatelli M, Mehling A, Natsch A, Pitchford JW, Prater D, Prieto P, Schepky A, Schuurmann G, Smirnova L, Toole C, van Vliet E, Weisensee D, Hartung T. Integrated testing strategies (ITS) for safety assessment. ALTEX. 2015;32(1):25–40.
35. Gomes C, Noçairi H, Thomas M, Collin JF, Ibanez F, Saporta G. Stacking prediction for a binary outcome. In COMPSTAT, 20th international conference on computational statistics, Limassol, Cyprus; 2012. p. 271–82.
36. Gomes CH, Nocairi H, Thomas M, Collin JF, Saporta G. A simple and robust scoring technique for binary classification. Artif Intell Res. 2014;3(1):52–8.
37. Urbisch D, Mehling A, Guth K, Ramirez T, Honarvar N, Kolle S, Landsiedel R, Jaworska J, Kern P, Gerberick F, Natsch A, Emter R, Ashikaga T, Miyazawa M, Sakaguchi H. Assessing skin sensitization hazard in mice and men using non-animal test methods. Regul Toxicol Pharmacol. 2015;71(2):337–51.
38. Bauch C, Kolle SN, Fabian E, Pachel C, Ramirez T, Wiench B, Wruck CJ, van Ravenzwaay B, Landsiedel R. Intralaboratory validation of four in vitro assays for the prediction of the skin sensitizing potential of chemicals. Toxicol In Vitro. 2011;25(6):1162–8.
39. Nukada Y, Ashikaga T, Miyazawa M, Hirota M, Sakaguchi H, Sasa H, Nishiyama N. Prediction of skin sensitization potency of chemicals by human cell line activation test (h-CLAT) and an attempt at classifying skin sensitization potency. Toxicol In Vitro. 2012;26:1150–60.

Human Peripheral Blood Monocyte Derived Dendritic Cells Assay for the Detection and Characterization of Sensitizers

23

Andreas Schepky, Hendrik Reuter, Jochen Kühnl, and Pierre Aeby

23.1 Principle of the Test Method and Scientific Basis

The various molecular and cellular events leading to a skin sensitization reaction have been analyzed in many details [1–4] and the complex interactions of a chemical with the different compartments of the immune system have been described: Briefly, to act as a skin sensitizer, a chemical must penetrate the skin barrier and reach the viable part of the epidermis. There, it must bind and modify endogenous skin proteins. Some chemicals react directly and others designated as pro-haptens, require prior conversion or activation through a metabolic activity.

The modified (haptenated) proteins initiate various intracellular signaling pathways within the viable part of the epidermis and activate surrounding skin dendritic cells (DCs). As a consequence, activated DCs internalize and process the haptenated proteins, up-regulate the expression of a set of surface proteins (e.g. CD86), secrete various cytokines and migrate from the epidermis to the draining lymph node. Once in the lymph node, they complete maturation and present fragments of the modified proteins to the adaptive immune system initiating a T-cell based, antigen-specific immune response. These key biological events have recently been documented in an OECD report on: "The Adverse

A. Schepky (✉) • H. Reuter • J. Kühnl
Beiersdorf AG, Hamburg, Germany
e-mail: Andreas.Schepky@Beiersdorf.com; Hendrik.Reuter@beiersdorf.com; Jochen.Kuehnl@beiersdorf.com

P. Aeby
Independant Consultant, Marly, Switzerland

C. Eskes et al. (eds.), *Alternatives for Dermal Toxicity Testing*, DOI 10.1007/978-3-319-50353-0_23

Outcome Pathway (AOP) for Skin Sensitisation Initiated by Covalent Binding to Proteins" [5].

Skin DCs thus play a pivotal role in orchestrating the immune response leading to the acquisition of contact dermatitis that is the clinical condition resulting from skin sensitization. The well-documented induction of CD86 expression on the surface of activated DCs [6] was rapidly perceived as an opportunity to detect the activated status of DCs and to develop DC-based *in vitro* assays for the detection of skin sensitizers [7–9].

DC or DC-like cells can be obtained from primary cells (cord or peripheral blood). However, the production of DC-like cells from primary cells was perceived as a difficult and sometimes inconsistent task [6, 10] and donor-dependent variability—even though embracing the biological variability—is known to affect the reproducibility of results. Hence, many research groups shifted their effort toward developing protocols based on cell lines (e.g. THP-1, U937, Mutz-3, etc.) (See [11]). However, such transformed cells have their own drawbacks: they carry severely altered genomes and may become genetically instable over prolonged culture period [12, 13]. In addition, their functional properties and metabolic capacity are not fully characterized and may be rather limited [14]. Moreover, differences exist between clones from different cell banks and recent findings have shown that many cell lines are contaminated with other cell types/lines [12]. Last but not least, their culture process must be strictly controlled in order to get consistent and reproducible results [15, 16].

Therefore, some laboratories pursued the development of *in vitro* protocols based on primary cells [7, 17, 18]. Recently, Reuter et al. published an optimized protocol based on human peripheral blood monocyte-derived dendritic cells (PBMDCs). Multiple culture parameters such as cytokine concentrations, incubation time, pooled vs. single donors or exposure conditions, (readout, and cytotoxicity) were re-assessed and a stable and reproducible protocol was proposed: Mononuclear cells isolated from buffy coat through Ficoll gradient density centrifugation and CD1a−/CD14+ selection are prepared to generate PBMDCs. The cells are suspended in a dedicated culture medium containing optimized concentrations of recombinant human granulocyte–macrophage colony-stimulating factor (rh GM-CSF) and interleukin-4 (rh IL-4). After 5 days of differentiation, the CD86 status of the cells is analyzed and if determined to be acceptable (<60%), the DC-like cells are seeded in 24 well plates. The test chemicals are then added and, after 48 h of exposure, the modulation of CD86 surface expression on PBMDCs is measured by flow cytometry (against a set of a negative (RPMI 1640 medium), positive (10 mM DNCB) and a functional (100 ng/mL LPS) controls. The induced modification in surface marker expression due to exposure to the test chemical is calculated as difference to untreated controls (ΔDCD86 = %CD86$^+$ cells in substance treated sample—%CD86$^+$ cells in untreated sample). A test item is considered as a skin sensitizer when the ΔDCD86 is >20% [17] (see Fig. 23.1).

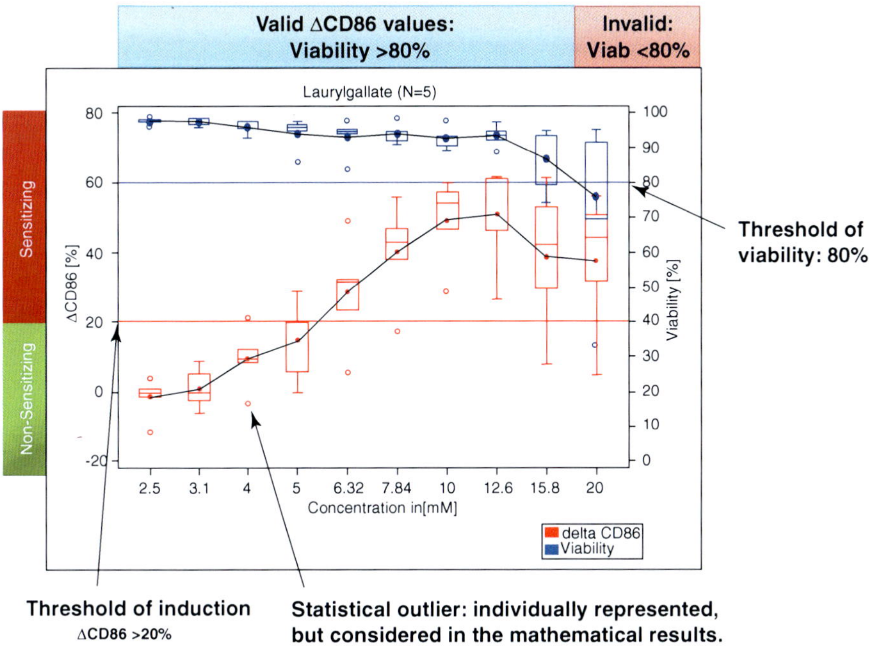

Fig. 23.1 Graphical representation of PBMDC results and explanation of important parameters. Test results are summarized graphically as shown above: The *red box-plots* indicate the ΔCD86 percentages at a given concentration. *Blue box-plots* indicate the percentage of cell viability at a given concentration. Statistical outliers are visualized as *circles*

23.2 Brief Description of the Protocol

The optimized test protocol is described in the [17] publication. The procedure is briefly described below:

23.2.1 Preparation of Human Monocytes

Mononuclear cell fractions are obtained from fresh buffy coats of random and anonymous human donors. Buffy coats are defined as the anti-coagulated blood sample fraction and contain mainly white blood cells and platelets. It is crucial that buffy coats are obtained by centrifugation and NOT by filtration. Moreover, the buffy coats must not be older than 24 h. Typically, around 5 10 × 107 cells can be isolated from one buffy coat (ca. 60 mL) using a two-step protocol. First, Ficoll density centrifugation (A) or Leucosep separation tube separation (B) are applied to obtain a monocyte-containing fraction.

A: Ficoll density centrifugation protocol:

- Add 20 mL Lymphocyte Separation Medium (LSM 1077) to each of 4 clean 50 mL Falcon tubes.
- Split the buffy coat into four clean 50 mL Falcon tubes (ca. 15 mL/tube.)
- Adjust volume to 30 mL with Hank's buffered salt solution (HBSS) + 2 mM EDTA and mix gently
- Carefully and slowly overlay the Ficoll with the buffy coat/Hank's-mixture
- Centrifuge for 35 min at 500 g and 20 °C (w/o break)

B: Leucosep separation tube protocol:

- Prepare four separation tubes according to manufacturer's instruction.
- Split blood into four clean 50 mL Falcon tubes (ca. 15 mL each.)
- Adjust volume to 30 mL with HBSS +2 mM EDTA and mix gently
- Carefully transfer blood/Hank's-mixture to separation tubes.
- Centrifuge according to manufacturer's instructions (Leucosep tubes for 15 min at 800 g and 20 °C (w/o break)).

Subsequent to A or B, carefully withdraw ca. ¾ of the supernatant using a pipette prior to transferring the remaining monocyte-rich cell fraction into two clean 50 mL Falcon tubes. Adjust the volume in each tube to 30–40 mL with HBSS (with EDTA) before washing the cells four times in HBSS, applying the following centrifugation protocol (all centrifugation steps are performed w/o brake!):

1. Centrifuge for 10 min at 600 g (4 °C), carefully discard supernatant. Resuspend the cells in 10 mL/tube prior to pooling cells from two Falcon tubes in one tube. Finally adjust the volume to 30 mL with Hank's (with EDTA).
2. Repeat centrifugation step (10 min at 500 g, 4 °C), carefully discard supernatant and resuspend cells in 30 mL Hank's (with EDTA)
3. Centrifuge for 10 min at 400 g (4 °C). Resuspend cells in 2–4 mL Hank's (with EDTA)

Following monocyte enrichment, the cells must be handled on ice at all times. Cells are counted, transferred into a clean 50 mL Falcon tube and the volume is adjusted to 30 mL with Hank's (with EDTA). A final centrifugation step for 5 min at 300 g (4 °C) (w/o brake) is carried out to collect the cells.

From the monocyte-enriched fraction, CD14+ monocytes are isolated by a positive selection process using anti-CD14-Ig coupled magnetic microbeads (Miltenyi Biotec) according to manufactor's instruction. Briefly, cells are resuspended in a calculated volume of MACS-buffer which results in a concentration of 10^8 cells in 800 µL (if fewer number of cells are obtained, use 800 µL as a minimal volume). MACS CD14 MicroBeads are then added to the cell suspension, followed by gently mixing prior to an incubation for 15 min at 4 °C. Subsequently, the volume is adjusted to 30 mL with MACS-buffer and the suspension is centrifuged for 10 min at 300 g (4 °C). Finally, cells are resuspended in MACS buffer (500 µL/10^8 cells).

23.2.2 Positive Selection of CD14+ Monocytes by Magnetic Separation

First, place a LS-column in magnetic field and equilibrate the column with 3 mL MACS-buffer. Carefully transfer the cell suspension into the column (for >10^9 cells use an additional column). Column is washed thrice each with 3 mL MACS-buffer. Take out the column off the magnetic field and elute the isolated CD14+ cells in 5 mL of MACS-buffer. Finally, adjust volume to 20 mL with MACS-buffer and collect the cells by centrifugation for 6 min at 400 g (4 °C). The purified human monocytes can be stored for up to 3 month at −80 °C. To this end, deep-freeze monocytes in cryo vials in RPMI with 10% FCS und 10% DMSO (final concentration 1×10^7 cells/mL).

23.2.3 Thawing and Differentiation of Human Monocytes into PBMDCs

Purified human monocytes are thawed, re-suspended in culture medium and their CD14 phenotype is controlled by FACS analysis. If the cells pass the acceptance criteria (>70% CD14+ cells), they are seeded at an initial cell density of 1×10^6 cells/ml RPMI medium (containing 10% FCS, 2 mM L-glutamine, 100 U/mL penicillin, 100 μg/mL streptomycin, 100 U/ml recombinant human (rh IL-4, 200 U/mL rh GM-CSF) in 6 well plates (2.5 mL per cavity). Monocytes are then incubated (at 37 °C; 5% CO_2) for 5 days to differentiate into PBMDCs.

23.2.4 Exposure to the Test Chemicals

23.2.4.1 Determination of the Test Concentration Range

Prior to final experiments, the chemical's cytotoxicity is analyzed in order to determine the adequate concentration range for the main experiments. To begin with, solubility in DMSO and water is determined up to a concentration of 1000 μM and the highest soluble concentration is used for a 1:2 dilution series with 6 concentrations. Secondly, the cytotoxicity is evaluated in initial screening experiments, exposing PBMDCs (48 h in 24-well plates) to a broad concentration range. Finally, the concentration range for the main experiments is defined based on the concentration that induces a 20% cytotoxicity (determined via 7-AAD-stainings [19] or a 20% induction of CD86 (against negative controls). Subsequently, a minimum of four lower and one higher test concentrations are determined by the „Decimal Geometric Concentration Series"([20]; also described in the ECVAM-INVITOXX protocol Nr. 78 for the 3T3 Neutral Red uptake assay). This determination of concentrations is sufficient to classify the majority of chemicals. However, if required, smaller concentration steps may be used to further substantiate initial results in a critical concentration range.

23.2.4.2 PBMDCs Exposure to the Test Chemicals

During the main experiment, PBMDCs are exposed to the defined range of test chemical concentrations. The test substance may be dissolved in RPMI medium or DMSO (Merck). The maximal DMSO concentration is 1.0% (v/v). Using 2× concentrated substance solution, the final exposure of cells is restricted to a maximal 0.5% DMSO concentration. Each chemical is tested in at least five independent experiments corresponding to five different donors. Briefly, the differentiated PBMDCs are seeded in 24-well plates and then exposed for 48 h (at 37 °C, 5% CO_2) to the five chosen test chemical concentrations. On day 7 (5 days of initial culture and 48 h exposure to the test chemical), cells are stained and analyzed for CD86 surface expression via flow cytometry (staining of 5×10^5 cells, measurement of at least 10,000 cells). Dead cells are detected using 7-AAD staining [19] and excluded from the CD86 analysis. For each donor, a negative control (RPMI 1640 medium), an isotype control (untreated cells only) and a functional control (100 ng/mL LPS) are included in the test set and analyzed in parallel. To obtain reproducible results, the data are normalized to the isocontrol (set to 1%). Acceptance criteria for the use of cells are determined by analyzing the negative control: Thresholds are <60% of CD86-positive cells and <8.5% of 7-AAD-positive cells. In addition, CD1a expression is also measured for the negative control.

23.2.4.3 Data Analysis and Prediction Model

A chemical is classified as a sensitizer if it induces a mean ΔCD86 > 20% (n = 5) at any concentration within the acceptable cytotoxicity domain (cytotoxicity <20%). This ΔCD86 reflects the increase in the proportion of CD86 expressing cells induced by the test chemical normalized to an untreated control. The acceptable relative cytotoxicity range is limited to ≤20% in order to avoid "danger signal"-dependent artifacts (due to e.g. damage-associated molecular pattern-dependent Toll Like Receptor activation). The EC20 value is calculated by (1) determination of the first concentration that induces a ΔCD86 expression of >20%, (2) interpolation of the CD86 expression between this concentration and the CD86 expression induced by the next-smaller concentration tested in the experiments (using the "trend" function of Excel that corresponds to a linear regression via method of least squares), (3) calculation of the concentration that induces a 20% increase of ΔCD86. Correspondingly, the CV80 values are calculated by (1) identification of the first concentration that results in >20% fraction of dead cells, (2) interpolation of the viability ratio between this concentration and the viability ratio of the next-smaller concentration that shows a cell viability of 80%.

23.3 Reliability and Relevance

23.3.1 Current (Pre)validation Status

Different versions of PBMDC-based test protocols have been published and evaluated for their individual capability to measure, analyze or predict the sensitizing properties of a wide range of chemicals [7, 21–23]. Recently, Reuter et al. optimized the sourcing, isolation, exposure and analysis procedures proposed by these different protocols. This work resulted in the publication of a stable and reproducible approach [17].

In a first phase, the performance of this protocol was evaluated with a test set comprising seven sensitizing and five non-sensitizing chemicals. All but one (benzalkonium chloride) were correctly classified (sensitizer/non-sensitizer). Moreover, information concerning the sensitizer category could be derived from the minimal test concentration inducing a positive response.

Its performance was then assessed in a ring study involving five laboratories [24]. The transferability and the predictivity were analyzed. A statistical evaluation of the results indicated that the test protocol could be successfully transferred to all participating laboratories (some of them without prior experience in deploying *in vitro* sensitization assays) and correctly predicted the sensitization potential of the tested chemicals (one non-sensitizer and six sensitizers). Karschuk et al. also successfully used this protocol as a platform for developing a method for the detection and characterization of photosensitizers [22].

23.3.2 Performance and Applicability of the Test Method

23.3.2.1 Reproducibility

Determination of inter-laboratory reproducibility is a challenging and critical endeavor on the way to validation and regulatory acceptance of any assay. The reproducibility of the PBMDC has been evaluated in a ring study comprising five laboratories. Following training in the lead laboratory at Beiersdorf, each participating laboratory was challenged to assess a set of seven chemicals (Table 23.1).

Table 23.1 List, characteristics and test concentrations of the chemicals included in the ring study test set

Chemical name (abbreviation)	Sensitizer category	Test concentrations
Sodium lauryl sulfate (SDS),	Non-sensitizer	100, 200, 300, 400, 500 and 600 μM
4-Hydroxybenzoic Acid (HA)	Non-sensitizer	5000, 6000, 7000, 8000, 9000 and 10,000 μM
Eugenol (EUG)	Weak sensitizer	50, 200, 350, 500, 650, 800, 950 and 1100 μM
Hydroxycitronellal (HCIT)	Weak sensitizer	350, 425, 500, 575, 650 and 725 μM
Alpha-Hexylcinnamaldehyde (HCA)	Moderate sensitizer	50, 60, 70, 80, 90 and 100 μM
$NiSO_4$	Moderate sensitizer	50, 100, 150, 200, 250 and 300 μM
2,4-Dinitrochlorobenzene (DNCB)	Extreme sensitizer	2.5, 5.0, 7.5, 10.0, 12.5 and 15.0 μM

Source: Reuter et al. [24]

23.3.2.2 Statistical Analysis

Inductive statistical results analysis of the inter-laboratory data (two-way ANOVA with interaction) confirmed the good inter-laboratory reproducibility of the evaluated protocol as no statistically relevant difference could be observed between the results obtained in the different laboratories (See Table 23.2).

23.3.2.3 Predictive Capacity

The predictive capacity of the test protocol has been evaluated after the finalization of the optimized protocol, [17], during the described ring study [24] and as part of a larger systematic evaluation of non-animal test methods [25].

During the ring study, seven chemicals (five sensitizers and two non-sensitizers) were analyzed in five different laboratories (see Table 23.3). Inductive statistics indicated that the hypothesis of relevant differences between substances is true: A statistically different EC20 value was obtained for each tested substances indicating that the assay was able to differentiate all tested chemicals. And last but not least, in most cases, the prediction model correctly estimated the sensitization hazard of the tested chemicals (See Table 23.3). A false positive result was obtained for SDS in

Table 23.2 Inductive statistical comparisons between laboratories across substances (n.s. indicates no statistically relevant difference)

EC20			
Comparison	Raw *p*-value	Adjusted *p*-value	Result
Lab 1 ~ Lead Lab	0.0078	0.0593	n.s.
Lab 1 ~ Lab 2	0.4058	0.9198	n.s.
Lab 1 ~ Lab 3	0.1299	0.5496	n.s.
Lab 1 ~ Lab 4	0.0951	0.4497	n.s.
Lead Lab ~ Lab 2	0.0667	0.3508	n.s.
Lead Lab ~ Lab 3	0.2472	0.7732	n.s.
Lead Lab ~ Lab 4	0.2561	0.7852	n.s.
Lab 2 ~ Lab 3	0.4949	0.9597	n.s.
Lab 2 ~ Lab 4	0.4263	0.9310	n.s.
Lab 3 ~ Lab 4	0.9367	1.0000	n.s.

Source: Reuter et al. [24]

Table 23.3 Classification obtained in the five different laboratories: +: Sensitizer; –: Non-Sensitizer; ±: Equivocal)

	SDS	HA	EUG	HCIT	HCA	NiSO4	DNCB
Lab Ref	±	–	±	+	+	+	+
Lab 1	±	–	+	+	+	+	+
Lab 2	–	–	+	+	+	+	+
Lab 3	+	–	+	+	–	+	+
Lab 4	–	–	+	+	Doubtful[a]	+	+

[a]Relevant increases in CD86 expression (>20%) were observed in 8 out of 9 donors
Source: Reuter et al. [24]

Table 23.4 Hazard classification for the ten chemical set (S: Sensitizer; NS: Non-sensitizer) (Source: Reuter et al. [25])

Test substance	CAS number	Reference result: potential	PBMDC
2-Mercaptobenzothiazole	149-30-4	S	S
4-Nitrobenzylbromide	100-11-8	S	S
Cinnamal	104-55-2	S	S
Lactic acid	50-21-5	NS	NS
Lauryl gallate	1166-52-5	S	S
Methyldibromoglutaronitrile	35691-65-7	S	NS
Phenyl benzoate	93-99-2	S	NS
Salicylic acid	69-72-7	NS	NS
Sodium lauryl sulphate	151-21-3	NS	NS
Tetramethyl thiuram disulphide	137-26-8	S	S
Accuracy			8/10

one laboratory (Lab 3). SDS is known as a human irritant and a false positive in the LLNA. Moreover, it lyses the test cells, releasing various cell components that may act as danger signals and activate the DC. On the other hand, the false negative result obtained with HCA in Lab 3 could have been due to a limited test concentration range.

Moreover, the PBMDC protocol entered a systematic and comparative evaluation of 16 *in vitro* test methods for skin sensitization safety assessment. This study was initiated and supervised by the task force skin tolerance of Cosmetics Europe with contributions from EURL-ECVAM. Each test method was challenged with a common set of ten substances (Table 23.4). Regarding the PBMDC, an overall accuracy of 8/10 was obtained (80% sensitivity, 100% specificity).

Prior to the aforementioned study by Cosmetics Europe, the PBMDC's predictive parameters and applicability domains were further evaluated in internal projects at Beiersdorf. Overall, the corresponding data of 61 chemicals were communicated in the final report, displaying an accuracy of 77.1% (sensitivity of 76.8%, specificity of 77.8%).

23.3.2.4 Applications and Limitations

- The described assay uses PBMDCs in suspension cell culture as test system. It is thus compatible with test items that can be solubilized in the culture medium at the desired test concentrations (concentrations ranging from non-toxic to cytotoxic). In case of negative results without toxic effects, the test item's solubility data should be reconsidered.
- A test item is considered as a sensitizer if the ΔDCD86 is >20%. Such a threshold-based prediction model has an inherent limitation: Very weak sensitizers may induce a ΔCD86 just around or below this threshold and may not be correctly classified as sensitizers.

- As PBMDCs are derived from primary cells, they are supposed to provide the adequate metabolic capacity for the activation and the detection of pro-haptens. Consistent with this assumption, all pro-haptens tested in the above-mentioned publication and during the ring study were detected and correctly classified as sensitizers.

On the other hand, chemicals known as pre-haptens need a non-enzymatic oxidation step to act as haptens [26]. Depending on their specific chemical properties, such molecules may not be detected in this assay.

23.3.2.5 Comparison to Human Data

The PBMDC assay is based on primary cells of human origin, hence, it is expected to deliver quality information relevant to the human *in vivo* situation. In fact, most of the results obtained with this *in vitro* test [17, 25] and Beiersdorf internal data are in agreement with the categories proposed by [27] according to human skin sensitization characteristics.

Dinitrochlorobenzene, in line with its subgroup 1 category ("Extensive evidence of contact allergy in relation to degree of exposure and size of exposed population") tested positive at very low concentrations with the PBMDC assay (positive at 7.5 μM [17] or at 11.1 μM [24].

Lauryl gallate and 2-mercaptobenzothiazole classified in Category 2 resp. 3 ("Frequent, resp. common cause of contact allergy") both tested positive (sensitizers) in the PBMDC assay [25]. Interestingly, eugenol (Category 3), is a pro-hapten requiring metabolic activation for conversion into a chemically reactive chemical [28] and was nevertheless correctly identified as a sensitizer in the PBMDC assay at concentrations <200 μM [17] or at 567 to >1100 μM [24].

Hydroxycitronellal classified in Category 4 ("Infrequent cause of contact allergy in relation to level of exposure"), was identified as a sensitizer at 425 μM [17] or at 431–643 μM [24].

An interesting and informative case is represented by hexylcinnamal, a known human sensitizer from the Category 5 subgroup ("A rare cause of contact allergy except perhaps in special circumstances"). This hapten often remains undetected as a sensitizer in cell line based *in vitro* assays [13, 29]. However, it was detected as a sensitizer with the PBMDC assay by 4 out of 5 laboratories during the ring study [24] at concentrations ranging from 51 μM to 77 μM. The fifth lab didn't classify it as a sensitizer due to the limited test concentration range evaluated.

On the other hand, benzalkonium chloride, also from the Category 5 subgroup, tested positive at a relatively low concentration (0.2 μM, [17]) indicating a possible over sensitivity of the PBMDC assay to surfactants.

Lactic acid, a well characterized non-sensitizer from Category 6 (Clinical data "essentially absent, with at least no systematic convincing evidence of contact allergy") tested negative in the PBMDC assay [25].

This list is certainly not exhaustive. Additional chemicals such as resorcinol (Category 4) or benzyl benzoate or benzyl salicylate (Category 5) have been evaluated with the PBMDC assay in Beiersdorf facilities (unpublished results) and were correctly classified as sensitizers.

23.4 Role in a Testing Strategy

To date, no single *in vitro* or *in silico* approach is able to predict the sensitizing property or potency of a new chemical without animals. Quality information will rely on a wise combination of various approaches [30, 31]. *In silico* methods may provide information on the physico-chemical properties, such as the presence of structural alerts, that allow for a read-across with similar chemicals. If the information is deemed as insufficient, *in vitro* methods should then be involved to confirm and expand this information. Chemistry-based protocols such as the Direct Peptide Reactivity Assay (DPRA), [32] allow the measurement of the binding capacity toward peptides. Information concerning the biological activity (e.g. cellular stress, DC activation) of the tested molecule can be obtained using cell based assays [15, 29, 33, 34]. Such assays are usually cell line-based (e.g. THP-1, U937, Mutz-3) and the obtained information is constrained by the inherent limitations of such cells (e.g. altered and instable genome, limited functional and metabolic properties). On the other hand, test protocols based on human PBMDCs do not suffer from the same limitation. Their role in a testing strategy would be to provide additional and/or confirmatory information, knowing that such information has been obtained with a test system based on DCs derived from primary, fully functional human cells. For example, a pro-hapten that needs to be modified by specific enzymes present in human cells may not exert it's full biological activity in the absence of the necessary metabolic activity [28, 35, 36]. Since the metabolic capacity of cell lines is often deficient or not fully characterized, additional and complementary information on the biological activity of such molecule (pro-hapten) may be gained through the use of PBMDCs based test protocols.

As an example, eugenol that is a known pro-hapten and a weak sensitizer [28], was successfully tested as a sensitizer by all laboratories participating in the above described ring study.

23.5 Perspectives from the Test Developer

23.5.1 Critical Steps in the Protocol

The optimized PBMDC protocol has been proven to be sufficiently stable for routine testing. Some key parameters that were optimized in order to obtain and ensure valid and reproducible data on primary cells should be highlighted: (1) Only buffy coats obtained by centrifugation (and not by filtration) are suitable, (2) A thorough quality control of the isolated monocytic cell fraction to be used for PBMDC generation must be performed. Only monocytes with a ratio of $CD14^+$ cells >70% should be used. (3) The basal CD86 expression of PBMDC to be used in the experiments should be <60%. This is important to exclude pre-activated PBMDCs and to ensure a sufficient dynamic range of the response. (4) Cytotoxic effects of the chemicals should be carefully evaluated in initial screening experiments to prevent unspecific activation due to damage-associated molecular patterns (e.g. HMGB1, heat-shock proteins, TLR-ligands) released by apoptotic and necrotic cells. False positive results are minimized by using test chemical concentrations which do not induce >20% of dead cells. Test concentrations inducing a higher proportion of dead cells should not be considered for the identification of sensitizers. Moreover, lytic compounds lead to the release of danger signals that activate PBMDCs, resulting in doubtful dead cell estimation. Thus, the assessment of cell lysis inducing compounds (e.g. surfactants, detergents) should be carefully evaluated.

23.5.2 Possible Protocol Adaptations

- Due to possible shortage in buffy coats of the required quality, the source of the mononuclear cells may have to be adapted: For example, monocytes isolated and differentiated from cord blood stem cells have been could be adapted to this protocol (unpublished data). However, such samples are difficult to obtain and the use of alternative source such as commercially available cells should be envisaged (using adaptation of the described protocol). The same is true for other approaches for obtaining the mononuclear fractions necessary for the assay. Variation in the preparation of the CD14 positive fraction should be acceptable as long as the DC acceptance criteria (see above) are met.

23.5.3 Challenges and Opportunities

- Automation of DC-isolation and assay
- 384 well format
- Integration in systemic tox assays (3D-models and SysToxChip)

23.6 Conclusions

The protocol described here is the result of almost 20 years of experience [7, 37] and has been recently re-evaluated and optimized [17]. It combines the use of human DCs generated from primary human cells with the measurement of CD86 expression as a marker of DC activation. These essential aspects of the protocol derive from a thorough analysis of the biological mechanisms leading to the acquisition of contact dermatitis: The central role of skin DCs and the up regulation of CD86 on the surface of activated DCs [38, 39]. Both aspects have been firmly established [6, 8, 40] and allowed the successful development of a variety of *in vitro* assays for detection of sensitizers [10]. However, unlike most cell based *in vitro* sensitization protocols published to date, this PBMDC assay rely on the use of human primary cell derived DCs. It is thus capable of delivering high quality information concerning the sensitizing properties of the tested chemicals relevant to the human *in vivo* situation. This assay will deploy its full potential when included in a wider test strategy including *in silico* or *in chemico* methods (see Sect. 5). It should be regarded as primary source of information concerning the DC activation properties of the tested molecule or as an addition and/or a confirmation to the results already obtained with a cell line based test system.

Its reproducibility and predictive capacity have been demonstrated after a final optimization phase [17] and during a ring study organized with five different laboratories ([24] and see pt. 3). As expected for an assay based on primary cells of human origin, the delivered information compared very well with corresponding human sensitization data [27], including for pro-haptens.

However, this assay has its inherent limitations: The physico-chemical properties (solubility, cytotoxicity, chemical activation) of the molecule to be tested should be evaluated and compatible with the assay.

– Keeping in mind these advantages and limitations, this protocol may represent a significant source of information in a test strategy. Its main role is to provide quality biological information on the DC activation properties of test chemical and, using the associated data analysis model, to predict some of its sensitization characteristics such as hazard and potency.

Acknowledgements We thank Silke Gerlach and Jochem Spieker for excellent technical assistance and valuable intellectual contributions in the course of assay development.

References

1. Aeby P, Ashikaga T, Bessou-Touya S, Schepky A, Gerberick F, Kern P, Marrec-Fairley M, Maxwell G, Ovigne JM, Sakaguchi H, Reisinger K, Tailhardat M, Martinozzi-Teissier S, Winkler P. Identifying and characterizing chemical skin sensitizers without animal testing: Colipa's research and method development program. Toxicol In Vitro. 2010;24:1465–73. doi:10.1016/j.tiv.2010.07.005.
2. Basketter D, Casati S, Gerberick GF, Griem P, Philips B, Worth A. Skin sensitisation. Altern Lab Anim. 2005;33(Suppl 1):83–103.
3. Maxwell G, Aeby P, Ashikaga T, Bessou-Touya S, Diembeck W, Gerberick F, Kern P, Marrec-Fairley M, Ovigne JM, Sakaguchi H, Schroeder K, Tailhardat M, Teissier S, Winkler P. Skin sensitisation: the Colipa strategy for developing and evaluating non-animal test methods for risk assessment. ALTEX. 2011;28:50–5.
4. Smith C, Hotchkiss SA. Allergic contact dermatitis. London: Taylor & Francis; 2001.
5. OECD. The Adverse Outcome Pathway for Skin Sensitisation Initiated by Covalent Binding to Proteins, OECD Publishing, Paris, 2014. doi:10.1787/9789264221444-en.
6. Aiba S, Terunuma A, Manome H, Tagami H. Dendritic cells differently respond to haptens and irritants by their production of cytokines and expression of co-stimulatory molecules. Eur J Immunol. 1997;27:3031–8. doi:10.1002/eji.1830271141.
7. Aeby P, Wyss C, Beck H, Griem P, Scheffler H, Goebel C. Characterization of the sensitizing potential of chemicals by in vitro analysis of dendritic cell activation and skin penetration. J Invest Dermatol. 2004;122:1154–64. doi:10.1111/j.0022-202X.2004.22402.x.
8. Degwert J, Steckel F, Hoppe U, Kligman LH. In vitro model for contact sensitization: I. Stimulatory capacities of human blood-derived dendritic cells and their phenotypical alterations in the presence of contact sensitizers. Toxicol In Vitro. 1997;11:613–8.

9. Hulette BA, Ryan CA, Gerberick GF. Elucidating changes in surface marker expression of dendritic cells following chemical allergen treatment. Toxicol Appl Pharmacol. 2002;182:226–33.
10. Basketter D, Maxwell G. In vitro approaches to the identification and characterization of skin sensitizers. Cutan Ocul Toxicol. 2007;26:359–73. doi:10.1080/15569520701622993.
11. Santos dos GG, Reinders J, Ouwehand K, Rustemeyer T, Scheper RJ, Gibbs S. Progress on the development of human in vitro dendritic cell based assays for assessment of the sensitizing potential of a compound. Toxicol Appl Pharmacol. 2009;236:372–82. doi:10.1016/j.taap.2009.02.004.
12. Geraghty RJ, Capes-Davis A, Davis JM, Downward J, Freshney RI, Knezevic I, Lovell-Badge R, Masters JRW, Meredith J, Stacey GN, Thraves P, Vias M, Cancer Research UK. Guidelines for the use of cell lines in biomedical research. Br J Cancer. 2014;111(6):1021–46. doi:10.1038/bjc.2014.166.
13. Sakaguchi H, Ryan C, Ovigne JM, Schroeder KR, Ashikaga T. Predicting skin sensitization potential and inter-laboratory reproducibility of a human Cell Line Activation Test (h-CLAT) in the European Cosmetics Association (COLIPA) ring trials. Toxicol In Vitro. 2010;24:1810–20. doi:10.1016/j.tiv.2010.05.012.
14. Skazik-Voogt C, Kühler K, Ott H, Czaja K, Zwadlo-Klarwasser G, Merk HF, Amann PM, Baron JM. Myeloid human cell lines lack functional regulation of aryl hydrocarbon receptor-dependent phase I genes. ALTEX. 2016;33(1):37–46.
15. Nelissen I, Selderslaghs I, Heuvel RVD, Witters H, Verheyen GR, Schoeters G. MUTZ-3-derived dendritic cells as an in vitro alternative model to CD34+ progenitor-derived dendritic cells for testing of chemical sensitizers. Toxicol In Vitro. 2009;23:1477–81. doi:10.1016/j.tiv.2009.08.022.
16. Python F, Goebel C, Aeby P. Comparative DNA microarray analysis of human monocyte derived dendritic cells and MUTZ-3 cells exposed to the moderate skin sensitizer cinnamaldehyde. Toxicol Appl Pharmacol. 2009;239:273–83. doi:10.1016/j.taap.2009.06.003.
17. Reuter H, Spieker J, Gerlach S, Engels U, Pape W, Kolbe L, Schmucker R, Wenck H, Diembeck W, Wittern KP, Reisinger K, Schepky AG. In vitro detection of contact allergens: development of an optimized protocol using human peripheral blood monocyte-derived dendritic cells. Toxicol In Vitro. 2011;25:315–23. doi:10.1016/j.tiv.2010.09.016.
18. Staquet MJ, Sportouch M, Jacquet C, Schmitt D, Guesnet J, Peguet-Navarro J. Moderate skin sensitizers can induce phenotypic changes on in vitro generated dendritic cells. Toxicol In Vitro. 2004;18:493–500. doi:10.1016/j.tiv.2003.12.005.
19. Schmid I, Krall WJ, Uittenbogaart CH, Braun J, Giorgi JV. Dead cell discrimination with 7-amino-actinomycin D in combination with dual color immunofluorescence in single laser flow cytometry. Cytometry. 1992;13:204–8. doi:10.1002/cyto.990130216.
20. Hackenberg U, Bartling H. Measurement & calculation in pharmacological laboratories with a special numbering system (WL24-system). Naunyn Schmiedebergs Arch Exp Pathol Pharmakol. 1959;235:437–63.
21. Aeby P, Sieber T, Beck H, Gerberick GF, Goebel C. Skin sensitization to p-phenylenediamine: the diverging roles of oxidation and N-acetylation for dendritic cell activation and the immune response. J Invest Dermatol. 2009;129:99–109. doi:10.1038/jid.2008.209.
22. Karschuk N, Tepe Y, Gerlach S, Pape W, Wenck H, Schmucker R, Wittern K-P, Schepky A, Reuter H. A novel in vitro method for the detection and characterization of photosensitizers. PLoS One. 2010;5:e15221. doi:10.1371/journal.pone.0015221.
23. Skazik C, Heise R, Ott H, Czaja K, Marquardt Y, Merk HF, Baron JM. Active transport of contact allergens in human monocyte-derived dendritic cells is mediated by multidrug resistance related proteins. Arch Biochem Biophys. 2011;508:212–6. doi:10.1016/j.abb.2011.01.013.
24. Reuter H, Gerlach S, Spieker J, Ryan C, Bauch C, Mangez C, Winkler P, Landsiedel R, Templier M, Mignot A, Gerberick F, Wenck H, Aeby P, Schepky A. Evaluation of an optimized protocol using human peripheral blood monocyte derived dendritic cells for the in vitro detection of senstizers: results of a ring study in five laboratories. Toxicol In Vitro. 2015;29:976–86. doi:10.1016/j.tiv.2015.03.021.

25. Reisinger K, Hoffmann S, Alépée N, Ashikaga T, Barroso J, Elcombe C, Gellatly N, Galbiati V, Gibbs S, Groux H, Hibatallah J, Keller D, Kern P, Klaric M, Kolle S, Kuehnl J, Lambrechts N, Lindstedt M, Millet M, Martinozzi-Teissier S, Natsch A, Petersohn D, Pike I, Sakaguchi H, Schepky A, Tailhardat M, Templier M, van Vliet E, Maxwell G. Systematic evaluation of non-animal test methods for skin sensitisation safety assessment. Toxicol In Vitro. 2014;29:259–70. doi:10.1016/j.tiv.2014.10.018.
26. Lepoittevin J-P. Metabolism versus chemical transformation or pro-versus prehaptens? Contact Dermat. 2006;54:73–4. doi:10.1111/j.0105-1873.2006.00795.x.
27. Basketter DA, Alépée N, Ashikaga T, Barroso J, Gilmour N, Goebel C, Hibatallah J, Hoffmann S, Kern P, Martinozzi-Teissier S, Maxwell G, Reisinger K, Sakaguchi H, Schepky A, Tailhardat M, Templier M. Categorization of chemicals according to their relative human skin sensitizing potency. Dermatitis. 2014;25:11–21. doi:10.1097/DER.0000000000000003.
28. Barratt MD, Basketter DA. Possible origin of the skin sensitization potential of isoeugenol and related compounds. (I). Preliminary studies of potential reaction mechanisms. Contact Dermat. 1992;27:98–104.
29. Python F, Goebel C, Aeby P. Assessment of the U937 cell line for the detection of contact allergens. Toxicol Appl Pharmacol. 2007;220:113–24. doi:10.1016/j.taap.2006.12.026.
30. Goebel C, Aeby P, Ade N, Alepee N, Aptula A, Araki D, Dufour E, Gilmour N, Hibatallah J, Keller D, Kern P, Kirst A, Marrec-Fairley M, Maxwell G, Rowland J, Safford B, Schellauf F, Schepky A, Seaman C, Teichert T, Tessier N, Teissier S, Weltzien HU, Winkler P, Scheel J. Guiding principles for the implementation of non-animal safety assessment approaches for cosmetics: skin sensitisation. Regul Toxicol Pharmacol. 2012;63:40–52. doi:10.1016/j.yrtph.2012.02.007.
31. Rovida C, Alépée N, Api AM, Basketter DA, Bois FY, Caloni F, Corsini E, Daneshian M, Eskes C, Ezendam J, Fuchs H, Hayden P, Hegele-Hartung C, Hoffmann S, Hubesch B, Jacobs MN, Jaworska J, Kleensang A, Kleinstreuer N, Lalko J, Landsiedel R, Lebreux F, Luechtefeld T, Locatelli M, Mehling A, Natsch A, Pitchford JW, Prater D, Prieto P, Schepky A, Schuurmann G, Smirnova L, Toole C, van Vliet E, Weisensee D, Hartung T. Integrated Testing Strategies (ITS) for safety assessment. ALTEX. 2014;32(1):25–40. doi:10.14573/altex.1411011.
32. Gerberick GF, Vassallo JD, Bailey RE, Chaney JG, Morrall SW, Lepoittevin JP. Development of a peptide reactivity assay for screening contact allergens. Toxicol Sci. 2004;81:332–43. doi:10.1093/toxsci/kfh213.
33. Ade N, Martinozzi-Teissier S, Pallardy M, Rousset F. Activation of U937 cells by contact sensitizers: CD86 expression is independent of apoptosis. J Immunotoxicol. 2006;3:189–97. doi:10.1080/15476910600978038.
34. Ashikaga T, Yoshida Y, Hirota M, Yoneyama K, Itagaki H, Sakaguchi H, Miyazawa M, Ito Y, Suzuki H, Toyoda H. Development of an in vitro skin sensitization test using human cell lines: the human Cell Line Activation Test (h-CLAT). I. Optimization of the h-CLAT protocol. Toxicol In Vitro. 2006;20:767–73. doi:10.1016/j.tiv.2005.10.012.
35. Nilsson AM, Bergstrom MA, Luthman K, Nilsson JL, Karlberg AT. A conjugated diene identified as a prohapten: contact allergenic activity and chemical reactivity of proposed epoxide metabolites. Chem Res Toxicol. 2005;18:308–16. doi:10.1021/tx049758c.
36. Smith Pease CK, Basketter DA, Patlewicz GY. Contact allergy: the role of skin chemistry and metabolism. Clin Exp Dermatol. 2003;28:177–83.
37. Reutter K, Jager D, Degwert J, Hoppe U. In vitro model for contact sensitization: II. Induction of IL-1beta mRNA in human blood-derived dendritic cells by contact sensitizers. Toxicol In Vitro. 1997;11:619–26.
38. Enk AH, Katz SI. Early events in the induction phase of contact sensitivity. J Invest Dermatol. 1992;99:39S–41S.
39. Kimber I, Pichowski JS, Betts CJ, Cumberbatch M, Basketter DA, Dearman RJ. Alternative approaches to the identification and characterization of chemical allergens. Toxicol In Vitro. 2001;15:307–12.
40. Tuschl H, Kovac R. Langerhans cells and immature dendritic cells as model systems for screening of skin sensitizers. Toxicol In Vitro. 2001;15:327–31.

VITOSENS™

24

Nathalie Lambrechts, Greet Schoeters, Rosette Van Den Heuvel, Hilda Witters, Inge Nelissen, and Jef Hooyberghs

24.1 Principle of the Test Method and Scientific Basis

Although the mechanisms underlying skin sensitization are not yet fully understood, some aspects are well-known. For instance, upon hapten encounter, DC undergo the maturation process during which their phenotypical appearance changes in order to permit their migration to the local lymph nodes and activation of surrounding cells [1]. These transitions are characterized by the altered expression of cell surface markers on DC and secretion of cytokines. Similar changes occur after exposure of *in vitro* differentiated DC to chemical skin sensitizers.

Several studies evaluated cytokines as possible biomarkers for skin sensitization and tested their discriminating power [2]. In the CD34$^+$ progenitor-derived dendritic cell (CD34-DC) model, secretion of cytokines such as IL-1β, IL-6, IL-12 and TNF-α were studied. These markers showed differential expression levels after exposure to a limited set of sensitizers, but changes in cytokine production were not consistently related with exposure to an allergen [3, 4].

N. Lambrechts (✉) • R. Van Den Heuvel • H. Witters • I. Nelissen
Unit of Environmental Risk and Health, VITO NV, Boeretang 200, 2400 Mol, Antwerp, Belgium
e-mail: nathalie.lambrechts@vito.be

G. Schoeters
Unit of Environmental Risk and Health, VITO NV, Boeretang 200, 2400 Mol, Antwerp, Belgium

Department of Biomedical Sciences, University of Antwerp, Antwerp, Belgium

Dept of Environmental Medicine, Institute of Public Health, University of Southern Denmark, Odense, Denmark

J. Hooyberghs
Unit of Environmental Risk and Health, VITO NV, Boeretang 200, 2400 Mol, Antwerp, Belgium

Theoretical Physics, Hasselt University, Hasselt, Belgium

C. Eskes et al. (eds.), *Alternatives for Dermal Toxicity Testing*,
DOI 10.1007/978-3-319-50353-0_24

Probably the most studied endpoint in DC are surface markers. In the CD34-DC model a correct classification of 9 skin sensitizers and 2 irritants could be established based on changes in surface marker expression (CD86, CD83 and HLA-DR) [5]. All allergens tested induced a significant increase in at least one of the DC surface markers. In contrast, none of the irritants tested were able to significantly up-regulate membrane marker expression in exposed DC. It therefore could be concluded that the *in vitro* CD34-DC model has the capacity to distinguish between chemical sensitizers and irritants based on altered phenotypic characteristics. However, the dynamic range of a phenotypic response is limited [6].

Based on the above mentioned findings, we applied a new, holistic approach using microarray technology. The genome of DC was screened for differential expression markers after *in vitro* exposure to four skin sensitizers and two non-sensitizers. This resulted in the selection of a set of 13 genes as novel potential biomarkers [7]. The discriminating potential of these markers was confirmed by qPCR experiments on an extended data set of 21 chemicals. A final classification model based on two genes, Chemokine (C–C motif) receptor (CCR)2 and cAMP responsive element modulator (CREM was built, that predicts the sensitizing identity of a chemical (Fig. 24.1). This dichotomous prediction assay was named VITOSENS™ [8].

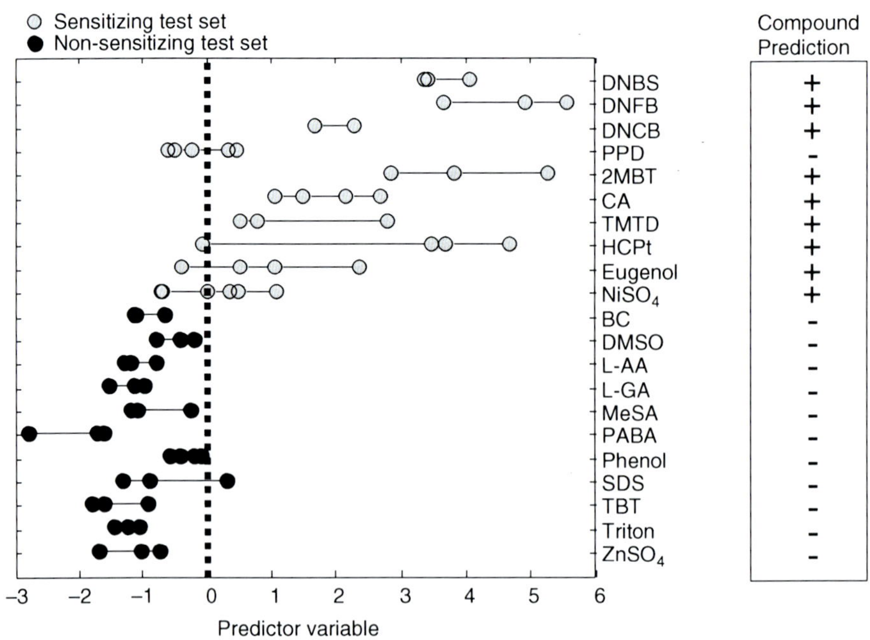

Fig. 24.1 Prediction of the VITOSENS™-assay based on induced CREM and CCR2 gene expression changes [8]

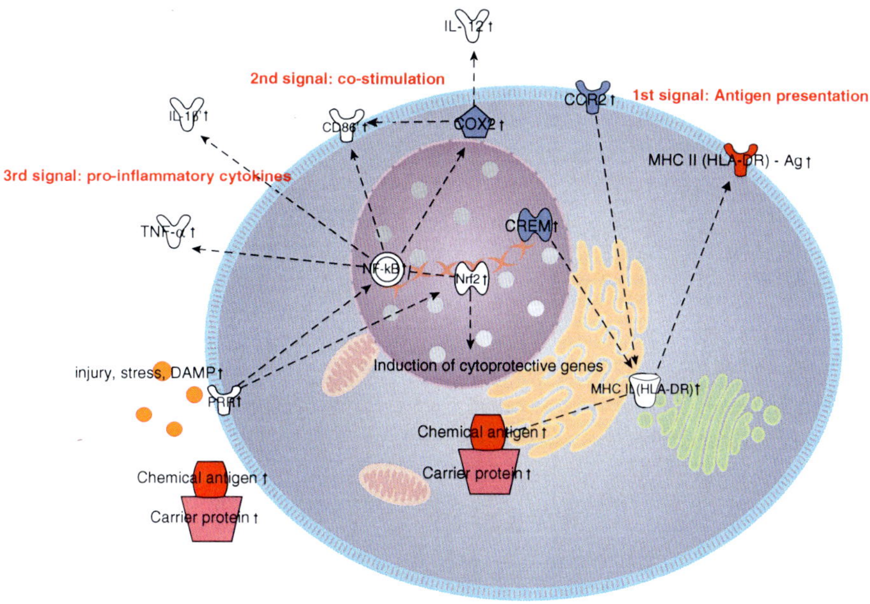

Fig. 24.2 Sensitization cascade hypothesized from mechanistic insights in VITOSENS™ (generated in Ingenuity Systems)

Based on the predictive gene expression markers, a solid mechanistic basis for the VITOSENS assay has been established. In current immunologic knowledge, it is stated that for sensitization to occur effector T cells should be activated by three signals generated by DC. First, the antigen encountered by DC needs to be presented on their surface by MHC molecules, secondly, by expression of co-stimulatory molecules on the DC surface, division and proliferation of T cells is activated, and thirdly, pro-inflammatory cytokines need to be secreted in order to induce T-cell polarization [9, 10] (Fig. 24.2). These signals can be applied to the screening mechanism of VITOSENS™.

CREM has been described to regulate the expression of MHC class II molecules [11]. This is an indication that the first signal of antigen presentation is present in the mechanism by which VITOSENS™ recognizes chemical sensitizers.

Besides CREM and CCR2, also cyclo-oxygenase (COX)2 gene expression was significantly differential between sensitizing versus non-sensitizing exposure in CD34-DC. By constructing a molecular network based on the VITOSENS™ markers using data-mining software, a relation was shown with other molecules that are associated with the skin sensitization process. CREM and COX2 appeared to be central hubs in this network [12].

A further link of CCR2 and COX2 in the functional cascade of DC activation was studied. The significantly differential protein expression of these molecules was established in sensitizer versus non-sensitizer exposed CD34-DC, and their protein expression pattern was confirmed to match the gene expression profile. Further functional investigation showed that counteracting the sensitizer-induced expression of CCR2 significantly influenced HLA-DR, an MHC II surface receptor that further strengthened incorporation of the first signal of antigen presentation.

Also the second signal of co-stimulation seemed to be activated when CD34-DC were exposed to sensitizers. This could be derived directly from the enhanced CD83 and CD86 surface expression in CD34-DC by chemical sensitizers, but also COX2 could be identified as a novel marker for this second signal. Inhibiting its protein activity after sensitizing exposure significantly reduced CD86 expression. COX2 is also involved in PGE2 synthesis, an eicosanoid that is known to polarize the T helper cell profile. Therefore, COX2 might also represent the third signal of cytokine secretion in VITOSENS™. Also, the differential over-expression of TNF-α indicated that pro-inflammatory cytokines play a role in the VITOSENS™ discrimination mechanism. Previous studies on CD34-DC revealed sensitizer-induced expression of TNF-α, IL-1β and IL-12 in the cellular assay [3, 5].

Further research showed that it was possible to rank skin sensitizers according to their potency by taking the chemical's inherent irritating capacity into account. Combining the induced gene expression changes of a skin sensitizer with the concentration that induced 20% cytotoxicity resulted in a significant correlation with *in vivo* potency predictions [13]. The lower the concentration of the chemical to induce a defined level of cellular damage, the more potent it is for sensitization induction. *In vivo* similar findings have been observed; concurrent application of allergens and irritants promotes a stronger clinical response than when applied separately [14, 15].

A possible explanation for the above described phenomena could be attributed to the danger hypothesis [16]. Briefly, this theory suggests that the distinction by which the immune system reacts to certain molecules is not based on self and non-self, but rather on their potential danger properties. More specifically, APC are suspected to be activated by alarm signals that emanate from stressed or injured tissues, without which no primary immune responses can occur [17]. This reasoning can easily be translated to the sensitization cascade. The usual outcome of antigen uptake is tolerance since the antigen-presenting DC will not properly express co-stimulatory molecules in the absence of inflammatory triggers. Consequently, the induced T-cell response is mainly characterized by cell division, but very little differentiation to effector cells [10]. However, in the presence of danger-related molecules, such as for instance 'pathogen-associated molecular-pattern molecules' (PAMPs), 'damage-associated molecular-pattern molecules' (DAMPs), or cell injury like cytotoxicity, DC will be activated to express the two other required signals of co-stimulation and pro-inflammatory cytokine secretion. As a consequence, recognition of danger molecules is suspected to determine the size and longevity of an immunological response [18].

At the level of the VITOSENS™ gene markers, a distinct expression pattern between LPS and DNFB exposure could be observed as well. The response of both CREM and CCR2, the two genes that react most discriminating upon sensitizing exposure of CD34-DC as for DNFB, was not affected by exposure to LPS. Being a PAMP, LPS is considered to be a sensitizing adjuvant or a molecule that lowers sensitization thresholds by operating via Toll-like receptor signaling in activating signals 2 and 3 [10]. LPS induced a significant induction of co-stimulatory molecules CD86 and CD83 in CD34-DC, while HLA-DR, needed for antigen presentation, was not significantly altered [19]. This was opposite to the results obtained after exposure to the chemical sensitizer DNFB that significantly enhanced the expression of HLA-DR, as well as of CD86 and CD83. Although these observations suggest that LPS is not linked to enhanced antigen presentation in *in vitro* cultured CD34-DC, this hypothesis should be considered with some care, as *in vitro*-derived DC already display high expression of MHC-II surface receptors without being exposed to maturation stimuli [20].

Altogether, the above findings indicate that VITOSENS™ only responds to sensitizing agents that encompass all three signals needed for the onset of sensitization: induction of inflammatory cytokines and surface molecules, as well as presentation of histocompatibility complexes. These signals are designated as so-called key events at the cellular level in the adverse outcome pathway (AOP) on skin sensitization published by the OECD (OECD N° 168, [21, 22]). Furthermore, the metabolism pathway, an event relating to chemical structure and properties, is also present in the CD34-DC model. After all, three pro-haptens (e.g. eugenol, geraniol and cinnamic alcohol) have all been ranked correctly according to their skin sensitizing potency [13].

24.2 Performance and Applicability of the Test Method

24.2.1 Reproducibility

With the VITOSENS™ assay data on 32 chemicals have been published [8, 13, 23]. Of this selection eight were re-tested in the blinded Cosmetics Europe ring trial: dihydroquinone, methyldibromoglutaronitrile, cinnamic alcohol, dinitrochlorobenzene, 2-mercaptobenzothiazole, cinnamaldehyde, tetramethylthiuram disulphide and sodium lauryl/dodecyl sulphate. For all these chemicals consistent predictions were made concerning both sensitizing hazard as well as potency.

24.2.2 Predictive Capacity

For the 32 chemicals evaluated in the VITOSENS™ assay, a specificity of 100% (13/13), a sensitivity of 94.7% (18/19) and an overall concordance of 96.9% (31/32) was calculated. This set of compounds contains chemicals that can be immediately haptenated, but also pre/pro-prohaptens are present, being chemicals that require prior activation either metabolic or chemical (Lepoittevin). Predictions of paraphenylene diamine, (*p*PD), have not been consistently positive as should be expected from a strong sensitizer. In the paper published by Hooyberghs et al. [8], it was predicted a non-sensitizer by VITOSENS™, while *in vivo* studies (both human and animal) and recent in house experiments as well as the blinded CE ring trial clearly show a sensitizing potential (see also Fig. 24.1). A possible explanation for an occasional false negative result is the fact that *in vitro* *p*PD is acetylated in cytosol with consequent loss of sensitization potency [24], while *in vivo* atmospheric oxidation prevents further acetylation thereby retaining the sensitizing capacity [25]. When the oxidized metabolite of pPD, Bandrowski's Base, was applied to VITOSENS, also positive identification of its sensitizing character was obtained. A compound that was consistently erroneously predicted as a non-sensitizer by VITOSENS™ is phenyl benzoate, a challenging compound for *in vitro* assays, as it was misclassified as a non-sensitizer by six test methods in the CE ring trial [23]. By combining both the induced gene expression changes of CREM and CCR2, with the 20% cytotoxic concentration in a robust multiple linear regression analysis a potency value was modelled that closely fitted the LLNA potency data, and this over the entire range from weak to extremely sensitizing chemicals (robust multiple linear regression with spearman rank-correlation coefficient: 0.91) [13].

24.2.3 Applications and Limitations

Delineating the group of chemicals for which a test is applicable in order to make predictions for new compounds is of great importance for future validation of assays such as VITOSENS™. Extending the so far limited chemical data set of VITOSENS™ is a first important step in this process. However, as was described higher, additional compounds should be selected with care, covering multiple reactivity classes, and not only sensitizers versus harmless or irritating compounds.

One observation concerning applicability domain that has been made involves testing non-cytotoxic compounds. As stated above, the mechanism by which VITOSENS™ discriminates is based on the recognition of three signals (24.1). The second and third signals can only take place if there is some degree of danger present, or in the case of chemical insult, if there is some degree of cytotoxicity. This hypothesis coincides with the observation that non-cytotoxic chemicals are often classified as false negatives in the VITOSENS™ assay.

To set off this applicability gap, experiments have been conducted in which an external danger signal was added to the sensitizer applied at a non-cytotoxic concentration. The results indicate that such a modification of the protocol can aid in the positive identification of non-cytotoxic sensitizers. Further, this observation will be helpful when exploring safety assessment of mixtures.

For further delineation of the applicability domains of VITOSENS™, characterization of the metabolic capacity of the CD34-DC model is essential. Being primary dendritic cells, CD34-DC have considerable metabolic capacity compared to immortalized cell lines. As such, they may also serve as a complementary read-out for assessment of pre-haptens.

As primary human cells, CD34-DC are able to closely mimic the induction phase of sensitization by DC in the skin, however their isolation and culturing procedure is intensive and their use depends on the availability of cord blood. Comparing the response of VITOSENS™ genes in CD34-DC versus THP-1-derived cell lines suggested that CD34-DC represent not only the most relevant model due to their primary origin, but also that they are the most capable cells in our set-up [26]. Primary cells are highly representative of the human standard [6], which is an important advantage considering that a fundamental aspect of *in vitro* assay development is the relationship between the test object and the heterogeneous human population [27].

24.2.4 Comparison to Human Data

VITOSENS™ was first developed using a training set of 21 chemicals (Table 24.1). For these compounds, concordance with human data was investigated [8]. Out of 21 chemical, the sensitizing capacity of 16 were described in humans: seven were identified as sensitizers, eight compounds as non-sensitizers, and Benzalkonium chloride (BC) is an irritating compound, but nonetheless cases have been described in which humans are reported to be sensitized after exposure. However, symptoms of irritating exposure may sometimes be confused with sensitizing exposure and therefore the classification of BC as sensitizing may not be objective [28].

Table 24.1 Set of 21 chemical compounds used for training of the VITOSENS™assay, and their response in animal assays and humans

Chemical	Cas-n°	Abbr.	Effect	LLNA	Modified LLNA	GPMT/ BT	Human
Dinitrobenzenesulfonic acid	885-62-1	DNBS	+	+			
Dinitrofluorobenzene	70-34-8	DNFB	+	+			
Dinitrochlorobenzene	97-00-7	DNCB	+	+	+	+	+
p-phenylenediamine	106-50-3	PPD	+/−	+		+	+
2-mercaptobenzothiazole	149-30-4	2MBT	+	+	+	+	+
Cinnamaldehyde	104-55-2	CA	+	+		+	+
Tetramethylthiuram disulfide	137-26-8	TMTD	+	+	+	+	+
Ammonium hexachloroplatinate IV	16919-58-7	HCPt	+	+	+		
Eugenol	97-53-0	eugenol	+	+	+	+	+
Nickel Sulfate	10101-97-0	$NiSO_4$	+	−		+	+
Benzalkonium Chloride	8001-54-5	BC	−	−	+	−	+
Dimethylsulfoxide	67-68-5	DMSO	−	−		−	
L-Ascorbic Acid	50-81-7	L-AA	−				−
L-Glutamic Acid	56-86-0	L-GA	−				−
Methyl salicylate	119-36-8	MeSA	−	−	−	−	−
p-Aminobenzoic Acid	150-13-0	PABA	−	−		−	−
Phenol	108-95-2	phenol	−	−			−
Sodium Lauryl/Dodecyl Sulphate	151-21-3	SDS	−	+	−	−	−
Tributyltin Chloride	1461-22-9	TBT	−			−	
Triton X-100	9002-93-1	triton	−	+	−	−	−
Zinc sulphate	7733-02-0	$ZnSO_4$	−	+	−		−

+, sensitizing; −, non-sensitizing; empty cell, not available. Effect: a priori sensitizing character, LLNA: local lymph node assay, Modified: a modified protocol of LLNA, GPMT/BT: Guinea Pig Maximization Test/Buehler Patch Test, Human: conclusions as reported by ECETOC (if available)

24.3 Brief Description of the Protocol

VITOSENS™ is developed as an *in vitro* assay that is able to identify skin sensitizing chemicals [8]. The assay is based on dendritic cells derived from CD34+ progenitor cells (CD34-DC) obtained from human cord blood. The response of CD34-DC to chemical exposure is assessed at the level of gene expression: by means of qPCR measurements of the expression fold-change of a set of genes. This fold-change is induced in the exposed sample versus solvent control sample. The current protocol evaluates the expression of the genes Chemokine (C–C motif) receptor 2 (CCR2) and cAMP responsive element modulator (CREM) after 6 h exposure to a chemical concentration that yields around 20% cell death (IC20). The latter is measured by propidium iodide (PI) staining and flow cytometry. The resulting fold-changes are combined by a weighted average into a predictor variable that should be positive for sensitizing compounds and negative for non-sensitizers. The VITOSENS™ assay has been designed as a dichotomous classifier: a test substance is classified as either sensitizing or non-sensitizing [8].

Briefly, the IC20 is determined in a dose range experiment on cells derived from 1 donor in 96-well plates by propidium iodide staining and flow cytometry. Then, an additional donor sample is exposed for 6 h, after which part of the cells are collected for RNA extraction and subsequent expression analysis of CREM and CCR2 using qPCR. After 24 h, the remainder of the cells are collected and the chemical-induced cell death in this sample is measured using PI. The cells from this sample are exposed to a dilution series (n = 8) of the test item including the estimated IC20. For each of these conditions both gene expression and cell death data are gathered. This procedure allows accurate selection of the exposure concentration as the one approximating as close as possible 20% cell death and this specific for the individuals that will be analyzed for gene expression read-out. Hence, in the past, the IC20 was defined in advance by experiments on three separate donor samples. Due to both the primary origin of the cells and technical variability, the pre-selected exposure concentration sometimes resulted in considerable variance in cell death induction between donor samples and consequently influenced the prediction outcome. The new approach takes into account the aspect of individual variability and turns it into an opportunity to approximate and optimize the prediction with an individualized response.

Next, a biologically distinct donor sample is exposed for further analyses after 6 h and 24 h to no more than three different concentrations also containing the estimated IC20. If the gene expression response of both donors did not coincide, an additional third donor sample is exposed using an analogous protocol. For all tested donors applies that if no cell damage is induced, the highest soluble concentration is added to the cells, with a maximum of 500 μg/mL. This maximum concentration is applied since in the past we observed that at high concentrations (>500 μg/mL) toxic effects may occur that are no longer representative of a danger signal (see Fig. 24.3).

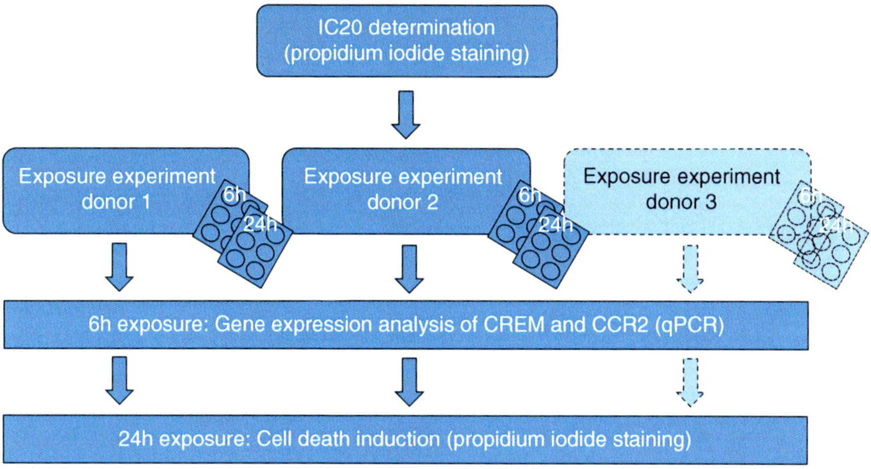

Fig. 24.3 Experimental set-up of VITOSENS™

24.4 Role in a Testing Strategy

Currently, the public opinion on alternative testing is that no single measurement will be sufficient to predict the sensitizing potency of the complex underlying biology of this biological process [29]. On the other hand, skin sensitization is a well-developed test case in an integrated testing strategy, due to a fairly good understanding of this biology. This is perhaps best illustrated by AOP for skin sensitization that the OECD have developed, thereby describing the existing biological knowledge as the linkage between a molecular initiating event and an adverse outcome at the individual or population level (OECD N° 168, [21, 22]).

In a number of recent publications, different testing strategies for skin sensitization are proposed [30–32]. Generally, it is stated that for simple hazard classification the existing information (*in vitro*, *in vivo* and/or human data) combined with *in silico* data (read-across, QSARs) may be sufficient. However, if more information is needed, for instance to define threshold levels for triggering human reactions, or to distinguish between skin or respiratory mechanisms, the inclusion of more complex and mechanistically relevant assays may be needed [33]. In the latter strategy, VITOSENS™ may be a valuable asset. As was described in 24.1 this assay reflects the key events 3 and 4 of the AOP (presentation of haptenated protein by dendritic cells resulting in activation and proliferation of T cells. It is this mechanistic underpinning in the primary cell-based assay that complements other more high-throughput dendritic cell assays as they more often represent only the second and third signal needed for T cell activation and lack the specificity of antigen presentation (first signal) (Fig. 24.2). Often, *in vitro* assays tend to have an unequal distribution of sensitizers versus non-sensitizers in their training set, and of these negative compounds, a strong majority does not induce cytotoxicity [34, 35], thus no danger and therefore no co-stimulation. The fact that VITOSENS™ does comprise the antigen signal and the danger signal, renders this assay highly specific for sensitizers, but also prone to false negative classification: as was explained in 24.3.3, non-cytotoxic sensitizers may not be identified as such. In an integrated testing strategy, this may be solved by using complementary assays with a different mode of action due to a different position in the sensitization cascade, and a higher sensitivity.

24.5 Perspectives from the Test Developer

24.5.1 Critical Steps in the Protocol

Since the cell model of the VITOSENS™ assays consists of primary cells, their generation and differentiation require some skills and time. However, as was said in 24.2, transfer of the protocol has been successfully performed in the past. To this end, the protocol is not thought to contain any critical aspects.

24.5.2 Possible Protocol Adaptations

As was described above, the applicability of the assay is currently confined to cytotoxic chemicals. To rectify this, addition of an adjuvant to the exposure in order to induce the danger signal can be considered. This would be in line with *in vivo* assays such as GPMT where Freund's complete adjuvant and/or SLS is co-administered to the animal, rendering this assay more suited for evaluating weak sensitizers [36]. In the LLNA, several different vehicles are used, including acetone: olive oil (AOO), DMSO, and propylene glycol, and they also have been shown to augment the LLNA response of certain chemicals [37].

24.5.3 Challenges and Opportunities

Besides adapting the VITOSENS protocol towards assessment of non-cytotoxic chemicals, also screening of substances that share the biological aspects of allergenicity, such as respiratory LMW chemicals and even high-molecular weight substances like pollen, should be investigated. Hence, nowadays respiratory sensitizers cannot be identified as such, neither by animal testing, nor by alternative assays [38]. Preliminary experiments on the subject have been conducted but further research is required to optimize the hazard and risk classification of respiratory sensitizers.

From the results obtained with mixture testing, it appears that the magnitude of the response of the VITOSENS™ markers is susceptible to danger-related signals.

The accomplished results indicate that VITOSENS™ allows *in vitro* evaluation of mixture effects, or of sensitization-modifying compounds. Further research into the matter may also contribute to more insights and optimized testing strategies for sensitizing potency.

In analyzing the individual response of cord blood donors, a tendency has been observed in the induction profile of CREM by a sensitizer versus a non-sensitizer. It was observed that the donor effect on the expression of CREM was much smaller than the effect induced by the compound, therefore these inter-individual differences do not compromise the predictions. However, these indicative findings raise new research questions; e.g. can VITOSENS™ contribute to predicting sensitization on an individual level? Can no or minimal-effect levels be derived? Can product efficacy be optimized by developing individualized anti-allergic cosmetics/therapeutics? These questions may be explored within new research goals on personalized medicine and health.

24.6 Conclusions

As is the general demand of legislation, our society should head towards chemical hygiene and this preferably in an animal-sparing approach. However, up to now, no alternative testing method has been validated to replace the LLNA or GPMT. Since *in vitro* assays only cover a specific step in the complex biological cascade of skin sensitization as it occurs *in vivo*, it is unlikely that a single method will be able to substitute animal tests [39]. This is especially true for assessing the risk imposed by exposure to skin sensitizing chemicals. Instead of dissecting all elements of the immunobiological response of skin sensitization using distinct tests, integration of the various data elements has been proposed [29, 33, 40].

Each assay in such an integrated approach should meet at least the EURL ECVAM criteria for entering pre-validation, including its mechanistic relevance, predictive capacity and evidence of the reproducibility of the method [41, 42]. Bearing this in mind, VITOSENS™ has good prospects to become an alternative-screening test of possible skin sensitizers. The assay has a good predictive capacity on both hazard and potency of skin sensitizers, but should be further underpinned by assessing more chemicals. Further, both the cell model of primary DC, and the novel sensitization-related markers contribute to the mechanistic relevance of VITOSENS® as a disease-based assay for skin sensitization. The obtained findings suggest that the discriminating mechanism of VITOSENS® is comprising the three signals that are required to evoke sensitization, thereby rendering this assay highly specific for sensitizers, and comprising key events of the AOP on skin sensitization. Furthermore, the metabolism pathway is also present in the CD34-DC model. VITOSENS™ may thus be a relevant assay to map the adverse outcome pathway for skin sensitization and this by a quantitative output that allows an indication of skin sensitizing potency.

References

1. Romani N, Holzmann S, Tripp CH, Koch F, Stoitzner P. Langerhans cells – dendritic cells of the epidermis. APMIS. 2003;111(7–8):725–40.
2. dos Santos GG, Reinders J, Ouwehand K, Rustemeyer T, Scheper RJ, Gibbs S. Progress on the development of human in vitro dendritic cell based assays for assessment of the sensitizing potential of a compound. Toxicol Appl Pharmacol. 2009;236(3):372–82.
3. De Smedt ACA, Van Den Heuvel RL, Zwi Berneman N, Schoeters GER. Modulation of phenotype, cytokine production and stimulatory function of CD34+-derived DC by NiCl2 and SDS. Toxicol In Vitro. 2001;15(4-5):319.
4. De Smedt AC, Van Den Heuvel RL, Van Tendeloo VF, Berneman ZN, Schoeters GE, Weber E, Tuschl H. Phenotypic alterations and IL-1beta production in CD34(+) progenitor- and monocyte-derived dendritic cells after exposure to allergens: a comparative analysis. Arch Dermatol Res. 2002;294(3):109–16.
5. De Smedt AC, Van Den Heuvel RL, Van Tendeloo VF, Berneman ZN, Schoeters GE. Capacity of CD34+ progenitor-derived dendritic cells to distinguish between sensitizers and irritants. Toxicol Lett. 2005;156(3):377–89.
6. Casati S, Aeby P, Basketter DA, Cavani A, Gennari A, Gerberick GF, Griem P, Hartung T, Kimber I, Lepoittevin JP, Meade BJ, Pallardy M, Rougier N, Rousset F, Rubinstenn G, Sallusto F, Verheyen GR, Zuang V. Dendritic cells as a tool for the predictive identification of skin sensitisation hazard. Altern Lab Anim. 2005;33(1):47–62.
7. Schoeters E, Verheyen GR, Nelissen I, Van Rompay AR, Hooyberghs J, Van Den Heuvel RL, Witters H, Schoeters GE, Van Tendeloo VF, Berneman ZN. Microarray analyses in dendritic cells reveal potential biomarkers for chemical-induced skin sensitization. Mol Immunol. 2007;44(12):3222–33.
8. Hooyberghs J, Schoeters E, Lambrechts N, Nelissen I, Witters H, Schoeters G, Van Den Heuvel R. A cell-based in vitro alternative to identify skin sensitizers by gene expression. Toxicol Appl Pharmacol. 2008;231(1):103–11.
9. Vocanson M, Hennino A, Rozieres A, Poyet G, Nicolas JF. Effector and regulatory mechanisms in allergic contact dermatitis. Allergy. 2009;64(12):1699–714.
10. Willart MA, Lambrecht BN. The danger within: endogenous danger signals, atopy and asthma. Clin Exp Allergy. 2009;39(1):12–9.
11. Moreno CS, Beresford GW, Louis-Plence P, Morris AC, Boss JM. CREB regulates MHC class II expression in a CIITA-dependent manner. Immunity. 1999;10(2):143–51.
12. Lambrechts N, Vanheel H, Hooyberghs J, De Boever P, Witters H, Van Den Heuvel R, Van Tendeloo V, Nelissen I, Schoeters G. Gene markers in dendritic cells unravel pieces of the skin sensitization puzzle. Toxicol Lett. 2010;196(2):95–103.
13. Lambrechts N, Vanheel H, Nelissen I, Witters H, Van Den Heuvel R, Van Tendeloo V, Schoeters G, Hooyberghs J. Assessment of chemical skin sensitizing potency by an in vitro assay based on human dendritic cells. Toxicol Sci. 2010;116(1):122–9.
14. Grabbe S, Steinert M, Mahnke K, Schwartz A, Luger TA, Schwarz T. Dissection of antigenic and irritative effects of epicutaneously applied haptens in mice. Evidence that not the antigenic component but nonspecific proinflammatory effects of haptens determine the concentration-dependent elicitation of allergic contact dermatitis. J Clin Invest. 1996;98(5):1158–64.
15. Pedersen LK, Johansen JD, Held E, Agner T. Augmentation of skin response by exposure to a combination of allergens and irritants – a review. Contact Dermat. 2004;50(5):265–73.
16. Matzinger P. An innate sense of danger. Semin Immunol. 1998;10(5):399–415.
17. Gallucci S, Matzinger P. Danger signals: SOS to the immune system. Curr Opin Immunol. 2001;13(1):114–9.
18. Noble A. Do we have memory of danger as well as antigen?. Trends in Immunology 2009;30 (4):150–156
19. Lambrechts N, Nelissen I, Van TV, Witters H, Van Den Heuvel R, Hooyberghs J, Schoeters G. Functionality and specificity of gene markers for skin sensitization in dendritic cells. Toxicol Lett. 2011;203(2):106–10.
20. Romani N, Clausen BE, Stoitzner P. Langerhans cells and more: langerin-expressing dendritic cell subsets in the skin. Immunol Rev. 2010;234(1):120–41.

21. OECD. The adverse outcome pathway for skin sensitisation initiated by covalent binding to proteins – part 1. In: OECD environment, health and safety publications series on testing and assessment, vol. 168; 2012. p. 1–59. http://www.oecd.org/env/the-adverse-outcome-pathway-for-skin-sensitisation-initiated-by-covalent-bindingto-proteins-9789264221444-en.htm
22. OECD. The adverse outcome pathway for skin sensitisation initiated by covalent binding to proteins. Part 2: use of the AOP to develop chemical categories and integrated assessment and testing approaches. In: OECD environment, health and safety publications series on testing and assessment, vol. 168; 2012. p. 1–46. http://www.oecd.org/env/the-adverse-outcome-pathway-for-skin-sensitisation-initiated-by-covalent-bindingto-proteins-9789264221444-en.htm
23. Reisinger K, Hoffmann S, Alepee N, Ashikaga T, Barroso J, Elcombe C, Gellatly N, Galbiati V, Gibbs S, Groux H, Hibatallah J, Keller D, Kern P, Klaric M, Kolle S, Kuehnl J, Lambrechts N, Lindstedt M, Millet M, Martinozzi-Teissier S, Natsch A, Petersohn D, Pike I, Sakaguchi H, Schepky A, Tailhardat M, Templier M van VE, Maxwell G. Systematic evaluation of non-animal test methods for skin sensitisation safety assessment. Toxicol In Vitro. 2015;29(1):259–70.
24. Aeby P, Wyss C, Beck H, Griem P, Scheffler H, Goebel C. Characterization of the sensitizing potential of chemicals by in vitro analysis of dendritic cell activation and skin penetration. J Invest Dermatol. 2004;122(5):1154–64.
25. Goebel C, Sieber T, Göttel O, Chassot L, Gerberick F, Aeby P. N-Acetylation of aromatic amine hair dyes antagonizes haptenization. Toxicol Lett. 2007;172:S31–2.
26. Lambrechts N, Verstraelen S, Lodewyckx H, Felicio A, Hooyberghs J, Witters H, Van Tendeloo V, Van Cauwenberge P, Nelissen I, Van Den Heuvel R, Schoeters G. THP-1 monocytes but not macrophages as a potential alternative for CD34+ dendritic cells to identify chemical skin sensitizers. Toxicol Appl Pharmacol. 2009;236(2):221–30.
27. Andersen ME, Krewski D. Toxicity testing in the 21st century: bringing the vision to life. Toxicol Sci. 2009;107(2):324–30.
28. Basketter DA, Gerberick GF, Kimber I. Strategies for identifying false positive responses in predictive skin sensitization tests. Food Chem Toxicol. 1998;36(4):327–33.
29. Jowsey IR, Basketter DA, Westmoreland C, Kimber I. A future approach to measuring relative skin sensitising potency: a proposal. J Appl Toxicol. 2006;26(4):341–50.
30. Bauch C, et al. Putting the parts together: combining in vitro methods to test for skin sensitizing potentials. Regul Toxicol Pharmacol. 2012;63(3):489–504.
31. Jaworska J, et al. Bayesian integrated testing strategy to assess skin sensitization potency: from theory to practice. J Appl Toxicol. 2013;33:1353–64.
32. Van der Veen JW, et al. Evaluating the performance of integrated approaches for hazard identification of skin sensitizing chemicals. Regul Toxicol Pharmacol. 2014;69:371–9.
33. Rovida C. Integrated Testing Strategies (ITS) for safety assessment. ALTEX. 2014;32(1):25–40.
34. Ashikaga T, et al. Database of h-CLAT (cell-based skin sensitization test) for clarification of applicability domain. Toxicol Lett. 2008;180S:S95.
35. Emter R, Ellis G, Natsch A. Performance of a novel keratinocyte-based reporter cell line to screen skin sensitizers in vitro. Toxicol Appl Pharmacol. 2010;245(3):281–90.
36. Andersen KE. Guinea pig maximization test: effect of type of Freund's complete adjuvant emulsion and of challenge site location. Dermatosen Beruf Umwelt. 1985;33(4):132–6.
37. Anderson SE, Siegel PD, Meade BJ. The LLNA: a brief review of recent advances and limitations. J Allergy (Cairo). 2011;2011:424203.
38. Arts JH, Mommers C, de Heer C. Dose-response relationships and threshold levels in skin and respiratory allergy. Crit Rev Toxicol. 2006;36(3):219–51.
39. ECHA. Chapter R.7a: Endpoint specific guidance. In: Guidance on information requirements and chemical safety assessment. 2008.
40. Basketter DA, Kimber I. Updating the skin sensitization in vitro data assessment paradigm in 2009. J Appl Toxicol. 2009;29(6):545–50.
41. Balls M. Validation of alternative tests in the European Union. Curr Probl Dermatol. 1995;23:265–74.
42. Hartung T, et al. A modular approach to the ECVAM principles on test validity. Altern Lab Anim. 2004;32(5):467–72.

The SENS-IS Assay

25

Françoise Cottrez, Elodie Boitel, Claude Auriault,
and Hervé Groux

25.1 Principle of the Test Method and Scientific Basis

Skin sensitization results from an interaction between a reactive chemical (hapten) with self-proteins [1]. That interaction induces modification in the protein that is thus recognized as a foreign antigen by the adaptive immune system, more precisely by $CD8^+$- and $CD4^+$-specific T cells [2]. In parallel, danger signals are induced that trigger the adaptive immune response [3]. The test method described here results from a careful analysis of the gene expression modification induced by these protein modifications in skin cells. In order to detect minute modification of the gene expression pattern, a multi-target quantitative real-time polymerase chain reaction (qPCR) approach was chosen [4].

As a prerequisite, a thorough data mining and literature review on skin biology and inflammation allowed to identify and select relevant genes and biological pathways. A further selection based on qPCR measurements of the modulation of gene expression during the *in vivo* sensitization processes either on mice (LLNA) or humans (blisters) led to the compilation of a comprehensive panel of *in vivo* skin-derived sensitization biomarkers. The selected genes include already identified markers such as the ARE family [5] and others not yet associated with the sensitization process (the so-called SENS-IS gene subset). The expression of these unique sets of genes was then measured by qRT-PCR on reconstituted human skin models (Episkin™) [6] exposed to various sensitizers and non-sensitizers. The data set obtained from these experiments was then used to further refine the selected genes set and to develop a prediction model [7].

The finalized test protocol (SENS-IS protocol) uses reconstituted 3D human epidermis (Episkin™) [6] as the test system and a qPCR analysis of the expression of a selected set of biomarkers. Briefly, set dilutions of the test item (50, 10, 1, and 0.1%)

F. Cottrez • E. Boitel • C. Auriault • H. Groux (✉)
ImmunoSearch, Les Cyclades, Chemin de Camperousse, 06130 Grasse, France
e-mail: hgroux@immunosearch.fr

C. Eskes et al. (eds.), *Alternatives for Dermal Toxicity Testing*,
DOI 10.1007/978-3-319-50353-0_25

are applied onto the skin model for 15 min and washed out [6]. The tissues are then incubated for 6 h and processed, and 65 specific biomarkers (subdivided into three subgroups, see below) are analyzed by qPCR. The expression of each gene is normalized to the expression of three housekeeping genes. A positive overexpression is noted when fold increase (ΔE) above vehicle controls is >1.25. Combined with the prediction model, this test protocol proposes a classification of the sensitizing potency of the tested chemicals into four categories similar to the one used by ECETOC with the LLNA (weak, moderate, strong, and extreme) [8].

In order to ensure reproducibility over different batches of reconstituted epidermis, the prediction model measures the number of overexpressed genes and not their expression level. This approach offers an overall picture of the reactions of the epidermis exposed to the test chemical and minimizes the influence of batch to batch variation of the skin models.

The set of 65 genes has been subdivided in three subgroups: two sets of sensitization and one set of irritation biomarkers.

The first set of sensitization biomarkers includes genes involved in the antioxidant (ARE genes) or redox responses and has been designated as the "Redox" subgroup. These 17 genes are predominantly induced by chemicals that bind to cysteine.

The second set includes sensitization biomarkers that are not related to the redox pathways and has been designated as the "SENS-IS" subgroup. These 21 genes are predominantly induced by chemicals that bind to lysine.

The third set includes 24 biomarkers related to the irritation process and has been designated as the "Irritation" subgroup. If a test chemical induces more than 20 genes from that set, it is considered as over-irritating and the corresponding results discarded. A lower test item concentration is then analyzed.

If a test chemical (at any given concentration) induces more than seven genes from either the "Redox" or "SENS-IS" group, it is classified as a sensitizer.

The sensitization potency is determined using four categories, weak, moderate, strong, and extreme, based on the lowest concentration required to induce a positive test according to the following scheme:

Positive at 0.1%, 1%, 10%, or 50% classifies the test item as "extreme," "strong," "moderate," or "weak," respectively.

25.2 Current Validation Status

The reproducibility (intra- and inter-laboratory), predictive capacity, and transferability of the SENS-IS assay have been evaluated in three different laboratories [9]. Overall, the results indicated a successful transfer in the participating laboratories. The reproducibility and the predictivity have been considered as very satisfactory. A pre-submission dossier has also been proposed to the ECVAM. After data review, the EURL ECVAM has invited the developers to submit a full validation dossier. However, the assay being protected by a patent request, the further evaluation of the validation dossier is pending on the OECD approval of the business model and license agreement to third parties.

25.3 Performance and Applicability of the Test Method

25.3.1 Reproducibility

The assay relies on the number of overexpressed genes in each of three biomarker subgroups to predict the eventual sensitization characteristics of a test item (see principle of the test method and scientific basis). The reproducibility of this assay was thus evaluated using the variability in the number of overexpressed genes in the different groups after exposure to a given chemical rather than simply measuring concordance of prediction.

The reproducibility was calculated using data obtained in three separate experiments. The chemical test set was based on the list of 27 chemicals published by the Sens-it-iv consortium (see Fig. 25.1). This list contains 13 skin sensitizers: 2,4-dinitrochlorobenzene (DNCB), cinnamaldehyde, tetramethyl thiuram disulfide (TMTD), resorcinol, oxazolone, glyoxal, 2-mercaptobenzothiazole (MBT), 2-bromo-2-(bromomethyl) glutaronitrile (BBMG), 4-nitrobenzylbromide (4NBB), and three pro-haptens (isoeugenol, eugenol, and cinnamic alcohol) and one pre-hapten (paraphenylene diamine (PPD)). This set was completed by five respiratory sensitizers, diphenylmethane diisocyanate (4′ MDI), trimellitic anhydride (TMA), ammonium hexachloroplatinate (AHCP), hexamethylene diisocyanate (HMDH), and glutaraldehyde, and nine negative controls, salicylic acid, phenol, glycerol, lactic acid, chlorobenzene, p-hydroxybenzoic acid, benzaldehyde, diethyl phthalate, and octanoic acid. This chemical set covers all the categories from the ECETOC

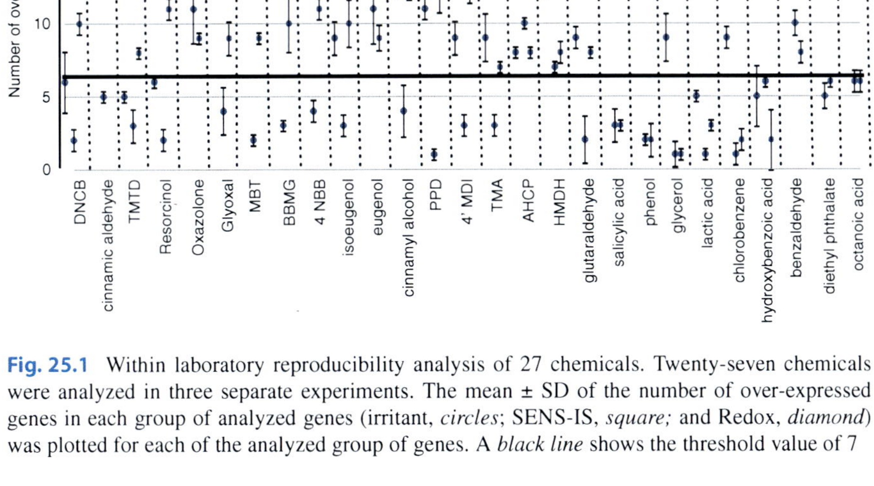

Fig. 25.1 Within laboratory reproducibility analysis of 27 chemicals. Twenty-seven chemicals were analyzed in three separate experiments. The mean ± SD of the number of over-expressed genes in each group of analyzed genes (irritant, *circles*; SENS-IS, *square;* and Redox, *diamond*) was plotted for each of the analyzed group of genes. A *black line* shows the threshold value of 7

potency category proposal (extreme, strong, moderate, and weak). The chemicals have also an even distribution between liquid and solid forms and for the solvent used in the assay (9 DMSO, 8 olive oil, and 10 water).

The mean ± SD of the number of over-expressed genes in each subgroup of analyzed genes was plotted for each of the analyzed group of genes ("irritant" subgroup, circles; "SENS-IS" subgroup, square; and "Redox" subgroup, diamond; see Fig. 25.1). A black line indicates the threshold value of seven over-expressed genes necessary in either the "SENS-IS" or "Redox" group of genes for classifying a test item as a sensitizer. All sensitizers included in the test set reached this threshold value (see Fig. 25.1) and were correctly classified as sensitizers in all three experiments. All the non-sensitizer (with the exception of benzaldehyde which was classified as a sensitizer) did not reach the threshold value (seven) in either the SENS-IS or Redox subgroups and were thus correctly classified as non-sensitizers. Moreover, as shown in Fig. 25.1, the small size of the standard deviations (error bars) and the absence of outliers indicate a very good reproducibility.

To analyze the transferability and the between laboratory reproducibility, the assay was transferred to two other laboratories. The results obtained in the lead laboratory and Lab 2 are presented here.

The same batches of the 16 chemicals included in the test set (see Table 25.1) were analyzed under blinded conditions. The set included three non-sensitizers (propylene glycol, dimethyl sulfoxide (DMSO), and sodium lauryl sulfate (SLS) (although SLS is detected as a weak sensitizer by the LLNA), one extreme sensitizer (DNCB), six strong sensitizers (propyl gallate, cinnamaldehyde, methylchloroisothiazolinone (MCI), potassium dichromate (PDIC), BBMG, 2-aminophenol), five moderate sensitizers (geraniol, resorcinol, trans-anethol, eugenol, diethyl sulfate), and one weak sensitizer (limonene). Considering these five reactivity classes, the within laboratory reproducibility (WLR) was 93% for the lead laboratory (MCI was detected as an extreme sensitizer in one replicate and a strong sensitizer in two other replicates). The WLR was 100% for Lab 2 resulting in an average WLR of 96.5% for the two laboratories! Moreover, any difference in the categorization of a test item was never greater than one class.

The two laboratories obtained an identical classification (sensitizers versus non-sensitizers) for all tested chemicals resulting in a between laboratory reproducibility (BLR) value of 100% (see Table 25.1). Moreover, only one chemical (propyl gallate) was categorized in a different reactivity class (strong versus moderate) by the two laboratories. The calculated BLR value taking into account the five reactivity classes was thus 93%.

25.3.2 Predictive Capacity According to Cooper Statistics and Comparison to Human Data

Cooper statistics was computed on the results obtained with the 41 chemicals already included in the WLR and BLR evaluation (see Table 25.2) and tested in at

Table 25.1 Analysis of reproducibility and transferability based on prediction value over five classes in two different laboratories

	Chemicals	Propylene glycol	DMSO	SLS	DNCB	Propyl gallate	Cinnamaldehyde	MCI	PDIC	BBMG	2-Aminophenol	Geraniol	Resorcinol	Trans-anethol	Eugenol	Diethyl sulfate	Limonene
Immunosearch	Repro1	NS	NS	NS	EXT	Strong	Strong	Strong	Strong	Strong	Strong	MOD	MOD	MOD	MOD	MOD	Weak
	Repro2	NS	NS	NS	EXT	Strong	Strong	EXT	Strong	Strong	Strong	MOD	MOD	MOD	MOD	MOD	Weak
	Repro3	NS	NS	NS	EXT	Strong	Strong	Strong	Strong	Strong	Strong	MOD	MOD	MOD	MOD	MOD	Weak
	WLR 5 class	3/3	3/3	3/3	3/3	3/3	3/3	2/3	3/3	3/3	3/3	3/3	3/3	3/3	3/3	3/3	3/3
Lab 2	Repro1	NS	NS	NS	EXT	MOD	Strong	Strong	Strong	Strong	Strong	MOD	Strong	MOD	MOD	MOD	Weak
	Repro2	NS	NS	NS	EXT	MOD	Strong	Strong	Strong	Strong	Strong	MOD	MOD	MOD	MOD	MOD	Weak
	Repro3	NS	NS	NS	EXT	MOD	Strong	Strong	Strong	failed	Strong	MOD	MOD	MOD	MOD	MOD	Weak
	WLR 5 class	3/3	3/3	3/3	3/3	3/3	3/3	3/3	3/3	3/3	3/3	3/3	3/3	3/3	3/3	3/3	3/3
BLR		Y	Y	Y	Y	N	Y	Y	Y	Y	Y	Y	Y	Y	Y	Y	Y

The indicated 16 chemicals were tested blindly three times in two different laboratories. The two laboratories made their evaluation according to the prediction model as indicated. The within laboratory reproducibility (WLR) for each laboratory and each chemical is indicated. The between laboratory reproducibility (BLR) is noted in the last line

Table 25.2 Analysis of the predictive capacity of the SENS-IS assay

Name	CAS	Vehicle used	Physical form	LLNA classification	Human classification	SENS-IS classification
2,4 Dinitrochlorobenzene	97-00-7	DMSO	SOL	EXT	1[a]	EXT
2,4,6-Trinitrobenzenesulfonic acid	2508-19-2	Water	SOL	EXT	No data	EXT
4-Nitrobenzylbromide	100-11-8	DMSO	SOL	EXT	No data	EXT
Oxazolone	15646-46-5	OO	SOL	EXT	+[b]	EXT
Propyl gallate	121-79-9	DMSO	SOL	STRONG	2[a]	STRONG
2-Aminophenol	95-55-6	OO	SOL	STRONG	2[a]	STRONG
Methylchloroisothiazolinone	26172-55-4	Water	LIQ	STRONG	1[a]	STRONG
Potassium dichromate	7778-50-9	Water	SOL	STRONG	1[a]	STRONG
2-Bromo-2-(bromomethyl) glutaronitrile	35691-65-7	DMSO	SOL	STRONG	2[a]	STRONG
Glutaraldehyde	111-30-8	Water	LIQ	STRONG	2[a]	STRONG
p-Phenylendiamine	106-50-3	Water	SOL	STRONG	1[a]	STRONG
Cinnamaldehyde	104-55-2	OO	LIQ	STRONG	2[a]	STRONG
Geraniol	106-24-1	OO	LIQ	MOD	4[a]	MOD
Trans anethole	4180-23-8	DMSO	LIQ	MOD	+	MOD
Diethyl sulfate	64-67-5	OO	LIQ	MOD	+	MOD
Tetramethyl thiuram disulfide	137-26-8	DMSO	SOL	MOD	3[a]	MOD
Diphenylmethane diisocyanate (MDI)	101-68-8	Water	SOL	MOD	+	MOD
Glyoxal	107-22-2	Water	LIQ	MOD	2[a]	MOD
2-Mercaptobenzothiazole	149-30-4	DMSO	SOL	MOD	3[a]	MOD
Isoeugenol	97-54-1	OO	LIQ	MOD	2[a]	MOD
Maléic anhydride	108-31-6	Water	SOL	MOD	+	MOD
Eugenol	97-53-0	OO	LIQ	WEAK	3[a]	MOD
Cinnamyl alcohol	104-54-1	OO	SOL	WEAK	3[a]	WEAK
Resorsinol	108-46-3	Water	SOL	WEAK	4[a]	MOD
Limonen	5989-54-8	DMSO	LIQ	WEAK	5[a]	WEAK
Alpha-hexylcinnamaldehyde	101-86-0	OO	LIQ	WEAK	5[a]	WEAK

Ammonium hexachloroplatinate	16919-58-7	Water	SOL	WEAK	No data	WEAK
Hexamethylenediisocyanate	822-06-0	OO	LIQ	WEAK	+	WEAK
Trimellitic anhydride	552-30-7	Water	SOL	MOD	+	WEAK
Sodium lauryl sulfate	151-21-3	Water	SOL	WEAK (FP)	6[a]	NON
Salicylic acid	69-72-7	DMSO	SOL	NON	6[a]	NON
Phenol	108-95-2	Water	SOL	NON	6[a]	NON
Glycerol	56-81-5	Water	LIQ	NON	6[a]	NON
Lactic acid	50-21-5	Water	LIQ	NON	6[a]	NON
Chlorobenzene	108-90-7	DMSO	LIQ	NON	+	NON
Propylene glycol	57-55-6	Water	LIQ	NON	5[a]	NON
Dimethylsulfoxide	67-68-5	Water	LIQ	NON	6[a]	NON
Benzaldéhyde	100-52-7	Water	LIQ	NON	5[a]	MOD
Diethyl phthalate	84-66-2	DMSO	LIQ	NON	6[a]	NON
Octanoic acid	124-07-2	OO	LIQ	NON	6[a]	NON
p-Hydroxybenzoic acid	99-96-7	DMSO	SOL	NON	–	NON

[a][10]
[b][11]

Table 25.3 Concordance of prediction according to the LLNA assay

		LLNA		
		Total	S	NS
		41	29	12
SENS-IS	S	30	29	1
	NS	11	0	11
		Accuracy	Sensitivity	Specificity
		97.5%	100%	91.6%

Table 25.4 Concordance of prediction according to the four classes of the LLNA assay or the five classes for human data

LLNA class	Extreme	Strong	Moderate	Weak	NS	Total
Extreme	4					4
Strong		8				8
Moderate			9	2	1	12
Weak			1	5		6
NS					11	11
Total	4	8	10	7	12	41
Concordance: 90.2%						

least two independent experiments. The results obtained with the SENS-IS assay were compared to published human and LLNA data (Tables 25.3 and 25.4).

The SENS-IS assay correctly predicted 40/41 substances (sensitizer versus non-sensitizer) when compared to LLNA data. The obtained cooper statistics values (Table 25.3) are sensitivity 100%, specificity 91.6%, and overall accuracy 97.5%. A false positive result was observed with one chemical (benzaldehyde). However, one should note that allergic reactions, although rare, have been described in human for benzaldehyde [12]. As the SENS-IS assay can categorize sensitizers in four categories, we also analyzed the concordance of prediction according to the four classes of the LLNA assay (Table 25.4), and an overall accuracy of 90.2% was obtained.

The comparison of the results obtained with the SENS-IS assay to human data was more challenging due to the lack of comprehensive human databases using a classification similar to the one proposed by ECETOC for the LLNA. The recent extensive work published by Basketter et al. [10] uses six classes, making a comparison with the five SENS-IS classes difficult. Moreover, as an attempt to regroup chemicals into similar risk categories for the human population, the authors introduced notions of level of exposure, frequency, and concentration of the chemicals. A direct comparison to potency categories predicted by *in vitro* assays not taking into account these parameters is thus quite difficult. Sensitization to nickel is a good example of this issue. Nickel sulfate is classified as a weak to non-sensitizer in the LLNA (as well as in the SENS-IS assay, data not shown). However, it is classified by Basketter et al., as a category 2 sensitizer due to its frequent and widespread uses. This discrepancy in the criteria for category assignment makes it difficult to compare the side-by-side potency predicted by *in vitro* or *in vivo* assays and the category system used by Basketter et al.

In an attempt to make comparison and because that approach has already been successfully used with historic LLNA data, we considered that (Basketter et al.) human category 1 can be extreme or strong, human category 2 can be strong or moderate, human category 3 can be moderate or weak, and human category 4 can be moderate or weak. The human category 5 chemicals are difficult to assign because these represent either very weak or non-sensitizer according to the LLNA and SENS-IS. Nevertheless, using this categorization, concordance prediction against human data was 96.6% (see Table 25.4).

25.3.3 Applications and Limitations

The SENS-IS protocol was developed and optimized to provide a comprehensive *in vitro* assay that allows the testing of all types of chemicals, natural products, and mixtures under conditions close to the human situation. The assay uses a reconstituted 3D epidermis as the test system and thus mimics normal skin interaction with a chemical and is compatible with a wide variety of exposure methods. The test item can be dissolved and applied in the usual vehicles such as water, buffers, and DMSO. However, the main advantage of using a reconstituted 3D epidermis is that other exposure method such as cosmetic bases can also be envisaged. This allows taking into account all the possibilities for the chemical(s) to penetrate through the skin barrier, to interact with the keratinocytes, and to get information on the risk associated with a chemical in a wide variety of vehicles or cosmetic bases.

3D reconstituted epidermis also provides a metabolic activity very similar to normal human skin [13] that should be sufficient to activate most pro-haptens, thus allowing their detection.

In order to provide specific and relevant information, the test system (reconstituted 3D epidermis) should not be overstressed. This means that the applied solution/mixture should not induce extreme cytotoxic effects. This aspect is described in the test protocol and controlled through the monitoring of the "Irritation" gene subgroup (see above).

Nevertheless, weak to very weak sensitizers with a strong irritation or cytotoxic potential represent very challenging chemicals since they may inflict heavy damages to the skin model before inducing measurable sensitizing signals.

The use of 3D reconstituted skin enables also the testing of complex finished products. That opportunity could be a help to minimize the risk for the end user or before any human test.

25.4 Brief Description of the Protocol

25.4.1 Test System

The SENS-IS assay is based on the Episkin™ 3D reconstituted human epidermis (RHE) as the test system. Although other skin models have been successfully used, the final version of the protocol is based on the Episkin™ model. These models are obtained by culturing adult human keratinocytes on a collagen substrate in conditions which permit their terminal differentiation and the reconstruction of an epidermis with a functional horny layer. After 3 days of immersed culture conditions, the epidermis is airlifted during 10 days allowing differentiation and formation of a horny layer. The human keratinocytes come from mammary samples obtained from healthy consenting donors during plastic surgery. HIV 1 and 2, B and C, hepatitis tests are carried out on the donor bloods as well as verification of the bacteriological and fungal sterility of the cells and absence of mycoplasma. The reconstructed human epidermis expresses the major differentiation markers (filaggrin and involucrin in granular cell layers, transglutaminase I and keratin 10 in supra basal cell layers, and loricrin in upper granular cell layers), as well as expressing the basement membrane markers (type IV collagen, integrin alpha 6, integrin beta 4, antigen BP, laminin I, and laminin V). Free fatty acids and ceramides are detected in the lipid profile. The ultrastructural features show secretion and normal arrangement of bilayered lipid content into the intercellular spaces of the cornified cell layers (formation of normal permeability barrier).

25.4.2 Procedure

The different steps of the protocol are clearly described in Fig. 25.2. Each test substance (vehicle, test item, negative and positive controls) is topically applied on the Episkin™ model for 15 min at room temperature. The test substance is then rinsed

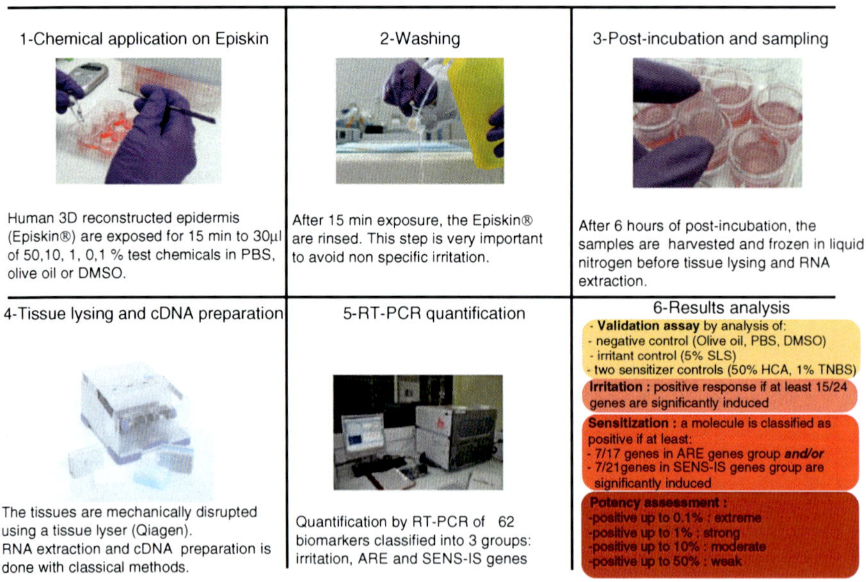

Fig. 25.2 Protocol of the SENS-IS assay

with phosphate-buffered saline (PBS). Epidermis is then transferred to fresh medium and incubated at 37 °C for 6 additional hours. Tissues are then lysed in RNA lysis buffer, and RNA is prepared using the extraction system (affinity column) from Qiagen according to the manufacturer's instruction. cDNA is then synthesized for RT-PCR expression measurement of the three sets of selected genes.

A specific prediction model using the number of upregulated biomarkers ($\Delta E > 1.25$; see below) allows classifying the test item as a sensitizer or a non-sensitizer. Moreover, the sensitizing category (extreme /strong/moderate/weak) can be predicted according to the dose effect relationship (see below). An experimental run comprises at least two independent experiments to take into account eventual variations in the RHE.

25.4.3 End Point Measurement

Samples quantification is performed as an absolute quantification analysis using the second derivative maximum method, as set up using the algorithm developed by ROCHE.

The second derivative maximum method identifies the Cp (curve point) of a sample as the point where the sample's fluorescence curve turns sharply upward. This turning point corresponds to the maximum of the second derivative of the amplification curve. Thus, this method is called "Second Derivative Maximum Method." The big advantage of this method is that it requires little user input. The software performs the calculation automatically.

25.4.4 Data Analysis

Data are analyzed through an automated process based on Excel file macros. All computational steps are performed without user's input. One only needs to transfer the Cp values from the PCR apparatus to the corresponding cells of the Excel spreadsheet.

The first step checks the internal reproducibility of the PCR analysis. Three separate PCR analyses for the 65 tested genes are performed for both the phosphate-buffered saline (PBS) and olive oil (OO) samples. The data variability is measured and the run is rejected if the variation exceeds a fixed value.

Three housekeeping genes (GUSB, NONO and B2M) are routinely measured in order to normalize the amount of ARN in the different samples for the comparison with the vehicle controls. The fold increased over negative control is then measured by the $2^{\Delta\Delta Cp}$ method routinely used for quantitative PCR measurement.

The modulation of the expression of a particular gene is then determined by measuring the expression level for the analyzed test item and dividing it by the results obtained for the corresponding negative control (ΔE). A gene is considered as over-induced (positive) if its ΔE is >1.25.

25.4.5 Acceptance Criteria

To evaluate the acceptance criteria applied to the results, for every batch of Episkin™ analyzed, six different controls are used:

- Three vehicle controls, phosphate buffer saline (PBS), olive oil (OO), and dimethyl sulfoxide (DMSO)
- One irritant control, 5% sodium lauryl sulfate (SLS)
- Two positive sensitizer controls, 50% alpha-hexylcinnamaldehyde and 1% trinitrobenzenesulfonic acid (TNBS)

Multiple acceptance criteria are applied.

1. To address tissue damage caused by an excessive concentration of corrosive or very irritant chemicals, the Cp value of the HSPAA1 gene is measured: all the cDNAs with a HSPAA1 Cp value >110% that of the control OO or PBS are a sign of unacceptable cell death. The run is discarded and the tested chemical will be tested at lower concentrations.
2. To address an excessive irritation or cytotoxicity that may cause the unspecific induction of sensitizer specific genes, the number of over-expressed genes from the "Irritation" subgroup is measured. If more than 20 irritation genes are over-expressed ($\Delta E > 1.25$) and if more than 7 genes in the "SENS-IS" or "Redox" groups of genes are over-expressed, the result is considered as false positive and rejected.
3. Analysis of the negative controls: three vehicle (negative) controls are used, OO, PBS, and dimethyl sulfoxide (DMSO). If these controls induce the over-expression ($\Delta E > 1.25$) of >7 genes in the "SENS-IS" or "Redox" group of genes, the run is rejected.
4. Analysis of the irritation control: sodium lauryl sulfate (SLS) at 5% is used as a negative sensitization control. If 5% SLS induces the over-expression ($\Delta E > 1.25$) of more than seven genes in the "SENS-IS" or "Redox" group of genes, the run is rejected.
5. Sensitization controls: alpha-hexylcinnamaldehyde (HCA) and trinitrobenzenesulfonic acid (TNBS) are used as positive sensitization controls. If 50% HCA or 1% TNBS do not induce the over-expression ($\Delta E > 1.25$) of more than seven genes in either the "SENS-IS" or "Redox" group of genes, the run is rejected.

25.4.6 Prediction Model

A test substance at a given concentration is considered a sensitizer if it induces the upregulation ($\Delta E > 1.25$) of at least seven genes among the "SENS-IS" and "Redox" subgroups within the acceptable cytotoxicity domain (see acceptance criteria).

To predict the sensitizing potency category of the test item, the lowest concentration classifying it as a sensitizer (as defined above) is noted. If the chemical is

Table 25.5 Concordance of prediction according to the 6 classes for human data

Human class	1	2	3	4	5	6	Total
Extreme	1						1
Strong	3	5					8
Moderate		2	3	2	1		8
Weak			1		2		3
NS					2	8	10
Total	4	7	4	2	5	8	30
Concordance: 96.6%							

positive at 50%, it is considered a weak sensitizer, up to 10% it is considered a moderate sensitizer, up to 1% it is considered a strong sensitizer, and if it is positive at 0.1%, it is considered as an extreme sensitizer. Table 25.5 summarizes the SENS-IS prediction model. The number of positive genes ($\Delta E > 1.25$) in each group of genes and the corresponding sensitization hazard are indicated. Moreover, the potency category predicted using the concentration necessary for observing a positive response (weak, moderate, strong, extreme) is indicated in the last four rows.

25.5 Role in a Testing Strategy

The SENS-IS is an *in vitro* test method that can be used as a stand-alone or as part of a non-animal integrated test strategy for assessing the skin sensitization potential of chemicals (e.g., as a replacement of the LLNA). It is intended for use as a replacement of regulatory *in vivo* tests for skin sensitization hazard classification and labeling, relevant to current "REGULATION (EC) N° 1272/2008 on classification, labeling, and packaging of substances and mixtures".

The use of a reconstituted human epidermis providing skin penetration and metabolic properties similar to the *in vivo* situation combined with quantitative data generated by qPCR analysis of the expression of a relevant panel of genes makes the SENS-IS approach a promising alternative for risk assessment and safety prediction and quantitative risk assessment calculation (for definition of maximum use level). It covers most biological steps described in the OECD document "The Adverse Outcome Pathway for Skin Sensitisation…" [13]:

Step 1: "Skin penetration" is covered due to the use of a reconstituted epidermis with a horny layer as a test system.

Step 2: "Detection of …electrophilic substance directly or via auto-oxidation of metabolism": Skin models have been shown to replicate most of the characteristics of native skin regarding metabolism capabilities [14].

Steps 3–4: The test system should provide the necessary activity for the covalent modification of epidermal proteins, and the readout system can discriminate between binding onto cysteine residues (inducing the expression of genes from the "Redox" subgroup) and binding onto lysine residues (expression of genes from the "SENS-IS" subgroup).

Steps 5 and 6: Addressing the activation of epidermal keratinocytes and dendritic cells. The SENS-IS assay analyzes the gene expression pattern in skin keratinocytes. However, dendritic cells (DC) are not included in the proposed skin model. Assays analyzing DC activation and/or T cell activation would thus represent perfect partners in a testing strategy. However the chemical should be handled and presented to the dendritic cells and T cells in a way similar to the SENS-IS assay in order not to loose on one end what could be gain on the other. These would require the development of a complex model incorporating the skin, dendritic cells, and T cells.

25.6 Perspectives from the Test Developer: Challenges and Opportunities/Conclusions

The most challenging chemicals for the SENS-IS assay would be highly irritating or cytotoxic compounds that have a weak to very weak sensitizing activity. Such chemicals may damage the skin model before inducing measurable sensitization signals. However, a peptide-binding assay using a sensitive set of peptides should be able to detect this type of chemicals.

References

1. Martin SF, Esser PR, Weber FC, Jakob T, Freudenberg MA, Schmidt M, Goebeler M. Mechanisms of chemical-induced innate immunity in allergic contact dermatitis. Allergy. 2011;66(9):1152–63. doi:10.1111/j.1398-9995.2011.02652.x.
2. Nosbaum A, Vocanson M, Rozieres A, Hennino A, Nicolas J-F. Allergic and irritant contact dermatitis. Eur J Dermatol. 2009;19(4):325–32. doi:10.1684/ejd.2009.0686.
3. Freudenberg MA, Esser PR, Jakob T, Galanos C, Martin SF. Innate and adaptive immune responses in contact dermatitis: analogy with infections. G Ital Dermatol Venereol. 2009;144(2):173–85.
4. Moreilhon C, Gras D, Hologne C, Bajolet O, Cottrez F, Magnone V, et al. Live Staphylococcus aureus and bacterial soluble factors induce different transcriptional responses in human airway cells. Physiol Genomics. 2005;20(3):244–55. doi:10.1152/physiolgenomics.00135.2004.
5. Ade N, Leon F, Pallardy M, Peiffer J-L, Kerdine-Romer S, Tissier M-H, et al. HMOX1 and NQO1 genes are upregulated in response to contact sensitizers in dendritic cells and THP-1 cell line: role of the Keap1/Nrf2 pathway. Toxicol Sci. 2009;107(2):451–60. doi:10.1093/toxsci/kfn243.
6. Cotovio J, Grandidier M-H, Portes P, Roguet R, Rubinstenn G. The in vitro skin irritation of chemicals: optimisation of the EPISKIN prediction model within the framework of the ECVAM validation process. ATLA. 2005;33(4):329–49.
7. Cottrez F, Boitel E, Auriault C, Aeby P, Groux H. Genes specifically modulated in sensitized skins allow the detection of sensitizers in a reconstructed human skin model. Development of the SENS-IS assay. Toxicology in Vitro. 2015;29(4):787–802. ISSN 0887-2333.
8. Loveless SE, Api A-M, Crevel RWR, Debruyne E, Gamer A, Jowsey IR, et al. Potency values from the local lymph node assay: application to classification, labelling and risk assessment. Regul Toxicol Pharmacol. 2010;56(1):54–66. doi:10.1016/j.yrtph.2009.08.016.
9. Cottrez F, Boitel E, Ourlin J-C, Peiffer J-L, Fabre I, Henaoui I-S, Mari B, Vallauri A, Paquet A, Barbry P, Auriault C, Aeby P, Groux H. SENS-IS, a 3D reconstituted epidermis based model

for quantifying chemical sensitization potency: Reproducibility and predictivity results from an inter-laboratory study. Toxicology in Vitro. 2016;32:248–260. ISSN 0887-2333.
10. Basketter DA, Alépée N, Ashikaga T, Barroso J, Gilmour N, Goebel C, et al. Categorization of chemicals according to their relative human skin sensitizing potency. Dermatitis. 2014;25(1):11–21. doi:10.1097/DER.0000000000000003.
11. Basketter DA, Lea LJ, Cooper K, Stocks J, Dickens A, Pate I, et al. Threshold for classification as a skin sensitizer in the local lymph node assay: a statistical evaluation. Food Chem Toxicol. 1999;37(12):1167–74.
12. Andersen A. Final report on the safety assessment of benzaldehyde. Int J Toxicol. 2006;25(Suppl 1):11–27. doi:10.1080/10915810600716612.
13. OECD (2014), The Adverse Outcome Pathway for Skin Sensitisation Initiated by Covalent Binding to Proteins, OECD Publishing, Paris. http://dx.doi.org/10.1787/9789264221444-en.
14. Hewitt NJ, Edwards RJ, Fritsche E, Goebel C, Aeby P, Scheel J, et al. Use of human in vitro skin models for accurate and ethical risk assessment: metabolic considerations. Toxicol Sci. 2013;133(2):209–17. doi:10.1093/toxsci/kft080.

The SenCeeTox® Assay

26

Donald Keller, Caroline Bauch, and Prakash Patel

26.1 Principle of the Test Method and Scientific Basis

Allergic contact dermatitis (ACD) is a common adaptive, delayed type allergy experienced by about 15–20% of the population [1, 2]. ACD is an adverse reaction to chemicals in cosmetic products, cleaning products, medication, etc. [1, 3]. Thus the detection of the sensitizing potential of ingredients or finished products is critical across the pharmaceutical, personal care, and medical device industries. With recent enforcement of testing and marketing ban on the European Market with the seventh amendment of the European Cosmetic Directive and increasing interest of reducing or replacing animal test systems, the use of *in vitro* test methods became popular [4, 5]. Models for sensitization profiling must be applicable across a broad spectrum of chemical classes, e.g., aldehydes, anhydrides, alcohols, aromatic compounds, personal care formulations, plant extracts, medical device extracts, and materials with different physicochemical properties like hydrophobicity or hydrophilicity. The model also needs to be sensitive enough to correctly identify sensitizing compounds from non-sensitizing compounds or irritants and if possible to provide a prediction of potency of the test article [6–8].

Sensitization is the development of ACD and is triggered by repeated contact to allergens and the activation of several cell types in the skin and immune system. Although chemical allergens, also called haptens, are antigenic, due to their small molecular structure they are not immunogenic and thus not recognized by the immune system. In order to obtain immunogenicity chemical allergens require to form conjugates with skin proteins [9–11]. Chemical allergens have electrophilic properties or, in the case of pro- and pre-haptens can obtain it by either autooxidation or enzymatic activation. Electrophilic properties allow the formation of

D. Keller (✉) • C. Bauch • P. Patel
Cyprotex, 4717 Campus Drive, Kalamazoo, MI 49008, USA
e-mail: d.keller@cyprotex.com

C. Eskes et al. (eds.), *Alternatives for Dermal Toxicity Testing*,
DOI 10.1007/978-3-319-50353-0_26

stable conjugates with nucleophilic residues in proteins [12]. Formation of covalent bonds between proteins and chemical allergens is called haptenization. The presence of these "haptenated" proteins or the chemical itself in the skin activates keratinocytes and dendritic cells [13, 14]. Activated dendritic cells migrate out of the skin towards the local lymph node where they present chemical-protein-conjugate to naïve immune cells, T cells [15]. Every naïve T cell carries a unique T cell receptor. If dendritic cells present a matching antigen to the T cell, activation of T cells to specific T cells is triggered and with following encounter of the chemical allergen the typical symptoms of the allergic contact dermatitis are caused by the specific T cells at the site of contact [11, 16, 17].

The SenCeeTox® assay is a multi-parametric approach covering several important steps of the skin sensitization process, namely protein reactivity, keratinocyte activation alongside cell viability. Changes in gene expression of keratinocytes and protein reactivity are important hallmarks of the sensitization response. In the SenCeeTox® assay protein reactivity of the test material is measured in a cell-free approach using glutathione (GSH) as a model peptide. Electrophilic test material will form conjugates with GSH with the free, nucleophilic cysteine residue [18]. Activation of keratinocytes is assessed using either the human keratinocyte cell line HaCaT or the reconstituted human skin model supplied by MatTek, EpiDerm™. The latter system offers the possibility to test finished products as the HaCaT cell based approach is only applicable for testing of single ingredients. Activation of keratinocytes is monitored by detection of changes in gene expression using qRT-PCR. A set of 11 genes underlying the Keap-1/Nrf-2 or EpRE/ARE pathways were selected to assess the activation of keratinocytes after exposure to test material [18, 19]. The Nrf-2/Keap-1 signaling pathway is known as cellular sensor to oxidative stress also caused by electrophilic compounds and initiates gene expression of the so-called Phase 2 enzymes to compensate cell stress [20–22]. In unstressed cells the proteins Keap-1 and Nrf-2 are dimerized and thus inactive. Keap-1 acts as a sensor for electrophiles with its free cysteine residues. Reactive molecules bind to Keap-1 causing conformational changes leading to the dissociation and translocation of Nrf-2 into the nucleus. Nrf-2 is a transcription factor and binds to a specific genetic element, the Electrophilic/Aromatic Response Element (EpRE/ARE) [19]. The EpRE/ARE element is located in the promotor region of numerous genes like NADPH-quinone oxidoreductase 1 (NQO1), Aldoketoreductase (AKR1C2), Interleukin 8 (IL8), Aldehyde dehydrogenase 3A (ALDH3A), Heme-oxygenase 1 (HMOX1), Glutamate cysteine ligase catalytic subunit C (GCLC), Metallothionein 1A (MT1A), Metallothionein 2A (MT2A), Thioredoxin (TXN), v-maf avian musculoaponeurotic fibrosarcoma oncogene homolog F (MAFF). With the increased expression of those enzymes the cells are able to respond to xenobiotic assault with the ultimate aim of turning the xenobiotic material into non-reactive molecules that are easy to remove from the cell [18]. It was shown that genes controlled by the xenobiotic response element (XRE) such as cytochrome P450 (CYP1A1) are also induced. Cytochrome P450 induction can be particularly informative as it is often induced in situations where sensitizers need to be metabolically activated to gain full electrophilicity [18].

In addition to activation of genes to allow cells to respond to xenobiotic molecules and the caused cellular stress, the NRF-2/Keap-1 pathway has been shown to interact with a pathway facilitating the induction of the allergic response. After exposure to skin sensitizers activated keratinocytes show an increased expression of CCL5, also called RANTES, and TNFα. CCL5 is a chemokine attracting immature dendritic cells and memory T cells [23].

In addition cell viability is measured by either lactose dehydrogenase (LDH) leakage when using EpiDerm™ or by testing for 3-(4,5-dimethylthiazol-2-yl)-2,5-diphenyltetrazolium bromide (MTT) reduction when using HaCaT cells. Cell viability is a crucial factor to assess alongside any cell activation. Increased cytotoxicity can activate expression of the Nrf-2/Keap-1 signaling pathway a cellular defense mechanism and could result in false positive results.

Data of the gene responses were processed by calculating the gradient and intercept of the linear regression. A pattern recognition approach was used to calculate predicted LLNA EC3 values in mM. Alongside the gene response data the IC_{50} for cell viability, GSH percent depletion, and *in silico* molecular descriptors, e.g., molecular weight, LogP which are generated from the SMILES representation using the RDKit software (www.rdkit.org) if available. The predicted LLNA EC (mM) is then converted to LLNA EC3 (%) and compounds are classified into the five potency categories as given in [24].

26.2 Brief Description of the Protocol

a. Dosing and test material preparation. For pure test compounds, stock solutions of 500 mM in appropriate vehicle were prepared fresh and diluted in media on the day of dosing. An initial experiment to asses viability limits was done when using HaCaT cells across a broad concentration range. Results of this assessment were then used to determine a range for the main study and gene expression analysis. A six point dilution series was prepared in the appropriate vehicle. Vehicles suitable for this test system include DMSO and Ethanol at 0.5% final concentration in media.
b. Finished products and extracts can be applied neat or diluted in a compatible vehicle if tested in the EpiDerm™. Vehicles tested in this model include PBS, water, olive oil, sesame oil, propylene glycol, 4:1 acetone: olive oil, up to 30% ethanol and up to 30% DMSO.
c. Glutathione reactivity. For the GSH reactivity assay a reaction mix of the following reagents and final concentrations was prepared: 20 mM sodium phosphate buffer, pH 7.4, 20 mM test material (pure test material), and 100 μM GSH. Extracts or finished products were added neat to the reaction mix. Samples were prepared in 96-well deep well plates and each test material was assessed in triplicates. The samples and reaction mix were incubated for 24 h at room temperature on a plate shaker [19, 25]. The reaction was stopped by adding metaphosphoric acid (2.8% final concentration) to each well and mixed for 2 min on a plate shaker at room temperature protected from light. 50 μL of reaction mix was transferred to a new

clear 96-well plate. To each well 10 μL of 1.5 M triethanolamine was added and incubated for additional 2 min on a plate shaker. Per well 150 μL GSH assay buffer containing 0.7 mM 5,5′-dithiobis-2-nitrobenzoic acid (DTNB), 0.25 mM NADPH, and 0.6 U/mL glutathione reductase was added. The plate was shaken for an additional 5 min and absorbance was read at 415 nm using an absorbance plate reader [26]. To assess the percent of GSH depletion, the average absorbance of the triplicates of each sample was divided by the average of vehicle control samples.

d. Tissue/Cell culture conditions. Human Keratinocyte (HaCaT) cells were cultured in Dulbecco's Minimal Essential Medium (DMEM) supplemented with 10% Fetal Bovine Serum (FBS) and 100 U/mL penicillin and 100 μg/mL streptomycin. Cells were maintained at 37 °C, 5% CO_2 with humidity [27]. For the assay cells were seeded into 96-well culture plates at a density of 12,000 cells per well in 100 μL of culture medium and allowed to equilibrate for 48 h at 37 °C, 5% CO_2 with humidity.
e. The reconstructed, three dimensional tissue culture model Epiderm™ was obtained from MatTek (Ashlands, Massachusetts). For tissue treatment prior to dosing vendor protocols were followed. In brief, tissues were received and transferred from shipping agar to pre-warmed media. Tissues were allowed to equilibrate for at least 1 h after which the media is changed and test articles applied.
f. Exposure of Tissue/Cells to test material. Prior to dosing, dosing solutions were prepared by diluting the test material vehicle stocks 1:200 with assay media. The media of the HaCaT cells was removed and replaced with dosing solutions. Each compound and concentration was dosed in triplicates. The plates are returned to the incubator for 24 h at 37 °C and 5% CO_2.
g. Dosing solutions of pure chemicals made to dose EpiDerm™ were prepared by diluting the test material vehicle stocks 1:200 with DPBS. The final vehicle concentrations were kept at 0.5%. Tissues were treated by applying 30 μL on the apical side of each tissue. Each compound and concentration was dosed in triplicates. The plates are returned to the incubator for 24 h at 37 °C and 5% CO_2.
h. Harvest and RNA isolation. After a 24 h incubation time of the cells or 3D tissues RNA was isolated using the QIAGEN's RNeasy 96 kit using the vacuum/centrifugation protocol as described by the manufacturer. In brief, tissue lysates are mixed with ethanol and transferred to a 96-well plate containing silica to bind the RNA. RNase free DNase is applied in the wells and digestion of DNA allowed to proceed for 15 min. RNA to rebound to the plates with buffer and then washed twice. The isolation plates are centrifuged for 10 min at 5600 × *g* to remove excess buffer. RNA is eluted from the isolation plates by applying 60 μL of RNase-free water and centrifuged for 4 min at 5600 × *g*. The elution is done twice to maximize RNA yield. The RNA concentration was quantified by using a Nanodrop ND-8000 spectrophotometer at the wavelength 260 and 280 nm.
i. Gene expression analysis. To assess the relative abundance of mRNA for 11 target genes normalization was performed to housekeeping genes. qRT-PCR was performed using QuantiTect Primer assays obtained from QIAGEN, Inc. (Germantown, Maryland).

The following target genes were assessed:

1. NADPH-quinone oxidoreductase 1 (NQO1)
2. Aldoketoreductase (AKR1C2)
3. Thioredoxin (TXN)
4. Interleukin 8 (IL8)
5. Aldehyde dehydrogenase 3A (ALDH3A)
6. Heme-oxygenase 1 (HMOX1)
7. Musculoaponeurotic fibrosarcoma (MafF)
8. Glutamate cysteine ligase catalytic subunit C (GCLC)
9. Cytochrome P450 1A1 (CYP1A1)
10. Metallothionein 1 (MT1)
11. Metallothionein 2 (MT2).

The four housekeeping gene primers used were also obtained from QIAGEN, Inc.:

a. 18s-ribosomal RNA (RRN18S)
b. Glyceraldehyde 3-phosphate dehydrogenase (GAPDH)
c. Transferrin receptor protein 1 (TFRC)
d. β2 microglobulin (B2M).

Experiments for qRT-PCR were conducted using QIAGEN's QuantiTect SYBR Green Kit (QIAGEN, Germantown, MD) and a Roche LightCycler 480 (Roche Applied Science, Mannheim, Germany). Reactions were loaded with approximately 50 ng of input RNA. Standard curves were made using serial diluted RNA from non-treated tissues. The one-step qRT-PCR was run using the following settings: reverse transcription was 1 cycle of 50 °C for 30 min followed by 95 °C for 15 min. Amplification and signal acquisition was set for 45 cycles of 94 °C for 15 s, 55 °C for 20 s, and 72 °C for 20 s. A melting curve was generated by 1 cycle of 95 °C for 5 s followed by 65 °C for 1 min.

qRT-PCR data were analyzed using LightCycler® 480 software version 1.5.0.39 (Roche Diagnostics Corp., Indianapolis, IN). For data analysis standard curve slope, error and efficiency, and Cp values within 3 cycles of 4 vehicle controls were recorded. Each test samples was analyzed using the Dixon-Q test and potential data point outliers were identified and removed from further analysis. Data were exported into a Microsoft Excel spreadsheet and normalized to vehicle control samples. Normalized housekeeping gene expression was presented in a box and whisker plot. The control gene which showed the lowest variation was selected and the ratio with normalized gene of interest values was calculated.

26.2.1 Viability

Viability of HaCaT cells was assessed using the MTT assay [28]. Briefly, 3-(4,5-dimethylthiazo-2-yl)-2,5-diphenyl-tetrazolium bromide (MTT) was dissolved in HaCaT cell media to a concentration of 0.5 mg/mL. Dosing media was removed and MTT media added to the cells and incubated for an additional 3 h at 37 °C under 5% CO_2 with humidity. Then MTT media was removed and cells were lysed by adding 200 μL isopropanol to each well and shaking for 10 min. The absorbance was read at 570 nm and 650 nm on a plate reader and values were normalized to the vehicle control samples. Gene expression data of compound concentrations reducing cell viability below 50% viability were not considered in further evaluations, as cytotoxicity can increase the false positive response.

Viability of 3D tissue models was assessed by analyzing the lactate dehydrogenase (LDH) released into the media using the CytoTox-ONE Homogeneous Membrane Integrity Assay from Promega (Madison, WI). 50 μL of the media was transferred into a new 96-well plate and 50 μL of CytoTox-one reagent was added. The mixture was incubated for 10 min at room temperature, then 25 μL of Stop Solution was added and mixed for 2 min on a plate shaker. Fluorescence signal was read with an excitation wavelength of 560 nm and an emission wavelength of 590 nm on a Biotek Synergy plate reader. Values were normalized to tissues treated with 10% Triton-X 100 (St. Louis, MO) as a control for 100% loss of cell viability after 24 h. Gene expression data of compound concentrations reducing cell viability below 50% viability were not considered in further evaluations, as cytotoxicity can increase the false positive response.

26.3 Initial Validation Study

The SenCeeTox® assay was initially set up using the human keratinocyte cell line HaCaT. Experiments to identify optimum conditions of incubation time and cell seeding density were performed. The seeding density was optimized to 12,000 cells per well in a 96-well plate and an incubation time over 48 h prior to dosing to allow the growth to confluence. Gene expression changes were monitored at 6, 24, and 48 h to determine optimal dosing time. Glutathione reactivity was also tested at 6 and 24 h with and without microsomes to determine the best set of conditions. Ultimately 24 h was chosen as the optimal time point for both, GSH reactivity and induction of gene expression. Proof of concept was shown using a test set of 39 known sensitizers and non-sensitizers of various potency classes according to the LLNA [19]. These compounds covered a large chemical space and included many fragrance and personal care materials, as well as pesticide, fungicides, and antimicrobials; and industrial catalysts, solvents, and raw materials. The vast majority of the compounds were soluble in aqueous media although some were limited in the concentrations which could be tested due to solubility. The data from this test set was used to develop a gated algorithm to analyze GSH reactivity, gene expression, and cell viability. The model was trained to translate the input data into a Predicted Toxicity Index (PTI) which was shown to correlate with LLNA potency predictions [18]. The assay and algorithm were challenged with additional three sets of compounds, each consisting of 20 blinded test compounds. Each set contained non-sensitizers and weak to moderate, strong, or extreme sensitizers [18].

To improve the test system and its applicability for testing finished products, extracts, or mixtures the test method was transferred to use reconstructed, three dimensional tissue culture models (3D RHE models) [23]. For this purpose two different RHE models were evaluated for their suitability, SkinEthic™ (RHE, Leon, France) and MatTek (Epiderm™, Ashland, Massachusetts) [23]. Both models were generated using normal human keratinocytes grown in transwell plates and both show normal differentiation markers for human skin. In addition 3D RHE models provide similar pattern of cell types and cell morphology as compared to human skin including multiple cell layers [23]. Additional advantage of 3D RHE models over to 2D cell based assay is the ability of apical surface treatment of the models, which is more relevant to the *in vivo* situation. Apical surface treatment allows the testing of finished products, extracts, e.g., of medical devices, and the use of the hydrophobic vehicles like sesame oil, olive oil, or acetone: olive oil mixtures. If required ethanol and DMSO can be used in concentrations up to 50% without impacting cell viability (see Fig. 26.1).

The 3D RHE models were tested using a test set of ten compounds previously assessed in the SenCeeTox® HaCaT validation. Both the Episkin™ and Epiderm models were obtained and challenged with the same set of compounds. Positive prediction of the sensitizers was determined in both models with only 2 of 10 placed in the incorrect potency category [23]. While both models seemed equal in terms of predictivity, the Epiderm model was chosen due to an ease in shipping logistics. Work with the Episkin™ samples was complicated due to high background oxidative stress [23] probably due to X-Ray scans carried out in shipping.

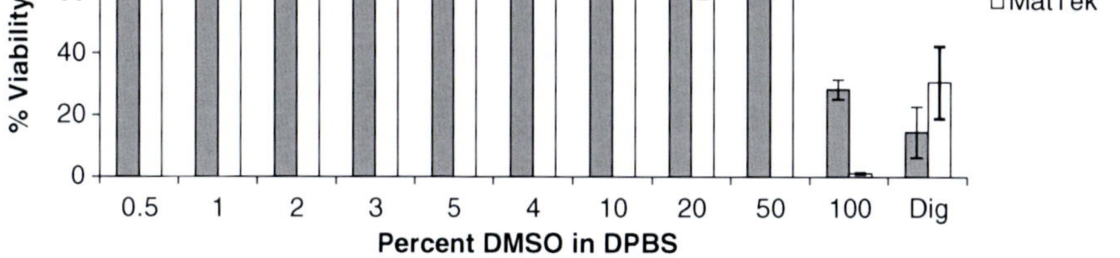

Fig. 26.1 Effect of increased DMSO on viability of 3D RHE models: SkinEthic™ (*grey*) and MatTek (*white*). 3D RHE models were treated with increasing concentration of DMSO in DPBS over 24 h. Cell viability was assessed using the LDH release assay. Data represent $n = 8$

Recent review of all the data generated over the course of the past 4 years allowed replacement of the gated algorithm with a more robust statistical model. The dose dependent gene expression data published by McKim et al. [18] was processed by calculating the IC_{50} values from the MTT cell viability assay, and calculating the gradient and intercept of $\log_{10}$(gene response) versus $\log_{10}$(dose) for each of the 11 genes in the assay. A model was constructed by pattern recognition by combining the processed gene expression data with percent of GSH depletion and molecular descriptors provided by the RDkit (www.rdkit.org) which were calculated from the SMILES representation of the compounds. Further on a regression model was constructed to predict LLNA EC3 (%) using statistical pattern recognition modelling software developed at Cyprotex [29] automatically constructing various tenfold cross-validated models like the General Linear Model, Random Forest, and Neural Net. The final model chosen was a General Linear Model which gave the highest cross-validated R^2 value against LLNA EC3 (mM) and also the lowest error in the prediction of skin sensitizing potency when the calculated LLNA EC3 (mM) was converted to LLNA EC3 (%) as given in [19].

26.4 Current Validation Status

The applicability of both HaCaT cells and 3D RHE models was shown to have a good correlation to *in vivo* data. To show transferability of the test method to other laboratories, an inter-laboratory validation study was initiated with the Flemish Institute for Technological Research (VITO). Scientists at VITO were trained on the test methods and proficiency of the assay was shown by assessment of four compounds (p-benzoquinone, cinnamic aldehyde, 2-aminophenol, and benzoic acid). The validation study was performed by testing ten blinded compounds at VITO with comparison to previous data generated at CeeTox/Cyprotex. The experiments were run using EpiDerm tissues from MatTek. The data showed a correlation to initial data generated by CeeTox/Cyprotex. Both labs were able to correctly identify all sensitizers and to assign potency categories with ±1 category of LLNA potency categories. Data are shown in Table 26.1.

An additional validation study was performed within a study initiated by Cosmetic Europe to determine suitable alternatives to animal testing for skin sensitization [34]. Cosmetic Europe provided a set of ten blinded compounds to all participating laboratories. Results were returned to Cosmetics Europe for further analysis and comparison of the several different test methods. The results indicated that SenCeeTox® provided 90% concordance with animal testing ([34], Table 26.2). The data were also re-analyzed using the pattern recognition methods. All sensitizers were correctly identified with some mismatch in the potency predictions. Two out of the three non-sensitizers were misclassified as moderate sensitizers. Both test compounds have higher irritant effects as acids, which could explain the false positive response.

Table 26.1 Comparison of SenCeeTox® data generated at CeeTox/Cyprotex and VITO compared to LLNA and human patch test data [30]

Test article name (Unblinded)	CeeTox/ Cyprotex		VITO		LLNA potency category	Human patch test
	PTI	PPC	PTI	PPC		
p-benzoquinone (+ control)	7	Extreme	6	Strong	Extreme	+
1-chloro-2,4-dinitrobenzene	6	Strong	7	Extreme	Extreme	+
metol	6	Strong	5	Moderate	Strong	
2-mercaptobenzothiozol	6	Strong	5	Moderate	Moderate	+
Isoeugenol	6	Strong	6	Strong	Moderate	+
2-hydroxyethylmethacrylate	4	Moderate	4	Moderate	Moderate	
2-hydroxyethylacrylate	6	Strong	6	Strong	Moderate	+
2,3-butanedione	3	Weak	3	Weak	Weak	
Eugenol	3	Weak	3	Weak	Weak	+
Glycerol	0	Non-sensitizer	1	Non-sensitizer	Non-sensitizer	–
Lactic acid	3	Weak	1	Non-sensitizer	Non-sensitizer	+
Benzoic acid (– control)	3	Weak	1	Non-sensitizer	Non-sensitizer	

PTI predicted toxicity index, *PPC* predicted potency category. Summarizes results of the validation studies at CeeTox/Cyprotex and VITO compared to the LLNA and human data. LLNA and human data were taken from published sources [24, 30–33]

Table 26.2 Results of ten compound set from Cosmetics Europe

Chemical tested for Cosmetic Europe	Prediction based on [18]	Prediction based on regression model	Potency expected
4-Nitrobenzylbromide	Strong	Moderate	Extreme
Methyldibromoglutaronitrile	Strong-extreme	Strong	Strong
2-Mercaptobenzothiazole	Moderate	Strong	Moderate
Cinnamal	Moderate	Moderate	Moderate
Tetramethyl thiuram disulfide	Moderate	Strong	Moderate
Salicylic acid	Non-sensitizer	Moderate	Non-sensitizer
Lactic acid	Non-sensitizer	Moderate	Non-sensitizer
Sodium lauryl sulfate	Weak	Non-sensitizer	Non-sensitizer
Phenyl benzoate	Weak	Moderate	Weak
Lauryl gallate	Weak	Moderate	Strong

26.5 Performance and Applicability of the Test Method

26.5.1 Reproducibility

With every experiment of the SenCeeTox® assay, one negative control (benzoic acid or glycerol, both non-sensitizers) and one positive control (p-benzoquinone, a

Table 26.3 Potency prediction of the SenCeeTox® model against assigned LLNA categories

	Actual potency				
Predicted potency	Non	Weak	Moderate	Strong	Extreme
Non	15	1	0	0	0
Weak	7	16	3	0	0
Moderate	6	3	29	0	0
Strong	0	0	2	9	0
Extreme	0	0	0	1	5

strong sensitizer) are run alongside the test material and solvent controls. In six separate runs, the assay predicted the correct potency for benzoic acid, glycerol, and p-benzoquinone in 56%, 75%, and 100% of the cases, respectively.

26.5.2 Predictive Capacity

The confusion matrix of the actual versus predicted skin sensitizing categories of the 97 compounds tested in the assay is given in Table 26.3. The correct potencies were assigned with 76% accuracy, with most incorrect predictions being within one potency category of the actual category. When the non-weak and strong-extreme categories are pooled, the accuracy rate for the non-weak, moderate, and strong-extreme categories were 81%, 85%, and 100% respectively.

26.5.3 Applications and Limitations

The SenCeeTox® assay provides accurate prediction of the level of potency (non-sensitizer to extreme) for single compounds whose structure is known, as it combines *in vitro* sensitization data with physicochemical data from RDKit molecular descriptors calculated *in silico*.

There are various limitations of the SenCeeTox® method. The method was optimized on the 100 compounds tested using the HaCaT cell line [18] and transferred to use in the 3D model. The transfer was performed using a test set of ten compounds and showed a high concordance [23]. Nevertheless, more reference compounds need to be tested through the 3D approach to prove the validity of the model on this test system.

Another limitation is that the model demonstrated best predictivity with the inclusion of RDKit molecular descriptors, which needs structural information in the form of SMILES. If extracts, finished products or mixtures are tested, the application of the molecular descriptors is untested so far. It is unknown if potential sensitizers have additive effects if cells or skin is exposed to mixtures. To accommodate this, another model without inclusion of RDKit descriptors was constructed, although it needs to be considered that this approach gives less accurate results.

When the non-weak and strong-extreme categories are pooled, the accuracy rate for the non-weak, moderate, and strong-extreme categories were reduced to 65%, 68%, and 53%, respectively.

26.5.4 Role in a Testing Strategy

It has been shown and highly discussed that the replacement of the *in vivo* test methods for skin sensitization by *in vitro* test methods is more likely to be achieved using a test battery or tiered test approach [6, 19, 35]. The discussed potential approaches cover endpoints like peptide reactivity, cell activation of either keratinocytes and/or dendritic cells, and the use of *in silico* analysis of structures.

The SenCeeTox® method already combines peptide reactivity using GSH as model peptide, cell activation by assessment of relevant gene expression, and cell viability. Thus, the SenCeeTox® can be seen as a small test battery itself. It has been shown that no further improvement of the predictively of the SenCeeTox® assay can be achieved combining with methods like DPRA and/or KeratinoSens (Poster presentation at IVTS 2014, SOT 2015: [36, 37]). A proposed strategy SenCeeTox® for the utilization of the SenCeeTox® assay may be as a late stage method for testing finished products to rate their sensitizing potential if multiple compounds with potential skin sensitizing effect have been combined.

26.6 Perspectives from the Test Developer

26.6.1 Critical Steps in the Protocol

As with any model system the results are only as good as the information going in and so every piece of data must be as reliable as possible. For SenCeeTox®, there are two critical steps that must be carefully performed and considered.

The first is the glutathione assay, which is a colorimetric assay based on reduction of oxidation of GSH and reduction of 5,5′-dithiobis-2-nitrobenzoic acid (DTNB) to TNB which absorbs at 412 nm. As a result of the high concentration of GSH present in the reactions, the color development occurs fairly quickly and the rate of color development can vary depending on the enzyme lot. As such the linear range must be firmly established and reliable metrics for absorbance of the vehicle and negative controls should be established. Percent depletion of the positive control should also be standardized so that each run can be shown based on the vehicle, negative and positive controls. It is also advisable that a moderate sensitizer such as cinnamic aldehyde be run. By running a compound, which is not expected to deplete all or none of the GSH, confidence that the data is in the linear range is further raised.

The second critical step is the analysis of the PCR data. The data must be carefully collected, normalized, and examined for outliers to obtain the most reliable data possible. In order to normalize the data correctly it needs to be considered that the vehicles do not show gene induction and the control gene should not be affected by the use of different vehicles. The assay is run using four control genes and with the help of a box and whiskers plot the genes with the lowest variability is determined. Alternatively, the mean of three control genes with the lowest variability can be used to normalize the data. Either approach is acceptable, but needs to be applied consistently throughout one experiment.

26.6.2 Possible Protocol Adaptations

With the variability of using either HaCaT cells or RHE models, the applicability of the SenCeeTox® method covers a broad range of test materials from single ingredients to finished products. One adaption to this has been developed to predict photosensitization. For this the test materials are dosed with a predefined dose of UVA radiation. Photosensitizers such as TSCA were shown to increase their effect on gene expression and GSH reactivity upon UVA exposure. Known sensitizers such as para-benzoquinone and non-sensitizers remain unchanged with and without UVA exposure. To prove the correct performance of the UVA exposure during the assay an appropriate control needs to be selected.

The use of gene expression as an endpoint of sensitization response allows the assessment in changes of other genes as well. Genes involved in any number of processes can be monitored for changes.

26.7 Challenges and Opportunities

It has only been in the last year that a statistical bioinformatics model was developed for the SenCeeTox® assay. The model was produced from a large amount of HaCaT data with a small number of compound having been assessed using the 3D model. The goal for the future is to test more compounds in the 3D model to develop a more robust bioinformatics model specific to the test system. The hope is to continue to challenge and refine the model.

26.8 Conclusions

The SenCeeTox® model provides a robust assessment of skin sensitization potential. The use of multiple endpoints (GSH depletion, LDH leakage, molecular descriptors, and changes in gene expression) generates a wealth of data, which is analyzed by a model trained with over 100 compounds. The model has been shown to have a high level of accuracy, sensitivity, and specificity. The model is applicable over a wide chemical space because of the use of 3D skin tissue models. Pure compounds, medical device extracts, cosmetic formulations, and non-aqueously soluble materials can be assessed in the assay. A LLNA EC3 value can be predicted which can generate predicted potency category. The assignment of a potency category is critical for risk assessment and definition of exposure limits, especially in the cosmetic industry.

The SenCeeTox® method covers a broad range of endpoints and biochemical processes relevant for the skin sensitization process. The assay benefits of parallel assessment of potential peptide reactivity, keratinocyte activation, and impact on cell viability. Considering the time and cost involved in the SenCeeTox® assay it is best suited to determine the potency of flagged sensitizers or finished products especially if compared to methods like DPRA or KeratinoSens assay. However, SenCeeTox®'s ability to predict potency using 3D tissue models may be a primary means of screening compound which have low aqueous solubility which are required for several of the current sensitization assays.

References

1. Thyssen JP, Linneberg A, Johansen JD. The epidemiology of contact allergy in the general population-prevalence and main findings. Contact Dermatitis. 2007;57:287–99.
2. Zug KA, Warshaw EM, Fowler JF Jr, Maibach HI, Belsito DL, Pratt MD, Sasseville D, Storrs FJ, Taylor JS, Mathias CG. Patch test results of the North American contact dermatitis group 2005-2006. Dermatitis. 2009;20:149–60.
3. Kimber I, Basketter DA, Gerberick GF, Dearman RJ. Allergic contact dermatitis. Int Immunopharmacol. 2002;2:201–11.
4. 2003/15/EC. Directive 2003/15/EC of the European Parliament and the Council of 27 February 2003 amending Council Directive 76/768/EEC on the approximations of laws of the Member States relating to cosmetic products. Off J Eur Union. 2003;L66:26–35.
5. EU 1223/2009. Regulation (EC) No 1223/2009 of the European parliament and of the council. Off J Eur Union. 2009;L342:59–209.
6. Jaworska J, Harol A, Kern PS, Gerberick GF. Integrating non-animal test information into an adaptive testing strategy – skin sensitization proof of concept case. ALTEX. 2011;28:211–25.
7. Jowsey IR, Basketter DA, Westmoreland C, Kimber I. A future approach to measuring relative skin sensitising potency: a proposal. J Appl Toxicol. 2006;26:341–50.
8. Mehling A, Eriksson T, Eltze T, Kolle SN, Ramirez T, Teubner W, van Ravenzwaay B, Landsiedel R. Non-animal test methods for predicting skin sensitization potentials. Arch Toxicol. 2012;86:1273–95.
9. Dupuis G, Benezra C. Allergic contact dermatitis to simple chemicals: a molecular approach. New York: Marcel Dekker; 1982.
10. Landsteiner K, Jacobs J. Studies on the sensitization of animals with simple chemical compounds. J Exp Med. 1936;64:625–39.
11. Weltzien HU, Moulon C, Martin SF, Padovan E, Hartmann U, Kohler J. T cell immune response to haptens. Structural models for allergic and autoimmune reaction. Toxicology. 1996;107:141–51.
12. Roberts DW, Lepoittevin JP. Hapten-protein interactions. In: Lepoittevin JP, Basketter DA, Goossens A, Karlbert A-T, editors. Allergic contact dermatitis: the molecular basis. Berlin: Springer; 1998. p. 81–111.
13. Kim BS, Miyagawa F, Cho Y-H, Bennett CL, Clausen BE, Katz SI. Keratinocytes function as accessory cells for presentation of endogenous antigen expressed in the epidermis. J Investig Dermatol. 2009;129:2805–17.
14. Kimber I, Cumberbatch M. Dendritic cells in cutaneous immune response to chemical allergens. Food Chem Toxicol. 1992;117:137–46.
15. Aiba S, Terunuma A, Manome H, Tagami H. Dendritic cells differently respond to haptens and irritants by their production of cytokines and expression of co-stimulatory molecules. Eur J Immunol. 1997;27:3031–8.
16. Banchereau J, Steinman RM. Dendritic cells and the control of immunity. Nature. 1998;392(6673):245–52.

17. Banchereau J, Briere F, Caus C, Davoust J, Lebecque S, Liu YJ, Pulendran B, Palucka K. Immunobiology of dendritic cells. Annu Rev Immunol. 2000;18:767–811.
18. McKim JM Jr, Keller DJ, Gorski JR. A new *in vitro* method for identifying chemical sensitizers combining peptide binding with ARE/EpRE-mediated gene expression in human skin cells. Cutan Ocul Toxicol. 2010;29(3):170–97.
19. Natsch A, Emter R. Skin sensitizers induce antioxidant response element dependent genes: application to the *in vitro* testing of the sensitization potential of chemicals. Toxicol Sci. 2008;102(1):110–9.
20. Dinkova-Kostova AT, Holtzclaw WD, Kensler TW. The role of Keap1 in cellular protective responses. Chem Res Toxicol. 2005;18(12):1779–91.
21. Wakabayashi N, Dinkova-Kostova AT, Holtzclaw WD, Kang MI, Kobayashi A, Yamamoto M, Kensler TW, Talalay P. Protection against electrophile and oxidant stress by induction of the phase 2 response: fate of cysteines of the Keap1 sensor modified by inducers. Proc Natl Acad Sci U S A. 2004;101(7):2040–5.
22. Wang W, Jaiswal AK. Nuclear factor Nrf2 and antioxidant response element regulate NRH:quinone oxidoreductase 2 (NQO2) gene expression and antioxidant induction. Free Radic Biol Med. 2006;40(7):1119–30.
23. McKim JM Jr, Keller DJ, Gorski JR. An *in vitro* method for detecting chemical sensitization using human reconstructed skin models and its applicability to cosmetic, pharmaceutical and medical device safety testing. Cutan Ocul Toxicol. 2012;31(4):292–305.
24. Kimber I, Basketter DA, Butler M, Gamer A, Garrigue JL, Gerberick GF, Newsome C, Steiling W, Vohr HW. Classification of contact allergens according to potency: proposals. Food Chem Toxicol. 2003;41(12):1799–809.
25. Gerberick GF, Vassallo JD, Bailey RE, Chaney JG, Morrall SW, Lepoittevin J-P. Development of a peptide reactivity assay for screening contact allergens. Toxicol Sci. 2004;81:332–43.
26. Griffith OW. Determination of glutathione and glutathione disulfide using glutathione reductase and 2-vinylpyridine. Anal Biochem. 1980;106(1):207–12.
27. Chein CW, Chiang MC, Ho IC, Lee TC. Association of chromosomal alterations with arsenite-induced tumorigenicity of human HaCaT keratinocytes in nude mice. Environ Health Perspect. 2004;112(17):1704–10.
28. Mosmann T. Rapid colorimetric assay for cellular growth and survival: application to proliferation and cytotoxicity assays. J Immunol Methods. 1983;16:55–63.
29. Krstajic D, Buturovic LJ, Leahy DE, Thomas S. Cross-validation pitfalls when selecting and assessing regression and classification models. J Chem. 2014;6(1):10.
30. Clippinger A, Keller D, McKim JM, Witters H, Van Rompay AR. Inter-laboratory validation of an in vitro method to classify skin sensitizers. In: Society of toxicology 53rd annual meeting. Society of toxicology. Phoenix Convention Center, Phoenix. 24–27 Mar 2014. Poster presentation. 2014.
31. Basketter DA, Lea LJ, Cooper K, Stocks J, Dickens A, Pate I, Dearman RJ, Kimber I. Threshold for classification as a skin sensitizer in the local lymph node assay: a statistical evaluation. Food Chem Toxicol. 1999;37(12):1167–74.
32. Natsch A, Emter R, Ellis G. Filling the concept with data: integrating data from different in vitro and in silico assays on skin sensitizers to explore the battery approach for animal-free skin sensitization testing. Toxicol Sci. 2009;107(1):106–21.
33. Gerberick GF, Ryan CA, Kern PS, Schlatter H, Dearman RJ, Kimber I, Patlewicz GY, Basketter DA. Compilation of historical local lymph node data for evaluation of skin sensitization alternative methods. Dermatitis. 2005;16(4):157–202.
34. Reisinger K, Hoffmann S, Alépée N, Ashikaga T, Barroso J, Elcombe C, Gellatly N, Galbiati V, Gibbs S, Groux H, Hibatallah J, Keller D, Kern P, Klaric M, Kolle S, Kuehnl J, Lambrechts N, Lindstedt M, Millet M, Martinozzi-Teissier S, Natsch A, Petersohn D, Pike I, Sakaguchi H, Schepky A, Tailhardat M, Templier M, van Vliet E, Maxwell G. Systematic evaluation of non-animal test methods for skin sensitisation safety assessment. Toxicol In Vitro. 2015;29(1):259–70.

35. Bauch C, Kolle SN, Ramirez T, Eltze T, Fabian E, Mehling A, Teubner W, van Ravenzwaay B, Landsiedel R. Putting the parts together: combining in vitro methods to test for skin sensitizing potentials. Regul Toxicol Pharmacol. 2012;63(3):489–504.
36. Bauch C, Walker P, Patel P, Thomas S, Keller D, Dilworth C. Comparison and combination of in silico and in vitro skin sensitization assays to predict skin sensitization potential and potency. In: Poster session at: in vitro toxicology society, annual IVTS meeting, Nov 10–11, AstraZeneca, Alderley Park, Cheshire, UK. 2014.
37. Bauch C, Jamin A, Willoughby JA, Keller DJ, Patel P, Thomas S, Dilworth C, Walker P. Combination of cheminformatics and in vitro assays to predict skin sensitization potential and potency. In: session at: Society of toxicology, 54th annual meeting and ToxExpo™, Mar 22–26, San Diego, California, USA. 2015.

GARD: Genomic Allergen Rapid Detection

27

Malin Lindstedt, Kathrin Stephanie Zeller, Henrik Johansson, and Carl Borrebaeck

27.1 Principle of the Test Method and Scientific Basis

Genomic allergen rapid detection (GARD) is an assay designed to predict the ability of chemical compounds to induce skin sensitization in humans. Consequently, GARD is intended to serve as an accurate tool for hazard identification and risk assessment of chemicals. GARD measures transcriptional levels of selected genes, called the GARD prediction signature (GPS), in a derivative of the dendritic cell-like human myeloid cell line MUTZ-3 following a chemical stimulation. The identified transcripts of the GPS have been associated with recognition of xenobiotic substances and innate immune response signalling, both of which lead to maturation and activation of dendritic cells (DCs). Mechanistically, GARD is linked to key event 3 "Activation of DCs", as defined by OECD's adverse outcome pathways for skin sensitization, including "Step 5: Biochemical pathways related to skin sensitization" and "Step 6: Immune recognition of chemical allergens and maturation of DCs" [1]. In the GARD assay, cells are exposed to test chemicals at non-toxic concentrations or at concentrations, which result in 90% relative cell viability (RV90). Targeting concentrations with low cytotoxicity compared to other DC-based assays has the clear advantage that the measured immunological responses are not concealed by molecular pathways associated with toxicity or necrosis.

The transcriptome of the stimulated cells is analysed with NanoString nCounter technology (NanoString). The measured endpoints are the transcribed mRNA levels of 196 genes, collectively referred to as the GPS. All 196 genes are measured simultaneously, using NanoString digital counts of expressed transcripts. Based on these

M. Lindstedt (✉) • K.S. Zeller • C. Borrebaeck
Department of Immunotechnology, Lund University, Medicon Village (Building 406), S-22183 Lund, Sweden
e-mail: malin.lindstedt@immun.lth.se

H. Johansson
SenzaGen AB, Medicon Village (Building 401), S-22381 Lund, Sweden

C. Eskes et al. (eds.), *Alternatives for Dermal Toxicity Testing*,
DOI 10.1007/978-3-319-50353-0_27

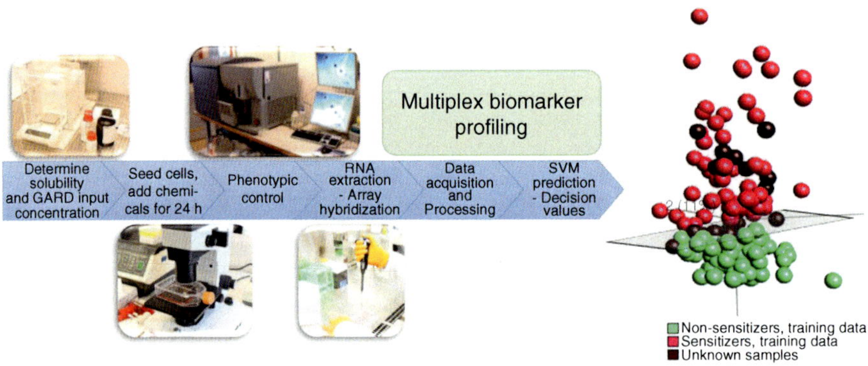

Fig. 27.1 GARD assay—schematic overview of workflow

raw data, decision values are generated *in silico* with a support vector machine (SVM), using an open-source and easy-to-use software. The SVM, previously trained on the predictive biomarker signature (the GPS), is applied to assign a decision value to each sample. In so-called prediction calls, this decision value is then in turn used to classify the samples as either sensitizers or non-sensitizers. The classification of a test substance is based on three biological replicates. A schematic overview is presented in Fig. 27.1.

The development of *in vitro* tests was initiated under the EU FP6 Programme-funded project "Novel Testing Strategies for *In Vitro* Assessment of Allergens, Sens-it-iv" (2005–2010), dedicated to develop and optimize test strategies that could reduce or replace animal testing for sensitization. For initial assay development and to investigate the potential of using a myeloid human cell line as test basis for evaluation of chemical sensitizers, a panel of 38 reference compounds was used for stimulation. Protein expression and transcriptional profiles using genome arrays of stimulated cells were measured. While a number of protein markers, including CD86, were considered insufficient for accurate classification of chemical sensitizers, the GPS set of 196 informative genes was identified as potent predictor of sensitizers [2–4]. The initial transcriptome analysis of stimulated cells was intended to complement the simpler flow cytometry-based assays in search for novel biomarkers. However, the need for additional assays exhibiting a higher predictive accuracy became increasingly apparent. The idea of applying the entire GPS as a predictive instrument took form, and a testing strategy and protocols were published [5].

27.2 Current Validation Status

The functionality of the assay was demonstrated by the publication of data from three separate pre-validation studies, of which two were performed using entirely blinded samples. The predictive performance of the assay was estimated to 89% [6]. The GARD assay, offered by SenzaGen AB, has been reviewed by all OECD WNT (the Working Group of the National Coordinators of the Test Guidelines Programme) task force members and included in the OECD Test Guideline Programme (TGP) work plan with the TGP No 4.106. Thus, the formal validation process for GARD is now ongoing.

27.3 Performance and Applicability of the Test Method

27.3.1 Reproducibility

Within-laboratory reproducibility (WLR) was assessed using estimation of concordance of predictions between different experiments. Two to three biological replicates based on different cell batches were investigated routinely. Thus, biological (e.g. cell cycle and growth rate) and technical variations (e.g. RNA isolation, probe hybridization) should be included in order to provide appropriate data for WLR calculations. GARD classifications are concordant between experiments in 92% of available cases. Notably, these calculations were only performed with compounds that have been repeatedly used for cell stimulations, i.e. in more than one experiment.

27.3.2 Predictive Capacity

Assessment of predictivity was performed using three test sets of in total 37 compounds [6]. Only the first of these sets was selected in-house, as set #2 and #3 were run as blinded samples, selected by third parties. The rationale and criteria applied for selection of test set #1 were the following: (1) sensitizers should not have been used for model training, (2) sensitizers should range from weak to extreme (assessed by LLNA), and (3) non-sensitizers should not have been used for model training, with one exception for internal control and normalization purposes. Compounds of test set #2 and #3 were selected by third parties, according to the following criteria: (1) they should include "standard" sensitizers, normally used in assay evaluation, (2) they should include chemicals that are known to be inaccurately classified by LLNA and/or other *in vitro* assays (e.g. SDS, benzalkonium chloride), and (3) they should include compounds that are generally accepted as extremely difficult to accurately classify by *in vitro* assays (e.g. surfactants, emulsifiers). Using 37 chemicals (of which 26 chemicals were blinded) for in-house pre-validation and estimation of prediction performance, the Cooper statistics were as follows: accuracy 92%, sensitivity 92% and specificity 92%. However, as being part of a proof-of-concept study, the first (non-blind) test set was used for optimization of the prediction model cut-off (PMCO). Applying the optimized PMCO on the 26 blind chemicals in test set #2 and #3, the prediction performance was estimated to be as follows: accuracy 89%, sensitivity 89% and specificity 88%. Notably, these blinded test substances included some challenging compounds ([6], as described above). Predictive performance is calculated by comparing GARD to LLNA classifications, except for chemicals that are known to be misclassified by LLNA in the context of clinical experience.

For certain compounds false predictions were received. For instance, 4-chloroaniline appeared as a false negative in the GARD assay. This chemical is also frequently misclassified by the LLNA, whereas it often appears as a sensitizer in guinea pig tests [7, 8]. To the best of our knowledge, human reference data is not available. Further, phenyl benzoate (false negative in GARD) is classified as a weak sensitizer by the LLNA and is assigned to the human potency category 3 according to Basketter et al. [9]. Category 3 contains substances that typically need a "substantial degree of exposure" in order to cause sensitization in 0.01–0.1% of subjects exposed; in lower concentrations they may not sensitize at all or only few individuals. Notably, phenyl benzoate is also published as a false negative in the KeratinoSens assay [10], which in this context may support a certain relevance of data produced using assays based on human cells.

27.3.3 Applications and Limitations

The advantage of GARD compared to other assays for prediction of chemical skin sensitizers is the use of a multivariate readout, enabling mechanistic-based classifications of test substances. Indeed, models for prediction of sensitizing potency are currently being established, which will provide further details in addition to the binary classifications. Finally, the global readout allows for the simultaneous assessment of multiple biological endpoints by applying the same test system but with different biomarker prediction signatures. Prediction models for chemical respiratory sensitizers have been demonstrated [11], and yet another for protein respiratory sensitizers is in development.

The GARD test is considered to be applicable on any chemical compound that to some degree is soluble in cell media, with or without dilution in DMSO. Experimental experience, so far, has only revealed one compound (oxazolone) that has been unable to be evaluated due to solubility issues. Furthermore, being based on a biological system, little information regarding the chemicals to be tested is required. Experiments investigating mixtures are currently being carried out and have so far shown promising results. The metabolic capacity of cell lines, i.e. the relevance of the cell line in the investigation of pre- and pro-haptens, has been discussed in several contexts. However, in this case, both pre- and pro-haptens were tested during the data-driven GARD assay development. Examples include geraniol, ethylenediamine and resorcinol (pro-haptens), as well as 1,4-phenylenediamine and isoeugenol (pre-haptens). Both pre- and pro-haptens have accurately been classified with GARD, with examples also including cinnamic alcohol, hydroquinone and lauryl gallate [2, 6].

One limitation of the GARD assay, as it was used during assay development, was the requirement of expensive equipment or access to core facilities providing such equipment, most notably Affymetrix GeneChip-associated workstations. Consequently, a technology platform transfer was recently carried out and validated in-house [12]. Initially, several state of the art transcriptome quantification technologies were evaluated and compared with respect to simplicity of protocols, sample throughput capacity and resource effectiveness. Of importance, the generated data was also correlated to historical array data, ensuring comparable detection of transcriptional profiles [13]. Upon the successful technology platform transfer, protocols are today simpler and the number of steps requiring personnel expertise has been drastically reduced. Further limitations of the GARD assay include the somewhat high costs per chemical if only few substances (<10) are tested at a time. However, this is currently mitigated by the use of so-called GARD campaigns, in which a high number of chemicals (~30) are assayed simultaneously.

A limiting and confounding factor, based on the experimental data collected with the GARD test so far, is the biological experiment-to-experiment variability, which causes a shift in all measured transcript levels. This in turn can cause the SVM model to produce unreliable decision values if the PMCO for each batch of experiments is not closely monitored. These limitations are currently being minimized with an SVM calibration, using benchmarked controls. On the other hand, it should be noted that GARD, because it is based on DNA hybridization techniques, is not affected by a number of issues that are frequently described as limitations in assays based on merely a spectrophotometric/fluorescent readout. Consequently, autofluorescent compounds and dyes can be tested by GARD. Of note, compounds regularly described as "problematic", e.g. certain surfactants, emulsifiers and volatiles, have also been successfully classified by GARD.

The previous focus has been to provide a reliable tool for classification in the context of REACH and the Cosmetic Directive. Present and ongoing development envisages the inclusion of product candidates from pharmaceutical and food industry, where GARD by its information-rich readout is expected to stand its ground as well.

27.4 Brief Description of the Protocol

27.4.1 Cell Culture

The MUTZ-3 derivative is cultured in maintenance medium, prepared by supplementing α-MEM (Thermo Scientific) with 20% FCS (Gibco Invitrogen); GM-CSF (Miltenyi Biotec) is added freshly to the medium at every cell split to a final concentration of 40 ng/mL. Media supplemented with both GM-CSF and FCS are referred to as complete medium. All cell work should be performed under sterile conditions. All centrifugations are performed at 1200 rpm, 5 min, 4 °C. All incubations are performed at 37 °C and 5% CO_2. Working stocks of cultures should not be grown for more than 20 passages or 2 months after thawing. Cells should be maintained at 200,000 cells/mL. The medium is changed every 3–4 days or when cell density exceeds 500,000 cells/mL.

27.4.2 Phenotypic Quality Control

Prior to any chemical stimulation, cells are quality controlled by phenotypic analysis. This is done to ensure that the cells are maintained in an immature state and to detect phenotypic drift. PBS (Thermo Scientific) with 1% (w/w) BSA (Saveen & Werner), sterilized by filtration, is used for all washing and staining steps for flow cytometry. All incubations are performed at 4 °C. Transfer 200,000 cells to FACS tubes, wash cells twice and add 50 µL wash buffer. Stain cells by adding the mouse antihuman antibodies as indicated to each tube; (1) Isotype controls mouse polyclonal anti-IgG1-FITC and anti-IgG1-PE (both BD Pharmingen), (2) anti-CD86-FITC and anti-CD54-PE (both BD Pharmingen), (3) anti-HLA-DR-FITC and anti-CD80-PE (both BD Biosciences), (4) anti-CD34-FITC (BD Bioscience) and anti-CD14-PE (Dako) and (5) anti-CD1a-FITC (Dako) and propidium iodide (PI, BD Pharmingen). Incubate 15 min, wash and resuspend in 200 µL wash buffer. Analyse the samples on FACSCanto II (BD Bioscience) or an equivalent instrument. During analysis, use tube 1 for setting gates and quadrants. Exclude dead cells and cell debris by setting a gate in the FSC/SSC scatterplot, see Fig. 27.2a. Set quadrants by the isotype controls, see Fig. 27.2b. Figure 27.2c provides an example for the positive fraction of cells (%) for CD34 and CD14. Acceptance criteria for each phenotypic marker are shown in Table 27.1.

27.4.3 Establishing the GARD Input Concentration

All test compounds should be stored according to instructions from the supplier, in order to ensure the stability of the compounds. All test compounds should be dissolved in sterile water if soluble, otherwise in DMSO (Sigma Aldrich). For cell stimulations, chemicals should be dissolved in their appropriate solvent as 1000x stocks of target in well concentration for establishment of GARD input concentration. The GARD input concentration of a test compound is decided according to the following decision tree: (1) chemicals that induce cytotoxicity should be used

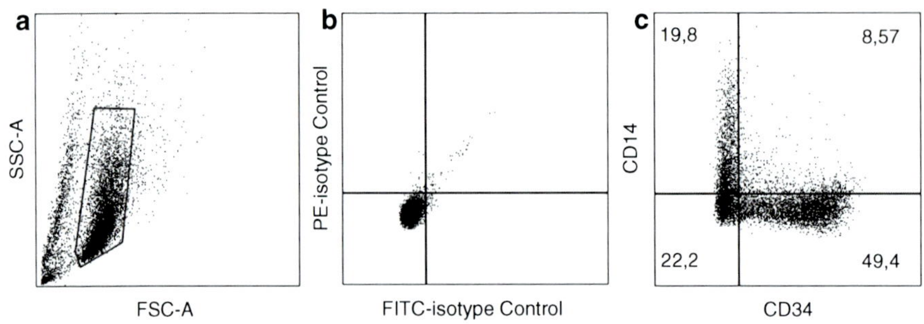

Fig. 27.2 Setting gates and quadrants and recording the fraction of positive cells for phenotypic markers, as described in Sec. 27.4.2

Table 27.1 Acceptance criteria for phenotypic markers

Phenotypic biomarker	Accepted range of positive cells (%)
CD86	10–40
CD54	>95
HLA-DR	>60
CD80	<10
CD34	35–70
CD14	5–50
CD1a	10–60
Propidium iodide	<15

The accepted range of phenotypic biomarker expression is based on observations made in the developing laboratory during assay development. Variations within these ranges are to be considered normal

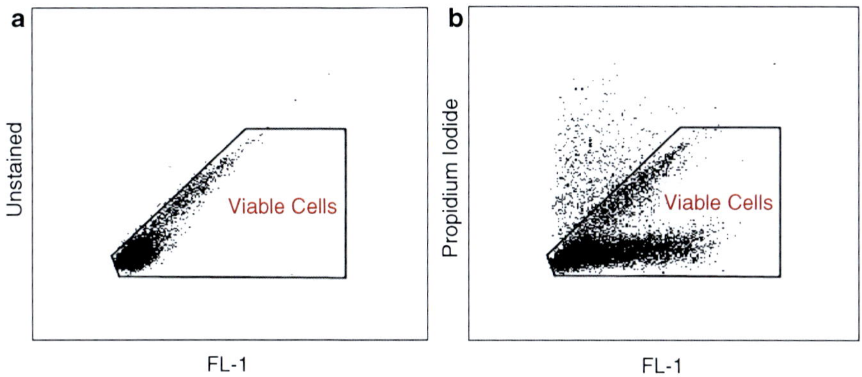

Fig. 27.3 Gating for determining the viability of stimulated cells

at the concentration that induces a relative viability of 90% (RV90), (2) chemicals that are not cytotoxic are used at 500 μM, and (3) chemicals that are not cytotoxic and not soluble to 500 μM are used at the highest soluble concentration in cell media. To determine the decision criteria and the GARD input concentration for a test compound, cell stimulations are performed using a titration range of concentrations, in duplicates, ranging from 1 to 500 μM. Harvest the chemical-stimulated cells, wash, resuspend in 50 μL wash buffer and add 1 μL PI. Incubate at 4 °C for 10 min. Wash, resuspend in 200 μL wash buffer and analyse the samples with FACSCanto II. During analysis, use unstimulated unstained samples to set a gate for viable and dead cells, respectively. Applying the preset gate on remaining tubes, record the positive fraction of cells (%) in the gate for viable cells for each concentration (Fig. 27.3).

Once the viability (fraction of cells in %) for the entire titration range of a test compound has been recorded, the relative viability for each sample is calculated according to the equation below, where Rv is the relative viability of the sample, V_S is the viability of the sample in % and V_C is the viability of the unstimulated (PI-stained) control sample.

$$Rv = \frac{V_S}{V_C} \bullet 100$$

For each concentration of the titration range, calculate the mean value of the replicate samples. Once the mean relative viability for each concentration has been calculated, the RV90 concentration is identified as the concentration that yields 90% relative viability. If an RV90 value exists within the titration range, this is used as the GARD input concentration. If an RV90 value does not exist within the titration range, the maximum concentration of the titration range is used as GARD input concentration. This concentration should be either 500 μM or the maximum soluble concentration of the test compound.

27.4.4 Positive, Negative and Benchmark Control(s)

A set of controls is run in biological triplicates (Table 27.2). The purpose of the benchmark controls is twofold: Firstly, they are used to calibrate the prediction model by identifying the optimal prediction model cut-off (PMCO) for binary classification of test substances. Secondly, the positive control (DNCB) and the negative control (DMSO) serve as acceptance criteria. The positive and negative control must be accurately classified as positive and negative, respectively, for a test to be accepted as valid.

27.4.5 Cell Stimulations for the Purpose of mRNA Isolation

Once the GARD input concentration for each chemical is established, three independent biological experiments should be performed. Also, each chemical stimulation should be performed in triple wells, including benchmark controls. For mRNA preparation, cells are harvested into RNase-free Eppendorf tubes. Collect

Table 27.2 List of controls

Compound	Purpose	Vehicle	GARD input concentration (μM)
Dinitrochlorobenzene	Benchmark/positive control	DMSO	4
Diethylmaleate	Benchmark (high)		
2-Hydroxyethylacrylate	Benchmark (high)	Water	100
2-Mercaptobenzothiazole	Benchmark (medium)	DMSO	250
Phenyl benzoate	Benchmark (low)	DMSO	200
2-Ethylhexyl acrylate	Benchmark (low)		
Benzakonium chloride	Benchmark (low)	DMSO	3
Chlorobenzene	Benchmark (negative)	DMSO	100
1-Butanol	Benchmark (negative)	DMSO	500
Unstimulated cells	Benchmark (negative)	–	–
DMSO	Benchmark/negative control	–	0.1% (v/v)

During the publication process of this book, protocol optimization in the developing laboratory has drastically reduced the number of benchmarks used. Current GARD SOPs utilizes three (3) benchmark controls. Authors' remark duplicate samples of each stimulation, 1 mL of cell suspension in each tube. Centrifuge tubes at 1200 rpm, 4 °C, 5 min. Remove supernatant carefully by aspiration. Add 500 μL of TRIzol™ reagent (Ambion) to each cell pellet. Vortex samples for 30 s and rapidly freeze samples at −20 °C. For flow cytometry analyses, cells are harvested into FACS tubes in duplicates, stained with PI and analysed as described above when establishing the GARD input concentration. For cytotoxic compounds inducing an RV90 value, acceptance criterion for each sample is a relative viability of 85–95%.

27.4.6 mRNA Isolation

Total RNA, including mRNA, is isolated from the TRIzol™ samples using commercially available kits and reagents (Direct-zol RNA MiniPrep, Zymo Research, Cat.No R2052), according to protocols provided by the supplier. Eluted RNA samples are stored at −80 °C until NanoString CodeSet hybridization and mRNA quantification.

27.4.7 Digital Quantification of mRNA Transcripts

The endpoint measurements of GARD are the mRNA quantifications of the GPS, using NanoString nCounter technology. All steps following mRNA isolation are performed according to NanoString standard protocols, using a custom CodeSet comprising probes for the GPS, as described by the supplier. For quality control of RNA, the RNA 6000 Nano Kit is used with a Bioanalyzer 2100 (Agilent).

27.4.8 Data Analysis

Scanning of the NanoString arrays leads to raw data in the form of .RCC-files. All data analysis from this point is performed in R, an open-source statistical programming language and software freely available for download at www.r-project.org. The arrays are preprocessed, normalized and annotated as benchmark controls or unknown test samples. Historical reference data are used to train an SVM model. The benchmark controls are used for SVM calibration in order to identify the PMCO, which is subsequently applied to classify test substances. In order to meet the acceptance criteria, the positive control (DNCB) should be accurately classified as a sensitizer, and the negative control (DMSO) should be accurately classified as a non-sensitizer after SVM calibration using the benchmark controls. For binary predictions, the prediction model is defined as "If the mean calibrated SVM DV from three biological triplicates of any test substance is greater than (>) zero (0), the test substance is classified as a sensitizer. Consequently, if the mean calibrated SVM DV is <0, the test substance is classified as a non-sensitizer".

27.5 Role in a Testing Strategy

The GARD test contributes significantly as part of an integrated testing strategy (ITS) for the assessment of chemical skin sensitizers for the purpose REACH- and/or Cosmetic Directive-associated testing. Of note, the GARD test has also shown promising performance to deliver potency prediction.

27.6 Perspectives from the Test Developer

27.6.1 Critical Steps in the Protocol

Provided that a laboratory is familiar with the techniques involved, the amount of training required is minimal for an assay transfer. Receiving laboratories need to have routines for the maintenance of mammalian cell lines. If the complete assay is to be transferred, the receiving lab needs to be experienced and fully equipped with NanoString equipment. Another option is to perform the cell stimulations in the receiving lab and to send mRNA samples to the test method developers as the acting core facility.

27.6.2 Challenges and Opportunities

As of June 2016, GARD should be considered a functional, internally validated assay. Current development of the assay is related to the following areas: (1) applicability/functionality domain of chemical respiratory sensitizers, (2) applicability/functionality domain of protein sensitizers and (3) potency assessment.

We are also currently working on defining subgroups among sensitizers based on differential activation and/or inhibition of signalling and metabolic pathways. Indeed, we have shown differential engagement of canonical pathways associated with different chemical reactivities of sensitizers [14]. In addition, GARD has been pre-validated in-house, in a proof-of-concept study demonstrating the functionality of the assay [6].

27.7 Conclusions

GARD is a functional, internally validated test platform, based on chemical-induced transcriptional regulation, which results in acquisition of large amounts of data. This, in turn, results in the ability to contribute with information on the adverse outcome pathways for skin sensitization, the potency of chemicals and an increased mechanistic understanding of sensitization. The information content is one of the distinguishing features of GARD, apart from being based on dendritic cells, a key cellular component of the human immune system.

References

1. OECD. ENV/JM/MONO(2012)10/PART1. The adverse outcome pathway for skin sensitisation initiated by covalent binding to proteins. 2012.
2. Johansson H, Albrekt AS, Lindstedt M, Borrebaeck CAK. A genomic biomarker signature can predict skin sensitizers using a cell-based in vitro alternative to animal tests. BMC Genomics. 2011;12:399.
3. Johansson H, Lindstedt M. Prediction of skin sensitizers using alternative methods to animal experimentation. Basic Clin Pharmacol. 2014;1(1):110–7.
4. Lindstedt M, Borrebaeck CAK. Pattern rules: biomarker signatures for sensitization as an alternative to animal testing. Biomark Med. 2011;5(6):809–11.
5. Johansson H, Albrekt AS, Borrebaeck CAK, Lindstedt M. The GARD assay for assessment of chemical skin sensitizers. Toxicol In Vitro. 2013;27:1163–9.
6. Johansson H, Rydnert F, Kuehnl J, Schepky A, Borrebaeck CAK, Lindstedt M. GARD in-house validation – a proof of concept. Toxicol Sci. 2014;139(2):362–70.
7. Haneke KE, Tice RR, Carson BL, Margolin BH, Stokes WS. ICCVAM evaluation of the murine local lymph node assay. Data analyses completed by the national toxicology program interagency center for the evaluation of alternative toxicological methods. Regul Toxicol Pharmacol. 2001;34(3):274–86.
8. World Health Organization. Concise international chemical assessment document 48. 4-CHLORO-ANILINE, ISBN 9241530480. Geneva. 2003.
9. Basketter DA, Alépée N, Ashikaga T, Barroso J, Gilmour N, Goebel C, Hibatallah J, Hoffmann S, Kern P, Martinozzi-Teissier S, Maxwell G, Reisinger K, Sakaguchi H, Schepky A, Tailhardat M, Templier M. Categorization of chemicals according to their relative human skin sensitizing potency. Dermatitis. 2014;25(1):11–21.
10. EURL ECVAM. Summary sheet on the KeratinoSens assay prepared by EURL ECVAM. Update 14 Feb 2014. p. 4(34).
11. Forreryd A, Johansson H, Albrekt AS, Borrebaeck CAK, Lindstedt M. Prediction of chemical respiratory sensitizers using GARD, a novel in vitro assay based on a genomic biomarker signature. PLoS One. 2015;2015:11.
12. Forreryd A, Zeller KS, Johansson H, Lindstedt M. From genome-wide arrays to tailor-made biomarker readout – progress towards routine analysis of skin sensitizing chemicals with GARD. Toxicol In Vitro. 2016;37:178–88.
13. Forreryd A, Johansson H, Albrekt AS, Lindstedt M. Evaluation of high throughput gene expression platforms using GARD – an assay for prediction of skin sensitization based on a genomic biomarker signature. BMC Genomics. 2014;15(1):379.
14. Albrekt AS, Johansson H, Borrebaeck CAK, Lindstedt M. Differentially regulated signaling pathways in MUTZ-3 cells stimulated with skin sensitizers. BMC Pharmacol Toxicol. 2014;15:5.

Proteomics Testing for Sensitization Potency Using the Sensiderm™ TMT-SRM 10-Plex Assay

28

Petra Budde, Hans-Dieter Zucht, and Ian Pike

28.1 Rationale and Basic Concepts for the Analysis of Proteomic Response Signatures of Dendritic Cellular Models for Dermal Toxicity Testing

There is a pressing and continued need for alternative assays that replace animal testing in toxicological evaluation of cosmetic ingredients [1]. Current alternative testing approaches to predict the sensitization potential of chemicals involve *in silico* approaches, peptide reactivity assays as well as *in vitro* tissue and cell culture models [2]. Given the diversity of the physical and chemical properties of skin sensitizers, it is reasonable to assume that multiple cellular pathways will play a role in response, and so a combination of different assays and molecular markers will be required for predicting the sensitization potency for a growing number of substances used as cosmetic ingredients. State-of-the-art proteomics analytical technologies hold the potential to provide a more comprehensive picture of molecular and cellular changes that accompany skin sensitization than more restricted genomic methods. By following unbiased discovery approaches, it is possible to interrogate the entire proteome including post-translational modifications that regulate the activity of stress response pathways. Subsequently, statistical methods are used to determine minimal panels of peptide or protein changes that predict a particular phenotype or in this case assign chemicals into one of three groups, non-sensitizer, sensitizer or irritant.

In order to discover protein biomarker signatures related to one of the key events of allergic skin sensitization, which is the activation of dendritic cells, we have previously performed unbiased proteomic discovery experiments for a range of chemical sensitizers in human skin sensitization cell models [3]. All test chemicals used

P. Budde • H.-D. Zucht • I. Pike (✉)
Proteome Sciences PLC, Hamilton House, Mabledon Place, London WC1H 9BB, UK
e-mail: ian.pike@proteomics.com

C. Eskes et al. (eds.), *Alternatives for Dermal Toxicity Testing*,
DOI 10.1007/978-3-319-50353-0_28

Table 28.1 The table summarizes gene name as well as protein identifiers such as IPI-ID, Uniprot ID, Swissprot ID and common protein name

Gene ID	Uniprot ID	Protein	Pathway
G6PD	P11413	*Glucose-6-phosphate 1-dehydrogenase*	Oxidative stress response Nrf2-regulated gene [17, 18]
PGD	P52209	*6-phosphogluconate dehydrogenase, decarboxylating (PGD)*	Oxidative stress response Nrf2-regulated gene [17, 18]
HSPA8	P11142	*Isoform 1 of heat shock cognate 71 kDa protein*	Stress response and danger signal Immune system activation [19, 20]
MPO	P05164	*Isoform H17 of myeloperoxidase (light and heavy)*	Peroxidase enzyme involved in intracellular pathogen killing [21]
S100A4	P26447	*Protein S100-A4*	Mediates inflammatory response via TLR4 [22] Candidate gene in allergy [23]
S100A8 S100A9	P05109 P06702	*Protein S100-A8* *Protein S100-A9*	Calprotectin (S100A8/S100A9) heterodimer Activation of innate and inflammatory immune response via TLR4 [24]
SLC3A2	B4E2Z3	*4F2 heavy chain (CD98)*	Dendritic cell antigen Role in adaptive immunity and amino acid transport [25]
SOD1	P00441	*Superoxide dismutase [Cu-Zn]*	Cytosolic enzyme Removal of superoxide radicals [26]
TMSL3/ TMSB4X	P62328	*Thymosin beta-4-like protein 3*	Multifunctional regenerative peptide Prevents oxidative stress [27]

in the model-building phase were provided by the European Integrated Project *Sens-it-iv*, which was finished in March 2011 [4, 5]. This approach has led to the discovery of a panel of ten proteins expressed in the human dendritic cell model MUTZ-3, which has subsequently been developed into the multiplex 10-marker protein multiplexing assay, referred to as Sensiderm™ TMT®-SRM 10 plex assay or shortly "Sensiderm" assay. As shown in Table 28.1, the ten protein targets were selected to provide information on the activation of different biological pathways.

The Sensiderm assay uses the MUTZ-3 cell line that has been extensively evaluated as a surrogate for dermal dendritic cells (DC) in the context of skin sensitization testing [6–9]. DCs are highly relevant during skin sensitization processes because they physiologically internalize, process and transport antigens to the lymph node inducing T cell differentiation and proliferation. Whilst primary DCs isolated from living donors or *in vitro* differentiated DCs can be used in assay systems, the widespread use of primary cells in standardized screening assays is limited by donor variations and difficulties to obtain sufficient quantities of cells. Thus, human cell lines with dendritic-like properties such as the MUTZ-3 cell line are used in the Sensiderm assay. MUTZ-3 cells were isolated from a patient with acute myelomonocytic leukaemia and show cytokine-dependent proliferation and survival [10, 11].

The intended use of the assay is to discriminate chemical sensitizers and non-sensitizers based on pathway-specific biomarker proteins, where the pathways interrogated are related to antigen presentation, the NF-kappa B cascade, leukocyte migration, activation of oxidative stress, inflammatory response, energy metabolism, activation of systems belonging to the innate immunity like the TLR4 system and danger signals. The multitude of analytical targets was selected to provide measures, which apparently are triggered simultaneously upon the action of a sensitizer. The algorithm creates a combined potency score by combining this complementing information. The added value and upside potential of the Sensiderm test is that it provides particular information by its individual protein measures, which allows a deeper scientific interpretation of the results and the clustering of similar compounds. It has been shown that it is possible to differentiate chemical classes using mathematical gradation methods such as partial least squares regression discriminant analysis (PLS-DA) or principal component (PCA) analysis (see Fig. 28.1) [12].

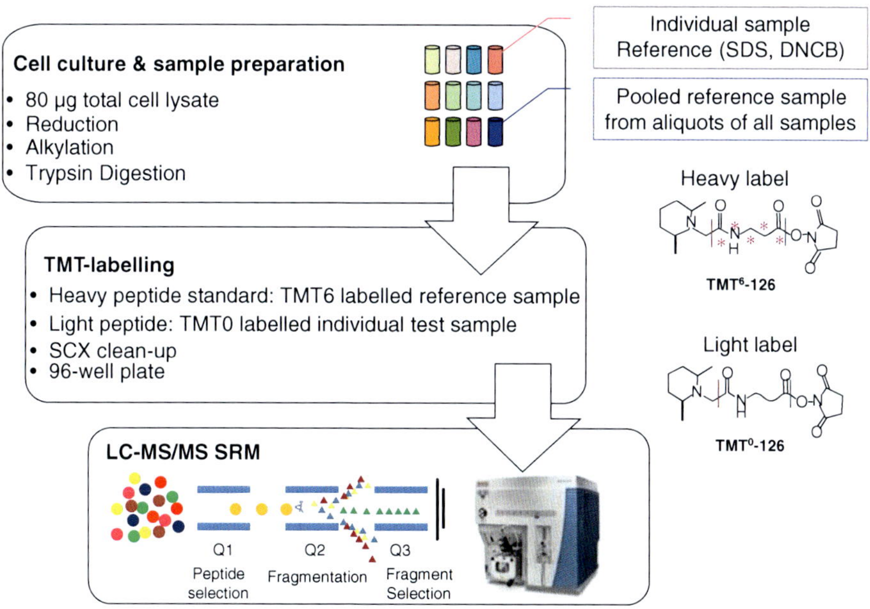

Fig. 28.1 Test principle for the analysis of exposed cell culture cells in the Sensiderm assay: Proteins from treated Mutz-3 cells are extracted, cleaved into peptides and chemically labelled with the weight labelling reagent (TMT®) used to tag case and control samples with different masses. During the measurement in a triple stage mass spectrometer, the peptides are measured using mass detection filters (Q1–Q3), which are highly specific

The analytical technology behind the Sensiderm assay is based on the targeted quantitative mass spectrometry of protein-specific peptides. The mass spectrometry method uses multiplexed selected reaction monitoring (SRM) on a triple-stage mass spectrometer coupled to high-performance liquid chromatography (HPLC). This technology provides highly accurate measures of protein concentration, with a typical precision less than 5% relative standard deviation. High assay accuracy and specificity is achieved by calibrating the assay with molecular standards and by measuring a highly specific mass spectrometric fingerprint of individual peptides. This method is therefore superior to common immunological detection methods of proteins. The Sensiderm assay allows the analysis of cell lysates using the typically small protein amounts derived from Petri dishes with a typical throughput of several hundred samples per week and single instrument. The specificity and sensitivity of the SRM method, as well as the ability to multiplex the measurement of proteins in parallel, has made this method now widely used in clinical and toxicology laboratories [13, 14]. In order to make the cellular protein content accessible to the mass spectrometer, a dedicated sample preparation procedure is required. For this purpose, the cellular protein extract is digested into small peptides with the protease trypsin, which are then labelled with the mass spectrometric detection and quantification reagent called Tandem Mass Tags® (TMT®) as shown in Fig. 28.1. TMT reagents are sets of isobaric and isotopic compounds, which introduce a sample-specific label, so that multiple samples can be mixed and measured in a single MS analysis [15] This simultaneous measurement of case and control samples keeps the measurement and classification errors very low.

In the Sensiderm assay we measure the abundance of ten tryptic peptides derived from digests of MUTZ-3 cells exposed to the test compound with the levels of the same ten peptides from MUTZ-3 cells exposed to a control buffer. Digests of cells treated with control compound are labelled with a TMTzero tag (TMT^0-126; Thermo Scientific, Germany) that has no heavy atom substitutions whilst the test compound cell digests are labelled with a TMT 6-plex tag (TMT^6-126; Thermo Scientific, Germany) that has five heavy atom substitutions. The two digests are then mixed and analysed by the programmed TMT-SRM method, and the ratio of the ion intensities for the ten peptides between treated and control cells is determined. Threshold values for each of the ten peptides are then used to create the potency score.

Considering the mathematical model used in the Sensiderm assay (Fig. 28.2), the assay yields a potency score, which is used as a classifier for sensitizers by combining the quantitative measures from ten target proteins by means of linear regression.

$$Y(x_1,\ldots,x_n) = a_0 + \sum_{i=1}^{m} a_i f_i$$

Fig. 28.2 Linear regression model for potency prediction: The individual protein measures (f_i) are weighted according to their importance (a_i) and summed up yielding the sensitization score (Y)

The calibration of the model is done using skin sensitizers and respiratory sensitizers as proposed by Casati et al. [4] relative to cells exposed to non-sensitizing or inert compounds. Measures of the test are provided as a ratio of protein expression between the exposed cells and cells grown in control medium containing solvent. The linear regression model is fitted using the set of potency data of the training chemicals. The model has to be fitted for each experimental round correcting for inter-assay variation. A threshold of the computed score can be varied after analysing the corresponding sensitivity and specificity. New chemicals gain only a high score when the majority of the metabolic pathways have been activated simultaneously.

An added value of the Sensiderm assay is the possibility for a post-hoc interpretation of the data. Besides the computation of the potency score, the data can be analysed using multivariate chemometric methods such as partial least squares regression, principal component analysis (PLS-DA, PCA, (reviewed in [16])). These multivariate methods allow an interpretation of the underlying latent variables in the multivariate space. Hence, the combination of the quantitative readout of ten protein targets allows creating a model, which is able to reflect a chemical's sensitization potency as well as to cluster components according to similarities in the individual pathway responses. Figure 28.3 gives an example of data analysis using a PLS-DA model using the Sensiderm analytes separating the compounds DNCB, TMA, SDS and solvent.

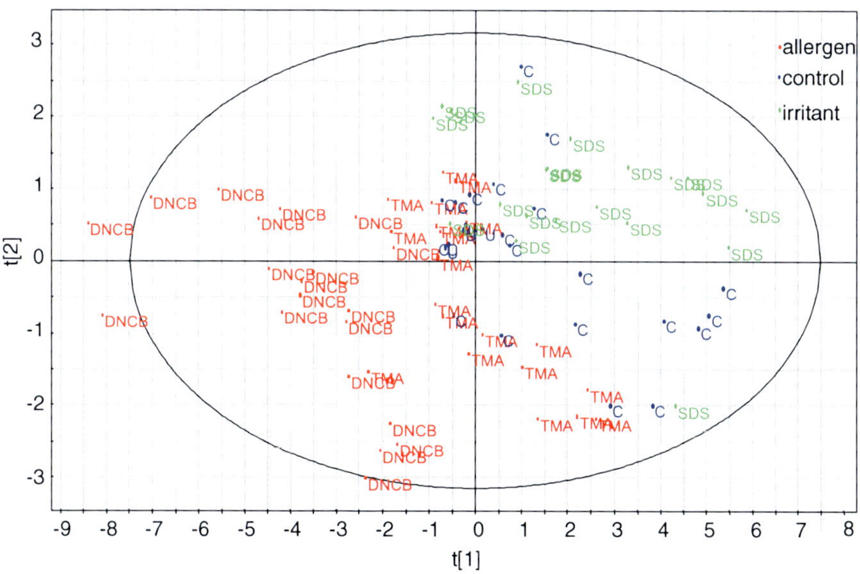

Fig. 28.3 A visualisation of the statistical computation of the so called score plot of a PLS-DA model using Sensiderm data. PLS-DA reduces all measurements to fewer components, which values can be plotted in such *x*/*y* diagrams. A good separation of those scores computed for each sample type (here *red* = samples treated with sensitizer, *blue* = controls, *green* = irritants) illustrates the utility of the markers measured that they altogether allow a classification of samples and clustering of chemicals

The analytical targets of the Sensiderm assay have been discovered during a FP6 funded EU project called *Sens-it-iv* ("Novel testing strategies for *in vitro* assessment of chemicals"). Within this research programme, proteomic investigation of MUTZ-3 cells using quantitative mass spectrometry profiling was performed. Putative biomarker targets have been selected from thousands of measured proteins after experimental replication and rigorous statistical testing for prioritizing the most relevant analytical targets.

The proteins measured in the Sensiderm 10-plex mass spectrometric assay are shown in Table 28.1.

28.2 Current (Pre)Validation Status

The test system, which has been derived as a result of the Sens-it-iv project, has been refined to a route of pre-validation. The testing protocols have been transferred into standard operating procedures (SOPs) guided by a quality management system according to ISO 9001:2008. SOPs include cell culture conditions, sample preparation procedures, quantitative mass spectrometry and data analysis procedures.

Assay transfer procedures have been performed between two analytical laboratories concerning mass spectrometry as well as the provision of cells and their incubation procedures with testing chemicals. A set of testing chemicals has been investigated in a blinded and non-blinded fashion [2]. Further characterization efforts are currently ongoing to complement data from a larger set of chemicals.

28.3 Performance and Applicability of the Test Method

The performance of the test method has been investigated under the participation of the COLIPA initiative for the systematic evaluation of non-animal test methods for skin sensitization [2] in a blinded fashion. From the ten testing chemicals, the testing concordance was 6/6 scoring correctly three sensitizers and three non-sensitizers. The remaining chemicals were not scored unambiguously. Due to the low number of chemicals, sensitivity and specificity cannot yet be computed. Further validation is therfore required.

The test method's applicability allows for the testing of a large number of chemical classes as long as they can be applied on the MUTZ-3 cell culture system. The analytical test format is currently a laboratory-developed test (LDT) available as analytical service from Proteome Sciences plc (London, UK). For future widespread testing, the analytical method can be easily transferred requiring a simple method transfer and some analyte specific reagents (ASRs) for MUTZ-3 cell culture.

28.4 Brief Description of the Protocol

The Sensiderm protocol includes the following procedures:

- Pre-analysis of toxicity of the chemical compounds towards MUTZ-3 cells, as well as solubility analyses
- Incubation of MUTZ-3 cells with the novel compound in cell culture dishes together with a multitude of other chemicals in replicates
- Harvesting of cells, washing of cells and cell lysis
- Extraction of proteins, trypsinization of proteins to obtain peptides and labelling of the peptides with a mass spectrometric label
- Injection of the peptide mixtures in an LC-MS instrument and measurement of selective reaction monitoring mode (SRM-LC-MS) on a triple quadrupole mass spectrometer
- Data preprocessing to yield raw data of the target peptides as ratios treated versus untreated cells
- Combination of peptide measures to protein measures
- Linear regression of the data using the reference chemicals to yield the ten weights for the linear equation
- Computation of a score using the linear equation and setting thresholds
- Exploration of data using multivariate statistics

All chemicals are initially diluted in DMSO. Growth assays with increasing dosages of test compounds in biological duplicates are performed to assess the maximum non-toxic dosage of the compound, which is used in the subsequent risk assessment TMT-SRM assay. The highest compound concentration that retains >80% cell viability will be used in the final assay for testing purposes. Calibration is performed using DNCB, SDS and DMSO as control chemicals.

The assay system is based on the stimulation of differentiated MUTZ-3 cells, which have previously been shown to be able to reflect the initial triggering mechanisms of sensitization events. In brief, MUTZ-3 cells are propagated in conditioned medium (MEM alpha plus 10% medium from 5637 cells). Chemical exposure experiments are performed in 6-well plates. Cells are exposed for 24 h to non-toxic concentrations of the test chemicals (cell viability >80%). After 24 h the cells are washed and harvested by addition of a lysis buffer. Measurement shall be performed at least in triplicates.

In order to enable multiplex quantification of protein targets, proper sample processing steps are required prior to mass spectrometry. The SRM analytical procedure can be performed in any laboratory possessing a triple quadrupole mass spectrometer coupled to a micro reversed-phase high-performance liquid chromatography (RP-HPLC) unit.

Each cell lysate (100 μg protein per sample) is digested with sequencing grade trypsin. The resulting tryptic peptides of the test sample are then labelled with a mass spectrometric reagent referred to as heavy isotope-doped tandem mass tag reagents (TMT6-126; Thermo Scientific, Germany), and peptides of the control sample digest are labelled with the light version of the reagent (TMT^0-126; Thermo Scientific, Germany).

The SRM method is performed on a TSQ Vantage (Thermo Scientific, Germany) triple quadrupole mass spectrometer coupled using an Accela 1250 microLC (Thermo Scientific, Germany) with a flow rate of 100 μL/min. Reverse-phase separations are carried out using a 1 × 50 mm Hypersil Gold C18 1.9 μm column (Thermo Scientific, Germany). Solvent A is LC-MS grade water with 0.1% (v/v) formic acid, and solvent B is LC-MS grade acetonitrile with 0.1% (v/v) formic acid. Peptides were separated over a gradient from 5% to 60% B over 15 min.

The MS data are first analysed using specific mass spectrometry software such as Pinpoint™ (Thermo Scientific, Germany) to calculate peptide transition ratios from the heavy- and light-labelled peptides. The ratios for each protein within a technical replicate are averaged.

A linear regression model is fitted using the training data obtained with the reference compounds, to fit the parameters for each protein to model the potency of each training chemical. Alternatively, the chemicals are encoded by their class membership, and the model is trained to be able to predict sensitizers and non-sensitizers. PLS-DA analysis or cluster analysis can be performed using the ten protein measures for each chemical to cluster compounds based on similarity.

28.5 Role in a Testing Strategy

Allergic contact sensitization is the result of a complex sequence of chemical and biological events. Therefore, it seems obvious that the currently available *in vitro* tests have been optimized to model specific key events of hapten formation, skin penetration (key event 1), keratinocyte and dendritic cell (DC) activation (key events 2 and 3) and antigen presentation (key event 4). Consequently, for each hapten and depending on its chemical properties, a battery of different assays will be employed to characterize a novel chemical [2]. The Sensiderm assay utilizes the MUTZ-3 cell line as a cell culture surrogate of DC activation upon exposure to skin sensitizing chemicals. Sensiderm is based on ten pathway-specific biomarker proteins and therefore complements single marker assays of DC maturation such as myeloid U937 skin sensitisation test (MUSST) and modified MUSST (mMUSST). Both assays measure CD86 expression in the myeloid cell line U937 as an early DC maturation marker and primarily focus on prediction of hazard potential. Since the Sensiderm assay has a quantitative readout of protein expression, it holds potential to be used for skin sensitization potency evaluation. As Sensiderm investigates pathway-specific markers, it can be expected that the method is especially well suited to provide further insight into sensitization mechanisms by which (new) chemical sensitizers induce skin sensitization [2].

28.6 Perspectives from the Test Developer

Quantitative mass spectrometry is an evolving technology, which offers extraordinary high analytical specificity and analytical precision. Currently, the demand for instrumentation and technological skills limits its accessibility. However, the flexibility of mass spectrometry is determined by the fact that the development of analytical targets is mainly defined by setting up physical instrument settings tuned specifically for the peptide analytes. We have developed the Sensiderm assay as a relative quantitative assay using a control treated cell digest and calculating the ratio of protein expression for ten key DC activation pathways. This is possible by using the isotopic TMT reagents and allows use of post-translational modified peptides. However, it would also be possible to develop this into an absolute quantitative assay by using synthetic peptides which can be purchased from many service providers either with pre-incorporated heavy atom substitutions or which can be labelled with isotopic TMT reagents.

28.7 Critical Steps in the Protocol

The protocol contains a few critical steps defined by cell culture and chemical toxicity of test compounds. It is required to adhere to cell culture and incubation protocols and measurement conditions. In addition, physiological variability requires that test compounds are tested with the same lot of cells, which is used for the new compounds to be able to correct for systematic bias. Initial setup of cell culture requires the investigation if the deployed cells are sufficiently responsive towards calibration chemicals.

28.8 Possible Protocol Adaptations

The protocol has certain degrees of freedom for putative protocol adjustments: The number of training chemicals and number of replicates can be adopted to the respective needs for potency prediction. In a post-hoc cluster analysis, a large number of training chemicals (up to 100) can be used to identify novel chemicals with similar pathway induction. If needed, the flexible design of mass spectrometric assays allows including further analytical targets representing complementary biochemical pathways induced by novel chemical sensitizers. Cell lines can be adapted if the analytical targets are expressed in an appropriate fashion in those cells.

28.9 Challenges and Opportunities

The opportunity to complement a testing strategy with a set of proteomic analytical targets provides a chance for a more "holistic" view of sensitization thereby increasing sensitivity of the testing as well as providing scientific evidence for events of sensitization. The computed summary score reflecting chemical potency is complemented with an additional source of information. The individual pathway readouts help for later interpretation of experimental data to find interpretations for unexpected behaviour *in vivo*. A large database of apparent chemical potency can be built providing highly informative data of reactivity classes and clusters leading to a more sophisticated categorisation of chemicals based on mode of action, potency and bioavailability. Multiplex analysis fits better the demand of a risk minimizing screening test since a series of "sensors" are combined for the screening procedure. Validation of the testing method requires funding for analytical work, to create this valuable data repository.

28.10 Conclusions

The ultimate goal of *in vitro* toxicity testing is to create surrogates of measures obtained using *in vivo* testing models. Although animal models encompass inherently more aspects of sensitization, a major shortcoming is their imperfect correlation with human sensitization test. A major advantage of multi-analyte *in vitro* tests is that they create reproducible and robust readouts. As these cell lines are of human origin, their response to chemical sensitizers may be more relevant for modelling human allergic contact dermatitis. The development and setup of a novel multiplex assay principle utilizing proteomic signatures have required a series of necessary consecutive steps during the design and validation process of the assay. After setting up of the biological model (MUTZ-3), the successful research led to the discovery of relevant biological protein signatures reflecting the stimulation of different sensitization-related pathways. The final assay development utilized those analytical targets providing a quantitative assay. With this data mathematical models are used for modelling the sensitization potential and classifiers. Further validation studies will need to be performed to conclude the development cycle.

The Sensiderm proteomic profile offers for the first time, a sensitization analysis method which deploys a multiplex proteomic procedure with a large number of analytes. Whilst the number of analytes is currently only ten, this still offers the means for a proper interpretation of biological pathway data for a post-hoc analysis of suspicious chemicals supporting the decision-making process.

References

1. Ashton R, De Wever B, Fuchs HW, Gaca M, Hill E, Krul C, Poth A, Roggen EL. State of the art on alternative methods to animal testing from an industrial point of view: ready for regulation? ALTEX. 2014;31(3):357–63.
2. Reisinger K, Hoffmann S, Alépée N, Ashikaga T, Barroso J, Elcombe C, Gellatly N, Galbiati V, Gibbs S, Groux H, Hibatallah J, Keller D, Kern P, Klaric M, Kolle S, Kuehnl J, Lambrechts N, Lindstedt M, Millet M, Martinozzi-Teissier S, Natsch A, Petersohn D, Pike I, Sakaguchi H, Schepky A, Tailhardat M, Templier M, van Vliet E, Maxwell G. Systematic evaluation of non-animal test methods for skin sensitisation safety assessment. Toxicol In Vitro. 2015;29(1):259–70.

3. Weltzien H-U, Corsini E, Gibbs S, Lindstedt M, Borrebaeck CAK, Budde P, Schulz-Knappe P, Thierse H-J, Martin SF, Roggen EL. Safe cosmetics without animal testing? Contributions of the EU Project Sens-it-iv. J Verbr Lebensm. 2009;4(Suppl 2):S41–8.
4. Casati S, Aeby P, Kimber I, Maxwell G, Ovigne JM, Roggen E, Rovida C, Tosti L, Basketter D. Selection of chemicals for the development and evaluation of in vitro methods for skin sensitisation testing. Altern Lab Anim. 2009;37(3):305–12.
5. Rovida C, Martin SF, Vivier M, Weltzien HU, Roggen E. Advanced tests for skin and respiratory sensitization assessment. ALTEX. 2013;30(2):231–52.
6. Hitzler M, Bergert A, Luch A, Peiser M. Evaluation of selected biomarkers for the detection of chemical sensitization in human skin: a comparative study applying THP-1, MUTZ-3 and primary dendritic cells in culture. Toxicol In Vitro. 2013;27(6):1659–69.
7. Johansson H, Lindstedt M, Albrekt AS, Borrebaeck CA. A genomic biomarker signature can predict skin sensitizers using a cell-based in vitro alternative to animal tests. BMC Genomics. 2011;12:399.
8. Nelissen I, Selderslaghs I, Heuvel RV, Witters H, Verheyen GR, Schoeters G. MUTZ-3-derived dendritic cells as an in vitro alternative model to CD34+ progenitor-derived dendritic cells for testing of chemical sensitizers. Toxicol In Vitro. 2009;23(8):1477–81.
9. Python F, Goebel C, Aeby P. Comparative DNA microarray analysis of human monocyte derived dendritic cells and MUTZ-3 cells exposed to the moderate skin sensitizer cinnamaldehyde. Toxicol Appl Pharmacol. 2009;239(3):273–83.
10. Hu ZB, Ma W, Zaborski M, MacLeod R, Quentmeier H, Drexler HG. Establishment and characterization of two novel cytokine-responsive acute myeloid and monocytic leukemia cell lines, MUTZ-2 and MUTZ-3. Leukemia. 1996;10(6):1025–40.
11. Masterson AJ, Sombroek CC, De Gruijl TD, Graus YM, van der Vliet HJ, Lougheed SM, van den Eertwegh AJ, Pinedo HM, Scheper RJ. MUTZ-3, a human cell line model for the cytokine-induced differentiation of dendritic cells from CD34+ precursors. Blood. 2002;100(2):701–3.
12. Palermo G, Piraino P, Zucht HD. Performance of PLS regression coefficients in selecting variables for each response of a multivariate PLS for omics-type data. Adv Appl Bioinform Chem. 2009;2:57–70.
13. Strathmann FG, Hoofnagle AN. Current and future applications of mass spectrometry to the clinical laboratory. Am J Clin Pathol. 2011;136(4):609–16.
14. Titz B, Elamin A, Martin F, Schneider T, Dijon S, Ivanov NV, Hoeng J, Peitsch MC. Proteomics for systems toxicology. Comput Struct Biotechnol J. 2014;11(18):73–90.
15. Končarević S, Lößner C, Kuhn K, Prinz T, Pike I, Zucht HD. In-depth profiling of the peripheral blood mononuclear cells proteome for clinical blood proteomics. Int J Proteom. 2014;2014:129259.
16. Wold S. Chemometrics; what do we mean with it, and what do we want from it? Chemom Intell Lab Syst. 1995;30(1):109–15. doi:10.1016/0169-7439(95)00042-9.
17. Thimmulappa RK, Mai KH, Srisuma S, Kensler TW, Yamamoto M, Biswal S. Identification of Nrf2-regulated genes induced by the chemopreventive agent sulforaphane by oligonucleotide microarray. Cancer Res. 2002;62(18):5196–203.
18. Natsch A. The Nrf2-Keap1-ARE toxicity pathway as a cellular sensor for skin sensitizers – functional relevance and a hypothesis on innate reactions to skin sensitizers. Toxicol Sci. 2010;113(2):284–92.
19. Tamura Y, Torigoe T, Kutomi G, Hirata K, Sato N. New paradigm for intrinsic function of heat shock proteins as endogenous ligands in inflammation and innate immunity. Curr Mol Med. 2012;12(9):1198–206.
20. Yusuf N, Nasti TH, Huang CM, Huber BS, Jaleel T, Lin HY, Xu H, Elmets CA. Heat shock proteins HSP27 and HSP70 are present in the skin and are important mediators of allergic contact hypersensitivity. J Immunol. 2009;182(1):675–83.
21. Arnhold J, Flemmig J. Human myeloperoxidase in innate and acquired immunity. Arch Biochem Biophys. 2010;500(1):92–106.
22. Cerezo LA, Remáková M, Tomčik M, Gay S, Neidhart M, Lukanidin E, Pavelka K, Grigorian M, Vencovský J, Šenolt L. The metastasis-associated protein S100A4 promotes the

inflammatory response of mononuclear cells via the TLR4 signalling pathway in rheumatoid arthritis. Rheumatology (Oxford). 2014;53(8):1520–6.

23. Bruhn S, Fang Y, Barrenäs F, Gustafsson M, Zhang H, Konstantinell A, Krönke A, Sönnichsen B, Bresnick A, Dulyaninova N, Wang H, Zhao Y, Klingelhöfer J, Ambartsumian N, Beck MK, Nestor C, Bona E, Xiang Z, Benson M. A generally applicable translational strategy identifies S100A4 as a candidate gene in allergy. Sci Transl Med. 2014;6(218):218ra4.
24. Petersen B, Wolf M, Austermann J, van Lent P, Foell D, Ahlmann M, Kupas V, Loser K, Sorg C, Roth J, Vogl T. The alarmin Mrp8/14 as regulator of the adaptive immune response during allergic contact dermatitis. EMBO J. 2013;32(1):100–11.
25. Cantor JM, Ginsberg MH. CD98 at the crossroads of adaptive immunity and cancer. J Cell Sci. 2012;125(Pt 6):1373–82.
26. Milani P, Gagliardi S, Cova E, Cereda C. SOD1 transcriptional and posttranscriptional regulation and its potential implications in ALS. Neurol Res Int. 2011;2011:458427.
27. Goldstein AL, Hannappel E, Sosne G, Kleinman HK. Thymosin β4: a multi-functional regenerative peptide. Basic properties and clinical applications. Expert Opin Biol Ther. 2012;12(1):37–51.

In Vitro Dendritic Cell-Based Test for Skin Sensitizers Identification and Potency Estimation

29

Bruno Miguel Neves, João Demétrio Martins, Isabel Ferreira, Ana Silva, and Maria Teresa Cruz

29.1 Principle of the Test Method and Scientific Basis

Skin sensitization resulting in allergic contact dermatitis is the outcome of numerous complex interactions at molecular, cellular and tissue levels. Due to the significant level of complexity associated with the pathophysiology of skin sensitization, it has become clear that no single *in vitro* test will likely be adequate for hazard characterization and potency estimation. Rather, the integration of results obtained from different assays that cover the different phases of the skin sensitization process is currently underlined as the most promising approach (reviewed in [1–3]). Studies focused in the pathophysiology of skin allergy have pointed out key molecular events triggered by skin allergens that are crucial for the development of the so-called adverse outcome pathways (AOPs) [4]. AOPs are novel tools in toxicology and human risk assessment that provide a clear-cut mechanistic representation of critical toxicological effects that span over different layers of biological organization. An elaborate AOP framework has been proposed by OECD for chemical-induced skin sensitization that can serve a number of ubiquitous purposes, including the establishment of (quantitative) structure-activity relationships, the development of alternative paradigms for hazard characterization and the elaboration of prioritization strategies [5, 6]. Given that low molecular weight chemicals

B.M. Neves (✉) • J.D. Martins • I. Ferreira • M.T. Cruz (✉)
Centre for Neuroscience and Cell Biology, University of Coimbra, 3004-517 Coimbra, Portugal

Faculty of Pharmacy, Center for Neuroscience and Cell Biology, University of Coimbra, 3000-548 Coimbra, Portugal
e-mail: neves_bruno@sapo.pt; trosete@ff.uc.pt

A. Silva
Centre for Neuroscience and Cell Biology, University of Coimbra, 3004-517 Coimbra, Portugal

C. Eskes et al. (eds.), *Alternatives for Dermal Toxicity Testing*,
DOI 10.1007/978-3-319-50353-0_29

(haptens) are too small to be directly recognized by the immune system, the first key molecular event evoked during skin sensitization is their reaction with proteins. Therefore, chemical allergens are either naturally protein reactive or are rapidly metabolized to protein-reactive species. The second molecular occurrence is the induction of stress responses and xenoinflammation by haptenated proteins through the formation of reactive oxygen species (ROS) and damage-associated molecular patterns (DAMPs) such as purines and hyaluronic acid. DAMPs positively modulate internalization of the haptenated proteins by dendritic cells (DC) inducing their maturation. Then, as the third postulation, DC process the conjugates and subsequently migrate to the draining lymph nodes where the antigen is presented to responsive T lymphocytes (reviewed in [4, 7]). This flow of events, in which DC have a leading role, prompted the development of *in vitro* DC-based assays gathered on the rational that changes triggered by allergens in DC phenotype can be differentiated from those induced by irritants. This assumption derives from the unique capacity of DC to convert environmental signals encountered at the skin into a receptor expression (MHC class II molecules, co-stimulatory molecules CD83 and CD86, chemokine receptors CXCR4 and CCR7) and soluble mediator release profile that will stimulate T lymphocytes. Importantly, signal transduction cascades precede changes in surface marker expression and cytokine/chemokine secretion, implying that all DC phenotypic modifications are a consequence of a signal transduction profiles that is specifically triggered by sensitizers and not by irritants (reviewed in [8]). All the above statements constituted the scientific basis of the herein presented test that describes a predictive assay based on the analysis and integration of gene expression and signal transduction profiles resulting from the exposure of a mouse foetal skin-derived dendritic cell line (FSDC) to chemicals. These results were previously published by Neves and colleagues [9]. Furthermore, the challenge of classifying sensitizers according to their potency is also addressed. The potency of chemicals *in vivo* is determined by a complex conjunction of factors such as the bioavailability, the peptide reactivity and the cytotoxicity/irritancy. Therefore, and as an update to the test developed to detect skin sensitizers, the feasibility to predict the sensitizers' potency using *in vitro*-generated data and *in silico*-calculated parameters (namely, the skin permeation coefficient and the electrophilicity index) is also disclosed and highlighted in this chapter.

29.2 Performance and Applicability of the Test Method

29.2.1 Reproducibility

29.2.1.1 Skin Sensitizers' Identification

The developed *in vitro* predictive test is based on the analysis and integration of gene expression and intracellular signalling profiles of chemical-exposed skin-derived dendritic cells. The chemicals used in this work were selected from the set proposed by the European Centre for the Validation of Alternative Methods

(ECVAM) as the reference chemicals to be used in method development phase in order to facilitate the comparison with other tests being developed [10] (Table 29.1). In a first approach, cells were treated with four known sensitizers (DNFB, OXA, PPD, Ni) and two non-sensitizers (SDS and BC), and the effects on the expression of 20 candidate genes (correlated with ACD pathophysiology and/or DC maturation) and on the activation of pro-inflammatory signalling pathways (MAPKs, PI3K/Akt and NF-κB) were analysed by real-time RT-PCR and Western blot, respectively (Table 29.2). The genes *Trxr1, Hmox1, Nqo1* and *Cxcl10* and the signalling pathways p38 and JNK MAPKs were identified as good predictor variables and used to construct a dichotomous classifier. In order to graphically represent the profiles of treated cells considering all the genes studied and the effects on the activation of p38 and JNK MAPKs, a principal component analysis was performed. As can be seen in Fig. 29.1, sensitizers (red) and non-sensitizers (green) are completely separated, allowing a clear discrimination between the two groups. It can also be observed that with exception of Ni1- and PPD1-treated samples, biological replicates group together, indicating a high reproducible response of FSDC to chemical treatment (Fig. 29.1). Posteriorly, to validate the developed classifier, 12 additional chemicals (test set) were tested, comprising 9 sensitizers and 3 non-sensitizers in a chemical-coded blind assay (Table 29.3).

Table 29.1 Panel of chemicals tested and classification according to their skin sensitizing capacity determined in the local lymph node assay (LLNA). (Reproduced from [9])

Compound name	Category	CAS No.	Solvent	EC_{30} (μM)
Training compounds				
1-Fluoro-2,4-dinitrobenzene (DNFB)	Extreme sensitizer	70-34-8	DMSO	9
Oxazolone (OXA)	Extreme sensitizer	15646-46-5	DMSO	410
1-4-Phenylenediamine (PPD)	Strong sensitizer	106-50-3	H_2O	400
Nickel sulphate (NI)	Moderate sensitizer	10101-97-0	H_2O	270
Sodium dodecyl sulphate (SDS)	Non-sensitizer	151-21-3	H_2O	195
Benzalkonium chloride (BC)	Non-sensitizer	63449-41-2	H_2O	1.8
Test compounds				
4-Nitrobenzylbromide (4-NB)	Extreme sensitizer	100-11-8	DMSO	6
Methyldibromo glutaronitrile (MDGN)	Strong sensitizer	35691-65-7	DMSO	13
Isoeugenol (ISO)	Moderate sensitizer	97-54-1	DMSO	342
Glyoxal (GLX)	Moderate sensitizer	107-22-2	H_2O	820
2-Mercaptobenzothiazole (MBT)	Moderate sensitizer	149-30-4	DMSO	410
Cinnamaldehyde (CIN)	Moderate sensitizer	104-55-2	DMSO	215
Tetramethyl thiuram disulphide (TMTD)	Moderate sensitizer	137-26-8	DMSO	16
Eugenol (EUG)	Weak sensitizer	97-53-0	DMSO	468
Cinnamyl alcohol (CIN AL)	Weak sensitizer	104-54-1	DMSO	1740
2,4-Dichloronitrobenzene (DCNB)	Non-sensitizer	611-06-3	DMSO	79
Lactic acid (LA)	Non-sensitizer	50-21-5	H_2O	6120
Salicylic acid (SA)	Non-sensitizer	69-72-7	Ethanol	594

EC_{30} values were calculated for every compound as the concentration that induced 30% reduction on cell viability based on linear regression analysis of data from the 24 h MTT reduction assay

Table 29.2 Effect of training compounds on the transcription of selected genes. (Reproduced from [9])

	Mean $\log_2$ of fold changes ± SD					
	Sensitizers				Non-sensitizers	
Gene	DNFB	OXA	PPD	Ni	SDS	BC
Cd40	0.19 ± 1.35	0.68 ± 0.62	−0.06 ± 0.90	0.00 ± 0.25	0.22 ± 0.28	0.04 ± 0.23
Cd86	−0.2 ± 1.37	−0.19 ± 0.28	0.19 ± 0.24	−0.03 ± 0.28	−0.26 ± 0.32	−0.01 ± 0.13
Ccr7	0.28 ± 1.32	2.12 ± 1.95	−1.01 ± 0.87	−0.32 ± 0.48	−0.49 ± 0.60	−0.57 ± 1.23
Cxcr4	−0.26 ± 0.25	−0.05 ± 0.21	−0.46 ± 0.42	0.13 ± 0.30	0.06 ± 0.30	0.19 ± 0.41
Trx	0.15 ± 0.33	0.06 ± 0.29	0.31 ± 0.48	0.19 ± 0.39	−0.08 ± 0.51	−0.01 ± 0.27
Trxr1	1.23 ± 0.35**	0.93 ± 0.19*	1.70 ± 0.25**	0.49 ± 0.21*	−0.01 ± 0.12	0.33 ± 0.19*
Hmox1	2.74 ± 0.86**	1.97 ± 0.34**	4.19 ± 0.65**	2.24 ± 0.39**	0.32 ± 0.49	0.46 ± 0.11*
Nqo1	1.54 ± 0.40**	0.81 ± 0.27*	3.54 ± 0.79*	2.78 ± 0.42**	−0.24 ± 0.30	−0.13 ± 0.14
Il17f	1.95 ± 1.33*	0.12 ± 1.09*	1.37 ± 0.31**	0.68 ± 0.58	0.32 ± 0.74	0.51 ± 0.28*
Il1b	0.77 ± 0.70	−0.16 ± 0.16	0.59 ± 1.23	−0.27 ± 1.66	1.17 ± 0.97	0.19 ± 0.60
Tnfa	−0.70 ± 0.23	−0.67 ± 0.38	−1.06 ± 1.05	−1.13 ± 1.53	−0.93 ± 0.83	−0.93 ± 1.21
Csf3	0.73 ± 1.05	−0.27 ± 0.38	−1.00 ± 0.17	−0.02 ± 0.78	−0.54 ± 0.61	−1.08 ± 0.55
Cxcl10	−1.92 ± 1.15*	−0.78 ± 0.08	−2.83 ± 0.98***	−1.24 ± 0.27***	0.34 ± 0.14*	0.59 ± 0.29
Ccl2	−0.07 ± 0.69	−0.41 ± 0.52	−0.25 ± 1.23	−0.89 ± 1.07	−0.58 ± 0.53	−0.33 ± 0.01*
Ccl3	0.11 ± 0.27	0.67 ± 0.34	0.25 ± 0.47	0.53 ± 0.07	−1.58 ± 0.66	0.94 ± 0.19
Ccl4	−1.05 ± 0.60	−0.90 ± 0.44	−0.25 ± 0.28	−0.32 ± 1.03	−0.92 ± 0.47	−0.20 ± 0.20
Ccl5	−0.52 ± 0.62	−0.66 ± 0.11	−0.30 ± 0.26	−0.14 ± 0.32	−0.18 ± 0.50	−0.17 ± 0.17
Ccl17	2.26 ± 1.46	0.46 ± 0.48	0.81 ± 1.09	1.29 ± 0.73	0.49 ± 0.76	0.03 ± 0.54
Cxcl2	−0.06 ± 0.67	−0.19 ± 0.10	−1.17 ± 1.08	0.39 ± 0.47	−0.06 ± 0.86	−0.23 ± 0.21

FSDC were exposed to the EC_{30} concentration of each chemical, for 6 h, followed by transcription analysis of selected genes by real-time RT-PCR. Results are presented as mean $\log_2$ values of fold changes relatively to control (cells treated with the respective vehicle) for 3–5 independent experiments and their corresponding standard deviation. Student's *t*-test was used to determine significant variation relatively to controls (0). $^{*}p < 0.05$, $^{**}p < 0.01$, $^{***}p < 0.001$

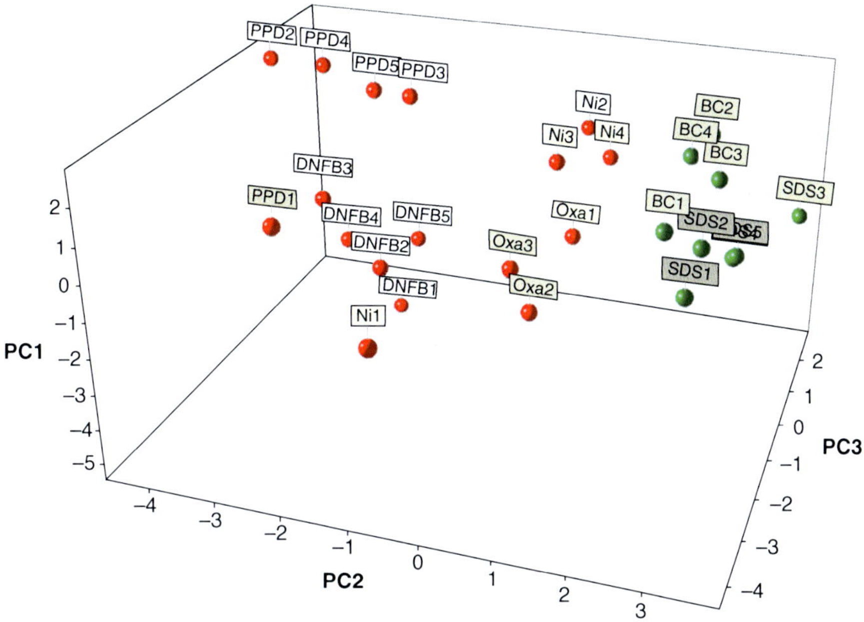

Fig. 29.1 3D graphical representation of principal component analysis of transcripts and signalling pathways investigated after FSDC exposure to chemicals of the training set (reproduced from Neves et al. [9]). mRNA levels were analysed by real-time RT-PCR after cell treatment for 6 h, and phosphorylated levels of JNK and p38 MAPKs were assessed by Western blot following a 2 h exposure. Similarities in the profiles induced by sensitizers (*red*) and non-sensitizers (*green*) were investigated using principal component analysis in the software Genex (MultiD Analyses AB, Göteberg, Sweden)

29.2.1.2 Skin Sensitizers' Potency Estimation

The feasibility to predict the sensitizers' potency using *in vitro*-generated data and several *in silico*-calculated descriptors was also performed. Multiple linear regression analysis of data resulted in an optimized model with just two explanatory variables: the concentration that induced 30% cytotoxicity (EC30) and the electrophilicity index (Table 29.4). The information added by the skin permeation coefficient, although theoretically relevant, did not increase the performance of the model and was therefore not included. Consequently, the only variable prone to experimental variation is EC30, which, however, has demonstrated a high reproducibility between biological replicates and experiments performed at different days and by different operators.

29.2.2 Predictive Capacity

29.2.2.1 Skin Sensitizers' Identification

Considering the total 18 chemicals tested, the developed classifier showed an accuracy of 94% (17/18 compounds correctly classified), a sensitivity of 92% (12/13

Table 29.3 Effect of test compounds on the selected predictor variables. (Reproduced from [9])

Test compounds	Predictor variables					
	Cxcl10	*Hmox1*	*Nqo1*	*Trxr1*	p-p38 MAPK	p-JNK
Sensitizers						
CIN	−1.10 ± 0.53	5.39 ± 1.27***	4.54 ± 0.80**	2.08 ± 0.24***	3.36 ± 1.05**	3.66 ± 0.40***
GLX	−0.54 ± 0.18	4.90 ± 0.96**	0.34 ± 0.71	1.02 ± 0.31*	4.39 ± 0.56**	3.96 ± 0.43***
EUG	1.06 ± 0.35*	2.45 ± 0.56*	1.39 ± 0.67	0.76 ± 0.22**	0.55 ± 0.29*	0.87 ± 0.37**
ISO	0.34 ± 1.54	3.49 ± 1.20**	2.80 ± 1.00*	1.12 ± 0.36**	1.14 ± 0.46*	1.69 ± 0.89*
TMTD	−0.93 ± 0.62	3.53 ± 0.46***	4.92 ± 1.20*	1.50 ± 0.40**	2.69 ± 1.07*	3.97 ± 0.42***
4-NB	−0.90 ± 1.15	1.40 ± 0.75*	1.31 ± 0.34*	1.17 ± 0.60	0.98 ± 1.50	3.41 ± 0.49***
MDGN	−0.29 ± 0.36	1.29 ± 0.16***	1.76 ± 0.01**	0.56 ± 0.18**	1.44 ± 0.94	2.14 ± 0.99*
CIN AL	−1.32 ± 1.57	4.75 ± 0.89**	3.32 ± 0.28**	1.62 ± 0.15***	0.66 ± 1.02	2.51 ± 0.47**
MBT	−0.17 ± 0.65	1.04 ± 0.30*	0.66 ± 0.25*	0.34 ± 0.04**	0.21 ± 0.59	0.71 ± 0.51*
Non-sensitizers						
SA	0.18 ± 0.20	0.77 ± 0.25**	−1.15 ± 1.21	−0.04 ± 0.18	−0.47 ± 0.98	0.11 ± 0.26
LA	6.66 ± 1.14**	0.87 ± 0.34*	−0.03 ± 0.54	−0.01 ± 0.29	0.58 ± 1.26	1.23 ± 0.40**
DCNB	0.49 ± 0.34	1.72 ± 018*	−0.20 ± 0.44	0.17 ± 0.14	−1.10 ± 1.26	0.13 ± 0.41
Non-allergenic immunogen						
LPS	10.65 ± 1.26**	0.90 ± 0.82	0.56 ± 0.80	0.99 ± 0.70	1.20 ± 0.30*	1.13 ± 0.42*

FSDC were exposed to EC_{30} concentration of each chemical followed by analysis of transcription of *Cxcl10*, *Hmox1*, *Nqo1* and *Trxr1* genes by real-time RT-PCR (6 h exposure) and analysis of activation of p38 MAPK and JNK signalling pathways (2 h exposure) by Western blot. Results are presented as mean $\log_2$ values of fold changes relatively to control (cells treated with the respective vehicle) for 3–5 independent experiments and their corresponding standard deviation. Student's *t*-test was used to determine significant variation relatively to controls (0). $^*p < 0.05$, $^{**}p < 0.01$, $^{***}p < 0.001$

Table 29.4 Summary of multiple linear regression model for potency estimation

	Dependent variable	Explanatory variables		
Chemicals	LLNA EC3 (%)[a]	EC_{30} (μM)		ω (eV)
OXA	0.003	410		1.83
DNFB	0.03	9		3.26
4-NB	0.05	6		2.74
PPD	0.16	400		2.55
MDGN	0.9	13		2.28
ISO	1.2	342		0.74
GLX	1.4	820		3.23
MBT	1.7	410		1.24
CIN	3	215		2.18
TMTD	5.2	16		1.09
EUG	13	468		0.59
CIN AL	21	1740		1.14
Model summary				
R	R^2		Significance	
0.85	0.72		0.003	
Coefficients				
	Unstandardized coefficients	Standardized coefficients	Significance	Collinearity statistics (VIF)
Constant	5.456			
EC_{30}	8.863	0.664	0.006	1.064
ω	−2.674	−0.386	0.065	1.064

[a]Sensitizers' potency ranking according their threefold stimulation of lymph node proliferation (EC3) in the LLNA [11]

sensitizers identified) and a specificity of 100% (5/5 non-sensitizers identified) (Fig. 29.2). The classifier was able to identify pro- and pre-haptens (PPD and eugenol, respectively) and metal sensitizers (Ni) and could also discriminate sensitizers from non-allergenic immunogens, namely, lipopolysaccharide (LPS). The only chemical misclassified was the moderate sensitizer MBT, with a predictive score (Psc) of −0.2 very close to the cut-off established.

29.2.2.2 Skin Sensitizers' Potency Estimation

A strong correlation between predicted values and the correspondent *in vivo* LLNA EC3 data over a wide range of magnitudes (from weak to extreme sensitizers) was obtained (Pearson correlation coefficient $r = 0.85$, $p < 0.001$, $n = 12$) (Fig. 29.3).

29.2.3 Applications and Limitations

The FSDC cell line proved to be metabolically competent since the pre- and prohaptens eugenol and PPD, respectively, were correctly classified as skin sensitizers. Both tests (skin sensitizer identification and subsequent potency estimation) could be applied to a synthetic chemical, a natural-occurring chemical or a botanically fragrance; drug mixtures were not evaluated. Extremely poor water-soluble chemicals

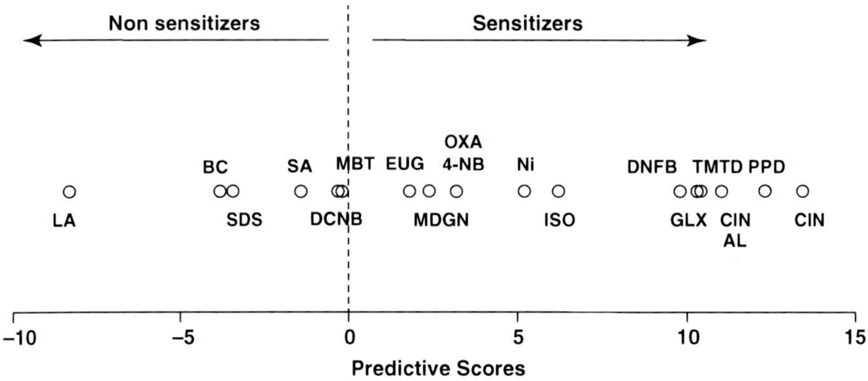

Fig. 29.2 Classification of chemicals according to their calculated predictive scores (reproduced from Neves et al. [9]). A dichotomous classifier was constructed following linear discriminant analysis of the effects of training chemicals over the identified predictor variables: phospho-p38 MAPK, phospho-JNK and the genes *Cxcl10*, *Trxr1*, *Hmox1* and *Nqo1*. Two linear discriminant functions were obtained: a function for calculation of the score of the chemical as a sensitizer (sensitizer score) and another for calculation of the score as a non-sensitizer (non-sensitizer score). For classification of a new chemical, its effects on the predictor variables are quantified and then used as entries on the functions. According to this model, a chemical is classified as sensitizer if the difference between sensitizer score and non-sensitizer sore (predictive score) is greater than 0

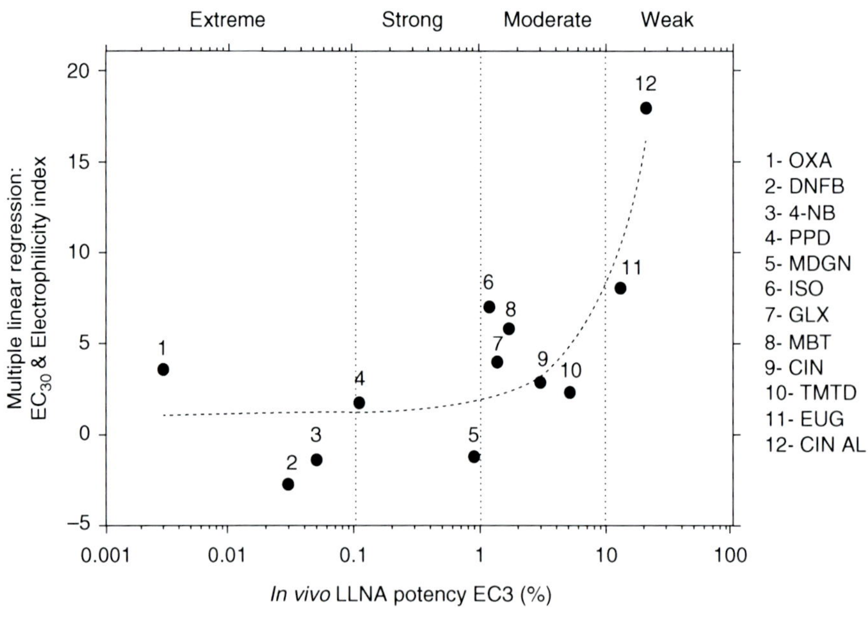

Fig. 29.3 Potency estimation and correlation with the *in vivo* LLNA EC_3 values. Multiple linear regression analysis of *in vitro*-generated data and several *in silico*-calculated descriptors resulted in a statistically significant model ($p = 0.003$) with two explanatory variables: the concentration that induced 30% cytotoxicity (EC_{30}) and the electrophilicity index (ω) of chemicals. Estimated values were plotted against correspondent *in vivo* LLNA EC3 values (%) and correlation was analysed. Pearson coefficient correlation: $r = 0.85$, $p < 0.001$, $n = 12$. Dashed line indicates linear regression curve fit: $r^2 = 0.72$, $p < 0.001$. It assumes a curve shape as x axis is in log scale

constitute a limitation for both the identification and potency estimation of skin sensitizers. Indeed, OXA, an extreme sensitizer [12] reveals to be underestimated concerning potency prediction, indicating some limitations of the proposed model. This can in part be due to the unexpected elevated concentrations of OXA that were required to induce minimal toxicity in the FSDC cell line. Although the potency prediction model also encompasses the possible contribution of the electrophilicity index, which can be regarded as a measure of reactivity (ω), the cytotoxicity descriptor EC30 has a much greater impact on the correlation as shown by its high standardized beta coefficient. Accordingly, in other cell-based tests that use *in vitro*-assessed cytotoxicity as a parameter, oxazolone is identified as a clear outlier, being its sensitizing potency always underestimated [13, 14]. Additionally, metals were excluded from the potency estimation test. Truly, metal sensitizers induce sensitization through mechanisms that are not similar to those evoked by the vast majority of organic chemicals. Rather than covalently binding to proteins, metals form coordinate complexes with cell surface proteins like MHC, thus inducing sensitization in a protein/peptide-independent manner [15, 16]. Therefore, descriptors such as electrophilicity index are not adequate for potency prediction of metals and were excluded from this assay.

29.2.4 Comparison to Human Data

As depicted in Table 29.5, most of the results obtained with the developed test are in agreement with the results obtained by Basketter and colleagues [17] according to human skin sensitization characteristics. The only chemical misclassified was MBT belonging to human class 3.

Table 29.5 Comparison to human data

Compound name	LLNA classification	Human class[a]	Skin sensitizers identification
1-Fluoro-2,4-dinitrobenzene (DNFB)	Extreme sensitizer	1	Sensitizer
Oxazolone (OXA)	Extreme sensitizer	NA	Sensitizer
1-40-Phenylenediamine (PPD)	Strong sensitizer	1	Sensitizer
Nickel sulphate (NI)	Moderate sensitizer	2	Sensitizer
Sodium dodecyl sulphate (SDS)	Non-sensitizer	6	Non-sensitizer
Benzalkonium chloride (BC)	Non-sensitizer	5	Non-sensitizer
4-Nitrobenzylbromide (4-NB)	Extreme sensitizer	NA	Sensitizer
Methyldibromo glutaronitrile (MDGN)	Strong sensitizer	2	Sensitizer
Isoeugenol (ISO)	Moderate sensitizer	2	Sensitizer
Glyoxal (GLX)	Moderate sensitizer	2	Sensitizer
2-Mercaptobenzothiazole (MBT)	Moderate sensitizer	3	*Misclassified*
Cinnamaldehyde (CIN)	Moderate sensitizer	2	Sensitizer
Tetramethyl thiuram disulphide (TMTD)	Moderate sensitizer	3	Sensitizer
Eugenol (EUG)	Weak sensitizer	3	Sensitizer
Cinnamyl alcohol (CIN AL)	Weak sensitizer	3	Sensitizer
2,4-Dichloronitrobenzene (DCNB)	Non-sensitizer	NA	Non-sensitizer
Lactic acid (LA)	Non-sensitizer	6	Non-sensitizer
Salicylic acid (SA)	Non-sensitizer	6	Non-sensitizer

NA not available

[a]Accordingly to Basketter et al. [17]

29.3 Current (Pre)Validation Status

29.3.1 Skin Sensitizers' Identification

The classifier obtained with assays using a group of training compounds (consisting of four known sensitizers and two non-sensitizers) was internally validated by two independent operators. For that, 12 additional chemicals, comprising 9 sensitizers and 3 non-sensitizers, were tested (test set) in a chemical-coded blind assay. In the blind assay, a first operator coded the new 12 chemicals with alphabet letters, performed cell viability assays, prepared cell extracts and executed RNA extraction. Posteriorly, a second operator performed Western blots and real-time RT-PCRs and finally used the obtained results in the previously developed classifier to classify the samples as sensitizers or non-sensitizers. The identity of samples was revealed at the end of the process and the performance of the model was assessed by comparing the predicted classification with LLNA classification data.

29.4 Brief Description of the Protocol

Part of the optimized test protocol herein presented is described in the manuscript of Neves and colleagues [9].

29.4.1 Cell Culture

The mouse foetal skin-derived dendritic cell line (FSDC) is a skin dendritic cell precursor with antigen-presenting capacity that was characterized as having a surface phenotype consistent with immature myeloid DC progenitors (H-2d.b+, CD54+, MHCI+, MHCII+, CD11c+, CD44high, CD11bhigh, CD86+, B220−, CD3−, CD4−, CD8−) [18]. Cells were cultured in endotoxin-free Iscove's Modified Dulbecco's Medium (IMDM) (Sigma-Aldrich, St. Louis, MO, USA) supplemented with 10% (v/v) foetal bovine serum (Invitrogen, Paisley, UK), 1% (w/v) glutamine, 3.02 g/L sodium bicarbonate, 100 μg/mL streptomycin and 100 U/mL penicillin, in a humidified incubator with 5% CO_2/95% air, at 37 ° C. Experiments were performed with cells between passage no. 5 and 35 after each stock thawing.

29.4.2 Experimental Design

Optimization of experimental conditions and identification of possible discriminative biomarkers were initially performed with a set of six chemicals designated as training set. This training set included four known sensitizers (DNFB, PPD, Ni and OXA) and two irritants (SDS and BC). Sensitizers that integrate the training set were chosen to encompass their chemical-associated heterogeneity. DNFB and

OXA are extreme sensitizers chemically characterized as an SN2Ar electrophile and an acyl transfer agent, respectively, PPD is a pro-hapten classified as a strong sensitizer that reacts hypothetically as a pro-Michael acceptor, and Ni is a moderate/weak metal sensitizer [10]. The concentrations that cause 30% cytotoxicity (EC30) were determined for each of these chemicals and used in subsequent experiments. The effect of chemicals on the activation of MAPKs, PI3K/Akt and NF-κB signalling pathways was analysed by Western blot at several time-points (5, 15, 30, 60 and 120 min), and the expression of 20 candidate genes was analysed by real-time RT-PCR at 6 and 24 h post-cell treatment. Obtained data, quantitatively expressed as fold change relatively to untreated cells, was subjected to linear discriminant analysis allowing the identification of variables with strong discriminative power. A dichotomous classifier was then constructed based on Fisher's classification coefficients of the selected variables. Finally, the performance of the classifier was evaluated in a blind assay by testing a set of 12 additional chemicals.

29.4.3 MTT Reduction Assay for Assessment of Cytotoxicity

Since a certain level of cytotoxicity is required for effective DC activation [19], for each chemical, the concentration that induced 30% cytotoxicity (EC30) was determined, which was used along the subsequent experiments. FSDC were exposed for 24 h to several concentrations of chemicals, in a dose response experiment, and analysed for viability by the reduction of the tetrazolium bromide salt, 3-(4,5-dime thylthiazol-2-yl)-2,5-diphenyl tetrazolium bromide (MTT) [20]. EC30 values were calculated by linear regression of obtained data. The solvent used to dissolve each chemical and EC30 values are presented in Table 29.1. For chemicals dissolved in dimethyl sulphoxide (DMSO) (Sigma-Aldrich, St. Louis, MO, USA) and ethanol, the maximum final concentrations of solvents were 0.5% and 1%, respectively.

29.4.4 Cell Treatment

FSDC were plated in 12-well microplates at 1×10^6 cells/well in a final growth media volume of 2 mL. They were then incubated with the EC30 concentration of each chemical of the training set, during 5, 30, 60 or 120 min for cell lysate preparation or 6 and 24 h for RNA extraction. In experiments with chemicals of the test set, only the optimal time points determined for the training set were used.

29.4.5 Cell Lysate Preparation and Western Blot Analysis

To obtain the lysates, cells were washed in ice-cold PBS and harvested in RIPA lysis buffer (50 mM Tris-HCl (pH 8.0), 1% Nonidet P-40, 150 mM NaCl, 0.5%

sodium deoxycholate, 0.1% SDS, 2 mM EDTA and 1 mM DTT) freshly supplemented with protease and phosphatase inhibitor cocktails (Roche, Mannheim, Germany). The nuclei and the insoluble cell debris were removed by centrifugation at 4 °C, at 12,000 × *g* for 10 min. Protein concentration was determined using the bicinchoninic acid method, and the cell lysates were denatured at 95 °C, for 5 min, in sample buffer (0.125 mM Tris pH 6.8; 2%, w/v SDS; 100 mM DTT; 10% glycerol and bromophenol blue) for posterior use in Western blot analysis.

After electrophoretic separation in SDS-PAGE gels, proteins were transferred to polyvinylidene difluoride membranes (Millipore Corporation, Bedford, MA, USA). The membranes were blocked with 5% (w/v) fat-free dry milk in Tris-buffered saline containing 0.1% (v/v) Tween 20 (TBS-T), for 1 h, at room temperature and blots incubated overnight at 4 °C with the primary antibodies against the different studied proteins diluted 1:1000 in TBS-T. After washing for 25 min with TBS-T, membranes were incubated for 1 h at room temperature with alkaline phosphatase-conjugated anti-rabbit or anti-mouse antibodies (1:20,000) (GE Healthcare, Chalfont St. Giles, UK). The immune complexes were detected by membrane exposure to the ECF reagent for 5 min, followed by scanning for blue excited fluorescence on the typhoon imager (GE Healthcare). The generated signals were analysed using ImageQuant TL software. To test whether similar amounts of protein were loaded for each sample, the membranes were stripped and reprobed with antibodies to total JNK1 (R&D Systems, Minneapolis, MN, USA), ERK1/2, p38 MAPKs, and Akt (Cell Signaling Technologies, Danvers, MA, USA). Blots were then developed with alkaline phosphatase-conjugated secondary antibodies and visualized by enhanced chemifluorescence.

29.4.6 Analysis of Gene Expression by Real-Time RT-PCR

Total RNA was isolated from cells with TRIzol reagent, according to the manufacturer's instructions, and the concentration was determined by OD260 measurement using a NanoDrop spectrophotometer. RNA quality was inspected for the absence of degradation or genomic DNA contamination using Experion RNA Std-Sens Chips in the Experion automated microfluidic electrophoresis system, and samples were stored in RNA Storage Solution at −80 °C until use.

After retro-transcription, real-time RT-PCR reactions were performed, in duplicate for each sample, on a Bio-Rad MyCycler iQ5 [21]. Gene expression changes were analysed using the built-in iQ5 optical system software. The results were normalized using *Hprt1* as reference gene. This gene was experimentally determined with Genex software (MultiD Analyses AB, Göteberg, Sweden) as the most stable for the treatment conditions used. Primer sequences were designed using Beacon Designer software version 7.7 (Premier Biosoft International, Palo Alto, CA, USA) and thoroughly tested.

29.4.7 Calculation of Real-Time RT-PCR Results

Because the real-time RT-PCR results are presented as ratios of chemical-treated samples over untreated (control) cells, a two-base logarithmic transformation was used to make observations symmetric and closer to a normal distribution. If x represents the fold change of a gene in one sample, then the two-base logarithmic transformation ($\log_2(x)$) is $\ln(x)/\ln(2)$. Therefore, fold changes of 2 and 0.5 correspond to mean $\log_2$ values of 1 and −1, respectively.

29.4.8 Construction of the Predictive Model

Signaling pathways and gene expression signatures of chemicals were used to construct a dichotomous classifier. Briefly, the quantitative values of variables, expressed as $\log_2$ of fold changes relatively to untreated cells, were submitted to linear discriminant analysis. Two Fisher's linear discriminant functions were obtained: a function for calculation of the score of the chemical as a sensitizer (sensitizer score) and another for calculation of the score as a non-sensitizer (non-sensitizer score). Functions assume the following formula:

Sensitizer score = (X1 × x1 s) + (X2 × x2 s) + (X3 × x3 s) + (Xn × xn s)… + k s.

Non-sensitizer score = (X1 × x1 ns) + (X2 × x2 ns) + (X3 × x3 ns) + (Xn × xn ns)… + k ns.

where Xn is the predictor variable expressed as mean $\log_2$ of fold change, xn s and xn ns are the classification coefficients from Fisher's linear discriminant functions of sensitizers and non-sensitizers and k s and k ns are the constants from each function.

Finally, a predictive score (Psc) is calculated as the difference between sensitizer score and non-sensitizer sore. A chemical is classified as sensitizer if the calculated Predictive score is greater than 0. IBM SPSS Statistics 20 software package was used for all statistical analysis.

29.4.9 Estimation of Sensitizers' Potency

The correlation between LLNA EC3 potency classification data and several *in vitro* and *in silico* parameters was evaluated by multiple linear regression analysis. This allows defining a model where a dependent variable (EC3 value) is predicted given a set of explanatory variables. The generic model is represented by Eq. (29.1):

$$Y = \beta 0 + \beta 1X1 + \beta 2X2 + \ldots + \beta pXp \tag{29.1}$$

where Y is the dependent variable, β0 is the constant term and β1 to βp are the coefficients relating the p explanatory variables to the variable of interest.

As possible explanatory variables, we tested *in vitro*-generated data regarding *Hmox1*, *Nqo1*, *Trxr1* and *Cxcl10* expression levels, phospho-p38MAPK and phospho-JNK levels, as well as sensitizer's cytotoxicity (EC30). Additionally, several *in silico*-calculated parameters that are known to reflect chemical characteristics, which in turn determine their potency *in vivo*, were analysed. Such parameters include the capacity of chemicals to permeate the skin, represented by skin permeation coefficient (kp), and their reactivity, represented by electrophilicity index (ω). The skin permeation coefficient is a mathematical estimation of the rate at which a chemical penetrates the stratum corneum and reaches the viable cells within the epidermis. It is calculated based on chemical molecular weight and octanol/water partition coefficient (Log P). The calculations were made with Skin Permeation Calculator (http://www.cdc.gov/niosh/topics/skin/skinpermcalc.html) according to the modified Robinson model [22]. Despite the great chemical heterogeneity, a common feature of skin sensitizers is their intrinsic electrophilicity (with the exception of metals and thiols) or their potential to be metabolized to electrophilic compounds. This feature determines their ability to covalently bind to skin proteins forming sensitizing complexes, which has been demonstrated to be correlated with sensitizers' potency [23]. Thus, as an indicative measure of biological reactivity of chemicals, their electrophilicity index was calculated.

The electrophilicity index is a quantum chemical descriptor that quantifies the global electrophilic nature of a molecule within a relative scale [24] and is frequently used in the development of quantitative structure activity relationship (QSAR) models [25].

Electrophilicity index was calculated according to Eq. (29.2) proposed by Parr and colleges [24]:

$$\begin{aligned} &\text{Electrophilicity index}(\omega) \\ &= \left[(\text{EHOMO} + \text{ELUMO})/2\right]2/\left[(\text{ELUMO} - \text{EHOMO})^{*}2\right] \end{aligned} \tag{29.2}$$

where EHOMO and ELUMO are the one-electron energies of the highest occupied and lowest unoccupied molecular orbitals, respectively. All computational calculations of the one-electron energies based on chemical structures were performed using Spartan'10 software (Wavefunction, Inc., CA, USA) and the B3LYP/6-31G(d) level of theory.

29.5 Role in a Testing Strategy

Integrated testing strategies (ITS) enable to significantly increase the use of non-testing information for regulatory decision-making and thus to minimize the need for animal testing. This is in conformity with the new REACH paradigm appealing to move away from extensive standard testing to a more intelligent, substance-tailored approach. The envisaged decision theory framework includes alternative methods such as chemical and biological read-across, *in vitro* results, *in vivo* information on analogues, qualitative and quantitative structure-activity relationships and thresholds of toxicological concern, although the way towards this goal remains unclear. Concerning skin sensitization, alternative approaches currently under development and evaluation are designed to address and model the key biological mechanisms of the induction phase of skin sensitization, namely [4]:

- The ability of the chemical to penetrate the skin and reach the site of haptenation (skin bioavailability).
- The covalent binding of the chemical to the skin proteins (haptenation).
- The ability of the haptenated chemical to evoke xenoinflammation and subsequent cytotoxicity triggered on epidermal keratinocytes and Langerhans cells (tissue stress/damage).
- The activation, maturation and migration of dendritic cells (DC) from the skin to the regional lymph nodes and presentation of the antigens to T cells with subsequent proliferation of specific T cells.

These key biological mechanisms are in line with the recently developed and endorsed adverse outcome pathway for skin sensitization by the Organisation for Economic Co-operation and Development (OECD) [5, 6]. Having this in mind, a future approach for skin sensitization identification will be surely an integrative platform encompassing several readouts for the different phases of skin sensitization tied to adverse outcomes of regulatory concern. The test herein dissected could represent a valuable tool reflecting the interaction of DCs with chemicals at different cellular and molecular compartments with several readouts, which include cytotoxicity/xenoinflammation, intracellular signalling pathways as well as genes codifying detoxifying proteins and chemokines involved in DC migration. We believe that integration of several readouts while increasing the complexity of the test could indeed increase its strength. In our perspective, the concomitant use of *in chemico*-based tests (such as the direct peptide reactivity assay, [11]) and the herein presented test could give valuable information about the skin sensitizing potential of chemicals, allowing the measurement of their binding capacity towards peptides and their subsequent biological activity, thus presenting potential of being included in a decision-making process. Furthermore, the European implementation of the Globally Harmonized System of Classification and Labelling (GHS) (amendment to CLP regulation EC no. 1272/2008) foresees classifying allergens into category 1A (strong sensitizers) or category 1B (other skin sensitizers), which is in effect for single chemicals since 2010 and come into force in 2015 for mixtures. Therefore, in addition to being able to determine whether or not a chemical is a sensitizer, it is equally important to determine its potency in order to establish a maximum safe concentration for human exposure. Thus, in an attempt to update the developed labelling test, we further unravel a method for classifying sensitizers according to their potency based on their cytotoxicity and electrophilicity index, two key biological mechanisms involved in skin sensitization. A putative role of the herein described approach in a testing strategy is depicted in Fig. 29.4.

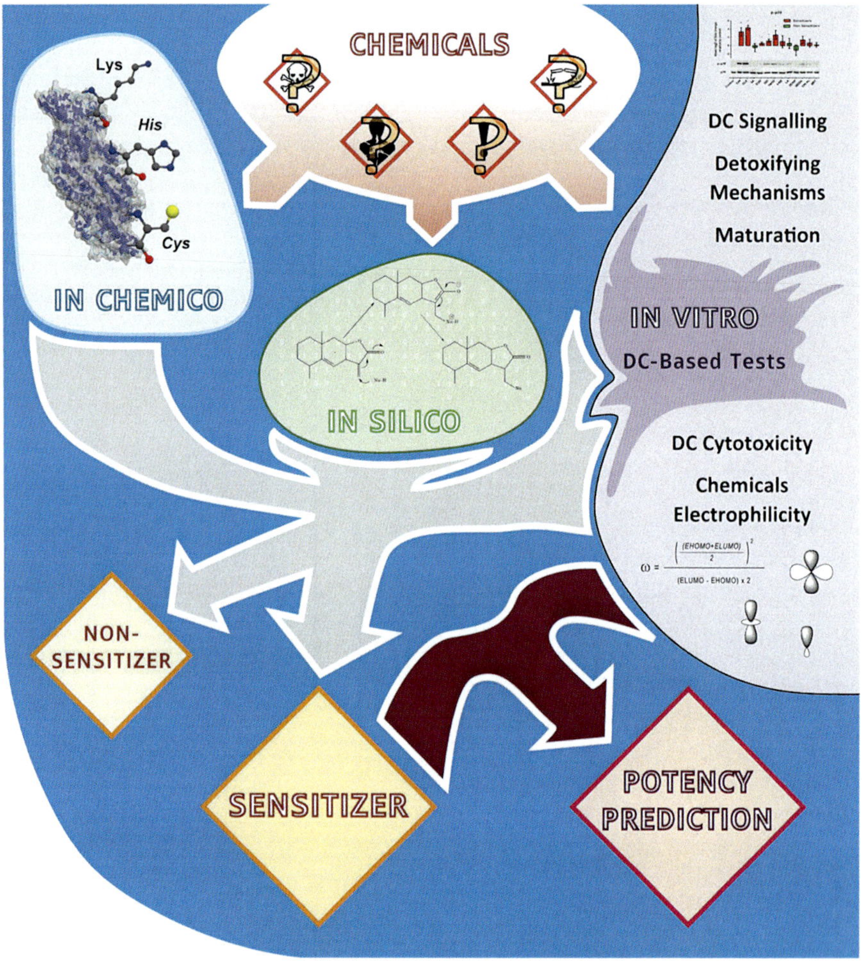

Fig. 29.4 Putative role of the dendritic cell-based approach in a testing strategy. The concomitant use of *in chemico* and/or *in silico*-based tests and the herein presented DC-based approach could give valuable information about the skin sensitizing potential of chemicals, allowing the measurement of their binding capacity towards peptides and their subsequent biological activity, thus presenting potential of being included in a decision making process (*DC* dendritic cells)

29.6 Perspectives from the Test Developer

29.6.1 Critical Steps in the Protocol

A major critical step in the protocol is the determination of the concentrations of chemicals that cause 30% cytotoxicity. As the test relies on the exposure of cells to chemicals in aqueous medium, it is therefore limited by the solubility of the compounds and to a certain extent to their lipophilicity. Highly lipophilic chemicals may not effectively interact with cells and have their effects underestimated. Although metabolically and phenotypically stable, FSDC cells should be preferentially used up to passage 40 and routinely checked for mycoplasma contamination. The adequate responsiveness of cells could be checked by including in the assays DNFB and SDS as positive and negative controls, respectively.

29.6.2 Possible Protocol Adaptations

The test makes use of the mouse DC-like cell line FSDC; however, other mammalian DC surrogates may be used. From our preliminary data, the approach described for the test, including the discriminatory variables, could be transposed to human DC surrogates such as THP1 cells. The use of cells from human origin may hypothetically improve the performance of the test particularly if results were checked against human data rather than LLNA. Additionally, several discriminatory parameters (activation of intracellular signalling pathways) are semi-quantitatively determined by Western blot, a laborious and time-consuming technique. This could represent a drawback to the transposition of the test to large-scale screening. A possible step to further optimize the model will be the incorporation of quantitative and high-throughput techniques replacing Western blot. Accordingly, assessment of phosphorylated JNK and p38 MAPK could be performed by flow cytometry or even by suspension array technologies such as Luminex®. Concerning the estimation of sensitizers' potency, the number of chemicals tested was very low; therefore, it is envisaged to increase the panel of skin sensitizers tested.

29.6.3 Challenges and Opportunities

The promising ongoing tests addressing the key steps in the sensitization process (protein reactivity, Nrf2/ARE activation as well as dendritic cell activation) point to the replacement, in a near future, of animal testing through the integration of the various readouts in a combinatory platform. Future developments will need to focus on potency estimation. The better way to integrate the already known pathophysiological mechanisms and adverse outcome pathway behind skin sensitization towards the development of a prediction model to assess the potency of skin sensitizers will be a challenge. The predictive test herein presented is a prototype with putative value for this challenging issue, through a biphasic approach that first identifies skin sensitizers and afterwards estimates their potency (Fig. 29.4).

29.7 Conclusions

The dendritic cell-based approach discussed in this chapter is founded on key elements of the adverse outcome pathway for skin sensitization. More specifically, it is based on the analysis and integration of several functional mechanistic endpoints that reproduce the interaction of DCs with chemicals at different cellular and molecular levels, including cytotoxicity evocation (EC30), intracellular signalling pathways (JNK and p38 MAPKs) activation as well as modulation of genes codifying detoxifying proteins (*Trxr1, Hmox1, Nqo1*) and chemokines involved in DC migration (*Cxcl10*). From the total 18 compounds tested, 17 were correctly classified, representing a concordance of 94%, with a sensitivity of 92% and a specificity of 100%. Among the panel of chemicals tested, metals and pro- and pre-haptens were correctly classified, accordingly with corresponding human sensitization data. Interestingly, the model was also able to properly classify the non-sensitizer but immunogenic *Toll-like* receptor 4 agonist lipopolysaccharide (LPS), thus reflecting some specificity of the triggered toxicity pathways attributable to skin sensitizers. Highly lipophilic and extremely poor water-soluble chemicals may not effectively interact with cells, and their effects may be underestimated, which constitutes a limitation of the test. This assay can be combined with *in chemico* and/or *in silico* data in order to maximize the predictive power (Fig. 29.4).

Importantly, the feasibility to predict the potency of sensitizers using *in vitro*-generated data and several *in silico*-calculated descriptors was also analysed. A strong correlation with LLNA EC3 values was obtained (Pearson correlation coefficient $r = 0.85$, $p < 0.001$, $n = 12$).

Taken together, a biphasic approach could be designed to provide the identification and classification of skin sensitizing chemicals having putative potential to be further included in integrated testing strategies.

Acknowledgements This work is funded by FEDER funds through the Operational Programme Competitiveness Factors—COMPETE, "ToxVitroTesting", Quadro de Referência Estratégico Nacional (QREN)—Empresas em co-promoção project ref.: 38977 and national funds by FCT—Foundation for Science and Technology under the project grant number PTDC/SAU-OSM/099762/2008 and strategic project UID/NEU/04539/2013. João Demétrio Martins had a FCT grant number SFRH/BD/73065/2010.

References

1. Basketter D, Alépée N, Casati S, et al. Skin sensitisation-moving forward with non-animal testing strategies for regulatory purposes in the EU. Regul Toxicol Pharmacol. 2013;67:531–5. doi:10.1016/j.yrtph.2013.10.002.
2. Bauch C, Kolle SN, Ramirez T, et al. Putting the parts together: combining in vitro methods to test for skin sensitizing potentials. Regul Toxicol Pharmacol. 2012;63:489–504. doi:10.1016/j.yrtph.2012.05.013.
3. Rovida C. Integrated Testing Strategies (ITS) for safety assessment. ALTEX. 2014;32:25–40. doi:10.14573/altex.1411011.
4. Reisinger K, Hoffmann S, Alépée N, et al. Systematic evaluation of non-animal test methods for skin sensitisation safety assessment. Toxicol In Vitro. 2015;29:259–70. doi:10.1016/j.tiv.2014.10.018.
5. OECD. OECD guideline for the testing of chemicals No. 168, part 1. In: The adverse outcome pathway for skin sensitization initiated by covalent binding proteins. Paris: Organisation for Economic Cooperation and Development; 2012. Accessible at: http://www.oecd.org/officialdocuments/publicdisplaydocumentpdf/?cote=env/jm/mono(2012)10/part1&doclanguage=en.
6. OECD. OECD guideline for the testing of chemicals No. 168, part 2. In: The adverse outcome pathway for skin sensitisation initated by covalent binding to proteins. Paris: Organisation for Economic Cooperation and Development; 2012. Accessible at: http://www.oecd.org/officialdocuments/publicdisplaydocumentpdf/?cote=env/jm/mono(2012)10/part2&doclanguage=en.

7. Roggen EL. In vitro approaches for detection of chemical sensitization. Basic Clin Pharmacol Toxicol. 2014;115:32–40. doi:10.1111/bcpt.12202.
8. Neves BM, Gonçalo M, Figueiredo A, et al. Signal transduction profile of chemical sensitisers in dendritic cells: an endpoint to be included in a cell-based in vitro alternative approach to hazard identification? Toxicol Appl Pharmacol. 2011;250:87–95. doi:10.1016/j.taap.2010.10.003.
9. Neves BM, Rosa SC, Martins JD, et al. Development of an in vitro dendritic cell-based test for skin sensitizer identification. Chem Res Toxicol. 2013;26:368–78. doi:10.1021/tx300472d.
10. Casati S, Aeby P, Kimber I, et al. Selection of chemicals for the development and evaluation of in vitro methods for skin sensitisation testing. Altern Lab Anim. 2009;37:305–12.
11. Gerberick GF, Vassallo JD, Bailey RE, et al. Development of a peptide reactivity assay for screening contact allergens. Toxicol Sci. 2004;81:332–43. doi:10.1093/toxsci/kfh213.
12. Loveless SE, Ladies GS, Gerberick GF, et al. Further evaluation of the local lymph node assay in the final phase of an international collaborative trial. Toxicology. 1996;108:141–52. doi:10.1016/0300-483X(95)03279-O.
13. Dos Santos GG, Spiekstra SW, Sampat-Sardjoepersad SC, et al. A potential in vitro epidermal equivalent assay to determine sensitizer potency. Toxicol In Vitro. 2011;25:347–57. doi:10.1016/j.tiv.2010.10.008.
14. Lambrechts N, Vanheel H, Nelissen I, et al. Assessment of chemical skin-sensitizing potency by an in vitro assay based on human dendritic cells. Toxicol Sci. 2010;116:122–9. doi:10.1093/toxsci/kfq108.
15. Gamerdinger K, Moulon C, Karp DR, et al. A new type of metal recognition by human T cells: contact residues for peptide-independent bridging of T cell receptor and major histocompatibility complex by nickel. J Exp Med. 2003;197:1345–53. doi:10.1084/jem.20030121.
16. Lu L, Vollmer J, Moulon C, et al. Components of the ligand for a Ni++ reactive human T cell clone. J Exp Med. 2003;197:567–74. doi:10.1084/jem.20021762.
17. Basketter DA, Alépée N, Ashikaga T, et al. Categorization of chemicals according to their relative human skin sensitizing potency. Dermatitis. 2014;25:11–21. doi:10.1097/DER.0000000000000003.
18. Girolomoni G, Lutz MB, Pastore S, et al. Establishment of a cell line with features of early dendritic cell precursors from fetal mouse skin. Eur J Immunol. 1995;25:2163–9. doi:10.1002/eji.1830250807.
19. Hulette BC, Ryan CA, Gildea LA, Gerberick GF. Relationship of CD86 surface marker expression and cytotoxicity on dendritic cells exposed to chemical allergen. Toxicol Appl Pharmacol. 2005;209:159–66. doi:10.1016/j.taap.2005.03.019.
20. Mosmann T. Rapid colorimetric assay for cellular growth and survival: application to proliferation and cytotoxicity assays. J Immunol Methods. 1983;65:55–63. doi:10.1016/0022-1759(83)90303-4.
21. Neves BM, Cruz MT, Francisco V, et al. Differential roles of PI3-Kinase, MAPKs and NF-κB on the manipulation of dendritic cell Th1/Th2 cytokine/chemokine polarizing profile. Mol Immunol. 2009;46:2481–92. doi:10.1016/j.molimm.2009.05.021.
22. Wilschut A, ten Berge WF, Robinson PJ, McKone TE. Estimating skin permeation. The validation of five mathematical skin permeation models. Chemosphere. 1995;30:1275–96. doi:10.1016/0045-6535(95)00023-2.
23. Chipinda I, Ajibola RO, Morakinyo MK, et al. Rapid and simple kinetics screening assay for electrophilic dermal sensitizers using nitrobenzenethiol. Chem Res Toxicol. 2010;23:918–25. doi:10.1021/tx100003w.
24. Parr RG, Szentpály L, Liu S. Electrophilicity index. J Am Chem Soc. 1999;121:1922–4. doi:10.1021/ja983494x.
25. Enoch SJ, Roberts DW, Cronin MTD. Mechanistic category formation for the prediction of respiratory sensitization. Chem Res Toxicol. 2010;23:1547–55. doi:10.1021/tx100218h.

An *In Vitro* Human Skin Test for Assessing Adverse Immune Reactions and Sensitization Potential

30

Anne Dickinson, Xiao Nong Wang, and Shaheda Ahmed

30.1 Principle of the Scientific Basis of the Test Method

The skin explant test was developed originally as a predictive test [1] for graft-versus-host disease (GvHD), a severe systemic complication developing in patients post-allogeneic bone marrow transplantation. GvHD involves recognition of foreign target (patient) tissue by the incoming donor T cells causing activation, proliferation and cytokine release. For predicting GvHD, the skin explant test involves sensitization of donor cells to patient cells in a mixed lymphocyte reaction. After 7 days of co-culture, the sensitized donor cells are then added to a small skin biopsy from the patient *in vitro*. After 3 days of co-culture of the skin plus activated T cells, the skin is assessed for damage by conventional histopathology. The assay has been used for over 20 years, mainly in the laboratory of Dickinson for both predicting GvHD and understanding its immunobiology [2–6]. It has also been used in the clinic to predict GvHD and modify therapy based on the assay results [7–10]. The predictive outcome of the test is based on a histopathological score (grades I–IV) according to the severity of damage or lesions exhibited in the skin [11]. This allogeneic skin explant test (between patient and donor) was modified as an autologous skin explant test for assessment of sensitizing potential by utilizing skin tissue and cells from the same individual and

A. Dickinson (✉) • S. Ahmed
Alcyomics, Ltd., Bulman House, Regent Centre, Gosforth, Newcastle-upon-Tyne NE3 3LS, UK

Haematological Sciences, Institute of Cellular Medicine, Newcastle University, Newcastle-upon-Tyne NE2 4HH, UK
e-mail: anne.dickinson@ncl.ac.uk

X.N. Wang
Haematological Sciences, Institute of Cellular Medicine, Newcastle University, Newcastle-upon-Tyne NE2 4HH, UK

C. Eskes et al. (eds.), *Alternatives for Dermal Toxicity Testing*,
DOI 10.1007/978-3-319-50353-0_30

developed to mimic immunological responses that occur *in vivo*. This is achieved by priming dendritic cells with the chemical of interest followed by T cell activation and cytokine release, similar to that observed in an allergic contact dermatitis reaction in the skin. The end point of the test is again histopathological damage, which is novel, and the grading of histopathological changes in the skin can allow the determination of relative sensitizing potency. Potency is important to determine a safe usage level of a compound and determined by quantifying the minimum dose level of a compound that is required to initiate a sensitization response. Other alternative approaches of sensitization testing have not been suitable for assessment of relative potency.

A wide variety of chemicals are able to cause allergic sensitization. This can take a variety of forms, including skin sensitization resulting in allergic contact dermatitis and sensitization of the respiratory tract culminating in rhinitis and asthma.

There are a number of alternative methods for characterization of sensitizing potential which have been proposed, ranging from *in silico* models [12, 13] to methods using various cell culture systems [14–16]. For the latter assays, the endpoints measured are commonly induced changes in expression of specified membrane determinants or cytokine production [17–20]. However, none of the suggested alternative methods have made use of human skin explants (rather than 3D skin equivalent models) to predict sensitization potential.

Here we describe a novel human skin explant test as a possible alternative approach to the use of animal models.

30.2 Current Pre-validation Status

We have tested 44 chemicals based on their sensitizing and non-sensitizing potential as classified by the published mouse local lymph node assay (LLNA) [21, 22] evaluation (Sens-it-iv website (http://www.sens-it-iv.eu)). A wide range of positive and negative chemicals were used for the study including respiratory sensitizers and pre-/pro-haptens. A set of 3 respiratory sensitizers, 4 pre-/pro-haptens, 14 sensitizers, 21 non-sensitizers and 2 LLNA misclassified compounds were tested. All chemicals were purchased at the highest possible purity (>85%) from Sigma-Aldrich. Exposure concentration for each chemical was selected after performing a cell viability test. Several concentrations for each chemical were tested with healthy volunteer peripheral blood mononuclear cells, and the highest concentration demonstrating a minimum of 80% cell viability was selected.

Results included measurement of T cell proliferation, interferon gamma (IFNγ) release and grades of histopathological damage and compared with data for the LLNA as well as, where available, human sensitization data [23]. We have also determined if the observed responses to T cell priming was antigen specific. In addition to chemical sensitizers and non-sensitizers, we have used the test to assess adverse immune reactions to nine clinical relevant monoclonal antibodies and assessed the skin test results with reactions observed in the clinic. This has included an analogue of TGN1412, the CD28-super-agonist, which caused the Northwick Park incident in 2006. In these later tests, the assay was modified to assess the role of total peripheral blood cell populations as a replacement to dendritic cell activation alone.

30.3 Performance and Applicability of the Test Method

30.3.1 Reproducibility

The test is highly reproducible. The same two compounds (dinitrochlorobenzene (DNCB) and Triton-x) were initially validated in six separate experiments using peripheral blood and skin samples from six healthy volunteers and then used as positive and negative controls, respectively, in all subsequent experiments with a consistent positive or negative response rate of greater than 95%. In addition the test is easily transferable from laboratory to laboratory and between research staff due to standardization of protocols and methodology.

30.3.2 Predictive Capacity

The results show a high predictive capacity for identification of sensitizers and non-sensitizers. The results were compared with the LLNA results for 44 compounds, including 22 sensitizers and 22 non-sensitizers and gave a 95% specificity, 95% sensitivity and 95% concordance with a correlation coefficient of 0.9. In addition, monoclonal antibody reactions in the skin test correlated to 96% with those observed in the clinic by way of either injection site rashes or systemic allergic reactions. The test could also have predicted the systemic adverse reactions observed with TGN1412, since severe grades III–IV reactions were observed in over 90% of the tests.

30.3.3 Applications and Limitations

The test has been applied to identify known and unknown sensitizers and non-sensitizers, monoclonal antibody versus biosimilar compounds and the detection of IgE responses. It has also been used to detect responses to small-molecule compounds, such as antibiotics. One of the main limitations of the test is that occasionally high concentrations tested in animal models are too toxic to human peripheral blood lymphocytes, and therefore a direct comparison cannot be made. In addition, the compounds must be soluble in either water or solvent, and to date the test cannot be used for absorption assays. The test does not detect anti-idiotype antibody- or anti-drug antibody (ADA)-type responses.

The tests' main advantages include that it can be used to detect potency responses to chemicals not detected by animal models, e.g. nickel sulphate, and can be modified for efficacy testing of immunomodulatory compounds. The T cell proliferation and IFNγ assays can also be used as screening tests prior to the skin test.

30.3.4 Comparison to Human Data

Certain chemicals and therapeutic monoclonal antibodies have human response data recorded, and these results can be compared with the skin explant test *in vitro* data with a high degree of concordance (correlation coefficient 0.91). For example, nickel sulphate shows a negative response in the LLNA but is positive in man and the skin explant test; conversely, propylene glycol is also a weak positive in man and the skin explant test but negative in the mouse LLNA assay.

30.4 Brief Description of the Protocol

All healthy volunteer samples are taken with consent and Local Research Ethics Committee approval. From each volunteer, peripheral blood and two skin biopsies (taken from the abdominal region) are used for each skin explant test. Peripheral blood mononuclear cells (PBMC) are separated by density-gradient centrifugation using the Lymphoprep™ method and treated with the compound of interest to assess viability. $CD14^+$ monocytes are selected using the MACS® technology (Miltenyi Biotec). Monocyte-derived dendritic cells (MoDC) are then generated as described by Kvistborg and colleagues [24] with some modification. The MoDCs are then co-cultured with autologous lymphocytes containing T cells. During this incubation period, T cell proliferation and IFNγ release are measured. After incubation, the activated cells are then added to autologous skin explants and cultured, after which time the skin is paraffin embedded, sectioned and stained with haematoxylin and eosin. Histopathological evaluation of each skin explant is performed blindly by two independent histopathologists and histological damage graded from I to IV according to the Lerner criteria [1, 11] as follows: grade 0 represents no observable damage to skin keratinocytes, grade I displays mild vacuolization of basal cells, grade II displays vacuolization of basal cells and evidence of dyskeratotic bodies, grade III displays sub-epidermal cleft formation at the dermal epidermal junction, and grade IV displays complete epidermal separation (Fig. 30.1). Based on previous results using this grading system, grades 0 and I are regarded as negative results and a histopathological grade of II or greater as a positive result. Controls consist of skin explants cultured with medium alone (background control) or with autologous lymphocytes (negative control) or with third-party mixed lymphocyte responder allogeneic lymphocytes (positive control) as previously described [8].

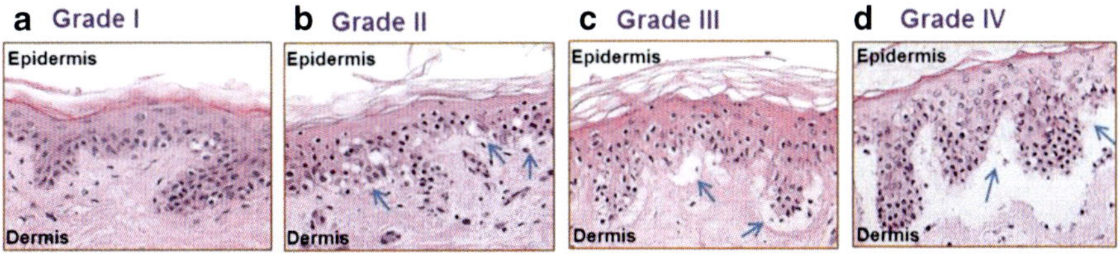

Fig. 30.1 Histopathological changes in each grade. (**a**) *Grade I* normal skin negative control. (**b**) *Grade II* dyskeratotic bodies and vacuolization. (**c**) *Grade III* sub-epidermal cleft formation. (**d**) *Grade IV* complete separation of the dermis and epidermis. Histopathological assessment of the samples is performed blindly and a grade assigned to each sample. Grading criteria can be seen in figure. Grade II or above is considered a positive result. Skimune® has shown an excellent correlation with the LLNA with 95% overall correlation (95% sensitivity and 95% specificity)

30.4.1 [^{3}H]-Thymidine T cell Proliferation Assay and IFNγ Release Screening Assays

To measure the induction of a T cell proliferative response during co-culture with MoDCs exposed to chemicals, MoDCs are cultured with the test chemical, then harvested and washed. Cells are then co-cultured in RPMI media (with supplements) with autologous lymphocytes (CD14$^-$ fraction) in triplicate for 5 days. A negative control (untreated MoDC co-cultured with autologous lymphocytes) and a positive control (untreated MoDC co-cultured with allogeneic lymphocytes) are included in each assay. After 5 days, supernatants are collected for IFNγ analysis, before [^{3}H]-Thymidine addition for 16–18 h. Cells are harvested and subsequently counted using a β-scintillation counter. Data is interpreted using Prism GraphPad software (V5).

Interferon-γ (IFNγ) levels are quantified in supernatants collected from the T cell proliferation assays using a multiplex cytometric bead array (CBA) flex set following the manufacturer's instructions (BD). Data are acquired using a FACS Canto (BD Biosciences) and analysed using FCAP software (BD Biosciences).

30.4.2 Data and Statistical Analysis

For each compound, tests were performed using samples from a minimum of 4–6 healthy volunteers. Data from skin explant tests are recorded as graded scores from I to IV as described previously. A compound is regarded as a sensitizer if 75% or more of the tests are positive and regarded as a weak sensitizer if 60–74% of the tests are positive. Conversely a compound is regarded as a non-sensitizer if 60–100% of the tests are negative. Compounds giving inconsistent results are further tested at various dose response ranges.

The data for the T cell proliferation assays are given as stimulation indices (SI) and are representative of the level of T cell proliferation (cpm) in response to exposure to the test compound. To determine the SI, the background value (cells plus medium alone) is subtracted from the test value and calculated as a percentage increase compared to the negative control (untreated cells co-cultured with autologous lymphocytes). As reported for the LLNA and peer reviewed by the Interagency Coordinating Committee on the Validation of Alternative Methods (ICCVAM), the cut-off SI value of 3 is considered to be a positive response. The data for IFNγ secretion is given as a fold increase. The fold increase is representative of the level of IFNγ secretion by cells in response to the test compound and is calculated in a similar manner to the SI. The cut-off value of a threefold increase is considered to be a positive response. Statistical analysis is carried out using Mann-Whitney U tests or one-way ANOVA using Prism GraphPad software V5. Correlation coefficients are determined using SPSS Statistics 7. Chi-squared analysis is carried out to compare the observed and expected results for sensitizers and non-sensitizers compared to those reported for the LLNA assay or for human sensitization data as published in ICCVAM LLA potency evaluation report.

30.5 Role in Testing Strategy

The mechanism of action of the skin explant test fits with the components of the key events of the sensitization process including dendritic cell activation and T cell proliferation responses, which are critical paths in the adverse outcome pathways (AOP) (Fig. 30.2).

30.6 Perspectives from the Test Developer

30.6.1 Critical Steps in the Protocol

Critical steps include (1) the assessment of optimal concentrations of the compounds to be used, using viability testing, (2) dendritic cell generation, (3) T cell proliferation and (4) cytokine release. Appropriate controls for the compounds to be tested need to be taken into careful consideration, e.g. isotype controls for monoclonal antibody testing, appropriate positive and negative sensitizers and/or allergens. As way of illustration, we have added several case studies as illustrated below.

30.6.1.1 Case Study 1

A blinded study was carried out to investigate the effect of various antibody formulations on normal skin in the presence and absence of autologous lymphocytes under serum-free conditions using the modified skin explant assay. Three anonymized antibodies were tested (antibody A, antibody B and antibody C) at a concentration of 0.1 μg/mL. Peripheral blood mononuclear cells (PBMCs) were isolated from whole blood and incubated with the test monoclonal antibody and

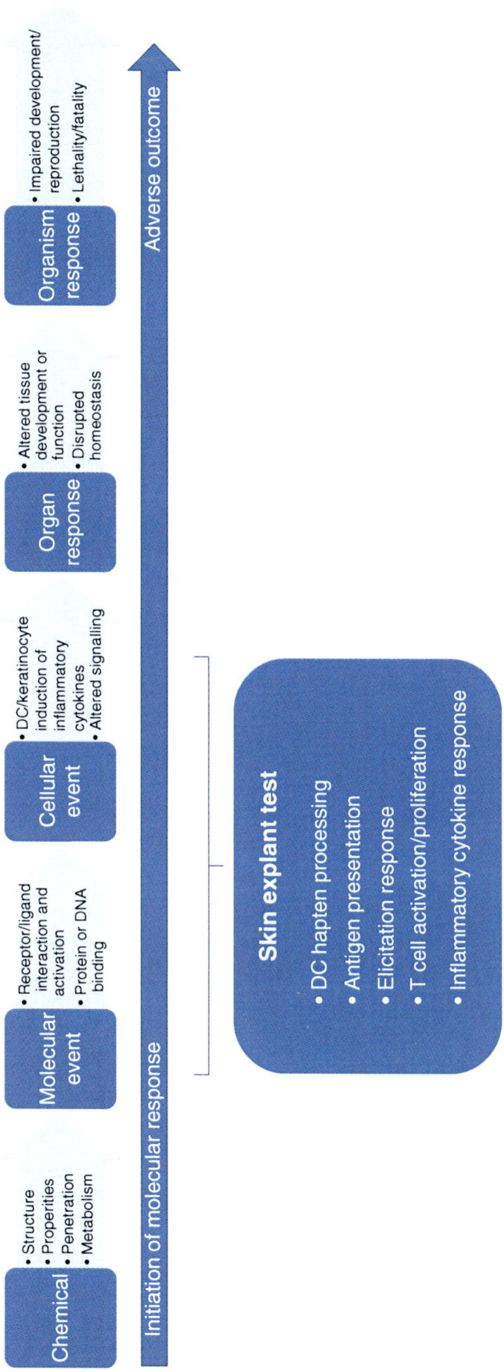

Fig. 30.2 Adverse outcome pathway and the skin explant assay as a sensitization predictive tool

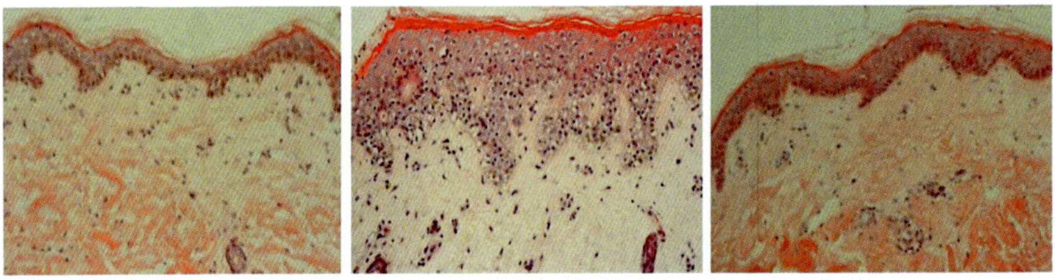

Fig. 30.3 Histological analysis of responses to monoclonal antibodies. Antibodies A and C showing a grade I (*negative*) response and antibody B showing a grade II (*positive*) response

an autologous skin biopsy. After 3 days, the skin was removed from each well, placed in formalin, paraffin embedded, sectioned and routinely stained for histopathology.

Slides were graded for histological damage from grades 0 to IV relative to the level of damage observed. Skin explant results and respective images are given in Fig. 30.3. Slides graded II or above were considered to be positive for damage. The results showed antibody A gave a negative (grade I) response. Antibody B gave a positive grade II response where histological damage was observed at the dermis/epidermis junction. Antibody C gave a negative (grade I) response.

Compounds when uncoded were as follows: Antibody A was a negative control; antibody B was a new drug which had caused a rash in phase I clinical trials and was withdrawn. Antibody C was a modified version of antibody B. The results show the modified version of the new drug (antibody B) does not have the potential to cause an adverse reaction.

30.6.1.2 Case Study 2

Hydrolysed wheat-based proteins (HWP) are a common ingredient in the formulation of cosmetics. More recently these proteins have been implicated in the induction of rare but severe allergic reactions [25, 26]. The effects of two wheat protein products (compound X and compound Y) were tested using the modified skin explant assay to determine any potential sensitizing characteristics (Fig. 30.4), and levels of IgE antibody production were measured in cell culture supernatants using the enzyme-linked immunosorbent assay (ELISA) method. The HWP compounds were tested at 1% concentration. Compounds were incubated overnight with dendritic cells derived from whole blood and then further incubated with autologous T cells. The primed T cells were then further incubated with autologous skin. The results show compound X displayed no damage in skin histopathology (Fig. 30.4b) and low IgE antibody levels (Fig. 30.4d) indicating this compound was unlikely to cause a hypersensitivity reaction. Compound Y (Fig. 30.4c) gave a positive response showing a high degree of histopathological damage (grade III) and elevated IgE antibody levels (Fig. 30.4d) indicating that this compound was recognized as an allergen and had the potential to cause an adverse reaction.

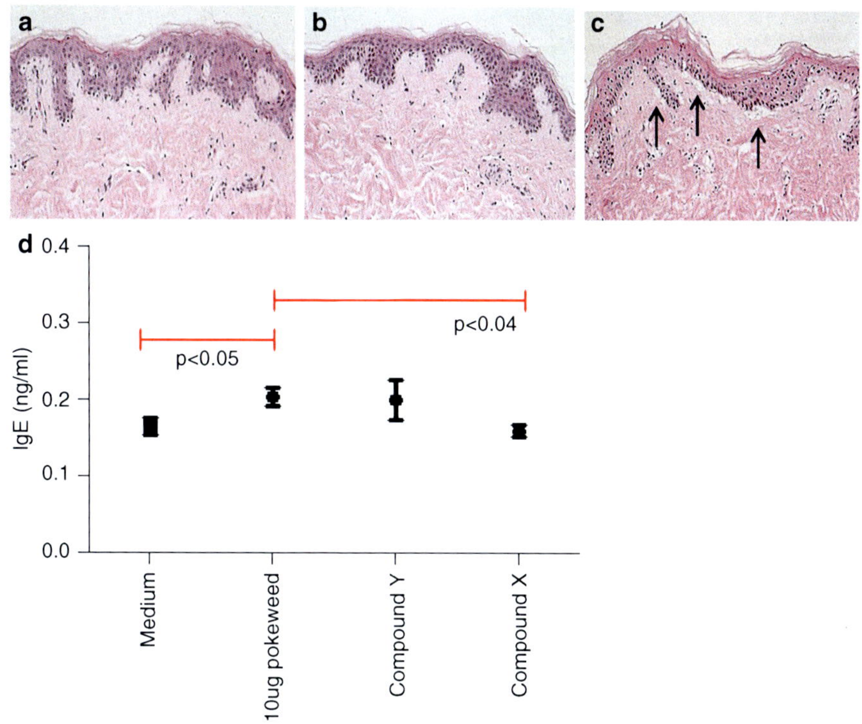

Fig. 30.4 Immunogenic responses to wheat-based proteins. (**a**) Negative control; skin incubated in medium alone showing a grade I negative response with an intact epidermis/dermis junction. (**b**) Compound X showing a grade I negative response. (**c**) Compound Y showing a positive grade III response; *arrows* indicate damage to the tissue and the formation of clefts on the epidermis/dermis junction. (**d**) IgE antibody levels measured in cell culture supernatants showing a significant increase in IgE levels in the positive control, Pokeweed-stimulated cells in comparison to the negative control, cells in medium only and to compound X. IgE levels in response to compound Y was similar to that observed in response to the positive control

30.6.2 Possible Protocol Adaptions

Adaptations of the test can include the use of T cell subsets, e.g. regulatory T cells or naive T cells to further understand mode of action of the compounds and further extension of cytokine analysis using flow cytometry or ELISA.

30.6.3 Challenges and Opportunities

The use of the assay for absorption studies is a challenge, but the readout could be used alongside standard toxicology assessments. Compounds need to be soluble and currently the test is not used for topical creams or ointments.

30.7 Conclusions

The skin explant test has been trademarked as Skimune® and has been used commercially since 2009 to:

1. Assess the efficacy of immunomodulatory compounds (anti-Fas monoclonal antibody [5] and anti-TNF antibody [27] using the assay developed for predicting GvHD as previously described [8]
2. Determine the sensitizing potential/adverse reactions of biosimilars compared to monoclonal antibodies
3. Assess the allergen potential of cosmetic, e.g. wheat-based products
4. Assess numerous chemicals and small-molecule compounds for sensitization or adverse reactions [23].
5. Assess the safety of certain cellular therapy products [28]

Skimune® is a highly sensitive test which can be use preclinically to assess the safety and efficacy of therapeutics [29] reducing the need for extensive animal testing and has been shown to correlate not only with animal models of sensitivity but also human outcome data. Greater use of the test would enable potential drugs to be eliminated earlier in the drug development pathway saving costs to the pharmaceutical industry and reducing animal testing.

References

1. Vogelsang GB, Hess AD, Berkman AW, Tutschka PJ, Farmer ER, Converse PJ, Santos GW. An in vitro predictive test for graft versus host disease in patients with genotypic HLA-identical bone marrow transplants. N Engl J Med. 1985;313:645–50.
2. Dickinson AM, Sviland L, Dunn J, Carey P, Proctor SJ. Demonstration of direct involvement of cytokines in graft-versus-host reactions using an in vitro human skin explant model. Bone Marrow Transplant. 1991;7:209–16.
3. Dickinson AM, Wang XN, Sviland L, Vyth-Dreese FA, Jackson GH, Schumacher TN, Haanen JB, Mutis T, Goulmy E. In situ dissection of the graft-versus-host activities of cytotoxic T cells specific for minor histocompatibility antigens. Nat Med. 2002;8(4):410.
4. Jarvis M, Schulz U, Dickinson AM, Sviland L, Jackson G, Konur A, Wang XN, Hromadnikova I, Kolb HJ, Eissner G, Holler E. The detection of apoptosis in a human in vitro skin explant assay for graft versus host reactions. J Clin Pathol. 2002;55:127–32.
5. Ruffin N, Ahmed SS, Osorio LM, Wang XN, Jackson GH, Collin MP, Ekre HP, Chiodi F, Dickinson AM. The involvement of epithelial Fas in a human model of graft versus host disease. Transplantation. 2011;91:946–51.
6. Wang XN, Haniffa MA, Holtick U, Collin MP, Jackson G, Hilkens CM, Holler E, Edinger M, Hoffmann P, Dickinson AM. Regulatory T-cell suppression of CD8+ T-cell-mediated graft-versus-host reaction requires their presence during priming. Transplantation. 2009;88:188–97.
7. Dickinson A, Hromadnikova I, Sviland L, Jackson G, Taylor P, Vavrinec J, Sedlacek P, Cermakova M, Stary J, Vitek A, Sajdova J, Proctor S. Use of a skin explant model for predicting GVHD in HLA-matched bone marrow transplants – effect of GVHD prophylaxis. Bone Marrow Transplant. 1999;24:857–63.
8. Dickinson AM, Sviland L, Carey P, Reid MM, Hamilton PJ, Pearson AJ, Proctor SJ. Skin explant culture as a model for cutaneous graft-versus-host disease in humans. Bone Marrow Transplant. 1988;3:323–9.

9. Sviland L, Dickinson AM. A human skin explant model for predicting graft-versus-host disease following bone marrow transplantation. J Clin Pathol. 1999;52:910–3.
10. Sviland L, Dickinson AM, Carey PJ, Pearson AD, Proctor SJ. An in vitro predictive test for clinical graft-versus-host disease in allogeneic bone marrow transplant recipients. Bone Marrow Transplant. 1990;5:105–9.
11. Lerner KG, Kao GF, Storb R, Buckner CD, Clift RA, Thomas ED. Histopathology of graft-vs.-host reaction (GvHR) in human recipients of marrow from HL-A-matched sibling donors. Transplant Proc. 1974;6:367–71.
12. Chaudhry Q, Piclin N, Cotterill J, Pintore M, Price NR, Chretien JR, Roncaglioni A. Global QSAR models of skin sensitisers for regulatory purposes. Chem Cent J. 2010;4(Suppl 1):S5.
13. Dimitrov SD, Low LK, Patlewicz GY, Kern PS, Dimitrova GD, Comber MH, Phillips RD, Niemela J, Bailey PT, Mekenyan OG. Skin sensitization: modeling based on skin metabolism simulation and formation of protein conjugates. Int J Toxicol. 2005;24:189–204.
14. Ashikaga T, Yoshida Y, Hirota M, Yoneyama K, Itagaki H, Sakaguchi H, Miyazawa M, Ito Y, Suzuki H, Toyoda H. Development of an in vitro skin sensitization test using human cell lines: the human Cell Line Activation Test (h-CLAT). I. Optimization of the h-CLAT protocol. Toxicol In Vitro. 2006;20:767–73.
15. Faulkner L, Martinsson K, Santoyo-Castelazo A, Cederbrant K, Schuppe-Koistinen I, Powell H, Tugwood J, Naisbitt DJ, Park BK. The development of in vitro culture methods to characterize primary T-cell responses to drugs. Toxicol Sci. 2012;127:150–8.
16. Python F, Goebel C, Aeby P. Assessment of the U937 cell line for the detection of contact allergens. Toxicol Appl Pharmacol. 2007;220:113–24.
17. Azam P, Peiffer JL, Chamousset D, Tissier MH, Bonnet PA, Vian L, Fabre I, Ourlin JC. The cytokine-dependent MUTZ-3 cell line as an in vitro model for the screening of contact sensitizers. Toxicol Appl Pharmacol. 2006;212:14–23.
18. Christiansen J, Farm G, Eid-Forest R, Anderson C, Cederbrant K, Hultman P. Interferon-gamma secreted from peripheral blood mononuclear cells as a possible diagnostic marker for allergic contact dermatitis to gold. Contact Dermatitis. 2006;55:101–12.
19. Dhabhar FS, Satoskar AR, Bluethmann H, David JR, McEwen BS. Stress-induced enhancement of skin immune function: a role for gamma interferon. Proc Natl Acad Sci U S A. 2000;97:2846–51.
20. Miyazawa M, Ito Y, Yoshida Y, Sakaguchi H, Suzuki H. Phenotypic alterations and cytokine production in THP-1 cells in response to allergens. Toxicol In Vitro. 2007;21:428–37.
21. Kimber I, Hilton J, Weisenberger C. The murine local lymph node assay for identification of contact allergens: a preliminary evaluation of in situ measurement of lymphocyte proliferation. Contact Dermatitis. 1989;21:215–20.
22. Zaghi D, Maibach HI. The local lymph node assay compared with the human maximization test as an indicator of allergic potency in humans using patch test clinic populations. Cutan Ocul Toxicol. 2009;28:61–4.
23. Ahmed SS, Wang XN, Fielding M, Kerry A, Dickinson II, Munuswamy R, Kimber I, Dickinson AM. An in vitro human skin test for assessing sensitization potential. J Appl Toxicol. 2016;36:669–84.
24. Kvistborg P, Boegh M, Pedersen AW, Claesson MH, Zocca MB. Fast generation of dendritic cells. Cell Immunol. 2009;260:56–62.
25. Kobayashi T, Ito T, Kawakamo H, Fuzishiro K, Hirano H, Okubo Y, Tsuboi R. Eighteen cases of wheat allergy and wheat-dependent exercise-induced urticarial/anaphylaxis sensitized by hydrolyzed wheat protein in soap. Int J Dermatol. 2015;54(8):302–5.
26. Lauriere M, Pecquet C, Bouchez-Mahiout I, Snegaroff J, Bayrou O, Raison-Peyron N, Vigan M. Hydrolysed wheat proteins present in cosmetics can induce immediate hypersensitivities. Contact Dermatitis. 2006;54(5):283–9.
27. Dickinson A, Sviland L, Jackson G, Dunn J, Stephens S, Proctor SJ. Monoclonal anti TNFα suppresses graft versus host disease reactions in an in vitro skin explant model. Cytokine. 1994;6(2):141–6.

28. Qesari M, Richter A, Ogonek J, Mischak-Weissinger E, Wang XN, Dickinson AM. Cytomegalovirus-Specific T Cells Isolated by IFN-gamma Secretion Assay Do Not Induce Significant Graft-Versus-Host Reactions In Vitro. Transplantation. 2016;100:2352–61.
29. Ogese MO, Ahmed SS, Alferivic A, Betts CJ, Dickinson AM, Failkner L, French N, Gibson A, Hirschfield GM, Kammuller M, Meng X, Martin SF, Musette P, Norris A, Pirmohamed M, Park BK, Purcell AW, Spraggs CF, Whritenour J, Naisbitt D. New approaches to investigate drug-induced hypersensitivity. Chem Res Toxicol. 2017;30:239–59.

The Human T Cell Priming Assay (hTCPA) 31

Philipp R. Esser and Stefan F. Martin

31.1 Introduction

Allergic contact dermatitis (ACD) as an occupational skin disease causes high socio-economic costs due to the lack of causative treatments [1]. Up to now, the only possibility for sensitized individuals is to symptomatically treat their inflamed skin with corticosteroids or non-steroidal anti-inflammatory drugs—and eventually to avoid contact with the causative chemical. This often necessitates a change in professions, making careful determination of the sensitizing potential of new chemicals mandatory. However, animal testing for the skin sensitizing potential of chemical compounds is considered unacceptable. The European Directive 86/609/EEC aims to reduce the number of animals used for research and other purposes and to promote alternatives to animal testing. In addition, the seventh amendment to the EU Cosmetics Directive prohibits animal testing for skin sensitization potential of chemicals including a complete marketing ban for products that contain ingredients tested on animals. Moreover, the Regulation on Registration, Evaluation, Authorisation and Restriction of Chemicals (REACH) also aims at a reduction of animal testing. Hence, there is an urgent need for the development of *in vitro* alternatives.

The previous gold standard for testing the skin sensitizing potential of chemicals was the local lymph node assay (LLNA) (OECD Test Guideline 429). This *in vivo* assay addresses the proliferation of cells in the lymph nodes of mice after repeated application of a test chemical onto their ear skin [2]. In order to develop *in vitro* alternatives to the LLNA, the immunological mechanisms underlying the ACD have to be considered, and key events of the sensitization process must be addressed

P.R. Esser • S.F. Martin (✉)
Allergy Research Group, Department of Dermatology, Medical Center-University of Freiburg, Faculty of Medicine, University of Freiburg, Freiburg, Germany
e-mail: philipp.esser@uniklinik-freiburg.de; stefan.martin@uniklinik-freiburg.de

C. Eskes et al. (eds.), *Alternatives for Dermal Toxicity Testing*,
DOI 10.1007/978-3-319-50353-0_31

in *in vitro* assays. Given the complexity of this process, the development of integrated testing strategies (ITS) that combine different assays is the current aim.

Immunologically, contact dermatitis can be of two different types. The irritant contact dermatitis (ICD) is an eczematous skin reaction previously attributed solely to toxic effects of chemicals. However, ICD seems to involve activation of the innate immune system. In contrast, ACD is defined by its activation of both the innate and the adaptive immune system. Here, the initial activation of innate immune responses is necessary to facilitate the activation and migration of skin dendritic cells to the draining lymph node and, subsequently, the efficient priming of antigen-specific T cells. As a classical type IV allergy, ACD results in erythema and eczema formation driven by the cytotoxic effects of different T cell subsets. While in the murine model of ACD—the contact hypersensitivity (CHS) model—CD8+ T cells seem to play a dominant role [3–6], IL-17 producing Th17/Tc17 cells can also be found. Additionally, Th22 cells can be detected during ACD reactions in the human system [7].

Several approaches have been proposed to set up a test system for the *in vitro* identification of the potential allergenicity and the allergenic potency of chemical compounds, mostly analysing various parameters depending on the activation of innate immune responses.

Over the last few years, it has become clear that most probably no single test will be sufficient to provide a standalone assay to determine the sensitizing potential of a chemical. Although in February 2015 the Direct Peptide Reactivity Assay (DPRA) and the ARE-Nrf2 Luciferase Test Method (e.g. Keratinosens™) and more recently the human Cell Line Activation Test (h-CLAT) have been adopted by the OECD as Test No. 442C, 422D and 442E, respectively, in all cases, the use of these assays as part of an IATA (integrated approach to testing and assessment, IATA) was recommended. To this end, the combination of assays addressing different mechanistic aspects of the ACD might prove to be most useful, and different approaches have been evaluated recently [8, 9].

31.2 The hTCPA Principle

We have developed a standard operating procedure (SOP) for a human T cell priming assay in close collaboration with the group of J.F. Nicolas and M. Vocanson in Lyon that enables the differentiation between contact sensitizers and irritants. For this assay, fresh human peripheral blood is initially separated into CD14+ monocytes and CD14- cells. While the CD14- cells are stored until 6–8 days, the monocytes are differentiated with GM-CSF and IL-4 over 5–7 days to immature monocyte-derived dendritic cells (MDDCs). Subsequently, they are incubated with the test chemical for 24 h at a concentration previously determined to result in a cytotoxicity of ~20%. To facilitate a full maturation of the MDDCs regardless of the chemical that is used, the TLR4 agonist LPS is added during this incubation step. On day 7, the CD14- fraction is further purified by depletion of CD25+/CD56+/CD45RO+ and non-T cells to contain only naive T cells. The unbound chemical is removed by washing, and the MDDCs are plated into 96 wells together with the autologous naive T cells. While the optimal MDDC/T cell ratio has to be determined for the specific setup, a ratio of 1:10 usually works best in our hands. After 2 days, IL-7 and IL-15 are added to the culture, after 4 days IL-2, IL-7 and IL-15 are added, and after 6 days the three cytokines are added once again in order to allow for a most efficient T cell priming and proliferation. After 9–10 days of this priming phase, the T cells are restimulated with autologous MDDCs that were either treated with the same chemical as before (antigen specific restimulation), left untreated (background control) or have been treated with an irrelevant chemical (specificity control). As a positive control, a fraction of the T cells can be stimulated by addition of PMA/Ionomycin. Depending on the readout, we usually analyse the T cell activation 6 h after the restimulation step in a multiparametric flow cytometry analysis and detection of intracellular cytokines like IFN-γ or TNF-α. However, other readouts/readout systems can easily be chosen like detection of the cytotoxic activity of the primed T cells via CD107a staining, proliferation via CFSE/Ki67 analysis as well as ELISPOT assays for cytokine and Granzyme B measurement.

31.3 Recent Efforts to Optimize Antigen Detection: Obstacles to Overcome

One of the problems arising with the set-up of a T cell-based assay is the limitation in cell numbers arising from the need to use primary autologous MDDCs and naive T cells from the same blood donor to avoid unspecific activation. This limitation makes a T cell-based assay at best a medium-throughput assay that cannot compete with an assay set-up, for example, with cell lines. However, we have been able to enhance the number of chemicals that can be tested in one approach due to two changes in the original protocol. One change is the use of CD19+ B cells as antigen-presenting cells for the restimulation step. Since the priming of the naive T cells can only be achieved with mature MDDCs, the use of B cells for the restimulation of already primed T cells allows saving the precious MDDCs for the initial step. In addition, the use of alternative readout methods like the ELISPOT to detect IFN-γ or TNF-α production instead of a FACS-based analysis allows reducing the necessary number of DC/T cells by about tenfold. First experiments comparing the efficiency of the ELISPOT-based detection of antigen-specific T cells with the FACS analysis showed comparable results. However, whether or not the important information that can be gained by the multiparametric FACS analysis and that is not provided by the ELISPOT is needed will have to be evaluated in the future.

Another important issue with human T cell priming assays is their ability to detect not only strong sensitizers but also moderate/weak sensitizers with a sufficient sensitivity. While historical protocols developed to analyse innate and adaptive immune responses *in vitro* used unfractionated human PBMC or murine lymph node cells in the presence or absence of antigens or hapten-loaded MDDCs as antigen-presenting cells, recent protocols have been improved regarding their cellular composition. One approach to enhance the efficiency of lymphocyte reactions was, for example, the use of specific antigen-presenting cell populations modified with haptens. This way the detection of sensitizers like oxazolone that gave no response in assays with unfractionated lymphocytes more closely reflected the *in vivo* situation where a strong ear swelling response was observed in the CHS model [10]. Especially work by Vocanson et al. and our own results have shown that the presence of CD25+ or CD56+ regulatory T cells hampers the antigen-specific priming of naive T cells. This effect was also described for the induction of antigen-specific responses to viral proteins [11] and is known to reduce the sensitivity of a T cell priming assay [12]. In addition, Vocanson et al. have recently reported that CD1a^{low} monocyte-derived dendritic cells inhibited T cell activation, while usage of CD1a^{high} MDDCs for the priming enhanced T cell activation [13].

Therefore, using advanced protocols where not only CD45RO+ memory T cells but also CD25+ and CD56+ cells are depleted [14, 15] in combination with a protocol using CD1a^{low} depleted MDDCs for priming should allow to significantly increase the sensitivity of future T cell priming assays.

31.4 Allergenic Potency Determination with the hTCPA: An Opportunity?

A big aim in terms of hazard identification and risk management is the set-up of an assay that not only allows differentiation between sensitizers and non-sensitizers but also enables the classification of contact sensitizers according to their potency as is possible in the LLNA via EC3 values. Several factors may be involved in the determination of the sensitizing potency *in vivo*. Empirically, the first step in determining whether or not a sensitizer will be strong or weak is its ability to cross the skin barrier and the depth of penetration into the skin that can be reached by the chemical. Another factor is the ability of the chemical to induce the generation of a pro-inflammatory cytokine milieu and the activation of an innate immune response. Without the activation and full maturation of DCs, there will be no priming of naive T cells in the draining lymph nodes. This is an effect particularly to be taken into account for weak allergens—as soon as these are combined with either irritants or mixed with other weak sensitizers, the sensitizing potency of the weak sensitizer is enhanced due to an increased inflammatory immune reaction [16–18]. Moreover, the efficiency of the activation of counter-regulatory mechanisms like the activation of ICOS+ regulatory T cells can determine the strength of the sensitizing potential of a given chemical [19].

In this respect, the hTCPA in its current form may not be the optimal assay for potency assessment. As mentioned before, efforts were made to set up an assay that is as sensitive as possible—i.e. all kinds of cells that may dampen the T cell priming were removed, thus possibly removing one of the factors determining allergenic potency. The activation of the MDDCs has been optimized by addition of pro-inflammatory factors like TNF-α or LPS resulting in a full activation of the DCs. In addition, a cytokine supplementation strategy has been worked out to allow for enhanced priming even with weak allergens. This seems to be possible without an increase in background T cell activation. However, all of these optimization steps that improve the sensitivity of the hTCPA remove the factors co-determining the potency of a contact sensitizer *in vivo*. This precludes potency assessment.

31.5 Conclusions

Although further optimisation and evaluation of the most sensitive readout for the human T cell priming assay are still ongoing, this assay may be of relevance as a third-line assay/final validation step in a IATA. The hTCPA addresses the crucial step in the sensitization to contact allergens, i.e. the priming of antigen-specific T cells. The major advantage of the hTCPA protocol is the testing for antigen specificity by an antigen-specific restimulation step—i.e. this assay provides the only protocol, where not only the activation of innate immune cells like the MDDCs but also the extreme specificity of the T cell receptor is taken into account. Chemicals that might provide false-positive results in other assays due to their inherent ability to activate innate immune reactions will not be able to elicit the priming and antigen-specific recognition by a T cell. This makes the hTCPA an assay, that—though not allowing to perform a high-throughput screening of substances—will allow to identify without any doubt the sensitizing potential of a limited number of crucial chemicals. In addition, a comparable protocol has already been successfully implemented to characterize primary T cell responses to drugs [20].

References

1. Peiser M, Tralau T, Heidler J, Api AM, Arts JHE, Basketter DA, English J, Diepgen TL, Fuhlbrigge RC, Gaspari AA, et al. Allergic contact dermatitis: epidemiology, molecular mechanisms, in vitro methods and regulatory aspects. Cell Mol Life Sci. 2011;69:763–81.
2. Kimber I, Dearman RJ, Basketter DA, Ryan CA, Gerberick GF. The local lymph node assay: past, present and future. Contact Dermatol. 2002;47:315–28.
3. Kehren J, Desvignes C, Krasteva M, Ducluzeau MT, Assossou O, Horand F, Hahne M, Kägi D, Kaiserlian D, Nicolas JF. Cytotoxicity is mandatory for CD8(+) T cell-mediated contact hypersensitivity. J Exp Med. 1999;189:779–86.
4. Martin S, Lappin MB, Kohler J, Delattre V, Leicht C, Preckel T, Simon JC, Weltzien HU. Peptide immunization indicates that CD8+ T cells are the dominant effector cells in trinitrophenyl-specific contact hypersensitivity. J Invest Dermatol. 2000;115:260–6.
5. Martin SF, et al. T-cell recognition of chemicals, protein allergens and drugs: towards the development of in vitro assays. Cell Mol Life Sci. 2010;67:4171–84.
6. Vocanson M, Hennino A, Rozieres A, Cluzel-Tailhardat M, Poyet G, Valeyrie M, Benetiere J, Tedone R, Kaiserlian D, Nicolas J-F. Skin exposure to weak and moderate contact allergens induces IFN[gamma] production by lymph node cells of CD4+ T-cell-depleted mice. J Invest Dermatol. 2008;129:1185–91.
7. Cavani A, Pennino D, Eyerich K. Th17 and Th22 in skin allergy. In: Ring J, Darsow U, Behrendt H, editors. Chemical immunology and allergy. KARGER: Basel; 2012. p. 39–44.
8. Reisinger K, Hoffmann S, Alépée N, Ashikaga T, Barroso J, Elcombe C, Gellatly N, Galbiati V, Gibbs S, Groux H, et al. Systematic evaluation of non-animal test methods for skin sensitisation safety assessment. Toxicol In Vitro. 2015;29:259–70.

9. Van der Veen JW, Rorije E, Emter R, Natsch A, van Loveren H, Ezendam J. Evaluating the performance of integrated approaches for hazard identification of skin sensitizing chemicals. Regul Toxicol Pharmacol. 2014;69:371–9.
10. Robinson MK. Optimization of an in vitro lymphocyte blastogenesis assay for predictive assessment of immunologic responsiveness to contact sensitizers. J Invest Dermatol. 1989;92:860–7.
11. Mattarollo SR, Yong M, Gosmann C, Choyce A, Chan D, Leggatt GR, Frazer IH. NKT cells inhibit antigen-specific effector CD8 T cell induction to skin viral proteins. J Immunol. 2011;187:1601–8.
12. Goubier A, Vocanson M, Macari C, Poyet G, Herbelin A, Nicolas J-F, Dubois B, Kaiserlian D. Invariant NKT cells suppress CD8+ T-cell-mediated allergic contact dermatitis independently of regulatory CD4+ T cells. J Invest Dermatol. 2013;133:980–7.
13. Vocanson M, Achachi A, Mutez V, Cluzel-Tailhardat M, Varlet BL, Rozières A, Fournier P, Nicolas J-F. Human T cell priming assay: depletion of peripheral blood lymphocytes in CD25(+) cells improves the in vitro detection of weak allergen-specific T cells. EXS. 2014;104:89–100.
14. Dietz L, Esser PR, Schmucker SS, Goette I, Richter A, Schnölzer M, Martin SF, Thierse H-J. Tracking human contact allergens: from mass spectrometric identification of peptide-bound reactive small chemicals to chemical-specific naive human T-cell priming. Toxicol Sci. 2010;117:336–47.
15. Richter A, Schmucker SS, Esser PR, Traska V, Weber V, Dietz L, Thierse H-J, Pennino D, Cavani A, Martin SF. Human T cell priming assay (hTCPA) for the identification of contact allergens based on naive T cells and DC–IFN-γ and TNF-α readout. Toxicol In Vitro. 2013;27:1180–5.
16. Bonefeld CM, Nielsen MM, Rubin IMC, Vennegaard MT, Dabelsteen S, Gimenéz-Arnau E, Lepoittevin J-P, Geisler C, Johansen JD. Enhanced sensitization and elicitation responses caused by mixtures of common fragrance allergens. Contact Dermatitis. 2011;65:336–42.
17. Johansen JD, Skov L, Volund A, Andersen K, Menné TJ. Allergens in combination have a synergistic effect on the elicitation response: a study of fragrance-sensitized individuals. Br J Dermatol. 1998;139:264–70.
18. De Jong WH, Tentij M, Spiekstra SW, Vandebriel RJ, Van Loveren H. Determination of the sensitising activity of the rubber contact sensitisers TMTD, ZDMC, MBT and DEA in a modified local lymph node assay and the effect of sodium dodecyl sulfate pretreatment on local lymph node responses. Toxicology. 2002;176:123–34.
19. Vocanson M, Rozieres A, Hennino A, Poyet G, Gaillard V, Renaudineau S, Achachi A, Benetiere J, Kaiserlian D, Dubois B, et al. Inducible costimulator (ICOS) is a marker for highly suppressive antigen-specific T cells sharing features of TH17/TH1 and regulatory T cells. J Allergy Clin Immunol. 2010;126:280–289.e7.
20. Faulkner L, Martinsson K, Santoyo-Castelazo A, Cederbrant K, Schuppe-Koistinen I, Powell H, Tugwood J, Naisbitt DJ, Park BK. The development of in vitro culture methods to characterize primary T-cell responses to drugs. Toxicol Sci. 2012;127:150–8.

Promising Test Systems Beyond the Current Status 32

Erwin L. Roggen

32.1 Introduction

The new technologies described in the previous sections are to facilitate animal-free toxicity testing in the area of skin sensitization. While the presented methods are being evaluated, refined for industrial and regulatory applicability, and build into approaches for testing and assessment of substances, new methods or variants of existing methods addressing specific issues of the parent method are emerging.

32.2 Emerging Tools for Animal-Free Assessment of Skin Sensitization: A Non-Exhaustive List

32.2.1 Bioavailability: The Chemical has the Capability of Reach the Viable Cell Layers

With the methodologies available, efforts should now go to the development and implementation of methods for quantification of compound disposition in the skin, to obtain information on kinetics, potential tissue bioaccumulation, and actual exposure at cellular level. Currently only a very limited number of readouts such as colorimetric assays or histology are used for the analysis of tissue models in toxicology. However, several new developments have the potential of providing more detailed information about the fate of a chemical following exposure of, e.g., the skin barrier. Here, three such examples are described.

E.L. Roggen
3Rs Management and Consulting ApS, Kongens Lyngby, Denmark
e-mail: 3rsmc.eu@gmail.com

C. Eskes et al. (eds.), *Alternatives for Dermal Toxicity Testing*,
DOI 10.1007/978-3-319-50353-0_32

32.2.1.1 Impedance Spectroscopy

Groeber et al. [1] applied a nondestructive technology to analyze the epithelial barriers based on impedance spectroscopy. Epithelial tissues such as skin are impermeable to electrical charged particles and thus show a significant resistance. This resistance is usually employed in testing strategies as the transepithelial electrical resistance (TEER) value. Although impedance spectroscopy is using a similar approach, here a full impedance spectrum is determined and used to gain the specific electrical characteristics such as the resistance and the capacitance of the biological components only. Reconstructed human epidermis (RHE) shows characteristic impedance spectra in a frequency range between 1 Hz and 100 kHz, which are comparable to the spectra of isolated human epidermal samples. From these spectra, electrical parameters of the RHEs such as the capacity and the resistance are extracted. These parameters change significantly during the epidermal differentiation and are useful for quantification of the effects of a mechanical and chemical disruption of the epidermal integrity. Impedance spectroscopy shows a sufficient sensitivity to detect a transient decreased resistance caused by 2-propanol, which is classified as a non-irritant by MTT assays. Furthermore, the method can detect in which layer a substance is affecting an electrical barrier.

32.2.1.2 Raman Spectroscopy

To allow the use in downstream applications, non-invasive technologies for biological sample characterization are highly relevant for biomedical research [2]. It was demonstrated that living cells can be monitored by non-invasive Raman spectroscopy, which is based on inelastic scattering of photons at molecules. Raman spectra can be collected using the BioRam® system (CellTool GmbH) with a nondestructive 785 nm diode laser. This system combines Raman spectroscopy with digital microscopy. All molecules within the laser focus contribute to a sum spectrum resulting in a unique biochemical fingerprint. Spectral data are analyzed using multivariate data analysis. Published data show that Raman spectroscopy is a suitable method to investigate if vital primary cells obtained from surgical specimen are suitable for 3D test system generation or should be discarded due to tumor-specific characteristics.

32.2.1.3 High-Content Imaging (HCI)

HCI platforms are alternative fluorescent imaging techniques that have evolved to better predict toxicological effects through *in vitro* assays and have in some instances been shown to predict toxicity with a high accuracy [3]. This is in contrast to, e.g., cytotoxicity assays that often are criticized for a low concordance with human toxic effects, particular in view of potential multifactorial effects of a single compound. HCI was developed to measure multiple endpoints in one assay, thus more accurately predicting complex and multifactorial mechanisms of toxicity [4].

32.2.2 Haptenation: The Chemical Reacts Covalently with a "Carrier Protein"

In vitro studies suggest that the specificity of the covalent modification is time and dose dependent and that the target proteins *in vitro* become more general and less discriminative over time and with increasing concentrations of the chemical [5]. By applying modern proteomic technologies together with a target peptide containing all amino acids, "allergen-protein interaction assay (APIA) intends to address these issues by profiling all amino acid-specific allergen–peptide interactions" [6]. Moreover, potentially crucial allergen-specific Cys-modifications can be qualitatively monitored by mass spectrometry and confirmed by a dual peptide approach. Assay conditions chosen mimic the distinct human epidermal reactivity compartments of the skin surface (pH 5.5), stratum basale (pH 6.8), and typical physiological conditions (pH 7.4). An extreme as well as a moderate human contact sensitizer produced Cys-specific mass shifts, whereas a skin irritant did not. While this approach is an important step toward a better understanding of haptenation of proteins, further experimentation is required to substantiate the promising results.

Several variants of the DPRA have emerged, each addressing specific issues related to the technology. It was observed that peptide depletion as measured in the DPRA may be due to either adduct formation or peptide oxidation. Natsch and Gfeller [7] modified the assay to generate a more detailed characterization of the reactivity of a molecule by simultaneous determination of peptide depletion, peptide oxidation (dimerization), adduct formation, and thiol reactivity. Highly reactive molecules are further discriminated with a kinetic measure. The assay was validated on 80 chemicals. The majority of the sensitizers formed detectable peptide adducts, but many sensitizers were also able to catalyze peptide oxidation. Whereas adduct formation was only observed for sensitizers, this oxidation reaction was also observed for two non-sensitizing fragrance aldehydes, indicating that peptide depletion might not always be regarded as sufficient evidence for rating a chemical as a sensitizer.

Yamamoto et al. (2015) further improved the applicability of their amino acid derivative reactivity assay (ADRA) making it possible to assess reactivity at a 100-fold lower concentration of the test substance [8]. The predictive accuracy of the "ADRA-dilutional method" (ADRA-DM) for skin sensitization (90%) was higher than for the ADRA and DPRA. Furthermore, the ADRA-DM does not have the restrictions on test compound solubility which is considered a major problem with the DPRA.

The glutathione depletion assay builds upon the understanding that glutathione (GSH) is the most prominent antioxidant in cells and the cofactor of an important set of enzymes involved in the skin metabolic clearance system, glutathione *S*-transferases (GST) [9]. Jacquoilleot et al. [10] described an LC–MS (liquid chromatography–mass spectroscopy) method to measure GSH and its disulfide form (GSSG) in HaCaT cells and a 3D reconstructed human epidermis (RHE) model. In this assay, the level of GSH in HaCaT cells treated with a single nontoxic dose of dinitrochlorobenzene was shown to increase. Cells treated with 1-chloro-2,4-dinitrobenzene (DNCB) and 1-fluoro-2,4-dinitrobenzene (DNFB) repleted GSH to levels similar to untreated control cells within 24 h, 1-bromo-2,4-dinitrobenzene (DNBB) seemed to prevent such a repletion and appeared to be the most toxic compound in all assays. These data show that GSH depletion and repletion occur rapidly in skin cells and emphasize the importance of conducting kinetic studies when performing *in vitro* experiments exploring skin sensitization. Activation of the nuclear factor E2-related factor 2 (Nrf2) pathway was observed with all compounds within 2 h, and at concentrations less than 10 μM.

32.2.3 Inflammation: Innate Recognition Followed by Activation of Innate Immunity

As yet, it is not properly understood which subtle balance between danger signals and/or intracellular interactions are decisive with respect to initiating Th-1 (e.g., in contact dermatitis)- or Th-2-mediated responses (e.g. in asthma) mediated immune phenotypes. One way of acquiring mechanistic information is by application of -omics approaches. The genomic allergen rapid detection (GARD) test provides information about the mechanisms driving sensitization and potency in the dendritic cell (DC)-like MUTZ-3 cell. Similar mechanistic information is now being acquired now for also for keratinocytes.

Assessing chemical sensitizers using a specific gene signature: Saito et al. [11] developed an *in vitro* epidermal sensitization assay (EpiSensA) using reconstructed human epidermis, RhE model, which is expected to have broader applicability domain rather than existing *in vitro* assays. Microarray analysis identified five genes related to cellular stress responses being upregulated after treatment with skin sensitizers and not with a non-sensitizer. When assessed for predictivity using a small number of substances, the applied genes exhibited a high predictive accuracy. All tested pre-/pro-haptens were correctly predicted by both ATF3 and DNAJB4. These preliminary results suggested that the EpiSensA has the potential of becoming a new tool for assessing skin sensitization. The analysis of chemical-induced changes in gene expression by the HaCaT human keratinocyte cell revealed ten genes that accurately discriminated sensitizers and non-sensitizers, including irritants. An algorithm was developed to compare changes in gene regulation of chemicals of unknown class to that induced by chemicals of known class. A chemical was assigned the most predominant class indicated by these algorithms [12].

32.2.4 Dendritic Cell Activation: From Innate Responses to Dendritic Cell (DC) Maturation

Most of the advanced methods for assessing DC activation rely on detection of membrane markers describing phenotypic changes related to DC activation and maturation. In addition to phenotypic markers (e.g., CD54, CD86), interleukin (IL)-8 has been suggested as an activation-related marker that can discriminate skin sensitizers and non-sensitizers [13–15]. Inspired by these observations, Takahashi et al. [16] established a stable THP-1-derived IL-8 reporter cell line (THP-G8) which was capable of discriminating most of the tested sensitizers and non-sensitizers. Interestingly, pretreatment with N-acetylcysteine suppressed the increase triggered by the sensitizers suggesting Cys–Cys interaction as the driving force behind cell activation.

32.2.5 Dendritic Cell Migration: Translating the Message into Specific Actions

The molecular mechanisms driving migration of DC to and from peripheral tissues were reviewed [17]. Fibroblasts play a key role both as advisors helping the KCs and Langerhans cells (LCs) to discriminate irritants from sensitizers, which in many cases are irritants themselves, and as guides helping the LCs out of the epidermis into the dermis and further toward lymphatic vessels [18]. Using a full-thickness tissue-engineered skin model containing fully functional MUTZ-3-derived LCs (MUTZ-LC), the MUTZ-LCs were demonstrated to mature and to acquire the ability to migrate toward C–X–C motif ligand (CXCL)12 and C–C motif ligand (CCL)19/21 in a comparable manner with primary LCs in skin explants [19].

The acquired knowledge has resulted in a DC-migration assay which is based on carboxyfluorescein succinimidyl ester (CFSE)-labeled MUTZ-3 cells. The discriminating feature of the assay is that irritant-induced migration is CCL5 dependent, while sensitizer-induced migration is CXCL12 dependent. The readout of the test is the ratio between migration toward CXCL12 or to CCL5 [20].

While the preliminary data on 12 chemicals are promising (no misclassification), further evaluation performed with more chemicals is required. The test is also expensive and rather complicated which may hamper its application by industry. More work is required to refine the test to make it more attractive for industrial use.

32.3 Concluding Summary

While the most mature testing methods are currently evaluated and/or validated by industry and regulatory authorities, several of these methods are already improved and refined to better meet industrial and regulatory applicability. In addition, some new approaches are emerging. To what extent these new and modified tools represent an added value to the currently emerging approaches for testing and assessment remains to be substantiated though.

References

1. Groeber F, Engelhardt L, Egger S, Werthmann H, Monaghan M, Walles H, Hansmann J. Impedance spectroscopy for the non-desctructive evaluation of in vitro epidermal models. Pharm Res. 2014;32(5):1845–54. doi:10.1007/s11095-014-1580-3.
2. Brauchle E, Schenke-Layland K. Raman spectroscopy in biomedicine – non-invasive in vitro analysis of cells and extracelluilar matrix components in tissues. Biotechnol J. 2013;8(3):288–97. doi:10.1002/biot.201200163.
3. O'Brien PJ, Irwin W, Diaz D, Howard-Cofield E, Krejsa CM, Slaughter MR, Gao B, Kaludercic N, Angeline A, Bernardi P, Brain P, Hougham C. High concordance of drug-induced human hepatotoxicity with in vitro cytotoxicity measured in a novel cell-based model using high content screening. Arch Toxicol. 2006;80(9):580–604.
4. Tolosa L, Pinto S, Donato MT, Lahoz A, Castell JV, O'Connor JE, Gómez-Lechón MJ. Development of a multiparametric cell-based protocol to screen and classify the hepatoxicity potential of drugs. Toxicol Sci. 2012;127:187–98.
5. Manchanda T, Hess D, Dale L, Ferguson SG, Rieder MJ. Haptenation of sulfonamide reactive metabolites to cellular proteins. Mol Pharmacol. 2002;62:1011–26.
6. Dietz L, Kinzebach S, Ohnesorge S, Franke B, Goette I, Koenig-Gressel D, et al. Proteomic allergen-peptide/protein interaction assay for the identification of human skin sensitizers. Toxicol In Vitro. 2013;27:1157–62. doi:10.1016/j.tiv.2012.08.013.
7. Natsch A, Gfeller H. LC-MS-based characterization of the peptide reactivity of chemicals to improve the in vitro prediction of the skin sensitization potential. Toxicol Sci. 2008; 106(2):464–78.

8. Yamamoto Y, Tahara H, Usami R, Kasahara T, Jimbo Y, Hioki T, Fujita M. A novel in chemico method to detect skin sensitizers in highly diluted reaction conditions. J Appl Toxicol. 2015;35:1348–60.
9. Aptula AO, Patlewicz G, Roberts DW, Schultz TW. Non-enzymatic glutathione reactivity and in vitro toxicity: a non-animal approach to skin sensitization. Toxicol In Vitro. 2006;20:239–47.
10. Jacquoilleot S, Sheffield D, Olayanju A, Sison-Young R, Kitteringham NR, Naisbitt DJ, Aleksic M. Glutathione metabolism in the HaCaT cell line as a model for the detoxification of the model sensitizers 2,4-dinitrohalobenzenes in human skin. Toxicol Lett. 2015;237:11–20.
11. Saito K, Nukada Y, Takenouchi O, Miyazawa M, Sakaguchi H, Nishiyama N. Development of a new in vitro skin sensitization assay (epidermal sensitization assay; EpiSensA) using reconstructed human epidermis. Toxicol In Vitro. 2013;27:2213–24.
12. van der Veen JW, Pronk TE, van Loveren H, Ezendam J. Applicability of a keratinocyte gene signature to predict skin sensitizing potential. Toxicol In Vitro. 2013;27:314–22.
13. Nukada Y, Miyazawa M, Kosaka N, Ito Y, Sakaguchi H, Nishiyama N. Production of IL-8 in THP-1 cells following contact allergen stimulation via mitogen-activated protein kinase activation or tumor necrosis factor-alpha production. J Toxicol Sci. 2008;33:175–85.
14. Python F, Goebel C, Aeby P. Assessment of the U937 cell line for the detection of contact allergens. Toxicol Appl Pharmacol. 2007;220:113–24.
15. Toebak MJ, Gibbs S, Bruynzeel DP, Scheper RJ, Rustemeyer T. Dendritic cells: biology of the skin. Contact Dermatitis. 2009;60:2–20.
16. Takahashi T, Kimura Y, Saito R, Nakajima Y, Ohmiya Y, Yamasaki K, Aiba S. An in vitro test to screen skin sensitizers using a stable THP-1-derived IL-8 reporter cell line, THP-G8. Toxicol Sci. 2011;124(2):359–69.
17. Villablanca EJ, Russo V, Mora JR. Dendritic cell migration and lymphocyte homing imprinting. Histol Histopathol. 2008;23:897–910.
18. Ouwehand K, Scheper RJ, de Gruijl TD, Gibbs S. Epidermis-todermis migration of immature Langerhans cells upon topical irritant exposure is dependent on CCL2 and CCL5. Eur J Immunol. 2012;40:2026–34.
19. Ouwehand K, Spiekstra SW, Waaijman T, Scheper RJ, de Gruijl TD, Gibbs S. Langerhans cells derived from a human cell line in a full-thickness skin equivalent undergo allergen-induced maturation and migration. Tech Adv. 2011;90:1028–33.
20. Rees B, Spiekstra SW, Carfi M, Ouwehand K, Williams CA, Corsini E, McLeod JD, Gibbs S. Inter-laboratory study of the in vitro dendritic cell migration assay for identification of contact allergens. Toxicol In Vitro. 2011;25:2124–34.

Part IV

UV-Induced Effects (Phototoxicity and Photoallergy)

33 Overview on the Current Status of Available Test Methods and Additional Promising Methods for Assessing UV-Induced Effects

Lorena Rigo Gaspar, Camila Martins Kawakami, and Carolina Gomes Benevenuto

33.1 Photosafety Requirements

The assessment of photo-induced toxicity is required when chemical preparations are expected or intended to be used on sunlight-exposed skin, i.e., dermal drugs or other products topically applied to the skin such as sunscreens or multifunctional cosmetic formulations. However, photo-induced toxicity of systemic exposure to some drugs has also been reported [1].

Photo-induced toxicity depends on the chemical concentration in the skin at the time of light exposure, which in turn depends on a variety of factors such as bioavailability. In systemic exposure, the factors are plasma concentration, perfusion of the tissue, partitioning from vascular to interstitial and cellular compartments, and binding, retention, and accumulation of the chemical in the tissue. The duration of exposure depends on clearance rates as reflected by half-lives in plasma and tissue. Taken together, these parameters define the mean residence time of the photoreactive chemical in tissue [1].

A cosmetic chemical or a pharmaceutical should be tested for photo-induced toxicity when it significantly absorbs ultraviolet/visible (UV/VIS) radiation in the 290–700 nm range, promotes reactive oxygen species (ROS) generation, and has high skin penetration or partition into the skin or eyes when applied orally or intravenously. Since the vehicle influences skin penetration, it also represents an important parameter to take into account in photosafety evaluation. Furthermore, if the chemical has low photostability or contains phototoxic impurities, it should also be tested for photo-induced toxicity [1].

L.R. Gaspar (✉) • C.M. Kawakami • C.G. Benevenuto
School of Pharmaceutical Sciences of Ribeirão Preto, University of São Paulo, Ribeirão Preto, São Paulo, Brazil
e-mail: lorena@fcfrp.usp.br; camilakawakami@usp.br; carolbenevenuto@gmail.com

C. Eskes et al. (eds.), *Alternatives for Dermal Toxicity Testing*,
DOI 10.1007/978-3-319-50353-0_33

Photo-induced toxicity is based on the photosensitization process as described below, on the light, which is normally UV/VIS radiation, and on the chemical that represents the photosensitizer.

33.2 Photosensitization

The photosensitization process is observed when a photosensitizer, i.e., a molecule that interacts with a nontoxic dose of UV/VIS radiation (chromophore), can induce some effects in a substrate. When this substrate is the skin, photosensitization can induce three main processes: phototoxicity, photoallergy, and photogenotoxicity. This chapter addresses only phototoxicity and photoallergy effects, since photogenotoxicity assays were considered oversensitive and are not currently considered useful for clinical relevance of human pharmaceuticals [1].

33.2.1 Photosensitizers

There are some known topical photosensitizers used in cosmetics, such as UV filters, some fragrances (bergamot oil and musk ambrette), as well as the antimicrobial agent triclosan and the nonsteroidal anti-inflammatory agent ketoprofen. Among the systemic photosensitizers, there are some antibiotics such as sulfonamides and neuroleptics such as chlorpromazine and promethazine [2].

33.2.2 Mechanisms Involved in Photosensitization

The photosensitizer absorbs a photon of UV/VIS radiation and reaches the excited state.

Photosensitizers have two systems of electronically excited states, the singlet and triplet states. The singlet state is usually short-lived but may undergo intersystem crossing to a longer lifetime triplet state. The triplet state of the photosensitizer can subsequently react via two major pathways: (a) by electron or hydrogen transfer (free radical) processes (Type I reaction), which may or may not require oxygen and can generate superoxide, hydrogen peroxide, and hydroxyl radicals, or (b) by energy transfer (typically) to oxygen (Type II reaction), to form excited-state singlet oxygen.

The relative contributions of the Type I and Type II processes depend on the features from the photosensitizer and from the substrate, on the reaction conditions (e.g., solvent, pH, concentrations of photosensitizer, substrate, and oxygen) and, in some cases, on whether the photosensitizer absorbs light into its first or second absorption band.

Following excitation, photosensitizers may also undergo photolysis/photodegradation, which result in photoproducts that may be toxic. Photodegradation reactions can be present in both singlet and triplet excited states.

The determination of the UV/VIS absorption spectrum should precede any testing for photosensitizing properties in biological systems *in vitro* or *in vivo*, since chemicals with absorption in the 300–750 nm range usually reach the singlet excited state, which is an essential feature for photosensitizers. The chemical may undergo intersystem crossing to the longer lifetime triplet state, which can lead to the photosensitized oxidations. Therefore, effective photosensitizers are usually those providing a high absorption in the 300–750 nm range and also high yield of a long-lived triplet state [3].

33.3 Phototoxicity

Phototoxicity represents an acute toxic response observed after a first exposure of, e.g., skin to a photosensitizer, with subsequent exposure to light (commonly UV/VIS radiation). Phototoxicity can also occur following systemic administration of the photosensitizer.

33.3.1 Clinical Signs of Phototoxicity

The signs of phototoxicity are usually related to exaggerated sunburn, and it takes minutes to hours after UV exposure for the reactions to occur [2, 4].

33.3.2 Recommended Tiered Strategy for *In Vitro* Phototoxicity Evaluation

There is a recommended tiered strategy for phototoxicity evaluation in order to maximize the potential of each test and to reduce needless evaluations. The strategy includes both nonbiological and biological assays.

33.3.2.1 Nonbiological Assays

An UV/VIS absorption spectrum of the test chemical must be determined before biological testing is considered, since a test chemical with a molar extinction coefficient of less than 1000 L mol^{-1} cm^{-1} is unlikely to be photoreactive.

Since generation of ROS after UV/VIS irradiation is the most important mechanism for chemically induced phototoxicity, the ROS assay was recently included in this tiered strategy for the determination of the phototoxic potential of chemicals [5–7].

In this photosafety strategy, UV/VIS spectral analysis should be employed as the first screening to indicate a potential phototoxic risk by measuring the molar extinction coefficient (MEC). Next, as a second step, the ROS assay is recommended to assess the photoreactivity of the chemical. Chemicals that are non-photoreactive in the ROS assay would not require further testing since a negative result indicates a very low probability of phototoxicity. On the other hand, for photoreactive, weakly photoreactive, and inconclusive results, it is generally recommended, as the next step of testing, the use of an *in vitro* test system such as the 3T3 neutral red uptake phototoxicity test (3T3 NRU-PT), which is a monolayer fibroblast culture test [1, 5, 6].

33.3.2.1.1 Calculation of the Molar Extinction Coefficient

According to Lambert–Beer law, absorption is related to the molar extinction coefficient, the length of solution the light passes through (cm), and the concentration of the solution in mol/L. The solutions of each chemical should be analyzed by UV spectrophotometry and used in the following formula to calculate the molar extinction coefficient:

$$\varepsilon = \frac{A}{l\,c}$$

where

A = absorbance

l = length of solution the light passes through

c = concentration of the chemical

If the result shows a molar extinction coefficient higher than 1000 L mol^{-1} cm^{-1}, [7], the next step should be performed, as recommended by the European Authority [8].

33.3.2.1.2 ROS Assay to Examine the Photoreactivity of Chemicals

The ROS assay was designed for the identification of ROS produced by photoreactive chemicals after exposure to UV/VIS light based on the two following major pathways, electron/hydrogen transfer (Type I reaction) that produces superoxide anion (SA) and energy transfer to oxygen (Type II reaction) that produces singlet oxygen (SO) [1, 5, 6].

In the ROS assay, it is important to ensure that irradiation conditions satisfy the recommended criteria, a goal that can be met by using proficiency chemicals (positive and negative controls). SO generation is measured by spectrophotometric monitoring of *p*-nitrosodimethyl aniline (RNO) bleaching, followed by decreased RNO absorbance at 440 nm. In this reaction, the imidazole ring captures the generated SO, resulting in the formation of a trans-annular peroxide intermediate, which is capable of inducing the bleaching of RNO, as follows [5]:

$$SO + \text{imidazole} \rightarrow [\text{peroxide intermediate}] \rightarrow \text{ oxidized imidazole}$$
$$[\text{peroxide intermediate}] + RNO \rightarrow RNO + \text{products}$$

On the other hand, SA generation is determined by the observation of the reduction of nitroblue tetrazolium (NBT) at 560 nm by SA via a one-electron transfer reaction, yielding partially reduced (2 e^-) monoformazan (NBT+) as a stable intermediate, as follows:

$$SA + NBT \rightarrow O_2 + NBT^+$$

As demonstrated by validation studies, all phototoxic reference chemicals that produced conclusive results were identified as photoreactive, resulting in a sensitivity of 100% for predicting phototoxicants and a false-negative rate of 0%. However, the test can result in some false positives [5, 9].

33.3.2.2 Biological Assays

The 3T3 NRU-PT is the first step of the biological assays conducted to evaluate the phototoxic potential of a test chemical. The 3T3 NRU-PT is considered a standalone test for negative results due to its high sensitivity (100%) for the identification of absence of phototoxic potential [10].

If a positive result is obtained, a follow-up testing should be performed to obtain data with models that better reflect the human situation, such as 3D skin models. The human 3D reconstructed skin model phototoxicity test (H3D-PT) contains a barrier system (*stratum corneum*), which takes into account the bioavailability of the test chemical for the verification of positive results obtained with the 3T3 NRU-PT. These 3D models involve a skin barrier similar to *in vivo* human skin, with metabolically competent skin cells. This method is also a good choice for test chemicals having limited solubility (lipophilic) and for testing formulations such as dermal drugs and cosmetic preparations [1, 11, 12].

Finally, the third step is the confirmatory photopatch test to be applied to a group of volunteers and to be performed using the first non-phototoxic concentration determined in the H3D-PT [1, 11].

33.3.2.2.1 3T3 NRU Phototoxicity Test

Rationale

The *in vitro* 3T3 NRU phototoxicity test is based on a comparison of the cytotoxicity of a chemical when tested in the presence and in the absence of exposure to a non-cytotoxic dose of simulated solar light.

Protocol

The phototoxicity test should be performed according to the test guideline 432 of the Organisation for Economic Cooperation and Development (OECD TG 432) and based on the INVITTOX protocol 78 [10, 13]. The 3T3 fibroblasts are dispensed in two 96-well plates and incubated for 24 h (7.5% CO_2, 37 °C). Plates are washed with 150 μL DPBS (Dulbecco's phosphate-buffered saline), and eight different concentrations of the test chemicals are applied in sextuplicate to the 96-well plates. After 1-h incubation with the test chemicals (7.5% CO_2, 37 °C), the ultraviolet A (+UVA) plate is irradiated (total dose = 5 J cm^{-2}, usually 1.7 mW cm^{-2} for 50 min) with UVA radiation, and the -UVA plate is kept in a dark box. Plates are washed with 150 μL DPBS and incubated (7.5% CO_2, 37 °C).

The day after, plates are washed with 150 μL DPBS and neutral red medium is added to each well. After 3 h, plates are washed and a desorb solution (ethanol/acetic acid/water) is added to each well. The +UVA and -UVA plates are analyzed with a microtiter plate reader at 540 nm. The data are analyzed using the Phototox

Software version 2.0 for the evaluation of phototoxic potential. This software is freely available at the OECD website.

The software calculates the mean photo effect (MPE) and the photoirritation factor (PIF). The MPE is a statistical comparison of dose–response curves obtained with (+UV) and without (−UV) UV irradiation and test substances. The PIF is a factor generated by comparing two equally effective cytotoxic concentrations (IC_{50}) of the test chemical obtained in the absence (−UV) and in the presence (+UV) of a non-cytotoxic irradiation with UVA/VIS [10].

According to the OECD TG 432, the test chemical is predicted as phototoxic if the MPE is higher than 0.15 or if a PIF is higher than 5. A prediction as "probably phototoxic" is obtained if $0.1 < MPE < 0.15$ or if $2 < PIF < 5$. Finally, a $MPE < 0.1$ and a $PIF < 2$ predicts non-phototoxic effects. Use of MPEs is preferable when obtaining unusual dose–response curves and should be used instead of PIF if no IC_{50} values can be derived [10].

Quality Control

In order to check the quality of 3T3, the following parameters should be controlled. At 5 J cm^{-2}, the cell viability of +UVA plates should be at least 80% of that of non-irradiated cells.

Some MPE and PIF ranges for reference chemicals (positive and negative controls) are reported in OECD TG 432. These substances should be tested prior to the testing of test chemicals, and at least one positive control should be included in each batch of experiments to guarantee the reliability of the results.

The UV sensitivity of cells should be regularly tested. Both xenon arcs and (doped) mercury-metal halide light sources are used. However, all solar simulators should be suitably filtered to attenuate highly cytotoxic ultraviolet B (UVB) wavelengths [10].

Limitations

One of the limitations of the test is that only hydrophilic substances can be analyzed; however, solubility is not considered to be a relevant limitation since most phototoxins are positive in the 3T3 NRU-PT at low concentrations (0.01–10 μg/mL). For chemicals with topical exposure (e.g., cosmetic ingredients), the H3D-PT (INVITTOX Protocol 121) could be used in a second tier testing in order to account for biokinetics/bioavailability which are poorly modeled in the 3T3 NRU-PT. Although the method does not allow the identification of the exact phototoxic mechanism, it has the advantage of potentially detecting photogenotoxins and photoallergens [9, 14, 15].

One drawback of 3T3 NRU-PT is the detection of UVB photosensitizers. The major UVB phototoxins are detected in the assay since, even with the use of the OECD TG 432 recommended filter to attenuate UVB radiation, low amounts of UVB radiation can still reach the cells. In addition, the clinical impact of UVB absorbers is significantly lower than that of UVA absorbers, especially for systemic drugs. Alternatively, *in vitro* skin models and the photo red blood cells assay (RBC-PT), which better tolerate UVB, could be considered [9, 15].

Recommendations

The prediction model of this test is based on the OECD TG 432 and is described in the INVITTOX protocol 78 [10, 13]. Despite its relevance and robustness, it must be followed without deviations including no use of different cell lines instead of Balb/c 3T3 clone 31 fibroblasts, no changes in the number of cells/well, incubation period, UV source, irradiation time, and the conditions of UV exposure. The vital dye cannot be altered, i.e., the tetrazolium dye (3-(4,5-dimethylthiazol-2-yl)-2,5-diphenyltetrazolium bromide – MTT) should not be used instead of neutral red, since some substances have the ability to reduce MTT, directly leading to false predictions.

The plates should be irradiated with the lid on, and consequently the UV irradiance should be measured with the lid on. The plates should be ventilated with a fan to prevent water condensation under the lid and to guarantee that the right doses are reaching the cells (Fig. 33.1).

Regulatory Aspects

The 3T3 NRU-PT was formally validated for phototoxicity assessment and adopted for regulatory purposes by the European Union (in 2000), by the OECD (in 2004) and more recently by the ICH S10 (ICH Harmonised Tripartite Guideline: European Union, USA and Japan) as a guideline for photosafety evaluation of pharmaceuticals.

Considering that fibroblasts are highly sensitive to UV radiation, some false-positive predictions may be obtained, a problem that has been solved with the use of the recommended tiered strategy involving 3D skin models. However, if a test chemical does not show a phototoxic potential in this test, as described earlier, this is the only assay required to determine that the substance has no phototoxic potential [15].

33.3.2.2.2 Three-Dimensional Skin Models for Phototoxicity Assays

The most used and pre-validated skin models for phototoxicity assays are the EpiDerm™ Reconstructed human Epidermis (RhE) from MatTek and the Episkin™ and SkinEthic™ models from EPISKIN™ S.A. Some studies have demonstrated a

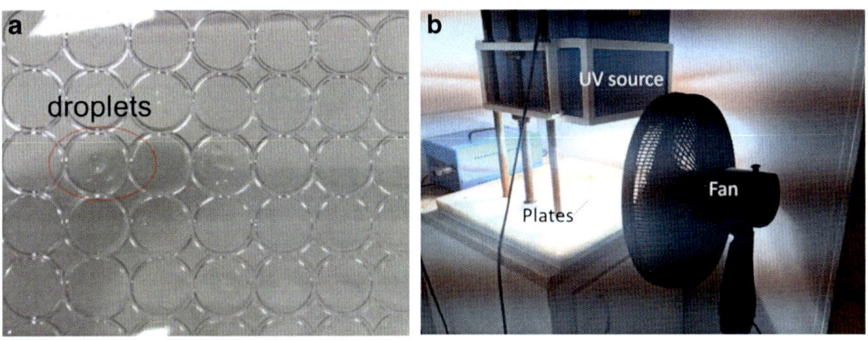

Fig. 33.1 Droplet formation (**a**) should be avoided with the use of a fan (**b**)

good correlation of H3D-PT performed in these skin models and known phototoxic and non-phototoxic chemicals (see Chap. 35).

Rationale

This assay was designed to detect the phototoxic potential of test chemicals by using a three-dimensional human epidermis model.

Similarly to 3T3 NRU-PT, the test is based on a comparison of the cytotoxicity of a chemical tested with and without additional exposure to a nontoxic dose of UVA/VIS. Cytotoxicity is expressed as reduction of mitochondrial conversion of MTT to formazan, determined 1 day after chemical treatment and UVA exposure [11].

Protocol

One hour before starting the assay, tissues are transferred to 6-well plates with assay medium and the medium is changed. Five concentrations of the chemical are topically applied onto two tissues per concentration (i.e., one vehicle control + five concentrations = 12 tissues), and the tissues are incubated overnight (18–24 h) with the test chemicals. A second set of 12 tissues is treated identically. The five different concentrations will be used to determine the first non-phototoxic concentration. Plates are incubated overnight.

The next day, +UVA tissues are exposed to 6 J cm^{-2} UVA (usually 1.7 mW cm^{-2} for 60 min), and the -UVA tissues are kept in a dark box for the same period of time. Tissues are then rinsed with PBS and transferred to new 6-well plates with fresh medium and incubated overnight.

The next day, tissues are incubated for 3 h with MTT-medium. Tissues are then rinsed with PBS, and the formazan is extracted with isopropanol. Optical density is determined at 570 nm in a spectrophotometer and cell viability is calculated for each tissue as % of the corresponding irradiated (+UVA) or non-irradiated (−UVA) vehicle control [14].

Advantages of H3D–PT over 3 T3 NRU-PT

The assay allows application of test materials topically to the stratum corneum, which has two main consequences: (1) It mimics the *in vivo* situation and thus may allow to predict phototoxic potency of test materials applied at usage concentrations; (2) poorly soluble test chemicals can be used, such as oils and powders, as well as the formulations that will be employed under actual conditions of use, since the vehicle also influences skin penetration and consequently the photosafety of the substances.

There are some clear cases that show the need for this assay as a second step in the tiered testing strategy for photosafety assessment. In particular, the presence of a barrier system (*stratum corneum*) allows to take into consideration the bioavailability of the test chemical and the verification of positive results obtained with the 3T3 NRU-PT. One example is the UV filter avobenzone, which shows a phototoxic potential in 3T3 NRU-PT that is not confirmed by H3D-PT [16, 17], due to its low skin penetration [18]. Another example is the fragrance ingredient bergamot oil, in which the 3T3 NRU phototoxic potential was confirmed by H3D-PT and by human photo patch tests. The authors also observed the influence of different solvents used on the phototoxicity potential [11].

Limitations

Extrapolation of *in vitro* results to the human effects may be performed only to a limited extent, due in part to the higher permeability of the experimental skin tissues as compared to human skin *in vivo* [17]. Thus, a precautionary factor of 10 would be recommended when approaching testing on human volunteers based on the outcomes of *in vitro* H3D phototoxicity test methods [9].

Regulatory Aspects

Although the validation of H3D-PT has not been completed, previous studies show its high sensitivity (87%), specificity (93%), and accuracy (93%) [19]. Thus, ICH recently recommended it in S10 guidelines as a second step of biological assays for the photosafety assessment of pharmaceuticals [1].

33.3.2.2.3 Clinical Tests for Phototoxicity: Photopatch Test

When a photo-induced effect is suspected, a photopatch test is indicated for diagnosis. However, to produce consistent results, human photopatch tests need to be standardized.

A group of interested European contact dermatologists/photobiologists met in 2004 aiming to arrive at a consensus statement on the methodology, test materials, and interpretation of photopatch testing [20].

The photopatch test is mainly used for the assessment of photoallergic reactions, although it can also be used, with some protocol modifications, for the assessment of phototoxic reactions.

General Protocol

There are different protocols for the photopatch test so that there is a need for standardization.

However, the test is generally based on the following steps: a sample of each formulation is applied once to the protected area of the subject's back skin. The patch is removed after 24 h and the test sites are assessed and irradiated with UVA radiation. The area adjacent to the site of application and the spot of the control patch with saline is irradiated and used as control. Readings are taken immediately after irradiation and also after 24, 48, and 72 h if a positive reaction is observed.

The selection of concentrations used is based on the first non-phototoxic concentration determined by the H3D-PT using a margin of safety factor of 10 [9, 11, 20].

33.3.2.2.4 And What about Phototoxicity Evaluation in Animals?

The phototoxicity evaluation in animals fails in some aspects, such as the fact that it has not been validated and the quality of the available animal data is poor. However, animal testing could be used for systemic drugs since it provides relative tissue to plasma concentration ratios, tissue residence time, and the potential for retention and accumulation into the skin.

33.4 Photoallergy

Photoallergic reactions are delayed hypersensitivity responses (type IV), which require the specific sensitization of an individual to a photosensitizer.

33.4.1 Clinical Signs of Photoallergy

Photoallergic reactions can occur in response to topical or systemic chemicals. The most common clinical manifestation of photoallergic reactions is photoallergic contact dermatitis (PACD). PACD is a hypersensitive delayed reaction mediated by T cells and is induced by a photoallergen. UVA, UVB, and even VIS radiation can be involved in the development of PACD. Differentiation between photoallergic and phototoxic reactions is often difficult because most drugs are able to cause both photoallergic and phototoxic reactions, and both are very similar in terms of clinical and histological features [21].

Photoallergy has a lower incidence than phototoxicity. The main differences between them are that photoallergic reactions require a smaller amount of chemical and a lower dose of light; in addition, photoallergy requires more time and more than one contact with the test chemical to develop.

33.4.2 Mechanisms

Two main mechanisms have been proposed for the conversion of a chemical (hapten) to an immunologically active substance (photoallergen) [21]:

1. A photosensitizer absorbs light and reaches the excited state (as described before). When the electron returns to the ground state, it releases enough energy to conjugate the photosensitizer with a protein, generating an antigen.
2. A photosensitizer absorbs light, reaches the excited state, and releases energy, forming a stable photoproduct, which is a hapten. When this hapten is conjugated with a protein, it forms a complete antigen.

The immunological reaction is then similar to the one occurring in allergic contact dermatitis, during what is called a "sensitization phase." Briefly, after the generation of the new antigen by one of the two mechanisms described above, dermis dendritic cells or epidermis Langerhans cells capture the antigen, migrate to the lymph node, and present this antigen to the T cells.

In a second exposure, which is called the "elicitation phase," T cells may proliferate and initiate an inflammatory response that is responsible for the clinical signs of photoallergy.

33.4.3 *In Vitro* Tests for the Assessment of Photoallergy

In vitro tests are only able to predict the sensitization phase of allergy. However, while the binding of an excited chemical to proteins is a prerequisite for photoallergy, at present no validated *in vitro* method is available to predict photoallergy effects [15].

Oeda et al. [22] described few reported *in vitro* tests to assess photosensitization, such as the modified h-CLAT (human cell line activation test; the original h-CLAT is already validated for skin sensitization: TG 442E; [23]) and the dendritic cell activation test [24–26]; the assessment of IL-18 by keratinocytes [27], the apoptose assay using HaCaT [28], the *in vitro* photo-binding test to human serum albumin [29], and the photo-SH/NH2 [22], which assess changes of cell-surface thiols and/or amines as biomarkers.

33.4.4 Clinical Tests for Photoallergy: Photopatch Test

The photopatch test for photoallergy is based on induction, rest, and challenge phases to correctly provide time and stimuli to the immunological system.

33.4.4.1 General Protocol

The protocol is based on the application of a patch containing the test chemical for a period of time, for example, twice a week for 3 weeks. The patch is removed 24–48 h after the last application, and the area is immediately evaluated and irradiated with UVA.

Next, a rest period of some days follows the induction period, when no patches are applied and no irradiation is performed. In the challenge phase, patches of each sample and control are applied to the back of the subjects in a different area, i.e., where no patches had been applied before.

The patch is then removed after 24 h of contact with the skin, and the test sites are irradiated with UVA. Readings are taken immediately after irradiation and thereafter [30].

33.4.4.2 Considerations and Recommendations

According to the European Multicentre Photopatch Study Taskforce consensus methodology, the recommended dose is 5 J cm^{-2} UVA. The UVA meters should be calibrated [20].

All reactions obtained in the photopatch test should be graded according to the International Contact Dermatitis Research Group (ICDRG) system.

33.4.4.3 Epidemiological Studies

A review of specific allergens that induce PACD, recently published by Victor et al. [31], suggested that sunscreens and antimicrobial agents were the most frequent allergens eliciting PACD. A decrease in PACD caused by fragrances and an increase of number of reactions to drugs was also observed. Benzophenone-3, avobenzone, and octocrylene are some UV filters detected in individuals with a photosensitivity disorder.

However, these skin reactions have not been detected in compatibility testing or photopatches of finished cosmetic products [32] recommended by the European Commission Scientific Committee on Consumer Safety (SCCS).

It has long been debated whether sunscreens increase or decrease skin cancer risk. Two recently published systematic reviews [33, 34] indicate that the incidence of basal cell carcinoma and melanoma skin cancer is no longer increasing in younger age groups of some countries. These results may indicate an early effect of broad-spectrum sunscreen (containing avobenzone and other UVA filters) use and sun protection programs, leading to a beneficial reduction of intermittent solar exposure.

Today, sunscreens are becoming safer since awareness of photostability/photo-instability of sunscreen products have arguably improved modern sunscreens. However, even a photo-unstable and probably phototoxic combination (detected as positive with the 3T3 NRU-PT but not confirmed with the H3D-PT) [16] reduced the incidence of both melanoma and non-melanoma skin cancers in a population-based controlled clinical trial performed with the population of Nambour in south-east Queensland, Australia [35, 36].

33.5 Conclusion

The assessment of photo-induced toxicity, which includes phototoxic and photoallergic evaluations, is recommended when chemical preparations are intended to be applied on sunlight-exposed skin.

The recommended tiered testing strategy for phototoxic evaluation, including determination of MEC, the ROS assay, the 3T3 NRU-PT, and the H3D-PT *in vitro* test methods, allows the identification of phototoxicity potential without animal testing. On the other hand, no validated *in vitro* method is yet available for predicting photoallergic reactions.

Finally, considering that photoallergy has a lower incidence as compared to phototoxicity, the use of the tiered strategy involving 3T3 NRU-PT and H3D-PT for phototoxicity prediction is considered relevant for clinical prediction of human photosafety [1, 12].

References

1. ICH. Photosafety evaluation of pharmaceuticals S10. 2013. http://www.ich.org/fileadmin/Public_Web_Site/ICH_Products/Guidelines/Safety/S10/S10_Step_4.pdf. Accessed 8 Jul 2016.
2. Wilm A, Berneburg M. Photoallergy. J Dtsch Dermatol Ges. 2015;13(1):7–12.
3. Spielmann H, Muller L, Averbeck D, Balls M, Brendler-Schwaab S, Castell JV, Curren R, de Silva O, Gibbs NK, Liebsch M, Lovell WW, Merk HF, Nash JF, Neumann NJ, Pape WJW, Ulrich P, Vohr HW. The second ECVAM workshop on phototoxicity testing. ATLA. 2000;28:777–814.
4. Gould JW, Mercurio MG, Elmets CA. Cutaneous photosensitivity diseases induced by exogenous agents. J Am Acad Dermatol. 1995;33(4):551–73.

5. JaCVAM. Reactive oxygen species (ROS) assay to examine photoreactivity of chemicals. 2013. http://www.jacvam.jp/files/doc/02_03/02_03_E2.pdf. Accessed 8 Jul 2016.
6. Onoue S, Hosoi K, Wakuri S, Iwase Y, Yamamoto T, Matsuoka N, Nakamura K, Toda T, Takagi H, Osaki N, Matsumoto Y, Kawakami S, Seto Y, Kato M, Yamada S, Ohno Y, Kojima H. Establishment and intra–/inter-laboratory validation of a standard protocol of reactive oxygen species assay for chemical photosafety evaluation. J Appl Toxicol. 2013;33:1241–50.
7. Henry B, Foti C, Alsante K. Can light absorption and photostability data be used to assess the photosafety risks in patients for a new drug molecule? J Photochem Photobiol B. 2009;96:57–62.
8. EMA. Questions and answers on the "Note for guidance of photosafety testing". 2011. http://www.ema.europa.eu/ema/. Accessed 8 Jul 2016.
9. Ceridono M, Tellner P, Bauer D, Barroso J, Alépée N, Corvi R, De Smedt A, Fellows MD, Gibbs NK, Heisler E, Jacobs A, Jirova D, Jones D, Kandárová H, Kasper P, Akunda JK, Krul C, Learn D, Liebsch M, Lynch AM, Muster W, Nakamura K, Nash JF, Pfannerbecker U, Phillips G, Robles C, Rogiers V, Van de Water F, Liminga UW, Vohr HW, Wattrelos O, Woods J, Zuang V, Kreysa J, Wilcox P. The 3T3 neutral red uptake phototoxicity test: practical experience and implications for phototoxicity testing- the report of an ECVAM-EFPIA workshop. Regul Toxicol Pharmacol. 2012;63(3):480–8.
10. OECD. OECD guidelines for the testing of chemicals, test no. 432: in vitro 3T3 NRU phototoxicity test. 2004. http://www.oecd.org/. Accessed 8 Jul 2016.
11. Kejlová K, Jírová D, Bendová H, Kandárová H, Weidenhoffer Z, Kolárová H, Liebsch M. Phototoxicity of bergamot oil assessed by in vitro techniques in combination with human patch tests. Toxicol In Vitro. 2007;21(7):1298–303.
12. SCCS. SCCS/1564/15: the SCCS notes of guidance for the testing of cosmetic ingredients and their safety evaluation 9th revision, adopted by SCCS during the 11th plenary meeting of 29 September 2015 (revised version of 25 Apr 2016). 2016.
13. Liebsch M, Spielmann H. INVITTOX protocol No. 78: 3T3 NRU phototoxicity assay. European Commission DB-ALM, ECVAM. 2002. https://eurl-ecvam.jrc.ec.europa.eu. Accessed 8 Jul 2016.
14. Liebsch M. INVITTOX protocol No. 121: EpiDerm™ Phototoxicity assay. European Commission DB-ALM, ECVAM. 1999. https://eurl-ecvam.jrc.ec.europa.eu. Accessed 8 Jul 2016.
15. Liebsch M, Spielmann H, Pape W, Krul C, Deguercy A, Eskes C. UV-induced effects. Altern Lab Anim. 2005;33:131–46.
16. Gaspar LR, Tharmann J, Maia Campos PM, Liebsch M. Skin phototoxicity of cosmetic formulations containing photounstable and photostable UV-filters and vitamin a palmitate. Toxicol In Vitro. 2013;27(1):418–25.
17. Kandárová H. Evaluation and validation of reconstructed human skin models as alternatives to animal tests in regulatory toxicology. Dissertation, Freie Universit ät Berlin, 2006.
18. Freitas JV, Praça FSG, Bentley MVLB, Gaspar LR. *Trans*-resveratrol and beta-carotene from sunscreens penetrate viable skin layers and reduce cutaneous penetration of UV-filters. Int J Pharm. 2015;484(1–2):131–7.
19. Liebsch M, Traue D, Barrabas C, Spielmann H, Gerberick GF, Cruse L, Diembeck W, Pfannenbecker U, Spieker J, Hottzhutter H, Brantom P, Aspin P, Southee J. Prevalidation of the Epiderm™ phototoxicity test. In: Clark DG, Lisansky SG, Macmillan R, editors. Alternatives to animal testing II. Brussels: COLIPA; 1999. p. 160–7.
20. Bruynzeel DP, Ferguson J, Andersen K, Gonçalo M, English J, Goossens A, Holzle E, Ibbotson SH, Lecha M, Lehmann P, Leonard F, Moseley H, Pigatto P, Tanew A. Photopatch testing: a consensus methodology for Europe. J Eur Acad Dermatol Venereol. 2004;18(6):679–82.
21. Dean JH, Luster MI, Munson AE, Kimber I, editors. Immunotoxicology and immunopharmacology. New York: Raven Press; 1994.
22. Oeda S, Hirota M, Nishida H, Ashikaga T, Sasa H, Aiba S, Tokura Y, Kouzuki H. Development of an in vitro photosensitization test based on changes of cell-surface thiols and amines as biomarkers: the photo-SH/NH_2 test. J Toxicol Sci. 2016;41(1):129–42.

23. OECD. OECD guideline for the testing of chemicals No. 442E: in vitro skin sensitisation: human cell line activation test (h-CLAT). Paris, France: Organisation for Economic Cooperation and Development. 2016. Available at http://www.oecd.org/env/testguidelines.
24. Hino R, Orimo H, Kabashima K, Atarashi K, Nakanishi M, Kuma H, Tokura Y. Evaluation of photoallergic potential of chemicals using THP-1 cells. J Dermatol Sci. 2008;52(2):140–3.
25. Hoya M, Hirota M, Suzuki M, Hagino S, Itagaki H, Aiba S. Development of an in vitro photosensitization assay using human monocyte-derived cells. Toxicol In Vitro. 2009;23(5):911–8.
26. Karschuk N, Tepe Y, Gerlach S, Pape W, Wenck H, Schmucker R, Wittern K, Schepky A, Reuter H. A novel in vitro method for the detection and characterization of photosensitizers. PLoS One. 2010;23:e15221.
27. Galbiati V, Bianchi S, Martínez V, Mitjans M, Corsini E. NCTC 2544 and IL-18 production: a tool for the in vitro identification of photoallergens. Toxicol In Vitro. 2014;28:13–7.
28. Kurita M, Shimauchi T, Kobayashi M, Atarashi K, Mori K, Tokura Y. Induction of keratinocyte apoptosis by photosensitizing chemicals plus UVA. J Dermatol Sci. 2007;45:105–12.
29. Lovell WW, Jones PA. Evaluation of mechanistic in vitro tests for the discrimination of photoallergic and photoirritant potential. Altern Lab Anim. 2000;28(5):707–24.
30. Bryden AM, Moseley H, Ibbotson SH, Chowdhury MM, Beck MH, Bourke J, English J, Farr P, Foulds IS, Gawkrodger DJ, George S, Orton DI, Shaw S, McFadden J, Norris P, Podmore P, Powell S, Rhodes LE, Sansom J, Wilkinson M, van Weelden H, Ferguson J. Photopatch testing of 1155 patients: results of the U.K. multicentre photopatch study group. Br J Dermatol. 2006;155(4):737–47.
31. Victor FC, Cohen DE, Soter NA. A 20-year analysis of previous and emerging allergens that elicit photoallergic contact dermatitis. J Am Acad Dermatol. 2010;62:605–10.
32. Benevenuto CG, Guerra LO, Gaspar LR. Combination of retinyl palmitate and UV-filters: phototoxic risk assessment based on photostability and in vitro and in vivo phototoxicity assays. Eur J Pharm Sci. 2015;68:127–36.
33. Erdmann F, Lortet-Tieulent J, Schüz J, Zeeb H, Greinert R, Breitbart EW, Bray F. International trends in the incidence of malignant melanoma 1953–2008 are recent generations at higher or lower risk? Int J Cancer. 2013;132(2):385–400.
34. Xiang F, Lucas R, Hales S, Neale R. Incidence of nonmelanoma skin cancer in relation to ambient UV radiation in white populations, 1978–2012: empirical relationships. JAMA Dermatol. 2014;150(10):1063–71.
35. Green A, Williams G, Neale R, Hart V, Leslie D, Parsons P, Marks GC, Gaffney P, Battistutta D, Frost C, Lang C, Russell A. Daily sunscreen application and betacarotene supplementation in prevention of basal-cell and squamous-cell carcinomas of the skin: a randomised controlled trial. Lancet. 1999;354(9180):723–9.
36. Green AC, Williams GM, Logan V, Strutton GM. Reduced melanoma after regular sunscreen use: randomized trial follow-up. J Clin Oncol. 2011;29(3):257–63.

34 Reactive Oxygen Species Assay for Evaluating Phototoxicity Potential

Hajime Kojima, Kazuhiro Hosoi, and Satomi Onoue

34.1 Principle and Scientific Basis of the Test Method

Drug-induced phototoxicity is elicited by exposure to sunlight after pharmaceutical substances are administered either topically or systemically to the skin and/or eyes. This undesirable side effect is an impediment to drug discovery and development, and substantial efforts have been made to avoid drug-induced phototoxic reactions. Effective methodologies for evaluating the phototoxic potential of test substances have been developed over the past few years, and screening strategies have also been proposed for predicting *in vivo* phototoxic reactions. European and US regulatory agencies as well as the International Conference on Harmonisation of Technical Requirements for Registration of Pharmaceuticals for Human Use (ICH) have published guidelines for predicting and avoiding drug-induced phototoxicity during the early stages of pharmaceutical development. The guidelines indicate the requirements for assessing photosafety of test substances on the basis of their photochemical behaviors and recommend phototoxic assessment tools for aiding new drug development. A number of phototoxic screening systems based on the pathogenesis of drug-induced phototoxicity have also been proposed, and some of them have already been applied to the phototoxic evaluation of new drug entities in drug discovery and development [1].

H. Kojima (✉)
National Institute of Health Sciences (NIHS), Tokyo 158-8501, Japan
e-mail: h-kojima@nihs.go.jp

K. Hosoi
Research and Development Center, Santen Pharmaceutical, Co., Ltd., Osaka 530-8552, Japan
e-mail: kazuhiro.hosoi@santen.co.jp

S. Onoue
The School of Pharmaceutical Sciences, University of Shizuoka, Shizuoka Prefecture, Shizuoka 422-8002, Japan
e-mail: onoue@u-shizuoka-ken.ac.jp

C. Eskes et al. (eds.), *Alternatives for Dermal Toxicity Testing*,
DOI 10.1007/978-3-319-50353-0_34

34.2 Brief Description of the Protocol

The reactive oxygen species (ROS) assay was developed by Onoue and Tsuda [2] and is a high-throughput, high-performance system for predicting the phototoxic potential of pharmaceutical substances. This assay employs multiwell plates and a quartz reaction container, the advantages of which include reduced sample volumes, improved assay productivity, and highly uniform irradiation. Singlet oxygen is measured in an aqueous solution by spectrophotometrically monitoring the bleaching of RNO (*p*-nitrosodimethylaniline) at 440 nm using imidazole as a selective acceptor of singlet oxygen. Samples containing the test chemical (2–200 μM), RNO (50 μM), and imidazole (50 μM) in 20 mM NaPB (sodium phosphate buffer) are mixed in a tube. Two hundred microliters of the sample are transferred to a well in a clear, non-treated, flat-bottomed plastic 96-well plate. The plate is subjected to measurement of absorbance at 440 nm using a microplate spectrophotometer. The plate is fixed in the reaction container with a quartz cover and then is irradiated with simulated sunlight for 1 h. A 1500-W xenon arc lamp is used as a solar simulator. The irradiation tests were carried out at 25 °C with an irradiance of ca. 2.0 mW/cm^2 as determined using a calibrated UVA detector. After agitation on a plate shaker, UV absorbance at 440 nm is measured. Superoxide is determined using samples containing the 2–200 μM test chemical and 50 μM nitroblue tetrazolium (NBT) in 20 mM of NaPB, which are then irradiated with the simulated sunlight for 1 h. The reduction in NBT is measured by the increase in absorbance at 560 nm, just as with the singlet oxygen determination.

34.3 Reliability and Relevance

34.3.1 Validation Status

The Japan Pharmaceutical Manufacturers Association (JPMA) and Japanese Center for the Validation of Alternative Methods (JaCVAM) initiated a study to validate intra- and interlaboratory variability and transferability of the ROS assay method in order to incorporate this assay for photoreactivity testing of drug candidates into an ICH framework [3, 4]. A validation management team (VMT) coordinated the evaluation of the robustness of the ROS assay at three facilities using a standardized protocol with Atlas and Seric solar simulators to assess 27 phototoxic and 19 non-phototoxic chemicals. The intra- and interlaboratory reproducibility and transferability of the method were confirmed to be high prior to assessing predictivity. Although problems with the experimental procedure arising from limited solubility had a major impact on assay performance and applicability, results show the capacity to classify a balanced set of 41 test chemicals with a high degree of accuracy and no false negatives.

34.3.2 Performance and Applicability of the Test Method

Significant effort went into establishing well-defined classification criteria based on ROS assay endpoints, thereby maximizing the applicability domain and assay performance to achieve a sensitivity of 100% (20 chemicals/20 chemicals), specificity of 80.0% (12 chemicals/15 chemicals), positive predictivity of 86.4% (19 chemicals/22 chemicals), negative predictivity of 100% (12 chemicals/12 chemicals), and accuracy of 91.2% (31 chemicals/34 chemicals), as shown in Tables 34.1 and 34.2. These results support the routine use of the validated ROS assay protocol in preclinical drug screening for phototoxic potential. The VMT proposed a standard ROS assay protocol with defined data and analytical methods based on a validation study in which the two different solar simulators were used.

Table 34.1 Results of ROS assay validation studies

Lab. No.	Solar simulator	Sensitivity	Specificity	Positive predictive ratio	Negative predictive ratio	Accuracy
1	Atlas	100% (21/21)	75.0% (12/16)	84.0% (21/25)	100% (12/12)	89.2% (33/37)
2	Suntest	100% (18/18)	71.4% (10/14)	81.8% (18/22)	100% (10/10)	87.5% (28/32)
3	CPS/CPS+	100% (20/20)	88.2% (15/17)	90.9% (20/22)	100% (15/15)	94.6% (35/37)
4	Seric	100% (19/19)	71.4% (10/14)	82.6% (19/23)	100% (10/10)	87.9% (29/33)
5	SXL-	100% (19/19)	50.0% (6/12)	76.0% (19/25)	100% (6/6)	80.6% (25/31)
6	2500 V2	100% (19/19)	73.3% (11/15)	82.6% (19/23)	100% (11/11)	88.2% (30/34)
7		100% (18/18)	69.2% (9/13)	81.8% (18/22)	100% (9/9)	87.1% (27/31)

Table 34.2 Within-laboratory reproducibility of ROS assay validation studies

		Test substances used by validation studies			
		Phototoxicant		Non-phototoxicant	
Results	Positive	Suntest CPS [#1] 21 [#2] 18 [#3] 20	SXL-2500 V2 [#4] 19 [#5] 19 [#6] 19 [#7] 18	Suntest CPS [#1] 4 [#2] 4 [#3] 2	SXL-2500 V2 [#4] 4 [#5] 6 [#6] 4 [#7] 4
	Negative	Suntest CPS [#1] 0 [#2] 0 [#3] 0	SXL-2500 V2 [#4] 0 [#5] 0 [#6] 0 [#7] 0	Suntest CPS [#1] 12 [#2] 10 [#3] 15	SXL-2500 V2 [#4] 10 [#5] 6 [#6] 11 [#7] 9

34.3.3 Independent Peer Review

JaCVAM convened an independent scientific peer panel to review the validation of the ROS assay in accordance with established international criteria [5]. The panel met initially in February and again in August 2013, in Tokyo, Japan. At their initial meeting, the panel considered two international validation study reports and a proposed ROS assay protocol. The panel subsequently reviewed updated versions of the ROS assay protocol and the validation study reports as revised by the VMT. This report summarizes the panel's final evaluation and conclusions.

The panel concluded that the reproducibility and predictivity of the ROS assay is sufficient to support its use in an integrated photosafety testing and decision strategy for drug research and development. In this strategy, negative results in the ROS assay would not require further testing in animals or other tests, while positive, weakly positive, and inconclusive results would undergo further testing using an *in vitro* test method, such as the 3T3 phototoxicity assay [6]. The panel also concluded that use of the ROS assay could potentially provide significant savings in time, cost, and reduced animal use for photosafety assessments. Thus, incorporating the ROS assay into a photosafety testing strategy is expected to significantly reduce the overall number of substances that would require additional testing in the *in vitro* 3T3 Phototoxicity Assay or further testing in animals [7].

34.4 Role in a Testing Strategy

The ICH S10 Guideline outlines further details on when photosafety testing is warranted and on possible assessment strategies. In this guideline, photoreactivity tests using chemical assays are described as follows.

If a drug developer chooses to assess photoreactivity of new drug candidates, the assay should be qualified using pharmaceutical agents under appropriate conditions to demonstrate assay sensitivity. Data suggests that the ROS assay [8] has high sensitivity for direct prediction of *in vivo* phototoxicants. One problematic aspect, however, is low specificity, which results in a high percentage of false positives. Nevertheless, as long as a test concentration of 200 μM is achieved, a negative result in a properly conducted ROS assay indicates a very low probability of phototoxicity, and a positive result at any concentration is merely a flag for follow-up assessment [9].

34.5 The Developer's Perspective

1. The solar simulator should be equipped with an appropriate temperature control unit or fan since ROS production is affected by temperature.
2. The proficiency chemicals listed in the test method protocol can be used for transferability check and improvement in the assay skill [10].

3. Each lab should develop historical positive and negative control value acceptance ranges that can be used to determine the acceptability of an individual test.
4. It is necessary to use appropriate reference chemicals when qualifying solar simulators other than the two used in the validation studies.

34.5.1 Photodegradation

The ROS assay assesses chemical photoreactivity in a non-biological system and therefore might overpredict phototoxicity potential, because it does not assess the direct interaction of chemicals with biological tissues. Also, it might overestimate phototoxicity potential, because some chemicals might never achieve sufficient concentration in the skin for phototoxic reactions to occur or might undergo photodegradation. Accordingly, a positive result in the ROS assay is generally construed to be a flag for further evaluation in a photosafety testing strategy. Seto et al. have developed a modified assay that partially resolves this issue [11].

34.5.2 Solubility

The applicability domain of the ROS assay is currently restricted to only those chemicals that meet the solubility criteria outlined in the recommended protocol. Precipitation, coloration, or other interference at both 20 and 200 renders a chemical incompatible with the ROS assay, and results are judged to be inconclusive. When precipitation or coloration is observed only at 200 μM, judgment can be made based on 20 μM. For regulatory purposes, the stability of the test chemical in the reaction mixture both before and after light exposure must be confirmed, when results at 20 μM are used for judgment of a non-photoreactive chemical for which no further phototoxicity testing is necessary. In the ICH test guideline, unfortunately, if a test concentration of 20 μM cannot be achieved, a positive result at any concentration is a flag for follow-up assessment. Onoue et al. are currently developing a modified assay to partially resolve this issue [4, 8, 11]. Insoluble chemicals in the reaction mixtures are not suitable for testing with the ROS assay and may be able to be tested by ROS assay with addition of solubilizing enhancers, such as Tween 20 and bovine serum albumin (BSA), in the reaction mixtures as follow-up assays.

34.5.3 Prediction Model

This assay is useful for drugs and other pure chemicals only. In order to resolve this issue, Nishida et al. developed a prediction model for unknown substances based on molecular weight [12]. Limited information is currently available on the applicability of the ROS assay to multi-constituent substances/mixtures.

34.6 Conclusion

The ROS assay has been proposed for use as a component in an integrated photosafety testing strategy to evaluate the phototoxicity potential of pharmaceuticals and other test substances. The reproducibility and predictivity of the ROS assay is sufficient to support its use in an integrated photosafety testing and decision strategy for drug research and development. In this strategy, negative results in the ROS assay would not require further testing in animals or other tests; positive, weakly positive, and inconclusive results would proceed to the next level of testing in an *in vitro* test system such as the 3T3 Phototoxicity Assay [6], as accepted by ICH S10 in 2013. The use of the ROS assay could potentially provide significant savings in time, cost, and reduced animal use for photosafety assessments.

References

1. Onoue S, Igarashi N, Yamada S, Tsuda Y. High-throughput reactive oxygen species (ROS) assay: an enabling technology for screening the phototoxic potential of pharmaceutical substances. J Pharm Biomed Anal. 2008;46(1):187–93.
2. Onoue S, Tsuda Y. Analytical studies on the prediction of photosensitive/phototoxic potential of pharmaceutical substances. Pharm Res. 2006;23(1):156–64.
3. Onoue S, Hosoi K, Wakuri S, Iwase Y, Yamamoto T, Matsuoka N, Nakamura K, Toda T, Takagi H, Osaki N, Matsumoto Y, Kawakami S, Seto Y, Kato M, Yamada S, Ohno Y, Kojima H. Establishment and intra–/inter-laboratory validation of a standard protocol of reactive oxygen species assay for chemical photosafety evaluation. J Appl Toxicol. 2013;33(11):1241–50.
4. Onoue S, Hosoi K, Toda T, Takagi H, Osaki N, Matsumoto Y, Kawakami S, Wakuri S, Iwase Y, Yamamoto T, Nakamura K, Ohno Y, Kojima H. Intra–/inter-laboratory validation study on reactive oxygen species assay for chemical photosafety evaluation using two different solar simulators. Toxicol In Vitro. 2014;28(4):515–23.
5. OECD. Environment, health and safety publications series on testing and assessment no. 34: guidance document on the validation and international acceptance. 2005.
6. OECD. Guidelines for the testing of chemicals section test no. 432: in vitro 3T3 NRU phototoxicity test. 2004.
7. Japanese Center for the Validation of Alternative Methods. Peer review panel evaluation of the reactive oxygen species (ROS) photosafety assay. 2013. Retrieved from http://www.jacvam.jp/files/news/ROS_Peer_Review_Report_FINAL16OCT13.pdf.
8. Onoue S, Yamauchi Y, Kojima T, Igarashi N, Tsuda Y. Analytical studies on photochemical behavior of phototoxic substances: effect of detergent additives on singlet oxygen generation. Pharm Res. 2008;25:861–8.
9. International Conference on Harmonisation of Technical Requirements for Registration of Pharmaceuticals for Human Use (ICH). S10 photosafety evaluation of pharmaceuticals. 2013. Retrieved from http://www.pmda.go.jp/ich/s/step4_s10_e.pdf.
10. Japanese Center for the Validation of Alternative Methods. Reactive oxygen species (ROS) assay to examine photoreactivity of chemicals. 2013. Retrieved from http://www.jacvam.jp/files/news/ROS_protocol_v3.1_130920_clean.pdf.
11. Seto Y, Kato M, Yamada S, Onoue S. Development of micellar reactive oxygen species assay for photosafety evaluation of poorly water-soluble chemicals. Toxicol In Vitro. 2013;27(6):1838–46.
12. Nishida H, Hirota H, Seto Y, Suzuki G, Kato M, Kitagaki M, Sugiyama M, Kouzuki H, Onoue S. Non-animal photosafety assessment approaches for molecular weight unknown substance using photochemical approach. Regul Toxicol Pharmacol. 2015;72:578–85.

35 The EpiDerm™ Phototoxicity Test (EpiDerm™ H3D-PT)

Helena Kandarova and Manfred Liebsch

35.1 Principle of the Test Method and Scientific Basis

Phototoxicity is defined as a toxic response from a substance applied to the body which is either elicited or increased (apparent at lower dose levels) after exposure to light, or that is induced by skin irradiation after systemic administration of a substance [1]. Identification of the phototoxic potential of topically or systemically applied test chemicals is a crucial step in the safety assessment of cosmetic and pharmaceutical compounds absorbing UV and visible light. According to the European Medicines Agency [2] and to the SCNFP [3] all test chemicals absorbing significant amount of UV light should be investigated for their phototoxic potency.

For ethical reasons, testing of unknown or newly synthesised test chemicals directly in human volunteers is impossible and testing of cosmetic compounds in animals is not an option in the EU [4]. Therefore, before considering any human patch studies, all relevant available *in vitro* methods should be used for the complex evaluation of the toxicological profile of the compound. Knowledge about the phototoxic potency and skin penetration properties of the test compound are crucial steps in this procedure.

The first step to determine the phototoxicity of UV absorbing compound is its evaluation in the validated 3T3 Neutral Red Uptake Phototoxicity Test (3T3 NRU PT) adopted as the OECD TG 432 [1]. In case of negative (non-phototoxic) result, due to high sensitivity of this *in vitro* method, the test chemical can be considered as 'photo-safe' [5]. However, if the result is positive, the test chemical should be further evaluated in a tiered testing strategy to avoid (1) false positive classification

H. Kandarova (✉)
MatTek In Vitro Life Science Laboratories, Mlynske Nivy 73, 821 05 Bratislava, Slovakia
e-mail: hkandarova@mattek.com

M. Liebsch
Retired Head of Unit, "Centre of Alternative Methods to Animal Experiments—ZEBET" at the BfR, Berlin, Germany

C. Eskes et al. (eds.), *Alternatives for Dermal Toxicity Testing*,
DOI 10.1007/978-3-319-50353-0_35

from the sensitive 3T3 NRU PT or (2) to determine safety margins for the test chemical use in a case of a true phototoxin (this consideration should be applicable only for pharmaceuticals where risk-benefit approach can be used).

In vitro reconstituted human skin models are increasingly being investigated for their usability in hazard identification and safety testing, because their organotypic structure with a functional stratum corneum allows for assessment of bioavailability of topically applied test chemicals [5]. An *in vitro* phototoxicity test using human reconstructed epidermis model EpiDerm™ (EpiDerm™ H3D–PT) has been developed and pre-validated almost 20 years ago [6, 7].

The EpiDerm™ H3D–PT assay is designed to detect the phototoxic potential and phototoxic potency of topically applied chemicals and formulations. Since the assay allows application of test materials to the air exposed surface (stratum corneum), it mimics the *in vivo* situation and thus may allow to predict phototoxic potency of test materials applied in usage concentrations. The test is based upon a comparison of the cytotoxicity of a chemical when tested with and without additional exposure to a non-toxic dose of UVA and visible light. Cytotoxicity is expressed as reduction of mitochondrial conversion of MTT to formazan, determined 1 day after chemical treatment and UVA exposure.

35.2 Current Validation Status

The EpiDerm™ H3D PT test was pre-validated by ECVAM in 1999 [7]. Three laboratories experienced in phototoxicity testing participated in the pre-validation study. ZEBET at the BfR acted as Lead testing laboratory participating in all three phases of the pre-validation. Procter & Gamble (P&G) acted as laboratory 2 participating in phases II and Ill and the Beiersdorf AG (BDF) joined the project in phase III (blind trial). The study was sponsored by ECVAM through a contract with Microbiological Bioservices (MB, Stirling, UK) [7].

The outcome of the pre-validation was a fully standardised protocol that was able to reliably predict the phototoxicity of the selected compounds. Later on, ECVAM sponsored a feasibility study on the prediction of the phototoxicity potential of the topically applied compounds [8]. Promising results were reported in several papers linked to this feasibility project [9–11] and in some other studies [12–17], however, no formal validation study has been initiated so far.

Despite the fact that this method is not fully validated, it is accepted by ICH as a useful component of the testing strategy for pre-clinical testing of topically applied pharmaceuticals [18].

35.3 Performance and Applicability of the Test Method

The EpiDerm™ H3D PT was developed and designed not only to predict acute phototoxicity (i.e. hazard identification) but also to estimate phototoxic potential of chemicals and formulations (i.e. risk assessment).

Based on the situation and testing strategy, the EpiDerm™ H3D PT can be used either as:

1. A method for the prediction and confirmation of phototoxicity of topically applied phototoxins in combination with 3T3 NRU PT, or
2. A test that can be used to identify false positive classification from the sensitive 3T3 NRU PT, or
3. An assay to determine safety margins for the phototoxic test chemical in pre-clinical phase of testing for topically applied pharmaceuticals where risk-benefit approach can be used.

35.3.1 Reproducibility

The between laboratory reproducibility of the EpiDerm™ H3D PT test has been assessed in the pre-validation Phase II that led to some improvement of the SOP and final evaluation was conducted in Phase III by testing ten coded reference chemicals in two independent runs [7]. Table 35.1 shows very good reproducibility of the prediction between the three laboratories.

The analysis of the data revealed that Tetracycline was not bioavailable under the conditions of the experiments, since sesame oil was used as a solvent, although the compound is water soluble. This error in selection of the solvent led to the false negative result. False positive classifications were only obtained

Table 35.1 Phase III—Data from Blind Trial between three laboratories (ten compounds)

Chemical	Run	Class	Beiersdorf	P&G	ZEBET
Chlorpromazine	1	PT	PT	PT	PT
	2	PT	PT	PT	PT
Acridine hydrochloride	1	PT	PT	PT	PT
	2	PT	PT	PT	PT
Bergamot oil	1	PT	PT	PT	PT
	2	PT	PT	PT	PT
Neutral red	1	PT	PT	PT	PT
	2	PT	PT	PT	PT
Tetracycline free base	1	PT	NPT	NPT	PT
	2	PT	NPT	NPT	PT
Penicillin G	1	NPT	NPT	NPT	NPT
	2	NPT	NPT	PT	NPT
Lauryl sulphate sodium	1	NPT	NPT	NPT	NPT
	2	NPT	NPT	NPT	NPT
Octyl salicylate (S13)	1	NPT	NPT	NPT	NPT
	2	NPT	NPT	NPT	NPT
4-Methylbenzylidene camphor (S60)	1	NPT	NPT	NPT	NPT
	2	NPT	NPT	NPT	NPT
Octyl methoxycinnamate (S28)	1	NPT	NPT	NPT	NPT
	2	NPT	NPT	NPT	PT

in two single assays with the chemicals Penicillin G and Octyl methoxycinnamate (S28). In both cases the result was just 0.2–9.3% beyond the classification borderline.

35.3.2 Predictive Capacity

The predictive capacity of the final version of the EpiDerm™ H3D PT was assessed during the validation phase III (Table 35.1) resulting in a sensitivity of 86.7% and a specificity of 93.3%. Sensitivity of the dataset was, however, affected by the use on inappropriate solvent for one of the compound tested (Tetracycline). This error in the selection of solvent led to the false negative result in the pre-validation.

In follow-up studies conducted by several research groups (See Table 35.2) many other compounds have been assessed using the original ZEBET's protocol or protocol with minor modifications introduced by the testing laboratories. MB Research Laboratories conducted a large in-house validation study and presented the data at several international conferences in form of posters (e.g. [16]). Unfortunately, until now the available data have not been summarised into a published review, therefore Table 35.2 may serve as a database of chemicals tested in the EpiDerm™ H3D PT over the past 20 years.

35.3.3 Applications and Limitations

The EpiDerm™ H3D PT was developed and designed to predict acute photo-irritation and photo-irritation potential of topically applied compounds. For systemically applied compounds, protocols are still under development and evaluation. It is known that systemic phototoxins, if applied topically on the *stratum corneum* may be under-predicted by all 3D models.

The EpiDerm™ H3D PT tests can be used as an alternative to the 3T3 NRU PT, or as a second tier to identify false positive or equivocal classifications for topically applied compounds. It has been reported [19, 20] that some photoallergens that are not acute photoirritants *in vivo* (for example, coumarin, 6-methyl-coumarin, musk ambrette) are classified correctly as negative by the skin model phototoxicity tests, while they are positive in the 3T3-NRU-PT. Thus the combination of 3T3 NRU PT and H3D PT could possibly be used for identification of photo-sensitisers.

One limitation of this assay is the possible interference of the test chemicals with the MTT measurement. A coloured test chemical or one that directly reduces MTT (and thereby mimics dehydrogenase activity of the cellular mitochondria) may interfere with the MTT measurement. However, these test chemicals are a problem only if at the time of the MTT test (i.e. 18–24 h after test chemical exposure) sufficient amounts of the test chemical are still present on (or in) the tissues. In case of this unlikely event, the (true) metabolic MTT reduction and the

Table 35.2 Fifty compounds tested in the follow-up studies by several testing laboratories worldwide over the past 20 years

#	Chemical	Class	*In vivo*	3T3 NRU PT prediction	EpiDerm™ prediction		
					ECVAM protocol–development and pre-validation	MB Research Laboratories validation[g]	Other studies using ECVAM protocol
1	Chlorpromazine	Phenothiazine derivate	PT	PT	PT[a,b]	PT	PT[h,i,j]
2	Promethazine	Phenothiazine derivate	PT	PT	PT[a]	PT	PT[j]
3	Bergamot oil	Fragrance/food additive	PT	PT	PT[a,b]	PT	PT[c,e,h,i,j]
4	5-Methoxypsoralen	Psoralen compound	PT	PT	PT[a]	PT	
5	8-Methoxypsoralen	Psoralen compound	PT	PT	PT[a]	PT	PT[c]
6	Anthracene	component of coal-tar	PT	PT	PT/NPT[a]	PT	PT[c,i]
7	Acridine	component of coal-tar	PT	PT	PT[b]	PT	PT[c]
8	Neutral red	colour	PT	PT	PT[b]	PT	
9	Tetracycline	pharmaceutical	PT	PT	PT/NPT[b]		
10	Rose Bengal	Fragrance	PT	PT		PT	PT[c,j]
11	Ichthammol	Pharmaceutical	PT	PT			PT[d]
12	Orange oil	Fragrance/food additive	PT	NPT/PT			PT[f]
13	Lemon oil	Fragrance/food additive	PT	NPT/PT			PT[f]
14	5-Aminolevulinic acid	Phenothiazine derivate	PT	PT		PT	
15	7-Methylcoumarin	Fragrant organic compound	PT	PT		PT	
16	Amiodarone	component of coal-tar	PT	PT		PT	
17	Bithionol	veterinary medicine	PT	PT		PT	
18	Norfloxacin	Pharmaceuticum	PT	PT		PT	
19	Protoporphyrin IX	Pharmaceuticum	PT	PT		PT	
20	TetraChlorSalicylAnilide	Pharmaceuticum	PT	PT		PT	
22	Deterpenated Lemon	Fragrance/food additive	NPT/PT	NPT/PT			PT[f]

(continued)

Table 35.2 (continued)

23	*Litsea cubeba*	Fragrance/food additive	NPT	PT			NPT[f,g]
24	Ichthyol pale	Pharmaceutical	NPT	PT			NPT[d]
25	Bergamot oil -purified	Fragrance	NPT	NPT			NPT[c,i,j]
26	BM-DBM, Eusolex9020 (Avobenzone)	UVA filter, cosmetics	NPT	PT			NPT[c,e,h,i]
27	PABA	UV-filter	NPT	NPT	NPT[a]		NPT[i]
28	Benzophenone-3	UV-filter	NPT	NPT	NPT[a]		
29	Methoxycinnamate	UV-filter	NPT	NPT	NPT[a]		
30	Mexoryl SX	UV-filter	NPT	PT/ NPT	NPT[a]		NPT[j]
31	Penicillin G	pharmaceutical	NPT	NPT	NPT[b]		NPT[j]
32	SDS	surfactant	NPT	NPT	NPT[b]	NPT	NPT[j]
33	Octyl salicylate	UV-filter, cosmetics	NPT	NPT	NPT[b]		
34	4-Methylbenzylidene camphor	UV-filter, cosmetics	NPT	NPT	NPT[b]		
35	Octyl methoxycinnamate	UV-filter, cosmetics	NPT	NPT	NPT[b]		NPT[c]
36	Musk ambrette	Fragrance	Photoallergen	PT	NPT[a]		
37	6-Methylcoumarin	Fragrance	Photoallergen	PT	NPT[a]		
38	Benzalkonium Chloride	surfactant	NPT	NPT		NPT	
39	DMSO	Organic solvent	NPT	NPT		NPT	
40	EtOH	Solvent	NPT	NPT		NPT	
41	Eucalyptus oil	essential oil/food additive	NPT	NPT		NPT	
42	Hexachlorophene	pharmaceuticum	NPT	NPT		NPT	
43	L-Histidine	Essential amino acid	NPT	NPT		NPT	
44	SDS	surfactant	NPT	NPT		NPT	
45	Sulisobenzone	UV-filter, cosmetics	NPT	NPT			
46	Coumarin	fragrant org. Compound	NPT	NPT			
47	Cinnamaldehyde	fragrant org. Compound	NPT	NPT			NPT[i]

48	TiO_2 nano-form: Eusolex T-2000	UV-filter, cosmetics	NPT[h,i]
49	TiO_2 nano-form: P25 AEROXID	Catalysis, nanomaterial	NPT[i]
50	TiO_2 nano-form: TIG-115	Catalysis, nanomaterial	NPT[i]

PT phototoxic, *NPT* non-phototoxic
[a]Liebsch et al. [6]
[b]Liebsch et al. [7]
[c]King et al. [25]
[d]Jırova et al. [12]
[e]Kejlova et al. [10]
[f]Kejlová et al. [11]
[g]Pratt et al. [16]
[h]Kandarova [9]
[i]Liskova [15]
[j]Sohn et al. [17]

contribution by a coloured test material or (false) direct MTT reduction by the test material can be quantified by a special procedure described in detail in the SOP of the assay [21].

35.3.4 Comparison to Human Data

Since there are a number of good clinical reports on human topical and systemic phototoxic events available, direct comparison to the results from the EpiDerm™ H3D PT was possible. Based on the currently available experimental data *in vitro*, the EpiDerm™ H3D PT correlates well with the acute human responses of topical phototoxins. It may, however, under-predict phototoxic effects of systemic photo-toxins or weak photo-sensitisers.

35.4 Brief Description of the Protocol: Experimental Procedure

35.4.1 Reconstituted Human Skin Model

The reconstructed tissue model EpiDerm™ (MatTek, Ashland, USA and MatTek IVLSL, Bratislava, Slovakia) consists of normal, human-derived epidermal keratinocytes which have been cultured to form a multilayered, highly differentiated model of the human epidermis. It consists of organised basal, spinous and granular layers, and a multilayered stratum corneum containing intercellular lamellar lipid layers arranged in patterns analogous to those found *in vivo* [22].

The epidermal cells are taken from healthy volunteers that are negative to HIV and Hepatitis. The EpiDerm™ tissues (surface 0.63 cm^2) are cultured on specially prepared cell culture inserts and shipped to customers as kits, containing 24 tissues on shipping agarose together with the necessary amount of culture media and 6-well plates. In addition the MTT kit (containing MTT concentrate, diluent, extractant, PBS and 24-well plate) can be provided by MatTek.

The EpiDerm™ System is manufactured according to defined quality assurance procedures compliant to GMP and ISO 9001:2008 process. All biological components of the epidermis and the culture medium are tested by the manufacturer for viral, bacterial, fungal and mycoplasma contamination. Barrier properties of each manufactured tissue lot are controlled by the manufacturer. Upon request, MatTek provides detailed information about ET_{50} experiment with Triton X-100 (1%) (chemical recommended as the penetration functional quality control of the barrier properties of the model by the OECD TG 431 and TG 439), information of tissue viability (MTT test), together with historical database of results.

The appropriate handling procedures for biological materials should be followed. It is recommended to wear gloves during handling with the skin model and kit components. After use, the epidermis, the material and all media in contact with it should be decontaminated prior to disposal (e.g. using 10% bleach or special containers).

35.4.2 Materials

All material required for the conductance of the EpiDerm™ H3D PT is summarized in Table 35.3. The basic EPI- 200-PHO kit contains 24 units of standard EpiDerm™ model (EPI-200) embedded in transporting agar, bottle of the assay medium, sterile 6 and 24-well plates small amount of DPBS and one vial of control material that can be used to test barrier properties of the EpiDerm™ model (Triton X-100). In addition, MatTek also offers kit for conducting MTT-assay that contains MTT-concentrate, MTT-diluent and extracting solution. Further details on the material, equipment and reagents required for the EpiDerm™ Phototoxicity Test can be found in Table 35.3.

Table 35.3 Material required for the conductance of the EpiDerm™ H3D PT

#	Description	Detail
(A) Material provided by MatTek Corporation with standard EPI-200-PHO Kit		
1	One sealed 24-well plate containing 24 inserts of EpiDerm™ embedded in transporting agar	EPI-200, 0.6 cm^2
2	Sterile 6-well plates used during the assay	4 pieces
3	Sterile 24-well plates used for MTT assay	2 pieces
4	One bottle of DPBS	125 mL
5	One bottle assay medium, EPI-100-ASY	50 mL
6	One vial, containing the positive control chemical—1% Triton	10 mL
7	Protocol for photo-irritation test	1 piece
(B) MTT-100 Assay Kit Components (ordered separately)		
1	One vial containing MTT concentrate (5 mg/mL)	2 mL
2	One vial MTT diluent (DMEM based culture medium)	8 mL
3	One bottle containing extracting solution isopropanol	60 mL
(C) Additional material and equipment needed		
1	Sterile Dulbecco's PBS (DPBS) without Ca^{2+} and Mg^{2+} (e.g. PAN or Biochrom)	2 L
2	Sterile, sharp blunt-edged forceps	
3	Positive displacement pipette for application of semi-solid test materials	30 μL
4	Mortar and pestle for grinding of granular solids	
5	Sterile disposable pipettes, pipette tips	20-200 μL 200–1000 μL
6	Bulb headed Pasteur pipettes—for spreading of test substances	NaCl weight: 25 mg
7	Beakers—for washing and collecting DPBS	200 mL
8	Parafilm	
9	Sterile cotton tip swabs	
10	Laminar flow hood—for work under sterile conditions	
11	Humidified incubator	37 °C, 5% CO_2, 95% RH
12	96-well plate photometer equipped with filter 570 nm	
13	Laboratory balance	
14	Plate shaker	
15	Stop-watches	
16	Wash bottle	500 mL
17	Vortex	

(continued)

Table 35.3 (continued)

#	Description	Detail
18	Sterile H_20 and sesame seed oil (pharmaceutical grade)	Solvents
(D) Irradiation equipment		
1	Sun simulator	For example, UV-sun simulator type SOL 500 Dr.Hönle GmbH
2	Any appropriate, adjustable and stable tripod	For holding the SOL 500 simulator
3	UVA-radiometer (ideally 2 to be used for calibration)	Dr. Hönle or UVX Radiometer UVP Jena
4	Filter to cut off emitted UVB	For example, H1 filter from Hönle

35.4.3 Protocol Steps

This protocol is closely following the ZEBET SOP published in 1997 [23], however, it includes minor modifications that reflect the lessons learnt after the pre-validation and may contribute to obtain improved outcomes for some test materials.

35.4.3.1 Calibration of the Solar Simulator

Before using the irradiation equipment, standardisation and calibration of the Solar Simulator should take place. New metal halide burners should be burned for ~100 h prior to first use to achieve a stable emittance. According to Dr. Hönle (supplier of the recommended Solar Simulator SOL 500) the burner has a shelf life in which the spectrum is stable for at least 800 h. Recording of lamp usage hours is, therefore, recommended. Extended use is only acceptable if the emitted energy spectrum can be checked.

1. Mount the Solar Simulator (e.g. SOL 500), equipped with a H1-filter, on any appropriate stable tripod allowing fine-adjustment of the exposure distance.
2. Adjust SOL 500 to a distance of about 60 cm.
3. Switch the Lamp on, wait at least 15 min and measure irradiance through the lid of a cell culture plate using the calibrated UVA-radiometer, equipped with a UVA-sensor.
4. Adjust distance of SOL 500 to achieve an irradiance of 1.7 mW/cm^2 of UVA (the resulting dose of UVA will be 1 J/cm^2 per 10 min. Exposure time).
5. According to the number of plates to be exposed concurrently, check the exposure area for equal distribution of irradiance: A range of 1.6–1.8 mW/cm^2 is acceptable. Important: a maximum difference of 1.5 and 1.9 mW/cm^2 can be accepted, if positions of the plates with low and high irradiance are changed after half time of the irradiation (30 min) is reached.
6. Calibration of the SOL 500/SOL 3 shall be checked as described above each time before performing a phototoxicity assay.

In case measurements with the UV radiometer reveal unexpected results, either the metal halide burner may have reached the end of its shelf life, or the radiometer is de-calibrated due to various reasons. In this case, a second reference radiometer shall be used for cross check.

Main EpiDerm™ Phototoxicity Test

DAY 0: Tissue conditioning

Before any testing on the viable reconstructed human tissues is performed, it is recommended to perform the evaluation of the test substance for interference with the measured endpoint (MTT assay). This procedure is described in detail in the SOP that is provided together with the testing kit by MatTek.

1. EPI-200-PHO kits are shipped from MatTek facilities in the USA and Slovakia (EU) every Monday.
2. Upon receipt of the shipment, examine all kit components for integrity. If there is a concern call MatTek immediately.
3. Record all information about supplied material into the Methods Documentation Sheet (MDS).
4. Place the DPBS into the refrigerator (5 ± 3 °C) and the vial containing the MTT concentrate in the freezer (−20 ± 5 °C).
5. Let the assay medium reach room temperature (20-25 °C). Do not pre-heat to 37 °C.
6. Pipette 0.9 mL of the assay medium into each well of sterile 6-well plates (for 24 inserts prepare four 6-well plates).
7. Under sterile conditions, open the plastic bag containing the 24-well plate with epidermal tissues. Under a sterile airflow, remove the sterile gauze and carefully (using sterile forceps) take out each insert containing the epidermal tissue. Remove any remaining agarose that adheres to the outer sides of the insert by gentle blotting on the sterile filter paper or gauze, and place the tissues in the empty, sterile 24-well plate.
8. Perform visual inspection of the inserts within the next 5 min. Record any tissue defects and excess moisture on the surface. Do not use tissues with defects or tissues with excessive moisture on the surface.
9. Dry the surface of the tissues with a sterile cotton tip swab and transfer tissues to a 6-well plate pre-filled with 0.9 mL medium. Place the plates for 60 ± 5 min into the incubator (37 ± 1 °C, 5 ± 1% CO_2, 95% RH).
10. At the end of the first (60 min) pre-incubation period, exchange the medium in the 6-well plates.
11. Place the plates back into incubator for overnight pre-incubation (18–24 h, 37 ± 1 °C, 5 ± 1% CO_2, 95% RH).
12. Place the rest of the assay medium into the refrigerator (5 ± 3 °C) and the vial containing the MTT concentrate in the freezer (−20 ± 5 °C).
13. Prepare and sterilise all devices which will be used in the assay.

Table 35.4 Ranges of solubility

Descriptive term	Range of solubility	% (w/v)	Category
Very soluble	>1000 mg/mL	>100.00	1
Freely soluble	>100–1000 mg/mL	>10.00	2
Soluble	>30–100 mg/mL	>3.00	5
Sparingly soluble	>10–30 mg/mL	>1.00	4
Slightly soluble	>1–10 mg/mL	>0.10	5
Very slightly soluble	>0.1–1 mg/mL	>0.01	6
Practically insoluble	0.1 me/nil and lower	<0.01	7

As a basic recommendation, poorly water soluble test materials (category 5–7) should be tested dissolved or suspended in sesame oil. Water soluble test materials (category 1–4) shall be tested dissolved in water. If suspensions are tested, use appropriate techniques for preparing, e.g. a homogeniser or vortex

Table 35.5 Concentration series

Vehicle %	(w/v) %	(w/v) %	(w/v) %	(w/v) %	(w/v) %
Oil	10	3.16	1	0.316	0.1
Water	1	0.316	0.1	0.0316	0.01

Note: According to ZEBET's experience, the series for oil fits for many test materials. Materials dissolved in water pass the stratum corneum more quickly. If they are, in addition, highly cytotoxic, the concentration series may have to be shifted to a lower range in a second experiment

DAY 1: Exposure

1. Place all devices, solution and chemicals necessary for the test into the sterile hood.
2. For each test chemical, prepare a series of five concentrations in the vehicles to which the test materials have its best solubility.
3. The recommended solutions/vehicles for the EpiDerm™ Phototoxicity test are water and sesame seed oil (pharmaceutical grade, pretested in the EpiDerm™ for phototoxicity and cytotoxicity). Affinity to an optimal solvent can be evaluated using the ranges of solubility shown in Table 35.4.

Concentration Series

1. Prepare five concentrations of the test material. Where possible, the highest concentration of a test material should show cytotoxicity in non-irradiated tissues. Since many test chemicals are likely to absorb UV they can act as UV-filter. Therefore, the highest test concentration should not exceed 10% (see Table 35.5).
2. Remove the pre-equilibrated, 6-well plates from the incubator and exchange the medium in all of the 6-well plates.
3. Label all 6-well plate lids with the test material codes or names. Per each test chemical use one EpiDerm™ EPI-200-PHO kit of 24 tissues.
4. Twelve tissues are used in the (−UVA) cytotoxicity part and 12 in the (+UVA) phototoxicity part of the test. Both parts of the test are dosed identically and

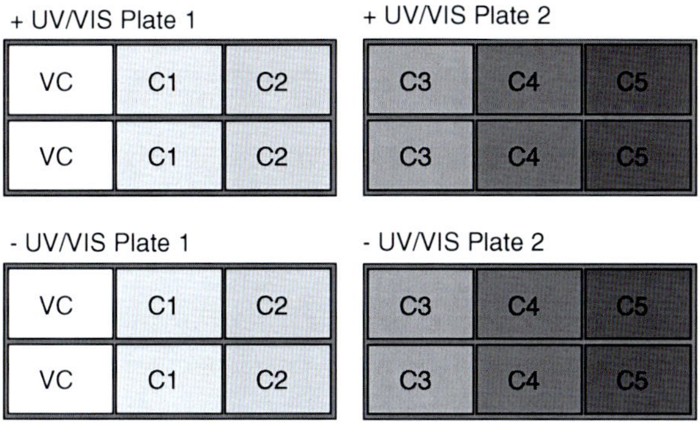

Fig. 35.1 Plate design for the phototoxicity experiment. To be on the safe side, position the lowest concentration (C1) beside the vehicle control (VC)

include Vehicle Control (VC, $N = 2$) and 5 concentrations of the test chemical (C1–C5, $N = 2$).

5. Mark lids and plates to prevent transposition errors as shown in Fig. 35.1.

Application of Test Sample

Note: The Original ZEBET SOP advices to use a patch technique for dosing of materials dissolved in oil (20 μL on a sterile patch). Experience has shown, however, that this technique may sometimes lead to false negative outcomes so that it is therefore recommended not to use this technique in the updated versions of the protocol.

1. Dose the EpiDerm™ tissue topically using standard or positive displacement pipette.
   a. Solutions in H_20: apply 50 μL atop the EpiDerm™ tissue and gently spread in necessary with bulb headed Pasteur pipette.
   b. Solutions in oil: apply 25 μL atop the EpiDerm™ tissue and gently spread using bulb headed Pasteur pipette. If the spreading is not sufficient, consider to apply a nylon mesh (circular shape) atop the tissue as an additional spreading tool. The mesh is supplied by MatTek.
   c. Any other vehicles: if a material is poorly soluble in water as well as in sesame seed oil and there is risk that suspension may lead to limited bioavailability, the next solvent of choice is Ethanol. The volume of EtOH should not exceed 25 μL, since higher volumes will lead into cytotoxicity. Another solvent that could be considered is a mixture of Acetone and Olive oil (4:1) also with a maximum volume of 25 μL. It is advised that Untreated Controls ($N = 2$) should be included in addition to Vehicle Controls to evaluate the possible cytotoxicity of an alternative solvent. If a solvent decreases tissue viability by >25%, it should not be used for the experiments.
2. Once all tissues have been dosed, cover the plates with the lids and incubate overnight (18–24 h) at 37 °C, 5% CO_2 and 95% RH.

DAY 2: Irradiation

1. Remove 6-well plates from the incubator.
2. Transfer the tissues into the new 6-well plates pre-filled with 0.9 mL of DPBS.
3. Irradiate the +*UVA*-plates (covered with lids) for 60 min with 1.7 mW/cm^2 *of UVA* (=6 J/cm^2 *of UVA, measured through the lid*) at room temperature. Ventilate with fan to prevent condensation under the lid.
4. Place the -*UVA*-plates in the dark box at room temperature.
5. While tissues are irradiated, prepare appropriate amount of new 6-well plates pre-filled with 0.9 mL of fresh assay medium per well and pre-warm in the incubator.
6. After UVA irradiation is completed, use wash bottle with sterile DPBS and rinse each insert of the +UVA-plates and –UVA-plates. About 20 washes are needed to remove the materials properly from the tissue surface.
7. Transfer all washed inserts to the fresh media in new plates prepared in step 5. After all inserts are washed, do not forget to carefully dry the surface of each tissue with a sterile cotton tipped swab. In case that traces of the chemical are still present on the surface, try to remove it with the sterile wetted cotton swab. You may evaluate visually tissue surface under a dissecting stereoscope.
8. Incubate the tissues overnight (18–24 h) at 37 °C, 5% CO_2, 95% RH.

DAY 3: MTT Viability Test

1. Prior to the MTT assay, label a sufficient number of 24-well plates.
2. Prepare MTT medium (1 mg/mL) from frozen concentrate and pipette 300 μL of the MTT medium in each well.
3. Remove inserts from the 6-well plates, blot the bottom of the inserts and transfer them into the 24-well plates, pre-filled with 0.3 mL of MTT (1 mg/mL). Place the plates in the incubator (37 ± 1 °C, 5 ± 1% CO_2, 95% RH), record the start time of MTT incubation in the MDS and incubate for 3 *h* ± 5 *min*.
4. After MTT incubation is complete, gently blot the tissue on the absorbing paper and transfer inserts into new 24-well plates.
5. Immerse the inserts by gently pipetting 2 mL of isopropanol (extractant solution) into each insert. The level will rise above the upper edges of the insert, thus completely covering the tissues from both sides.
6. Seal the 24-well plates (e.g. with Parafilm or place into a sealable plastic bag) to inhibit extractant evaporation. Record start time of extraction in the MDS and extract formazan for at least 2 h at room temperature with gentle shaking on a plate shaker (120–200 rpm).
7. As an alternative, overnight extraction is also possible. Seal plates as described above and extract at room temperature in the dark, without shaking. Before using the extracts, shake for at least 15 min on plate shaker.
8. After the extraction period is complete, pierce the inserts with an injection needle (~gauge 20, ~0.9 mm diameter) and allow the extract to run into the well

from which the insert was taken. Afterwards the insert can be discarded. Before transferring the extract to 96-well plates pipette up and down 3× until the extractant solution is homogenous.

9. For each tissue, transfer 2 × 200 μL aliquots of the blue formazan solution into a 96-well flat bottom microtiter plate according to the fixed plate design given in spreadsheet. Use isopropanol as blanks.
10. Read the OD in a 96-well plate spectrophotometer using a wavelength between 540 and 595 nm, preferably at 570 nm, *without using a reference filter.*

35.4.4 Test Data

A blank, password protected MS EXCEL workbook *EpiDerm™-PHO-SPREAD.XLS* can be provided by MatTek. A copy should be made before the first data entry. The workbook consists of two single spreadsheets named: IMPORT and SPREAD. Data files of optical densities (ODs) generated by the microplate reader (without blank subtraction) are copied from the reader software to the Windows Clipboard and then pasted into the first spreadsheet of the EXCEL workbook. The blank corrections, calculation of results and statistical parameters are done automatically in the second part of the workbook. Use the fixed 96-well plate design as specified in the SOP and spreadsheet provided by MatTek.

After data entry, the spreadsheet performs the following calculations:

1. For each individual tissue treated with a test substance (TS) and the vehicle control (VC) the individual relative tissue viability is calculated according to the following formulas:
   Relative viability TS (%) = [OD TS/Mean of OD VC] × 100
   Relative viability VC (%) = [OD VC/Mean of OD VC] × 100
2. For each Test Substance and Vehicle Control, the mean relative viability of the two individual tissues is calculated and used for classification according to the Prediction Model.
3. The spreadsheet shows a graph of the results (% of relative viability ± Difference).

   Difference is calculated as:

% Viability Tissue 1 − % Viability Tissue 2.

35.4.5 Data Interpretation Procedure (Prediction Model)

According to the pre-validation outcome, a chemical is predicted to have a phototoxic potential if one or more test concentrations of the (+UVA) part of the experiment reveal a decrease in viability exceeding 30% when compared with identical concentrations of the (−UVA) part of the experiment. Some more conservative laboratories opted for a 25% cut-off.

Prediction of phototoxicity is supported if, in addition, the (+UVA) induced reduction in tissue viability shows a dose–response relationship.

35.4.6 Assay Quality Controls

35.4.6.1 Assay Acceptance Criterion 1: Vehicle Control

The *absolute OD* of the vehicle control (VC) tissues in the MTT-test is an indicator of tissue viability obtained in the testing laboratory after shipping and storing procedures and under specific conditions of use.

The assay meets the acceptance criterion if the mean OD570 of the VC tissues is >0.8 and ≤2.8.

35.4.6.2 Assay Acceptance Criterion 2: Positive Control—Chlorpromazine (CPZ)

The original ZEBET SOP did not include a mandatory PC testing for each run. When the assay is newly established it is recommended to perform a full experiment with five concentrations of Chlorpromazine dissolved in H_20 ranging from 0.001% up to 0.1%. This test should be repeated on a regular basis. A dose dependent reduction of cell viability occurring only in the UVA-irradiated tissues shall be observed between 0.00316% and 0.0316%.

If CPZ shall be included in each assay, e.g. due to the GLP study, it is advised to use the concentration 0.01% of CPZ in water. Typical data received with 0.01% of CPZ are shown in Fig. 35.2.

The assay will meet the acceptance criterion if the difference between irradiated and non-irradiated part of the experiment will exceed 30%.

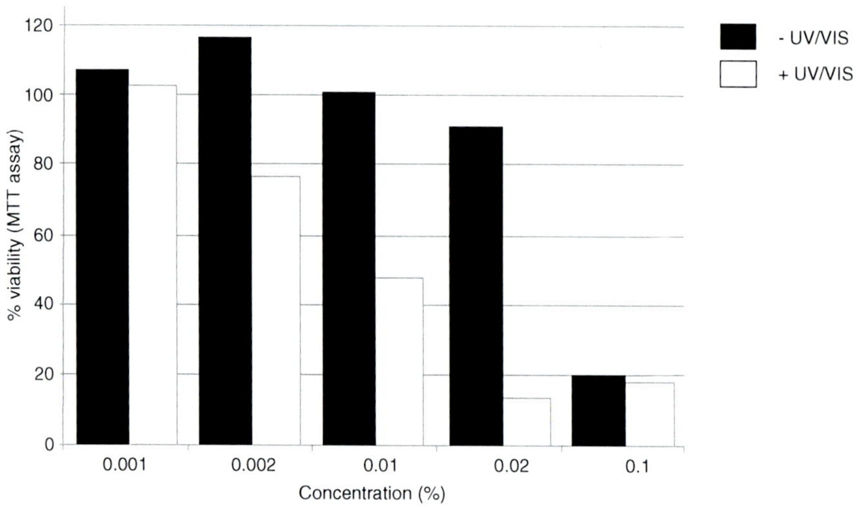

Fig. 35.2 Positive control (Chlorpromazine) reference data—ZEBET SOP, 1997

35.4.6.3 Assay Acceptance Criterion 3: Difference in Viability between Tissues Duplicates (Diff)

The difference in viability between tissue couples that are treated identically should not exceed 30%. According to the historical database of ZEBET the mean difference between untreated tissue duplicates is 9% ± 7%.

35.4.6.4 Additional Quality Measures: UVA-Sensitivity of the Epi-200 Tissues

A UVA-sensitivity experiment should be performed once the test is newly set up in a laboratory. If the UVA-sensitivity of the tissues is within the acceptance range this type of experiment should be repeated in greater intervals (e.g. once every 6 months). Brief description of the method and expected outcome is given in the paragraph below.

1. Pre-incubate 24 tissues (37 °C, 5% CO_2, 95% RH) according to the SOP.
2. Adjust irradiance of the SOL 500 to 1.7 mW/cm^2 of UVA (measure through plate lid!).
3. For UV irradiation, transfer 21 tissues into two 12-well plates filled with 0.5 mL DPBS per well.
4. Prepare another set of two 12-well plates with 0.5 mL DPBS per well and transfer the 3 tissues serving as non-irradiated controls into one of these plates. Place the plates in a dark box at room temperature.
5. Start irradiation of the 21 tissues through the lid of the 12-well plates. Use a fan to prevent H_20 condensation under the lids. Every 30 minu (=3 J/cm^2) transfer three tissues from the irradiation site to the dark box. The resulting dose series is 3, 6, 9, 12, 15, 18, 21 J/cm^2.
6. After the irradiation, transfer inserts into the fresh assay media and incubate tissues overnight (18–24 h) at 37 °C, 5% CO_2, 95% humidity.
7. Determine tissue viability using the MTT assay and calculate tissue viability of the irradiated triplicated compared to the non-irradiated tissues (set to 100% viability).

There shall be no reduction of viability exceeding 20% up to 6 J/cm^2. Per ZEBET's SOP, the historical ID50 UVA is in the range of ~12–18 J/cm^2 of UVA (See Fig. 35.3). More recent measurements [15] indicate that ID50 UVA is in the range of 18–24 J/cm^2 of UVA. This shift is most likely related to the improved transport conditions and reduced transport time.

35.5 Role in a Testing Strategy

The EpiDerm™ H3D PT tests can be used as an alternative to 3T3 NRU PT, or as a second tier to identify 3T3 NRU PT false positive or equivocal classifications for topically applied compounds [5, 24]. It has been reported [19, 20] that some photoallergens that are not acute photoirritants *in vivo* (for example, coumarin,

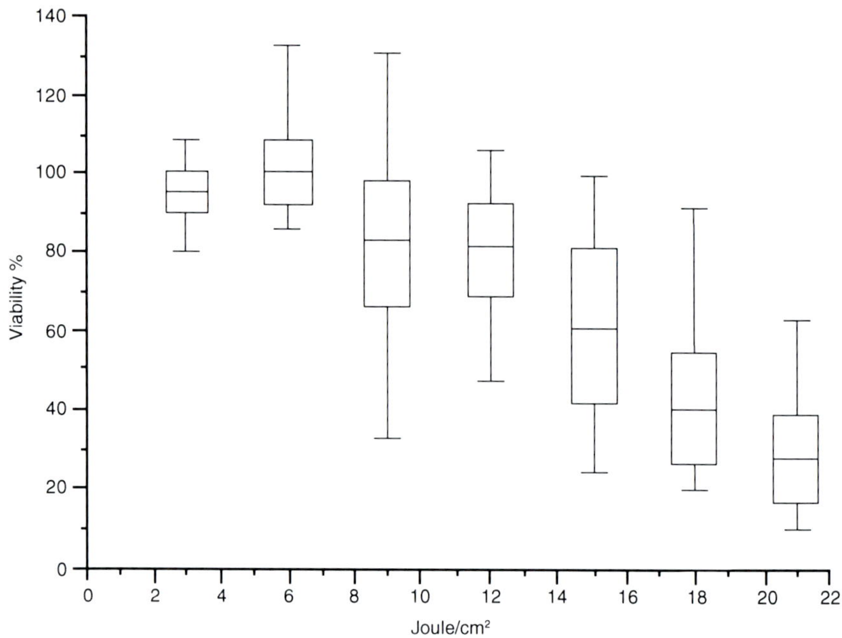

Fig. 35.3 UVA-Sensitivity of the Epi-200 Tissues—ZEBET SOP, 1997

6-methyl-coumarin, musk ambrette) are classified correctly as negative by the skin model phototoxicity tests, while they are positive in the 3T3-NRU-PT. Thus the combination of 3T3 NRU PT and H3D PT could possibly be used for identification of photo-sensitisers.

The updated ICH guideline S10 recommends the use of the 3D models as an additional test for the assessment of topically applied pharmaceuticals, especially those materials having a limited solubility, or those predominantly absorbing in the UVB range of the spectra, which may present issues in the very sensitive 3T3 NRU PT.

35.6 Perspectives from the Test Developer

35.6.1 Critical Steps in the Protocol

The EpiDerm™ H3D-PT is a robust and easy to perform method that utilises readily available laboratory equipment. The test can be performed by most laboratory personnel, provided that care is taken during the critical steps of exposure and washing.

Unequal spreading of a test material and use of inappropriate solvent may lead to false negative outcomes or high variability. Insufficient washing of the UV absorbing materials and particles may lead to the variability of the data. Omission of aseptic techniques and use of non-sterile tools may lead to contamination.

Special care should be taken when calibrating the solar simulator and assessing the tissues UVA-sensitivity. It is important to understand the output of the radiometers (UVA-radiometers or UVA + visible light radiometers), since they may provide different readings depending on their bandwidth. In some cases, literature reports information about a tolerance of tissues up to 50–60 J/cm^2, which is most probably referring to the overall irradiation, i.e. UVA + visible light measurements.

Testing colorant materials (blue, deep red, violet) and MTT direct reducing materials may be challenging since they can interfere with the MTT endpoint. MatTek and other tissue model suppliers have developed procedures to deal with such materials. The procedures are described in detail in the SOPs provided by the tissue producer.

35.6.2 Possible Protocol Adaptations

The protocol for the topical application could be enhanced by the screening of the inflammatory mediators, e.g. IL-1α, IL-18. There is no specific protocol developed, however, statistically significant increase of IL-1α and IL-18 might be a sign of mild skin irritation or sensitisation (in the non-irradiated part of the experiment) and photo-irritation/photo-sensitisation in the irradiated part of the expedient.

Some attempts have been made toward developing systemic phototoxicity protocol (for systemically applied pharmaceuticals and food additives), however, these studies are not yet completed.

35.6.3 Challenges and Opportunities

The EpiDerm™ H3D-PT can be regarded as a valid and highly standardised test that could be ideally used as an adjunct test to the 3T3-NRU-PT or as an alternative method if 3T3-NRU-PT is not applicable (UVB-absorbance range, poor solubility of materials, questionable bioavailability). The method was pre-validated by ECVAM and further assessed in several studies by independent laboratories worldwide. An official, formal validation study, specifically designed for the role of this test as described above would help in further establishing its use.

35.7 Conclusions

The EpiDerm™ H3D PT has been developed and pre-validated almost 20 years ago. Since then the protocol underwent only minor modifications reflecting the knowledge acquired by the testing laboratories during the use of the EpiDerm™ model and the *in vitro* phototoxicity assay.

Most of the data obtained with the EpiDerm™ H3D PT are summarised in this chapter and provide an evidence, that despite lack of the formal validation, the assay is extremely robust and provides reliable and reproducible outcomes over 2 decades.

Since implementation of the H3D PT assays into the ICH S10 Guideline, the interest of scientists in *in vitro* phototoxicity testing is again increasing. Not only pharmaceutical, but also cosmetic and chemical industries would benefit from the implementation of this assay into the OECD TGs. Implementation of a 3D tissue based assay would enable to further deal with some false positive outcomes obtained for chemicals and cosmetic ingredients using the formally validated test methods based on a mouse cell line (OECD TG 432).

References

1. OECD. OECD guidelines for the testing of chemicals no. 432: in vitro 3T3 NRU phototoxicity test. Paris: Organisation for Economic Co-operation and Development; 2004. p. 15.
2. EMEA. Note for Guidance on Photosafety Testing (CPMP/SWP/398/01). London, UK: European Agency for the Evaluation of Medicinal Products, Committee for Proprietary Medicinal Products; 2002. p. 8. http://www.emea.eu.int.
3. SCCNFP. The Scientific Committee on Cosmetic Products and Non-Food Products Intended for Consumers: Notes of Guidance for Testing of Cosmetic Ingredients for their Safety Evaluation. 5th Revision, adopted by the SCCNFP during the plenary meeting of 20 October 2003 (SCCNFP/0690/03 Final). 102pp. Brussels, Belgium: DG SANCO/C/2, European Commission. 2003.
4. EC. Commission Regulation (EU) No. 1197/2013 of 25 November 2013 amending Annex III to Regulation (EC) No. 1223/2009 of the European Parliament and of the Council on cosmetic products; 2013. http://eur-lex.europa.eu/legal-content/EN/TXT/?uri=celex:32013R1197.
5. Liebsch M, Spielmann H, Pape W, Krul C, Deguercy A, Eskes C. UV-induced effects. ATLA. 2005;33(1):131–46.
6. Liebsch M, Barrabas C, Traue T, Spielmann H. Entwicklung eines in vitro Tests auf dermale Phototoxizität in einem Modell menschlicher Epidermis (EpiDerm™). ALTEX. 1997;14:165–74.
7. Liebsch M, Traue D, Barrabas C, Spielmann H, Gerberick GF, Cruse L, Diembeck W, Pfannenbecker U, Spieker J, Holzhütter HG, Brantom P, Aspin P, Southee J. Prevalidation of the EpiDerm phototoxicity test. In: Clark D, Lisansky S, Macmillan R, editors. Alternatives to animal testing II: proceedings of the second international scientific conference organised by the European Cosmetic Industry, Brussels, Belgium. Newbury: CPL Press; 1999. p. 160–6.
8. ECVAM. ECVAM feasibility study. Can the pre-validated EpiDerm in vitro phototoxicity test be upgraded to quantify phototoxic potency of topical phototoxins? Contract no. 1986-2002-09 F1 EDISP DE. 2002.
9. Kandarova H. Evaluation and validation of reconstructed human skin models as alternatives to animal tests in regulatory toxicology. PhD thesis; 2006. http://www.diss.fu-berlin.de/diss/receive/FUDISS_thesis_000000002248.
10. Kejlova K, Jirova D, Bendova H, Kandarova H, Weidenhoffer Z, Kolarova H, Liebsch M. Phototoxicity of bergamot oil assessed by in vitro techniques in combination with human patch tests. Toxicol In Vitro. 2007;21:1298–303.
11. Kejlová K, Jírová D, Bendová H, Gajdoš P, Kolářová H. Phototoxicity of essential oils intended for cosmetic use. Toxicol In Vitro. 2010;24(8):2084–9.
12. Jırova D, Kejlova K, Bendova H, Ditrichova D, Mezulanikova M. Phototoxicity of bituminous tars—correspondence between results of 3T3 NRU PT, 3D skin model and experimental human data. TIV. 2005;19:931–4.
13. Jones P, King A, Lovell W, Earl L. Phototoxicity testing using 3-D reconstructed human skin models. In: Clark D, Lisansky S, Macmillan R, editors. Alternatives to animal testing II: proceedings of the second international scientific conference organised by the European cosmetic Industry, Brussels, Belgium. Newbury, UK: CPL Press; 1999. p. 138–41.

14. Jones PA, King AV, Earl LK, Lawrence RS. An assessment of the phototoxic hazard of a personal product ingredient using in vitro assays. Toxicol In Vitro. 2003;17:471–80.
15. Liskova A. Evaluation of the phototoxic potential of selected compounds using the 3D reconstructed human epidermis model – EpiDerm. Diploma Thesis. SCHK, Bratislava, Reg. number FCHPT-104538-42749; 2016. 93.p.
16. Pratt LF, Cerven DR, DeGeorge GL. Validation of 3D Skin Model for Cosmetic, Chemical and Medical Device Phototoxicity Testing (EPARS). MB Research Laboratories. IVTIP Conference. 2011.
17. Sohn S, Ju JH, Son KH, Lee JP, Kim J, Lim CH, Hong SK, Kwon TR, Chang M, Kim DS, Yoon HS, Park KL. Alternative methods for phototoxicity test using reconstructed human skin model. ALTEX. 2009;26(Spec. Issue):136.
18. ICH S10. ICH harmonised tripartite guideline photosafety evaluation of pharmaceuticals S10, 2013. http://www.ich.org/fileadmin/Public_Web_Site/ICH_Products/Guidelines/Safety/S10/S10_Step_4.pdf.
19. Api AM. In vitro assessment of phototoxicity. In Vitro Toxicol. 1997;10:339–50.
20. Liebsch M, Döring B, Donelly TA, Logemann P, Rheins LA, Spielmann H. Application of the human dermal model Skin2 ZK 1350 to phototoxicity and skin corrosivity testing. Toxicol In Vitro. 1995;9:557–62.
21. MatTek. Phototoxicity Protocol For EpiDerm™ Model (EPI-200) (Draft). Version April 21, 2015; 2015, p. 20.
22. Cannon CL, Neal PJ, Southee JA, Kubilus J, Klausner M. New epidermal model for dermal irritancy testing. Toxicol In Vitro. 1994;8:889–91.
23. ZEBET. Phototoxicity protocol for use with EpiDerm™ model (EPI-200). Version November 5; 1997. p. 22.
24. Ceridono M, Tellner P, Bauer D, Barroso J, Alépée N, Corvi R, et al. The 3T3 neutral red uptake phototoxicity test: practical experience and implications for phototoxicity testing-the report of an ECVAM-EFPIA workshop. Regul Toxicol Pharmacol. 2012;63(3):480–8.
25. King A, Lovel W, Earl L. Phototoxicity tesing using 3D reconstructed human skin models. In: Clark D, Lisansky S, Macmillan R, editors. Alternatives to animal testing II: proceedings of the second international scientific conference organised by the European cosmetic Industry, Brussels, Belgium. Newbury, UK: CPL Press; 1999.

Part V

Skin Genotoxicity

Overview on Current Status

36

Stefan Pfuhler and Kerstin Reisinger

36.1 Status Overview

The accumulation of persistent DNA damage in somatic or germ cells is associated with a variety of adverse health outcomes like the development of cancer, neurodegenerative conditions, or reproductive and developmental ailments. Therefore, the assessment of the genotoxicity hazard of chemicals is a central requirement in many legislations which often decides on the fate of raw materials during product development. In addition, the assessment of genotoxicity is utilized for the cancer risk assessment of the respective chemical, and the presence or absence of such a hazard is used for classification and labeling in the EU and globally [1, 2].

Genotoxicity serves as a superordinate term defining the potential of a chemical to damage DNA. This can happen through direct interaction with DNA or indirectly through interaction with proteins, which are related to DNA integrity like the spindle apparatus or through interference with DNA repair and DNA organization (e.g., transferases, topoisomerases). Resulting damage is recognized by the cell and may be repaired or could lead to cell death. In both cases, the effect of the chemical does not manifest as a heritable change in DNA sequence. Agents damaging DNA are differentiated into three classes. Mutagens are chemicals which lead to a change in the nucleotide sequence through base pair substitutions and small insertions or deletions. Clastogens interfere with the chromosomal structure by inducing strand breaks that lead to loss or rearrangement of chromosome segments. Aneugens induce numerical chromosome aberrations inducing gain or loss of entire chromosomes or chromosome sets. These effects are not limited to chemicals but may also

S. Pfuhler (✉)
Procter and Gamble, Mason Business Centre, Mason, 45040 OH, USA
e-mail: pfuhler.s@pg.com

K. Reisinger
Henkel Beauty Care, Henkelstr. 67, 40589 Düsseldorf, Germany
e-mail: kerstin.reisinger@henkel.com

C. Eskes et al. (eds.), *Alternatives for Dermal Toxicity Testing*,
DOI 10.1007/978-3-319-50353-0_36

evolve in response to physical (e.g., UV- or X-rays) or biological (e.g., viruses) stimuli.

Due to the diverse nature of mechanisms, no single *in vitro* or *in vivo* assay is able to detect all types of genotoxins. Therefore, international guidelines for assessing the genotoxic potential recommend the use of test batteries to evaluate industry chemicals (Regulation, Evaluation, authorization of Chemicals; REACh, [3]), pharmaceuticals (International Conference on Harmonization; [4]), biocides [5], plant protection products [6], or cosmetics [7]. Since genotoxicity assays are seen as hazard identification assays, the route of exposure of a substance has historically not always been paid much attention to, but this is changing as witnessed by recent revisions of genotoxicity testing guidelines, e.g., OECD TG 474 [8] (*In vivo* mammalian micronucleus test) and OECD TG 489 [9] (*in vivo* mammalian alkaline comet assay). These now state that the anticipated route of human exposure should be considered when designing an assay. With this comes an increased focus on effects that may occur at the organ of first contact, which for dermally exposed substances is the skin.

Generally, test batteries foresee the use of *in vitro* methods first. For industrial chemicals, biocides, and cosmetics, the evaluation of genotoxicity can normally be finalized if all *in vitro* assays performed (usually 2 or 3) were negative. In the case of positive (unfavorable) results, follow-up experiments have to be conducted to investigate the relevance of the results from the standard *in vitro* battery. These are usually performed *in vivo*. For pharmaceuticals and plant protection products, *in vivo* tests are mandatory independent of the outcome of the respective *in vitro* test(s). In contrast, since 2013, testing of cosmetics ingredients in Europe relies on *in vitro* methods only [10].

It has been shown that the standard *in vitro* test batteries are sensitive for identifying *in vivo* genotoxins and rodent carcinogens. However, they exhibit a low specificity (ability to correctly identify non-carcinogens) especially when two or more tests are combined as required for the safety assessment in several industries [11, 12]. This high rate of "false" or "misleading" positives may lead to unnecessary *in vivo* follow-up testing according to legislations which allow for, or demand, animal experiments, whereas cosmetic ingredients may be unnecessarily excluded from marketing. Efforts have been undertaken to improve the predictive capacity of existing *in vitro* methods, and new methods were developed to supplement existing test batteries to allow for animal-free follow-up testing [13].

Several reasons have been identified for the aforementioned low specificity of classic *in vitro* genotoxicity tests. The assays often use rodent cell lines which do not have functional p53 regulation and therefore suffer from compromised cell cycle control. In addition, a variety of mutations accumulated during their widespread use in the last decades, resulting in an instable genome for some of the cell lines. Most of these assays simulate the liver as the major organ of xenobiotic metabolism to convert potential pro-mutagens into reactive metabolites. For this purpose, an external metabolizing system, normally liver lysate of Aroclor 1254 or phenobarbital-/β-naphthoflavone-treated rats, so-called S9 mix, is added to the *in vitro* assays. However, this approach has several drawbacks. The preparation of

S9 mix requires a break-up of cell structures, resulting in reduced overall metabolic capacity but specifically an imbalance between phase I oxidative metabolism and phase II conjugation reactions. The CYP450 phase I system is often responsible for activation/toxification of a compound, while phase II usually is a detoxification reaction. Phase II enzymes are underrepresented in S9, and cofactors for the reaction are also often lacking. Furthermore, S9 mix tends to exhibit cytotoxic effects itself, preventing longer incubation times, which may be required to identify certain genotoxins. Finally, S9 mix is prepared from treated rats, and hence does not represent normal (human) liver metabolism and is certainly not at all representative of dermal metabolism.

In order to address genotoxic effects in the skin as the first site of contact of products, e.g., cosmetics, household cleaners, plant protection products, and dermal pharmaceuticals, until recently, only *in vivo* assays were available. Prior to the development of reconstructed skin (RS) models, only these took the barrier function of the skin and its organ-specific xenobiotic metabolism into account. Available animal models that allow for genotoxicity testing in the skin include the transgenic rodent gene mutation assay (TRG), the comet assay, and the skin micronucleus test, all of which are described here in more detail. For rodent TRG assays, an updated OECD Testing Guideline was published in 2013 [14]. The principle of the method is based on the presence of multiple copies of a chromosomally integrated plasmid or phage shuttle vector in every cell of a transgenic rat or mouse, carrying reporter genes that are normally not expressed in mammals. These readout systems allow for the detection of various types of mutations. After the dermal exposure of test compounds, the skin and other organs of interest are isolated and the genomic DNA is prepared. Subsequently, the shuttle vectors are recovered, and the phenotype of the reporter gene is analyzed in a bacterial host in terms of mutant frequencies. TGR models for which sufficient data are available to permit evaluation include MutaMouse™ and Big Blue® rat or mouse, both using the lac-operon for the detection of mutagens.

The second method allowing for detection of DNA damage, also recently adopted by the OECD, is the *in vivo* comet assay [9]. Like for the TRG assays, the skin and other organs of interest can be isolated after the dermal exposure of test compounds. Here, a cell suspension is prepared from the organs, and DNA damage is investigated after subjecting the cells/nuclei to a protocol that enables visualization of DNA strand breaks. Advantages of this method are its ease of application and the broad spectrum of DNA damage that can be detected; disadvantages are that the assay may not be sensitive for certain types of DNA-damaging agents, e.g., DNA intercalators [15].

A third method for detecting DNA damage in the skin is the skin micronucleus test which was developed in order to evaluate the effects of clastogens and aneugens in rodent skin [16]. In contrast to the more broadly used main variant which evaluates either bone marrow or peripheral blood erythrocytes, usually after oral exposure [8], the skin micronucleus considers a different exposure route and therefore also different metabolism. The skin micronucleus test, which is not a guideline assay, may help detect highly reactive compounds with a short lifespan that may not reach the bone marrow. For this purpose, keratinocytes are isolated and analyzed after an incubation period of a total of 72 h and repeated exposure. The frequency of micronuclei, small extra nuclei which contain entire chromosomes or fragments of them, is subsequently recorded [17, 18].

36.2 Summary

All of the above-described *in vivo* assays are time and cost intensive, and animal experiments are generally banned for cosmetic industry since 2013 due to the 7th Amendment to the EU Cosmetic Directive. When looking for alternatives to *in vivo* assays, it is important to consider the recent developments and the learnings with regard to *in vitro* genotoxicity assays that were described earlier. A new assay would need to address shortcomings of the standard *in vitro* test methods, i.e., would be composed of human primary cells which eliminates the species barrier and uses p53 competent cells, allow a more realistic exposure scenario with the topical application of the compounds, and better reflect key properties of human skin-like barrier function and organ-specific metabolism. Such alternative assays for substances exposed via the dermal route do now exist and base on human 3D skin models, namely, the 3D Skin Comet assay and the reconstructed skin micronucleus assay (RSMN). They allow for identification of DNA lesions resulting from mutagens, clastogens, and aneugens and are composed of human primary cells which eliminate the species barrier and at the same time use p53 competent cells, thus resembling human skin in many ways. Relevant doses of compounds with different physical-chemical properties can be applied, and, contrary to testing in 2D cultures, compounds which do not penetrate the barrier cannot evoke any toxicological effect. Thus the bioavailability of a substance after dermal exposure is simulated more realistically. Furthermore, the organ-specific xenobiotic metabolism of human skin is appropriately considered by these models [19, 20].

Chapters 37 and 38 of this book continue with a description of both methodologies and elaborate on their intended use in current test batteries and on their regulatory status.

References

1. EU Regulation. EC. No 1272/2008 OF THE EUROPEAN PARLIAMENT AND OF THE COUNCIL on classification, labelling and packaging of substances and mixtures, amending and repealing Directives 67/548/EEC and 1999/45/EC, and amending Regulation (EC) No 1907/2006. Official Journal of the European Union L 353/1. 2008.
2. UN Globally Harmonized System. 4th ed., United Nation, New York Geneva. 2011. http://www.unece.org/fileadmin/DAM/trans/danger/publi/ghs/ghs_rev04/English/ST-SG-AC10-30-Rev4e.pdf.
3. EU Regulation. EC No. 1907/2006 OF THE EUROPEAN PARLIAMENT AND OF THE COUNCIL concerning the Registration, Evaluation, Authorisation and Restriction of Chemicals (REACH), establishing a European Chemicals Agency, amending Directive 1999/45/EC and repealing Council Regulation (EEC) No. 793/93 and Commission Regulation (EC) No. 1488/94 as well as Council Directive 76/769/EEC and Commission Directives

91/155/EEC, 93/67/EEC, 93/105/EC and 2000/21/EC. Official Journal of the European Union L 36/84. 2006.

4. ICH Guideline. GUIDANCE ON GENOTOXICITY TESTING AND DATA INTERPRETATION FOR PHARMACEUTICALS INTENDED FOR HUMAN USE S2(R1). Approval by the Steering Committee of S2(R1) under Step 4 and recommendation for adoption to the three ICH regulatory bodies (9 Nov 2011). http://www.fda.gov/downloads/Drugs/Guidances/ucm074931.pdf.
5. EU Regulation. EC No. 528/2012 of the European Parliament and of the Council of 22 May 2012 concerning the making available on the market and use of biocidal products. Official Journal of the European Union L 167. 2012.
6. EU Regulation. EC No. 1107/2009 of the European Parliament and of the Council concerning the placing of plant protection products on the market and repealing Council Directives 79/117/EEC and 91/414/EEC. Official Journal of the European Union L 309/1. 2009.
7. SCCS. ADDENDUM to the SCCS's Notes of Guidance (NoG) for the Testing of Cosmetic Ingredients and their Safety Evaluation, 8th Revision (SCCS/1501/12) 2014. http://ec.europa.eu/health/scientific_committees/consumer_safety/docs/sccs_o_156.pdf.
8. OECD. Guideline for the testing of chemicals, test no. 474: mammalian erythrocyte micronucleus test. 2014. http://www.oecd.org/env/test-no-474-mammalian-erythrocyte-micronucleus-test-9789264224292-en.htm.
9. OECD. Guidelines for the testing of chemicals, test no. 489: in vivo mammalian alkaline comet assay. 2014. http://www.oecd-ilibrary.org/environment/test-no-489-in-vivo-mammalian-alkaline-comet-assay_9789264224179-en;jsessionid=npkmv8v01k1s.x-oecd-live-02.
10. EU Regulation. EC No. 1223/2009 OF THE EUROPEAN PARLIAMENT AND OF THE COUNCIL of 30 November 2009 on cosmetic products. Official Journal of the European Union L 342/59.
11. Kirkland D, Aardema M, Müller L, Hayashi M. Evaluation of the ability of a battery of three in vitro genotoxicity tests to discriminate rodent carcinogens and non-carcinogens II. Further analysis of mammalian cell results, relative predictivity and tumour profiles. Mutat Res. 2006;608(1):29–42.
12. Kirkland D, Aardema M, Henderson L, Müller L. Evaluation of the ability of a battery of three in vitro genotoxicity tests to discriminate rodent carcinogens and non-carcinogens I. Sensitivity, specificity and relative predictivity. Mutat Res. 2005;584(1-2):1–256.
13. Pfuhler S, Kirst A, Aardema M, Banduhn N, Goebel C, Araki D, Costabel-Farkas M, Dufour E, Fautz R, Harvey J, Hewitt NJ, Hibatallah J, Carmichael P, Macfarlane M, Reisinger K, Rowland J, Schellauf F, Schepky A, Scheel J. A tiered approach to the use of alternatives to animal testing for the safety assessment of cosmetics: genotoxicity. A COLIPA analysis. Regul Toxicol Pharmacol. 2010;57(2-3):315–24.
14. OECD. Guideline for the testing of chemicals, test no. 488: transgenic rodent somatic and germ cell gene mutation assay. 2013. http://www.oecd-ilibrary.org/environment/test-no-488-transgenic-rodent-somatic-and-germ-cell-gene-mutation-assays_9789264203907-en.
15. Speit G, Kojima H, Burlinson B, Collins AR, Kasper P, Plappert-Helbig U, Uno Y, Vasquez M, Beevers C, De Boeck M, Escobar PA, Kitamoto S, Pant K, Pfuhler S, Tanaka J, Levy DD. Critical issues with the in vivo comet assay: a report of the comet assay working group in the 6th international workshop on Genotoxicity testing (IWGT). Mutat Res Genet Toxicol Environ Mol Mutagen. 2015;783:6–12.
16. Nishikawa T, Haresaku M, Adachi K, Masuda M, Hayashi M. Study of a rat skin in vivo micronucleus test: data generated by mitomycin C and methyl methanesulfonate. Mutat Res. 1999;444(1):159–66.
17. Morita T, MacGregor JT, Hayashi M. Micronucleus assays in rodent tissues other than bone marrow. Mutagenesis. 2011;26(1):223–30.
18. Nishikawa T, Haresaku M, Fukushima A, Nakamura T, Adachi K, Masuda M, Hayashi M. Further evaluation of an in vivo micronucleus test on rat and mouse skin: results with five skin carcinogens. MutatRes. 2002;513(1-2):93–102.
19. Hewitt NJ, Edwards RJ, Fritsche E, Goebel C, Aeby P, Scheel J, Reisinger K, Ouédraogo G, Duche D, Eilstein J, Latil A, Kenny J, Moore C, Kuehnl J, Barroso J, Fautz R, Pfuhler S. Use

of human in vitro skin models for accurate and ethical risk assessment: metabolic considerations. Toxicol Sci. 2013;133:209–17.

20. Wiegand C, Hewitt NJ, Merk HF, Reisinger K. Dermal xenobiotic metabolism: a comparison between native human skin, four in vitro skin test systems and a liver system. Skin Pharmacol Physiol. 2014;27:263–75.

Reconstructed Skin Micronucleus Assay (RSMN)

37

Stefan Pfuhler and Kerstin Reisinger

37.1 Principle of the Test Method and Scientific Basis

The micronucleus test has very widespread use in genetic toxicology testing since it can detect potentially clastogenic and aneugenic chemicals with high sensitivity (e.g. [1, 2]). OECD Testing Guidelines exist for both the *in vivo* version (OECD TG 474, [3]; first adopted in 1983) and for the *in vitro* micronucleus version, which was first adopted in 2010 (OECD TG 487, [4]). The micronucleus test is a genotoxicity test for the detection of small extra nuclei, i.e. micronuclei (MN), in the cytoplasm of interphase cells. Micronuclei may originate from acentric chromosome fragments or whole chromosomes that are unable to migrate to the poles during the anaphase stage of cell division, i.e. micronuclei represent damage that has been transmitted to daughter cells. Therefore, it is important to demonstrate that cell proliferation has occurred in both the control and treated cultures, and the extent of test chemical-induced cytotoxicity or cytostasis is assessed in all of the cultures that are scored for the presence of micronuclei [4]. Erythrocytes are the cells of choice for the *in vivo* MN, whereas cell lines or primary cells are suggested by the OECD TG for the *in vitro* micronucleus test (MNvit). It has been shown that cells of human origin like human lymphocytes or TK6 cells provide a better predictivity when compared to rodent cell lines [5, 6], suggesting that cells of human origin may be preferable for use with the MNvit. To allow for cell proliferation and the implementation of chromosome damage or other effects on cell cycle/cell division, cells are grown for a sufficient period of time during or after exposure to the test chemical.

S. Pfuhler (✉)
Procter and Gamble, Mason Business Centre, Mason, OH 45040, USA
e-mail: pfuhler.s@pg.com

K. Reisinger
Henkel Beauty Care, Henkelstr. 67, 40589 Düsseldorf, Germany
e-mail: kerstin.reisinger@henkel.com

C. Eskes et al. (eds.), *Alternatives for Dermal Toxicity Testing*,
DOI 10.1007/978-3-319-50353-0_37

Subsequently, interphase cells are harvested and analysed for the presence of micronuclei. Today's broad acceptance of the *in vitro* assay was supported by the development of the cytokinesis-block methodology [7] which allows easy identification of cells that have undergone nuclear division, thereby enabling accurate assessment of the appropriate cell population for quantifying micronuclei (MN).

In genotoxicity hazard identification the primarily used route of exposure for the *in vivo* micronucleus test is oral. While this makes sense for substance classes which humans are exposed to primarily via the oral route this seems less logical when addressing products, e.g. cosmetics and household cleaners, for which the skin is the first site-of-contact. Therefore, *in vivo* assays were developed that take skin-specific characteristics like its barrier function and organ-specific xenobiotic metabolism into account, i.e. the rodent skin micronucleus test [8]. In this assay keratinocytes are isolated after repeated daily exposure for 72 h, followed by analysis of the frequency of micronuclei [9, 10]. These approaches which were summarized by Morita et al. [11] utilized rat or mouse skin as a surrogate for human skin. Such animal methods are not acceptable for the use with cosmetic ingredients as described in Chap. 36. In conjunction with the '3R's', human reconstructed skin-based genotoxicity assays have been suggested as potential follow-up tools for positive results from standard genotoxicity assays [12, 13] and found support by international experts groups, e.g. the International Workshop on Genotoxicity testing (IWGT) [14] and the European Commission's Scientific Committee on Consumer Safety (SCCS) [15].

This methodology is based on the *in vivo* skin micronucleus method in that it involves dermal application of a compound two or more times, separated by 24 h intervals, isolation of keratinocytes and their analysis for the occurrence of micronuclei.

37.2 Current Validation Status

In 2005, in foresight of an upcoming animal testing ban, a task force initiated by the European Commission and led by the European Centre for the Validation of Alternative Methods (ECVAM) had recommended a new approach for the safety assessment of cosmetics which would include *in vitro* skin model-based assays as a potential replacement for *in vivo* approaches [13]. Shortly after that, a team of genotoxicity experts started developing and establishing the RSMN, utilizing MatTek's EpiDerm™ reconstructed human skin tissue models [12]. First proof-of-concept studies showed that dividing keratinocytes could be evaluated using the cytokinesis-block methodology with cytochalasin B, enabling scoring of micronucleus in binucleated (dividing) cells [12]. The background frequency of MN was shown to be low and reproducible in this model, and statistically significant increases in the frequency of micronucleated cells were induced by two model genotoxins [12]. Subsequently, a validation program was designed and funded by Cosmetics Europe and steered by its Genotoxicity Task Force with additional support from external experts [16]. Testing was extended to three US laboratories [17, 18] and then transferred to two European laboratories, which was of significance because the RS models were supplied from a US provider (MatTek, MA) [19]. As part of this process, two training workshops were held to standardize the protocol and harmonize scoring of micronuclei, both of which were subsequently described and published by Dahl et al. [20]. Next, three coded compounds (*N*-ethyl-*N*-nitrosourea (ENU), Mitomycin C (MMC) (both genotoxic carcinogens) and cyclohexanone (non-carcinogen and non-genotoxic)) were tested by three laboratories demonstrating reproducibility of the results within and between laboratories [19].

After having demonstrated good transferability and inter/intra-laboratory reproducibility, the number of chemicals tested in the RSMN, selected by external experts, was extended to 38. This set included true positive and true negative chemicals, which showed concordant results in historical *in vitro* and *in vivo* testing, as well as a sub-set of negative compounds, the so-called "misleading" positives, for which positive results were obtained *in vitro* that did not correlate with historical *in vivo* genotoxicity studies or carcinogenicity. The chemicals were selected to represent different chemical classes and modes of action and were provided to the three testing laboratories involved in a double-blinded fashion. Initial results suggest a high predictive capacity of this assay [21], with final validation data pending publication.

This is further supported by results generated by three research groups outside of the RSMN validation study. The first study compared the predictive capacity of the RSMN with that of a '2-dimensional' version of the *in vitro* micronucleus test, which was performed with normal human epidermal keratinocytes (NHEKs) [22]. These authors investigated eight chemicals that included three *in vivo* genotoxins and five *in vivo* non-genotoxins, all of which were giving false-positive results in the *in vitro* chromosomal aberration test with rodent cell lines. False-positive results were obtained with the NHEK cell assays, while the RSMN correctly predicted the *in vivo* results for both the *in vivo* genotoxins and non-genotoxins [22], thereby providing further support for the use of 3-dimensional, human skin-like tissue models. The second study presented results from testing a series of seven aromatic amines (five non-carcinogens and two carcinogens), all of which were correctly predicted by the RSMN ([23], submitted). In the third study, hydrogen peroxide (H_2O_2), a clastogen in 2D *in vitro* systems but non-genotoxic *in vivo*, and a non-genotoxic carcinogen, methyl carbamate, correctly produced negative results while two *in vivo* genotoxins produced the expected positive results in the RSMN [24]. Aside from testing chemicals, the RSMN has also been shown to work well with physical agents. Schmid et al. [25] irradiated EpiDerm™ reconstructed skin tissues with proton beams, a method used in laser accelerated cancer radiotherapy. Increasing doses of such laser-induced pulsed X-rays led to a linear increase in micronuclei in the tissues, thereby demonstrating excellent correlation between this DNA-breaking event and the biological response.

Some genotoxins require metabolic activation to exert genotoxic activity, so-called pro-mutagens, several of which were investigated in more detail by Aardema et al. DOI:10.1016/j.mrgentox.2012.08.009 [26]. Since the skin has been shown to have a very low phase 1 (often bioactivating) capacity [26], it was considered that these chemicals may require a longer incubation duration in order to generate sufficient levels of the ultimate genotoxin. Based on the result observed for one of the chemicals studied, 4-nitroquinoline-*n*-oxide (4NQO), it was recommended that a 48 h treatment is used for a first experiment, followed by a longer treatment period (72 h) if the outcome of the standard 48 h treatment was negative or questionable [26]. This practice would not cost extra time, since a negative result in the first 48 h experiment would need to be confirmed in a second experiment anyway. Interim results of the ongoing validation exercise do support this change from the initial 48 h exposure protocol and indicate that a 48/72 h regimen will help improve the sensitivity of the RSMN and the treatment schedule was therefore adapted accordingly.

The increase of the total exposure time to 72 h gets further support from the outcome of a thorough investigation of xenobiotic metabolizing enzymes (XMEs) in native human skin, RS, and monolayer cultures of skin cells [26], using both a proteomic approach and measurement of actual substrate metabolism. Although CYP 1 family enzymes are generally present only at low levels or even absent in skin, CYP1A seemes to be inducible by classic polycyclic aromatic hydrocarbon inducers like 2,3,7,8-tetrachlorodibenzo-p-dioxin or B[a]P [27]. The level of activity reached a peak 48 h after first exposure which supports the extension of substance exposure beyond 48 h. The studies, summarized by Hewitt et al. [26], showed that multiple other phase 1 enzymes were present at significant levels, such as alcohol dehydrogenases, aldehyde dehydrogenases, amine oxidases and epoxide hydrolases. Phase 2 enzymes like glutathione *S*-transferases, *N*-acetyltransferase 1 and UDP-glucuronosyltransferases were all readily measurable in whole skin and in 3D skin models at activity levels similar to those measured in the liver. These data support the view that skin tends to be more of a detoxification than a bioactivation organ and that the enzyme profile observed in human reconstructed skin models resembles the profile found in human native skin quite well [26].

In addition to protocol improvements and increasing the database of agents tested in the RSMN there also have been attempts for automation of the most time-consuming aspect of the assay—the manual evaluation (scoring) of the slides prepared from a single-cell suspension generated from the skin tissues. Aside from ongoing attempts to measure micronuclei using a flow-cytometric approach there has been some initial success with automated image analysis. Automated micronucleus detection using MetaSystems Metafer Slide Scanning Platform (Metafer) was compared with results from manual scoring and demonstrated concordance for the two model compounds scored [24].

Summarizing the above data and experience, it can be concluded that there is considerable evidence that the RSMN is a valuable new *in vitro* method for the assessment of genotoxicity of dermally exposed chemicals and drugs. Since its validation is already at an advanced stage it is suggested that the assay can be used for following-up positive or equivocal results generated in the standard *in vitro* genotoxicity tests, thereby serving as a direct replacment of animal studies.

37.3 Performance and Applicability of the Test Method

Reconstructed human skin based genotoxicity assays enable testing of dermally applied compounds that require hazard characterization. These models overcome limitations of 2D (submerged) monolayer or suspension cultures and allow for testing of lipophilic compounds as well as application of higher concentrations in cases where this is relevant for the in-use situation. In addition, they facilitate testing of particulate materials as shown by Willis et al. [28], which is discussed in more detail in Sect. 37.5.2.

The ongoing validation studies as well as published data [12, 17, 19, 22–24] cover a wide spectrum of chemical classes and the results generated to date indicate a high predictive capacity of the RSMN across chemical classes and therefore its broad applicability. While more than 50 chemicals have been tested so far it is recognized that its full domain of applicability has not been established, however, there is no indication that it would be any different than for the classical micronucleus assay, both *in vitro* and *in vivo*, which is accepted for use across substance classes. Notably the assay has also been shown to work with physical agents, i.e. radiation [25].

37.4 Brief Description of the Protocol

37.4.1 Study Design

The investigation of a test agent with the RSMN comprises a similar set of experiments than the standard OECD micronucleus test (OECD TG 487). The study design and performance was described in detail by Dahl et al. [20]. (1) First, an appropriate solvent is selected to dissolve the chemical before exposing the tissue, targeting a maximum concentration of 10 mg/100 μL or 10%, respectively. (2) A dose-range-finding experiment is designed to narrow-down the dose range and to enable a decision on the maximum test concentration which could be limited by: (a) the limit dose previously mentioned (10 mg/100 μL); (b) cytotoxicity or (c) solubility/precipitation of the test compound. Cytotoxicity of the test compound is determined by calculating the rate of binucleation (occurrence of cells that have undergone nuclear division since addition of cytochalasin B) relative to the control, as well as by determination of the relative proportion of living cells.

37.4.2 Experimental Design

An outline of the assay performance schedule is shown in Fig. 37.1. When the skin models arrive they are transferred from an agarose coated 24-well plate to a 6-well plate containing fresh medium. The models can be treated on the same day but ideally they are placed in an incubator overnight to recover from shipping. On the next day, the medium is replaced with medium containing cytochalasin B (cytoB) and the first

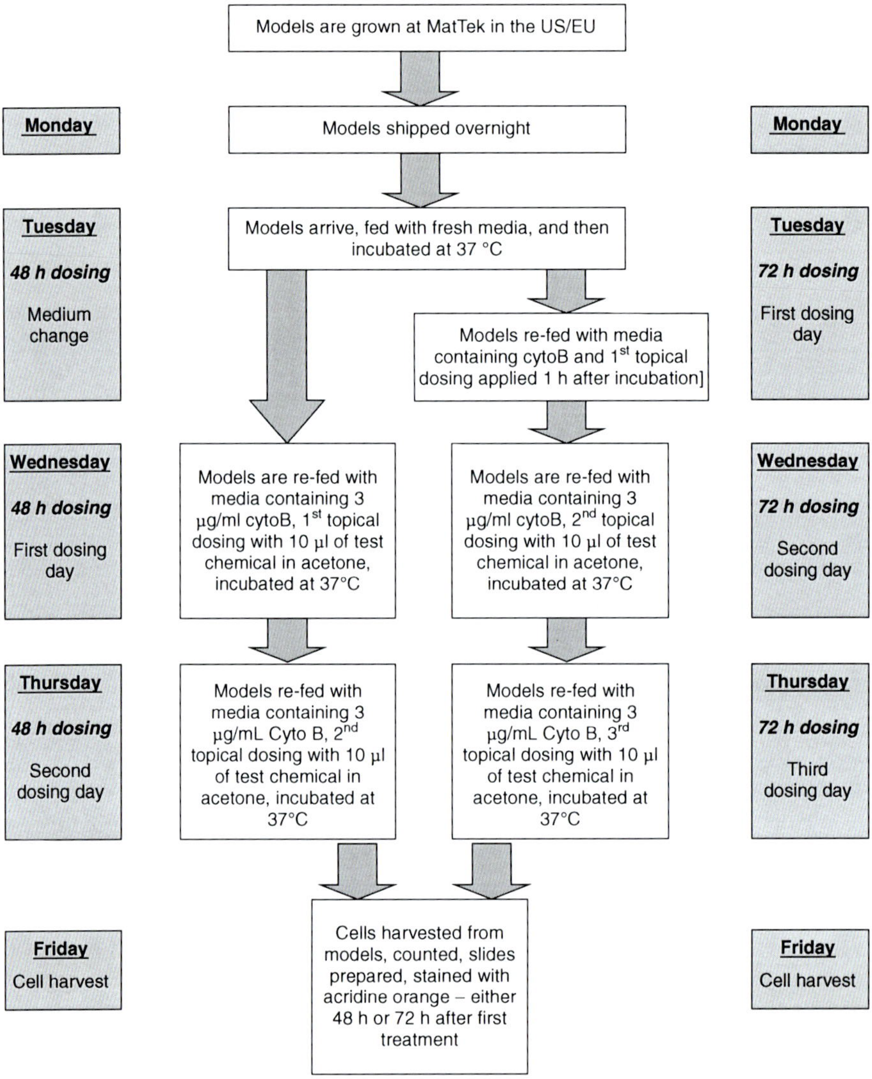

Fig. 37.1 An overview of the RSMN assay performance timeline (from [20], modified)

dose of test compound (normally in acetone, for alternatives refer to Sect. 37.5.1) is then applied to the upper side (stratum corneum) of the skin model. Twenty-four hours later, after replacing the medium again with fresh medium containing cytoB, the model is treated with a second dose of test compound. After 48 h of exposure, or after 72 h when longer treatment is desired for improved sensitivity as discussed in Sect. 37.2, cells from the basal layer and stratum spinosum of the models are harvested and prepared for analysis as described below (See Fig. 37.2 and Sect. 47.4.3; for resulting preparation see Fig. 37.2).

Dose-range selection for the definitive assay attempts including concentrations that reduce relative percent binucleation and/or relative live cell counts by 50 ± 10%

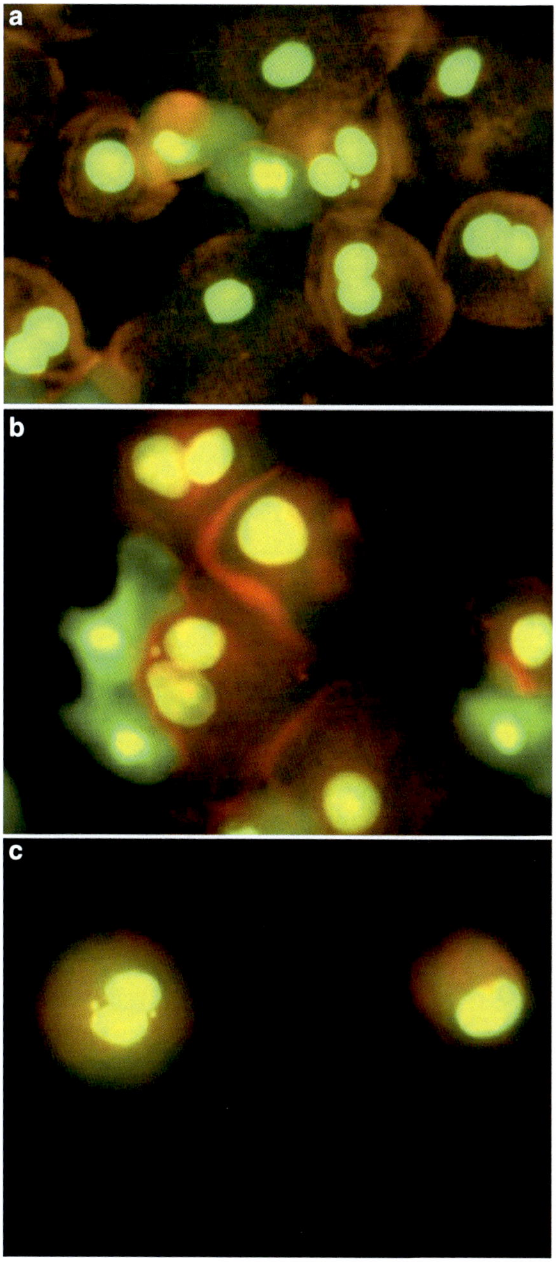

Fig. 37.2 Images of mono-and binucleated cells including cells positive for micronuclei. (**a**) mono- and binucleated cells from the stratum spinosum, all of which will be counted when determining the rate of cells that have undergone division (binucleation rate). (**b**) Picture includes cells stemming from the stratum corneum, identifiable by the cytoplasm staining green due to the lack of RNA. These are not counted when the binucleation rate is determined. (**c**) Binucleated cell with two micronuclei (on left)

(high cytotoxicity) and 30 ± 10% (intermediate cytotoxicity), and 10 ± 10% (low cytotoxicity), unless the test substance shows insignificant cytotoxicity up to the highest recommended test article concentration, or precipitation. The dose groups comprise at least two tissues, however, triplicate tissues are preferred. Ten microlitres of the solvent/test substance solution is topically applied to the centre of the EpiDerm™ tissue, assuring good spread of the solution across the entire surface of the skin models. Mitomycin C (MMC) is used as positive control. When establishing the assay it is recommended to include a negative (untreated) and a solvent control group. Untreated skin tissues do not need to be included anymore once sufficient solvent control data indicating that the solvent control has no impact on the background micronucleus rate of the tissues is available.

37.4.3 Cell Isolation and Micronucleus Test Procedure

Forty-eight or 72 h after the initial treatment, the skin models are trypsinized to obtain a single-cell suspension of keratinocytes from the basal layers, as described by Dahl et al. [20]. In brief, each tissue model insert is washed with buffer containing EDTA and then exposed to trypsin-EDTA solution for 10–15 min at 37 °C. After this initial trypsin incubation, the model is separated from the supporting membrane and transferred to a new well. The insert is thoroughly rinsed with trypsin-EDTA in the well to collect any remaining basal cells left on the supporting membrane. The skin model is then gently agitated to release additional attached basal cells from the detached model, with primarily stratum corneum remaining since it is resistant to further trypsinization. The single-cell suspension is then transferred to a conical tube containing warm medium with 10% fetal bovine serum to inactivate the trypsin. A sample of cell suspension is diluted with Trypan blue solution and counted using a haemocytometer to obtain a cell count and determine the proportion of live cells of each treatment compared to control. Others methods known to reliably obtain a live cell count can be used as well.

Once a single-cell suspension has been generated the protocol follows the steps as suggested by the OECD TG 487, also described by Dahl et al. [20]. The cell suspension is centrifuged and the cell pellet undergoes treatment with a hypotonic KCl solution before fixation with methanol/acetic acid fixative, followed by a second fixation step. At least two slides are prepared per skin tissue by gently dropping the cell suspension onto a clean, dry microscope slide. After the slides are completely dry, they are stained with acridine orange and can be analysed directly, or stored in the dark at 2–8 °C until analysis.

37.4.4 Analysis

All slides should be blind-coded before scoring and are then evaluated for binucleation and induction of micronuclei. The percentage of cells that have undergone division is determined by counting mononucleated, binucleated and multinucleated (>2 nuclei) cells, based on at least 500 cells per tissue (1000 per concentration).

Next, for all concentrations where the binucleation rate and the relative cell count are above 40% as compared to the concurrent solvent control, at least 500 cells per tissue (1000 per concentration) are evaluated for the occurrence of micronuclei. Aside from above scoring requirements the following validity criteria are applied for the assay (summarized from Dahl et al. [20]): (1) The yield of viable cells in the vehicle control should be higher than 5×10^4 cells per tissue. (2) The binucleation rate in each vehicle control tissue must be at least 25%. (3) The positive control must cause a statistically significant increase in the % MN compared with the average of the vehicle control tissues (one sided Fisher's exact test, $p < 0.05$). (4) Controls and each valid concentration must have at least two valid tissues per treatment. (5) The % relative viable cell count is 40% or more in each test article-treated tissue (no more than 60%), compared to the average of vehicle control tissues.

In addition to the statistical evaluation defining a positive or negative call, and in line with the procedures established for OECD TG 487, the following criteria are being looked at when deciding on the final outcome of a study. In brief, (a) dose-dependency of the response, (b) the statistical significance of the effect when compared to the solvent control, and (c) the magnitude of the effect in relation to the historical control data range. If all of these are positive, then the result of the study will be judged positive, if none of the three criteria is fulfilled the test item will be judged negative. In case one or two but not all of the criteria for a positive call are fulfilled for the test compound, expert judgment is required.

37.5 Perspectives from the Test Developer

37.5.1 Critical Steps in the Protocol

Since acetone (a volatile solvent with low viscosity) is a common solvent in the RSMN assay, care must be taken that the volumes pipetted are accurate and it is advisable to prepare the dosing solutions shortly before application and to cool the acetone to prevent evaporation of the dosing solution. The dosing solution is placed in the centre of the models, tilting the plate with a rotating movement to ensure that the whole surface of the model is covered by the dosing solution. Aside from the dosing there are two more critical protocol steps that require attention to detail—the cell separation and slide preparation. During cell separation it is important to assure a constant and reproducible procedure in order to achieve near complete separation of all viable keratinocytes from the remaining stratum corneum. This is essential since the total number of viable cells is an important cytotoxicity measure and this will also assure a good yield of scorable cells on the slides. To ensure a reproducible keratinocyte separation follow the steps as detailed by Dahl et al. [20] with one suggested difference: While the authors give a span of 10–15 min for the incubation of the tissues with trypsin-EDTA, we recommend the incubation to be 15 min to assist good cell separation. During slide preparation, cell fixation is another crucial step and will usually need some training/practice. During this step it is essential to avoid clumping of the cells which can be prevented by slowly adding (should take ~10–15 s) 3 mL of ice-cold, fresh MeOH/Acetic Acid (3:1,v/v) to the test tubes containing the cells in hypotonic KCl solution while mixing gently using a Vortex. After centrifugation and careful aspiration of the supernatant, the fixative is aspirated down to ~60 μL. A second fixation/centrifugation step using ice-cold MeOH/ Acetic Acid (40:1) will help improve the quality of the slides. Finally, the cell pellet is loosened by gently agitating the tube before the cell suspension is dropped onto pre-labelled slides using a glass Pasteur pipette. It is recommended to avoid overlap of cell drops on the slide and to prepare at least two slides per tissue.

The choice of solvents is another critical area since the choices of appropriate solvents for this assay that will penetrate but not damage the skin models are somewhat limited. Recommended solvents, limited to 10 μL/tissue, include acetone, ethanol, 3:1 ethanol/water, 4:1 acetone/olive oil and saline [20]. Solvents that will not immediately penetrate the skin, e.g. saline, should only be used if the other options are exhausted since the liquid will sit atop of the tissues for several hours which will likely affect the cells' oxygen supply leading to cellular stress [28]. This can also happen for chemicals that come out of solution after application, leading to precipitation of solid material on the tissues. For this reason it is recommended to avoid concentrations that cause extensive precipitation of the test substance.

The scoring process is also a critical protocol step and it is very important to assure proper training of the evaluator. Standards for the analysis of slides have been agreed on and were published recently [20]. In order to establish this assay in a laboratory, as proof of experimental competency in the RSMN, it is recommended to conduct a series of experiments providing a dose-response of positive controls, as well as low and reproducible micronucleus frequencies for non-treated and solvent exposed tissues. It is recommended that the slides are scored blind, without knowledge of the dose group. This phase aims at building proficiency and will also help build a historical database for the solvent, negative and positive controls.

37.5.2 Possible Protocol Adaptions

The micronucleus test can detect aneugenic as well as clastogenic effects, as outlined in Chap. 36. If mechanistic information is desired it is possible to differentiate aneugens from clastogens by checking micronuclei for the presence of whole chromosomes which will have a centromere, while micronuclei containing only chromosome fragments will usually not have a centromere (OECD TG 487). Centromere staining can be performed by using anti-kinetochore antibodies, FISH with pancentromeric DNA probes, or *in situ* labelling with pancentromere-specific primers, together with appropriate DNA counterstaining [2, 29].

Other foreseeable uses of the RSMN include the evaluation of the genotoxic potential of nano-sized particulate materials (nanomaterials (NM)). The idea of assessing NM or other solid, particulate materials that may come into contact with the skin using this skin-based method seems intriguing since these '3D' models possess a stratum corneum which exhibits an '*in vivo* like' barrier function. There is, however, only limited experience available for this type of use e.g, Willis et al. [28] doi: 10.1186/s12989-016-0161-5 and it remains to be seen how good the predictive capacity of the RSMN for these materials will be. More experience is needed especially in the context of the above described limitations of the assay, i.e. when precipitates from chemicals accumulate on the skin surface. During the validation phase such accumulation of precipitate has been associated with spurious increases in the micronucleus background frequency. Dose-selection will therefore be a key element in the successful adaptation of the RSMN for its use for particulate materials, including NM.

37.6 Conclusions

The RSMN assay combines the EpiDerm™ 3D reconstructed human skin model with a standard OECD Testing Guideline method, the *in vitro* micronucleus test, to provide a more realistic model for evaluating the genotoxic potential of dermally applied chemicals/products, such as cosmetics. The RSMN and 3D Skin Comet assays close a gap in the toolbox of *in vitro* genotoxicity assays since they have been shown to be advantageous for the evaluation of dermally exposed substances. The reconstructed skin models consist of human primary p53 competent cells, differentiating this approach from many *in vitro* genotoxicity assays, which are based on rodent cancer cell lines. The three-dimensional nature of the models supports their *in vivo*-like behaviour in terms of metabolism and barrier function and allows topical application of test compounds thereby mimicking the human exposure situation. From the results generated to date in independent research laboratories as well as during the ongoing validation of the RSMN it can be concluded that there is considerable evidence in support of the RSMN as a valuable new *in vitro* method for the assessment of genotoxicity of dermally exposed chemicals and drugs. Since its validation is already at an advanced stage it is suggested that the assay can be used for following-up positive or equivocal results generated in the standard *in vitro* genotoxicity tests, thereby filling a critical gap in the test battery.

References

1. Corvi R, Albertini S, Hartung T, Hoffmann S, Maurici D, Pfuhler S, van Benthem J, Vanparys P. ECVAM retrospective validation of in vitro micronucleus test (MNT). Mutagenesis. 2008;23(4):271–83.
2. Kirsch-Volders M, Sofuni T, Aardema M, Albertini S, Eastmond D, Fenech M, Ishidate M Jr, Kirchner S, Lorge E, Morita T, Norppa H, Surralles J, Vanhauwaert A, Wakata A. Report from the in vitro micronucleus assay working group. Mutat Res. 2003;540:153–63.
3. OECD. Guideline for the testing of chemicals, test no. 474: mammalian erythrocyte micronucleus test. 2014. http://www.oecdilibrary.org/environment/test-no-487-in-vitro-mammalian-cell-micronucleus-test_9789264224438-en;jsessionid=2nxqxkmd6f46d.x-oecd-live-02.
4. OECD. Guideline for the Testing of Chemicals, test no. 487: In vitro mammalian cell micronucleus test. 2014. http://www.oecdlibrary.org/docserver/download/9714561e.pdf?expires=1432204443&id=id&accname=guest&checksum=E96719F50FF0E66958D3C8A91B41B507.
5. Fowler P, Smith R, Smith K, Young J, Jeffrey L, Carmichael P, Kirkland D, Pfuhler S. Reduction of misleading ("false") positive results in mammalian cell genotoxicity assays. III: sensitivity of human cell types to known genotoxic agents. Mutat Res. 2014;767:28–36.

6. Fowler P, Smith K, Young J, Jeffrey L, Kirkland D, Pfuhler S, Carmichael P. Reduction of misleading ("false") positive results in mammalian cell genotoxicity assays. I. Choice of cell type. Mutat Res. 2012;742(1–2):11–25.
7. Fenech M, Morley AA. Cytokinesis-block micronucleus method in human lymphocytes: effect of in vivo ageing and low dose X-irradiation. Mutat Res. 1986;161:193–8.
8. Nishikawa T, Haresaku M, Adachi K, Masuda M, Hayashi M. Study of a rat skin in vivo micronucleus test: data generated by mitomycin C and methyl methanesulfonate. Mutat Res. 1999;444(1):159–66.
9. Nishikawa T, Haresaku M, Fukushima A, Nakamura T, Adachi K, Masuda M, Hayashi M. Further evaluation of an in vivo micronucleus test on rat and mouse skin: results with five skin carcinogens. Mutat Res. 2002;513(1–2):93–102.
10. Uno Y, Morita T, Luijten M, Beevers C, Hamada S, Itoh S, Ohyama W, Takasawa H. Micronucleus test in rodent tissues other than liver or erythrocytes: report of the IWGT working group. Mutat Res. 2015;783:19–22.
11. Morita T, MacGregor JT, Hayashi M. Micronucleus assays in rodent tissues other than bone marrow. Mutagenesis. 2011;26(1):223–30.
12. Curren RD, Mun GC, Gibson DP, Aardema MJ. Development of a method for assessing micronucleus induction in a 3D human skin model (EpiDerm™). Mutat Res. 2006;607:192–204.
13. Maurici D, Aardema M, Corvi R, Kleber M, Krul C, Laurent C, Loprieno N, Pasanen M, Pfuhler S, Phillips B, Sabbioni E, Sanner T, Vanparys P. Alternative (non-animal) methods for cosmetics testing: current status and future prospects. Genotoxicity/Mutagenicity. ATLA. 2005;33(Suppl 1):117–30.
14. Pfuhler S, Fellows M, van Benthem J, Corvi R, Curren R, Dearfield K, Fowler P, Frötschl R, Elhajouji A, Le Hégarat L, Kasamatsu T, Kojima H, Ouédraogo G, Scott A, Speit G. In vitro genotoxicity test approaches with better predictivity: summary of an IWGT workshop. Mutat Res. 2011;723:101–7.
15. SCCS. Position statement on genotoxicity/mutagenicity testing of cosmetic ingredients without animal experiments (SCCP/1212/09). 2009. http://ec.europa.eu/health/archive/ph_risk/committees/04_sccp/docs/sccp_s_08.pdf.
16. Pfuhler S, Kirst A, Aardema M, Banduhn N, Goebel C, Araki D, Costabel-Farkas M, Dufour E, Fautz R, Harvey J, Hewitt NJ, Hibatallah J, Carmichael P, Macfarlane M, Reisinger K, Rowland J, Schellauf F, Schepky A, Scheel J. A tiered approach to the use of alternatives to animal testing for the safety assessment of cosmetics: genotoxicity. A COLIPA analysis. Regul Toxicol Pharmacol. 2010;57(2–3):315–24.
17. Hu T, Kaluzhny Y, Mun GC, Barnett B, Karetsky V, Wilt N, Klausner M, Curren R, Aardema MJ. Intralaboratory and interlaboratory evaluation of the EpiDerm™ 3D human reconstructed skin micronucleus (RSMN) assay. Mutat Res. 2009;673:100–8.
18. Mun GC, Aardema MJ, Hu T, Barnett B, Kaluzhny Y, Klausner M, Karetsky V, Dahl EL, Curren RD. Further development of the EpiDerm™ 3D reconstructed human skin micronucleus (RSMN) assay. Mutat Res. 2009;673:92–9.
19. Aardema MJ, Barnett BB, Khambatta Z, Reisinger K, Ouedraogo-Arras G, Faquet B, Ginestet AC, Mun GC, Dahl EL, Hewitt NJ, Corvi R, Curren RC. International prevalidation studies of the EpiDerm™ 3D human reconstructed skin micronucleus (RSMN) assay: transferability and reproducibility. Mutat Res. 2010;701(2):123–31.
20. Dahl EL, Curren R, Barnett BC, Khambatta Z, Reisinger K, Ouedraogo G, Faquet B, Ginestet AC, Mun G, Hewitt NJ, Carr G, Pfuhler S, Aardema MJ. The reconstructed skin micronucleus assay (RSMN) in EpiDerm™: detailed protocol and harmonized scoring atlas. Mutat Res. 2011;720(1–2):42–52.
21. Pfuhler S, Fautz R, Ouedraogo G, Latil A, Kenny J, Moore C, Diembeck W, Hewitt NJ, Reisinger K, Barroso J. The cosmetics Europe strategy for animal-free genotoxicity testing: project status up-date. Toxicol In Vitro. 2014;28:18–23.
22. Yuki K, Ikeda N, Nishiyama N, Kasamatsu T. The reconstructed skin micronucleus assay in EpiDerm™: reduction of false-positive results - a mechanistic study with epigallocatechin gallate. Mutat Res. 2013;757(2):148–57.

23. Pfuhler S, Barnett B, Aardema MJ. Evaluation of aromatic amines in the EpiDerm™ 3D human reconstructed skin micronucleus (RSMN) assay. Mutat Res. 2017 (submitted).
24. Chapman KE, Thomas AD, Wills JW, Pfuhler S, Doak SH, Jenkins GJ. Automation and validation of micronucleus detection in the 3D EpiDerm™ human reconstructed skin assay and correlation with 2D dose responses. Mutagenesis. 2014;29(3):165–75.
25. Schmid TE, Dollinger G, Hable V, Greubel C, Zlobinskaya O, Michalski D, Molls M, Röper B. Relative biological effectiveness of pulsed and continuous 20 MeV protons for micronucleus induction in 3D human reconstructed skin tissue. Radiother Oncol. 2010;95(1):66–72.
26. Hewitt NJ, Edwards RJ, Fritsche E, Goebel C, Aeby P, Scheel J, Reisinger K, Ouédraogo G, Duche D, Eilstein J, Latil A, Kenny J, Moore C, Kuehnl J, Barroso J, Fautz R, Pfuhler S. Use of human in vitro skin models for accurate and ethical risk assessment: metabolic considerations. Toxicol Sci. 2013;133:209–17.
27. Götz C, Hewitt NJ, Jermann E, Tigges J, Kohne Z, Hübenthal U, Krutmann J, Merk HF, Fritsche E. Effects of the genotoxic compounds, benzo[a]pyrene and cyclophosphamide on phase 1 and 2 activities in EpiDerm™ models. Xenobiotica. 2012;42(6):526–37.
28. Reus AA, Reisinger K, Downs TR, Carr GJ, Zeller A, Corvi R, Krul CA, Pfuhler S. Comet assay in reconstructed 3D human epidermal skin models--investigation of intra- and interlaboratory reproducibility with coded chemicals. Mutagenesis. 2013;28(6):709–20.
29. Zijno P, Leopardi P, Marcon F, Crebelli R. Analysis of chromosome segregation by means of fluorescence in situ hybridization: application to cytokinesis-blocked human lymphocytes. Mutat Res. 1996;372(2):211–9.

3D Skin Comet Assay

38

Kerstin Reisinger and Stefan Pfuhler

38.1 Introduction

The comet assay allows the investigation of DNA damage in any cell type or tissue, which can be subjected to single cell isolation as it does not generally rely on proliferating cells. Therefore, the method is widely used in different areas like ecological and human monitoring or the analysis of DNA damage and repair. In addition, its versatile use in *in vivo* testing has been acknowledged to monitor effects of the first site of contact as well as organ-specific downstream effects, which are documented by the comet assay as DNA damage that may lead to clastogenic lesions or gene mutations. Recently, its increased recognition for regulatory testing led to the implementation of the *In Vivo* Mammalian Alkaline Comet Assay OECD Testing Guideline (OECD TG 489) [1].

Among the variety of *in vitro* models being subjected to the comet assay, several dermal test systems have been used to address the skin as the first site of contact for cosmetic ingredients, agrochemicals, and a growing number of pharmaceuticals. The following compilation concentrates on cell cultures of human origin, as they are of most relevance for human safety assessment. The first approaches utilized several 2D monolayer cultures, namely, primary keratinocytes (e.g., [2]), fibroblast (e.g., [3]), and melanocytes (e.g., [4]) or the HaCaT keratinocyte cell line (e.g. [5, 6]). The first study in which test compounds were applied on top of a 3D reconstructed skin model, namely, EpiSkin™ (SkinEthic™, France), was published in 2006 [7]. DNA damage was, however, not evaluated in the skin cells but in dendritic cells cultured in the medium below the skin model. Another approach published by Reus et al. [8] focused

K. Reisinger (✉)
Henkel AG & Co KGaA, Henkelstr. 67, 40589 Düsseldorf, Germany
e-mail: kerstin.reisinger@henkel.com

S. Pfuhler
Procter and Gamble, Mason Business Centre, Mason, OH 45040, USA
e-mail: pfuhler.s@pg.com

C. Eskes et al. (eds.), *Alternatives for Dermal Toxicity Testing*,
DOI 10.1007/978-3-319-50353-0_38

on the investigation of keratinocytes from *ex vivo* human skin. Twenty known genotoxins and non-genotoxins were applied atop of punch biopsies and were all predicted correctly.

Starting in 2007, a joint research project focused on a commercially available epidermal skin model, EpiDerm™ (MatTek, MA). After protocol transfer and optimization, three laboratories tested five coded genotoxic and non-genotoxic chemicals each by exposing the tissues from the top. They achieved a very high predictivity of >90% when considering the final call for a study, which comprised three independent experiments per chemical [9]. However, a relevant number of experiments had to be classified as invalid due to high values in the negative and solvent controls, accompanied by a suboptimal reproducibility. In consequence, efforts were undertaken to evaluate the suitability of commercially available full-thickness models, which consist of an epidermis and an underlying dermis connected by a basal membrane. It could be shown that the EpiDerm™ Full Thickness (EpiDerm™ FT; MatTek, MA, Fig. 38.1a) and the Phenion® Full-Thickness Skin Models (Phenion® FT; Henkel, Germany, Fig. 38.1b) were more reliable compared to the epidermal model initially used [10]. In a joint project comprising five European and US-American laboratories, EpiDerm™ FT and the Phenion® FT demonstrated lower and more consistent levels of background DNA damage in the negative and solvent controls as well as a dose-dependent increase in DNA migration after exposure with

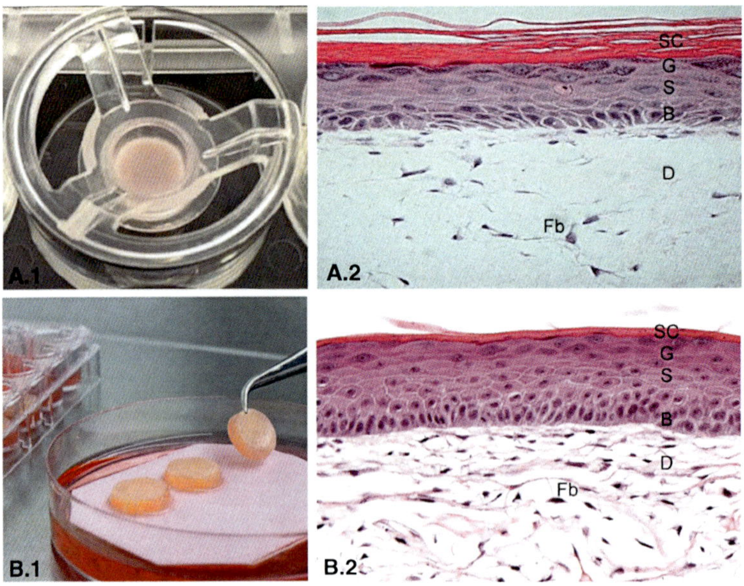

Fig. 38.1 Representative macroscopic views (A.1 and B.1) and cross sections (A.2 and B.2) of the EpiDerm™ FT (**a**) and the Phenion® Full-Thickness Skin Model (**b**). Hematoxylin and eosin-stained paraffin sections of both tissues reveal a fully differentiated epidermis containing distinct basal (B), spinous (S), granular layers (G), and a well-developed stratum corneum (SC). The collagen-based dermis (D) contains numerous primary dermal fibroblasts (Fb) (400X)

a positive control [10]. Furthermore, the rate of invalid experiments was negligible. In summary, this approach provides all advantages, which are linked with the use of 3D human reconstructed skin tissues, and adds the benefits of offering the possibility to assess the dermis as a second organ compartment. The tissues allow for topical application of compounds to mirror bioavailability relevant for the situation of use which is determined by the barrier function of the skin, mediated by the stratum corneum in the first instance (Fig. 38.1), and the organ- and species-specific xenobiotic metabolism. Furthermore, the skin models consist of p53 competent primary cells of human origin presuming normal cell cycle control. 3D tissues in general reflect cell-cell or cell-matrix interactions, as seen *in vivo*, more appropriately. The dermis, as an additional organ compartment, facilitates not only an intensive cross talk, pivotally contributing to the phenotype of the epidermis but also adds to the metabolic competency of the skin models [11]. Specifically, it could be shown that fibroblasts and keratinocytes cultured in 3D show a clear increased metabolic capacity compared to cells from the same donor propagated as 2D monolayer culture [11].

The prioritized tissues are commercially available in contrast to fresh *ex vivo* human skin, which in theory can also be used; however, it would be more difficult to handle and is not accessible globally. In consequence, the activities regarding the 3D Skin Comet assay, as further detailed below, concentrate on full-thickness skin models.

38.2 Principle of the Test Method and Scientific Basis

The comet assay methodology has first been introduced as single cell gel electrophoresis assay by Östling and Johanson [12]. Separated cells were imbedded in micro agarose gels. The cells were subsequently lysed with detergents under high salt conditions to degrade cellular and nuclear membranes and to liberate proteins like histones in order to prepare the DNA for subsequent electrophoresis. The method was further developed by Singh et al. [13] who introduced high alkaline conditions (pH > 13) during electrophoresis and a prior unwinding step, which allowed the detection of a broader range of DNA damage. After electrophoresis, which separates DNA according to size, the DNA is stained with an appropriate fluorescent dye to prepare for analysis with a full- or semiautomated image

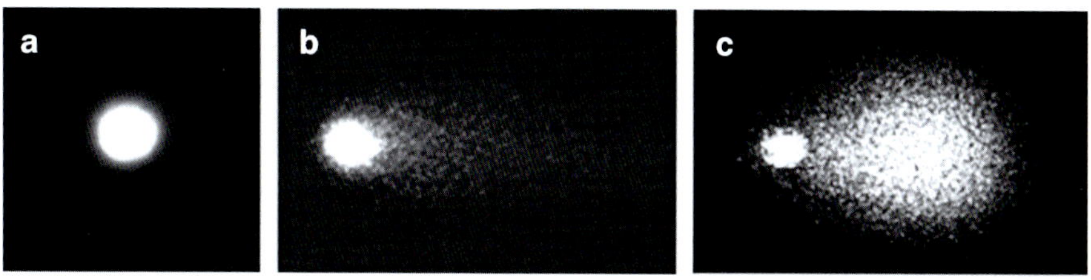

Fig. 38.2 A picture of (**a**) a comet representing normal non-fragmented DNA, which remains in the position of the nuclear DNA under the chosen electrophoresis conditions, while damaged DNA migrates toward the anode forming an increasing comet "tail" (**b**) and (**c**)

analyzer. The resulting structures appear in a comet-like shape with a head consisting of intact DNA (Fig. 38.2), which was not able to migrate under the used electrophoresis conditions due to its size, and a tail consisting of migrated DNA fragments and relaxed DNA loops, which appear after manifestation of strand breaks.

The high alkali condition allows for the detection of DNA double-strand breaks or single-strand breaks which may result from direct interaction of the test compound with the DNA or which are related to incomplete excision repair and alkali labile sites (OECD TG 489 [1]). In consequence, the modification by Singh enabled not only the detection of clastogenic DNA damage but also the identification of lesions which could be precursors of gene mutation events.

There are several parameters that are used to measure the extent of DNA that has migrated during electrophoresis and the distance it has traveled. Among these, the fluorescence intensity in the comet tail in comparison to the head (% tail DNA or % tail intensity) is of specific interest because it is considered linearly related to the DNA break frequency over a wide range of DNA damage [14]. This parameter has been recommended to assess DNA damage by the OECD TG 489 [1] and is also used for the methodology described here.

The comet assay in general is considered an indicator test since the DNA damage detected could be repaired or may be lethal to the cell resulting in nonpersistent effects. However, the strand breaks could also be fixed into mutations or chromosomal damage both resulting in permanent DNA damage of viable cells. In a recently published analysis of rodent carcinogens giving negative or equivocal results in the *in vivo* micronucleus test, the *in vivo* comet assay was positive for approximately 90% of these chemicals and was negative for nearly 80% of the noncarcinogens. Thus, the *in vivo* comet assay revealed a better predictivity in comparison to the *in vivo* rodent transgenic mutation assay or the *in vivo* UDS (Unscheduled DNA Synthesis Assay), [15]. Meanwhile the alkaline version of the comet assay, as described here, is the most widely used comet assay protocol, which has also been recommended for genotoxicity assessment by the International Workshop on Genotoxicity Testing (IWGT, formerly IWGPT; [16]).

38.3 Current Validation Status

3D skin assays have been flagged early for their potential to follow-up on positive results from *in vitro* genotoxicity assays, and consequently validation efforts have been suggested [17, 18]. Five European and US-American laboratories have since evaluated the within and between laboratory reproducibility of the 3D Skin Comet assay using full-thickness skin models. The generation of information on predictivity of the assay is in progress as 30 compounds, selected by external experts, are being investigated. The chemicals cover a balanced set of true positive and true negative chemicals, which showed concordant results in historical *in vitro* and *in vivo* testing, as well as a subset of compounds, the so-called irrelevant positives, for which positive results were obtained *in vitro* that did not correlate with historical *in vivo* genotoxicity or carcinogenicity studies. In addition, the chemicals represent different chemical classes and modes of action. For each of the selected chemicals, high-quality *in vitro* and *in vivo* data are available. However, only chemicals for which data from dermal *in vivo* studies exist can be used. This group of chemicals is limited and can therefore only support one standardized study. The ongoing ring trial is considering the respective validation standards; e.g., chemicals are tested double blinded, i.e., each chemical has an individual code, which differs between laboratories, though testing the same compound.

The ongoing study focuses on the investigation of the Phenion® FT using a so-called lean design. In the first phase, eight compounds were investigated by three laboratories. The data analysis showed that the reproducibility within and between the participating laboratories was sufficient to test each of the remaining 22 chemicals only in one laboratory during a second and last phase [10]. The validation will be finalized in 2017.

38.4 Performance and Applicability of the Test Method

In general, 3D skin models support testing of a great variety of compounds and compensate certain downsides of submerged 2D monolayer cultures. They allow for testing of lipophilic compounds and for application of higher concentrations if relevant for the situation of use. In addition, they facilitate testing of particulate materials although this has to be approached with caution as described below for the occurrence of precipitation.

Acetone or 70% ethanol (v/v) are used during the validation exercise. While tranfering the assay to other laboratories it has to be proven that these or other solvents do not disturb the air-liquid interface which is essential for a proper tissue cultivation over a period of 48 h [9]. Extensive precipitation of solids as well as small droplet of lipophilic liquids should be avoided as they may also disturb the air-liquid interface with the potential risk of causing false-positive results.

Taken these prerequisites into consideration, the results available from phase I of the validation indicate good reproducibility and predictivity comprising data of a pro-mutagen, a cross-linker, two direct-acting mutagens, as well as four compounds with an expected negative outcome. Colored substances were tested in parallel to the validation and did neither interfere with the tissue's integrity in the cultivation phase nor with DNA evaluation [19, 20].

Furthermore, skin models are increasingly used to investigate the impact of UV light on skin, which can directly modify DNA causing the formation of pyrimidine dimers as the major effect [21]. In consequence, 3D skin tissues have been used for the assessment of photoprotective compounds like UV filters [21]. For the general assessment of photogenotoxic effects, an epidermal model has successfully been used to assess the impact of UV light on DNA integrity with the comet assay [7]. After EpiSkin™ tissues were irradiated with UVA or solar-simulated light, keratinocytes were analyzed for DNA migration as a proof of concept for this approach. The use of 3D skin tissues is a relevant step forward in comparison to submerged 2D monolayers cultures. However, it should be noted that these nonstandard photo-genotoxicity approaches have not yet been investigated with regard to their predictive capacity and reproducibility within or between laboratories.

38.5 Brief Description of the Protocol

38.5.1 Study Design

The investigation of a specific compound with the 3D Skin Comet assay comprises a certain set of experiments similar to studies of standard *in vitro* genotoxicity assays (Fig. 38.3a). (1) First, an appropriate solvent is selected to dissolve the chemical before exposing the tissue, targeting a maximum concentration of 10 mg/100 μL or 10%, respectively. (2) The dose-range-finding experiment is designed to narrow down the dose range and especially to enable a decision on the maximum use concentration which could be limited by (a) the limit dose previously mentioned (10 mg/100 μL), (b) cytotoxicity, or (c) solubility/precipitation of the test compound. Cytotoxic effects of the test compound are measured in the form of intracellular adenosine triphosphate (ATP) concentration [22] and activity of adenylate kinase, which is released from the cells into the culture medium upon cell damage [23]. (3.1) Verification of a clear positive finding will usually not be required;

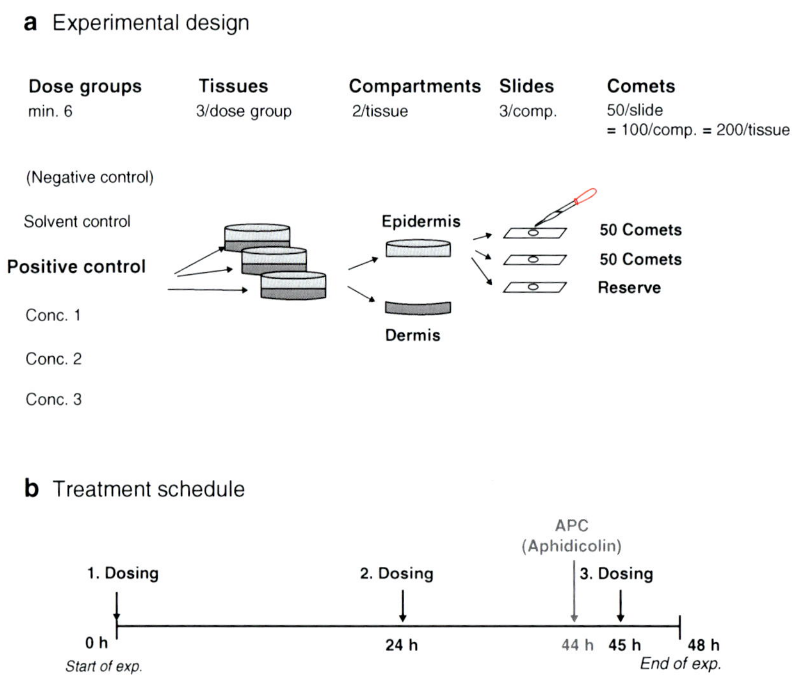

Fig. 38.3 Schematic of (**a**) the experimental design and (**b**) the treatment schedule. For details, please refer to Sect. 47.5. *comp.* compartment, *min* minimum, *exp.* experiment

however, a confirmatory second experiment should be added during the validation process. (3.2) In case the test item provides negative or inconclusive results, an additional test run should be performed using aphidicolin (APC), which is added 4 h before the end of the treatment period (Fig. 38.3b). APC, an inhibitor of DNA repair processes, was introduced into the protocol to improve the assay's sensitivity by accumulating excision repair-related strand breaks. This specific approach, which is outlined in paragraph 38.5.2 in more detail, has been shown to better reflect the assessment of pro-mutagens [24]. In the experiments complemented by APC, the pro-mutagen benzo(a)pyrene is used as positive control instead of methyl methane sulfonate (MMS) to prove the effectiveness of APC. (3.3) In case of non-concordant or equivocal findings in the APC experiment, a third test run with modified (usually tighter) concentration spacing is recommended.

38.5.2 Experimental Design

At least three concentrations of a test compound should be investigated in a test run complemented by a solvent control and a positive control (MMS, a direct-acting mutagen) group. The tissues are treated 48 h in total to ensure possible metabolic processing of the compound (see Fig. 38.3b). Twenty-four hours and 45 h after the first dosing, a second and third aliquot of the test compound is applied atop of the same tissue. Especially the latter time point is intended to capture damage, which may be subject to immediate DNA repair. To note: when establishing the assay, both negative (untreated) and solvent control groups should always be included. Once sufficient solvent control data is available which indicates that the solvent control has no impact on the background DNA damage of the tissues, untreated skin models do not need to be added anymore.

38.5.3 Cell Isolation and Comet Assay Procedure

At the end of the exposure period of 48 h, keratinocytes and fibroblasts are isolated in a tissue-specific procedure. The two compartments of the EpiDerm™ FT are separated using forceps before keratinocytes and fibroblasts are isolated separately by successive treatments with PBS, EDTA, and trypsin. The Phenion® FT is first incubated in thermolysin to allow for the degradation of the basal membrane and the subsequent separation of epidermis and dermis. Afterwards both cell types are isolated mechanically using a mincing procedure, similar to protocols used for the *in vivo* comet assay, leaving a mixture of cells and free nuclei. Subsequently, cells from both tissues are subjected to the same comet assay procedure in which cells/nuclei are first resuspended in low melting agarose (0.5%) which is then transferred onto glass slides. They are subjected to a lysis procedure overnight, which degrades cell and nuclear membranes by exposure to detergents. A high salt concentration removes proteins like histones. Afterwards, DNA strands are separated by high alkali conditions (pH >13) before DNA migrates to the anode during 30 min of electrophoresis using fresh high alkali buffer. Finally, the slides are neutralized and dried.

38.5.4 Analysis

After the experiment, four slides per skin model are subjected to analysis (two each for epidermis or dermis) as the evaluation of two slides per compartment and 50 comets per slide (i.e., 100 comets per cell type) was considered sufficient [25] (Fig. 38.3a). Before analysis, slides are randomized and stained with an appropriate fluorescence dye (e.g., SYBR Gold). The fluorescence intensity in the comet tail compared to the respective comet head is afterward analyzed semiautomated using a 200x magnification of a fluorescence microscope and comet assay image analysis software (sold by a variety of commercial providers, freeware is also available).

The 100 data points (2 × 50 comet measures/slide) per skin compartment (epidermis or dermis) are subjected to a variance-stabilizing transformation before they are summarized as median. Since each control or dose group is represented by three tissues, three medians are finally summarized as mean value for each dose or control group. These mean values are used for further statistical analysis since the skin tissue is considered the experimental unit.

Before evaluating the results for genotoxicity, the validity of an experiment is determined applying defined thresholds for % tail intensity for solvent and positive controls. Furthermore, the validity of a dose group is evaluated applying thresholds set for the two cytotoxicity measurements, which have been identified for the 3D Skin Comet assay, i.e., the intracellular concentration of ATP and the activity of adenylate kinase released into the culture medium. The latter criteria were established because DNA damage can be triggered by cellular toxicity, e.g., when cells go into apoptosis or necrosis. Therefore, as it is the case for other genotoxicity assays, increased DNA damage that occurs only in conjunction with strong cytotoxicity is not considered as biological relevant in the context of a genotoxicity assessment.

In the next step, an analysis of variance (ANOVA) of the solvent control and the dose groups is conducted. In case of a positive ANOVA, the statistical significance of an effect is analyzed by a pairwise comparison of the solvent control and single-dose groups using the Dunnett test. A test substance is considered to be genotoxic in the 3D Skin Comet assay if one or more concentrations produce a statistically significant increase in the percentage of tail DNA at concentrations that do not exceed the limits of cytotoxicity (for details, see [10]). In addition to the statistical evaluation, the following criteria, which are in line with procedures established for OECD TG 489 [1], need to be fulfilled to consider the biological relevance of effects observed. In brief, (a) the response has to be dose dependent, (b) at least one dose needs to be statistically significant different from the solvent control, and (c) at least one test group needs to be outside the historical control data range. If none of the three criteria is fulfilled, the test item is called negative. In case one or two but not all the criteria for a positive call are fulfilled, the test compound is considered negative, or equivocal, and further testing may considered (see also paragraph 38.5.1 on study design).

38.6 Perspectives from the Test Developer

38.6.1 Critical Steps in the Protocol

As with all variants of the comet assay, certain steps of the protocol need specific attention to support a high level of standardization and reproducibility within a laboratory. Altering sample preparation, electrophoresis conditions or microscope settings have been investigated and showed an effect on the DNA migration measured (e.g. [26]). These points have also been taken into account for the 3D Skin Comet assay protocol and are applied in the ongoing validation exercise. For example, trainings have been organized before the validation to ensure a proper implementation of the cell isolation procedures in the participating laboratories to avoid induced DNA damage caused by cell separation. To minimize between laboratory variability, the electrophoresis conditions were standardized by the use of a common commercially available electrophoresis chamber, uniform electrophoresis time, as well as voltage settings. In addition, standards for the analysis of slides have been agreed on and were published recently [9].

As with all other methods, laboratories should establish experimental competency in the 3D Skin Comet assay. Proof of competency could include a series of experiments providing low and reproducible % tail DNA values in non-treated or solvent exposed tissues. During the course of this proficiency phase, the laboratory should build a historical database of the solvent and negative controls. The same applies for the positive control using concentrations which induce a range of DNA damage that spans slight to clear DNA damage.

38.6.2 Possible Protocol Adaptations

The protocols for cell isolation and the comet assay procedure have been optimized and should provide high-quality data. However, the design of an experiment or an entire study might be optimized/modified further after evaluation of the entire data set generated in the ongoing validation exercise. A possible point for optimization could be the focus on one cell type to improve the throughput of the method. Protocol adaptations made after the first phase of the validation exercise included the rule that a negative control of untreated tissues will only be needed in case none of the two recommended solvents is used. This was decided after sufficient data became available showing that solvent and untreated controls did not differ in terms of their background DNA damage.

Apart from such data-driven changes, the standard protocol can be amended to gain mechanistic insights into DNA damage or to better reflect DNA damage induced by agents which cause cross-linking of DNA with DNA or protein. Such damage cannot be detected reliably with the standard experimental design [27]. Mitomycin C (MMC), as an example of a DNA-DNA cross-linker, generates covalent bonds between guanine nucleotides. These links between two DNA strands do not only suppress positive comet signals at higher doses but can even lead to a reduction of measurable strand breaks if compared to control. Therefore, a modified protocol that was developed for efficient detection of cross-linkers [28] was adapted to the Phenion® FT. Using this protocol, tissues were not only exposed to MMC but were co-treated with MMS to generate a high background level of single-strand breaks that allowed for efficient detection of a reduction in measurable DNA breaks [10].

The alkaline version of the comet assay not only enables the detection of strand breaks which are formed as direct effects of a treatment, it also allows detection of strand breaks which evolve in the course of excision repair processes in response to UV radiation, alkylating agents, or bulky adduct formation, to name a few. These strand breaks, set by specific enzymes to remove modified nucleotides or bases, can be short-lived. The incorporation of enzyme inhibitors to the comet assay protocol supports the accumulation of these DNA repair-related strand breaks thereby amplifying comet formation and increasing the sensitivity of the assay [29]. Two groups of inhibitors have successfully been included into comet assay protocols. The first group of inhibitors interferes with DNA repair and replication in general and contains, e.g., hydroxyurea (HU), which inhibits ribonucleotide reductase causing an imbalance in the nucleotide pool [30]. Cytosine arabinoside (ara-C) is incorporated into DNA during replication, after it was converted to cytosine arabinoside triphosphate, leading to chain termination and cell cycle arrest [31], and aphidicolin (APC) which was showing to inhibit DNA polymerases α and δ [29]. The latter one has been included in the 3D Skin Comet assay protocol, in case of negative findings in the first main experiment. APC is added 4 h before the experiment is terminated (44 h after the first treatment) (Fig. 38.3b). While strand breaks increase due to excision repair processes, the impact of this DNA repair enzyme inhibitor on cell proliferation is negligible. APC is added only for a limited period of time in which the keratinocytes of the stratum basale and stratum spinosum, the only proliferating cells in the tissues, are not affected due their low turnover compared to monolayer cultures. The marginal increase of % tail DNA in the solvent control after adding APC proved the suitability of the skin tissues for this approach which has been shown to better reflect the assessment of pro-mutagens, while the predictivity of non-genotoxins remained high with 100% [10, 24].

A second group of inhibitors, i.e., lesion-specific enzymes, can also be incorporated into the 3D Skin Comet assay to characterize DNA damage. These enzymes remove modified DNA bases leaving an apyrimidinic or apurinic (AP) site, which is subsequently converted to a single-strand break under high alkali conditions. In theory, any lesion for which a specific enzyme exists can be detected in this way. Hitherto, several enzymes, like 8-oxo-Gua DNA glycosylase (OGG1), have been identified to investigate oxidative DNA damage [32]. Alkylated nucleotides like 3-methyladenine can be identified by 3-methyladenine DNA gycosylase II (AlkA) [33], while uracil DNA glycosylase (UDG) has been shown to support the detection of uracil, as a miss-incorporated DNA base [34]. Furthermore, bulky adducts can be identified by applying uvrABC, an exonuclease derived from *E. coli* [35], whereas the enzyme T4 endonuclease V helps in detecting the dimerization of adjacent pyrimidine dimers which are observed as characteristic lesions induced by UV light [36]. The use of these lesion-specific enzymes has added value to the comet methodology in general and may add to the 3D Skin model approach in the future.

38.7 Conclusions

The 3D Skin Comet assay, together with the RSMN (Chap. 46), is considered to close a gap in the toolbox of *in vitro* genotoxicity assays since they have been shown to be advantageous for the evaluation of dermally exposed substances. The full-thickness models consist of human primary p53 competent keratinocytes and fibroblasts differentiating this approach from many *in vitro* genotoxicity assays, which are based on rodent cancer cell lines, some of them being p53 deficient. Furthermore the cultivation of keratinocytes and fibroblasts in a three dimensional environment not only supports the *in vivo*-like phenotype of the cells but also enables the topical application of compounds on top of the stratum corneum close to the situation of use. A validation study is ongoing, and the results generated so far indicate good reproducibility and predictivity of this method.

In parallel to the validation, the 3D Skin Comet assay has already been used to follow-up on unfavorable results from the standard genotoxicity *in vitro* test battery for regulatory testing of cosmetic ingredients [19, 20]. Successful validation is hoped to lead to wider regulatory acceptance that will include more product categories for which the dermal route is relevant for risk assessment.

References

1. OECD. Guidelines for the testing of chemicals, test no. 489: in vivo mammalian alkaline comet assay. 2014. http://www.oecd-ilibrary.org/environment/test-no-489-in-vivo-mammalian-alkaline-comet-assay_9789264224179-en;jsessionid=npkmv8v01k1s.x-oecd-live-02.
2. Tzung TY, Rünger TM. Assessment of DNA damage induced by broadband and narrowband UVB in cultured lymphoblasts and keratinocytes using the comet assay. Photochem Photobiol. 1998;67(6):647–50.
3. Emonet-Piccardi N, Richard MJ, Ravanat JL, Signorini N, Cadet J, Béani JC. Protective effects of antioxidants against UVA-induced DNA damage in human skin fibroblasts in culture. Free Radic Res. 1998;29(4):307–13.
4. Mouret S, Forestier A, Douki T. The specificity of UVA-induced DNA damage in human melanocytes. Photochem Photobiol Sci. 2012;11(1):155–62.
5. Wischermann K, Popp S, Moshir S, Scharfetter-Kochanek K, Wlaschek M, de Gruijl F, Hartschuh W, Greinert R, Volkmer B, Faust A, Rapp A, Schmezer P, Boukamp P. UVA radiation causes DNA strand breaks, chromosomal aberrations and tumorigenic transformation in HaCaT skin keratinocytes. Oncogene. 2008;27(31):4269–80.
6. Zeller A, Pfuhler S. *N*-acetylation of three aromatic amine hair dye precursor molecules eliminates their genotoxic potential. Mutagenesis. 2014;29(1):37–48.
7. Flamand N, Marrot L, Belaidi JP, Bourouf L, Dourille E, Feltes M, Meunier JR. Development of genotoxicity test procedures with Episkin, a reconstructed human skin model: towards

new tools for in vitro risk assessment of dermally applied compounds? Mutat Res. 2006;606(1–2):39–51.

8. Reus AA, Usta M, Krul CA. The use of ex vivo human skin tissue for genotoxicity testing. Toxicol Appl Pharmacol. 2012;261(2):154–63.
9. Reus AA, Reisinger K, Downs TR, Carr GJ, Zeller A, Corvi R, Krul CA, Pfuhler S. Comet assay in reconstructed 3D human epidermal skin models--investigation of intra- and inter-laboratory reproducibility with coded chemicals. Mutagenesis. 2013;28(6):709–20.
10. Reisinger K, Blatz V, Brinkmann JP, Downs TR, Fischer A, Henkler F, Krul C, Liebsch M, Luch A, Pirow R, Reus AA, Schulz M, Pfuhler S. Validation of the 3D Skin Comet assay using full thickness skin models: transferability and reproducibility. 2017.
11. Wiegand C, Hewitt NJ, Merk HF, Reisinger K. Dermal xenobiotic metabolism: a comparison between native human skin, four in vitro skin test systems and a liver system. Skin Pharmacol Physiol. 2014;27:263–75.
12. Östling O, Johanson KJ. Micorelectrophoretic study of radiation-induced DNA damaged in individual mammalian cells. Biochem Biophys Res Commun. 1984;123:291–8.
13. Singh NP, McCoy MT, Tice RR, Schneider EL. A simple technique for quantitation of low levels of DNA damage in individual cells. Exp Cell Res. 1988;175:184–91.
14. Hartmann A, Agurell E, Beevers C, Brendler-Schwaab S, Burlinson B, Clay P, Collins A, Smith A, Speit G, Thybaud V, Tice RR. Recommendations for conducting the in vivo alkaline comet assay. 4th international comet assay workshop. Mutagenesis. 2003;18:45–51.
15. Kirkland D, Speit G. Evaluation of the ability of a battery of three in vitro genotoxicity tests to discriminate rodent carcinogens and non-carcinogens III. Appropriate follow-up testing in vivo. Mutat Res. 2008;654(2):114–32.
16. Tice RR, Agurell E, Anderson D, Burlinson B, Hartmann A, Kobayashi H, Miyamae Y, Rojas E, Ryu JC, Sasaki YF. Single cell gel/comet assay: guidelines for in vitro and in vivo genetic toxicology testing. Environ Mol Mutagen. 2000;35:206–21.
17. Pfuhler S, Kirkland D, Kasper P, Hayashi M, Vanparys P, Carmichael P, Dertinger S, Eastmond D, Elhajouji A, Krul C, Rothfuss A, Schoening G, Smith A, Speit G, Thomas C, van Benthem J, Corvi R. Reduction of use of animals in regulatory genotoxicity testing: identification and implementation opportunities-report from an ECVAM workshop. Mutat Res. 2009;680:31–42.
18. SCCS. Position statement on genotoxicity/mutagenicity testing of cosmetic ingredients without animal experiments (SCCP/1212/09). 2009. http://ec.europa.eu/health/archive/ph_risk/committees/04_sccp/docs/sccp_s_08.pdf.
19. SCCS. Scientific Committee on consumer safety. Opinion on 2,6-Dihydroxyethylaminotoluene COLIPA no A138. SCCS/1563/15. Adopted on 25 Jun 2015. http://ec.europa.eu/health/scientific_committees/consumer_safety/docs/sccs_o_188.pdf.
20. SCCS (Scientific Committee on Consumer Safety). Opinion on Basic Brown 17 COLIPA no B007. SCCS/1531/14. Adopted on 24 Mar 2014, revision of 18 Jun 2014. http://ec.europa.eu/health/scientific_committees/consumer_safety/docs/sccs_o_152.pdf.
21. Marrot L, Belaïdi JP, Lejeune F, Meunier JR, Asselineau D, Bernerd F. Photostability of sunscreen products influences the efficiency of protection with regard to UV-induced genotoxic or photoageing-related endpoints. Br J Dermatol. 2004;151:1234–44.
22. Kangas L, Grönroos M, Nieminen AL. Bioluminescence of cellular ATP: a new method for evaluating cytotoxic agents in vitro. Med Biol. 1984;62(6):338–43.
23. Olsson T, Gulliksson H, Palmeborn M, Bergström K, Thore A. Leakage of adenylate kinase from stored blood cells. J Appl Biochem. 1983;5(6):437–45.
24. Brinkmann J, Stolpmann K, Trappe S, Otter T, Genkinger D, Bock U, Liebsch M, Henkler F, Hutzler C, Luch A. Metabolically competent human skin models: activation and genotoxicity of benzo[a]pyrene. Toxicol Sci. 2013;131(2):351–9.
25. Lovell DP, Omori T. Statistical issues in the use of the comet assay. Mutagenesis. 2008;23:171–82.
26. Ersson C, Möller L. The effects on DNA migration of altering parameters in the comet assay protocol such as agarose density, electrophoresis conditions and durations of the enzyme or the alkaline treatments. Mutagenesis. 2011;26(6):689–95.

27. Speit G, Kojima H, Burlinson B, Collins AR, Kasper P, Plappert-Helbig U, Uno Y, Vasquez M, Beevers C, De Boeck M, Escobar PA, Kitamoto S, Pant K, Pfuhler S, Tanaka J, Levy DD. Critical issues with the in vivo comet assay: a report of the comet assay working group in the 6th international workshop on Genotoxicity testing (IWGT). Mutat Res Genet Toxicol Environ Mol Mutagen. 2015;783:6–12.
28. Pfuhler S, Wolf HU. Detection of DNA-crosslinking agents with the alkaline comet assay. Environ Mol Mutagen. 1996;27:196–201.
29. Speit G, Schütz P, Hoffmann H. Enhancement of genotoxic effects in the comet assay with human blood samples by aphidicolin. Toxicol Lett. 2004;153(3):303–10.
30. Martin FL, Cole KJ, Orme MH, Grover PL, Phillips DH, Venitt S. The DNA repair inhibitors hydroxyurea and cytosine arabinoside enhance the sensitivity of the alkaline single-cell gel electrophoresis ('comet') assay in metabolically-competent MCL-5 cells. Mutat Res. 1999;445:21–43.
31. Gedik CM, Collins AR. The mode of action of 1-beta-D-arabinofuranosylcytosine in inhibiting DNA repair; new evidence using a sensitive assay for repair DNA synthesis and ligation in permeable cells. Mutat Res. 1991;254:231–7.
32. Collins AR. Measuring oxidative damage to DNA and its repair with the comet assay. Biochem Biophys Acta. 2014;1840:794–800.
33. Berdal KG, Johansen RF, Seeberg E. Release of normal bases from intact DNA by a native DNA repair enzyme. EMBO J. 1998;17:363–7.
34. Duthie SJ, McMillan P. Uracil misincorporation in human DNA detected using single cell gel electrophoresis. Carcinogenesis. 1997;18:1709–14.
35. Thomas DC, Husain I, Chaney SG, Panigrahi GB, Walker IG. Sequence effect on incision by (a)BC excinuclease of 4NQO adducts and UV photoproducts. Nucleic Acids Res. 1991;19:365–70.
36. Dizdaroglu M, Zastawny TH, Carmical JR, Lloyd RS. A novel DNA N-glycosylase activity of *E. coli* T4 endonuclear V that excises 4,6-diamino-5-foramidopyrimidive from DNA, a UV-radiation and hydroxyl radical-induced product of adenine. Mutat Res. 1996;362:1–8.

Role in a Testing Strategy

39

Kerstin Reisinger and Stefan Pfuhler

39.1 Role of the 3D Skin-Based Assays in a Testing Strategy

Due to the diverse nature of mechanisms involved in genotoxicity, it is known that no single assay is able to detect all classes and examples of genotoxic carcinogens. As a result, international guidelines for assessing the genotoxic potential of chemicals recommend the use of a battery of genotoxicity tests to detect gene, chromosome or genome mutations (for details please refer to Chap. 36).

In general, genotoxicity testing batteries foresee *in vitro* testing first. The predictivity of four assays often contributing to the initial *in vitro* battery was evaluated in terms of their predictive capacity for rodent carcinogenicity, namely, the bacterial reverse mutation test (Ames test); the *in vitro* micronucleus assay (MNT); the *in vitro* mammalian cell gene mutation test, i.e. Mouse lymphoma assay (MLA); and the *in vitro* mammalian chromosomal aberration assay (CA). Kirkland *et al.* [1, 2] showed that the sensitivity (i.e. ability to correctly predict rodent carcinogens) of the MNT was the highest of the tests analysed, although the database was much smaller compared to the other assays. The sensitivity increases to around 80% or higher when combining assays, but this, unfortunately, dramatically reduces the specificity (the ability to correctly identify *in vivo* non-genotoxic noncarcinogens) of such a battery. For instance, combining the Ames assay, which as a stand-alone test has a specificity of 74%, with two other tests decreases the specificity to as low as 5–23% [1].

Clearly, protection of consumers by employing sensitive tests to evaluate the potential genotoxicity of compounds is paramount. The use of tests with such low

K. Reisinger (✉)
Henkel AG & Co KGaA, Henkelstr. 67, 40589 Düsseldorf, Germany
e-mail: kerstin.reisinger@henkel.com

S. Pfuhler
Procter and Gamble, Mason Business Centre, Mason, OH 45040, USA
e-mail: pfuhler.s@pg.com

C. Eskes et al. (eds.), *Alternatives for Dermal Toxicity Testing*,
DOI 10.1007/978-3-319-50353-0_39

specificities, however, means that unacceptably high percentage of 'irrelevant-positive' results is generated. This can lead to unnecessary *in vivo* follow-up testing and/or trigger the need for performing mechanistic studies. Therefore, the complexity of risk assessments for compounds positive in standard *in vitro* genotoxicity assays increases as illustrated by the many publications (e.g. [3–5]), external workgroups and meetings devoted to this topic. In consequence, 'irrelevant-positive' *in vitro* results may not only trigger the use of large numbers of animals but also require extensive resources in regulatory agencies as well as industry. Because of the efforts involved in clarifying positive results from standard *in vitro* testing, companies often eliminate such ingredients from use, thereby losing potentially safe and useful compounds. For cosmetic ingredients, a positive result from a standard *in vitro* assay can trigger the loss of ingredients more directly since the 7th Amendment to the Cosmetics Directive of the European Commission prohibits the use of *in vivo* follow-up assays. In consequence, Cosmetics Europe, the European personal care association, has funded and driven projects aiming to address the lack of adequate alternatives to traditional *in vivo* tests and to help validate successful models.

The first pillar of the program aimed to optimize initial 'tier 1' testing and helped lowering the percentage of 'irrelevant-positive' results [6–9]. The second pillar of the program aims at establishing genotoxicity assays that are basing on reconstructed human skin models as 'tier 2' assays (direct replacement of animal studies). Since most cosmetics, as well as many other industrial chemicals, are exclusively or predominately in contact with the skin, assays using 3D human reconstructed skin models offer the potential for a more physiologically relevant approach to test effects after dermal exposure, as detailed in Chaps. 36, 37 and 38, namely, the reconstructed skin (RS) micronucleus assay (RSMN) and the 3D Skin Comet assay. 3D tissue constructs are therefore logical follow-up tools for standard '2D' genotoxicity assays because they allow for more natural cell-cell and cell-matrix interactions and show '*in vivo*-like' behaviour for key parameters such as cell viability differentiation, morphology, gene and protein expression and function [10].

The role of genotoxicity assays that are based on reconstructed skin models have been discussed by independent scientific committees. The 5th International Workshop on Genotoxicity Testing (IWGT), for example, discussed *in vitro* genotoxicity test approaches with better predictivity and concluded that these skin-based assays, once validated, will be useful to follow up on positive results from standard *in vitro* assays as they resemble the properties of human skin [11]. The latter statement was confirmed, with a focus on skin metabolism, by experimental data comparing native human skin with RS models. These experiments, also funded by Cosmetics Europe and performed at independent research facilities, confirm that RS models closely resemble metabolic properties of human skin (summarized in [12]). The Scientific Committee on Consumer Safety (SCCS), the independent expert panel mandated by the European Commission, provides guidance on testing of cosmetic ingredients including genotoxicity testing and reviews dossier submissions for ingredient categories that are regulated in the EU like colouring agents, preservatives and UV filters. The SCCS recently revised their genotoxicity testing guidelines to reflect progress made with the characterization and validation of the RS model-based assays [13]. In its so-called Notes of Guidance for the Testing of Cosmetic Ingredients and their Safety Evaluation, the SCCS calls these assays a 'good alternative to bridge the gap between *in vitro* and *in vivo* testing in terms of final hazard assessment' [13] and recommends using these as 'tier 2' assays to follow-up on unfavourable results from the *in vitro* standard test battery. Also, there already was an example where the 3D Skin Comet assay was used to support the safety of a hair dye [14]. Initial experiments with this dye in standard genotoxicity assays provided positive findings with the Ames test but found negative results in the standard *in vitro* micronucleus test. As no *in vivo* data could be generated to address the positive findings for gene mutation in the Ames test, due to the ban of animal studies in Europe, *in vitro* data were exclusively used to further address the Ames positive which included the 3D Skin Comet assay. This assay showed a lack of DNA damaging properties after application of the dye to the skin. The SCCS accepted these data as evidence that it does not have a genotoxic potential for the given dermal exposure scenario [14].

In order to obtain input on regulatory acceptance of both methods and potential obstacles independent of industry, EURL-ECVAM (European Union Reference Laboratory for alternatives to animal testing) is represented in the steering committee of both ongoing validation studies. The EURL-ECVAM input helps the validation team keep a high level of standardization during the studies and facilitates regular exchange with regulators from early project phases on which is hoped to promote broader regulatory acceptance of these assays as *in vitro* follow-up tools for unfavourable results of standard *in vitro* genotoxicity assays for dermally exposed substances.

References

1. Kirkland D, Aardema M, Müller L, Hayashi M. Evaluation of the ability of a battery of three in vitro genotoxicity tests to discriminate rodent carcinogens and non-carcinogens II. Further analysis of mammalian cell results, relative predictivity and tumour profiles. Mutat Res. 2006;608(1):29–42.
2. Kirkland D, Aardema M, Henderson L, Müller L. Evaluation of the ability of a battery of three in vitro genotoxicity tests to discriminate rodent carcinogens and non-carcinogens I. Sensitivity, specificity and relative predictivity. Mutat Res. 2005;584(1–2):1–256.
3. Pfuhler S, Kirst A, Aardema M, Banduhn N, Goebel C, Araki D, Costabel-Farkas M, Dufour E, Fautz R, Harvey J, Hewitt NJ, Hibatallah J, Carmichael P, Macfarlane M, Reisinger K, Rowland J, Schellauf F, Schepky A, Scheel J. A tiered approach to the use of alternatives to animal testing for the safety assessment of cosmetics: genotoxicity. A COLIPA analysis. Regul Toxicol Pharmacol. 2010;57(2–3):315–24.
4. Eastmond DA, Hartwig A, Anderson D, Anwar WA, Cimino MC, Dobrev I, Douglas GR, Nohmi T, Phillips DH, Vickers C. Mutagenicity testing for chemical risk assessment: update of the WHO/IPCS harmonized scheme. Mutagenesis. 2009;24:341–9.
5. Elespuru RK, Agarwal R, Atrakchi AH, Bigger CA, Heflich RH, Jagannath DR, Levy DD, Moore MM, Ouyang Y, Robison TW, Sotomayor RE, Cimino MC, Dearfield KL. Current and future application of genetic toxicity assays: the role and value of in vitro mammalian assays. Toxicol Sci. 2009;109:172–9.

6. Fowler P, Smith R, Smith K, Young J, Jeffrey L, Carmichael P, Kirkland D, Pfuhler S. Reduction of misleading ("false") positive results in mammalian cell genotoxicity assays. III: sensitivity of human cell types to known genotoxic agents. Mutat Res. 2014;767:28–36.
7. Fowler P, Smith R, Smith K, Young J, Jeffrey L, Kirkland D, Pfuhler S, Carmichael P. Reduction of misleading ("false") positive results in mammalian cell genotoxicity assays. II. Importance of accurate toxicity measurement. Mutat. Res. 2012;747(1):104–17.
8. Fowler P, Smith K, Young J, Jeffrey L, Kirkland D, Pfuhler S, Carmichael P. Reduction of misleading ("false") positive results in mammalian cell genotoxicity assays. I. Choice of cell type. Mutat Res. 2012;742(1–2):11–25.
9. Pfuhler S, Fautz R, Ouedraogo G, Latil A, Kenny J, Moore C, Diembeck W, Hewitt NJ, Reisinger K, Barroso J. The cosmetics Europe strategy for animal-free genotoxicity testing: project status up-date. Toxicol In Vitro. 2014;28:18–23.
10. Zeiger E, Gollapudi B, Aardema MJ, Auerbach S, Boverhof D, Custer L, Dedon P, Honma M, Ishida S, Kasinski AL, Kim JH, Manjanatha MG, Marlowe J, Pfuhler S, Pogribny I, Slikker W, Stankowski LF Jr, Tanir JY, Tice R, van Benthem J, White P, Witt KL, Thybaud V. Opportunities to integrate new approaches in genetic toxicology: an ILSI-HESI workshop report. Environ Mol Mutagen. 2015;56:277–85.
11. Pfuhler S, Fellows M, van Benthem J, Corvi R, Curren R, Dearfield K, Fowler P, Frötschl R, Elhajouji A, Le Hégarat L, Kasamatsu T, Kojima H, Ouédraogo G, Scott A, Speit G. In vitro genotoxicity test approaches with better predictivity: summary of an IWGT workshop. Mutat Res. 2011;723:101–7.
12. Hewitt NJ, Edwards RJ, Fritsche E, Goebel C, Aeby P, Scheel J, Reisinger K, Ouédraogo G, Duche D, Eilstein J, Latil A, Kenny J, Moore C, Kuehnl J, Barroso J, Fautz R, Pfuhler S. Use of human in vitro skin models for accurate and ethical risk assessment: metabolic considerations. Toxicol Sci. 2013;133:209–17.
13. SCCS. ADDENDUM to the SCCS's Notes of Guidance (NoG) for the Testing of Cosmetic Ingredients and their Safety Evaluation, 8th Revision (SCCS/1501/12) 2014. http://ec.europa.eu/health/scientific_committees/consumer_safety/docs/sccs_o_156.pdf.
14. SCCS. Scientific Committee on Consumer Safety. Opinion on Basic Brown 17 COLIPA No. B007. SCCS/1531/14. Adopted on 24 Mar 2014, revision of 18 Jun 2014. 2014. http://ec.europa.eu/health/scientific_committees/consumer_safety/docs/sccs_o_152.pdf.

Part VI

Other Exploratory Areas of Relevance

Progenitor Skin Cell Therapy and Evolution of Medical Applications

40

Lee Ann Applegate, Paris Jafari, Corinne Scaletta, Anthony de Buys Roessingh, Wassim Raffoul, and Nathalie Hirt-Burri

Abbreviations

AFIRM	Armed Forces Institute of Regenerative Medicine
BM-MSC	Bone marrow mesenchymal stem cell
DOD	Department of Defense
GMP	Good manufacturing Practice
GvHD	Graft versus host disease
MCB	Master cell bank
WCB	Working cell bank

40.1 Therapeutic Agent and Vaccine Development: Historical Uses with Cell Sources

Medical doctors and scientists have used defined tissue-derived cell lines since the 1930s for vaccine development. Tissues from different gestational stages were used for understanding cell biology and development and at the same time to

---

This article is dedicated to Sir Roger Moore (who was the dear Godfather of the Applegate Lab) and Lady Kristina for their continued support in the treatment of burned children, prevention and their rehabilitation.

L.A. Applegate (✉) • P. Jafari • C. Scaletta • W. Raffoul • N. Hirt-Burri
Department of Musculoskeletal Medicine, Plastic, Reconstructive and Hand Surgery, Unit of Regenerative Therapy, University Hospital of Lausanne, CHUV/UNIL, Chemin Croisettes 22, EPCR 02, 1066 Epalinges, Switzerland
e-mail: Lee.Laurent-Applegate@chuv.ch

A. de Buys Roessingh
Department of Pediatric Surgery, University Hospital of Lausanne, CHUV/UNIL, BH 10, 1011 Lausanne, Switzerland

C. Eskes et al. (eds.), *Alternatives for Dermal Toxicity Testing*,
DOI 10.1007/978-3-319-50353-0_40

develop therapeutic agents. In 1954, the Nobel Prize for Medicine was attributed to two American Immunologists who used human fetal progenitor cells for the development of the polio vaccine. The production of polio vaccine was historically produced using monkey cells, but in 1951 when the major polio epidemic occurred, it was necessary to find a solution for large-scale cell culture. Jonas Salk discovered the polio vaccine in 1952 but would not have been able to prove the efficacy without trying it on a large scale. The National Foundation for Infantile Paralysis (NFIP) was a charity created by President Roosevelt, who had been paralyzed by Polio, and they launched the largest field trial to inoculate 2 million children for which the blood was analyzed by the NFIP to see if they became immune. The monkey cells that were used in the cellular assays were mixed with the blood of the inoculated patients along with the polio virus, and the cells would be killed in the assay. As it would take so many monkeys (cells) to analyze the study, it was the first use of the famous HeLa cells in medicine. The cell line was derived from Henrietta Lacks, a patient at John's Hopkins with epidermoid carcinoma, and it was the first human cancer cell line to grow massively in culture in the laboratory and was accomplished by Dr. George Gey and his assistant Mary Kubichek [1]. Later, many other vaccines such as rubella, chicken pox, rabies, and hepatitis A vaccines were developed with the same revolutionary processing but with normal primary cells (non-transformed) from fetal tissues. Most interesting is that the original primary fetal cells cultured in the 1960s are still used today in the production of many vaccines. One of the first cell lines, the WI-38 (Wistar Institute 38) was developed by Leonard Hayflick in 1964 which has been known from the deposited cell source as the ATTC CCL-75 cells (American Type Culture Collection). Historically, this cell source came from normal fetal kidney tissue following voluntary pregnancy interruption which was a legal organ donation. It has been used for the production of the RA 27/3 vaccine against rubella for many years and in the last two decades integrated into a major campaign initiated by the World Health Organization (WHO) to eradicate rubella in developing countries. This campaign has successfully helped to decrease this disease and the handicaps associated [2].

There is another cell source that was developed 2 years later in 1966 from normal fetal lung tissue donated for medical use and is known as the MRC-5 (Medical Research Council 5). This cell source has produced vaccines for chickenpox, polio, smallpox, hepatitis A, and rabies. This same cell-banked source is still being used today for therapeutic product development [3].

As these cell sources have been very successful and their stability to be shown as remarkable, other fetal cell sources have also been introduced in the 1970s for pharmaceutical needs. The PER.C6 fetal cell line derived from retinal tissue was developed to promote adenovirus vector for gene therapy of developmental stage vaccines which include ebola virus, influenza, HIV, and Japanese encephalitis. Similarly, the HEK-293, human embryonic kidney cell line, was developed since these cells could easily be transfected and then used for drug development. Cell sources from human have been efficient for manufacturing of vaccines and also for the safety. Animal cellular sources could create potential allergy problems when used in production as has historically been shown when chicken embryos were used for primary tissue/cell sources for the production of measles, mumps, and rabies. Because of the continual problems of ovine protein presence in flu vaccines, Novartis adopted the cell-based production for influenza in the MDCK (Madin Darby Canine Kidney) cell line which came from the kidney of a cocker spaniel in 1958. Vaccines produced using this cell source has been available in Europe since 2007 and since 2009 in Switzerland. This canine cell line and the Vero green monkey cell lines are the only animal derived in FDA-approved vaccine production systems. There are few options for vaccines for rubella, chicken pox, and hepatitis A which are all produced by the 1960 human cell lines, WI-38 and MRC-5 fetal cell lines providing medically based evidence that human fetal cell sources can provide efficient and safe steps in manufacturing for universal use in therapeutic product development [4, 5].

40.2 Safety and Description of Cell Sources by Increased Detection

These fetal cell lines described above were developed under up-to-date processing at that time; however, techniques have had significant evolution for cell culture methods today. One of the main elements in cell culture that has evolved is changing from use of animal-derived products such as fetal bovine serum for nutritive support and porcine-derived trypsin for enzymatic release of cells during cellular passages. As many of the cell culture methods have evolved over time, it would be of high interest to develop new cell banks addressing these critical issues along with following strict criteria for organ donation and transplantation. Cell culture systems and events associated with cell growth can introduce contaminants and thus the safety of the final biological product. Viral contamination is of utmost concern during the processing, and the main sources of viral contaminants can include inherent cell line disposition, the raw materials in media composition for cell growth, personnel, environment, and equipment. Large battery of viral testing is therefore necessary to assure safe and effective therapeutics for human vaccines and biological medicines. Development and evolution of novel and increasingly sensitive technologies based on PCR assays has helped over the years.

There can be new circumstances of contamination, and this is why the techniques need to evolve. For instance, in 2010 a new contaminant was discovered using new technology termed massive parallel sequencing (MPS) or deep sequencing to discover the porcine circovirus (PCV). The source of this contaminant was most likely from trypsin batches during manufacture. Because most cell lines in the past have been exposed to this reagent, regulatory authorities require the screening of this and many other viruses when cell substrates are used in the manufacture of clinical material. Newer non-animal products have been implemented in recent years [6–9]. Cell culture systems thus require multiple steps of testing and quality control, and these will be dependent of the cell therapies, cell source chosen, and the manufacturing processing.

40.3 Technical Considerations of Cell Choices

The possibility to use cell culture systems to expand tissue and have cell populations that retain original tissue properties has revolutionized plastic surgery for the treatment of burn patients and provides enormous possibilities in the transplantation and tissue engineering field [10]. However, biopsies from the patient (autologous transplant) can take up to several weeks in culture in order to have enough cells for the first treatments. Therefore, other cell sources which are readily available could be of particular importance in the management of this severe trauma and to help save lives. One of the major challenges for assuring that more patients will benefit from cell-based therapies in the future will be the choice of the cell type and optimization of their isolation and proliferation to assure safety.

Tissue from both animal and human sources at all ages of development can be evaluated for cellular therapies. Each cell source has its advantages and disadvantages for each final cell type. Embryonic stem cells, adult and mesenchymal stem cells, and fetal progenitor cells can all be expanded in cell culture systems but differ highly in the complexity to obtain defined cell sources (Fig. 40.1).

Embryonic stem cells that are isolated from early-stage embryo are pluripotent and have an advantage over those cells from adult mesenchymal stem cells, which can differentiate only into a restricted number of cell lineages. However, cultures of both embryonic and adult stem cell types are technically very demanding because the amount of tissue to begin with is very low for embryonic stem cells (<100 cells) and isolation of adult mesenchymal stem cells from the tissue mass is difficult (only one stem cell for every 10^{4-5} cells in total adult tissue). Maintenance and expansion of stem cells in an undifferentiated state require the addition of many specific growth factors [11–14], and efficient culture of embryonic stem cells and some adult

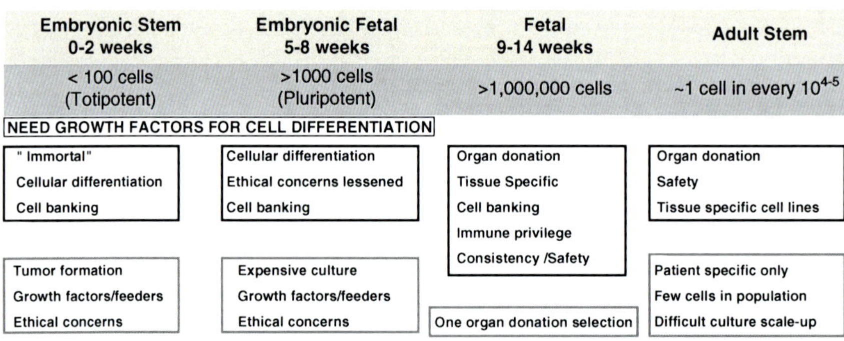

Fig. 40.1 Stem cell sources during development from embryonic to adult. Cells derived from tissue at the blastocyst stage (~100 cells) are embryonic stem cells which are totipotent. At 5–6 weeks of human development, the tissue is in the embryonic fetal stage (>1000 cells) which are pluripotent. Organ-specific cells are derived from 9 to 16 weeks of human fetal development, and these have a high number of cells ($>10^6$) for culture establishment. Adult stem cells can be found within all tissue types from 9 weeks of gestation, and these mesenchymal stem cells are pluripotent. Only one stem cell is found in every 10^4–10^5 cells in total-specific tissue

mesenchymal stem cells are not possible without feeder layers which is in some part responsible for the inconsistent colony cell growth. The necessity to use many exogenous growth factors as well as feeder layers to differentiate into specific cellular lineages are limiting factors for the scale-up of embryonic and adult stem cell cultures for clinical applications. There are other major issues with these stem cell types for security as the cells can dedifferentiate once placed into an *in vivo* environment and even develop into tumors. Many techniques involving cell cloning or encapsulation have been utilized to date for assuring delivery of correct cell populations (Fig. 40.1).

More specific aspects of each of the stem cell sources are related to their development features.

Embryonic stem cells are able to differentiate into a wide range of cell types, but in order to be able to do so, they have to be taken at an early age of the embryo (around 5 days after ovum fertilization). The *blastocyte* is created 5 days after ovum impregnation is created, and from this early structure embryonic stem cells can be obtained. These embryonic stem cells are "totipotent" until about eight cells (approximately 2 weeks postfertilization), and afterward they become "pluripotent." At this point, they cannot develop into another embryo, but they can develop in all different types of cells, and these cells are considered as an embryo until about 8 weeks after fertilization. The embryo is controlled under specific regulation for *in vitro* fertilization where in many countries a specific license is required to work with embryonic stem cells.

On the other end of the development scale are adult stem cells which have an advantage when used for autologous cell therapies (Fig. 40.2). Adult stem cells can be isolated from skin, hair follicles, adipose tissue, bone marrow, and all other tissues of the body. They have disadvantages for single-patient use and limited numbers of cells that may be stocked. Some of these adult cell sources can also be used for cell banking and are not new in medical treatments (i.e., bone marrow and adipose tissue). Bone marrow transplantation has been used for therapeutic purposes for over 40 years and cultured BM-MSC more recently for burns and scar

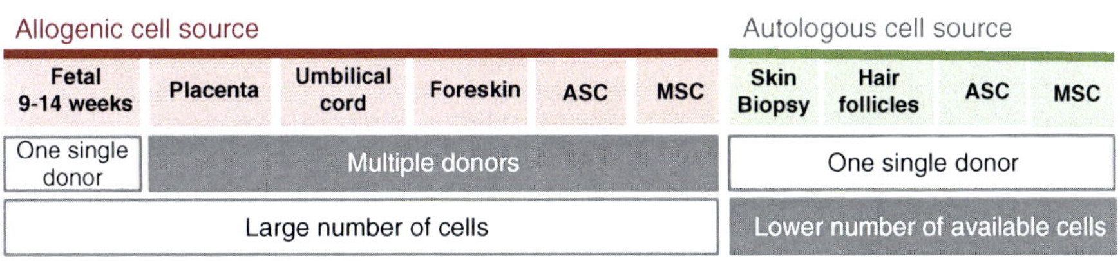

Fig. 40.2 Allogenic and autologous cell sources. Allogenic cell sources that are frequently used in the clinic for skin therapies are mesenchymal bone marrow stem cells (MSC), adipose-derived stem cells (ASC), foreskin cells, umbilical cord cells, placenta-derived cells, and from all fetal tissues at 9–14 weeks of development (i.e. skin, cartilage, bone, tendon, muscle, neurons, etc.) Autologous cell sources normally used in skin cell therapy are from patient skin, hair follicles of the patient (hair bulb), adipose tissue (ASC), and bone marrow (MSC). Allogenic cell sources have the advantage to be developed from one to several organ donations for very large numbers of cells for cell banking and multiple patient use

management. These advances in cell therapies are attributed to highly funded research through the US military, AFIRM program to develop new treatments using cell therapies and advanced plastic surgery. For instance, the DOD awarded 224.7 million dollars in 1997 to develop bone marrow-derived stem cells (BM-MSC) in extensive cell banking systems in collaboration with Osiris Therapeutics and Genzyme (www.osiris.com; www.genzyme.com) to first study radiation sickness and then used for meniscus cartilage regeneration. In their past cell banking procedures, adult donors (18–30 years) of bone marrow were isolated by density gradient, and stem cells were purified by adhesion to eliminate non-MSC cell sources. The resulting cell cultures from one donor were to make a stock of 10,000 doses of final product called Prochymal®. Unfortunately, in the early studies with cell-banked BM-MSCs, the cells were expanded over numerous passages that may have made the cells lose their effectiveness from over manipulation. The early use BM-MSCs in many clinical studies on graft vs host disease had not shown positive significant clinical results, but the lack of efficacy may be also related to patient populations chosen [15–17].

Other diseases (GvHD, Crohn's disease, cardiology, diabetes, pulmonary disease, multiple sclerosis, acute organ rejection, scleroderma, and arthritis) have also been investigated using BM-MSC cell therapy by Osiris [18]. BM-MSC cells are now being used for the first clinical trials on burns and wounds and to also look at combined surgical laser techniques with the stem cells in third degree burn scar treatments and management (http://med.miami.edu/news/miller-school-physician-scientists-receive-3-million-defense-grant-to-treat/). In optimizing the cell culture procedure, these allogenic BM-MSC cell sources could be efficient for treating burns and wounds and have readily available stocks of frozen cells.

More recently, ASC have been promoted as a more readily available autologous and allogenic cell source since adipose tissue can be extracted more easily and there are more ASC per gram of tissue than from bone marrow (Fig. 40.2).

Foreskin tissue from newborns is considered to be an organ donation and would be treated similar to operating room excess tissue which is destined to be destroyed once taken and is a readily available source for skin cells. Other allogenic cell sources include placenta with both amnion and chorion membranes, umbilical cord, and fetal tissues from 9 to 14 weeks of gestation (Fig. 40.2).

40.4 Organization of the Progenitor Cell Therapy Platform in Switzerland

Already in 1991, the Swiss Government allowed the Fetal Transplantation Platform to be registered with the Department of Public Health, and the program remained as such until 2007 when the new directions and law of transplantation took effect (Fig. 40.3). Since 2008, the Fetal Transplantation Platform was accepted for the development of "Clinical Grade Tissues for Musculoskeletal Bioengineering" and registered with Swissmedic (Federal Program for Registration of Human and Veterinary Medicines and Transplants). For this reason, there has been a transplantation program developed

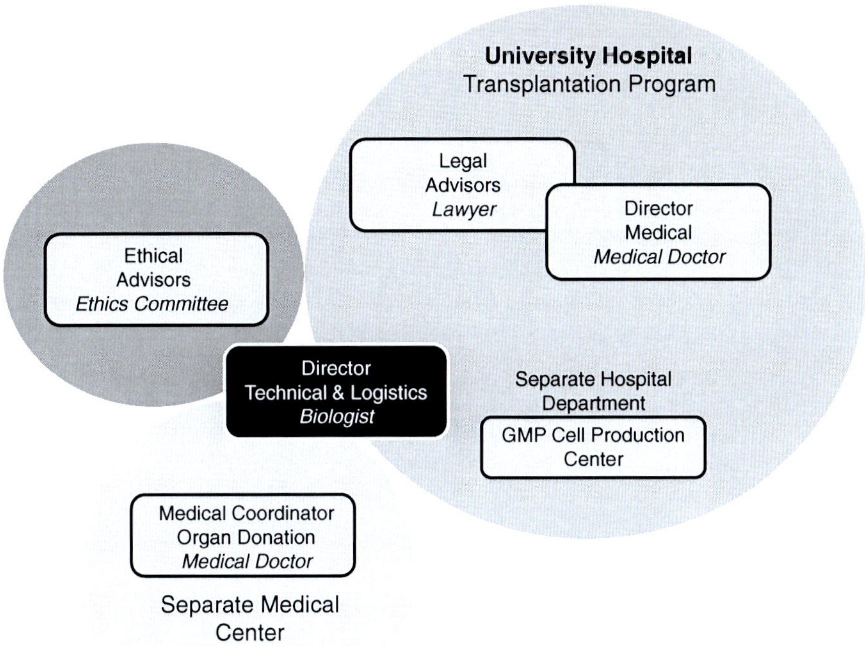

Fig. 40.3 Swiss progenitor cell platform. Organ donation program and cell bank development. Multiple fields of expertise are necessary to develop a successful platform for organ donations including lawyers to interpret regulatory issues of organ donations and defined therapeutic tissue and cellular products, biologists to assure the most appropriate cell choice and technical specifications, engineers for delivery and tissue engineering design, and medical doctors for donor screening, informed consent, and appropriate patient application of tissue and cellular products. Director of technological and logistics coordinates the program with the legal and medical directors and advisors. The medical coordinator for the organ donation is accomplished in a separate hospital facility to assure anonymous organ donation, tracing of samples with the laboratory reports and final assessment if the organ donation fulfills all inclusion criteria. The program and the director of technical and logistics along with the medical director assess the pathology and serology reports for inclusion, and parallel processing of tissue is accomplished in the GMP cell culture facility. Cell banks remain in quarantine for 3 months during the organ donation phase for retesting at 3 months followed by creation of the clinical cell banks for clinical trials

at the federal level in Switzerland that follows strict guidelines for organ donation and transplantation purposes. In brief, since fetal tissues have been central for many medical advances, it was thought to have a defined program for regulation that fit into the federal transplantation law in Switzerland. In fact, the legal framework in Switzerland could be regulated at different levels including constitutional law, federal law, application ordinances, and state Law and eventually applicable to international law, but it is mostly with the federal law regarding transplantation medicine that is applicable (http://www.bag.admin.ch/transplantation). An interdisciplinary approach for these types of programs is necessary to include expertise from lawyers (to interpret regulatory issues of organ donations and of defined therapeutic tissue and cellular products), biologists (to define appropriate cell choice and related technical issues), engineers (for delivery methods) and medicaio doctors (for screening, informed consent and appropriate patient use of cellular therapies) (Fig. 40.3).

This possibility led to the Swiss Fetal Cell Therapy Platform that was developed for musculoskeletal tissues including skin, muscle, bone, cartilage, tendon, and intervertebral disc [19–33] (Fig. 40.4). Organ donations were available through a defined program where tissue from pregnancy interruption was made accessible when the mother had to interrupt her pregnancy for medical grounds. Under full written and oral informed consent, the mother donor could be accepted into the program for organ donation. The mother donor was assessed for specific viruses and

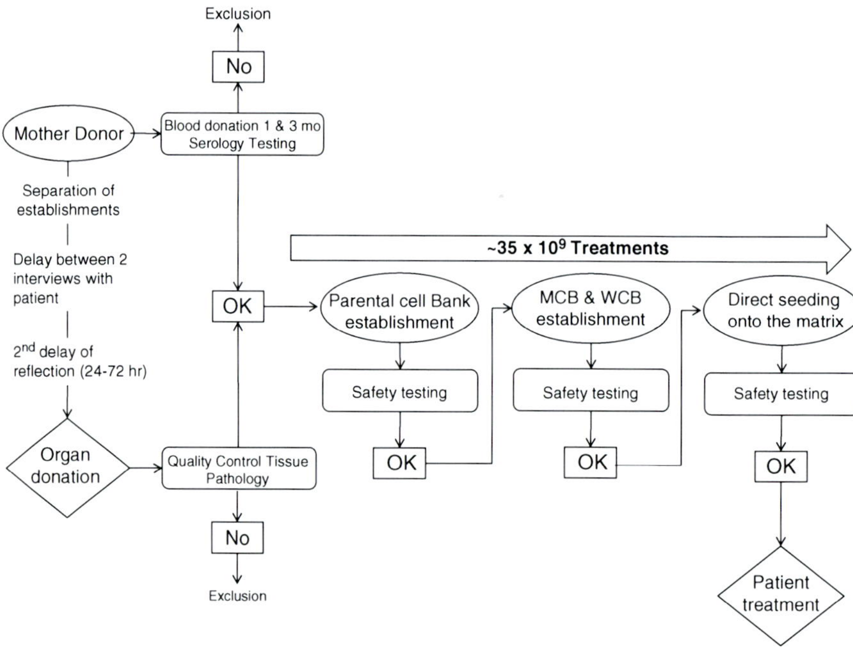

Fig. 40.4 Organ donation and potential for cell banking. One single organ donation enters into an extensive program for transplantation with all of the regulations in place for selection and testing. One single organ donation may produce first a parental cell bank (PCB) of primary cells. Tissue from skin (1–2 cm^2) or from bone, cartilage, tendon, and disc (0.2–0.5 mm^3) is enough to produce enough cells for further establishment of master and working cell banks (hundreds of vials each with 10×10^6 cells) that are frozen in liquid nitrogen (−165 °C) and can be kept for decades with high stability. From each of the master cell bank vials, it is possible to make one working cell bank containing hundreds of vials. Safety testing can be at each stage of development (patient to cell banking) before making final cell therapy treatments which can be as high as 35×10^9 for any given musculoskeletal tissue

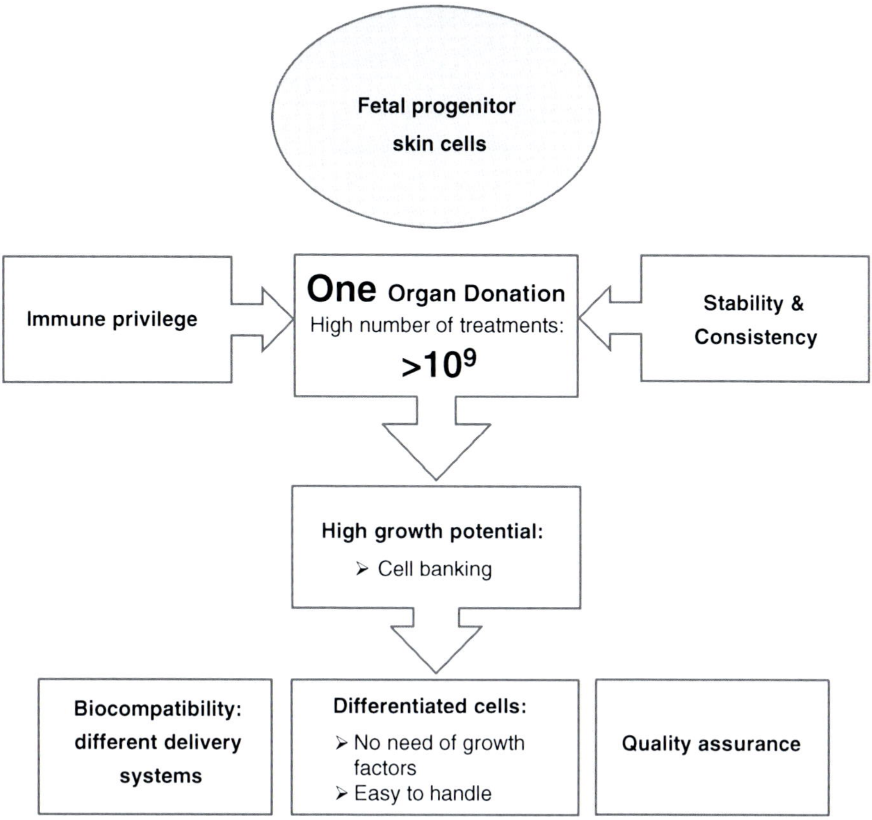

Fig. 40.5 Progenitor fetal cell advantages. From only one organ donation, stable sources of primary, diploid cells can be produced and stocked for long-term use for the clinic. Cells from individual tissues (i.e., skin, cartilage, bone, tendon, muscle, etc.) established in specific cell banking procedures are therefore easily screened for safety, have immune privilege, and are biocompatible with many matrixes making them ideal candidates for cell therapy and delivery to patients allowing millions of treatments from one cell stock due to the high growth potential and quality assurance assessment

infections by donating blood at the time of her pregnancy interruption and again after 3 months to assure that there was no seroconversion. This period of time also allowed the mother donor to change her mind regarding her participation in the program (Fig. 40.5). The tissue that had been donated could be put into primary culture immediately in the core facility for whole-cell bioprocessing developed in the University Hospital. The cells derived from the primary culture provided the parental cell banks necessary to submit to current good manufacturing practice (cGMP) production of cell banks similar to those processes used in vaccine production since the 1950s. A major advantage of allogenic cell sources is the potential of cell bank creation and out-scaling which could be used for multiple patient treatments (Fig. 40.4).

40.5 Cell Banking for Clinical Use

For cell banking procedures, they can be optimized for long-term storage and maximum future use. A "cell bank" is the stocked product of consistent cell cultures that are frozen into small vials that withstand long-term freezing in liquid nitrogen (−165 °C). A parental cell bank is from the very first cells derived from tissue. These are expanded and then frozen as the master cell bank (MCB) from which each vial can derive a working cell bank (WCB). Whole-cell bioprocessing and adaptable procedures to good manufacturing processes (GMP) make it possible to develop extensive MCB and WCB. These final stored cell banks allow thorough testing of the cells, for safety regarding sterility, pathogens, and adventitious agents and tumorigenicity. Once MCB are accomplished, WCB can be produced to establish individual batches of treatments for high numbers of patients (Fig. 40.4).

The development of master cell banks from cell sources provides a major advantage for the creation of a therapeutic biological agent. The least amount of cellular manipulation to provide enough cells for an extensive cell bank is also an advantage. It is important to mention that good manufacturing practices are imposed to assure that cell cultures are not contaminated not only by virus, bacteria, or molds but also by other cell lines. Even though this may seem difficult to imagine, it is more frequent in practice, and it is safer to accomplish isotyping to assure the cellular origin [34, 35].

Much effort has been made with allogenic adult stem cell sources particularly from BM-MSC and now adipose-derived stem cells (ASC) because they can be obtained with a less invasive technique and are 100 times more concentrated per ml of tissue compared to bone marrow [36]. Foreskin cells have also been successfully banked and used in a variety of final therapeutic products for skin cell therapies and tissue engineering.

There are other cell types coming from other developmental stages which may have more advantages than nondifferentiated adult cell sources which include fetal progenitor cells (Figs. 40.1 and 40.2). Unlike stem cells from either embryonic or adult sources, fetal progenitor cells are differentiated cells with high regeneration and low immunogenic properties [12, 37–40]. Since fetal cells are already differentiated and do not need to be directed, manipulated, or altered, the vast number of additional growth factors are not needed for cell culture and expansion. These cells are not known to dedifferentiate once placed into the *in vivo* environment since they are originally from defined tissues. They produce uniform primary cultures of cells from tissue explants without specific cell selection during tissue processing [11, 14, 24–26, 41].

Establishment of cell banks is a crucial step for musculoskeletal tissues and skin bioengineering of specific tissues. Fetal cells, because of their rapid growth and stability, are of high value, and clinical cell banks have been fully developed that can be used for patient treatment (Fig. 40.5).

Progenitor fetal cells adapt particularly well to biomaterials allowing efficient and simple delivery to the patient (Fig. 40.5). In our laboratory and in others, it has been shown that cells from donors (neonatal to adult) are not capable of efficient

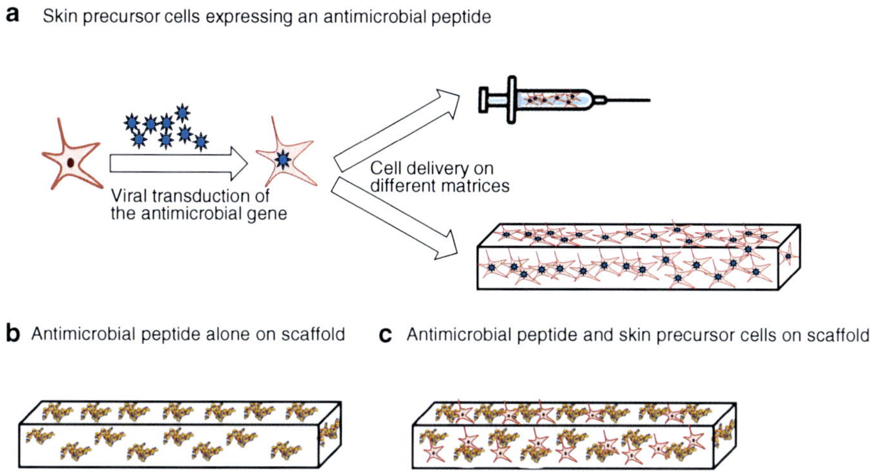

Fig. 40.6 Next-generation biological bandages with anti-infection properties. Bandages for skin repair can be developed using cellular therapies that have anti-infection activities directed toward bacteria implicated in burn and wound infections. (**a**) Antimicrobial peptides can be associated with cell therapy and implemented in different delivery modes which can have direct seeding of antimicrobial peptides alone (**b**) or in association with cell therapies (**c**) where the cells are modified as alternative strategies

integration into various biomaterials, and some biomaterials are in fact toxic to the cell. Ng, Khor, and Hutmacher [42] have emphasized that physical characteristics of scaffolds, such as porosity and mechanical stability, are important for withstanding cell contraction forces and assuring a homogenous distribution of cells throughout the scaffold. It is true that the scaffold is very important for tissue engineering, but the cell type is most probably the limiting factor. We have seen that other fetal progenitor cell types adapt well to various biomaterials [33, 42–44]. Progenitor skin cells have been used successfully in the clinic when associated with hemostatic dressings to date. This cell type can be easily integrated into these types of dressings to make biological bandages. Biocompatibility of the new formulations could easily be done with skin progenitor cells for screening purposes and also for development of biologically effective bandages. Importantly, future bandages could be directed for specific antimicrobial activity due to the variability of infectious agents in different patients such as *Pseudomonas aeruginosa* in burn patients (Fig. 40.6).

40.6 Regulatory Similarities and Differences in the World

The law on transplantation in Switzerland helps to encompass all of the legal and ethical issues and thus providing defined criteria for the medical researcher [19, 20]. The legal aspects have been addressed in an extensive review, and the program has also been described in detail elsewhere [19]. At later stages of development (9–12 weeks of gestation), the tissue is considered under transplantation programs as an organ donation and under the same regulation for adult stem cells. Fetal tissue can be considered to be an organ donation when from a voluntary pregnancy interruption and in Switzerland and many countries worldwide, there is legal availability of this medical procedure with trends remaining stable throughout the last two decades [19, 20, 45, 46]. In the United States, the importance of having access to legal pregnancy interruptions has been brought to attention in women's public health issues recently. As many as one in three women before the age of 45 seek this medical service. It has been proposed that abortion be integrated in primary care medicine and not only specialized clinics to assure the best health care [47].

Even though fetal cells are considered as adult stem cells, there can be certain considerations taken into account since this is associated with the voluntary interruption of pregnancy and to assure ethical use of available tissue donations. The specific platform begun in Switzerland in 2008 allowed for tracing all aspects of fetal tissue donation to assure complete transparency and respect of the Swiss laws and European regulations for tissue use (Fig. 40.7) [19, 20, 48–52]. These pathways have become more complex since the change of the directives and laws evolving from a linear pathway into one with multiple checks along the process for cell therapy use in the clinic (Fig. 40.7, left pathway before 2007; right pathway after 2007). Even though the pathways have intensified their complexity, the effort to develop progenitor cell banks would be worthwhile as these cell banks could be available for the next 100 years. In Israel, they have begun a program for embryonic stem cell banking to help address the ethical, scientific, and regulatory issues to serve in development of additional clinical-grade hESCs [8]. They have addressed ethical, scientific, and regulatory issues to pave the way to GMP feeder layer cell banking with foreskin, fetal, and umbilical cord tissues and their use for culture of clinical-grade embryonic stem cell banks to be used worldwide.

40.7 Adaptation to Regulatory Requirements

Because cellular therapies have had multiple changes in Regulatory worldwide, there have been some shifts to find alternative ways to reintroduce them into the clinic or to develop screening accessories for medical and cosmetic screening.

One of the pathways is the return of placental membrane products for use in the clinic. Since placental membranes are regulated as human cells, tissues or cellular and tissue-based products (HCT/P) under the 21 CFR part 1271 Section 361 of the Public Health Services (PHS) Act, they have a more simple process to be approved and do not need premarket approval. The downside is that they lack clinical data at the time they are used, and there are not randomized, controlled clinical data but only case studies to support their use. There are many products that are either cryopreserved or dehydrated in form for use [53]. Other applications are intensifying such as devices that can be used with individualized kits such as for autologous transplantation of BM-MSC. By having a centrifuge that is adapted to closed-system tubes for blood and bone marrow collection, these could be used directly at the treatment site for patient care (www.arteriocyte.com/). Other developments are

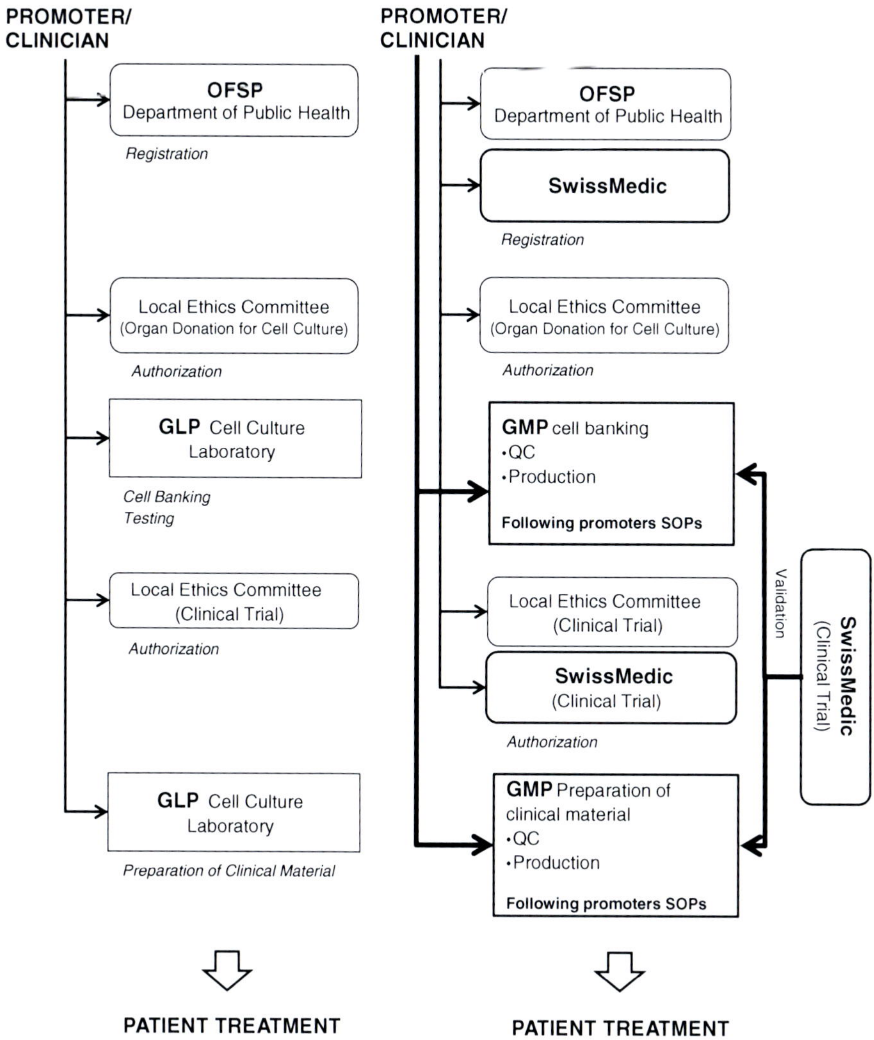

Fig. 40.7 Changes in regulatory for cellular therapies. Before 2007 (*left*), cell therapies were registered, and activities were accomplished in hospital standard laboratories following a linear pathway for clinical trial approval. After 2007 (*right*), administrative and quality control measures (legal, technical, and medical) along with new pathways and infrastructure (GCP and cGMP) are necessary for cell therapy use in the clinic and have thus created a complex pathway before new techniques finally arrive for patient use. *OFSP* Swiss Federal Office of Public Health (Office Fédérale de la Santé Publique)

biological scaffolds of extracellular matrix derived from porcine intestine to help patients cells integrate into new tissue formation and wound care [54].

AFIRM is also at the lead to bring foreign-developed techniques for direct use in clinical trials in the USA and has invested ~300 million dollars to top US universities and hospitals to develop new treatments using cell science and advanced plastic surgery which have included research on many musculoskeletal tissues (i.e., skin, muscle, cartilage, bone, and tendon). Efforts to make enough skin for severe burns have been shown to be possible in a technique developed in Australia. They have developed a device and associated kit that allows the take a 2-4 cm skin biopsy from the patient and process it in the operating room direcly within less than 1 h and then to apply the patient cells with a special sprayer device onto burns and wounds [55, 56]. Alternatively, foreskin tissue from circumcision has been assessed with this device to provide off-the-shelf cell sources ready for use. In a recent clinical study program, foreskin tissue cells which have been immortalized have been used for burns and wounds. These bioengineered cells make fresh artificial skin which has been compared to cryopreserved cadaver skin for traumatic wound cover and has been shown to be well tolerated in the clinic [57]. Importantly, these transformed cells near diploid keratinocytes have been shown to be pathogen-free and consistent and do not form tumors even though they spontaneously mutated. Other clinical trials that are funded by the DOD are concentrated adipose tissue and stem cells which also use innovative medical devices to assure easier regulatory pathways for stem cell therapy [58]. Many autologous and now allogenic cell therapies are being intensively investigated particularly for musculoskeletal tissue regeneration, and new techniques to stabilize end products such as freezing and lyophilization are being adapted to assist in logistics.

40.8 Future Recommendations of Hospital Developed and Applied Cell Therapies

It is a complex process in developing advanced cell therapy products for use in medicine. Advanced cell therapy products have to be developed taking into the consideration of many factors with patient safety being central. All other aspects with the cell choice and delivery are critical for success. Cell sources used in combination with delivery systems have been routinely used in the management of wounds. As more innovative types of dressings or bandages are evolving, there is particular scientific interest in the cell sources as well as the delivery system associated. Among the most used sources are allogenic and include acellular human cadaver dermis, human foreskin keratinocytes and fibroblasts, porcine skin, porcine small intestine mucosa, and cell lines established from human keratinocytes and human placenta and amnion membranes [53, 59–63].

Cell sources can be combined with many matrix types to allow easy delivery for patient use which may be nylon mesh, silicone, bovine collagen, porcine collagen, horse collagen, polyglycolic acid, or hyaluronic acid. The cell source choice will make a difference for healing efficiency due to the available growth factors. Early development paralleled to drug development in the search for one growth factor to do all of the necessary work for healing of the wound. As a single growth factor could ultimately be synthetically manufactured in a recombinant process, this would parallel the regulatory process of medicines easily. Although many single growth factors have been studied and tested, it could be questioned as to why use only one growth factor when cellular sources would provide so many factors at the same time. Some of the growth factors that have been of interest in wound healing include VEGF, sonic hedgehog (SHH), KGF-2, and PDGF-BB. The latter, platelet-derived growth factor, has been approved by the FDA for the use in neuropathic diabetic ulcers under the commercial name of REGRANEX gel showing approximately 20% improvement for wound healing after 20 weeks of topical treatment in aggressively debrided ulcers. There have been limitations to the total quantity in a lifetime for patients as post-marketing has presented results that overall cancer levels may be increased in patients having used this single growth factor at high dosages.

Cell sources providing lower dosages for individual growth factors but a multitude of different growth factors at very low dosages have been thought to be far better suited for wound healing compared to recombinant proteins.

Importantly, the physician needs to be aware of all of the surrounding issues in the development of biological products to assure the knowledge of the complexity on the ethical, legal, and political stakes due to the origin of the tissue and cell sources (organ donations) necessary for downline process development.

In the long run, technical aspects of working with cells will ultimately decide which cell choices are better adapted for future clinical practice.

The use of progenitor cells has been developed under regulations of a Federal Transplantation Program in Switzerland. In addition, these cell lines have been described and deposited in the European Protection Agency Cell Depository, Porton Down's. Thus, progenitor cells with their high expansion, simple culture conditions (do not require feeder layers or extensive growth factors for expansion which is a major reason for their consistency in scaling out), and low immunogenicity properties are ideal for whole-cell bioprocessing destined for cell therapy, tissue engineering, and medicinal products. Additionally, they have already been used in safety clinical phases I and II studies showing rapid and efficient tissue repair with minimal scarring. Overall development needs to keep organ donations and patient safety a priority to assist physicians with the use of these new emerging therapies for patient care (Fig. 40.3).

Bone marrow mesenchymal stem cells, foreskin cells, and autologous cells are the cell sources widely explored for cellular therapies to date with high financing through the DOD. Other cell sources that are stable and that can be easily stocked should be given the same opportunity to explore full potential (such as fetal progenitor and banked embryonic stem cells). Foreskin cells from newborns are an organ donation and are finally being widely used in tissue regeneration. Progenitor cells from fetal tissues have been shown to have more potent activity, can be stocked from only one single organ donation, and therefore are more easily screened for security and safety.

These cell types can be controlled through rigid federally registered programs of transplantation, can be deposited in cell depositories for research, and merit attention for further development of new cell therapies and advanced plastic surgery techniques. By developing consistent cell banks from only one organ donation, many of the risk factors can be eliminated for bringing safe and effective human cell-based therapies to the bedside.

References

1. Gey GO, Coffman WD, Kubicek MT. Tissue culture studies of the proliferative capacity of cervical carcinoma and normal epithelium. Cancer Res. 1952;12:264–5.
2. Banatvala JE, Brown DWG. Rubella. Lancet. 2004;363:1127–37.
3. Jacobs JP, Jones CM, Baille JP. Characteristics of a human diploid cell designated MRC-5. Nature. 1970;227:168–70.
4. Palache AM, Brands R, van Scharrenburg GJ. Immunogenicity and reactogenicity of influenza subunit vaccines produced in MCDH cells or fertilized chicken eggs. J Infect Dis. 1997;176:520–3.
5. Zimmerman RK. Ethical analyses of vaccines grown in human cell strains derived from abortion: arguments and internet search. Vaccine. 2004;22:4238–44.
6. Cavallo C, Cuomo C, Fantini S, Ricci F, Tazzari PL, Lucarelli E, Donati D, Facchini A, Lisignoli G, Fornasari PM, Grigolo B, Moroni L. Comparison of alternative mesenchymal stem cell sources for cell banking and musculoskeletal advanced therapies. J Cell Biochem. 2011;112:1418–30.
7. Moroni L, Fornasari PM. Human mesenchymal stem cells: a bank perspective on the isolation, characterization and potential of alternative sources for the regeneration of musculoskeletal tissues. J Cell Physiol. 2012;228:680–7.
8. Tannenbaum SE, Turetsky TT, Singer O, Aizenman E, Kirshberg S, Ilouz N, Gil Y, Berman-Zaken Y, Perlman TS, Geva N, Levy O, Arbell D, Simon A, Ben-Meir A, Shufaro Y, Laufer N, Reubinoff BE. Derivations of xeno-free and GMP-grade human embryonic stem cells- platforms for future clinical applications. PLoS One. 2012;7(6):ee35325.
9. Van der Valk J, Brunner D, De Smet K, Fex Svenningsen A, Honegger P, Knudsen LE, Lindl T, Noraberg J, Price A, Scarino ML, Gstraunthaler G. Optimization of chemicall defined cell culture media-replacing fetal bovine serum in mammalian *in vitro* methods. Toxicol In Vitro. 2010;24:1053–63.
10. Rheinwald JG, Green H. Serial cultivation of strains of human epidermal keratinocytes, the formation of keratinizing colonies from single cells. Cell. 1975;6:331–44.
11. de Buys Roessingh AS, Hohlfeld J, Scaletta C, Hirt-Burri N, Gerber S, et al. Development, characterization and use of a fetal skin cell bank for tissue engineering in wound healing. Cell Transplant. 2006;15:823–34.
12. Hohlfeld J, de Buys Roessingh A, Hirt-Burri N, Chaubert P, Gerber S, et al. Tissue-engineered fetal skin constructs for paediatric burns. Lancet. 2005;366:840–2.
13. Murry CE, Keller G. Differentiation of embryonic stem cells to clinically relevant populations: lessons from embryonic development. Cell. 2008;132:661–80.
14. Quintin A, Hirt-Burri N, Scaletta C, Schizas C, Pioletti DP, Applegate LA. Consistency and safety of fetal cell banks for research and clinical use. Cell Transplant. 2007;16:675–84.
15. Kebriaei P, Isola L, Bahceci E, Holland K, Rowley S, et al. Adult human mesenchymal stem cells added to corticosteroid therapy for the treatment of acute graft-versus-host disease. Biol Blood Marrow Transplant. 2009;15:804–11.
16. Mack GS. Osiris seals billion-dollar deal with Genzyme for cell therapy. Nat Biotechnol. 2009;27:106–7.
17. Newman RE, Yoo D, LeRoux MA, Danilkovitch-Miagkova A. Treatment of inflammatory diseases with mesenchymal stem cells. Inflamm Allergy Drug Targets. 2009;8:110–23.
18. Allison M. Genzyme backs Osiris, despite Prachymal flop. Nat Biotechnol. 2009;27:966–77.
19. Applegate LA, Hirt-Burri N, Scaletta C, Bauen J-F, Piolotti DP. Bioengineering of human fetal tissues for clinical use. In: Bioengineering: principles, methodologies and applications, Chapter 4. Hauppauge, NY: Nova Sciences Publishers; 2009. p. 1–19. isbn:978-1-60741-7620.
20. Applegate LA, Scaletta C, Hirt-Burri N, Raffoul W, Pioletti DP. Whole-cell bioprocessing of human fetal cells for tissue engineering of skin. Skin Pharmacol Physiol. 2009;22:63–73.
21. Applegate LA, Weber D, Simon J-P, Scaletta C, Hirt-Burri N, et al. Organ donation and whole-cell bioprocessing in the Swiss fetal progenitor cell transplantation platform. In: Organ donation and organ donors: issues, challenges and perspectives. Hauppauge, NY: Nova Publications; 2013.

22. Arvidson K, Abdallah BM, Applegate LA, Baldini N, Cenni E, et al. Bone regeneration and stem cells. J Cell Mol Med. 2011;15:718–46.
23. Darwiche S, Scaletta C, Raffoul W, Pioletti DP, Applegate LA. Epiphyseal chondroprogenitors provide a stable cell source for cartilage cell therapy. Cell Med. 2012;4:23–32.
24. Hirt-Burri N, de Buys Roessingh AS, Scaletta C, Gerber S, Pioletti DP, et al. Human muscular fetal cells: a potential cell source for muscular therapies. Pediatr Surg Int. 2008;24:37–47.
25. Hirt-Burri N, Scaletta C, Gerber S, Pioletti DP, Applegate LA. Wound-healing gene family expression differences between fetal and foreskin cells used for bioengineered skin substitutes. Artif Organs. 2008;32:509–18.
26. Hirt-Burri N, Ramelet A-A, Raffoul W, de Buys RA, Scaletta C, et al. Biologicals and fetal cell therapy for wound and scar management. ISRN Dermatol. 2011;2011:549870. doi:10.5402/2011/549870.
27. Krattinger N, Applegate LA, Pioletti DP, Caverzasio J. Regulation of proliferation and differentiatioon of human fetal bone cells. Eur Cell Mater. 2011;21:46–58.
28. Montjovent MO, Burri N, Mark S, Federici E, Scaletta C, et al. Fetal bone cells for tissue engineering. Bone. 2004;35:1323–33.
29. Montjovent M-O, Mathieu L, Schmoekel H, Silke M, Bourban P-E, et al. Repair of critical size defects in the rat cranium using ceramic-reinforced PLA scaffolds obtained by supercritical gas foaming. J Biomed Mater Res. 2007;83A:41–51.
30. Montjovent M-O, Silke M, Mathieu L, Scaletta C, Scherberich A, et al. Human fetal bone cells associated with ceramic reinforced PLA scaffolds for tissue engineering. Bone. 2008;42:554–64.
31. Pioletti DP, Montjovent MO, Zambelli P-Y, Applegate LA. Bone tissue engineering using foetal cell therapy. Swiss Med Wkly. 2006;136:557–60.
32. Quintin A, Schizas C, Scaletta C, Jaccoud S, Applegate LA, Pioletti DP. Plasticity of fetal cartilaginous cells. Cell Transplant. 2010;19:1346–57.
33. Tenorio DMH, Scaletta C, Jaccoud S, Hirt-Burri N, Pioletti DP, et al. Fetal bone cells in delivery systems for bone engineering. J Tissue Eng Regen Med. 2011;5:806–14.
34. Capes-Davis A, Theodosopoulos G, Atkin I, Drexler HG, Kohara A, et al. Check your cultures! A list of cross-contaminated or misidentified cell lines. Int J Cancer. 2010;127:1–8.
35. Torsvik A, Røsland GV, Svendsen A, Molven A, Immervoll H, et al. Spontaneous malignant transformation of human mesenchymal stem cells reflects cross-contamination: putting the research field on track. Cancer Res. 2010;70:6393–6.
36. Mosna F, Sensebé L, Krampera M. Human bone marrow and adipose tissue mesenchymal stem cells: a user's guide. Stem Cells Dev. 2010;19:1449–70.
37. Bhattacharya N. Fetal cell/tissue therapy in adult disease: a new horizon in regenerative medicine. Clin Exp Obstet Gynecol. 2004;31:167–73.
38. Montjovent MO, Bocelli-Tyndal C, Scaletta C, Scherberich A, Martin I, et al. In vitro characterization of immune-related properties of human fetal bone cells for potential tissue engineering applications. Tissue Eng Part A. 2009;15:1523–32.
39. Oster H, Wilson DI, Hanley NA. Human embryo and early fetus research. Clin Genet. 2006;70:98–107.
40. Quintin A, Schizas C, Scaletta C, Jaccoud S, Chapuis-Bernasconi C, et al. Human fetal spine as a source of cells for intervertebral disc regeneration. J Mol Cell Med. 2009;13:1–12.
41. Ramelet A-A, Hirt-Burri N, Raffoul W, Scaletta C, Pioletti DP, et al. Chronic wound healing by fetal cell therapy may be explained by differential gene profiling observed in fetal versus old skin cells. Exp. Gerontology. 2008;44:208–18.
42. Ng KW, Khor HL, Hutmacher DW. In vitro characterization of natural and synthetic dermal matrices cultured with human dermal fibroblasts. Biomaterials. 2004;25:2807–18.
43. Borcard F, Godinat A, Staedler D, Comas Blanco H, Dumont A-L, et al. Covalent cell surface functionalization of human fetal osteoblasts for tissue engineering. Bioconjug Chem. 2011;22:1422–32.
44. Krauss JF, Borcard F, Staedler D, Scaletta C, Applegate LA, et al. Functionalization of microstructured open-porous bioceramic scaffolds with human fetal bone cells. Bioconjug Chem. 2012;23:2278–90.

45. Addor V, Narring F, Michaud P-A. Abortion trends 1990–1999 in a Swiss region and determinants of abortion recurrence. Swiss Med Wkly. 2003;133:219–26.
46. Wyss D, Wirthner D, Renteria SC, De Grandi P. Les demandes d'interruption de grossesse de 1988 à 2002 au CHUV. Rev Med Suisse. 2004;2503:1–8.
47. Yanow S. It is time to integrate abortion into primary care. Am J Public Health. 2013;103:14–6.
48. EU. Setting standards of quality and safety for the donation, procurement, testing, processing, preservation, storage and distribution of human tissues and cells. In: Parliament E, editor. Directive 2004/23/EC; 2004.
49. EU. Implementing Directive 2004/23/EC of the European Parliament and of the Council as regards certain technical requirements for the donation, procurement and testing of human tissues and cells. In: Parliament E, editor. Directive 2006/17/EC, 2006.
50. EU. Implementing Directive 2004/23/EC of the European Parliament and of the Council as regards traceability requirements, notification of serious adverse reactions and events and certain technical requirements for the coding, processing, preservation, storage and distribution of human tissues and cells. In: Parliament E, editor. Directive 2006/86/EC, 2006.
51. PMP/ICH. Note for guidance on quality of biotechnological products: derivation and characterisation of cell substrates used for production of biotechnological/biological products. CPMP/ICH/294/95, 2001.
52. SwissMedics, Swiss Federal Council Transplantation Law, TxL; SR 81021, 2007.
53. Brantley JN, Verla TD. Use of placental membranes for the treatment of chronic diabetic foot ulcers. Adv Wound Care. 2015;4(9):545–59. doi:10.1089/wound.2015.0634.
54. Badylak SF, Freytes DO, Gilbert TW. Extracellular matrix as a biological scaffold material: structure and function. Acta Biomater. 2009;5:1–13.
55. Gravante G, Di Fede MC, Araco A, Grimaldi M, De Angelis B, et al. A randomized trial comparing ReCell system of epidermal cells delivery versus classic skin grafts for the treatment of deep partial thickness burns. Burns. 2007;33:966–72.
56. Wood FM, Giles N, Stevenson A, Rea S, Fear M. Characterisation of the cell suspension harvested from the dermal epidermal junction using a ReCell kit. Burns. 2012;38:44–51.
57. Centanni JM, Straseski JA, Wicks A, Hank JA, Rasmussen CA, et al. StrataGraft skin substitutes well-tolerated and is not acutely immunogenic in patients with traumatic wounds. Ann Surg. 2011;253:1–12.
58. Marra KG, Rubin JP. The potential of adipose-derived stem cells in craniofacial repair and regeneration. Birth Defects Res C Embryo Today. 2012;96:95–7.
59. Cazzell SM, Lange DL, Dickerson JE, Slade HB. The management of diabetic foot ulcers with porcine small intestine submucosa tri-layer matrix: a randomized controlled trial. Adv Wound Care. 2015;4(12):711–8. doi:10.1089/wound.2015.0645.
60. Duan-Arnold Y, Gyurdieva A, Johnson A, Uveges TE, Jacobstein DA, Danilkovitch A. Retention of endogenous viable cells enhances the anti-inflammatory activity of cryopreserved amnion. Advances Wound Care. 2015;4(9):523–33. doi:10.1089/wound.2015.0636.
61. Hart CE, Loewen-Rodriguez A, Lessem J. Dermagraft : use in the treatment of chronic wounds. Adv Wound Care. 2011; doi:10.1089/wound.2011.0282.
62. Soejima K, Shimoda K, Kashimura T, Yamaki T, Kono T, Sakurai H, Nakazawa H. Wound dressing material containing lyophilized allogeneic cultured cells. Cryobiology. 2013;66:210–4.
63. Zhou Y, Gan SU, Lin G, Lim YT, Masilamani J, Mustafa FB, Phua ML, Rivino L, Phan TT, Lee KO, Calne R, MacAry PA. Characterization of human umbilical cord lining-derived epithelial cells and tranplantation potential. Cell Transplant. 2011;20:1827–41.

Integrated Safety Strategy for the Development of Children's Cosmetic Products Using *In Vitro* and Clinical Methodologies

41

Andrezza di Pietro Micali Canavez, Talita M.T.P. Silveira, Natalia de Albuquerque Vita, Ana Cristina Weihermann, Caroline Radoski Neumann, Desirée Cigaran Schuck, Marcela Contador Baptista, Odivania Kruger, Carla Abdo Brohem, and Marcio Lorencini

41.1 Introduction

In today's society, it is increasingly difficult to clearly separate childhood and adolescence. Much has already been said about preteenager, an intermediate step between those two stages. Currently, children increasingly manifest early interest in adult clothing and habits, such as the use of makeup. The rapid development of this pattern has led the consumer market to offer a wide range of products of dubious quality, and in the absence of reliable product brands targeted at this audience, we have seen the use of items intended for adults. In this scenario, the provision of products intended for children's makeup has become increasingly important. Cosmetics should be formulated specifically for this audience to ensure their safety as much as possible.

Substances or mixtures intended for cleaning, perfuming, changing one's appearance, or providing protection are considered cosmetics. Their use is restricted to external parts of the human body (skin, hair, nails, lips, and the outside of the genitals) [1]. Studies show that on average, people use up to nine cosmetic products daily, and if the search is restricted to females, this number can rise to 15 products [2]. It is estimated that up to 10% of the world's population has some type of allergic reaction to cosmetics [3]. A study conducted by Wojciechowska et al. [4] showed that these reactions occur in 15% of cosmetic users. Regarding children, studies conducted by two different groups have shown that, on average, 45% of children develop atopic dermatitis when exposed to different cosmetic ingredients [5, 6].

A.P.M. Canavez • T.M.T.P. Silveira • N.A. Vita • A.C. Weihermann • C.R. Neumann
D.C. Schuck • M.C. Baptista • O. Kruger • C.A. Brohem • M. Lorencini (✉)
R&D Department, Grupo Boticário, São José dos Pinhais, PR, Brazil
e-mail: marciolo@grupoboticario.com.br

C. Eskes et al. (eds.), *Alternatives for Dermal Toxicity Testing*,
DOI 10.1007/978-3-319-50353-0_41

These results show that closer attention needs to be paid to the development of products for children's skin.

Several studies have shown that children's skin differs dramatically from adult skin in many ways. The characteristic fragility of children's skin is related to important differences: a pH-neutral skin surface, resulting in reduced defense against bacterial infection; reduced fat content due to the low quantity of sebaceous glands; thinner stratum corneum, epidermis, and dermis layers, increasing the permeability of the skin; and a higher concentration of proteoglycans, increasing the skin's water content [7, 8].

The 1980s marked the beginning of the development of alternative methods to animal use to obtain information about ingredients and formulations. This change occurred mainly due to differences in results between animal models and humans that represent the inherent differences between species. Alternative methods are primarily used in the evaluation of toxicological effects; however, as it is no longer ethically acceptable to use animals for cosmetic evaluation, these models are also used for efficacy evaluation. The biggest criticism of these alternative models is the fact that they have less complexity than the organism as a whole, and to overcome such difficulties, three-dimensional models, such as a reconstituted skin equivalent with greater complexity, are being produced.

In this study, we show the technical rationale for this process, in addition to *in vitro* and clinical tests conducted on a line of children's cosmetics (nail polishes, blushes, lipsticks, and lip glosses), their registration with the Brazilian regulatory agency (Agência Nacional de Vigilância Sanitária, ANVISA), and cosmetovigilance monitoring for 6 months after launch. This approach made it possible to register a children's line without using animal testing and to do so in an unprecedented manner in Brazil. Additionally, cosmetovigilance monitoring confirmed that care in the product development stage resulted in a product with minimal safety incidents, below the current average for cosmetics. Thus, this work shows that an animal-free approach to developing safe products is possible.

41.2 Regulatory and Safety Aspects in the Development and Registration of Children's Products

Brazil is one of the largest markets for children's cosmetics in the world. The use of personal care products, such as shampoos, conditioners, soaps, and beauty products, is already incorporated into the day-to-day lives of boys and girls of different age groups. This growing interest has attracted the attention of parents, doctors, and health authorities regarding the safety of these products. In Brazil, ANVISA, an autarchy linked to the Ministry of Health, is responsible for the registration of cosmetic products, including those for children's use.

With regard to makeup, which can have different pigments, every shade should be tested before being marketed to assess its potential for irritation, sensitization, and oral toxicity. In addition, ANVISA allows and encourages makeup to contain substances that have a bad (bitter) taste to prevent children from putting the product into their mouths. An essential requirement for children's makeup is to have low fixing strength and to be easily removed from the skin with water. For example, children's nail polishes are water-based and can be removed without the use of acetone or nail polish remover. Because children's nail polish has no solvent, its smell is very different from that of adult polish. Children's nail polish can also have bitter substances to prevent accidental ingestion. Lipsticks and lip glosses color the lips temporarily. As in other children's products, the formulas must include safe ingredients.

Children's cosmetic packaging must provide systems and dosing valves that allow the release of small amounts of the product and must not have sharp or dangerous ends. Furthermore, according to ANVISA, they must be free from toxic substances and cannot be delivered in aerosol form. The label must have specific safety instructions, including an indication of the age group on which the product can be used, along with guidance and warnings regarding its use. In small children, an adult must implement and oversee the use of the product [9].

Each of the ingredients used in the formulations is evaluated regarding toxicological data and restrictions of use and concentrations. Having collected preclinical information, for example, from *in vitro* studies, which are essential in determining the type and place of use, there is a need to demonstrate that cosmetic products are safe to use by conducting clinical trials on humans, which provide consumers with greater safety and less risk. In addition, the collection of preclinical information provides support for label warnings and guidance for the customer service department. Evaluation of cosmetic products on humans is not performed to investigate the potential risk but rather to confirm the safety of the finished product [10].

The challenge in developing the line presented here and the success of its registration was the fact that Brazilian law requires proof of children's makeup safety through animal studies (oral toxicity) because children tend to put products in their mouths. However, the Boticário Group has a policy of non-use of animals in its product and raw material testing. Therefore, it was decided to develop a technical and theoretical rationale to demonstrate the safety of children's products without the need for animal studies and to submit the dossier to CATEC (Câmara Técnica de Cosméticos, Technical Board of Cosmetics) to obtain permission to register with ANVISA without the need for animal testing.

During product development, in selecting products for the portfolio, the exclusion of "complex" formulations with many ingredients was considered to employ more safety data on the use of the products. Special attention was given to the development of nail polishes, with a water-based formula being chosen, as it is easily removed (with water and soap). The ingredients were in line with current Brazilian legislation, and the products did not contain prohibited raw materials. In addition, most of the ingredients used in these formulations were already used in products on the national and/or international markets aimed at children.

The fragrances used were within IFRA (International Fragrance Association) recommendations and were within the concentration range indicated for children's products. IFRA provides information about concentration, range, amount of use, and chemical composition and also provides comprehensive ingredient safety files prepared by the Research Institute for Fragrance Materials (RIFM) that also indicate possibly missing safety tests. A fragrance was chosen that has proven safety and that is also used in food products.

The characteristics of the raw materials were properly analyzed according to their nature, chemical structure and physicochemical characteristics, exposure level, toxicological profile, and in particular, the information provided in the suppliers' literature and MSDS (Material Safety Data Sheet), attesting the safety of these ingredients in the concentrations used. In addition to the previous items, the safety margins and LD50 (50% lethal dose) values resulting from historical experimental toxicity studies available in the literature were considered for each raw material used.

In terms of acute oral toxicity tests, required for registration confirmation [10], "most of the information required in evaluating the potential risk of a cosmetic product results from knowledge of the ingredients that make up its formula. It is they that can be directly responsible for any local and systemic effect". In other words, the oral toxicity of a product can be inferred from information regarding the oral toxicity of its ingredients. In the case of the formulations in question, the acute oral toxicity values accepted as limits for the selected ingredients were LD50 values between 1000 and 30,000 mg/kg (based on the results of historical experimental studies), values that are well above the quantities that can be absorbed in the case of accidental ingestion of the product.

With regard to the packaging of products, the use of small parts or "miniaturization" was avoided, thereby reducing the risk of accidental ingestion. In addition, attention given to the type of use, label warnings, and guidelines regarding adult supervision were all appropriate for the target market in question and contributed to product safety.

41.3 Alternative Methods to Animal Use in the Safety Evaluation of Cosmetic Products

In obtaining preclinical data, there are global efforts seeking to reduce, replace, and refine the use of animals in cosmetic safety and efficacy testing [11, 12]. The acceptance of such methods by regulatory agencies depends on proof of applicability of the techniques by a process of validation. Various international committees have the responsibility of validating *in vitro* methods that replace those performed on animals. These include the ICCVAM (Interagency Coordinating Committee on the Validation of Alternative Methods), ECVAM (European Centre for the Validation of Alternative Methods), BraCVAM (*Brazilian Center for the Validation of Alternative Methods*), and JaCVAM (*Japanese Center for the Validation of Alternative Methods*), among others. The mission of the OECD (*Organization for Economic Cooperation and Development*) is to promote policies that aim to improve the social and economic well-being of people around the world. The OECD proposes the three Rs principle—refinement, reduction, and replacement—as a form of rational use of animals in product testing and research. In addition, this agency is responsible for encouraging, evaluating, accepting, and regulating new methods and providing detailed guidelines of validated methodologies (OECD).

In vitro cytotoxicity studies were performed in this study according to the procedures described in OECD GD 129. This methodology measures the mouse embryonic fibroblast cell line (Balb/c 3T3) capacity to incorporate the neutral red marker (NRU, neutral red uptake). This marker is concentrated in the lysosomes of viable cells; upon contact with toxic substances, the cell loses its capacity to retain this marker. Thus, cells with low viability cannot retain the marker, while viable cells can retain it and turn into a reddish color, the absorbance of which is measured using a spectrophotometer [23]. In all, seven lipsticks, three blushes, three lip glosses, and three nail polishes were tested, all formulated specifically for children. None of the products tested showed cytotoxicity at the highest possible concentrations to be tested, making it impossible to determine their 50% inhibitory concentration (IC_{50}) values.

The phototoxic potential of these products was evaluated according to OECD TG 432 [13]. This test can efficiently identify substances that, *in vivo*, after ingestion or application to the skin, become toxic after sun exposure [13]. Zang and Dong (2005) [26] showed that this test has excellent predictive power regarding phototoxicity, anticipating toxic effects in humans with an assertiveness between 95 and 100%. Phototoxic substances are considered to be those that increase or cause toxicity upon exposure to light or solar radiation. In practical terms, the IC_{50} values of a cytotoxicity experiment with and without exposure to radiation are compared, and the difference obtained between the two experiments generates an index that expresses the capacity of the substance to absorb light energy and become toxic. Analyses were performed using Phototox software, which calculates the MPE (mean photo effect) index, which compares all points of the curve before and after irradiation. Using the generated MPE values, the samples were classified into three categories, as outlined in Table 41.1. In terms of phototoxicity, none of the products showed phototoxic potential, and all MPE values were below the 0.1 threshold.

In addition to the monolayer models, we also tested the product's potential for skin irritation using the reconstituted skin model, developed by the Boticário Group (Fig. 41.1), according to OECD TG 439 [14]. This test has become an alternative to skin irritation tests using animals and is currently widely used in the development of cosmetics [27]. Unlike monolayer tests, which require the tested active ingredient to be water soluble, this model can test any type of substance, including finished products. The use of equivalent skins follows a global trend of replacement and reduction of animal use, bringing research and preclinical testing more into line with ethical principles that preserve the dignity of animals [28]. Many models are internationally commercialized, such as those produced by MATEK and SkinEthic™.

Table 41.1 Classification of samples according to the MPE index

Factors	Non-phototoxic	Potentially phototoxic	Phototoxic
MPE	<0.1	0.1–0.15	≥ 0.15

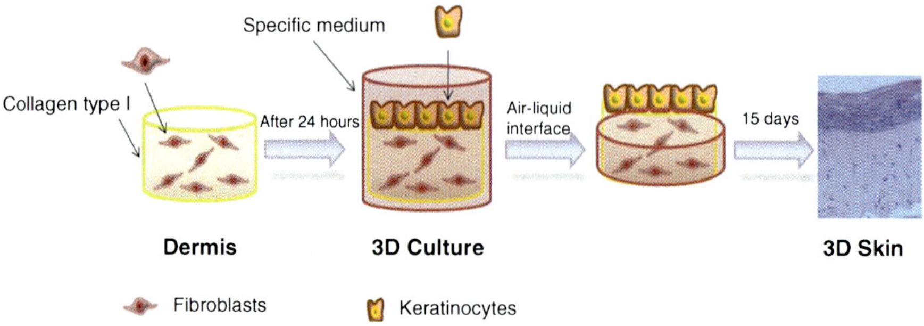

Fig. 41.1 Process for the construction of a three-dimensional reconstructed skin model

These models are not currently available in Brazil, which has led to the need to develop and implement our own equivalent skin model within the country.

The reconstructed skin model mimics human skin and maintains physiological characteristics that are lost in monolayer culture. Reconstructed skin model represents a dermal equivalent consisting of fibroblasts embedded in a type I collagen matrix and a mixture of reagents necessary for collagen polymerization. Above this layer, an epidermal equivalent is cultured consisting of keratinocytes, which remain in the air-liquid interface to differentiate into the different epidermis layers. After 10–12 days in the air-liquid interface, the skin equivalent is formed (Fig. 41.1). After this period, the product is applied topically onto the differentiated reconstructed skin model, and tissue viability is measured using the skin irritation test as described in OECD TG 439 [14]. In this protocol, we used a full-thickness model (epidermis and dermis) rather than skin equivalent models.

Using the reconstructed skin model, skin irritation evaluation can verify whether a substance has an irritant or nonirritant potential hazard. However, this model is limited in that it is unable to determine the degree of irritation (mild, medium, or severe). The following limits were used for test analysis: a substance was considered nonirritating when viability was above 50% in relation to the control, and a substance was considered an irritant when viability was below 50% compared to the control [21]. Faller and Bracher [15] demonstrated an excellent correlation between *in vivo* and *in vitro* irritation data using commercial EpiDerm™, EPISKIN™, and Cosmita reconstructed epidermis models. Using various correlation analyses and statistical models, they demonstrated that the 50% viability limit is a good predictor of irritant potential. Additionally, the skin was examined histologically to verify the differentiation of epidermal layers and to ensure tissue quality.

In the tests conducted after treatment, skins were selected for histological evaluation to determine differentiation and skin quality (Fig. 41.2). It was possible to verify formation of the epidermis in the equivalent model and to evaluate the effect of the controls on the skins. The 5% sodium dodecyl sulfate (SDS) positive control showed degradation of the epidermis, with separation of the dermis and epidermis, along with vacuoles in the epidermal cells, which indicate damage to the cellular structure. The controls without treatment and with phosphate-buffered saline (PBS)

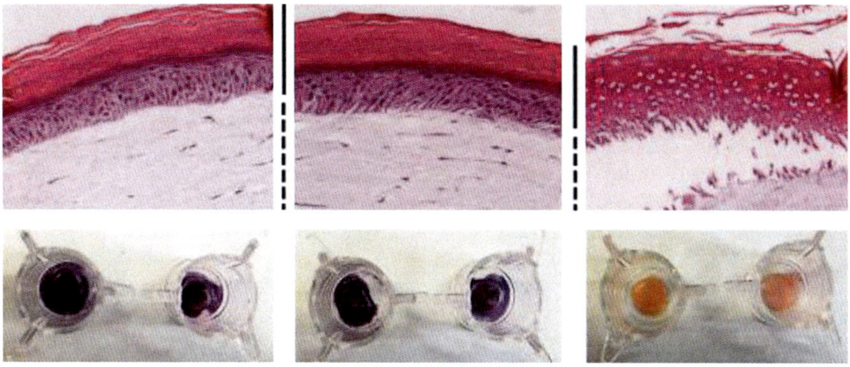

Fig. 41.2 Photos on the first row show hematoxylin- and eosin (HE)-stained histological sections of controls used in the skin irritation technique (10× magnification). The dashed line indicates the localization of the dermis equivalent; the solid line indicates the localization of the epidermis equivalent. Photos on the second row of the graphic show the equivalent skin after 3 h of incubation with MTT. The purplish color indicates viability, while white tissue indicates cell death

treatment were healthy, with all histological and cellular structures preserved, and were purplish in color, indicating MTT cell viability.

Regarding the MTT viability test results, all products tested in this model were found to be nonirritating, having viability values between 60 and 100%. As expected, the 5% SDS positive control had a mean viability of 4.82%, being irritating; the negative control, PBS, produced a mean viability of 96.3%, considered nonirritating.

41.4 Clinical Trials

In Brazil, the National Health Council (Conselho Nacional de Saúde) regulates research involving humans through Resolution 466/12 and constituted the National Research Ethics Committee (Comissão Nacional de Ética em Pesquisa, CONEP) to be responsible, among other things, for the registration of institutional research ethics committees. All research projects involving human subjects must comply with the recommendations of this resolution. All products must gain the approval of the ethics committee before being submitted to safety and efficacy tests. Furthermore, preclinical information and data provided must guarantee safe use prior to such evaluations (Resolution No. 466/12 National Health Council; CAAE N. 18225613.0.0000.5514; 20249413.2.0000.5514 and 24187713.0.0000.5514).

Clinical trials consist of two groups: compatibility studies and acceptability studies. Compatibility studies aim to demonstrate the safety of the finished product under maximized conditions, controlling the application area and quantity applied. This study, when indicated, should precede acceptability studies because, in this way, the risk of adverse reactions in those acceptability studies is minimized [10].

Compatibility studies demonstrate the absence of potential for irritation and sensitization caused by a particular product. According to the formulation, method of use, and whether the product is intended for sun exposure, the evaluation may be supplemented with phototests. Compatibility testing can be performed either with occlusive or semiocclusive applications or in open models [10].

Compatibility studies performed on children's products include evaluation of primary or cumulative skin irritation, dermal sensitization, photoirritation, and photosensitivity.

Evaluation of primary or cumulative skin irritation is performed for a single or repeated application, respectively, on the occlusive, semiocclusive, or open model, depending on the product to be tested. The first step consists of evaluating primary skin irritation, and the second step, evaluating cumulative skin irritation, with a longer application period (minimum of 3 weeks). The number of volunteers for this study was generally 50. The *International Contact Dermatitis Research Group* (ICDRG) scale was used to interpret results [10].

Dermal sensitization evaluation is divided into three stages: induction, rest, and challenge. This evaluation process serves to demonstrate the absence of allergic reactions by sensitization. This test includes a series of applications in occlusive or semiocclusive form, depending on the product. The number of volunteers is generally 50, and the results are evaluated according to ICDRG [10].

In evaluating photoirritation, one application is made in occlusive, semiocclusive, or open form on at least 25 volunteers to demonstrate the absence of potential irritation of a product applied to the skin when exposed to ultraviolet radiation. The results are also interpreted according to ICDRG [10].

In the photosensitization or photoallergy study, the form of application and the number of volunteers follow the parameters described above for photoirritation, except that repeated applications are made. The objective of this study is to demonstrate the absence of allergenic potential of a product applied to the skin when exposed to ultraviolet radiation. This test covers the phases of induction, rest, and challenge, and the results are interpreted according to ICDRG [10].

In the evaluated children's products, acceptability studies were also performed to confirm the absence of risk of primary and/or cumulative irritation and to capture feelings of discomfort reported by volunteers under normal or reasonably foreseeable usage conditions. These studies were not intended to confirm the absence of sensitizing risk (allergy). Acceptability protocols must take into account the usage conditions specified by the manufacturer, with standardized inclusion and exclusion criteria, where the only variable is use of the product [10, 16, 17]. This type of study lasts for 3 weeks and should be performed on a minimum of 30 volunteers. To utilize specific safety attributes, the study needs to be monitored by a specialist in addition to a dermatologist [10]. In the case of tests with children's cosmetic products, the "evaluated by a pediatrician" attribute may be used because the pediatrician also monitored the study phases.

To demonstrate safe use, the children's products were subjected to clinical trials (evaluating the irritant potential, sensitization, photoallergy and skin phototoxicity, and skin tolerability of the product in real usage conditions), which were first tested in adults, and after being approved, were adopted for a final stage (skin acceptability) for a children's audience, with pediatric monitoring.

In all analyses performed, no irritation, sensitization, or photosensitivity reactions were observed in any volunteer, and acceptability was 100% in adults and children. The dermatologist's evaluations concluded the following:

- No potential for primary skin irritation, potential cumulative skin irritation, or skin sensitization was observed.
- No potential for photoirritation or photosensitivity was observed.
- The products were demonstrated to be safe for human use in the area recommended by the manufacturer and did not cause irritation or sensitization reactions in the volunteer sample analyzed.
- As per recommendations from the Cosmetics Safety Evaluation Guide ([10]—2nd Edition), the products can be declared dermatologically tested.
- Products demonstrated to be safe for human use in the areas recommended by the manufacturer and did not cause irritation or sensitization reactions in the volunteer sample analyzed.
- Products can be described as "pediatrician evaluated."

Although not required, clinical compatibility and acceptability studies are highly recommended in the safety evaluation of cosmetics applied to the skin and external mucosa, as these applications can lead to unwanted local and systemic effects [29]. Local reactions include irritation, contact dermatitis, hives, or reaction upon exposure to sunlight. Since the prohibition of the use of animals in cosmetic safety checking, in addition to the low predictive value of the animal model, compatibility tests of finished products on humans can be seen as a scientifically and ethically more acceptable option [29]. It should be borne in mind that these clinical studies are conducted on a small sample population, and all previous safety analysis of the ingredients ensures a high degree of product safety. In the clinical trials conducted in this study, children took part only to the acceptability tests, after the compatibility tests were conducted on adults, always with pediatric monitoring, following the recommendations of the scientific committee of the European Commission, which follows a guideline describing the ethical use of human volunteers in safety tests involving the compatibility and acceptability of finished products [20].

41.5 Cosmetovigilance

RDC Resolution No. 332 of 01/12/2005 of ANVISA stipulates that manufacturers and/or importers of personal hygiene products, cosmetics, and perfumes in Brazil should implement a cosmetovigilance system. Cosmetovigilance is a process of monitoring the side effects caused by cosmetic products, which enables an assessment of risk to consumer health and guides the corrective actions that may be necessary [19]. In summary, this process includes recording of reports of occurrences of adverse events and their respective evaluations, recording of the measures taken to resolve the event and notification to ANVISA. In cases of complaints, the company must have a system in place for recording incoming contacts, undertaking an appropriate investigation, taking the necessary measures, and responding immediately to the claimant. Cosmetovigilance contributes to the safe and rational use of cosmetics and facilitates the construction of a database that can provide a source of information for health authorities and for future developments. Adverse events usually caused by cosmetics include itching, burning, erythema, allergic contact dermatitis, contact dermatitis caused by irritation, contact dermatitis due to phototoxicity, and cosmetic acne [25].

Cosmetovigilance complaint monitoring for a period of 6 months after the release of the products revealed three complaints, two related to "efficacy" and only one to "sensitivity." This number was very low, considering the high number of retail sales (201,872 products). None of the complaints were reported as serious, and there were no consumer accidents during the study period.

41.6 Integrated Strategy

The risk assessment approach based on existing data in conjunction with the tests performed proved to be comprehensive, as the tests were performed comprised both *in vitro* and clinical trials and were complemented by cosmetovigilance data, which showed no toxicity or adverse reactions to the products [22]. Multidisciplinary safety approaches can be advantageous, as they eliminate known hazardous ingredients, and the accuracy of data collected is confirmed both by *in vitro* tests and clinical trials (Fig. 41.3). Regarding cosmetovigilance, when comparing historical company data, not shown here, during the first year of launch of a product, there is a peak in the number of complaints; in this study, the data collected after 6 months of launch were very low in relation to historical averages, showing the importance of the integrated strategy adopted.

The approach to cosmetic risk evaluation should consider four criteria in the analysis process: (1) risk identification, (2) dose-response, (3) exposure assessment, and (4) risk characterization [18] (Fig. 41.3). In this work, for the identification of risks (1), historical data relating to carcinogenesis and other toxicological responses were considered, and raw materials having a potential hazard were excluded. Regarding dose-response (2), the concentrations of ingredients known to have potential toxic effects were considered, and only ingredients having a LD_{50} value higher than 1000 mg/kg were prioritized. Exposure assessment (3) considered the average exposure to the cosmetic product according to the Notes of Guidance for the Testing of Cosmetic Ingredients and Their Safety Evaluation from the Scientific Committee on Consumer Safety (SCCS), considering the magnitude, duration, and route of exposure [24]. The risks mapped out in steps (1), (2), and (3) were used to

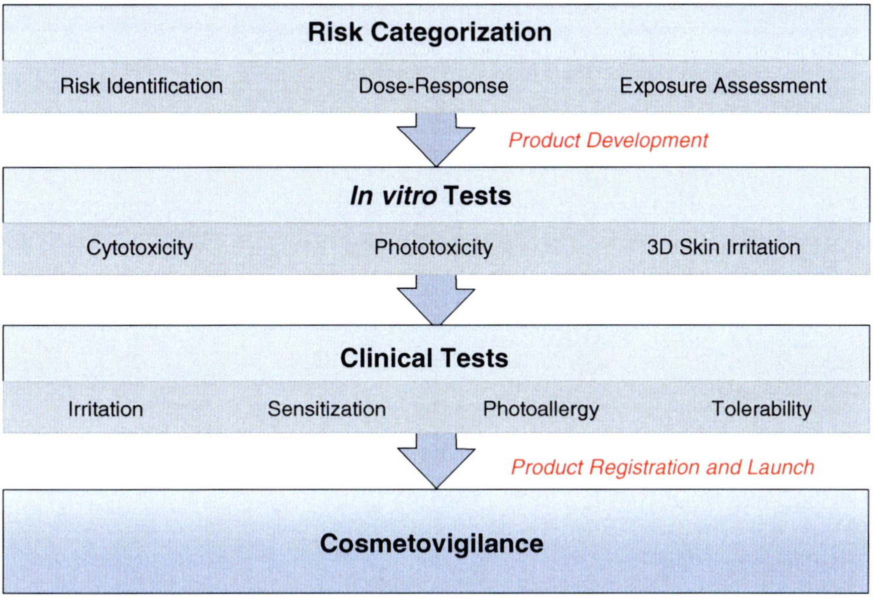

Fig. 41.3 Integrated safety strategy stages for the launch of children's products

construct the risk characterization rationale (4). Products developed from these theoretical rationales were then tested *in vitro*, following OECD recommendations, and confirmed the prior risk characterization based on existing data (Fig. 41.3). Following the *in vitro tests*, clinical trials reinforced and confirmed the appropriateness of the product development process and the *in vitro* safety testing results. Finally, cosmetovigilance represents a monitoring stage employed after product registration and launch. In the present study, it ensured that the consumer did not experience any unwanted or unexpected reactions and offered proof that a strategy of theoretical, *in vitro*, and clinical trials ensured the quality and safety of the commercialized children's cosmetic products.

41.7 Conclusion

To meet the needs of a growing market and to address ethical and product safety issues, a single evaluation or test does not meet all the necessary requirements. Thus, the theoretical rationale based on existing data, the *in vitro* (preclinical) evaluations, and the clinical trials of the developed formulations were able to produce safe products and enabled the products to be registered with the regulatory authority without the need for safety testing on animals. The results showed equivalence between the developed products and those on the market, i.e., they showed absence of potential phototoxic and irritating effects as well as negligible cytotoxicity.

References

1. EU. Regulation (EC) No. 1223/2009 of the European parliament and of the council of 30 November 2009 on cosmetic products (recast). 2009. OJ L 342/59, 22 December 2009.
2. Linda B, Sedlewicz BS. Cosmetic preservatives: friend or foe? Skin Med. 2005;4:8–100.
3. Wolf R, Wolf D, Tüzün B, Tüzün Y. Cosmetics and contact dermatitis. Dermatol Ther. 2001;14:181–7.
4. Wojciechowska M, Gocki J, Bartuzi Z. The occurrence of cosmetics side effects. In: Bartuzi Z, editor. Selected aspects of health care Bydgoszcz: Nicolaus Copernicus University; 2007. p. 524–8.
5. Giordano-Labadie F, Rancé F, Pellegrin F, Bazex J, Dutau G, Schwarze HP. Frequency of contact allergy in children with atopic dermatitis: results of a prospective study of 137 cases. Contact Dermatitis. 1999;40(4):192–5.
6. Rastogi SC, Johansen JD, Menné T, et al. Contents of fragrance allergens in children's cosmetics and cosmetic-toys. Contact Dermatitis. 1999;41(2):84–8.
7. Fluhr JW, Darlenski R, Taieb A, et al. Functional skin adaptation in infancy—almost complete but not fully competent. Exp Dermatol. 2010;19(6):483–92.
8. Shwayder T, Akland T. Neonatal skin barrier: structure, function, and disorders. Dermatol Ther. 2005;18(2):87–103.
9. ANVISA. Cartilha de Cosméticos Infantis. 2017 Disponível em: http://www.anvisa.gov.br/cosmeticos/material/cosmetico_infantil.pdf. Acessado em 06 July 2015.
10. ANVISA. Guia para avaliação de segurança de produtos cosméticos. Brasília, DF: Agência Nacional de Vigilância Sanitária. 2nd ed. 2012. Disponível em: http://www.anvisa.gov.br/cosmeticos/guia/index.htm. Acessado em 06 July 2015.
11. Eskes C, Sá-Rocha VM, Nunes J. Proposal for a Brazilian center on alternative test methods. ALTEX. 2009;26(4):303–6.
12. Kandárová H, Letašiová S. Alternative methods in toxicology: pre-validated and validated methods. Interdiscip Toxicol. 2011;4(3):107–13.
13. OECD. Guidelines for the Testing of Chemicals 432: In vitro 3T3 NRU phototoxicity test. OECD Organisation for Economic Cooperation and Development, Paris, France. 2004. Disponível em: http://www.oecd-ilibrary.org/environment/test-no-432-in-vitro-3t3-nru-phototoxicity-test_9789264071162-en. Acessado em 06 July 2015.
14. OECD. Guidelines for the Testing of Chemicals 439. In vitro Skin Irritation: Reconstructed Human Epidermis Test Method. Organisation for Economic Cooperation and Development, Paris, France. Originally adopted in 2010. 2013. Disponível em: http://www.oecd-ilibrary.org/environment/test-no-439-in-vitro-skin-irritation-reconstructed-human-epidermis-test-method_9789264203884-en. Acessado em 06 July 2015.
15. Faller C, Bracher M. Reconstructed skin kits: reproducibility of cutaneous irritancy testing. Skin Pharmacol Appl Ski Physiol. 2002;15(Suppl 1):74–91.
16. Jackson EM, Robillard NF. The controlled use test in a cosmetic product safety substation program. J Toxicol Cut Ocular Toxicol. 1982;1(2):117–32.
17. Waggoner WC. Clinical safety and efficacy testing of cosmetic. New York: Marcel Dekker; 1979.
18. Samet J. Risk assessment and child health. Pediatrics. 2004;113(4):952–6.
19. BRASIL. Ministério da Saúde. Agência Nacional de Vigilância Sanitária. Resolução RDC n° 332 de 01 de dezembro de 2005. As empresas fabricantes e/ou importadoras de Produtos de Higiene Pessoal Cosméticos e Perfumes, instaladas no território nacional deverão implementar um Sistema de Cosmetovigilância, a partir de 31 de dezembro de 2005. 2005. Disponível em: http://www.anvisa.gov.br/e-legis. Acessado em 06 July 15.
20. European Commission. European Commission scientific committee guidelines on the use of human volunteers in compatibility testing of finished cosmetic products. 2015. Available from: http://ec.europa.eu/health/ph_risk/committees/sccp/docshtml/sccp_out87_en.htm. Accessed in 06 July 2015.

21. Grisham JW, Smith GJ. Predictive and mechanistic evaluation to toxic responses in mammalian cell culture systems. Pharmacol Rev. 1984;36(2):151S–71S.
22. Kimber DA, Basketterb K, Bertholdc M, et al. Skin sensitization testing in potency and risk assessment. J Toxicol Sci. 2001;59:198–208.
23. OECD. Series on Testing and Assessment n. 129: Guidance document on using cytotoxicity tests to estimate starting doses for acute oral systemic toxicity tests. OECD Guideline for the Testing of Chemicals, Paris. 2010. Disponível em: http://www.oecd.org/env/ehs/testing/seriesontestingandassessmentpublicationsbynumber.htm. Acessado em 06 July 2015.
24. SCCS (Scientific Committee on Consumer Safety). The SCCS'S notes of guidance for the testing of cosmetic ingredients and their safety evaluation. 8th Revision. 2005.
25. Kohl L, Blondeel A, Song M. Allergic contact dermatitis from cosmetics-retrospective analysis of 819 patch-tested. Dermatology. 2002;204:334–7.
26. Zhang HW and Dong SX. Study on using in vitro 3T3 neutral red uptake to test phototoxicity. Wei Sheng Yan Jiu. 2005;34(5):628–630.
27. Hu T, Kaluzhny Y, Mun GC, Barnett B, Karetsky V, Wilt N, Klausner M, Curren RD, Aardema MJ. Intralaboratory and interlaboratory evaluation of the EpiDerm 3D human reconstructed skin micronucleus (RSMN) assay. 2009;673(2):100–8.
28. Welss T, Basketter DA, Schroder KR. In vitro skin irritation: facts and future. State of the art review of mechanisms and models. Toxicology in Vitro 2004;18(3):231–43.
29. SCCNFP Notes of Guidance for Testing of Cosmetic Ingredients for their Safety Evaluation. The Scientific Committee on Cosmetic Products and Non-Food Products Intended for Consumers SCCNFP/0321/00. Adopted by the SCCNFP during the plenary meeting of 24 October 2000.

High-Throughput Screening Techniques 42

Rico Brendtke, Bart De Wever, Florian Groeber,
Jan Hansmann, Freia Schmid, and Heike Walles

42.1 State of the Art of High-Throughput Screening

High-throughput screening (HTS) is a technique well known for the identification of hits in drug discovery [1–3]. For the selection of promising lead candidates, thousands of molecules have to be tested [4]. Therefore, the automation of the screening is essential to allow testing of 10,000–100,000 samples per day [5]. Automated HTS systems in drug discovery allow preparation, incubation

---

Rico Brendtke, Bart De Wever, Florian Groeber, Jan Hansmann, Freia Schmid and Heike Walles contributed equally to this work.

R. Brendtke
Department Tissue Engineering and Regenerative Medicine (TERM),
University Hospital Wuerzburg, Wuerzburg, Germany

Senetics Healthcare Group GmbH & Co. KG, Erlangen, Germany

B. De Wever (✉)
ATERA SAS, Nice, France
e-mail: bart@deweverconsulting.be

F. Groeber • F. Schmid
Translational Center Wuerzburg 'Regenerative Therapies in Oncology and Musculoskeletal Diseases', Fraunhofer Institute for Interfacial Engineering and Biotechnology (IGB), Wuerzburg, Germany

J. Hansmann • H. Walles
Department Tissue Engineering and Regenerative Medicine (TERM), University Hospital Wuerzburg, Wuerzburg, Germany

Translational Center Wuerzburg 'Regenerative Therapies in Oncology and Musculoskeletal Diseases', Fraunhofer Institute for Interfacial Engineering and Biotechnology (IGB), Wuerzburg, Germany

C. Eskes et al. (eds.), *Alternatives for Dermal Toxicity Testing*,
DOI 10.1007/978-3-319-50353-0_42

and analysis of many candidates simultaneously. The parallel testing of compounds generated through combinatorial chemical synthesis could significantly reduce the costs of drug development and results in substantial time saving for large numbers of samples [6]. Assays used for high-throughput drug discovery are mainly based on standard tests applied in biological and biochemical sciences, such as enzyme linked immunosorbent assays (ELISA), reporter gene assays or binding assays [5]. For increased sensitivity and reduced volume, biochemical assays are performed using scintillation proximity assays or fluorescence detection techniques, e.g. fluorescence resonance energy transfer, fluorescence correlation spectroscopy or homogeneous time resolved fluorescence. In addition, cell-based assays and ion channel techniques are applied in high throughput. In a first screening, the compounds are often tested as singlet, with low concentrations between 1 and 10 micromolar. If a positive result is generated, a secondary quantitative screening is performed and the IC_{50} is calculated [5].

In drug development, the majority of biological HTS applications are based on cell lines or primary cells cultured in 2D systems. To increase the amount of relevant information, 3D cultures have been successfully automated for HTS in drug development [7]. In contrast to 2D cultures, 3D systems exhibit higher *in vitro* to *in vivo* correlation and thus allow improved transferability to administration of a drug to human subjects. Hence, the combination of HTS techniques and 3D cell culture systems constitutes a promising research tool and can be applied for a variety of applications in biological and chemical sciences such as *in vitro* dermal toxicity testing.

In addition, programmes such as the Toxicology in the Twenty-First Century (Tox21), which is a federal collaboration of the Environmental Protection Agency (EPA) and the Food and Drug Administration (FDA), endorse the screening of thousands of chemicals for potential toxicity leading to an additional increasing need of *in vitro* models [8, 9]. The programme's aim is to rapidly and efficiently test whether certain chemical compounds have the potential to disrupt processes in the human body that may lead to adverse health effects. As these testing strategies require a significant increased number of test and *in vitro* toxicity testing is labour intensive and binds trained personnel, an HTS approach would be preferable. Automation of such *in vitro* assays could increase reproducibility and accuracy of measurements and reduces the error rate in comparison to manual processes and, therefore, enhances the quality of the generated data. Furthermore, an automated implementation leads to higher maximum capacities and is time efficient [9]. Nevertheless, in addition to automated testing, a sufficient number of complex 3D test systems must be available. Thus, there is also a need for a cost-effective, reproducible mass production of tissue models. When overcoming both low availability of 3D test systems and lack of devices for testing of 3D systems, HTS applications have great potential for meeting the challenges to alternative methods for animal testing and to become important components of modern toxicology testing strategies.

42.2 High-Throughput Production of *In Vitro* Models

42.2.1 High-Throughput Generation of Two-Dimensional Models

HTS requires a high amount of cells to generate sufficient numbers of both 2D and 3D models for the testing. Therefore, automation of the cell expansion of primary skin cells is the first step towards an application of HTS in dermal toxicity testing. The automation of cell culture processes such as the Cellerity™ (Tecan Trading AG, Switzerland) or the CompacT SelecT™ system (TAP Biosystems part of Sartorius AG, Germany) allows fully automated cell maintenance and expansion and provides the possibility to plate cells into a multiwell format for subsequent cell-based assays. Also the institution of commercially available reactor systems, e.g. from Eppendorf, BioSpherix, GE Healthcare Sciences or Pall Life Sciences, could improve the availability of cells (Eppendorf AG, Germany; BioSpherix Ltd., USA; GE Healthcare Sciences, Great Britain; Pall Life Sciences, USA). However, an effective expansion of primary human keratinocytes and dermal fibroblasts in microcarrier culture or bioreactors has already been shown [10, 11]. A current advantage of automated 2D cell cultures is that also different multiwell formats can be seeded with different cell types which supports a high compatibility to currently used systems for HTS in drug discovery and thus allows a good transferability of technologies for dermal toxicity testing into high-throughput processes. The KeratinoSens™ assay, for instance, is a non-animal skin sensitization test in a 96-well format.

However, 2D cell cultures constitute a rather artificial test system and can hardly be compared with native tissue, in which cells are embedded in a complex 3D microenvironment [12]. The unnatural rigid and flat substrates of 2D cell culture surfaces can alter cell metabolism and reduce functionality [13]. Moreover, these assays are not applicable to test hydrophobic substances as these are not soluble in physiological hydrophilic cell culture media. To overcome these drawbacks, great efforts are made to generate *in vivo*-like 3D cell cultures. However, the implementation of 3D cultures in an automated production to facilitate HTS faces several challenges due to more complex culture protocols and the use of nonstandard culture equipment.

42.2.2 High-Throughput Generation of Three-Dimensional Tissue Models

An automation of a 3D cell culture system was already realized by a collaboration of Tecan and TAP Biosystems. The two companies combined the Freedom EVO® liquid handling platform with the collagen-based 3D RAFT™ cell culture system, which led to an automated production of 3D collagen models and supports reproducible preparation of 3D cell cultures. TAP Biosystems established a 3D cell culture system based on a collagen matrix with a collagen content close to the native human skin tissue. The models could be manufactured either in 24- or 96-well formats and are suitable for oncology, toxicology, neuroscience and stem cell applications. The Freedom EVO® platform is composed of liquid handling devices and robotic arms and includes reagent, microplate cooling units and heated shaking devices. This configuration provides a fully automated production and culture of 3D RAFT™ models that could be used for a broad variety of cell biology applications (www.tecan.com; www.raft3dcellculture.com; www.tapbiosystems.com). In the field of toxicology research, a 3D liver model based on the RAFT™ system was established, providing higher functionality and a longer maintenance compared to a 2D culture [14]. Nevertheless, so far there is no application for dermal toxicity testing based on such a system as the standard configuration is lacking the possibility for culturing cells or tissues at the air-liquid interface, which is essential for the physiological epidermal differentiation.

However, 3D cornified epithelia such as the epidermis withstand mechanical or chemical damage to a greater extent than 2D cell layers. Epithelia cells cultured at the air-liquid interface mature into epithelial equivalents that resemble physiological properties, e.g. a histological architecture similar to the native human skin. Additionally, epithelial equivalents mimic a physiological barrier function impairing the penetration of toxic substances to viable cell layers, which is a crucial criteria, e.g. for *in vitro* irritation testing. Several test methods based on commercially available reconstructed human epidermis (RHE) have already received regulatory acceptance as a full replacement for *in vivo* skin irritation, and corrosion testing and other toxicity testing applications based on the use of 3D skin models, including sensitization, genotoxicity and phototoxicity, are currently in validation. Due to a raising demand, the availability of 3D skin models can become a major bottleneck in the replacement of animal tests. In addition to a higher availability, automation technology can increase the reproducibility of skin tissue engineering processes due to a higher degree of standardization. To date, manual production results in detectable variances in the RHE quality due to individual differences of the cell and tissue handling. The implementation of robotic systems conducting the entire production process ensures a reproducible quality of the models. Furthermore, an automated process produces a comprehensive data set of all manufacturing steps, which can be objectively analysed for quality control and assurance [15].

Based on the need for cost-effective, reproducible mass production of tissue models, the Fraunhofer-Gesellschaft developed a production plant, called 'Tissue Factory', which provides the possibility for the automated manufacturing of various kinds of human tissue models. The 'Tissue Factory' was a collaborative project of four Fraunhofer institutes coordinated by the Fraunhofer IGB. In a first phase, the facility was used to automate the production of the in-house developed Fraunhofer full-thickness human skin equivalent [16] which is manufactured using human keratinocytes seeded on top of a dermal equivalent consisting of fibroblasts embedded in a specific collagen scaffold. This allows the keratinocytes to differentiate into a multilayered epidermis with a stratified corneous layer (*stratum corneum*) exhibiting a barrier function to the penetration of topically applied substances. The production system of this first phase was organized in three modules. Focusing on high-throughput and maximized efficiency, a fully automated cell extraction module allows the isolation of primary keratinocytes and fibroblasts through a process chain comprising a combination of mechanical and enzymatic applications, while maintaining high cell vitality. Proliferation of primary human keratinocytes and fibroblasts to achieve the generation of a sufficient amount of cells is performed in the cell expansion module. The culture of primary skin cells, with a capacity of 500 flasks/bioreactors, was specifically adapted for the fully automated process. Finally, 3D skin equivalents are generated in the tissue culture module. Conducting this process without manual intervention is a challenging task with respect to process control and automation technology. It requires reliable handling and mixing of dispensed cells and other liquids with consistent time- and process-dependent properties. For instance, a reproducible application of high viscous liquids into tissue culture inserts requires precise positioning and dynamic control of active and passive dosing systems.

However, as regulatory guidelines currently endorse RHE tissue models to be used, the tissue factory was adapted in a second phase to meet the requirements for the production of such models. Hence, the automated production was adapted for the generation of the so-called 'open-source' reconstructed epidermis, based on an initial publication of Poumay et al. and further developed by the Henkel AG & Co. KGaA. The OS-REp is comprised of primary human epidermal keratinocytes which differentiate to a multilayered epidermis with a well-formed basal layer and a dense *stratum corneum* [17, 18]. The production plant ensures standardized and reproducible manufacturing with a monthly output of 2000–5000 epidermal equivalents. Whereas cell extraction and expansion are performed manually, the 'Tissue Factory' today can conduct all tissue reconstruction process steps fully automated (www.tissue-factory.com). Although the automated skin equivalent production has great potential for meeting the challenges of alternative dermal toxicity testing, classical automation approaches require substantial resources and lack a physiological continuous medium supply. As an alternative technology, bioreactor perfusion systems that mimic the *in vivo* nutrients supply can also be used.

42.2.3 Bioreactor-Based Generation of 3D Models

Bioreactor systems were firstly introduced for biotechnological applications. Per definition, a bioreactor is a system that converts biological processes. In addition to enzymatic and micro-bacterial applications, bioreactors are employed for processing plant and mammalian cells [19]. Commercial applications of bioreactors are, e.g. the culture of microorganisms for the production of substances such as amino acids or enzymes. Here, bioreactors allow robust concurrently adjustable process conditions in large-scale facilities [20–22]. In addition to these applications, bioreactor systems have been successfully employed in tissue engineering. Compared to conventional static tissue culture conditions, e.g. culture in multiwell plates, the advantage of bioreactor systems is that *in vivo*-like conditions can be mimicked [23]. With respect to dermal toxicity testing, this supports culturing the skin tissue under perfusion and convection and, thus, improved nutrients' supply and the skin tissue-specific culture at the air-liquid interface [24, 25]. Furthermore, bioreactors can be designed as closed systems. In combination with automatically controlled pumps that circulate defined volumes of media through models, this would allow the ability to perform middle- or long-term studies in toxicology.

Despite the advantages of bioreactor systems, their application in HTS is thwarted due to increased technical requirements compared to standard cell culture techniques. For maintaining controlled dynamic culture conditions in a tissue engineering process, additional equipment, e.g. pumps, sensors and feedback control systems, is required. In general, this can result in complex bioreactor embodiments exhibiting a need for extensive laboratory space, an increased prone to failure and a limited possibility for parallelization. In addition, a broad variety of bioreactor systems of different designs is currently available [26–29]. This demonstrates a lack of platform concepts and standards that is impairing the harmonization of bioreactor technology and limiting the transferability in industrial application [30].

Nevertheless, bioreactor systems are currently the only technology for generating complex, highly structured tissues composed of different cell types [31, 32]. Although simple RHE has been accepted by the Organization for Economic Co-operation and Development (OECD) to assess skin corrosion and irritation, more challenging endpoints might require more complex skin models which could be particularly useful to examine systemic effects of applied substances. Thus, the prediction level on reactions of human subjects might be enhanced. For instance, absorption and elimination kinetics can be studied, and thereby, the investigation of more than one drug at a time can be performed in order to identify drug/drug interactions. Complex skin models can be generated using decellularized native tissue containing the structure of the vascular system (BioVaSc®). This approach supports, e.g. the culture of a vascularized skin test model in combination with the dynamic conditions applied by the bioreactor [25]. The vascularization of the skin tissue can help to understand processes such as transdermal adsorption as well as the systemic availability of substances. Moreover, the system is applicable for the investigation of skin disease such as melanoma or psoriasis as the vasculature is one of the key components in the progression of these diseases [33, 34].

In addition, not only the skin but also intestinal and bladder test systems can be used for toxicity testing [35, 36]. Toxicological test applications, feasible via bioreactor-based testing, are toxicokinetics, dermal sensitization, repeated dose toxicity as well as carcinogenicity and reproductive toxicity [8, 37, 38]. Parallelization can be facilitated, when downscaling the total volume of a bioreactor. Due to significant efforts, it is possible to perform testing on a micro-bioreactor (μBR) chip with dimensions of just a few millimetres [39–42]. Researchers develop technologies that might be capable to build up a micro-electromechanical system (MEMS), where systemic effects can be investigated in a 'human-on-a-chip' perspective. The possibility to test effects not on a single organ but in complex systems comprised of different tissue models could help to investigate systemic effects *in vitro*. With this approach, potential benefits or harms can be investigated as shown previously especially in terms of drug development [43–45].

In conclusion, bioreactors allow controlled culture and testing conditions as well as a high level of robustness due to automated process steps of reproducible results [46]. However, it is challenging to employ high parallel processing within bioreactor systems. There are only a few studies reporting parallel bioreactor technology approaches and *in silico* simulations, where the system couples cell expansion and model generation in one device on a macroscopic scale [47–49]. Currently, research is performed to develop systems where many models can be cultured simultaneously [50–52]. These systems can prospectively support parallelized toxicity testing under controlled and standardized conditions [53, 54]. If successful, bioreactor systems might constitute, as already stated by the National Research Council (NRC) in 2007 [55], one solving strategy towards twenty-first century toxicology in the future.

42.3 High-Throughput Testing of *In Vitro* Models

A pivotal aspect of high-throughput testing is the assay that is used to predict toxic effects. To significantly increase the number of test chemicals in different concentrations that can be tested quantitatively, HTS has been proposed in different studies [9]. The applicability of the approach could be demonstrated by a publication of the US National Toxicology Program and the NIH National Chemical Genomics Center (NCGC) that assessed the toxic effects of a panel of 1408 chemicals in different human and rodent cells by automating a luminescent cell viability assay [56]. Moreover, a study of the European Commission Joint Research Centre demonstrated the successful automation of an assay to determine the acute oral toxicity by diligently implementing the manual test protocol, defined in an OECD guidance document, into a high-throughput test platform. However, the described approaches are currently limited to relative simple readouts such as cell viability nor do they employ 3D reconstructed tissues. A reason for this is that the current available test methods standardly use invasive techniques such as colorimetric assays or histology which are very difficult to combine with a high-throughput platform approach [57]. Pitfalls of these methods are that the test procedures are difficult to implement into a technical process [58] and that test samples are destroyed in the test procedure [59]. Hence, toxic effects need to be compared to controls which increase the needed number of tissue models. Additionally, the quality of a produced skin model batch can only be controlled via random samples. Especially for a continuous production process, as described in the bioreactor-based approaches, this is a major challenge as no in-process control can be used to monitor tissue formation. Hence, the process is dependent on a predefined protocol and no regulation is possible.

To overcome these pitfalls, nondestructive technologies can be employed to assess skin model reactions. Optical imaging systems that make use of the multiphoton effect techniques showing the highest spatial resolution seem very promising in this respect. In these systems, a femtosecond pulsed laser allows to excite a given sample with two or more photons. Hence, laser light with a near-infrared wavelength can be used to image a sample [60]. As light with a lower wavelength penetrates deeper into optical dense tissues such as the skin, information can be gained to a depth up to 1 mm [61]. Also this technique allows a label-free imaging due to the possibility to visualize the autofluorescence of endogenous substances such as nicotinamide adenine dinucleotide phosphate (NADPH), elastin or collagen [62]. Especially elastin and collagen can be detected specifically due to second harmonic effects. Due to the noncentrosym-3D structure, these biomolecules emit light with approximately the doubled energy (and thus doubled wavelength) than the light, which was used to excite the sample [63]. Using these systems, skin reactions could be investigated on a cellular level.

Besides imaging, advanced optical systems can also be used to investigate skin effects on a molecular level. In Raman spectroscopy, light is coupled into a sample where the light interacts with the present molecules and is scattered inelastically [64]. By counting the photons for each wavelength, a molecular fingerprint of a biological sample can be generated. Using this technology, different skin cell types can be identified [65] and different toxic reaction such as necrosis and apoptosis can be recognized. Furthermore, Raman spectra can be employed to analyse the extracellular matrix of tissues, which allows to investigate degradation processes in collagen [66] or differences in the molecular composition of the *stratum corneum* between human skin and *in vitro* skin models [67]. These findings were supported by standard invasive methods that showed that some aspects in the *stratum corneum* composition of reconstructed human epidermis and full-thickness skin models differ from the human skin *in vivo* [68].

Although Raman spectroscopy is a promising tool in the investigation of the molecular mechanisms behind skin toxicology, the long analysis times of 1 min per measurement restrict the broad application of the technology. To achieve much faster readouts, optical coherence tomography (OCT) can be used. Comparable to ultrasound measurements, OCT uses two light beams to generate optical cross-sections of a sample. Of these light beams, one is directed to the tissue sample and the second to a reference mirror. The combined reflected light from the two paths only forms an interference image if the working distance of both paths is matched. By adjusting the length in the reference path, the amplitude of the reflected light from the sample can be recorded depending on the depth [69]. Depending on the system used, OCT allows a spatial resolution down to 10 μm with measurement times of only a few seconds [69]. When employed to investigate the reconstructed human epidermis, OCT could discriminate between different epidermal layers and thus is a promising tool to assure the quality of commercially produced skin models. Accordingly, OCT was integrated into the first automated production facility of human skin models ('Tissue Factory') to ensure the quality of each produced model individually, without being dependant on random sampling.

In addition to optical systems, also electrical properties of the skin can be employed to assess skin models. Forming the interface between the human organism and the surrounding environment, the skin restricts the flow of electrically charged particles. Thus, the skin has a characteristic high trans-epithelial electrical resistance (TEER), which is usually measured with alternating current to avoid destructive effects of direct currents such as a polarization of cellular ions or the induction of electrolysis and heating in the used cell culture medium [70]. The potency of these electrical measurements to predict toxic effects on the skin was demonstrated *in vivo* [71] and by the European Union Reference Laboratory for alternatives to animal testing (EURL-ECVAM) validated TER method, in which the change of the electrical resistance of the *ex vivo* rat skin was used to assess skin corrosion [72]. However, these promising results could not be transferred to *in vitro* skin models as an unphysiological $CaSO_4$ solution was used in the TER method based on the *ex vivo* rat skin. Furthermore, no standardized measurement setup is available, and research is still dependant on commercially available systems that have been developed for 2D cell cultures. In contrast to these simple cell-based models, 3D skin models are composed of multiple layers with different electrical properties that result in high variability between measurements and currently restrict the applicability of TEER measurements in risk assessment [73, 74]. Moreover, TEER values are determined at one specific frequency only [75]. Due to the use of alternating currents, the electrical resistance or impedance is dependent on the applied frequency resulting in complex impedance spectra, which are specific for the tissue under investigation. Thus, a majority of information is lost in TEER measurements.

In a recent study, an experimental setup was presented that specially was designed to assess the electrical properties of 2D tissue constructs. In contrast to simple TEER measurements here, the impedance spectra between 1 Hz and 100 kHz of RHE was used to investigate the epidermal differentiation [76]. After the differentiation of the keratinocytes at the air-liquid interface, the RHE showed impedance spectra, which were comparable to the human skin *in vivo*. Employing mathematical modelling, electrical parameters such as the ohmic resistance and the capacitance could be extracted from the impedance spectra. Using this approach, the development of the epidermal barrier and the effects of different mechanical and chemical traumata on the models could be quantified. Most interestingly, when used as an additional endpoint in skin irritation testing, the method is sufficiently sensitive to detect the effect of the washing process and of non-irritants to the RHE. These results indicate that impedance spectroscopy might be applicable as a complementary endpoint in current skin toxicity testing.

42.4 Perspectives for Future Automation Approaches

Although automated platforms enable to produce considerably more skin models and a high-throughput strategy would allow to conduct more testing than in a manual process, significantly more resources are needed for the establishment of the technical processes. Being constructed for usually a single high-throughput test, most systems lack flexibility. Moreover, automated facilities need special cost-intensive peripheral equipment, such as liquid handling systems, incubators, centrifuges or shakers. Thus, before a test is automated, the cost of the automation should be carefully compared to its benefit. Especially if the expected demand for a specific test model or the test itself is moderate, classical automation approaches are too expensive for commercial use.

As an alternative, bioreactor technologies can help to reduce costs as here usually only fluidic systems are needed instead of expensive robotic systems. In addition to these systems, also new robotic devices that are able to use standard laboratory equipment can be used for laboratory automation. An interesting approach was presented by the Japanese company Yaskawa, which uses a dual-arm robotic system to automate even challenging laboratory processes. Much like a laboratory technician, the robotic system can use standard labware such as pipets and cellscrapers. Employing this system it could be demonstrated that the robot is able to perform even demanding cell culture processes [77]. In dermal toxicity testing, a pivotal part in standard operational procedures is the application of different test substances to the surface of skin models. Due to the different physico-chemical properties of the substances, this requires specially trained and experienced personnel, and so far no automated process has been proposed conducting this step. Due to the high flexibility of the dual-arm robotic system, different application scenarios could be programmed, which could allow an automated substance application in future high-throughput approaches.

A critical aspect of high-throughput testing is that these test methods need to be validated in order to receive regulatory acceptance to replace animal experimentation. As validation involves the testing in multiple independent laboratories, all employed testing platform need to have the same specifications. However, no international standard for lab automation is currently available. Thus, all participating laboratories need to comply with a single system before a validation study. Due to the significant cost for high-throughput testing systems, the available systems are usually adapted to the specific needs of an institution and usually differ significantly between one another. To an even greater extend, the 'Adam robotic scientist' system is using automation for the investigation of genomic analysis. Here, the system is not only conducting the experiments but is also able to decide autonomously which new experiments should be conducted [78]. Transferred to *in vitro* dermal toxicity testing, a future system could not only produce models and conduct assays on different toxicological endpoints but could also decide on the next steps in complex integrated testing strategies.

42.5 Conclusions

In this chapter, we have demonstrated that automated skin model production is feasible today, offering clear advantages over manual production, and can be achieved in a cost-effective way. In addition, since industry will continue to seek for better performing human test models, more 'complete' skin models including other cell types will surely be developed in the near future. Mass production of such complex tissue models is challenging using routine cell culture methods; hence, automation of tissue model manufacturing will soon show its benefits in industrial toxicity testing in the twenty-first century in general terms.

References

1. Russell WMS, Burch RL, Hume CW. The principles of humane experimental technique. 1959 [cited 2012 Sep 28]. http://altweb.jhsph.edu/pubs/books/humane_exp/addendum.
2. Mayr LM, Bojanic D. Novel trends in high-throughput screening. Curr Opin Pharmacol. 2009;9(5):580–8.
3. Smith A. Screening for drug discovery: the leading question. Nature. 2002;418:453–9.
4. Sundberg SA. High-throughput and ultra-high-throughput screening: solution- and cell-based approaches. Curr Opin Biotechnol. 2000;11(1):47–53.
5. Martis E. High-throughput screening: the hits and leads of drug discovery- an overview. J Appl Pharm Sci. 2011;01(01):02–10.
6. Mayr LM, Fuerst P. The future of high-throughput screening. Off J Soc Biomol Screen. 2008;13(6):443–8.
7. Rimann M, Angres B, Patocchi-Tenzer I, Braum S, Graf-Hausner U. Automation of 3D cell culture using chemically defined hydrogels. J Lab Autom. 2013;19(2):1–7.
8. Schmidt CW. TOX 21: new dimensions of toxicity testing. Environ Health Perspect. 2009;117(8):A348–53.
9. Shukla SJ, Huang R, Austin CP, Xia M. The future of toxicity testing: a focus on in vitro methods using a quantitative high-throughput screening platform. Drug Discov Today. 2010;15(23–24):997–1007.
10. Borg DJ, Dawson RA, Leavesley DI, Hutmacher DW, Upton Z, Malda J. Functional and phenotypic characterization of human keratinocytes expanded in microcarrier culture. J Biomed Mater Res A. 2009;88:184–94.
11. Kalyanaraman B, Boyce S. Assessment of an automated bioreactor to propagate and harvest keratinocytes for fabrication of engineered skin substitutes. Tissue Eng. 2007; 13(5):983–93.
12. Lee J, Cuddihy MJ, Kotov NA. Three-dimensional cell culture matrices: state of the art. Tissue Eng Part B Rev. 2008;14(1):61–86.
13. Cukierman E, Pankov R, Stevens DR, Yamada KM. Taking cell-matrix adhesions to the third dimension. Science. 2001;294(1998):1708–12.
14. Gieseck RL, Hannan NRF, Bort R, Hanley NA, Drake RA, Cameron GW, et al. Maturation of induced pluripotent stem cell derived hepatocytes by 3D-culture. PLoS One. 2014;9(1):e86372.
15. Kempner M. A review of cell culture automation. J Assoc Lab Autom. 2002;7(2):56–62.
16. Graeve T, Noll M. Three-dimensional skin model. Google Patents. 2002. https://www.google.de/patents/WO2001092477A3?cl=en.
17. Lemper M, De Paepe K, Rogiers V. Practical problems encountered during the cultivation of an open-source reconstructed human epidermis model on a polycarbonate membrane and protein quantification. Skin Pharmacol Physiol. 2013;27(2):106–12.
18. Poumay Y, Dupont F, Marcoux S, Leclercq-Smekens M, Hérin M, Coquette A. A simple reconstructed human epidermis: preparation of the culture model and utilization in in vitro studies. Arch Dermatol Res. 2004;296(5):203–11.
19. Williams JA. Keys to bioreactor selections. Chem Eng Prog. 2002;98(3):34–41.
20. Hambor JE. Bioreactor design and bioprocess controls for industrialized cell processing. BioProcess Int. 2012;10(6):22–33.
21. Placzek MR, Chung I-M, Macedo HM, Ismail S, Blanco TM, Lim M, et al. Stem cell bioprocessing: fundamentals and principles. J R Soc Interface. 2009;6(32):209–32.
22. Baudoin R, Griscom L, Prot JM, Legallais C, Leclerc E. Behavior of HepG2/C3A cell cultures in a microfluidic bioreactor. Biochem Eng J. 2011;53(2):172–81.
23. Krishnan V, Vogler EA, Sosnoski DM, Mastro AM. In vitro mimics of bone remodeling and the vicious cycle of cancer in bone. J Cell Physiol. 2014;229(4):453–62. doi:10.1002/jcp.24464.
24. Chaudhuri J. Special issue: design of bioreactor systems for tissue engineering. PRO. 2015;3(1):46–9.

25. Groeber F, Kahlig A, Loff S, Walles H, Hansmann J. A bioreactor system for interfacial culture and physiological perfusion of vascularized tissue equivalents. Biotechnol J. 2013;8(3):308–16. doi:10.1002/biot.201200160.
26. Martin I, Wendt D, Heberer M. The role of bioreactors in tissue engineering. Trends Biotechnol. 2004;22(2):80–6.
27. Chen H-C, Hu Y-C. Bioreactors for tissue engineering. Biotechnol Lett. 2006;28(18):1415–23. doi:10.1007/s10529-006-9111-x.
28. Korossis SA, Bolland F, Kearney JN, Fisher J, Ingham E. Bioreactors in tissue engineering. Top Tissue Eng. 2005;2(8):1–23.
29. Plunkett N, O'Brien FJ. Bioreactors in tissue engineering. Technol Health Care. 2011;19(1):55–69. doi:10.3233/THC-2011-0605.
30. Hansmann J, Groeber F, Kahlig A, Kleinhans C, Walles H. Bioreactors in tissue engineering—principles, applications and commercial constraints. Biotechnol J. 2013;8(3):298–307. doi:10.1002/biot.201200162.
31. Tocchio A, Tamplenizza M, Martello F, Gerges I, Rossi E, Argentiere S, et al. Versatile fabrication of vascularizable scaffolds for large tissue engineering in bioreactor. Biomaterials. 2015;45:124–31.
32. Ebrahimkhani MR, Young CL, Lauffenburger DA, Griffith LG, Borenstein JT. Approaches to in vitro tissue regeneration with application for human disease modeling and drug development. Drug Discov Today. 2014;19(6):754–62.
33. Bowcock AM, Krueger JG. Getting under the skin: the immunogenetics of psoriasis. Nat Rev Immunol. 2005;5(9):699–711. doi:10.1038/nri1689.
34. Mahabeleshwar GH, Byzova TV. Angiogenesis in melanoma. Semin Oncol. 2007;34(6):555–65.
35. Baker SC, Shabir S, Southgate J. Biomimetic urothelial tissue models for the in vitro evaluation of barrier physiology and bladder drug efficacy. Mol Pharm. 2014;11(7):1964–70. doi:10.1021/mp500065m.
36. Grabinger T, Luks L, Kostadinova F, Zimberlin C, Medema JP, Leist M, et al. Ex vivo culture of intestinal crypt organoids as a model system for assessing cell death induction in intestinal epithelial cells and enteropathy. Cell Death Dis. 2014;5:e1228. doi:10.1038/cddis.2014.183.
37. Osorio-Lozada A, Surapaneni S, Skiles GL, Subramanian R. Biosynthesis of drug metabolites using microbes in hollow fiber cartridge reactors: case study of diclofenac metabolism by Actinoplanes species. Drug Metab Dispos. 2008;36(2):234–40.
38. Lister PD. The role of pharmacodynamic research in the assessment and development of new antibacterial drugs. Biochem Pharmacol. 2006;71(7):1057–65.
39. Shah P, Vedarethinam I, Kwasny D, Andresen L, Dimaki M, Skov S, et al. Microfluidic bioreactors for culture of non-adherent cells. Sensors Actuators B Chem. 2011;156(2):1002–8.
40. Leclerc E, Sakai Y, Fujii T. Microfluidic PDMS (Polydimethylsiloxane) bioreactor for large-scale culture of hepatocytes. Biotechnol Prog. 2004;20(3):750–5. doi:10.1021/bp0300568.
41. Altmann B, Löchner A, Swain M, Kohal R-J, Giselbrecht S, Gottwald E, et al. Differences in morphogenesis of 3D cultured primary human osteoblasts under static and microfluidic growth conditions. Biomaterials. 2014;35(10):3208–19.
42. Marx U, Walles H, Hoffmann S, Lindner G, Horland R, Sonntag F, et al. "human-on-a-chip" developments: a translational cutting-edge alternative to systemic safety assessment and efficiency evaluation of substances in laboratory animals and man? Altern Lab Anim. 2012;40(5):235–57.
43. Esch MB, Smith AST, Prot J-M, Oleaga C, Hickman JJ, Shuler ML. How multi-organ microdevices can help foster drug development. Adv Drug Deliv Rev. 2014;69–70:158–69.
44. Bhatia SN, Ingber DE. Microfluidic organs-on-chips. Nat Biotechnol. 2014;32(8):760–72. doi:10.1038/nbt.2989.
45. Bhise NS, Ribas J, Manoharan V, Zhang YS, Polini A, Massa S, et al. Organ-on-a-chip platforms for studying drug delivery systems. J Control Release. 2014;190:82–93.
46. Mansbridge J. Commercial considerations in tissue engineering. J Anat. 2006;209(4):527–32. doi:10.1111/j.1469-7580.2006.00631.x.

47. Amanullah A, Otero JM, Mikola M, Hsu A, Zhang J, Aunins J, et al. Novel micro-bioreactor high throughput technology for cell culture process development: reproducibility and scalability assessment of fed-batch CHO cultures. Biotechnol Bioeng. 2010;106(1):57–67. doi:10.1002/bit.22664.
48. Spencer TJ, Hidalgo-Bastida LA, Cartmell SH, Halliday I, Care CM. In silico multi-scale model of transport and dynamic seeding in a bone tissue engineering perfusion bioreactor. Biotechnol Bioeng. 2013;110(4):1221–30. doi:10.1002/bit.24777.
49. Liu N, Zang R, Yang S-T, Li Y. Stem cell engineering in bioreactors for large-scale bioprocessing. Eng Life Sci. 2014;14(1):4–15. doi:10.1002/elsc.201300013.
50. Betts JPJ, Warr SRC, Finka GB, Uden M, Town M, Janda JM, et al. Impact of aeration strategies on fed-batch cell culture kinetics in a single-use 24-well miniature bioreactor. Biochem Eng J. 2014;82(0):105–16.
51. De Napoli IE, Scaglione S, Giannoni P, Quarto R, Catapano G. Mesenchymal stem cell culture in convection-enhanced hollow fibre membrane bioreactors for bone tissue engineering. J Membr Sci. 2011;379(1-2):341–52.
52. Misener R, Fuentes Garí M, Rende M, Velliou E, Panoskaltsis N, Pistikopoulos EN, et al. Global superstructure optimisation of red blood cell production in a parallelised hollow fibre bioreactor. Comput Chem Eng. 2014;71:532–53.
53. Leite SB, Teixeira AP, Miranda JP, Tostões RM, Clemente JJ, Sousa MF, et al. Merging bioreactor technology with 3D hepatocyte-fibroblast culturing approaches: improved in vitro models for toxicological applications. Toxicol In Vitro. 2011;25(4):825–32.
54. Gunness P, Mueller D, Shevchenko V, Heinzle E, Ingelman-Sundberg M, Noor F. 3D organotypic cultures of human HepaRG cells: a tool for in vitro toxicity studies. Toxicol Sci. 2013;133(1):67–78.
55. National Research Council. Toxicity testing in the 21st century: a vision and a strategy: Washington, DC, National Academies Press; 2007.
56. Xia M, Huang R, Witt KL, Southall N, Fostel J, Cho MH, et al. Compound cytotoxicity profiling using quantitative high-throughput screening. Environ Health Perspect. 2008;116(3):284–91.
57. Cotovio J, Grandidier M, Portes P, Roguet R, Rubinstenn G. The in vitro skin irritation of chemicals: optimisation of the EPISKIN prediction model within the framework of the ECVAM validation process. Altern Lab Anim. 2005;33(4):329–49.
58. Bouhifd M, Bories G, Casado J, Coecke S, Norlén H, Parissis N, et al. Automation of an in vitro cytotoxicity assay used to estimate starting doses in acute oral systemic toxicity tests. Food Chem Toxicol. 2012;50(6):2084–96.
59. Mahmud FA, Hastings GW, Martini M. Model to characterize strain generated potentials in bone. J Biomed Eng. 1988;10(1):54–6.
60. Schenke-Layland K, Riemann I, Damour O, Stock UA, König K. Two-photon microscopes and in vivo multiphoton tomographs - powerful diagnostic tools for tissue engineering and drug delivery. Adv Drug Deliv Rev. 2006;58(7):878–96.
61. Centonze V, White J. Multiphoton excitation provides optical sections from deeper within scattering specimens than confocal imaging. Biophys J. 1998;75(4):2015–24.
62. Skala MC, Riching KM, Bird DK, Gendron-Fitzpatrick A, Eickhoff J, Eliceiri KW, et al. In vivo multiphoton fluorescence lifetime imaging of protein-bound and free nicotinamide adenine dinucleotide in normal and precancerous epithelia. J Biomed Opt. 2007;12(2):024014.
63. Schenke-Layland K, Madershahian N, Riemann I, Starcher B, Halbhuber KJ, König K, et al. Impact of cryopreservation on extracellular matrix structures of heart valve leaflets. Ann Thorac Surg. 2006;81:918–26.
64. Brauchle E, Schenke-Layland K. Raman spectroscopy in biomedicine - non-invasive in vitro analysis of cells and extracellular matrix components in tissues. Biotechnol J. 2012; 8(3):1–10.
65. Pudlas M, Koch S, Bolwien C, Thude S, Jenne N, Hirth T, et al. Raman spectroscopy: a noninvasive analysis tool for the discrimination of human skin cells. Tissue Eng Part C. 2011;17(10):1027–40.

66. Votteler M, Carvajal Berrio DA, Pudlas M, Walles H, Stock UA, Schenke-Layland K. Raman spectroscopy for the non-contact and non-destructive monitoring of collagen damage within tissues. J Biophotonics. 2011;5(1):47–56.
67. Tfayli A, Piot O, Draux F, Pitre F, Manfait M. Molecular characterization of reconstructed skin model by Raman microspectroscopy: comparison with excised human skin. Biopolymers. 2007;87(4):261–74.
68. Thakoersing VS, van Smeden J, Mulder AA, Vreeken RJ, El Ghalbzouri A, Bouwstra JA. Increased presence of monounsaturated fatty acids in the stratum corneum of human skin equivalents. J Invest Dermatol. 2013;133(1):59–67.
69. Pierce MC, Strasswimmer J, Park BH, Cense B, de Boer JF. Advances in optical coherence tomography imaging for dermatology. J Invest Dermatol. 2004;123(3):458–63.
70. Yamamoto T, Yamamoto Y. Analysis for the change of skin impedance. Med Biol Eng Comput. 1977;15(3):219–27.
71. Nyrén M, Kuzmina N, Emtestam L. Electrical impedance as a potential tool to distinguish between allergic and irritant contact dermatitis. J Am Acad Dermatol. 2003;48(3):394–400.
72. Worth A. ECVAM protocol for rat skin transcutaneous electrical resistance: an in vitro assay for assessing dermal corrosivity. In Vitro. 2002;12(4):93–110.
73. Knott A, Koop U, Mielke H, Reuschlein K, Peters N, Muhr G, et al. A novel treatment option for photoaged skin. J Cosmet Dermatol. 2008;7(1):15–22.
74. Kandárová H. Evaluation and validation of reconstructed human skin models as alternatives to animal tests in regulatory toxicology. Berlin: Freie Universität Berlin; 2006. http://www.diss.fu-berlin.de/diss/receive/FUDISS_thesis_000000002248
75. Millipore Corporation. ERS-2 User Guide. 2009.
76. Groeber F, Engelhardt L, Egger S, Werthmann H, Monaghan M, Walles H, et al. Impedance spectroscopy for the non-destructive evaluation of in vitro epidermal models. Pharm Res. 2014;32(5):1–10.
77. Yaskawa. Motoman dual arm robot in biomedical cell. 2013. https://www.youtube.com/watch?v=Dq3zLFWjtYU.
78. King RD, Rowland J, Oliver SG, Young M, Aubrey W, Byrne E, et al. The automation of science. Science. 2009;324(5923):85–9.